STANDARD LEVEL

Mathematics

Analysis and Approaches

for the IB Diploma

IBRAHIM WAZIR
TIM GARRY

Published by Pearson Education Limited, 80 Strand, London, WC2R 0RL

https://www.pearson.com/international-schools

Copies of official specifications for all Pearson Edexcel qualifications may be found on the website: https://qualifications.pearson.com

Text © Pearson Education Limited 2019
Theory of Knowledge chapter authored by Ric Sims
Worked solutions authored by Matteo Merlo, Magdalena Stasiak, Dana Ungu and Ibrahim Wazir
Edited by Jim Newall and Sam Hartburn
Proofread by Penny Nicholson and Martin Payne
Indexed by Georgie Bowden
Designed by © Pearson Education Limited 2019
Typeset by © Tech-Set Ltd, Gateshead, UK
Original illustrations © Pearson Education Limited 2019
Illustrated by © Tech-Set Ltd, Gateshead, UK
Cover design by © Pearson Education Limited 2019

Cover images: Front: © **Getty Images**: Busà Photography
Inside front cover: **Shutterstock.com**: Dmitry Lobanov

The rights of Ibrahim Wazir and Tim Garry to be identified as the authors of this work have been asserted by them in accordance with the Copyright, Designs and Patents Act 1988.

First published 2019

24
IMP 13

British Library Cataloguing in Publication Data
A catalogue record for this book is available from the British Library

ISBN 978 1 292 26741 8

Printed by CPI Group (UK) Ltd, Croydon CR0 4YY

Acknowledgements

(Key: b-bottom; c-centre; l-left; r-right; t-top)

Getty Images: JPL/Moment/Getty Images 1, baxsyl/Moment/Getty Images 45, d3sign/Moment/Getty Images 71, Alberto Manuel Urosa Toledano/Moment/Getty Images 115, Franco Tollardo/EyeEm/Getty Images 149, Alberto Manuel Urosa Toledano/Moment/Getty Images 191, Domenico De Santo/Getty Images 235, Sebastian-Alexander Stamatis/Getty Images 307, Brasil2/ E+/Getty Images 349, Roc Canals Photography/Moment/Getty Images 399, Johner Images/Getty Images 431, Gabriel Perez/Moment/Getty Images 475.

All other images © Pearson Education

We are grateful to the following for permission to reproduce copyright material:

Text:

pages 536–537, Edge Foundation Inc.: What Kind of Thing Is a Number? A Talk with Reuben Hersh, Wed, Oct 24, 2018. Used with permission of Edge Foundation Inc.

Text extracts relating to the IB syllabus and assessment have been reproduced from IBO documents. Our thanks go to the International Baccalaureate for permission to reproduce its copyright.

Dedications

First and foremost, I want to thank my wife, friend and devotee, Lody, for all the support she has given through all of these years of my work and career. Most of that work occurred on weekends, nights, while on vacation, and other times inconvenient to my family. I could not have completed this effort without her assistance, tolerance and enthusiasm.

Most importantly, I dedicate this book to my four grandchildren, Marco, Roberto, Lukas and Sophia, who lived through my frequent absences from their events.

I would also like to extend my thanks to Catherine Barber, our Commissioning Editor at Pearson, for all her support, flexibility and help.

Ibrahim Wazir

In loving memory of my parents.

I wish to express my deepest thanks and love to my wife, Val, for her unflappable good nature and support – and for smiling and laughing with me each day. I am infinitely thankful for our wonderful and kind-hearted children – Bethany, Neil and Rhona. My love for you all is immeasurable.

Tim Garry

Contents

Introduction

IB Mathematics: Analysis and Approaches Standard Level **syllabus topics**

1. Number and Algebra
2. Functions
3. Geometry and Trigonometry
4. Statistics and Probability
5. Calculus

This textbook comprehensively covers all of the material in the syllabus for the two-year ***Mathematics: Analysis and Approaches Standard Level*** course of the International Baccalaureate (IB) Diploma Programme (DP). First teaching of this course starts in the autumn of 2019 with first exams occurring in May 2021. We, the authors, have strived to thoroughly explain and demonstrate the mathematical concepts and methods listed in the course syllabus.

Content

As you will see when you look at the table of contents, the five syllabus topics (see margin note) are fully covered, though some are split over different chapters in order to group the information as logically as possible. This textbook has been designed so that the chapters proceed in a manner that supports effective learning of the course content. Thus – although not essential – it is recommended that you read and study the chapters in numerical order. It is particularly important that you thoroughly review and understand all of the content in the first chapter, ***Algebra and function basics***, before studying any of the other chapters.

Other than the final two chapters (**Theory of knowledge** and **Internal assessment**), each chapter has a set of **exercises** at the end of every section. Also, at the end of each chapter there is a set of **practice questions,** which are designed to expose you to questions that are more 'exam-like'. Many of the end-of-chapter practice questions are taken from past IB exam papers. Near the end of the book, you will find answers to all of the exercises and practice questions. There are also numerous **worked examples** throughout the book, showing you how to apply the concepts and skills you are studying.

The Internal assessment chapter provides thorough information and advice on the required **mathematical exploration component**. Your teacher will advise you on the timeline for completing your exploration and will provide critical support during the process of choosing your topic and writing the draft and final versions of your exploration.

The final chapter in the book will support your involvement in the **Theory of knowledge** course. It is a thought-provoking chapter that will stimulate you to think more deeply and critically about the nature of knowledge in mathematics and the relationship between mathematics and other areas of knowledge.

eBook

Included with this textbook is an eBook that contains a digital copy of the textbook and additional high-quality enrichment materials to promote your understanding of a wide range of concepts and skills encountered throughout the course. These materials include:

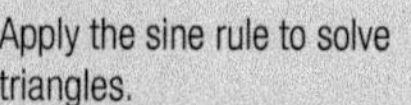

- Interactive ***GeoGebra*** **applets** demonstrating key concepts. Look out for the *GeoGebra* boxes and click on the icons in the eBook to open the applets in a new window.

- **Worked solutions** for all exercises and practice questions. These are downloadable PDFs located next to the exercises and practice questions themselves, and also with the short answers at the back of the book.

To access the eBook, please follow the instructions located on the inside cover.

Information boxes

As you read this textbook, you will encounter numerous boxes of different colours containing a wide range of helpful information.

Learning objectives

You will find learning objectives at the start of each chapter. They set out the content and aspects of learning covered in the chapter.

Learning objectives

By the end of this chapter, you should be familiar with...

- different forms of equations of lines and their gradients and intercepts
- parallel and perpendicular lines
- different methods to solve a system of linear equations (maximum of three equations in three unknowns)

Key facts

Key facts are drawn from the main text and highlighted for quick reference to help you identify clear learning points.

A function is **one-to-one** if each element y in the range is the image of exactly one element x in the domain.

Hints

Specific hints can be found alongside explanations, questions, exercises, and worked examples, providing insight into how to analyse/answer a question. They also identify common errors and pitfalls.

If you use a graph to answer a question on an IB mathematics exam, you must include a clear and well-labelled sketch in your working.

Notes

Notes include general information or advice.

Quadratic equations will be covered in detail in Chapter 2.

Examples

Worked examples show you how to tackle questions and apply the concepts and skills you are studying.

Example 1.5

Find x such that the distance between points $(1, 2)$ and $(x, -10)$ is 13 units.

Solution

$$d = 13 = \sqrt{(x-1)^2 + (-10-2)^2} \Rightarrow 13^2 = (x-1)^2 + (-12)^2$$

$$\Rightarrow 169 = x^2 - 2x + 1 + 144 \Rightarrow x^2 - 2x - 24 = 0$$

$$\Rightarrow (x-6)(x+4) = 0 \Rightarrow x - 6 = 0 \text{ or } x + 4 = 0$$

$$\Rightarrow x = 6 \text{ or } x = -4$$

How to use this book

This book is designed to be read by you – the student. It is very important that you read this book carefully. We have strived to write a readable book – and we hope that your teacher will routinely give you reading assignments from this textbook, thus giving you valuable time for productive explanations and discussions in the classroom. Developing your ability to read and understand mathematical explanations will prove to be valuable to your long-term intellectual development, while also helping you to comprehend mathematical ideas and acquire vital skills to be successful in the *Analysis and Approaches* SL course. Your goal should be understanding, not just remembering. You should always read a chapter section thoroughly before attempting any of the exercises at the end of the section.

Our aim is to support genuine inquiry into mathematical concepts while maintaining a coherent and engaging approach. We have included material to help you gain insight into appropriate and wise use of your GDC and an appreciation of the importance of proof as an essential skill in mathematics. We endeavoured to write clear and thorough explanations supported by suitable worked examples, with the overall goal of presenting sound mathematics with sufficient rigour and detail at a level appropriate for a student of SL mathematics.

For over 10 years, we have been writing successful textbooks for IB mathematics courses. During that time, we have received many useful comments from both teachers and students. If you have suggestions for improving this textbook, please feel free to write to us at globalschools@pearson.com. We wish you all the best in your mathematical endeavours.

Ibrahim Wazir and Tim Garry

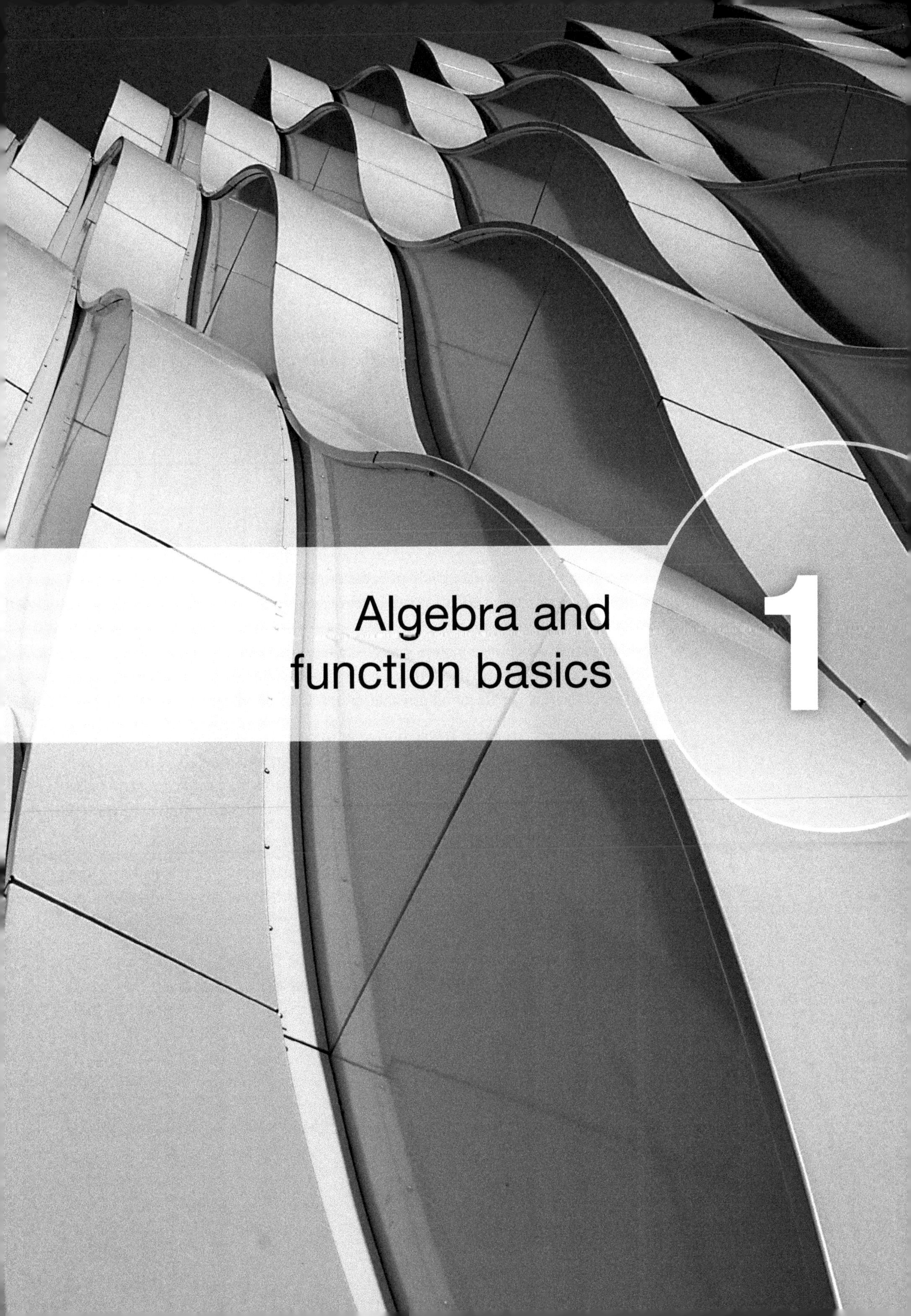

1 Algebra and function basics

1 Algebra and function basics

Learning objectives

By the end of this chapter, you should be familiar with...

- different forms of equations of lines and their gradients and intercepts
- parallel and perpendicular lines
- the concept of a function and its domain, range and graph
- mathematical notation for functions
- composite functions
- characteristics of an inverse function and finding the inverse function $f^{-1}(x)$
- transformations of graphs and composite transformations of graphs.

1.1 Equations and formulae

Equations, identities and formulae

You will encounter a wide variety of equations in this course. Essentially, an equation is a statement equating two algebraic expressions that may be true or false depending upon the value(s) substituted for the variable(s). Values of the variables that make the equation true are called **solutions** or **roots** of the equation. All of the solutions to an equation comprise the **solution set** of the equation. An equation that is true for all possible values of the variable is called an **identity**.

Many equations are often referred to as a **formula** (plural: formulae) and typically contain more than one variable and, often, other symbols that represent specific constants or **parameters** (constants that may change in value but do not alter the properties of the expression). Formulae with which you are familiar include: $A = \pi r^2$, $d = rt$, $d = \sqrt{(x_1 - x_2)^2 + (y_1 - y_2)^2}$ and $V = \frac{4}{3}\pi r^3$.

Whereas most equations that we encounter will have numerical solutions, we can solve a formula for one variable in terms of other variables – often referred to as changing the subject of a formula.

Example 1.1

(a) Solve for b in the formula $a^2 + b^2 = c^2$

(b) Solve for l in the formula $T = 2\pi\sqrt{\frac{l}{g}}$

(c) Solve for R in the formula $M = \frac{nR}{R + r}$

Solution

(a) $a^2 + b^2 = c^2 \Rightarrow b^2 = c^2 - a^2 \Rightarrow b = \pm\sqrt{c^2 - a^2}$

If b is a length then $b = \sqrt{c^2 - a^2}$

(b) $T = 2\pi\sqrt{\frac{l}{g}} \Rightarrow \sqrt{\frac{l}{g}} = \frac{T}{2\pi} \Rightarrow \frac{l}{g} = \frac{T^2}{4\pi^2} \Rightarrow l = \frac{T^2 g}{4\pi^2}$

(c) $I = \frac{nR}{R + r} \Rightarrow I(R + r) = nR \Rightarrow IR + Ir = nR$

$\Rightarrow IR - nR = -Ir \Rightarrow R(I - n) = -Ir$

$\Rightarrow R = \frac{Ir}{n - I}$

Note that factorisation was required in solving for R in part (c).

Equations and graphs

Two important characteristics of any equation are the number of variables (unknowns) and the type of algebraic expressions it contains (e.g. polynomials, rational expressions, trigonometric, exponential). Nearly all of the equations in this course will have either one or two variables. In this chapter we will only discuss equations with algebraic expressions that are polynomials. Solutions for equations with a single variable consist of individual numbers that can be graphed as points on a number line. The **graph** of an equation is a visual representation of the equation's solution set. For example, the solution set of the one-variable equation containing quadratic and linear polynomials $x^2 = 2x + 8$ is $x \in \{-2, 4\}$. The graph of this one-variable equation (Figure 1.1) is depicted on a one-dimensional coordinate system, i.e. the real number line.

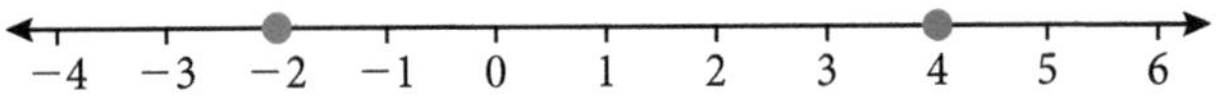

Figure 1.1 Graph of the solution set for the equation $x^2 = 2x + 8$

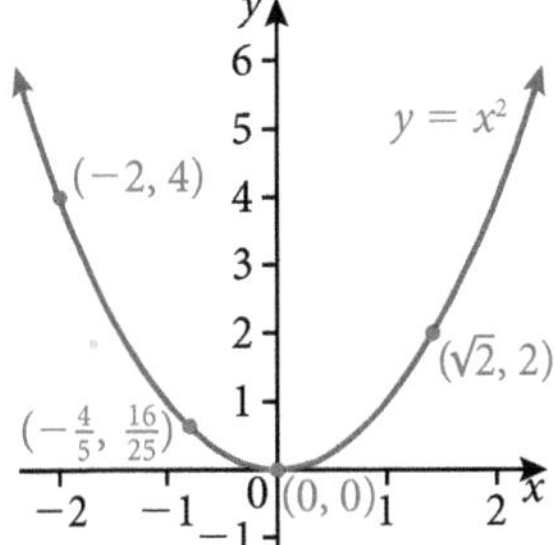

Figure 1.2 Graph of the solution set of the equation $y = x^2$

The solution set of a two-variable equation will be an **ordered pair** of numbers. An ordered pair corresponds to a location indicated by a point on a two-dimensional coordinate system, i.e. a **coordinate plane**. For example, the solution set of the two-variable **quadratic equation** $y = x^2$ will be an infinite set of ordered pairs (x, y) that satisfy the equation. Four ordered pairs in the solution set are shown in red in Figure 1.2. The graph of all the ordered pairs in the solution set forms a curve as shown in blue.

Quadratic equations will be covered in detail in Chapter 2.

Equations of lines

A one-variable **linear equation** in x can always be written in the form $ax + b = 0$, with $a \neq 0$, and it will have exactly one solution, namely $x = -\frac{b}{a}$. An example of a two-variable **linear equation in x and y** is $x - 2y = 2$. The graph of this equation's solution set (an infinite set of ordered pairs) is a **line** (Figure 1.3).

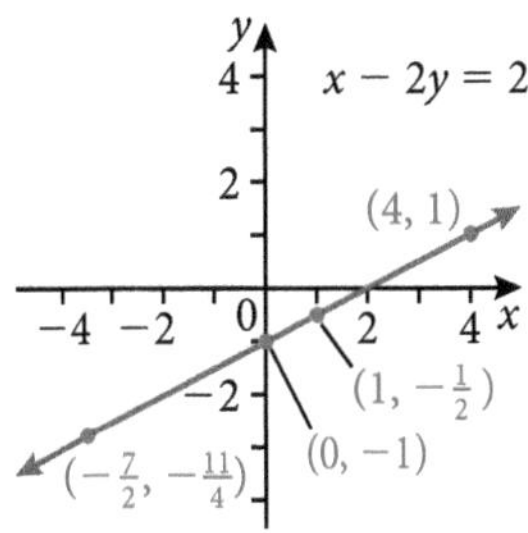

Figure 1.3 The graph of $x - 2y = 2$

The **slope** or **gradient**, m, of a non-vertical line is defined by the formula:

$$m = \frac{y_2 - y_1}{x_2 - x_1} = \frac{\text{vertical change}}{\text{horizontal change}}$$

Because division by zero is undefined, the slope of a vertical line is undefined. Using the two points $\left(1, -\frac{1}{2}\right)$ and (4, 1) we compute the slope of the line with equation $x - 2y = 2$ to be $m = \frac{1 - \left(-\frac{1}{2}\right)}{4 - 1} = \frac{\frac{3}{2}}{3} = \frac{1}{2}$

If we solve for y we can rewrite the equation in the form $y = \frac{1}{2}x - 1$

Note that the coefficient of x is the slope of the line and the constant term is the y-coordinate of the point at which the line intersects the y-axis, that is, the y-intercept. There are several forms for writing linear equations.

general form	$ax + by + c = 0$	every line has an equation in this form if both a and $b \neq 0$
slope-intercept form	$y = mx + c$	m is the slope; $(0, c)$ is the y-intercept
point-slope form	$y - y_1 = m(x - x_1)$	m is the slope; (x_1, y_1) is a known point on the line
horizontal line	$y = c$	slope is zero; $(0, c)$ is the y-intercept
vertical line	$x = c$	slope is undefined; unless the line is the y-axis, no y-intercept

Table 1.1 Forms for equations of lines

Most problems involving linear equations and their graphs fall into two categories: (1) given an equation, determine its graph; and (2) given a graph, or some information about it, find its equation.

For lines, the first type of problem is often best solved by using the slope-intercept form. For the second type of problem, the point-slope form is usually most useful.

Example 1.2

Without using a GDC, sketch the line that is the graph of each linear equation written in general form.

(a) $5x + 3y - 6 = 0$ (b) $y - 4 = 0$ (c) $x + 3 = 0$

Solution

(a) Solve for y to write the equation in slope-intercept form.

$$5x + 3y - 6 = 0 \Rightarrow 3y = -5x + 6 \Rightarrow y = -\frac{5}{3}x + 2$$

The line has a y-intercept of (0, 2) and a slope of $-\frac{5}{3}$

(b) The equation $y - 4 = 0$ is equivalent to $y = 4$, the graph of which is a horizontal line with a y-intercept of (0, 4)

(c) The equation $x + 3 = 0$ is equivalent to $x = -3$, the graph of which is a vertical line with no y-intercept; but, it has an x-intercept of $(-3, 0)$

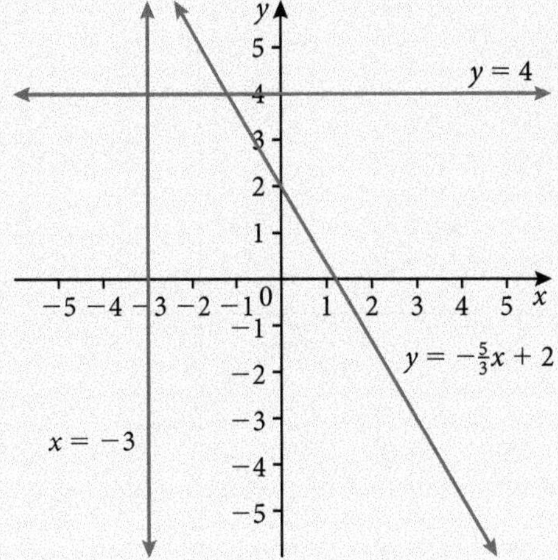

Example 1.3

(a) Find the equation of the line that passes through the point $(3, 31)$ and has a slope of 12. Write the equation in slope-intercept form.

(b) Find the linear equation in C and F knowing that $C = 10$ when $F = 50$, and $C = 100$ when $F = 212$. Solve for F in terms of C.

Solution

(a) Substitute $x_1 = 3$, $y_1 = 31$ and $m = 12$ into the point-slope form:

$$y - y_1 = m(x - x_1) \Rightarrow y - 31 = 12(x - 3) \Rightarrow y = 12x - 36 + 31$$
$$\Rightarrow y = 12x - 5$$

(b) The two points, ordered pairs (C, F), that are known to be on the line are $(10, 50)$ and $(100, 212)$. The variable C corresponds to the x variable and F corresponds to y in the definitions and forms stated above.

The slope of the line is $m = \dfrac{F_2 - F_1}{C_2 - C_1} = \dfrac{212 - 50}{100 - 10} = \dfrac{162}{90} = \dfrac{9}{5}$

Choose one of the points on the line, say $(10, 50)$, and substitute it and the slope into the point-slope form:

$$F - F_1 = m(C - C_1) \Rightarrow F - 50 = \frac{9}{5}(C - 10) \Rightarrow F = \frac{9}{5}C - 18 + 50$$
$$\Rightarrow F = \frac{9}{5}C + 32$$

The slope of a line is a convenient tool for determining whether two lines are parallel or perpendicular. The two lines shown in Figure 1.4 suggest that two distinct non-vertical lines are **parallel** if and only if their slopes are equal, $m_1 = m_2$.

The two lines shown in Figure 1.5 suggest that two non-vertical lines are perpendicular if and only if their slopes are negative reciprocals – that is, $m_1 = -\frac{1}{m_2}$, which is equivalent to $m_1 \cdot m_2 = -1$.

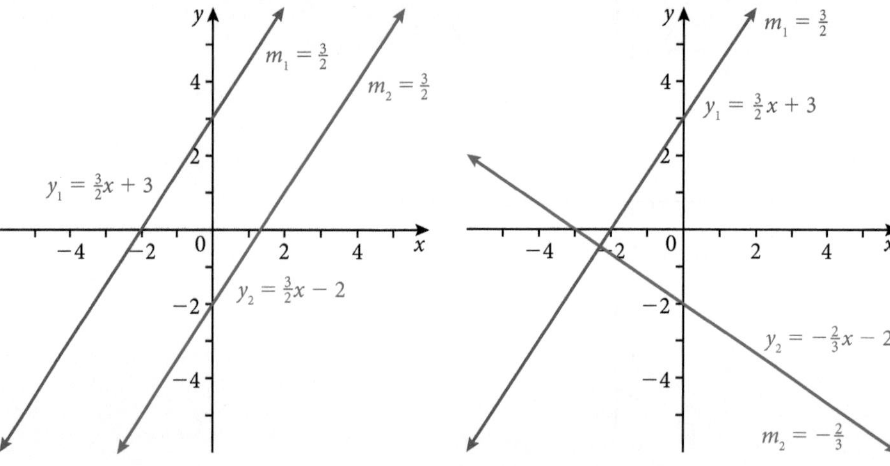

Figure 1.4 Parallel lines

Figure 1.5 Perpendicular lines

Online

See how the distance formula is an application of the Pythagorean theorem.

Distances and midpoints

Recall that absolute value is used to define the **distance** (always non-negative) between two points on the real number line. The distance between the points A and B on the real number line is $|B - A|$, which is equivalent to $|A - B|$.

The points A and B are the endpoints of a line segment that is denoted with the notation $[AB]$ and the length of the line segment is denoted AB. In Figure 1.6, the distance between A and B is $AB = |4 - (-2)| = |-2 - 4| = 6$.

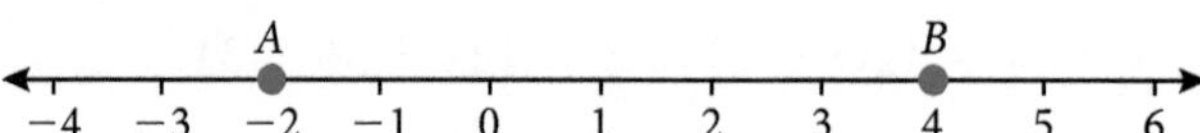

Figure 1.6 The length of the line segment $[AB]$ is denoted by AB

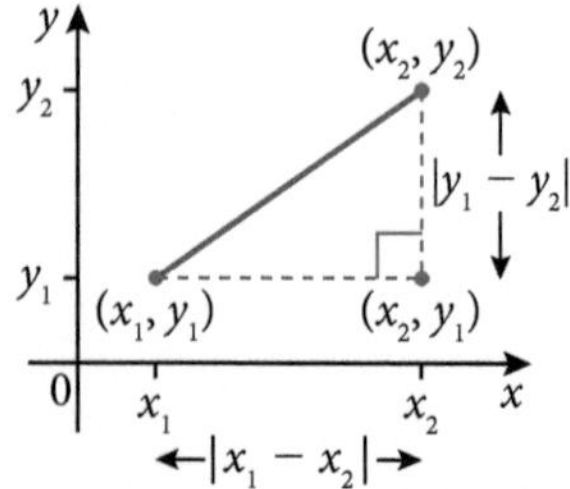

Figure 1.7 Distance between two points on a coordinate plane

We can find the distance between two general points (x_1, y_1) and (x_2, y_2) on a coordinate plane using the definition for distance on a number line and Pythagoras' theorem. For the points (x_1, y_1) and (x_2, y_2), the horizontal distance between them is $|x_1 - x_2|$ and the vertical distance is $|y_1 - y_2|$. As illustrated in Figure 1.7, these distances are the lengths of two legs of a right-angled triangle whose hypotenuse is the distance between the points. If d represents the distance between (x_1, y_1) and (x_2, y_2), then by Pythagoras' theorem $d^2 = |x_1 - x_2|^2 + |y_1 - y_2|^2$. Because the square of any number is positive, the absolute value is not necessary to give us the **distance formula** for two-dimensional coordinates.

The distance d between the two points (x_1, y_1) and (x_2, y_2) in the coordinate plane is $d = \sqrt{(x_1 - x_2)^2 + (y_1 - y_2)^2}$

The coordinates of the **midpoint** of a line segment are the average values of the corresponding coordinates of the two endpoints.

The midpoint of the line segment joining the points (x_1, y_1) and (x_2, y_2) in the coordinate plane is $\left(\frac{x_1 + x_2}{2}, \frac{y_1 + y_2}{2}\right)$

Example 1.4

(a) Show that the points $P(1, 2)$, $Q(3, 1)$ and $R(4, 8)$ are the vertices of a right-angled triangle.

(b) Find the midpoint of the hypotenuse of the triangle PQR.

Solution

(a) The three points are plotted and the line segments joining them are drawn in Figure 1.8. We can find the exact lengths of the three sides of the triangle by applying the distance formula.

$$PQ = \sqrt{(1-3)^2 + (2-1)^2} = \sqrt{4+1} = \sqrt{5}$$

$$QR = \sqrt{(3-4)^2 + (1-8)^2} = \sqrt{1+49} = \sqrt{50}$$

$$PR = \sqrt{(1-4)^2 + (2-8)^2} = \sqrt{9+36} = \sqrt{45}$$

$(PQ)^2 + (PR)^2 = (QR)^2$ because $(\sqrt{5})^2 + (\sqrt{45})^2 = 5 + 45 = 50 = (\sqrt{50})^2$

The lengths of the three sides of the triangle satisfy Pythagoras' theorem, confirming that the triangle is a right-angled triangle.

(b) QR is the hypotenuse. Let the midpoint of QR be point M. Using the midpoint formula, $M = \left(\frac{3+4}{2}, \frac{1+8}{2}\right) = \left(\frac{7}{2}, \frac{9}{2}\right)$. This point is plotted in Figure 1.8

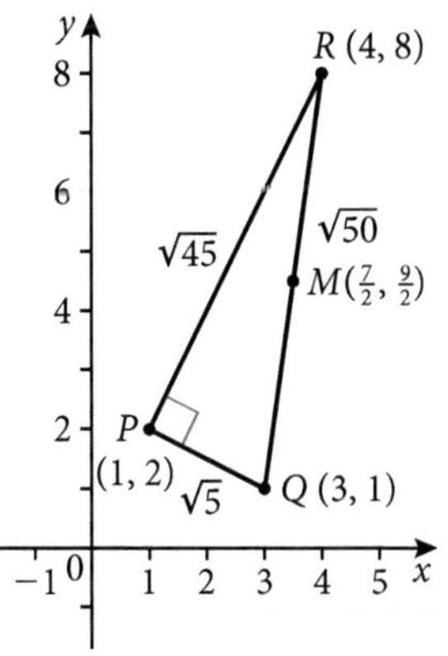

Figure 1.8 Diagram for Example 1.4

Example 1.5

Find x such that the distance between points $(1, 2)$ and $(x, -10)$ is 13 units.

Solution

$$d = 13 = \sqrt{(x-1)^2 + (-10-2)^2} \Rightarrow 13^2 = (x-1)^2 + (-12)^2$$

$$\Rightarrow 169 = x^2 - 2x + 1 + 144 \Rightarrow x^2 - 2x - 24 = 0$$

$$\Rightarrow (x-6)(x+4) = 0 \Rightarrow x - 6 = 0 \text{ or } x + 4 = 0$$

$\Rightarrow x = 6$ or $x = -4$ (see Figure 1.9)

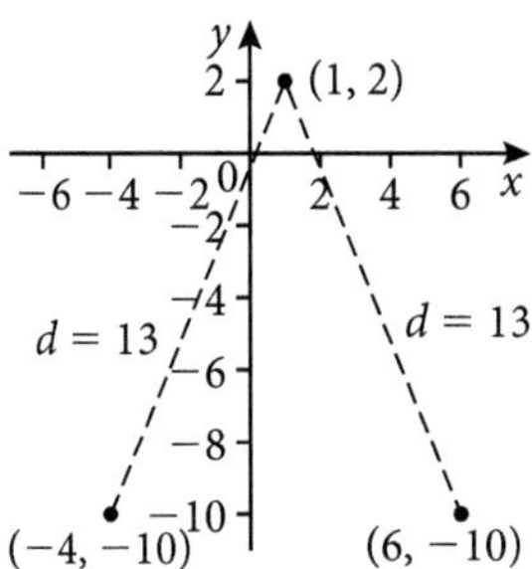

Figure 1.9 Graph for Example 1.5 showing the two points that are 13 units from (1, 2)

Exercise 1.1

1. Solve for the indicated variable in each formula.
   (a) $m(h - x) = n$, solve for x
   (b) $v = \sqrt{ab - t}$, solve for a
   (c) $A = \frac{h}{2}(b_1 + b_2)$, solve for b_1
   (d) $A = \frac{1}{2}r^2\theta$, solve for r
   (e) $\frac{f}{g} = \frac{h}{k}$, solve for k
   (f) $at = x - bt$, solve for t
   (g) $V = \frac{1}{3}\pi r^3 h$, solve for r
   (h) $F = \frac{g}{m_1 k + m_2 k}$, solve for k

2. Find the equation of the line that passes through the two given points. Write the line in slope-intercept form ($y = mx + c$), if possible.
   (a) $(-9, 1)$ and $(3, -7)$
   (b) $(3, -4)$ and $(10, -4)$
   (c) $(-12, -9)$ and $(4, 11)$
   (d) $\left(\frac{7}{3}, -\frac{1}{2}\right)$ and $\left(\frac{7}{3}, \frac{5}{2}\right)$
   (e) Find the equation of the line that passes through the point $(7, -17)$ and is parallel to the line with equation $4x + y - 3 = 0$. Write the line in slope-intercept form ($y = mx + c$).
   (f) Find the equation of the line that passes through the point $\left(-5, \frac{11}{2}\right)$ and is perpendicular to the line with equation $2x - 5y - 35 = 0$. Write the line in slope-intercept form ($y = mx + c$).

3. Find the exact distance between each pair of points and then find the midpoint of the line segment joining the two points.
   (a) $(-4, 10)$ and $(4, -5)$
   (b) $(-1, 2)$ and $(5, 4)$
   (c) $\left(\frac{1}{2}, 1\right)$ and $\left(-\frac{5}{2}, \frac{4}{3}\right)$
   (d) $(12, 2)$ and $(-10, 9)$

4. Find the value(s) of k so that the distance between the points is 5 units.
   (a) $(5, -1)$ and $(k, 2)$
   (b) $(-2, -7)$ and $(1, k)$

5. Show that the given points form the vertices of the indicated polygon.
   (a) Right-angled triangle: $(4, 0)$, $(2, 1)$ and $(-1, -5)$
   (b) Isosceles triangle: $(1, -3)$, $(3, 2)$ and $(-2, 4)$
   (c) Parallelogram: $(0, 1)$, $(3, 7)$, $(4, 4)$ and $(1, -2)$

1.2 Definition of a function

Many mathematical relationships concern how the value of one variable determines the value of a second variable. In general, suppose that the values of a particular **independent variable**, for example x, determine the values of a **dependent variable** y in such a way that for a specific value of x, a single value of y is determined. Then we say that y is a **function** of x and we write $y = f(x)$

(read y equals f of x) or $y = g(x)$, and so on, where the letter f or g represents the name of the function. For example:

- Period T is a function of length L: $T = 2\pi\sqrt{\frac{L}{g}}$
- Area A is a function of radius r: $A = \pi r^2$
- °F (degrees Fahrenheit) is a function of °C: $F = \frac{9}{5}C + 32$
- Distance d from the origin is a function of x: $d = |x|$

Other useful ways of representing a function include a graph of the equation on a **Cartesian coordinate system** (also called a rectangular coordinate system), a **table**, a **set of ordered pairs**, or a **mapping**.

A mapping illustrates how some values in the domain of a function are paired with values in the range of the function. Here is a mapping for the function $y = |x|$

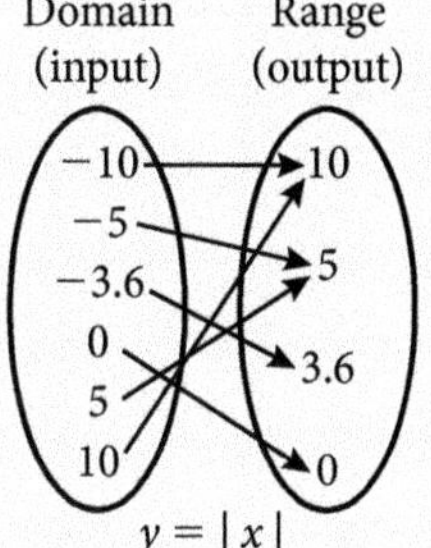

The largest possible set of values for the independent variable (the **input** set) is called the **domain**, and the set of resulting values for the dependent variable (the **output** set) is called the **range**. In the context of a mapping, each value in the domain is mapped to its **image** in the range. All of the various ways of representing a mathematical function illustrate that its defining characteristic is that it is a rule by which each number in the domain determines a unique number in the range.

The coordinate system for the graph of an equation has the independent variable on the horizontal axis and the dependent variable on the vertical axis.

A **function** is a correspondence (mapping) between two sets X and Y in which each element of set X corresponds to (maps to) exactly one element of set Y. The domain is set X (independent variable) and the range is set Y (dependent variable).

Not all equations represent a function. The solution set for the equation $x^2 + y^2 = 1$ is the set of ordered pairs (x, y) on the circle of radius equal to 1 and centre at the origin (see Figure 1.10). If we solve the equation for y, we get $y = \pm\sqrt{1 - x^2}$. It is clear that any value of x between -1 and 1 will produce two different values of y (opposites). Since at least one value in the domain (x) determines more than one value in the range (y), the equation does not represent a function. A correspondence between two sets that does not satisfy the definition of a function is called a **relation**.

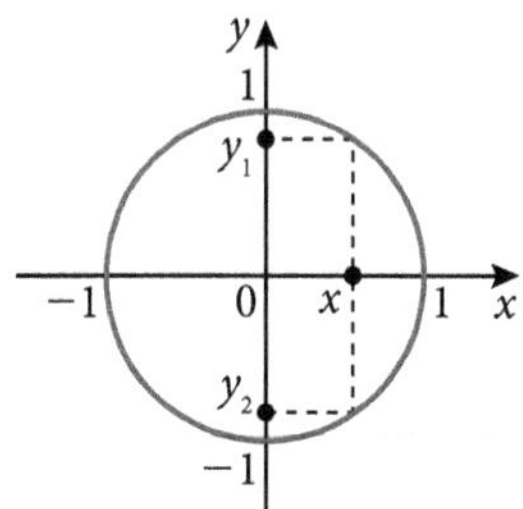

Figure 1.10 Graph of $x^2 + y^2 = 1$

Alternative definition of a function
A function is a relation in which no two different ordered pairs have the same first coordinate. A vertical line intersects the graph of a function at no more than one point (vertical line test).

For many physical phenomena, we observe that one quantity depends on another. The word function is used to describe this dependence of one quantity on another – that is, how the value of an independent variable determines the value of a dependent variable. A common mathematical task is to find how to express one variable as a function of another variable.

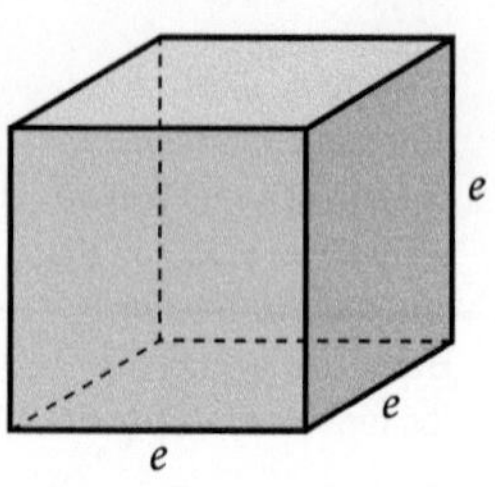

Figure 1.11 Cube for Example 1.6

Example 1.6

(a) Express the volume V of a cube as a function of the length e of each edge.

(b) Express the volume V of a cube as a function of its surface area S.

Solution

(a) V as a function of e is $V = e^3$

(b) The surface area of the cube consists of six squares each with an area of e^2. Hence, the surface area is $6e^2$; that is, $S = 6e^2$. We need to write V in terms of S. We can do this by first expressing e in terms of S, and then substituting this expression for e in the equation $V = e^3$.

$$S = 6e^2 \Rightarrow e^2 = \frac{S}{6} \Rightarrow e = \sqrt{\frac{S}{6}}.\text{ Substituting, } V = \left(\sqrt{\frac{S}{6}}\right)^3 = \frac{\left(S^{\frac{1}{2}}\right)^3}{\left(6^{\frac{1}{2}}\right)^3}$$

$$= \frac{S^{\frac{3}{2}}}{6^{\frac{3}{2}}} = \frac{S \cdot S^{\frac{1}{2}}}{6 \cdot 6^{\frac{1}{2}}} = \frac{S}{6}\sqrt{\frac{S}{6}}$$

V as a function of S is $V = \frac{S}{6}\sqrt{\frac{S}{6}}$

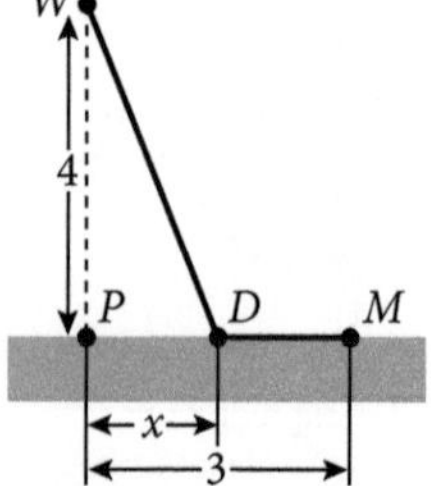

Figure 1.12 Diagram for Example 1.7

Example 1.7

An offshore wind turbine is located at point W, 4 km offshore from the nearest point P on a straight coastline. A maintenance station is at point M, 3 km along the coast from P. An engineer is returning by a small boat from the wind turbine. He sails to point D that is located between P and M at an unknown distance x km from point P. From there, he walks to the maintenance station. The boat sails at 3 km hr^{-1} and the engineer can walk at 6 km hr^{-1}. Express the total time T (hours) for the trip from the wind turbine to the maintenance station as a function of x (km).

Solution

To get an equation for T in terms of x, use the fact that time $= \frac{\text{distance}}{\text{rate}}$

We then have

$$T = \frac{\text{distance } WD}{3} + \frac{\text{distance } DM}{6}$$

The distance WD can be expressed in terms of x using Pythagoras' theorem.

$$WD^2 = x^2 + 4^2 \Rightarrow WD = \sqrt{x^2 + 16}$$

To express T in terms of only the single variable x, note that $DM = 3 - x$

Then the total time T can be written in terms of x by the equation

$$T = \frac{\sqrt{x^2 + 16}}{3} + \frac{3 - x}{6} \text{ or } T = \frac{1}{3}\sqrt{x^2 + 16} + \frac{1}{2} - \frac{x}{6}$$

Domain and range of a function

The domain of a function may be stated explicitly, or it may be implied by the expression that defines the function. For most of this course, we can assume that functions are real-valued functions of a real variable. The domain and range will contain only real numbers or some subset of the real numbers. The domain of a function is the set of all real numbers for which the expression is defined as a real number, if not explicitly stated otherwise. For example, if a certain value of x is substituted into the algebraic expression defining a function and it causes division by zero or the square root of a negative number (both undefined in the real numbers) to occur, that value of x cannot be in the domain.

The domain of a function may also be implied by the physical context or limitations that exist in a problem. For example, in both functions derived in Example 1.6 the domain is the set of positive real numbers (denoted by $\mathbb{R}^+$) because neither a length (edge of a cube) nor a surface area (face of a cube) can have a value that is negative or zero. In Example 1.7 the domain for the function is $0 < x < 3$ because of the constraints given in the problem. Usually the range of a function is not given explicitly and is determined by analysing the output of the function for all values of the input (domain). The range of a function is often more difficult to find than the domain, and analysing the graph of a function is very helpful in determining it. A combination of algebraic and graphical analysis is very useful in determining the domain and range of a function.

Online 

An interactive graph visually illustrates the domain and range of a given function.

Example 1.8

Find the domain of each function.

(a) $\{(-6, -3), (-1, 0), (2, 3), (3, 0), (5, 4)\}$

(b) Volume of a sphere: $V = \frac{4}{3}\pi r^3$

(c) $y = \frac{5}{2x - 6}$

(d) $y = \sqrt{3 - x}$

Solution

(a) The function consists of a set of ordered pairs. The domain of the function consists of all first coordinates of the ordered pairs. Therefore, the domain is the set $x \in \{-6, -1, 2, 3, 5\}$.

(b) The physical context tells us that a sphere cannot have a radius that is negative or zero. Therefore, the domain is the set of all real numbers r such that $r > 0$.

(c) Since division by zero is not defined for real numbers then $2x - 6 \neq 0$. Therefore, the domain is the set of all real numbers x such that $x \in \mathbb{R}$, $x \neq 3$.

(d) Since the square root of a negative number is not real, then $3 - x \geq 0$. Therefore, the domain is all real numbers x such that $x \leq 3$.

Example 1.9

Find the domain and range for the function $y = x^2$

Solution

Using algebraic analysis: Squaring any real number produces another real number. Therefore, the domain of $y = x^2$ is the set of all real numbers ($\mathbb{R}$). Since the square of any positive or negative number will be positive and the square of zero is zero, then the range is the set of all real numbers greater than or equal to zero.

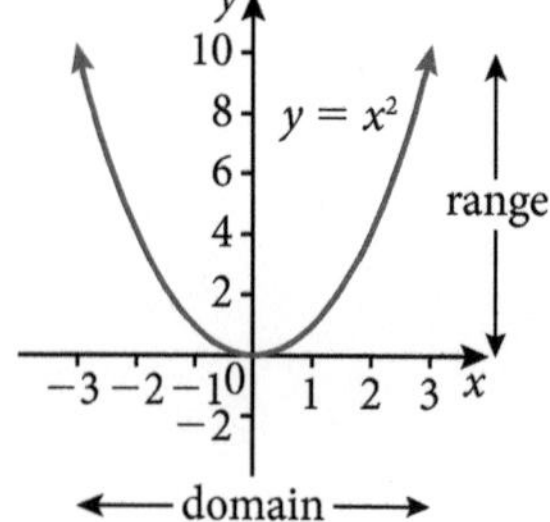

Figure 1.13 Graphical solution to Example 1.9

Using graphical analysis: For the domain, focus on the x-axis and scan the graph from $-\infty$ to $+\infty$. There are no gaps or blank regions in the graph and the parabola will continue to get wider as x goes to either $-\infty$ or $+\infty$. Therefore, the domain is all real numbers. For the range, focus on the y-axis and scan from $-\infty$ to $+\infty$. The parabola will continue to increase as y goes to $+\infty$, but the graph does not go below the x-axis. The parabola has no points with negative y coordinates. Therefore, the range is the set of real numbers greater than or equal to zero.

The inequality $2 \leqslant x < 5$ can also be written as $[2, 5[$. The number 2 is included, but 5 is not. When determining the domain and range of a function, use both algebraic and graphical analysis. Do not rely too much on using just one approach. For graphical analysis of a function, producing a graph on your GDC that shows all the important features is essential.

Description in words	Interval notation
Domain is any real number	Domain is $\{x : x \in \mathbb{R}\}$, or Domain is $x \in]-\infty, \infty[$
Range is any number greater than or equal to zero	Range is $\{y : y \geqslant 0\}$, or Range is $y \in]0, \infty[$

Table 1.2 Different ways of expressing the domain and range of $y = x^2$

Function notation

It is common practice to name a function using a single letter, with f, g and h commonly used. Given that the domain variable is x and the range variable is y, the symbol $f(x)$ denotes the unique value of y that is generated by the value of x.

Another notation – sometimes referred to as mapping notation – is based on the idea that the function f is the rule that maps x to $f(x)$ and is written $f{:}x \mapsto f(x)$. For each value of x in the domain, the corresponding unique value of y in the range is called the function value at x, or the image of x under f. The image of x may be written as $f(x)$ or as y. For example, for the function $f(x) = x^2$: '$f(3) = 9$', or 'if $x = 3$, then $y = 9$.'

Notation	Description in words
$f(x) = x^2$	The function f, in terms of x, is x^2; or, simply f of x equals x^2
$f{:}x \mapsto x^2$	The function f maps x to x^2
$f(3) = 9$	The value of the function f when $x = 3$ is 9; or, simply f of 3 equals 9
$f{:}3 \mapsto 9$	The image of 3 under the function f is 9

Table 1.3 Function notation

Example 1.10

Find the domain and range of the function $h{:}x \mapsto \dfrac{1}{x-2}$

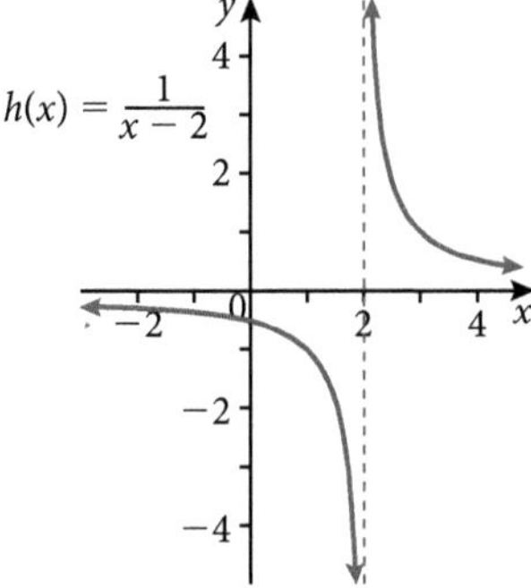

Figure 1.14 Diagram for Example 1.10

Solution

Using algebraic analysis: The function produces a real number for all x, except for $x = 2$ when division by zero occurs. Hence, $x = 2$ is the only real number not in the domain. Since the numerator of $\dfrac{1}{x-2}$ can never be zero, the value of y cannot be zero. Hence, $y = 0$ is the only real number not in the range.

Using graphical analysis: A horizontal scan shows a gap at $x = 2$ dividing the graph of the equation into two branches that both continue indefinitely with no other gaps as $x \to \pm\infty$. Both branches are **asymptotic** (approach but do not intersect) to the vertical line $x = 2$. This line is a **vertical asymptote** and is drawn as a dashed line (it is not part of the graph of the equation). A vertical scan reveals a gap at $y = 0$ (x-axis) with both branches of the graph continuing indefinitely with no other gaps as $y \to \pm\infty$. Both branches are also asymptotic to the x-axis. The x-axis is a **horizontal asymptote.**

Both approaches confirm that the domain and range for $h{:}x \mapsto \dfrac{1}{x-2}$ are:

domain: $\{x{:}x \in \mathbb{R}, x \neq 2\}$ or $x \in \,]-\infty, 2[\, \cup \,]2, \infty[$

range: $\{y{:}y \in \mathbb{R}, y \neq 0\}$ or $y \in \,]-\infty, 0[\, \cup \,]0, \infty[$

Example 1.11

Consider the function $f(x) = \sqrt{x+4}$

(a) Find:

(i) $f(7)$ (ii) $f(32)$ (iii) $f(-4)$

(b) Find the values of x for which f is undefined.

(c) State the domain and range of f.

Solution

(a) (i) $f(7) = \sqrt{7+4} = \sqrt{11} \approx 3.32$ (3 s.f.)

(ii) $f(32) = \sqrt{32+4} = \sqrt{36} = 6$

(iii) $f(-4) = \sqrt{-4+4} = \sqrt{0} = 0$

(b) $f(x)$ will be undefined (square root of a negative) when $x + 4 < 0$.

Therefore, $f(x)$ is undefined when $x < -4$.

(c) It follows from the result in (b) that the domain of f is $\{x{:}x \geqslant -4\}$.

The symbol $\sqrt{\ }$ stands for the **principal square root** that, by definition, can only give a result that is positive or zero. Therefore, the range of f is $\{y{:}y \geqslant 0\}$. The domain and range are confirmed by analysing the graph of the function.

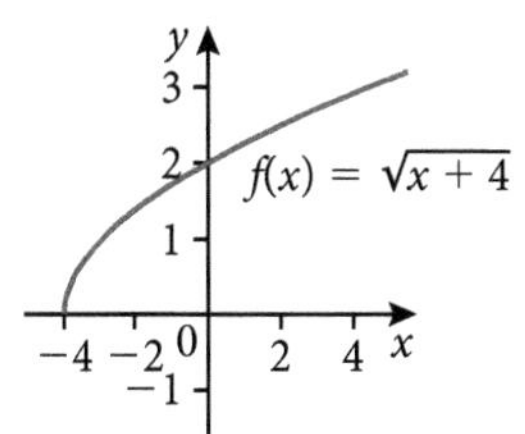

Figure 1.15 Graph for the solution to Example 1.11 (c)

As Example 1.12 illustrates, it is dangerous to completely trust graphs produced on a GDC without also doing some algebraic thinking. It is important to check that the graph shown is comprehensive (shows all important features), and that the graph agrees with algebraic analysis of the function, for example. where the function should be zero, positive, negative, undefined, or increasing/decreasing without bound.

Example 1.12

Find the domain and range of the function $f(x) = \dfrac{1}{\sqrt{9 - x^2}}$

Solution

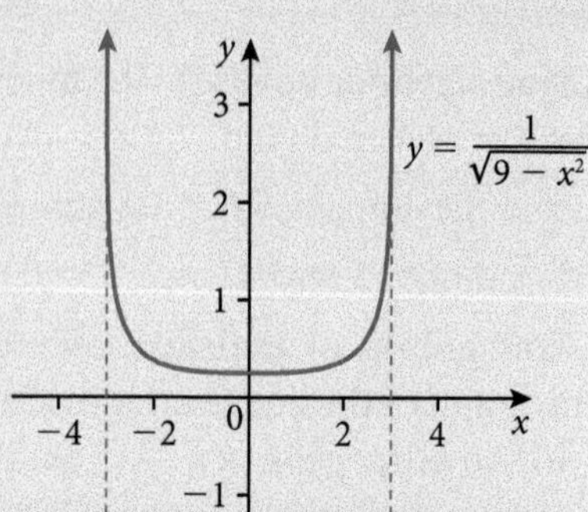

The graph of $y = \dfrac{1}{\sqrt{9 - x^2}}$ shown here, agrees with algebraic analysis indicating that the expression $\dfrac{1}{\sqrt{9 - x^2}}$ will be positive for all x, and is defined only for $-3 < x < 3$. Further analysis and tracing the graph reveals that $f(x)$ has a minimum at $\left(0, \dfrac{1}{3}\right)$. The graph on the GDC is misleading in that it appears to show that the function has a maximum value of approximately $y \approx 2.8037849$. Can this be correct? A lack of algebraic thinking and over-reliance on a GDC could easily lead to a mistake. The graph abruptly stops its curve upwards because of low screen resolution. Function values should get quite large for values of x a little less than 3, because the value of $\sqrt{9 - x^2}$ will be small, making the fraction $\dfrac{1}{\sqrt{9 - x^2}}$ large.

Using a GDC to make a table for $f(x)$ or evaluating the function for values of x very close to -3 or 3 confirms that as x approaches -3 or 3, y increases without bound – i.e. y goes to $+\infty$. Hence, $f(x)$ has vertical asymptotes of $x = -3$ and $x = 3$. This combination of graphical and algebraic analysis leads to the conclusion that the domain of $f(x)$ is $\{x : -3 < x < 3\}$, and the range of $f(x)$ is $\left\{y : y \geqslant \dfrac{1}{3}\right\}$

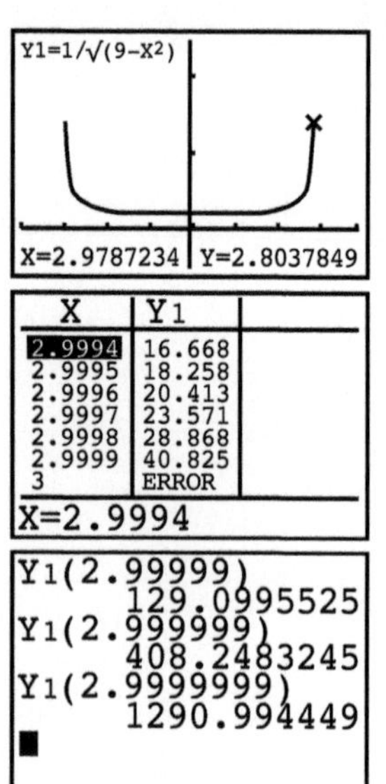

Figure 1.16 GDC screens for solution to Example 1.12

Exercise 1.2

1. **(i)** Match each equation to one of the graphs.

   (ii) State whether or not the equation represents any of the functions shown. Justify your answer. Assume that x is the independent variable and y is the dependent variable.

   (a) $y = 2x$ **(b)** $y = -3$ **(c)** $x - y = 2$

   (d) $x^2 + y^2 = 4$ **(e)** $y = 2 - x$ **(f)** $y = x^2 + 2$

   (g) $y^3 = x$ **(h)** $y = \dfrac{2}{x}$ **(i)** $x^2 + y = 2$

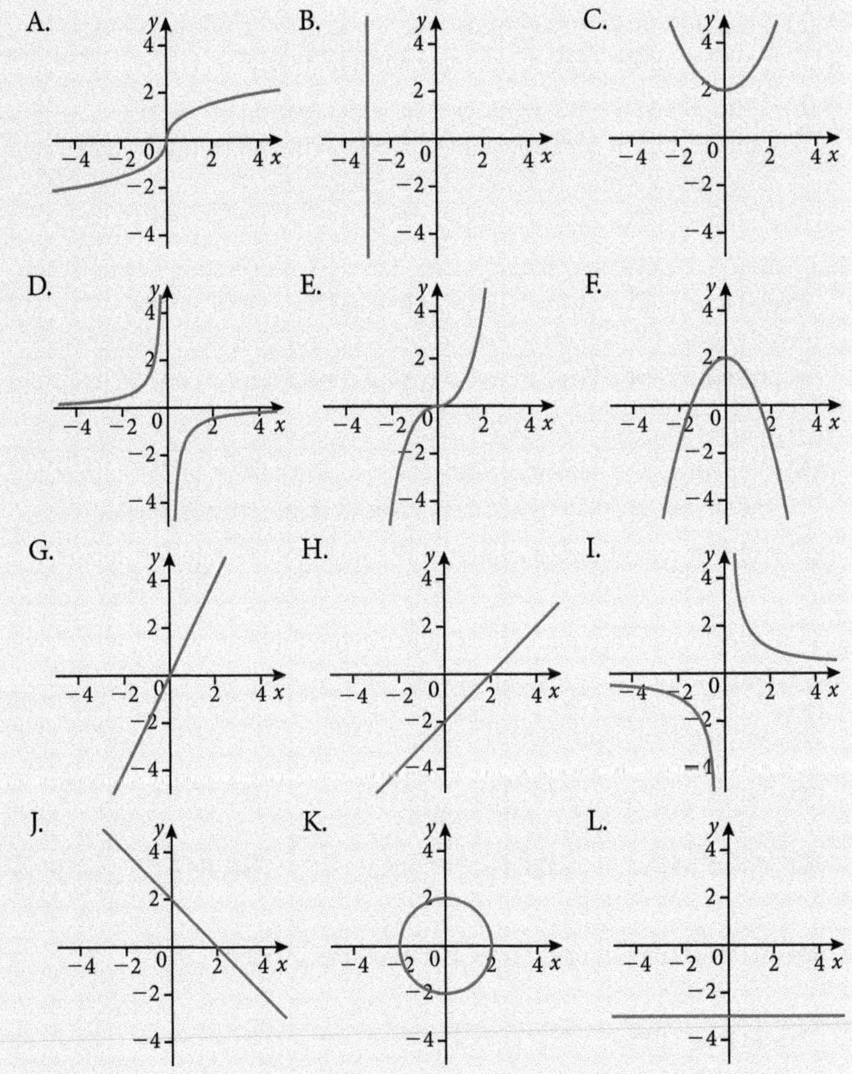

2. Express the area, A, of a circle as a function of its circumference, C.

3. Express the area, A, of an equilateral triangle as a function of the length, ℓ, of each of its sides.

4. A rectangular swimming pool with dimensions 12 metres by 18 metres is surrounded by a pavement of uniform width x metres. Find the area of the pavement, A, as a function of x.

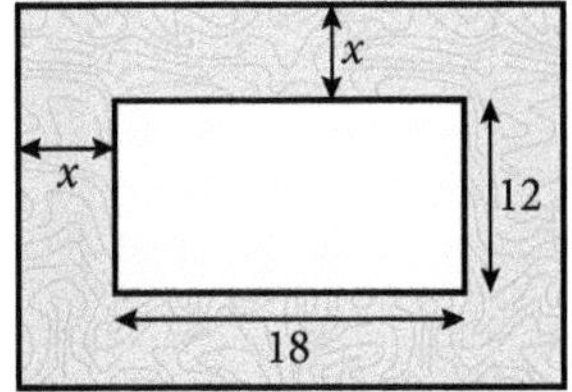

Figure 1.17 Diagram for question 4

5. In a right-angled isosceles triangle, the two equal sides have length x units and the hypotenuse has length h units. Write h as a function of x.

6. The pressure P (measured in kilopascals, kPa) for a particular sample of gas is directly proportional to the temperature T (measured in degrees kelvin, K) and inversely proportional to the volume V (measured in litres, L). With k representing the constant of proportionality, this relationship can be written in the form of the equation $P = k\frac{T}{V}$

(a) Find the constant of proportionality, k, if 150 L of gas exerts a pressure of 23.5 kPa at a temperature of 375K.

(b) Using the value of k from part **(a)** and assuming that the temperature is held constant at 375K, write the volume V as a function of pressure P for this sample of gas.

7. In physics, Hooke's law states that the force F (measured in newtons, N) needed to extend a spring by x units beyond its natural length is directly proportional to the extension x. Assume that the constant of proportionality is k (known as the spring constant for a particular spring).

(a) Write F as a function of x.

(b) A spring has a natural length of 12 cm and a force of 25 N stretches the spring to a length of 16 cm. Work out the spring constant k.

(c) What force is needed to stretch the spring to a length of 18 cm?

8. Find the domain of each of the following functions.

(a) $\{(-6.2, -7), (-1.5, -2), (0.7, 0), (3.2, 3), (3.8, 3)\}$

(b) Surface area of a sphere: $S = 4\pi r^2$

(c) $f(x) = \frac{2}{5}x - 7$ **(d)** $h: x \mapsto x^2 - 4$ **(e)** $g(t) = \sqrt{3 - t}$

(f) $h(t) = \sqrt[3]{t}$ **(g)** $f: x \mapsto \frac{6}{x^2 - 9}$ **(h)** $f(x) = \sqrt{\frac{1}{x^2} - 1}$

9. Do all linear equations represent a function? Explain.

10. Consider the function $h(x) = \sqrt{x - 4}$

(a) Find: **(i)** $h(21)$ **(ii)** $h(53)$ **(iii)** $h(4)$

(b) Find the values of x for which h is undefined.

(c) State the domain and range of h.

11. For each function below:

(i) find the domain and range of the function

(ii) sketch a comprehensive graph of the function, clearly indicating any intercepts or asymptotes.

(a) $f: x \mapsto \frac{1}{x - 5}$ **(b)** $g(x) = \frac{1}{\sqrt{x^2 - 9}}$ **(c)** $h(x) = \frac{2x - 1}{x + 2}$

(d) $p: x \mapsto \sqrt{5 - 2x^2}$ **(e)** $f(x) = \frac{1}{x} - 4$

1.3 Composite functions

Composition of functions

Consider the function in Example 1.11, $f(x) = \sqrt{x+4}$. When we evaluate $f(x)$ for a certain value of x in the domain, for example, $x = 5$, it is necessary to perform computations in two separate steps in a certain order.

$f(5) = \sqrt{5+4} \Rightarrow f(5) = \sqrt{9}$ Step 1: compute the sum of 5 + 4

$\Rightarrow f(5) = 3$ Step 2: compute the square root of 9

The **argument** of a function is the variable or expression on which a function operates. For example, the argument of $f(x) = x^3$ is x, the argument of $g(x) = \sqrt{x-3}$ is $x - 3$, and the argument of $y = 10^{2x}$ is $2x$.

Given that the function has two separate evaluation steps, $f(x)$ can be seen as a combination of two simpler functions that are performed in a specified order. According to how $f(x)$ is evaluated, the simpler function to be performed first is the rule of adding 4 and the second is the rule of taking the square root. If $h(x) = x + 4$ and $g(x) = \sqrt{x}$, then we can create (compose) the function $f(x)$ from a combination of $h(x)$ and $g(x)$ as follows:

$f(x) = g(h(x))$

$= g(x + 4)$ Step 1: substitute $x + 4$ for $h(x)$ making $x + 4$ the argument of $g(x)$

$= \sqrt{x+4}$ Step 2: apply the function $g(x)$ on the argument $x + 4$

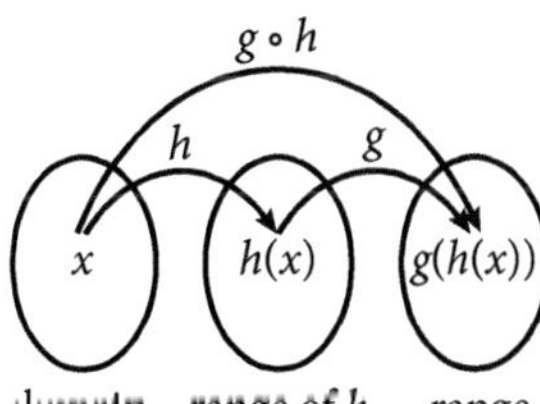

Figure 1.18 Mapping for composite function $g(h(x))$

We obtain the rule $\sqrt{x+4}$ by first applying the rule $x + 4$ and then applying the rule $\sqrt{x}$. A function that is obtained from simpler functions by applying one after another in this way is called a **composite function**. $f(x) = \sqrt{x+4}$ is the **composition** of $h(x) = x + 4$ followed by $g(x) = \sqrt{x}$. In other words, f is obtained by substituting h into g, and can be denoted in function notation by $g(h(x))$ – read 'g of h of x.'

The composition of two functions, g and h, such that h is applied first and g second is given by $(g \circ h)(x) = g(h(x))$. The domain of the composite function $g \circ h$ is the set of all x in the domain of h such that $h(x)$ is in the domain of g.

Start with a number x in the domain of h and find its image $h(x)$. If this number $h(x)$ is in the domain of g, we then compute the value of $g(h(x))$. The resulting composite function is denoted as $(g \circ h(x))$. See Figure 1.18.

Example 1.13

If $f(x) = 3x$ and $g(x) = 2x - 6$, find:

(a) (i) $(f \circ g)(5)$ (ii) Express $(f \circ g)(x)$ as a single function rule (expression).

(b) (i) $(g \circ f)(5)$ (ii) Express $(g \circ f)(x)$ as a single function rule (expression).

(c) (i) $(g \circ g)(5)$ (ii) Express $(g \circ g)(x)$ as a single function rule (expression).

The notations $(g \circ h)(x)$ and $g(h(x))$ are both commonly used to denote a composite function where h is applied first then followed by applying g. Since you are reading this from left to right, it is easy to apply the functions in the incorrect order. It may be helpful to read $g \circ h$ as 'g following h' to highlight the order in which the functions are applied. Also, in either notation, $(g \circ h)(x)$ or $g(h(x))$, the function applied first is closest to the variable x.

Solution

(a) (i) $(f \circ g)(5) = f(g(5)) = f(2 \cdot 5 - 6) = f(4) = 3 \cdot 4 = 12$

(ii) $(f \circ g)(x) = f(g(x)) = f(2x - 6) = 3(2x - 6) = 6x - 18$

Therefore, $(f \circ g)(x) = 6x - 18$

Check with result from (i): $(f \circ g)(5) = 6 \cdot 5 - 18 = 30 - 18 = 12$

(b) (i) $(g \circ f)(5) = g(f(5)) = g(3 \cdot 5) = g(15) = 2 \cdot 15 - 6 = 24$

(ii) $(g \circ f)(x) = g(f(x)) = g(3x) = 2(3x) - 6 = 6x - 6$

Therefore, $(g \circ f)(x) = 6x - 6$

Check with result from (i):$(g \circ f)(5) = 6 \cdot 5 - 6 = 30 - 6 = 24$

(c) (i) $(g \circ g)(5) = g(g(5)) = g(2 \cdot 5 - 6) = g(4) = 2 \cdot 4 - 6 = 2$

(ii) $(g \circ g)(x) = g(g(x)) = g(2x - 6) = 2(2x - 6) - 6 = 4x - 18$

Therefore, $(g \circ g)(x) = 4x - 18$

Check with result from (i): $(g \circ g)(5) = 4 \cdot 5 - 18 = 20 - 18 = 2$

It is important to notice that in parts (a)(ii) and (b)(ii) in Example 1.13, $f \circ g$ is not equal to $g \circ f$. At the start of this section, it was shown how the two functions $h(x) = x + 4$ and $g(x) = \sqrt{x}$ could be combined into the composite function $(g \circ h)(x)$ to create the single function $f(x) = \sqrt{x + 4}$. However, the composite function $(h \circ g)(x)$ (the functions applied in reverse order) creates a different function: $(h \circ g)(x) = h(g(x)) = h(\sqrt{x}) = \sqrt{x} + 4$. Since, $\sqrt{x} + 4 \neq \sqrt{x + 4}$ then $f \circ g$ is not equal to $g \circ f$. Is it always true that $f \circ g \neq g \circ f$? The next example will answer that question.

Example 1.14

Given $f{:}x \mapsto 3x - 6$ and $g{:}x \mapsto \frac{1}{3}x + 2$, find:

(a) $(f \circ g)(x)$ (b) $(g \circ f)(x)$

Solution

(a) $(f \circ g)(x) = f(g(x)) = f\left(\frac{1}{3}x + 2\right) = 3\left(\frac{1}{3}x + 2\right) - 6 = x + 6 - 6 = x$

(b) $(g \circ f)(x) = g(f(x)) = g(3x - 6) = \frac{1}{3}(3x - 6) + 2 = x - 2 + 2 = x$

Example 1.14 shows that it is possible for $f \circ g$ to be equal to $g \circ f$. You will learn in the next section that this occurs in some cases where there is a special relationship between the pair of functions. However, in general $f \circ g \neq g \circ f$.

Decomposing a composite function

In Examples 1.13 and 1.14, we created a single function by forming the composite of two functions. As with the function $f(x) = \sqrt{x+4}$, it is also important for us to be able to identify two functions that make up a composite function, in other words, to decompose a function into two simpler functions. When we are doing this it is very useful to think of the function that is applied first as the inside function, and the function that is applied second as the outside function. In the function $f(x) = \sqrt{x+4}$, the inside function is $h(x) = x + 4$ and the outside function is $g(x) = \sqrt{x}$.

Example 1.15

Each of these functions is a composite function of the form $(f \circ g)(x)$. For each, find the two component functions f and g.

(a) $h{:}x \mapsto \dfrac{1}{x+3}$ (b) $k{:}x \mapsto 2^{4x+1}$ (c) $p(x) = \sqrt[3]{x^2 - 4}$

Solution

(a) When we evaluate the function $h(x)$ for a certain x in the domain, we first evaluate the expression $x + 3$, and then evaluate the expression $\frac{1}{x}$. Hence, the inside function (applied first) is $y = x + 3$, and the outside function (applied second) is $y = \frac{1}{x}$. So the two component functions are $g(x) = x + 3$ and $f(x) = \frac{1}{x}$

(b) Evaluating $k(x)$ requires us to first evaluate the expression $4x + 1$, and then evaluate the expression 2^x. Hence, the inside function is $y = 4x + 1$, and the outside function is $y = 2^x$. The two composite functions are $g(x) = 4x + 1$ and $f(x) = 2^x$.

(c) Evaluating $p(x)$ requires us to perform three separate evaluation steps: squaring a number, subtracting four, and then taking the cube root. Hence, it is possible to decompose $p(x)$ into three component functions: $h(x) = x^2$, $g(x) = x - 4$ and $f(x) = \sqrt[3]{x}$. However, for our purposes it is best to decompose the composite function into only two component functions: $g(x) = x^2 - 4$, and $f(x) = \sqrt[3]{x}$.

Finding the domain of a composite function

It is important to note that in order for a value of x to be in the domain of the composite function $g \circ h$, two conditions must be met: (1) x must be in the domain of h, and (2) $h(x)$ must be in the domain of g. Likewise, it is also worth noting that $g(h(x))$ is in the range of $g \circ h$ only if x is in the domain of $g \circ h$. The next example illustrates these points – and also that, in general, the domains of $g \circ h$ and $h \circ g$ are not the same.

Example 1.16

Let $g(x) = x^2 - 4$ and $h(x) = \sqrt{x}$. Find:

(a) $(g \circ h)(x)$ and its domain and range

(b) $(h \circ g)(x)$ and its domain and range.

Solution

First, establish the domain and range for both g and h. For $g(x) = x^2 - 4$, the domain is $x \in \mathbb{R}$ and the range is $y \geqslant -4$. For $h(x) = \sqrt{x}$, the domain is $x \geqslant 0$ and the range is $y \geqslant 0$.

(a) $(g \circ h)(x) = g(h(x))$

$= g(\sqrt{x})$ To be in the domain of $g \circ h$, $\sqrt{x}$ must be defined for $x \Rightarrow x \geqslant 0$

$= (\sqrt{x^2}) - 4$ Therefore, the domain of $g \circ h$ is $x \geqslant 0$

$= x - 4$ Since $x \geqslant 0$, then the range for $y = x - 4$ is $y \geqslant -4$.

Therefore, $(g \circ h)(x) = x - 4$, and its domain is $x \geqslant 0$, and its range is $y \geqslant -4$

(b) $(h \circ g)(x) = h(g(x))$ $g(x) = x^2 - 4$ must be in the domain of $h \Rightarrow x^2 - 4 \geqslant 0 \Rightarrow x^2 \geqslant 4$

$= h(x^2 - 4)$ Therefore, the domain of $h \circ g$ is $x \leqslant -2$ or $x \geqslant 2$ and

$= \sqrt{x^2 - 4}$ with $x \leqslant -2$ or $x \geqslant 2$, the range for $y = \sqrt{x^2 - 4}$ is $y \geqslant 0$

Therefore, $(h \circ g)(x) = \sqrt{x^2 - 4}$, and its domain is $x \leqslant -2$ or $x \geqslant 2$, and its range is $y \geqslant 0$

Exercise 1.3

1. Let $f(x) = 2x$ and $g(x) = \dfrac{1}{x - 3}$, $x \neq 0$

Find the value of:

(a) $(f \circ g)(5)$ **(b)** $(g \circ f)(5)$

Find the function rule (expression) for:

(c) $(f \circ g)(x)$ **(d)** $(g \circ f)(x)$

2. Let $f: x \mapsto 2x - 3$ and $g: x \mapsto 2 - x^2$

Evaluate:

(a) $(f \circ g)(0)$ **(b)** $(g \circ f)(0)$ **(c)** $(f \circ f)(4)$

(d) $(g \circ g)(-3)$ **(e)** $(f \circ g)(-1)$ **(f)** $(g \circ f)(-3)$

Find the expression for:

(g) $(f \circ g)(x)$ **(h)** $(g \circ f)(x)$ **(i)** $(f \circ f)(x)$ **(j)** $(g \circ g)(x)$

3. For each pair of functions, find $(f \circ g)(x)$ and $(g \circ f)(x)$ and state the domain for each.

(a) $f(x) = 4x - 1, g(x) = 2 + 3x$ **(b)** $f(x) = x^2 + 1, g(x) = -2x$

(c) $f(x) = \sqrt{x+1}, g(x) = 1 + x^2$ **(d)** $f(x) = \frac{2}{x+4}, g(x) = x - 1$

(e) $f(x) = 3x + 5, g(x) = \frac{x-5}{3}$ **(f)** $f(x) = 2 - x^3, g(x) = \sqrt[3]{1-x^2}$

(g) $f(x) = \frac{2x}{4-x}, g(x) = \frac{1}{x^2}$

(h) $f(x) = \frac{2}{x+3} - 3, g(x) = \frac{2}{x+3} - 3$

(i) $f(x) = \frac{x}{x-1}, g(x) = x^2 - 1$

4. Let $g(x) = \sqrt{x-1}$ and $h(x) = 10 - x^2$. Find:

(a) $(g \circ h)(x)$ and its domain and range

(b) $(h \circ g)(x)$ and its domain and range.

5. Let $f(x) = \frac{1}{x}$ and $g(x) = 10 - x^2$. Find:

(a) $(f \circ g)(x)$ and its domain and range

(b) $(g \circ f)(x)$ and its domain and range.

6. Determine functions g and h so that $f(x) = g(h(x))$

(a) $f(x) = (x+3)^2$ **(b)** $f(x) = \sqrt{x-5}$ **(c)** $f(x) = 7 - \sqrt{x}$

(d) $f(x) = \frac{1}{x+3}$ **(e)** $f(x) = 10^{x+1}$ **(f)** $f(x) = \sqrt[3]{x-9}$

(g) $f(x) = |x^2 - 9|$ **(h)** $f(x) = \frac{1}{\sqrt{x-5}}$

7. Find the domain for:

(i) the function f **(ii)** the function g **(iii)** the composite function $f \circ g$

(a) $f(x) = \sqrt{x}, g(x) = x^2 + 1$ **(b)** $f(x) = \frac{1}{x}, g(x) = x + 3$

(c) $f(x) = \frac{3}{x^2-1}, g(x) = x + 1$ **(d)** $f(x) = 2x + 3, g(x) = \frac{x}{2}$

1.4 Inverse functions

Pairs of inverse functions

If we choose a number and cube it (raise it to the power of 3), and then take the cube root of the result, the answer is the original number. The same result would occur if we applied the two rules in the reverse order. That is, first take the cube root of a number and then cube the result; again, the answer is the original number.

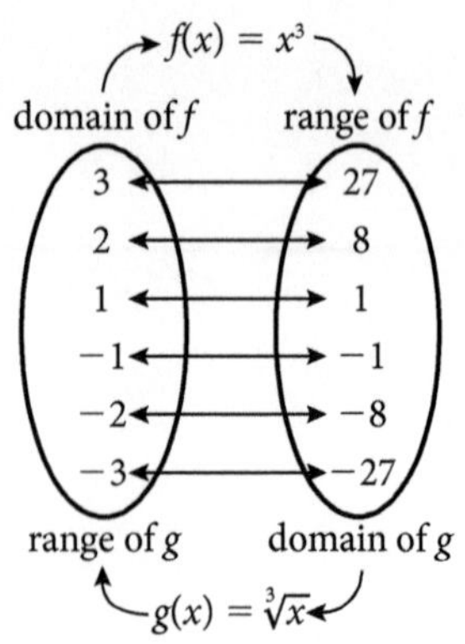

Figure 1.19 A mapping diagram for the cubing and cube root functions

Write each of these rules as a function with function notation. Write the cubing function as $f(x) = x^3$, and the cube root function as $g(x) = \sqrt[3]{x}$. Now, using what we know about composite functions and operations with radicals and powers, we can write what was described above in symbolic form.

Cube a number and then take the cube root of the result:

$$g(f(x)) = \sqrt[3]{x^3} = (x^3)^{\frac{1}{3}} = x^1 = x$$

For example, $g(f(-2)) = \sqrt[3]{(-2)^3} = \sqrt[3]{-8} = -2$

Take the cube root of a number and then cube the result:

$$f(g(x)) = \left(\sqrt[3]{x}\right)^3 = \left(x^{\frac{1}{3}}\right)^3 = x^1 = x$$

For example, $f(g(27)) = \left(\sqrt[3]{27}\right)^3 = (3)^3 = 27$

Because function g has this reverse (inverse) effect on function f, we call function g the **inverse** of function f. Function f has the same inverse effect on function g $[g(27) = 3$ and then $f(3) = 27]$, making f the inverse function of g. The functions f and g are inverses of each other. The cubing and cube root functions are an example of a pair of **inverse functions**. The mapping diagram for functions f and g (Figure 1.19) illustrates the relationship for a pair of inverse functions where the domain of one is the range for the other.

The composite of two inverse functions is the function that always produces the same number that was first substituted into the function. This function is called the **identity function** because it assigns each number in its domain to itself and is denoted by $I(x) = x$.

You should already be familiar with pairs of **inverse operations**. Addition and subtraction are inverse operations. For example, the rule of adding six $(x + 6)$, and the rule of subtracting six $(x - 6)$, undo each other. Accordingly, the functions $f(x) = x + 6$ and $g(x) = x - 6$ are a pair of inverse functions. Multiplication and division are also inverse operations.

Do not mistake the -1 in the notation f^{-1} for a power. It is not a power. If a superscript of -1 is applied to the name of a function, as in f^{-1} or $\sin^{-1}$, then it denotes the function that is the inverse of the named function (e.g. f or sin). If a superscript of -1 is applied to an expression, as in 7^{-1} or $(2x + 5)^{-1}$, then it is a power and denotes the reciprocal of the expression.

If f and g are two functions such that $(f \circ g)(x) = x$ for every x in the domain of g and $(g \circ f)(x) = x$ for every x in the domain of f, then the function g is the **inverse** of the function f. The notation to indicate the function that is the inverse of function f is f^{-1}. Therefore, $(f \circ f^{-1})(x) = x$ and $(f^{-1} \circ f)(x) = x$

The domain of f must be equal to the range of f^{-1}, and the range of f must be equal to the domain of f^{-1}.

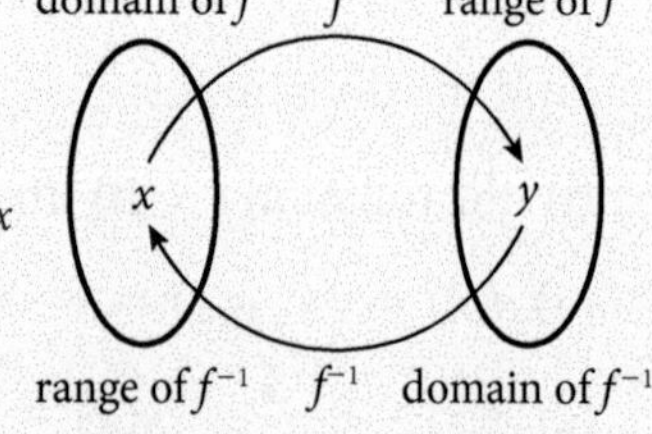

Remember that the notation $(f \circ g)(x)$ is equivalent to $f(g(x))$. It follows from the definition that if g is the inverse of f, then it must also be true that f is the inverse of g.

In general, the functions $f(x)$ and $g(x)$ are a pair of inverse functions if the following two statements are true:

1 $g(f(x)) = x$ for all x in the domain of f

2 $f(g(x)) = x$ for all x in the domain of g

For a pair of inverse functions, f and g, the composite functions $f(g(x))$ and $g(f(x))$ are equal. Remember that this is not generally true for an arbitrary pair of functions.

Example 1.17

Given $h(x) = \dfrac{x - 3}{2}$ and $p(x) = 2x + 3$, show that h and p are inverse functions.

Solution

Since the domain and range of both $h(x)$ and $p(x)$ is the set of all real numbers, then:

For any real number x, $p(h(x)) = p\left(\frac{x-3}{2}\right) = 2\left(\frac{x-3}{2}\right) + 3$

$= x - 3 + 3 = x$

For any real number x, $h(p(x)) = h(2x+3) = \frac{(2x+3)-3}{2} = \frac{2x}{2} = x$

Since $p(h(x)) = h(p(x)) = x$ then h and p are a pair of inverse functions.

It is clear that both $f(x) = x^3$ and $g(x) = \sqrt[3]{x}$ satisfy the definition of a function because for both f and g every number in its domain determines exactly one number in its range. Since they are a pair of inverse functions then the reverse is also true for both; that is, every number in its range is determined by exactly one number in its domain. Such a function is called a **one-to-one function**. The phrase one-to-one is appropriate because each value in the domain corresponds to exactly **one** value in the range, and each value in the range corresponds to exactly **one** value in the domain.

A function is **one-to-one** if each element y in the range is the image of exactly one element x in the domain.

The existence of an inverse function

Determining whether a function is one-to-one is very useful because the inverse of a one-to-one function will also be a function. Analysing the graph of a function is the most effective way to determine if a function is one-to-one. Let's look at the graph of the one-to-one function $f(x) = x^3$ shown in Figure 1.20. It is clear that as the values of x increase over the domain (from $-\infty$ to ∞), the function values are always increasing. A function that is always increasing, or always decreasing, throughout its domain is one-to-one and has an inverse function.

A function that is not one-to-one (always increasing or always decreasing) can be made so by restricting its domain.

The function $f(x) = x^2$ (Figure 1.21) is not one-to-one for all real numbers. However, the function $g(x) = x^2$ with domain $x \geqslant 0$ (Figure 1.22) is always increasing (one-to-one), and the function $h(x) = x^2$ with domain $x \leqslant 0$ (Figure 1.23) is always decreasing (one-to-one).

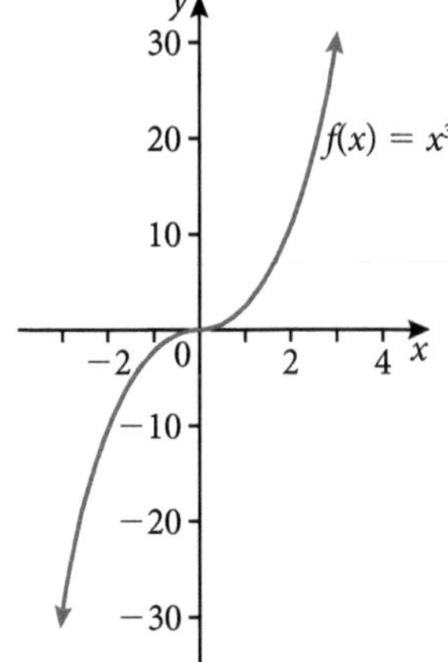

Figure 1.20 Graph of $f(x) = x^3$, which is increasing as x goes from $-\infty$ to ∞

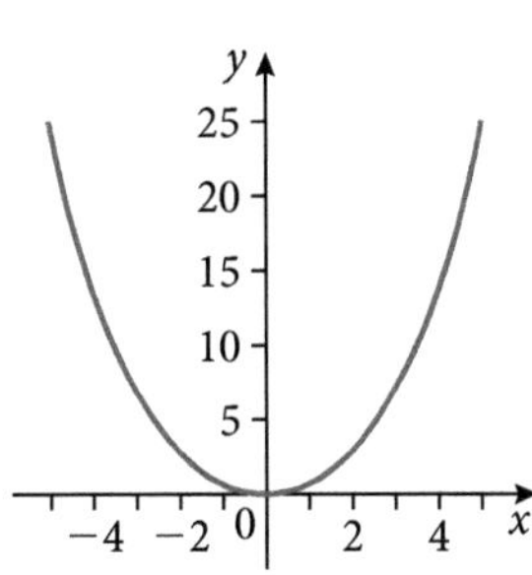

Figure 1.21 $f(x) = x^2$

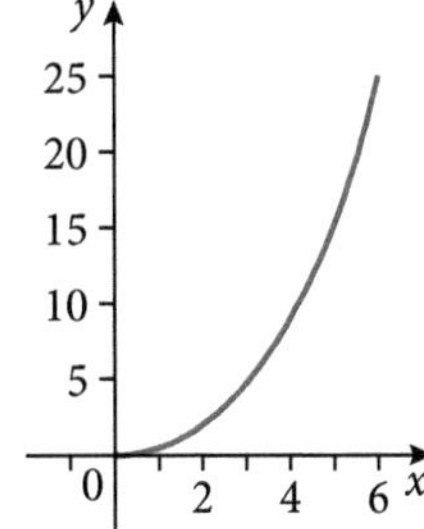

Figure 1.22 $g(x) = x^2, x \geqslant 0$

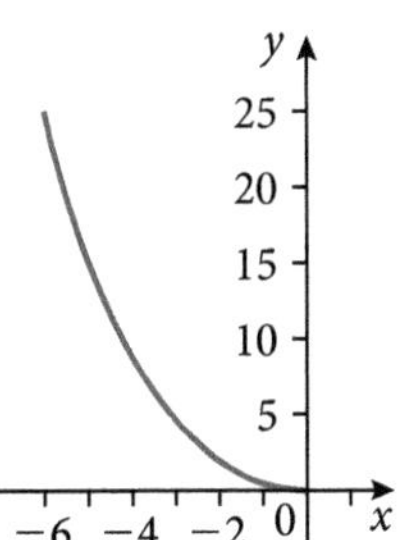

Figure 1.23 $h(x) = x^2, x \leqslant 0$

If a function f is always increasing or always decreasing in its domain (i.e. it is monotonic), then f has an inverse f^{-1}.

No horizontal line can pass through the graph of a one-to-one function at more than one point.

A function for which at least one element y in the range is the image of more than one element x in the domain is called a **many-to-one function**. Examples of many-to-one functions that we have already encountered are $y = x^2, x \in \mathbb{R}$ and $y = |x|, x \in \mathbb{R}$. As Figure 1.24 illustrates for $y = |x|$, a horizontal line exists that intersects a many-to-one function at more than one point. Thus, the inverse of a many-to-one function will not be a function.

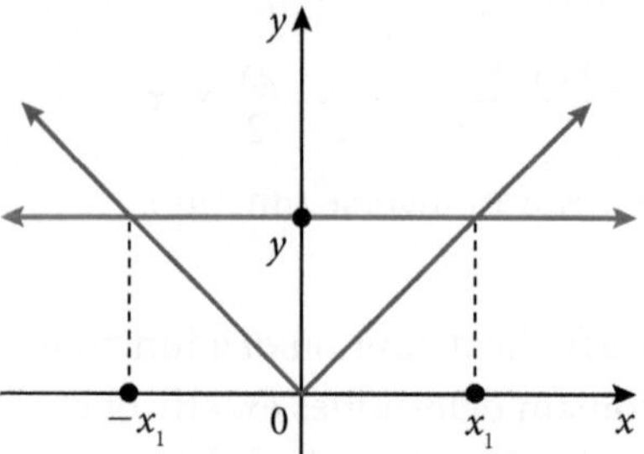

Figure 1.24 Graph of $y = |x|$; an example of a many-to-one function

Finding the inverse of a function

Example 1.18

The function f is defined for $x \in \mathbb{R}$ by $f(x) = 4x - 8$

(a) Determine if f has an inverse f^{-1}. If not, restrict the domain of f in order to find an inverse function f^{-1}

(b) Verify the result by showing that $(f \circ f^{-1})(x) = x$ and $(f^{-1} \circ f)(x) = x$

(c) Graph f and its inverse function f^{-1} on the same set of axes.

Solution

(a) Recognise that f is an increasing function for $(-\infty, \infty)$ because the graph of $f(x) = 4x - 8$ is a straight line with a constant slope of 4. Therefore, f is a one-to-one function and it has an inverse f^{-1}

(b) To find the equation for f^{-1}, start by switching the domain (x) and range (y) since the domain of f becomes the range of f^{-1} and the range of f becomes the domain of f^{-1}, as stated in the definition. Also, recall that $y = f(x)$.

$f(x) = 4x - 8$

$y = 4x - 8$	write $y = f(x)$
$x = 4y - 8$	interchange x and y (switch the domain and range)
$4y = x + 8$	solve for y (dependent variable) in terms of x (independent variable)

$y = \frac{1}{4}x + 2$

$f^{-1}(x) = \frac{1}{4}x + 2$ resulting equation is $y = f^{-1}(x)$

Verify that f and f^{-1} are inverses by showing that $f(f^{-1}(x)) = x$ and $f^{-1}(f(x)) = x$

$f\left(\frac{1}{4}x + 2\right) = 4\left(\frac{1}{4}x + 2\right) - 8 = x + 8 - 8 = x$

$f^{-1}(4x - 8) = \frac{1}{4}(4x - 8) + 2 = x - 2 + 2 = x$

This confirms that $y = 4x - 8$ and $y = \frac{1}{4}x + 2$ are inverses of each other. Here is a graph of this pair of inverse functions.

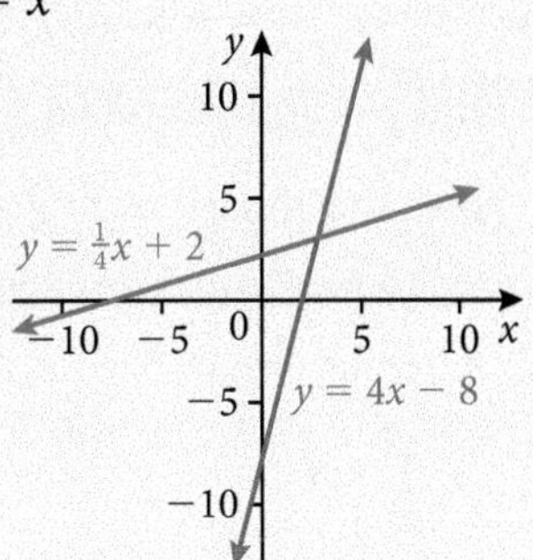

Online

Graph the inverse of an entered function and make a conjecture for the inverse function.

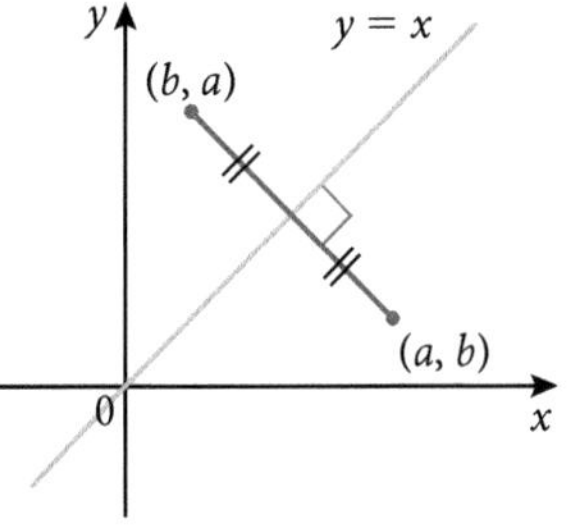

Figure 1.25 The point (b, a) is a reflection about the line $y = x$ of the point (a, b)

The method of interchanging domain (x) and range (y) to find the inverse function used in Example 1.18 also gives us a way for obtaining the graph of f^{-1} from the graph of f. Given the reversing effect that a pair of inverse functions have on each other, if $f(a) = b$ then $f^{-1}(b) = a$. Hence, if the ordered pair (a, b) is a point on the graph of $y = f(x)$, then the reversed ordered pair (b, a) must be on the graph of $y = f^{-1}(x)$. Figure 1.25 shows that the point (b, a) can be found by reflecting the point (a, b) about the line $y = x$. Therefore, the following statement can be made about the graphs of a pair of inverse functions.

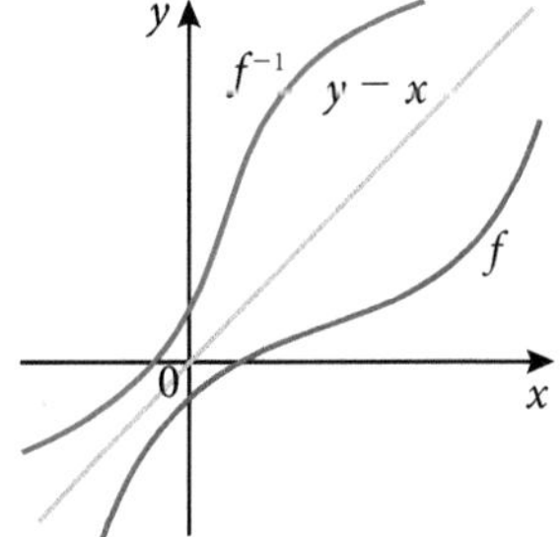

Figure 1.26 Graphs of f and f^{-1} are symmetric about the line $y = x$

The graph of f^{-1} is a reflection of the graph of f about the line $y = x$.

Example 1.19

Consider the function $f: x \mapsto \sqrt{x + 3}, x \geqslant -3$

(a) Determine the inverse function f^{-1}

(b) Find the domain of f^{-1}

Solution

(a) Following the steps for finding the inverse of a function gives:

$y = \sqrt{x + 3}$	replace $f(x)$ with y
$y^2 = x + 3$	solve for x in terms of y; squaring both sides
$x = y^2 - 3$	solve for x
$y = x^2 - 3$	interchange x and y
Therefore, $f^{-1}: x \mapsto x^2 - 3$	replace y with $f^{-1}(x)$

To find the inverse of a function f:

1. Determine if the function is one-to-one.
2. Replace $f(x)$ with y.
3. Solve for x in terms of y.
4. Interchange x and y.
5. Replace y with $f^{-1}(x)$.
6. The domain of f^{-1} is equal to the range of f, and the range of f^{-1} is equal to the domain of f.

(b) The domain explicitly defined for f is $x \geqslant -3$ and since the $\sqrt{\ }$ symbol stands for the principal square root (positive), then the range of f is all positive real numbers, i.e. $y \geqslant 0$. The domain of f^{-1} is equal to the range of f, therefore the domain of f^{-1} is $x \geqslant 0$.

Example 1.20

Consider the functions $f(x) = 2(x + 4)$ and $g(x) = \dfrac{1 - x}{3}$

(a) Find g^{-1} and state its domain and range.

(b) Solve the equation $(f \circ g^{-1})(x) = 2$

Solution

(a) $y = \dfrac{1 - x}{3}$ replace $f(x)$ with y

$x = \dfrac{1 - y}{3}$ interchange x and y

$3x = 1 - y$ solve for y

$y = -3x + 1$ solved for y

Therefore, $g^{-1}(x) = -3x + 1$ replace y with $g^{-1}(x)$

g is a linear function and its domain is $x \in \mathbb{R}$ and its range is $y \in \mathbb{R}$; therefore, for g^{-1} the domain is $x \in \mathbb{R}$ and the range is $y \in \mathbb{R}$.

(b) $(f \circ g^{-1})(x) = f(g^{-1}(x)) = f(-3x + 1) = 2$

$$2[(-3x + 1) + 4] = 2$$

$$-6x + 2 + 8 = 2$$

$$-6x = -8$$

$$x = \frac{4}{3}$$

Exercise 1.4

In questions 1–4, assume that f is a one-to-one function.

1. **(a)** If $f(2) = -5$, then what is $f^{-1}(-5)$?

(b) If $f^{-1}(6) = 10$, then what is $f(10)$?

2. **(a)** If $f(-1) = 13$, then what is $f^{-1}(13)$?

(b) If $f^{-1}(b) = a$, then what is $f(a)$?

3. If $g(x) = 3x - 7$, then what is $g^{-1}(5)$?

4. If $h(x) = x^2 - 8x$, with $x \geqslant 4$, then what is $h^{-1}(-12)$?

5. For each pair of functions, show algebraically and graphically that f and g are inverse functions by:

(i) verifying that $(f \circ g)(x) = x$ and $(g \circ f)(x) = x$

(ii) sketching the graphs of f and g on the same set of axes with equal scales on the x-axis and y-axis.

Use your GDC to assist in making your sketches on paper.

(a) $f{:}x \mapsto x + 6$; $g{:}x \mapsto x - 6$

(b) $f{:}x \mapsto 4x$; $g{:}x \mapsto \dfrac{x}{4}$

(c) $f{:}x \mapsto 3x + 9$; $g{:}x \mapsto \dfrac{1}{3}x - 3$

(d) $f{:}x \mapsto \dfrac{1}{x}$; $g{:}x \mapsto \dfrac{1}{x}$

(e) $f{:}x \mapsto x^2 - 2, x \geqslant 0$; $g{:}x \mapsto \sqrt{x + 2}, x \geqslant -2$

(f) $f{:}x \mapsto 5 - 7x$; $g{:}x \mapsto \dfrac{5 - x}{7}$

(g) $f{:}x \mapsto \dfrac{1}{1 + x}$; $g{:}x \mapsto \dfrac{1 - x}{x}$

(h) $f{:}x \mapsto (6 - x)^{\frac{1}{2}}$; $g{:}x \mapsto 6 - x^2, x \geqslant 0$

(i) $f{:}x \mapsto x^2 - 2x + 3, x \geqslant 1$; $g{:}x \mapsto 1 + \sqrt{x - 2}, x \geqslant 2$

(j) $f{:}x \mapsto \sqrt[3]{\dfrac{x + 6}{2}}$; $g{:}x \mapsto 2x^3 - 6$

6. Find the inverse function f^{-1} and state its domain.

(a) $f(x) = 2x - 3$

(b) $f(x) = \dfrac{x + 7}{4}$

(c) $f(x) = \sqrt{x}$

(d) $f(x) = \dfrac{1}{x + 2}$

(e) $f(x) = 4 - x^2, x \geqslant 0$

(f) $f(x) = \sqrt{x - 5}$

(g) $f(x) = ax + b, a \neq 0$

(h) $f(x) = x^2 + 2x, x \geqslant -1$

(i) $f(x) = \dfrac{x^2 - 1}{x^2 + 1}, x \leqslant 0$

(j) $f(x) = x^3 + 1$

7. Use your GDC to graph the function $f(x) = \dfrac{2x}{1 + x^2}, x \in \mathbb{R}$. Find three intervals for which f is a one-to-one function (monotonic) and hence, will have an inverse f^{-1} on the interval. The union of all three intervals is all real numbers.

8. Use the functions $g(x) = x + 3$ and $h(x) = 2x - 4$ to find the indicated value or the indicated function.

(a) $(g^{-1} \circ h^{-1})(5)$

(b) $(h^{-1} \circ g^{-1})(9)$

(c) $(g^{-1} \circ g^{-1})(2)$

(d) $(h^{-1} \circ h^{-1})(2)$

(e) $g^{-1} \circ h^{-1}$

(f) $h^{-1} \circ g^{-1}$

(g) $(g \circ h)^{-1}$

(h) $(h \circ g)^{-1}$

1.5 Transformations of functions

When analysing the graph of a function, it is often convenient to express a function in the form $y = f(x)$. As we have done throughout this chapter, we can refer to a function such as $f(x) = x^2$ by the equation $y = x^2$.

Even when you use your GDC to sketch the graph of a function, it is helpful to know what to expect in terms of the location and shape of the graph – and even more so, if you're not allowed to use your GDC for a particular question. In this section, you will look at how certain changes to the equation of a function can affect, or **transform**, the location and shape of its graph. You will investigate three different types of **transformations** of functions: how the graph of a function can be **translated, reflected** and **stretched** (or **shrunk**). Studying graphical transformations will help you to sketch and visualise many different functions efficiently. You will also take a closer look at two specific functions: the absolute value function, $y = |x|$, and the reciprocal function, $y = \frac{1}{x}$

Graphs of common functions

It is important to be familiar with the location and shape of a certain set of common functions. For example, from our previous knowledge about linear equations, we can determine the location of the linear function $f(x) = ax + b$. We know that the graph of this function is a line whose slope is a and whose y-intercept is $(0, b)$.

There are other important basic functions with which you should be familiar, for example, logarithmic and exponential functions, but you will learn about these in later chapters.

The eight graphs in Figure 1.27 represent some of the most commonly used functions in algebra. You should be familiar with the characteristics of the graphs of these common functions. This will help you predict and analyse the graphs of more complicated functions that are derived from applying one or more transformations to these simple functions.

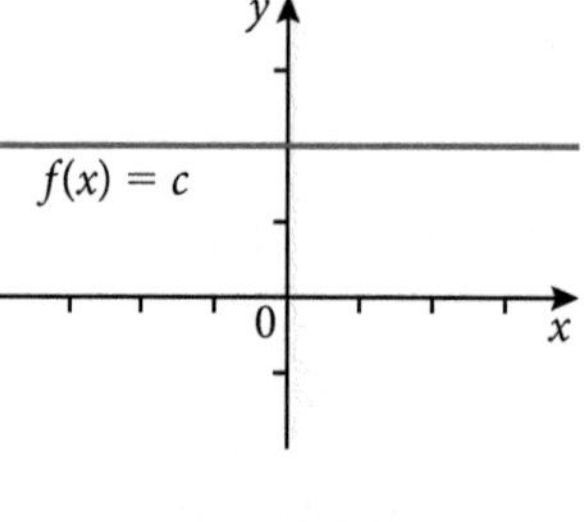

(a) Constant function

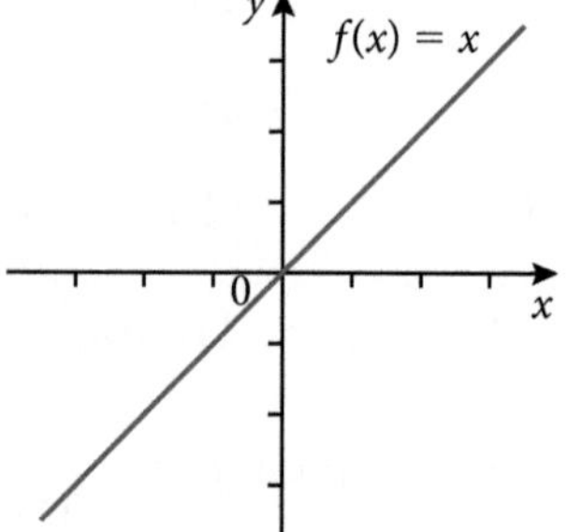

(b) Identity function

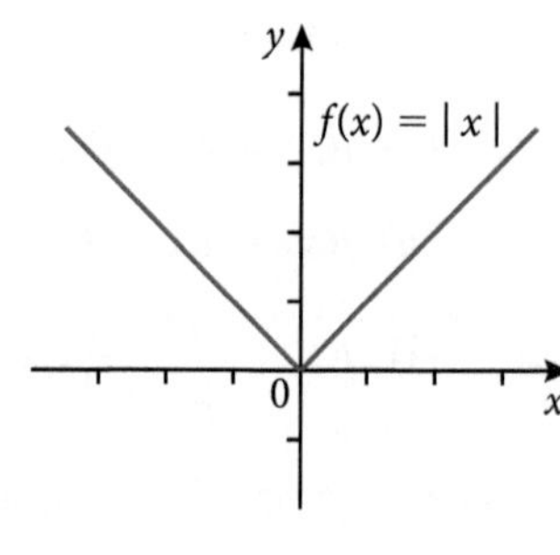

(c) Absolute value function

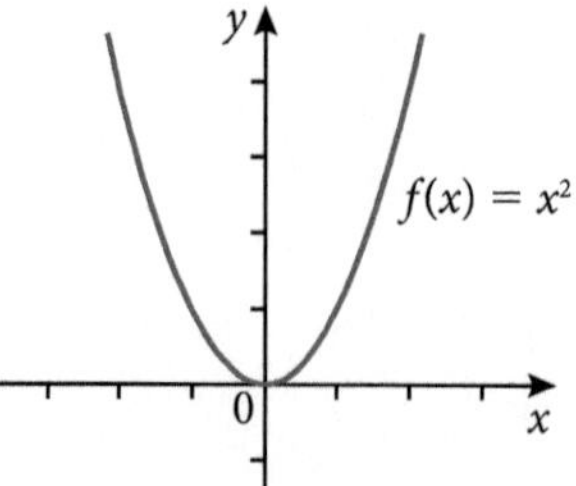

(d) Squaring function

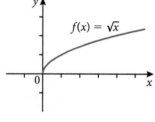

$f(x) = \sqrt{x}$

(e) Square root function

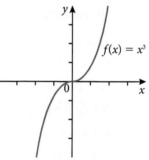

$f(x) = x^3$

(f) Cubing function

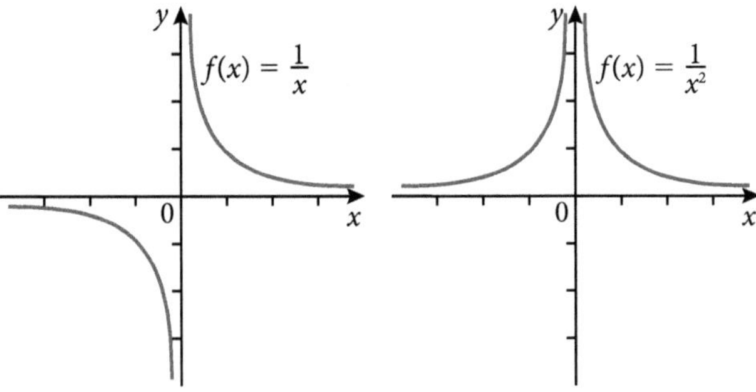

(g) Reciprocal function (h) Inverse square function

Figure 1.27 Graphs of common functions

We will see that many functions have graphs that are a transformation (translation, reflection or stretch), or a combination of transformations, of one of these common functions.

The word *inverse* can have different meanings in mathematics depending on the context. In section 1.4, 'inverse' is used to describe operations or functions that undo each other. However, 'inverse' is sometimes used to denote the **multiplicative inverse** (or **reciprocal**) of a number or function. This is how it is used in the names for the functions ***g*** and ***h*** shown above. The function in ***g*** is sometimes called the **reciprocal function**.

Vertical and horizontal translations

Use your GDC to graph each of these functions: $f(x) = x^2$, $g(x) = x^2 + 3$ and $h(x) = x^2 - 2$. How do the graphs of g and h compare with the graph of f? The graphs of g and h appear to have the same shape – it's only the location, or position, that has changed compared to f. Although the curves (parabolas) appear to be getting closer together, their vertical separation at every value of x is constant.

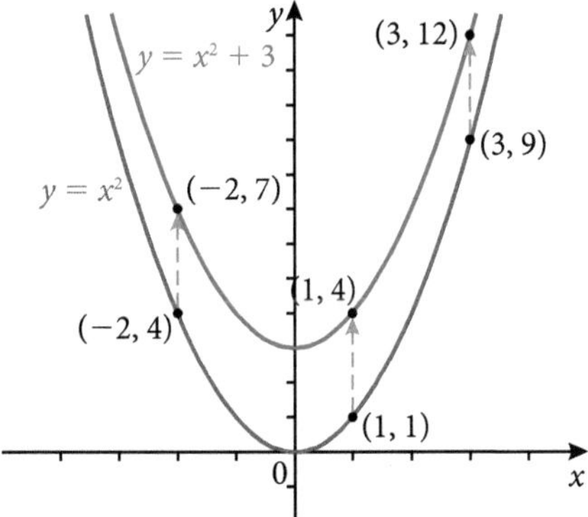

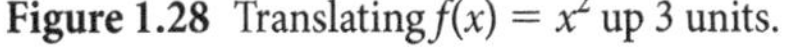

Figure 1.28 Translating $f(x) = x^2$ up 3 units.

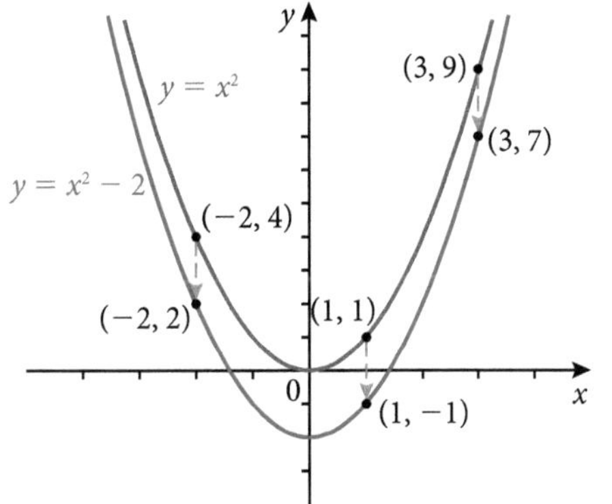

Figure 1.29 Translating $f(x) = x^2$ down 2 units.

Given $k > 0$:

- The graph of $y = f(x) + k$ is obtained by translating the graph of $y = f(x)$ up by k units.
- The graph of $y = f(x) - k$ is obtained by translating the graph of $y = f(x)$ down by k units.

As Figures 1.28 and 1.29 show, we can obtain the graph of $g(x) = x^2 + 3$ by translating (shifting) the graph of $f(x) = x^2$ **up** three units, and we can obtain the graph of $h(x) = x^2 - 2$ by translating the graph of $f(x) = x^2$ **down** two units.

Change function g to $g(x) = (x + 3)^2$ and change function h to $h(x) = (x - 2)^2$. Graph these two functions along with the original function $f(x) = x^2$ on your GDC. This time you can observe that the functions g and h can be obtained by a horizontal translation of f.

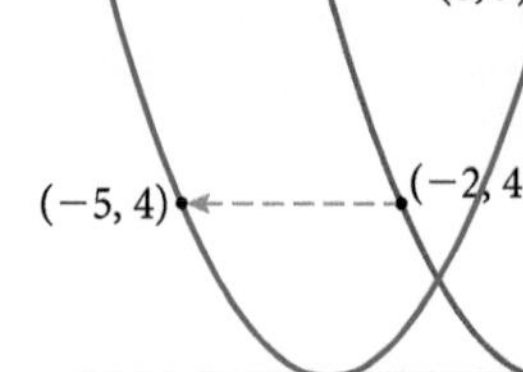

Figure 1.30 Translating $y = x^2$ left 3 units to produce $y = (x + 3)^2$

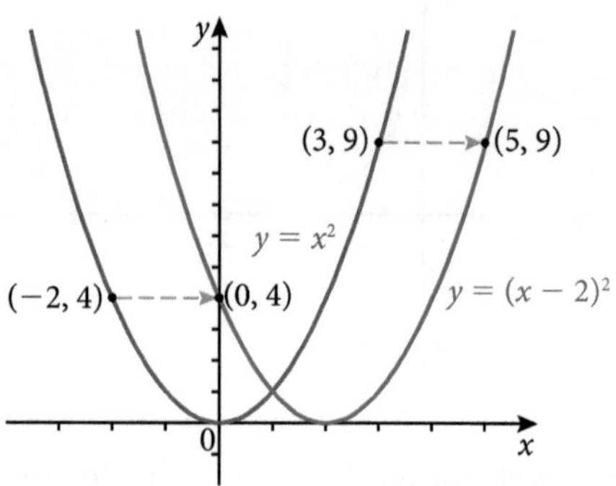

Figure 1.31 Translating $y = x^2$ right 3 units the graph of to produce the graph of $y = (x - 2)^2$

Given $k > 0$:

- The graph of $y = f(x - h)$ is obtained by translating the graph of $y = f(x)$ h units to the right.
- The graph of $y = f(x + h)$ is obtained by translating the graph of $y = f(x)$ h units to the left.

As Figures 1.30 and 1.31 show, we can obtain the graph of $g(x) = (x + 3)^2$ by translating the graph of $f(x) = x^2$ three units to the **left**, and we can obtain the graph of $h(x) = (x - 2)^2$ by translating the graph of $f(x) = x^2$ two units to the **right.**

In Example 1.21, if the transformations were performed in reverse order, that is, the vertical translation followed by the horizontal translation, we would get the same final graph (in part (b)) with the same equation. The order in which we apply both a vertical and horizontal translation on a function does not make any difference. The translations are commutative. However, as we will see further in the chapter, it can make a difference how other sequences of transformations are applied. In general, transformations are not commutative.

Example 1.21

The diagrams show how the graph of $y = \sqrt{x}$ is transformed to the graph of $y = f(x)$ in three steps. For each diagram, (a) and (b), give the equation of the curve.

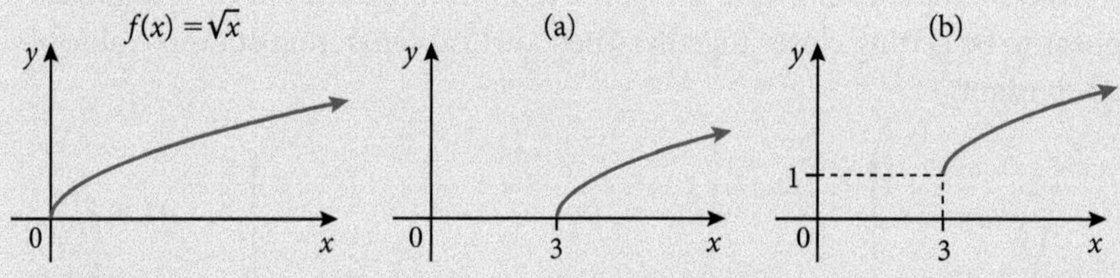

Solution

To obtain the graph in (a), the graph of $y = \sqrt{x}$ is translated three units to the right. To produce the equation of the translated graph, -3 is added inside the argument of the function $y = \sqrt{x}$.

Therefore, the equation of the curve graphed in (a) is $y = \sqrt{x - 3}$

To obtain the graph in (b), the graph of $y = \sqrt{x - 3}$ is translated up one unit. To produce the equation of the translated graph, $+1$ is added outside the function. Therefore, the equation of the curve graphed in (b) is

$y = \sqrt{x - 3} + 1$ or $y = 1 + \sqrt{x - 3}$

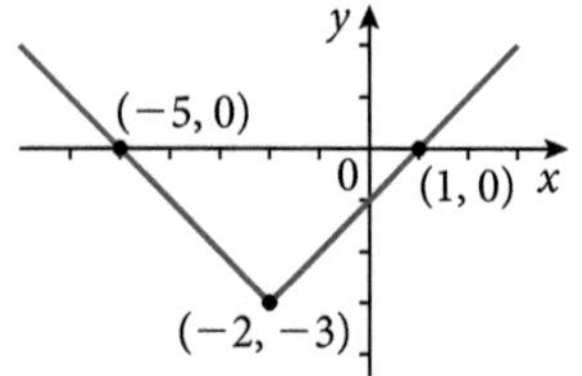

Figure 1.32 Diagram for Example 1.22

Example 1.22

Write the equation of the absolute value function shown by Figure 1.32.

Solution

The graph shown is exactly the same shape as the graph of the equation $y = |x|$ but in a different position. Given that the vertex is $(-2, -3)$, it is clear that this graph can be obtained by translating $y = |x|$ two units left and then three units down. When we move $y = |x|$ two units left we get the graph of $y = |x + 2|$. Moving the graph of $y = |x + 2|$ down three units down gives us the graph of $y = |x + 2| - 3$. Therefore, the equation of the graph shown is $y = |x + 2| - 3$

We would get the same result if we applied the translations in the reverse order.

Reflections

Use your GDC to graph the two functions $f(x) = x^2$ and $g(x) = -x^2$. The graph of $g(x) = -x^2$ is a reflection in the x-axis of $f(x) = x^2$. This certainly makes sense because g is formed by multiplying f by -1, causing the y-coordinate of each point on the graph of $y = -x^2$ to be the negative of the y-coordinate of the point on the graph of $y = x^2$ that has the same x-coordinate.

The expression $-x^2$ is potentially ambiguous. It is accepted to be equivalent to $-(x)^2$. It is not equivalent to $(-x)^2$. For example, if you enter the expression -3^2 into your GDC it gives a result of -9, *not* $+9$. The expression -3^2 is consistently interpreted as 3^2 being multiplied by -1. The same as $-x^2$ is interpreted as x^2 being multiplied by -1.

Figures 1.33 and 1.34 show that the graph of $y = -f(x)$ is obtained by reflecting the graph of $y = f(x)$ in the x-axis.

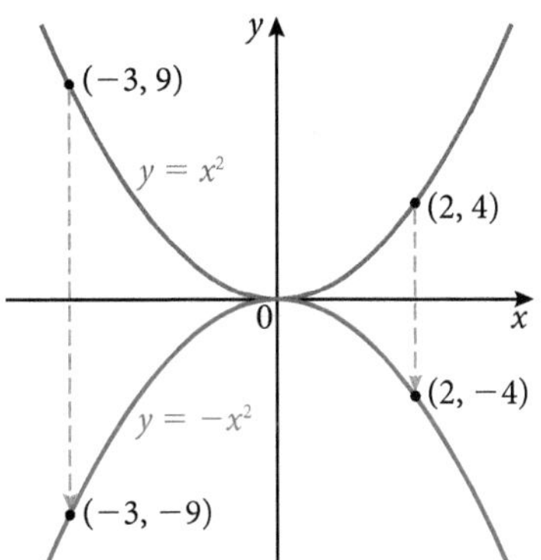

Figure 1.33 Reflecting $y = x^2$ in the x-axis

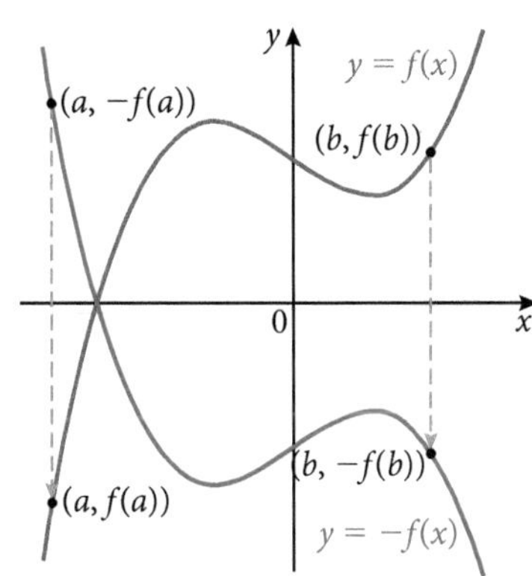

Figure 1.34 Reflecting $y = f(x)$ in the x-axis

Graph $f(x) = \sqrt{x - 2}$ and $g(x) = \sqrt{-x - 2}$. With $f(x) = x^2$ and $g(x) = -x^2$, g was formed by multiplying the entire function f by -1. However, for $f(x) = \sqrt{x - 2}$ and $g(x) = \sqrt{-x - 2}$, g is formed by multiplying the variable x by -1. In this case, the graph of $g(x) = \sqrt{-x - 2}$ is a reflection in the y-axis of $f(x) = \sqrt{x - 2}$. This makes sense if you recognise that the y-coordinate on the graph of $y = \sqrt{-x}$ will be the same as the y-coordinate on the graph of $y = \sqrt{x}$ if the value substituted for x in $y = \sqrt{-x}$ is the opposite of the x value in $y = \sqrt{x}$. For example, if $x = 9$ then $y = \sqrt{9} = 3$; and, if $x = -9$ then $y = \sqrt{-(-9)} = \sqrt{9} = 3$. Opposite values of x in the two functions produce the same y-coordinate for each.

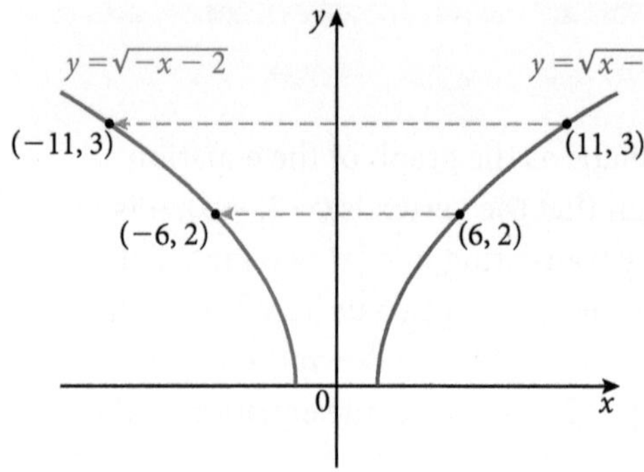

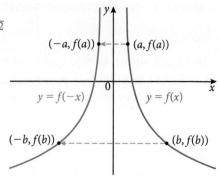

Figure 1.35 Reflecting $y = \sqrt{x-2}$ in the y-axis

Figure 1.36 Reflecting $y = f(x)$ in the y-axis

The graph of $y = -f(x)$ is obtained by reflecting the graph of $y = f(x)$ in the x-axis.
The graph of $y = f(-x)$ is obtained by reflecting the graph of $y = f(x)$ in the y-axis.

Figures 1.35 and 1.36 illustrate how the graph of $y = f(-x)$ is obtained by reflecting the graph of $y = f(x)$ in the y-axis.

Example 1.23

For $g(x) = 2x^3 - 6x^2 + 3$, find:

(a) the function $h(x)$ that is the reflection of $g(x)$ in the x-axis

(b) the function $p(x)$ that is the reflection of $g(x)$ in the y-axis.

Solution

(a) Knowing that $y = -f(x)$ is the reflection of $y = f(x)$ in the x-axis, then $h(x) = -g(x) = -(2x^3 - 6x^2 + 3) \Rightarrow h(x) = -2x^3 + 6x^2 - 3$ will be the reflection of $g(x)$ in the x-axis. We can verify the result on the GDC – graphing the original equation $y = 2x^3 - 6x^2 + 3$ in bold style.

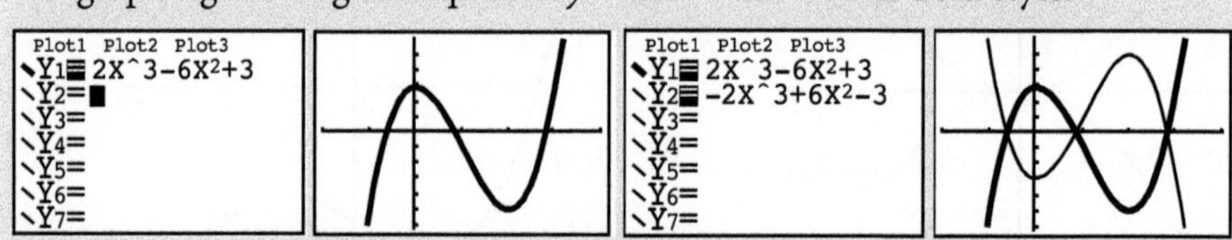

(b) Knowing that $y = f(-x)$ is the reflection of $y = f(x)$ in the y-axis, we need to substitute $-x$ in for x in $y = g(x)$.
Thus, $p(x) = g(-x) = 2(-x)^3 - 6(-x)^2 + 3 \Rightarrow p(x) = -2x^3 - 6x + 3$ will be the reflection of $g(x)$ in the y-axis. Again, we can verify the result on the GDC – graphing the original equation $y = 2x^3 - 6x^2 + 3$ in bold style.

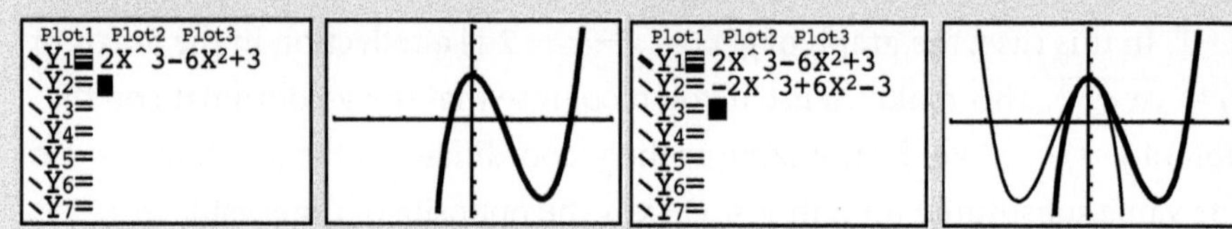

Non-rigid transformations: stretching and shrinking

Horizontal and vertical translations, and reflections in the x- and y-axes are called **rigid transformations** because the shape of the graph does not change – only its position is changed. **Non-rigid transformations** cause the shape of the original graph to change. The non-rigid transformations that you will study cause the shape of a graph to stretch or shrink in either the vertical or horizontal direction.

Vertical stretch or shrink

Graph the functions: $f(x) = x^2$, $g(x) = 3x^2$ and $h(x) = \frac{1}{3}x^2$. How do the graphs of g and h compare to the graph of f? Refer to figures 1.38 and 1.40. Clearly, the shape of the graphs of g and h is not the same as the graph of f. Multiplying the function f by a positive number greater than one, or less than one, has distorted the shape of the graph. For a certain value of x, the y-coordinate of $y = 3x^2$ is three times the y-coordinate of $y = x^2$. Therefore, the graph of $y = 3x^2$ can be obtained by **vertically stretching** the graph of $y = x^2$ by a factor of 3 (**scale factor** 3).

Likewise, the graph of $y = \frac{1}{3}x^2$ can be obtained by **vertically shrinking** the graph of $y = x^2$ by scale factor $\frac{1}{3}$

Figures 1.37 and 1.38 below show how multiplying a function by a positive number, a, greater than 1 causes a transformation in which the function stretches vertically by scale factor a. A point (x, y) on the graph of $y = f(x)$ is transformed to the point (x, ay) on the graph of $y = af(x)$.

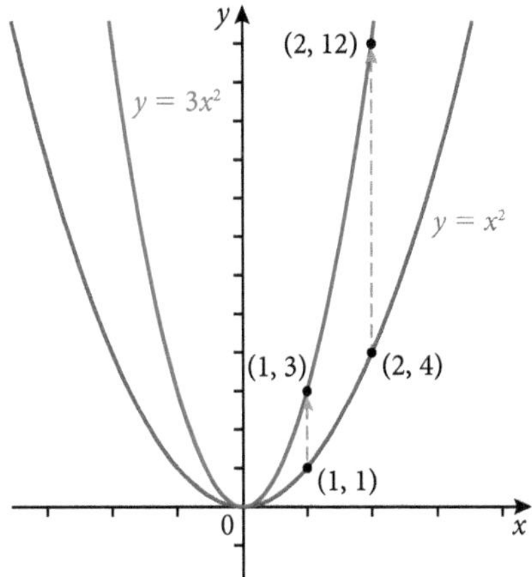

Figure 1.37 Vertical stretch of $y = x^2$ by scale factor 3

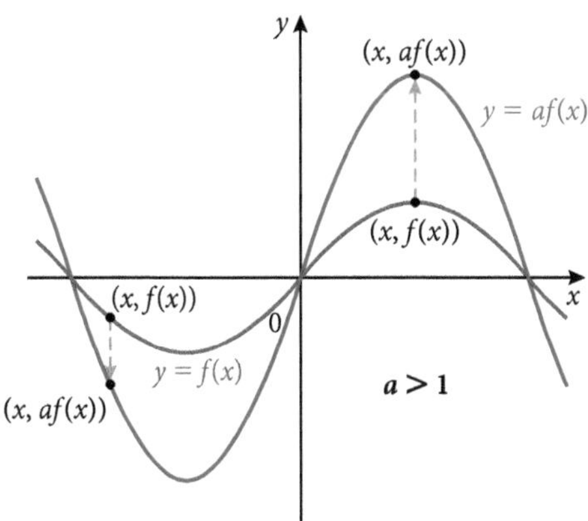

Figure 1.38 Vertical stretch of $y = f(x)$ by scale factor a when $a > 1$

Figures 1.39 and 1.40 below show how multiplying a function by a positive number, a, greater than 0 and less than 1 causes the function to shrink vertically by scale factor a. A point (x, y) on the graph of $y = f(x)$ is transformed to the point (x, ay) on the graph of $y = af(x)$.

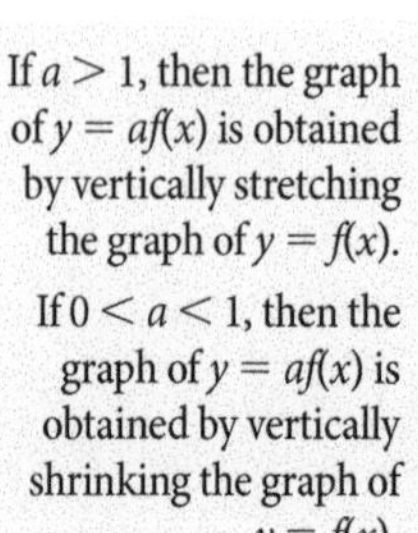

If $a > 1$, then the graph of $y = af(x)$ is obtained by vertically stretching the graph of $y = f(x)$.
If $0 < a < 1$, then the graph of $y = af(x)$ is obtained by vertically shrinking the graph of $y = f(x)$.

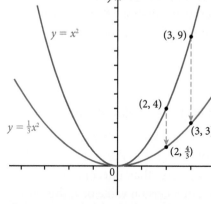

Figure 1.39 Vertical shrink of $y = x^2$ by scale factor $\frac{1}{3}$

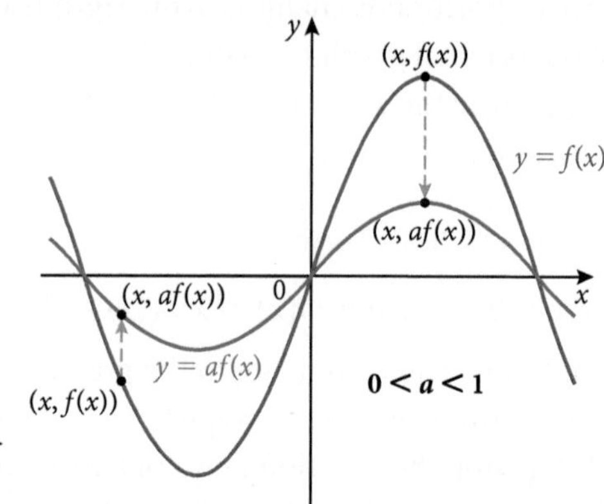

Figure 1.40 Vertical shrink of $y = f(x)$ by scale factor a, when $0 < a < 1$

Horizontal stretch or shrink

We will now investigate how the graph of $y = f(ax)$ is obtained from the graph of $y = f(x)$. Given $f(x) = x^2 - 4x$, find another function, $g(x)$, such that $g(x) = f(2x)$. We substitute $2x$ for x in the function f, giving $g(x) = (2x)^2 - 4(2x)$. For the purposes of our investigation, leave $g(x)$ in this form. On your GDC, graph these two functions, $f(x) = x^2 - 4x$ and $g(x) = (2x)^2 - 4(2x)$, using the indicated viewing window and graphing f in bold style (Figure 1.41).

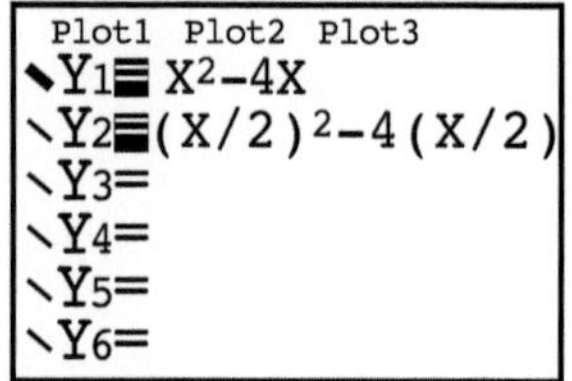

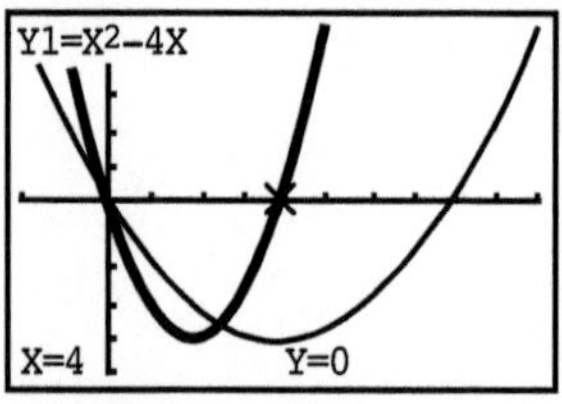

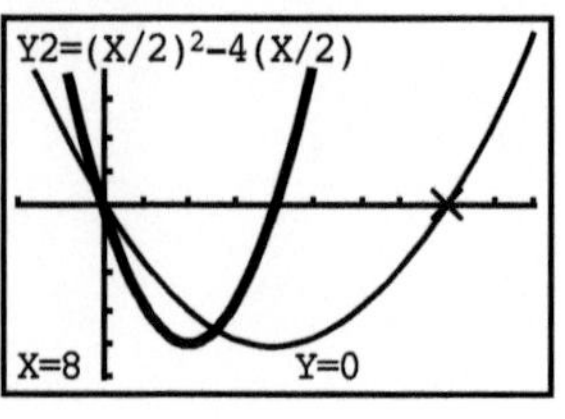

Figure 1.41 Graphs of $y = x^2 - 4x$ (in bold) and $y = \left(\frac{x}{2}\right)^2 - 4\left(\frac{x}{2}\right)$

Comparing the graphs of the two equations, we can see that $y = g(x)$ is not a translation or a reflection of $y = f(x)$. It is similar to the shrinking effect that occurs for $y = af(x)$ when $0 < a < 1$, except, instead of a vertical shrinking, the graph of $y = g(x) = f(2x)$ is obtained by horizontally shrinking the graph of $y = f(x)$. Given that it is a shrinking, the scale factor must be less than 1.

Consider the point $(4, 0)$ on the graph of $y = f(x)$. The point on the graph of $y = g(x) = f(2x)$ with the same y-coordinate and on the same side of the parabola is $(2, 0)$. The x-coordinate of the point on $y = f(2x)$ is the x-coordinate of the point on $y = f(x)$ multiplied by $\frac{1}{2}$. Use your GDC to confirm this for other pairs of corresponding points on $y = x^2 - 4x$ and $y = (2x)^2 - 4(2x)$ that have the same y-coordinate. The graph of $y = f(2x)$ can be obtained by horizontally shrinking the graph of $y = f(x)$ with scale factor $\frac{1}{2}$. This makes sense because if $f(2x_2) = (2x_2)^2 - 4(2x_2)$ and $f(x_1) = {x_1}^2 - 4x_1$ are to produce the same y-value, then $2x_2 = x_1$, and thus $x_2 = \frac{1}{2}x_1$. Figures 1.42 and 1.43 show how multiplying the x variable of a function by a positive number, a, greater than 1, causes the function to shrink horizontally by scale factor $\frac{1}{a}$. A point (x, y) on the graph of $y = f(x)$ is transformed to the point $\left(\frac{1}{a}x, y\right)$ on the graph of $y = f(ax)$.

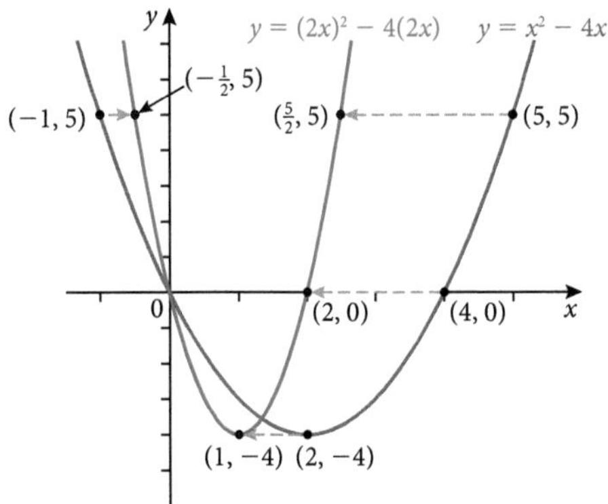

Figure 1.42 Horizontal shrink of $y = x^2 - 4x$ by scale factor $\frac{1}{2}$

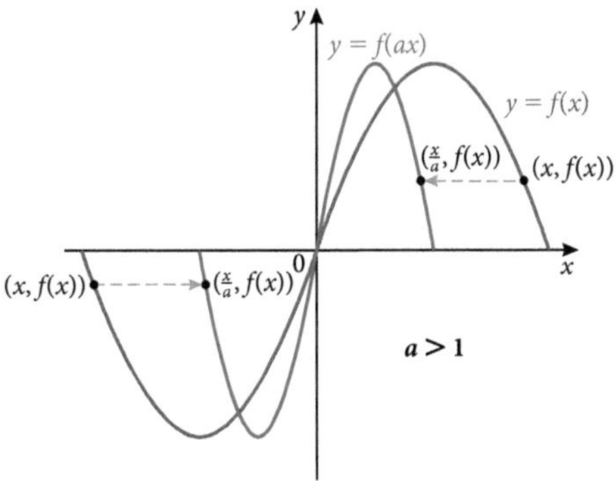

Figure 1.43 Horizontal shrink of $y = f(x)$ by scale factor $\frac{1}{a}$, $a > 1$

If $0 < a < 1$, then the graph of the function $y = f(ax)$ is obtained by a horizontal stretching – rather than a shrinking – of the graph of $y = f(x)$ because the scale factor $\frac{1}{a}$ will be a value between 0 and 1 if $0 < a < 1$. Now, letting $a = \frac{1}{2}$ and, again using the function $f(x) = x^2 - 4x$, find $g(x)$, such that $g(x) = f\left(\frac{1}{2}x\right)$.

Substitute $\frac{x}{2}$ for x in f, giving $g(x) = \left(\frac{x}{2}\right)^2 - 4\left(\frac{x}{2}\right)$. On your GDC, graph the functions f and g using the indicated viewing window with f in bold.

The graph of $y = \left(\frac{x}{2}\right)^2 - 4\left(\frac{x}{2}\right)$ is a horizontal stretching of the graph of $y = x^2 - 4x$ by scale factor $\frac{1}{a} = \frac{1}{\frac{1}{2}} = 2$. For example, the point $(4, 0)$ on $y = f(x)$ has been moved horizontally to the point $(8, 0)$ on $y = g(x) = f\left(\frac{x}{2}\right)$.

Figures 1.44 and 1.45 below show how multiplying the x variable of a function by a positive number, a, greater than 0 and less than 1, causes the function to stretch horizontally by scale factor $\frac{1}{a}$. A point (x, y) on the graph of $y = f(x)$ is transformed to the point $\left(\frac{1}{a}x, y\right)$ on the graph of $y = f(ax)$.

If $a > 1$, then the graph of $y = f(ax)$ is obtained by horizontally shrinking the graph of $y = f(x)$.

If $0 < a < 1$, then the graph of $y = f(ax)$ is obtained by horizontally stretching the graph of $y = f(x)$.

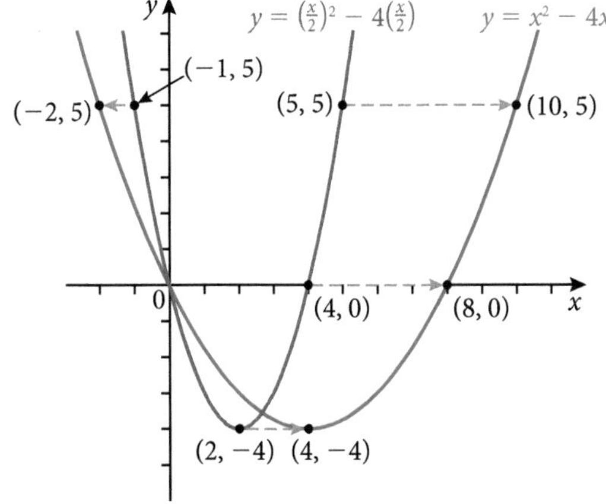

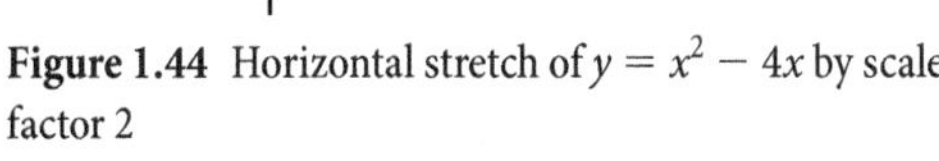

Figure 1.44 Horizontal stretch of $y = x^2 - 4x$ by scale factor 2

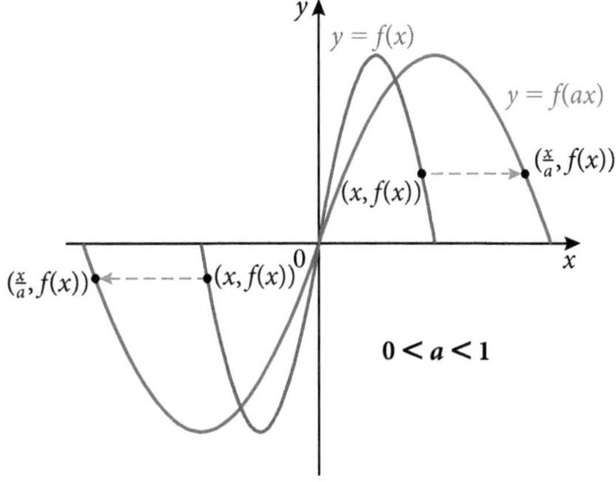

Figure 1.45 Horizontal stretch of $y = f(x)$ by scale factor $\frac{1}{a}$, $0 < a < 1$

Example 1.24

The graph of $y = f(x)$ is shown. Sketch the graph of each transformation.

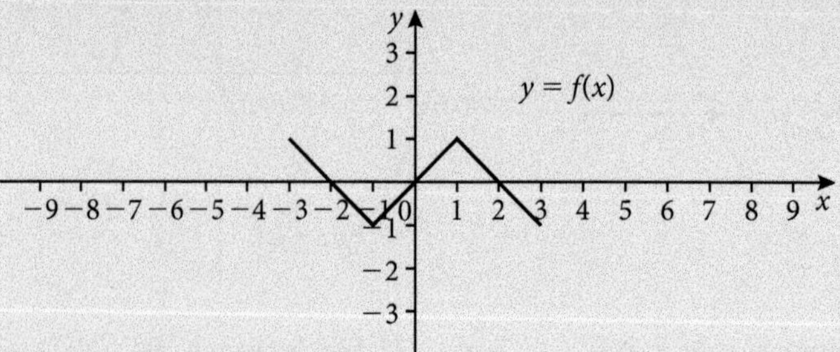

(a) $y = 3f(x)$

(b) $y = \frac{1}{3}f(x)$

(c) $y = f(3x)$

(d) $y = f\left(\frac{1}{3}x\right)$

Solution

(a) The graph of $y = 3f(x)$ is obtained by vertically stretching the graph of $y = f(x)$ with scale factor 3.

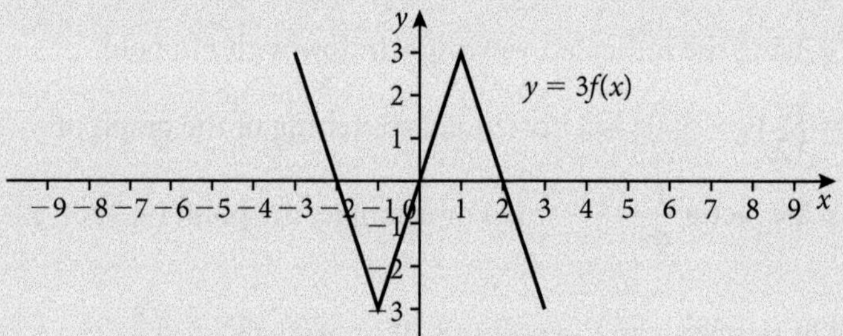

(b) The graph of $y = \frac{1}{3}f(x)$ is obtained by vertically shrinking the graph of $y = f(x)$ with scale factor $\frac{1}{3}$.

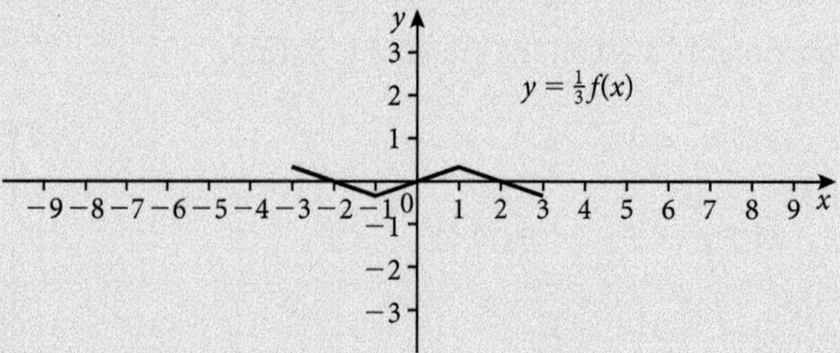

(c) The graph of $y = f(3x)$ is obtained by horizontally shrinking the graph of $y = f(x)$ with scale factor $\frac{1}{3}$.

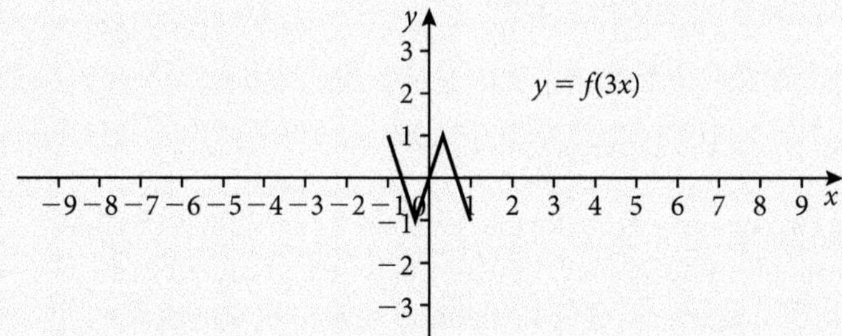

(d) The graph of $y = f\left(\frac{1}{3}\right)x$ is obtained by horizontally stretching the graph of $y = f(x)$ with scale factor 3.

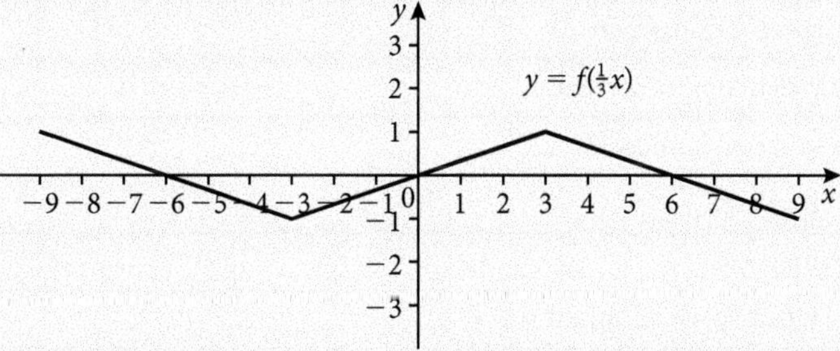

Example 1.25

Describe the sequence of transformations performed on the graph of $y = x^2$ to obtain the graph of $y = 4x^2 - 3$

Solution

Step 1: Start with the graph of $y = x^2$

Step 2: Vertically stretch $y = x^2$ by scale factor 4

Step 3: Vertically translate $y = 4x^2$ three units down

Step 1:

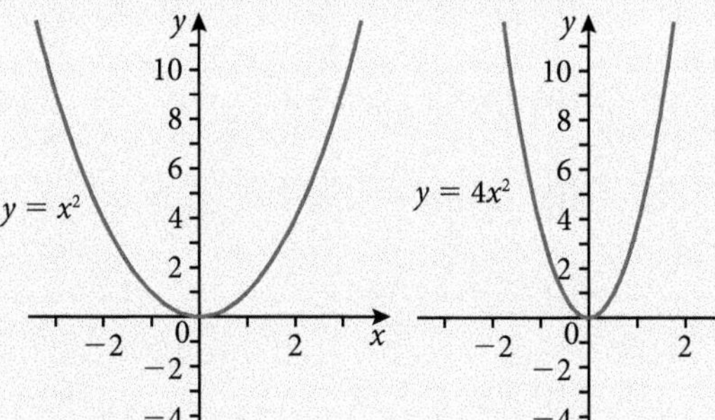

Step 2:

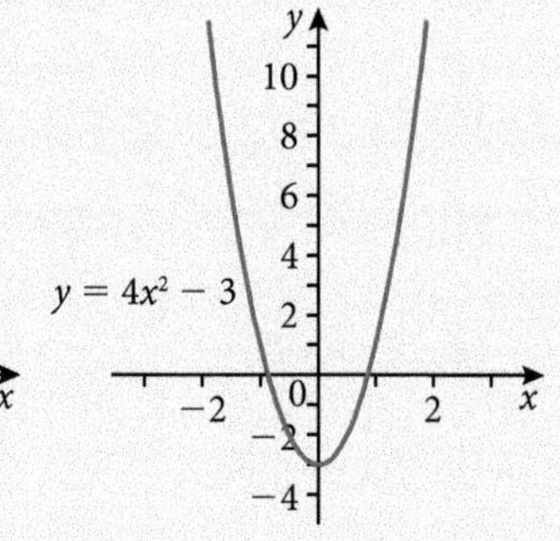

Step 3:

$y = 4x^2 - 3$

Online

Check your understanding of basic graph transformations.

Note that in Example 1.25, a vertical stretch followed by a vertical translation does not produce the same graph if the two transformations are performed in reverse order. A vertical translation followed by a vertical stretch would generate the following sequence of equations:

Step1: $y = x^2$ Step 2: $y = x^2 - 3$ Step 3: $y = 4(x^2 - 3) = 4x^2 - 12$

This final equation is not the same as $y = 4x^2 - 3$

In Table 1.4, assume that a, h and k are positive real numbers.

When combining two or more transformations, the order in which they are performed can make a difference. In general, when a sequence of transformations includes a vertical or horizontal stretch or shrink, or a reflection through the x-axis, the order may make a difference.

Transformed function	Transformation performed on $y = f(x)$
$y = f(x) + k$	vertical translation k units up
$y = f(x) - k$	vertical translation k units down
$y = f(x - h)$	horizontal translation h units right
$y = f(x + h)$	horizontal translation h units left
$y = -f(x)$	reflection in the x-axis
$y = f(-x)$	reflection in the y-axis
$y = af(x)$	vertical stretch ($a > 1$) or shrink ($0 < a < 1$)
$y = f(ax)$	horizontal stretch ($0 < a < 1$) or shrink ($a > 1$)
$y = \|f(x)\|$	portion of the graph of $y = f(x)$ below x-axis is reflected above the x-axis
$y = f(\|x\|)$	symmetric about the y-axis; portion right of the y-axis is reflected in the y-axis

Table 1.4 Summary of transformations on the graphs of functions

Exercise 1.5

1. Sketch the graph of f, without a GDC or plotting points, by using your knowledge of some of the basic functions shown at the start of the Section 1.5.

(a) $f{:}x \mapsto x^2 - 6$

(b) $f{:}x \mapsto (x - 6)^2$

(c) $f{:}x \mapsto |x| + 4$

(d) $f{:}x \mapsto |x + 4|$

(e) $f{:}x \mapsto 5 + \sqrt{x - 2}$

(f) $f{:}x \mapsto \dfrac{1}{x - 3}$

(g) $f{:}x \mapsto \dfrac{1}{(x + 5)^2} + 2$

(h) $f{:}x \mapsto -x^3 - 4$

(i) $f{:}x \mapsto -|x - 1| + 6$

(j) $f{:}x \mapsto \sqrt{-x + 3}$

(k) $f{:}x \mapsto 3\sqrt{x}$

(l) $f{:}x \mapsto \dfrac{1}{2}x^2$

(m) $f{:}x \mapsto \left(\dfrac{1}{2}x\right)^2$

(n) $f{:}x \mapsto (-x)^3$

2. Write the equation for each graph.

(a)

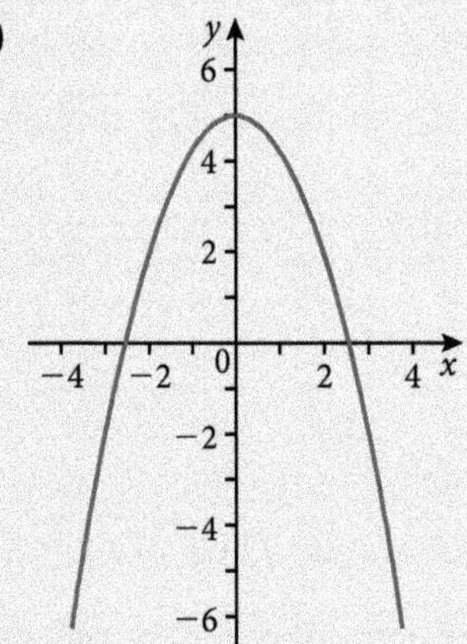

(b)

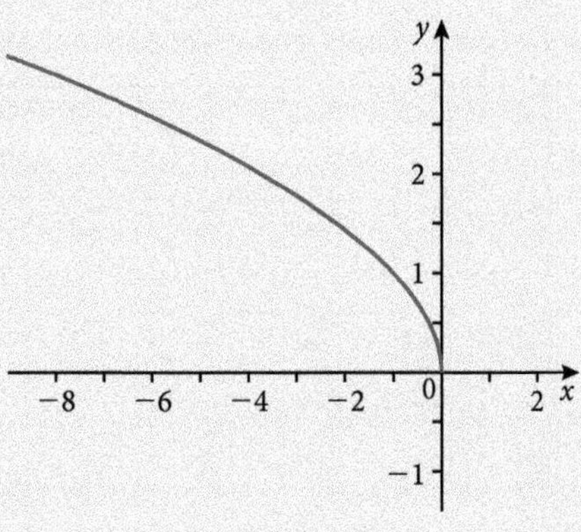

(c)

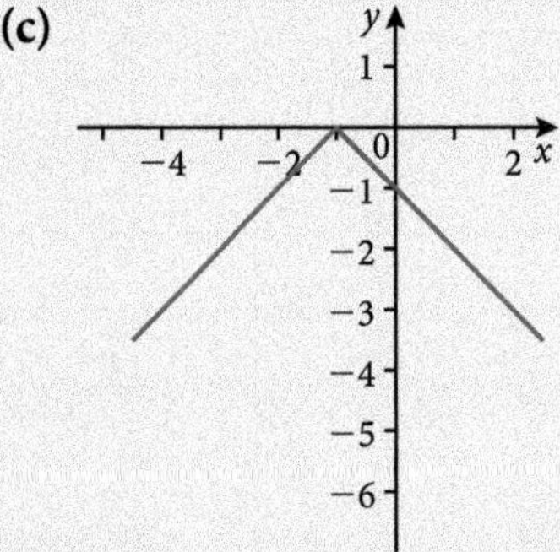

(d)

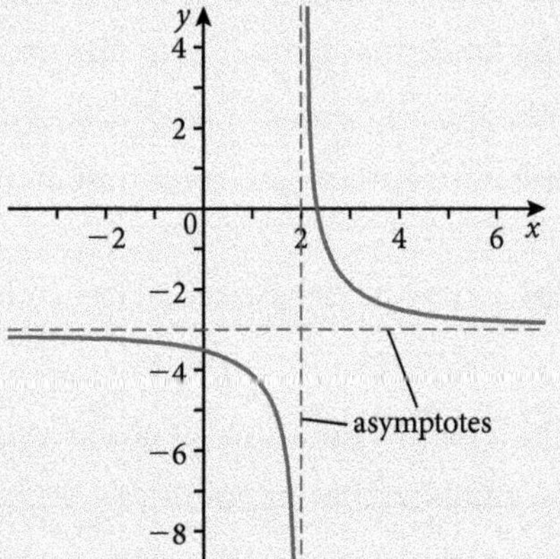

3. The graph of f is given. Sketch the graph of each transformed function.

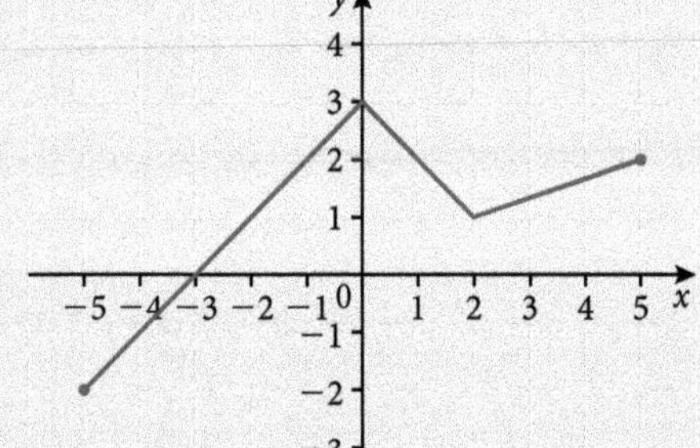

(a) $y = f(x) - 3$

(b) $y = f(x - 3)$

(c) $y = 2f(x)$

(d) $y = f(2x)$

(e) $y = -f(x)$

(f) $y = f(-x)$

(g) $y = 2f(x) + 4$

4. Specify a sequence of transformations to perform on the graph of $y = x^2$ to obtain the graph of the given function.

(a) $g:x \mapsto (x - 3)^2 + 5$

(b) $h:x \mapsto -x^2 + 2$

(c) $p:x \mapsto \frac{1}{2}(x + 4)^2$

(d) $f:x \mapsto [3(x - 1)]^2 - 6$

Chapter 1 practice questions

1. The functions f and g are defined as $f{:}x \mapsto \sqrt{x-3}$ and $g{:}x \mapsto x^2 + 2x$. The function $(f \circ g)(x)$ is defined for all $x \in \mathbb{R}$, except for the interval $]\,a, b\,[$.

   (a) Calculate the values of a and b (b) Find the range of $f \circ g$

2. Two functions g and h are defined as $g(x) = 2x - 7$ and $h(x) = 3(2 - x)$

   Find: (a) $g^{-1}(3)$ (b) $(h \circ g)(6)$

3. Consider the functions $f(x) = 5x - 2$ and $g(x) = \dfrac{4 - x}{3}$

   (a) Find g^{-1} (b) Solve the equation $(f \circ g^{-1})(x) = 8$

4. The functions g and h are defined by $g{:}x \mapsto x - 3$ and $h{:}x \mapsto 2x$

   (a) Find an expression for $(g \circ h)(x)$

   (b) Show that $g^{-1}(14) + h^{-1}(14) = 24$

5. The diagram shows the graph of $y = f(x)$. It has maximum and minimum points at $(0, 0)$ and $(1, -1)$, respectively.

   (a) Copy the diagram, and then add in the graph of $y = f(x + 1) - \frac{1}{2}$

   (b) Find the coordinates of the minimum and maximum points of $y = f(x + 1) - \frac{1}{2}$

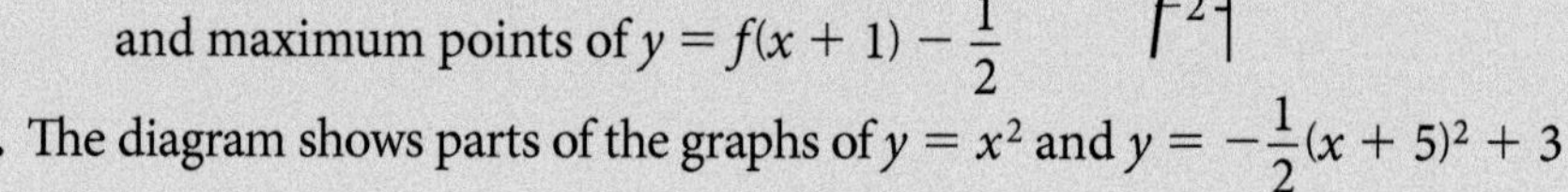

6. The diagram shows parts of the graphs of $y = x^2$ and $y = -\frac{1}{2}(x + 5)^2 + 3$

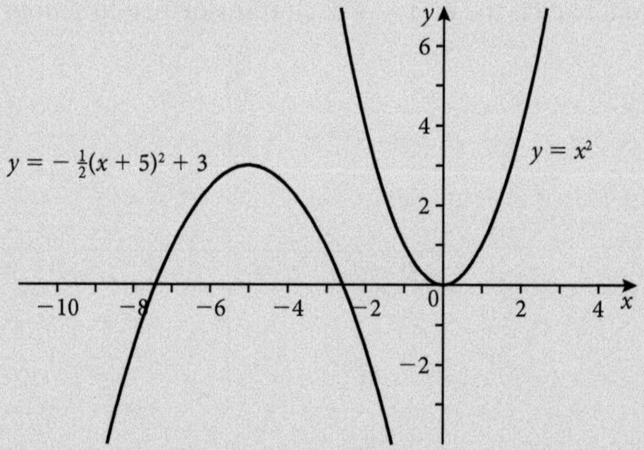

   The graph of $y = x^2$ may be transformed into the graph of $y = -\frac{1}{2}(x + 5)^2 + 3$ by these transformations:

   A reflection in the line $y = 0$, followed by a vertical stretch with scale factor k, followed by a horizontal translation of p units, followed by a vertical translation of q units.

   Write down the value of:

   (a) k (b) p (c) q

7. The function f is defined by $f(x) = \frac{4}{\sqrt{16 - x^2}}$, for $-4 < x < 4$

(a) Without using a GDC, sketch the graph of f.

(b) Write down the equation of each vertical asymptote.

(c) Write down the range of the function f.

8. Let $g{:}x \mapsto \frac{1}{x}, x \neq 0$

(a) Without using a GDC, sketch the graph of g.

The graph of g is transformed to the graph of h by a translation of 4 units to the left and 2 units down.

(b) Find an expression for the function h.

(c) (i) Find the x- and y-intercepts of h.

(ii) Write down the equations of the asymptotes of h.

(iii) Sketch the graph of h.

9. Consider $f(x) = \sqrt{x + 3}$

(a) Find:

(i) $f(8)$ (ii) $f(46)$ (iii) $f(-3)$

(b) Find the values of x for which f is undefined.

(c) Let $g{:}x \mapsto x^2 - 5$. Find $(g \circ f)(x)$.

10. Let $g(x) = \frac{x - 8}{2}$ and $h(x) = x^2 - 1$

(a) Find $g^{-1}(-2)$

(b) Find an expression for $(g^{-1} \circ h)(x)$

(c) Solve $(g^{-1} \circ h)(x) = 22$

11. Given the functions $f{:}x \mapsto 3x - 1$ and $g{:}x \mapsto \frac{4}{x}$, find the following:

(a) f^{-1}

(b) $f \circ g$

(c) $(f \circ g)^{-1}$

(d) $g \circ g$

12. (a) The diagram shows part of the graph of the function

$$h(x) = \frac{a}{x - b}$$

The curve passes through the point $A(-4, -8)$.

The vertical line MN is an asymptote.

Find the value of:

(i) a **(ii)** b

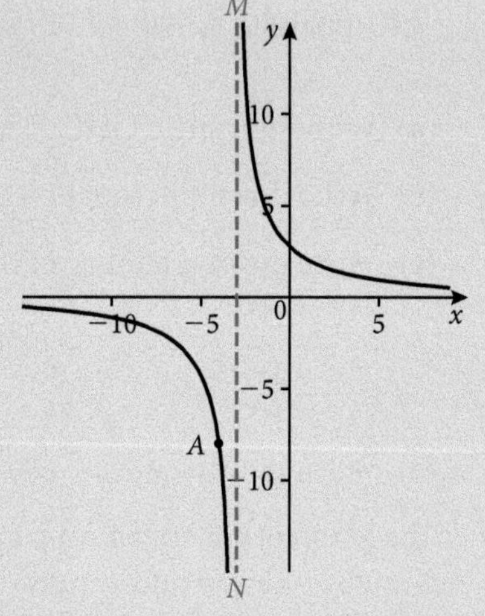

(b) The graph of $h(x)$ is transformed as shown in the diagram.

The point A is transformed to $A'(-4, 8)$.

Give a full geometric description of the transformation.

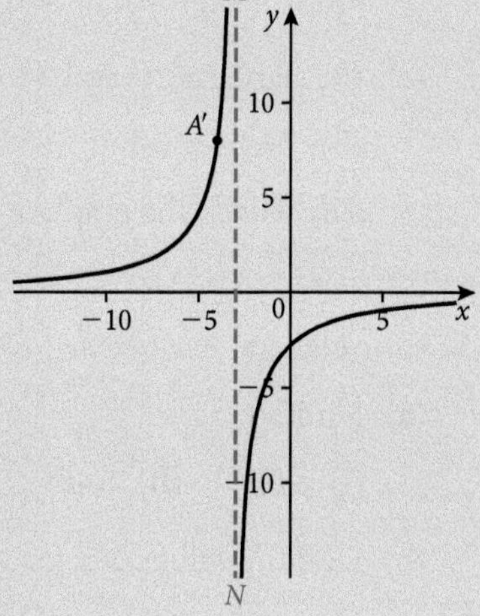

13. The graph of $y = f(x)$ is shown in the diagram.

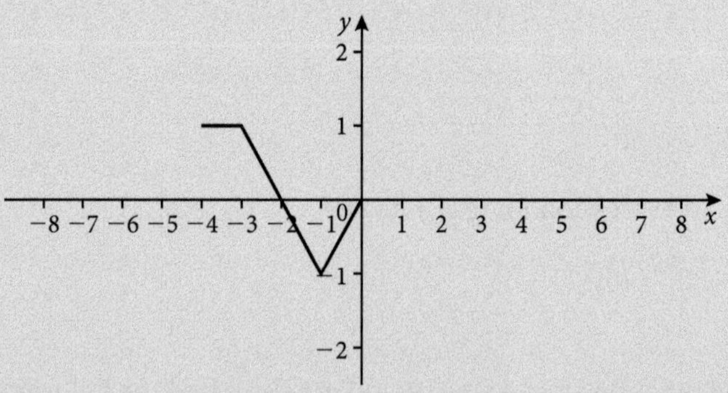

(a) Make two copies of the coordinate system as shown in the diagram but without the graph of $y = f(x)$. On the first diagram sketch a graph of $y = 2f(x)$, and on the second diagram sketch a graph of $y = f(x - 4)$

(b) The point $A(-3, 1)$ is on the graph of $y = f(x)$. The point A' is the corresponding point on the graph of $y = -f(x) - 1$. Find the coordinates of A'.

14. The diagram shows the graph of $y_1 = f(x)$. The x-axis is a tangent to $f(x)$ at $x = m$ and $f(x)$ crosses the x-axis at $x = n$.

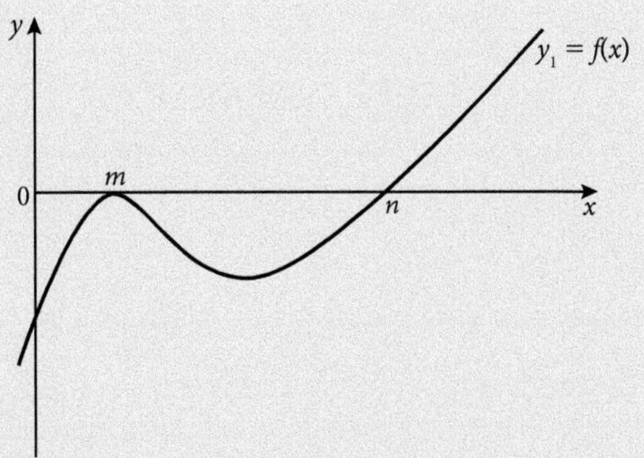

On the same diagram, sketch the graph of $y_2 = f(x - k)$, where $0 < k < n - m$ and indicate the coordinates of the points of intersection of y_2 with the x-axis.

15. Given functions $f: x \mapsto x + 1$ and $g: x \mapsto x^3$, find the function $(f \circ g)^{-1}$

16. If $f(x) = \frac{x}{x+1}$, for $x \neq -1$ and $g(x) = (f \circ f)(x)$, find:

(a) $g(x)$

(b) $(g \circ g)(2)$

17. Let $f: x \mapsto \sqrt{\frac{1}{x^2} - 2}$. Find:

(a) the set of real values of x for which f is real and finite

(b) the range of f.

18. The function $f: x \mapsto \frac{2x+1}{x-1}$, $x \in \mathbb{R}$, $x \neq 1$. Find the inverse function, f^{-1}, clearly stating its domain.

19. The one-to-one function f is defined on the domain $x > 0$ by

$$f(x) = \frac{2x-1}{x+2}$$

(a) State the range, A, of f.

(b) Obtain an expression for $f^{-1}(x)$, for $x \in A$.

20. The function f is defined by $f: x \mapsto x^3$

Find an expression for $g(x)$ in terms of x in each of the following cases:

(a) $(f \circ g)(x) = x + 1$

(b) $(g \circ f)(x) = x + 1$

21. **(a)** Find the largest set S of values of x such that the function $f(x) = \frac{1}{\sqrt{3 - x^2}}$ takes real values.

(b) Find the range of the function f defined on the domain S.

22. Let f and g be two functions. Given that $(f \circ g)(x) = \frac{x + 1}{2}$ and $g(x) = 2x - 1$, find $f(x - 3)$.

23. The diagram shows part of the graph of $y = f(x)$ that passes through the points A, B, C, and D.

Sketch, indicating clearly the images of A, B, C, and D, the graphs of

(a) $y = f(x - 4)$ **(b)** $y = f(-3x)$

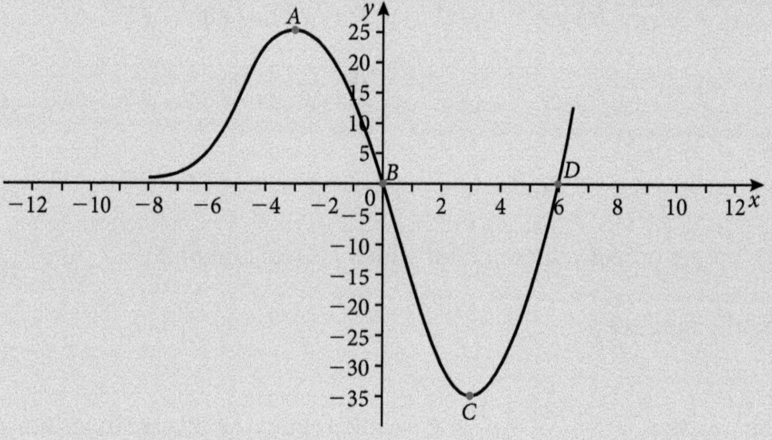

2 Functions, equations, and inequalities

2 Functions, equations, and inequalities

Learning objectives

By the end of this chapter, you should be familiar with...

- quadratic functions, and different forms in which to express them
- finding characteristics of a parabola: axis of symmetry, x-intercepts, and vertex
- solving quadratic equations and inequalities, both analytically and graphically
- applying the quadratic formula and the discriminant of a quadratic equation
- rational functions and their graphs, and identifying all asymptotes
- solving a variety of equations, both analytically and graphically.

This chapter will focus on polynomial functions (which includes quadratic functions) and rational functions. There are other function types that you need to be familiar with for this course. Chapter 4 will cover exponential functions and logarithmic functions, and Chapter 5 will focus on trigonometric functions. Along with polynomial and rational functions, this chapter will address solving polynomial equations, solving other types of equations and solving inequalities.

2.1 Quadratic functions

A **linear function** is a polynomial function of degree one that can be written in the general form $f(x) = ax + b$, where $a \neq 0$. Linear equations were briefly reviewed in Section 1.1. Any linear function will have a single solution (root) of $x = -\frac{b}{a}$. This is a formula that gives the zero of any linear equation.

If a, b, and c are real numbers, and $a \neq 0$, then the function $f(x) = ax^2 + bx + c$ is a **quadratic function.** The graph of $y = f(x)$ is called a **parabola.**

In this section, we will focus on **quadratic functions**, which are functions expressed in terms of a second-degree polynomial that can be written in the form $f(x) = ax^2 + bx + c$, where $a \neq 0$. You are probably familiar with the formula that gives the zeros of any quadratic polynomial; that is, the **quadratic formula.** We will also investigate other methods of finding zeros of quadratics and consider important characteristics of the graphs of quadratic functions.

Quadratic formula

The solution(s) to a quadratic equation in the form $ax^2 + bx + c = 0$ are given by
$x = \frac{-b \pm \sqrt{b^2 - 4ac}}{2a}$.

Every parabola is symmetric about a vertical line called its **axis of symmetry.** The axis of symmetry passes through a point on the parabola called the **vertex** of the parabola, as shown in Figure 2.1. If the leading coefficient, a, of the quadratic function is positive, then the parabola opens upward (concave up), and the y-coordinate of the vertex is a **minimum value** for the function. If the leading coefficient is negative, then the parabola opens downward (concave down), and the y-coordinate of the vertex is a **maximum value** for the function.

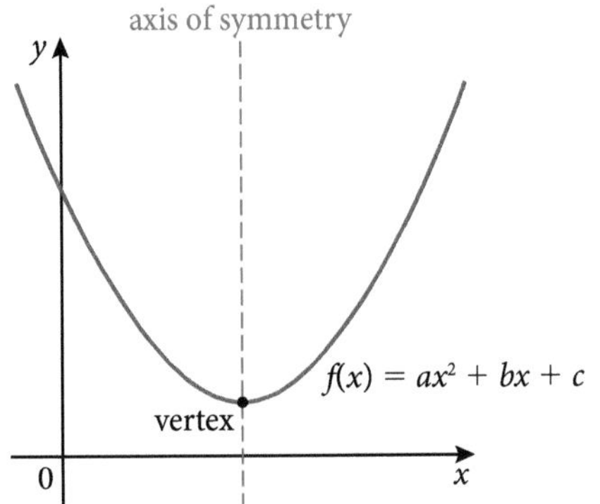

If $a > 0$ then the parabola opens upward.

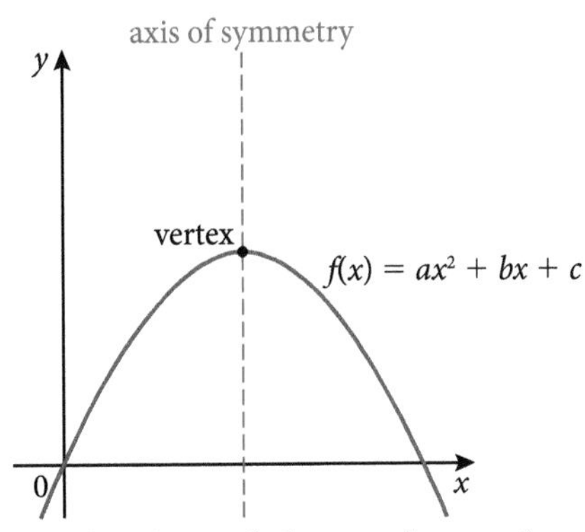

If $a < 0$ then the parabola opens downward.

Figure 2.1 'Concave up' and 'concave down' parabolas

The graph of $f(x) = a(x - h)^2 + k$

From Section 1.5, we know that the graph of the equation $y = (x + 3)^2 + 2$ can be obtained by translating $y = x^2$ three units to the left and two units up. As we are familiar with the shape and position of the graph of $y = x^2$ and we know the two translations that transform $y = x^2$ to $y = (x + 3)^2 + 2$, we can sketch the graph of $y = (x + 3)^2 + 2$, as shown in Figure 2.2.

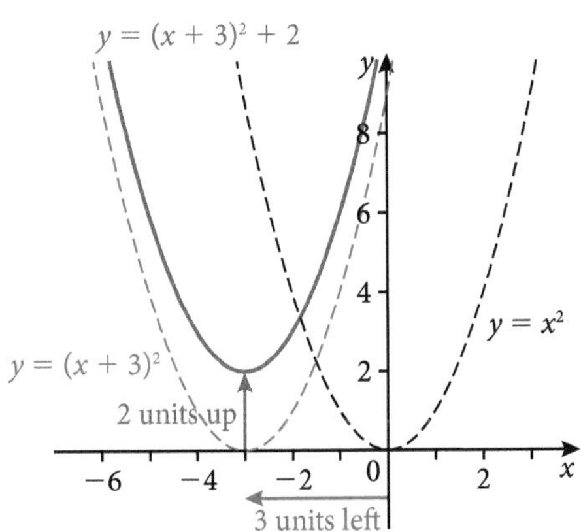

Figure 2.2 Translating $y = x^2$ to produce $y = (x + 3)^2 + 2$

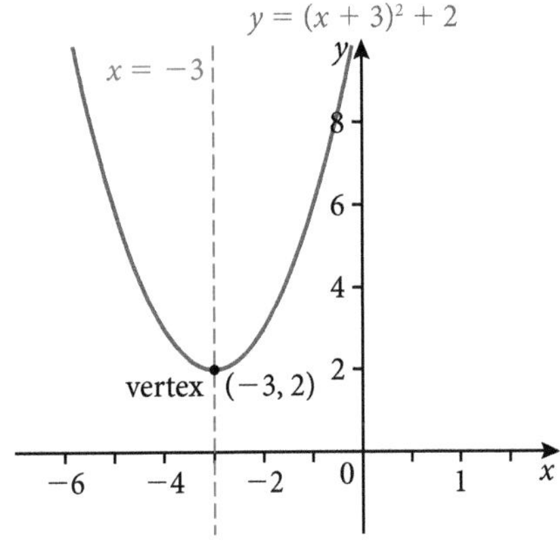

Figure 2.3 The axis of symmetry and the vertex of a parabola

We can also determine the axis of symmetry and the vertex of the graph. Figure 2.3 shows that the graph of $y = (x + 3)^2 + 2$ has an axis of symmetry of $x = -3$ and a vertex at $(-3, 2)$. The equation $y = (x + 3)^2 + 2$ can also be written as $y = x^2 + 6x + 11$. As we can identify the vertex of the parabola easily when the equation is written as $y = (x + 3)^2 + 2$, we often refer to this as the **vertex form** of the quadratic equation, and $y = x^2 + 6x + 11$ as the **general form** (or **expanded form**).

Online

See how changing constants in the vertex form of quadratic function affects the graph.

If a quadratic function is written in vertex form, that is $f(x) = a(x - h)^2 + k$, with $a \neq 0$, then the graph of f has an axis of symmetry of $x = h$ and a vertex at (h, k).

$f(x) = a(x - h)^2 + k$ is sometimes referred to as the standard form of a quadratic function.

Completing the square

When we are visualising and sketching quadratic functions, it is helpful to have them written in vertex form. We can convert a quadratic function written in general form into vertex form by completing the square. For any real number p, the quadratic expression $x^2 + px + \left(\frac{p}{2}\right)^2$ is the square of $\left(x + \frac{p}{2}\right)$. Convince yourself of this by expanding $\left(x + \frac{p}{2}\right)^2$. To complete the square, we add a constant to a quadratic expression to make it the square of a binomial. If the coefficient of the quadratic term (x^2) is $+1$, the coefficient of the linear term is p, and the constant term is $\left(\frac{p}{2}\right)^2$, then $x^2 + px + \left(\frac{p}{2}\right)^2 = \left(x + \frac{p}{2}\right)^2$ and the square is completed. Remember that the coefficient of the quadratic term (leading coefficient) must be equal to $+1$ before completing the square.

Example 2.1

Find the equation of the axis of symmetry and the coordinates of the vertex of the graph of $f(x) = x^2 - 8x + 18$ by rewriting the function in the form $f(x) = a(x - h)^2 + k$.

Solution

To complete the square and get the quadratic expression $x^2 - 8x + 18$ in the form $x^2 + px + \left(\frac{p}{2}\right)^2$, the constant term needs to be $\left(\frac{-8}{2}\right)^2 = 16$. We need to add 16, but also subtract 16, so that we are adding zero overall and not changing the original expression.

$f(x) = (x^2 - 8x + 16) - 16 + 18$ This effectively adds zero $(-16 + 16)$ to the right side.

$f(x) = x^2 - 8x + 16 + 2$ $x^2 - 8x + 16$ fits the pattern $x^2 + px + \left(\frac{p}{2}\right)^2$ with $p = -8$.

$f(x) = (x - 4)^2 + 2$ $x^2 - 8x + 16 = (x - 4)^2$

The axis of symmetry of the graph of f is the vertical line $x = 4$ and the vertex is at $(4, 2)$.

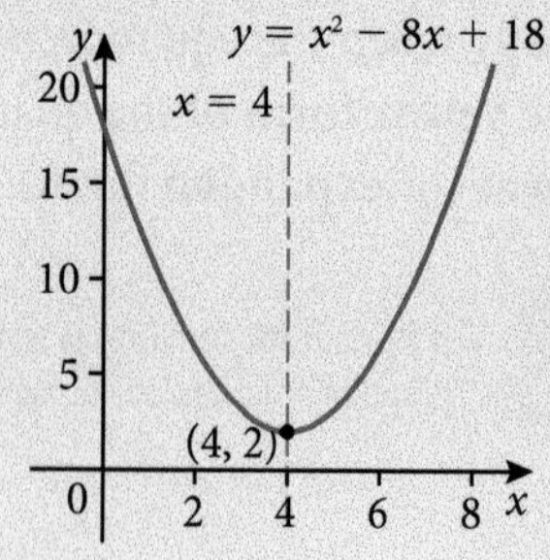

Example 2.2

For the function $g:x \mapsto -2x^2 - 12x + 7$:

(a) find the axis of symmetry and the coordinates of the vertex of the graph of g

(b) indicate the transformations that can be applied to $y = x^2$ to obtain the graph of g

(c) find the minimum or the maximum value.

Solution

(a) $g:x \mapsto -2\left(x^2 + 6x - \frac{7}{2}\right)$ — Factorise so that coefficient of quadratic term is $+1$.

$g:x \mapsto -2\left[(x^2 + 6x + 9) - 9 - \frac{7}{2}\right]$ — $p = 6 \Rightarrow \left(\frac{p}{2}\right)^2 = 9$; hence, add $+9 - 9$ (zero).

$g:x \mapsto -2\left[(x + 3)^2 - \frac{18}{2} - \frac{7}{2}\right]$ — $x^2 + 6x + 9 = (x + 3)^2$

$g:x \mapsto -2\left[(x + 3)^2 - \frac{25}{2}\right]$

$g:x \mapsto -2(x + 3)^2 + 25$ — Multiply through by -2 to remove outer brackets.

$g:x \mapsto -2(x - (-3))^2 + 25$ — Express in vertex form $g:x \mapsto a(x - h)^2 + k$.

The axis of symmetry of the graph of g is the vertical line $x = -3$ and the vertex is at $(-3, 25)$. The graph is shown in Figure 2.4.

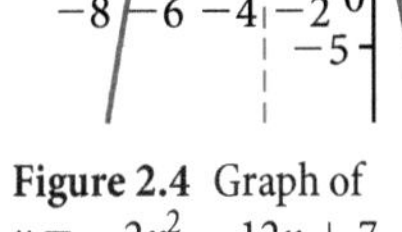

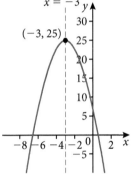

Figure 2.4 Graph of $y = -2x^2 - 12x + 7$

(b) Since $g:x \mapsto -2x^2 - 12x + 7 = -2(x + 3)^2 + 25$, we can obtain the graph of g by applying the following transformations (in the order given) on the graph of $y = x^2$:

1 horizontal translation of 3 units left
2 reflection in the x-axis (parabola opening down)
3 vertical stretch of factor 2
4 vertical translation of 25 units up.

(c) The parabola opens down because the leading coefficient is negative. Therefore, g has a maximum value and no minimum value. The maximum value is 25 (y-coordinate of vertex) at $x = -3$.

We can use the technique of completing the square to derive the **quadratic formula**. Example 2.3 derives a general expression for the axis of symmetry and the vertex of a quadratic function in the general form $f(x) = ax^2 + bx + c$ by completing the square.

Example 2.3

Find the axis of symmetry and the vertex for the general quadratic function $f(x) = ax^2 + bx + c$.

Solution

$f(x) = a\left(x^2 + \frac{b}{a}x + \frac{c}{a}\right)$ — Factorise so that the coefficient of x^2 term is $+1$.

$f(x) = a\left[\left(x^2 + \frac{b}{a}x + \left(\frac{b}{2a}\right)^2\right) - \left(\frac{b}{2a}\right)^2 + \frac{c}{a}\right]$ — $p = \frac{b}{a} \Rightarrow \left(\frac{p}{2}\right)^2 = \left(\frac{b}{2a}\right)^2$

$f(x) = a\left[\left(x + \frac{b}{2a}\right)^2 - \frac{b^2}{4a^2} + \frac{c}{a}\right]$ — $x^2 + \frac{b}{a}x + \left(\frac{b}{2a}\right)^2 = \left(x + \frac{b}{2a}\right)^2$

$f(x) = a\left(x + \frac{b}{2a}\right)^2 - \frac{b^2}{4a} + c$ — Multiply through by a.

$f(x) = a\left(x - \left(-\frac{b}{2a}\right)\right)^2 + c - \frac{b^2}{4a}$ — Express in vertex form $f(x) = a(x - h)^2 + k$.

The result in Example 2.3 leads to the following generalisation. For the graph of the quadratic function $f(x) = ax^2 + bx + c$, the axis of symmetry is the vertical line with the equation $x = -\frac{b}{2a}$ and the vertex has coordinates $\left(-\frac{b}{2a}, c - \frac{b^2}{4a}\right)$.

We can check the results from Example 2.2 using these formulae for the axis of symmetry and vertex. For the function $g: x \mapsto -2x^2 - 12x + 7$:

$$-\frac{b}{2a} = -\frac{-12}{2(-2)} = -3 \quad \Rightarrow \text{axis of symmetry is the vertical line } x = -3$$

$$c - \frac{b^2}{4a} = 7 - \frac{(-12)^2}{4(-2)} = \frac{56}{8} + \frac{144}{8} = 25 \quad \Rightarrow \text{vertex has coordinates } (-3, 25)$$

These results agree with the results from Example 2.2.

Zeros of a quadratic function

A specific value of x is a **zero** of a quadratic function $f(x) = ax^2 + bx + c$ if it is a solution (or **root**) to the equation $ax^2 + bx + c = 0$. The x-coordinate of any point(s) where f crosses the x-axis (y-coordinate is zero) is a **real zero** of the function.

A quadratic function can have no, one, or two real zeros, as Figure 2.5 illustrates. Finding all real zeros of a quadratic function requires you to solve quadratic equations of the form $ax^2 + bx + c = 0$. Although $a \neq 0$, it is possible for b or c to be equal to zero. There are five general methods for solving quadratic equations, as outlined in Table 2.1.

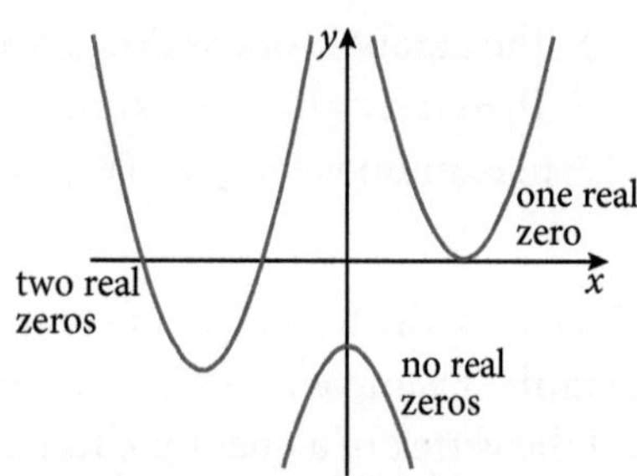

Figure 2.5 Different quadratic functions with different numbers of real zeros

Method	Examples
Square root If $x^2 = c$ and $c > 0$, then $x = \pm\sqrt{c}$	$x^2 - 25 = 0$ $x^2 = 25$ $x = \pm 5$ $(x + 2)^2 = 15$ $x + 2 = \pm\sqrt{15}$ $x = -2 \pm \sqrt{15}$
Factorising If $mn = 0$, then $m = 0$ or $n = 0$	$x^2 + 3x - 10 = 0$ $(x + 5)(x - 2) = 0$ $x = -5$ or $x = 2$ $x^2 - 7x = 0$ $x(x - 7) = 0$ $x = 0$ or $x = 7$
Completing the square If $x^2 + px + q = 0$, then $x^2 + px + \left(\frac{p}{2}\right)^2 = -q + \left(\frac{p}{2}\right)^2$, which leads to $\left(x + \frac{p}{2}\right)^2 = -q + \frac{p^2}{4}$ ... and then square root both sides (as above).	$x^2 - 8x + 5 = 0$ $x^2 - 8x + 16 = -5 + 16$ $(x - 4)^2 = 11$ $x - 4 = \pm\sqrt{11}$ $x = 4 \pm \sqrt{11}$
Quadratic formula If $ax^2 + bx + c = 0$, then $x = \frac{-b \pm \sqrt{b^2 - 4ac}}{2a}$	$2x^2 - 3x - 4 = 0$ $x = \frac{-(-3) \pm \sqrt{(-3)^2 - 4(2)(-4)}}{2(2)}$ $x = \frac{3 \pm \sqrt{41}}{4}$
Graphing Graph the equation $y = ax^2 + bx + c$ on your GDC. Use the GDC's graph analysis features to determine the x-coordinates of the point(s) where the parabola intersects the x-axis.	$2x^2 - 5x - 7 = 0$ GDC calculations reveal that the zeros are $x = \frac{7}{2}$ and $x = -1$ 1: Actions, 2: View, 3: Graph Entry/Edit, 4: Window/Zoom, 5: Trace, 6: Analyse Graph, 7: Table, 8: Geometry, 9: Settings... 1: Zero, 2: Minimum, 3: Maximum, 4: Intersection, 5: dy/dx, 6: Integral, 7: Bounded Area f1(x)=2·x²−5·x−7; 6.67; y; 1; x; -10; (-1,0); 1; (3.5,0); 10; -6.67

Table 2.1 Methods for finding zeros of quadratic functions

The quadratic formula and the discriminant

The expression beneath the radical sign in the quadratic formula, $b^2 - 4ac$, tells us whether the zeros of a quadratic function are real or not real (imaginary). Because it acts to 'discriminate' between the types of zeros, $b^2 - 4ac$ is called the **discriminant**. It is often labelled with the Greek letter Δ (upper case delta). The value of the discriminant can also indicate if the zeros are equal and if they are rational.

For the quadratic function $f(x) = ax^2 + bx + c$, $(a \neq 0)$ where a, b, and c are real numbers:

- If $\Delta = b^2 - 4ac > 0$, then f has two distinct real zeros, and the graph of f intersects the x-axis twice.
- If $\Delta = b^2 - 4ac = 0$, then f has one real zero (double root), and the graph of f intersects the x-axis once (i.e. it is tangent to the x-axis).
- If $\Delta = b^2 - 4ac < 0$, then f has two imaginary zeros (non-real), and the graph of f does not intersect the x-axis.
- In the special case when a, b, and c are integers and the discriminant is the square of an integer (a perfect square), the polynomial $ax^2 + bx + c$ has two distinct **rational zeros**.

If the zeros of a quadratic polynomial are rational (either two distinct rational zeros or two equal rational zeros) then the polynomial can be factorised. That is, if $ax^2 + bx + c$ has rational zeros then $ax^2 + bx + c = (mx + n)(px + q)$ where m, n, p, and q are rational numbers.

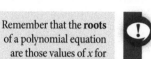

When the discriminant is zero, the solution of a quadratic function is $x = \frac{-b \pm \sqrt{b^2 - 4ac}}{2a} = \frac{-b \pm \sqrt{0}}{2a} = -\frac{b}{2a}$. This solution of $-\frac{b}{2a}$ is called a double zero (or root).
If a and b are integers, then the zero $-\frac{b}{2a}$ will be rational.

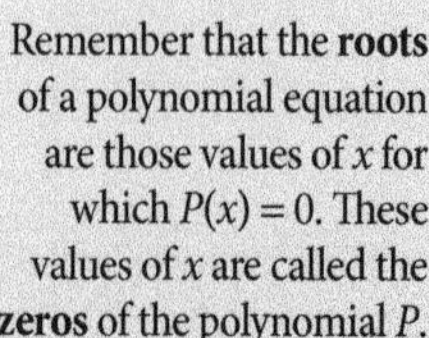

Remember that the **roots** of a polynomial equation are those values of x for which $P(x) = 0$. These values of x are called the **zeros** of the polynomial P.

Example 2.4

Use the discriminant to determine how many real roots each equation has. Confirm the result by graphing the corresponding quadratic function for each equation on your GDC.

(a) $2x^2 + 5x - 3 = 0$ (b) $4x^2 - 12x + 9 = 0$ (c) $2x^2 - 5x + 6 = 0$

Solution

(a) The discriminant is $\Delta = 5^2 - 4(2)(-3) = 49 > 0$.
Therefore, the equation has two distinct real roots.
This result is confirmed by the graph of the quadratic function $y = 2x^2 + 5x - 3$, which clearly shows that it intersects the x-axis twice. Since $\Delta = 49$ is a perfect square, the two roots are also rational and the quadratic polynomial $2x^2 + 5x - 3$ can be factorised:

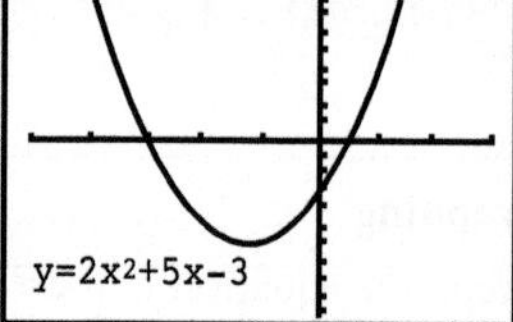

$2x^2 + 5x - 3 = (2x - 1)(x + 3) = 0$

The two rational roots are $x = \frac{1}{2}$ and $x = -3$

(b) The discriminant is $\Delta = (-12)^2 - 4(4)(9) = 0$. Therefore, the equation has one rational root (a double root). The graph on the GDC of $y = 4x^2 - 12x + 9$ appears to intersect the x-axis at only one point. We can be more confident with this conclusion by investigating further, for example by tracing or looking at a table of values on the GDC.

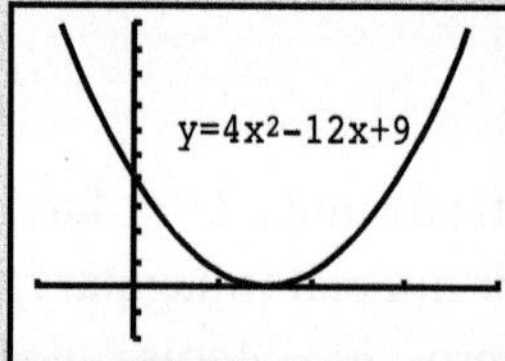

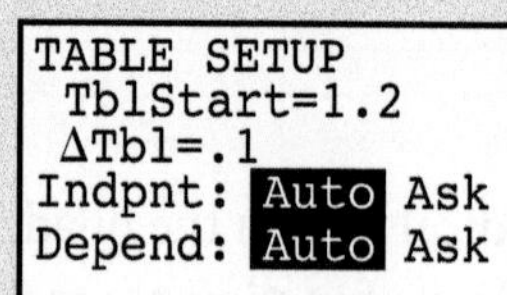

X	Y1
1.2	.36
1.3	.16
1.4	.04
1.5	0
1.6	.04
1.7	.16
1.8	.36

Y1=0

Since the root is rational ($\Delta = 0$) the polynomial $4x^2 - 12x + 9$ can be factorised:

$4x^2 - 12x + 9 = (2x - 3)(2x - 3) = \left[2\left(x - \frac{3}{2}\right)2\left(x - \frac{3}{2}\right)\right] = 4\left(x - \frac{3}{2}\right)^2 = 0$

There are two equal linear factors, which means there are two equal rational zeros, both equal to $\frac{3}{2}$ in this case.

(c) The discriminant is $\Delta = (-5)^2 - 4(2)(6) = -23 < 0$.
Therefore, the equation has no real roots.
This result is confirmed by the graph of the quadratic function $y = 2x^2 - 5x + 6$, which clearly shows that the graph does not intersect the x-axis.
The equation will have two imaginary (non-real) roots.

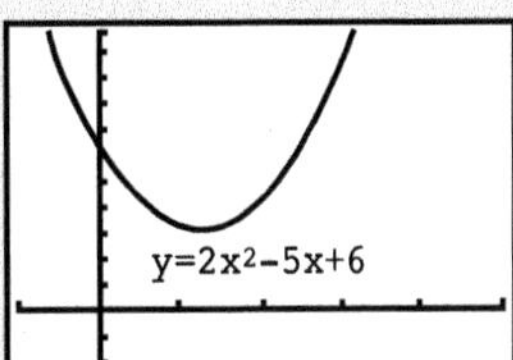

Example 2.5

For $4x^2 + 4kx + 9 = 0$, determine the value(s) of k so that the equation has:

(a) one real zero (b) two distinct real zeros (c) no real zeros.

Solution

(a) For one real zero, $\Delta = (4k)^2 - 4(4)(9) = 0$
$\Rightarrow 16k^2 - 144 = 0 \Rightarrow 16k^2 = 144 \Rightarrow k^2 = 9 \Rightarrow k = \pm 3$

(b) For two distinct real zeros, $\Delta = (4k)^2 - 4(4)(9) > 0$
$\Rightarrow 16k^2 > 144 \Rightarrow k^2 > 9 \Rightarrow k < -3 \text{ or } k > 3$

(c) For no real zeros, $\Delta = (4k)^2 - 4(4)(9) < 0$
$\Rightarrow 16k^2 < 144 \Rightarrow k^2 < 9 \Rightarrow k > -3 \text{ and } k < 3 \Rightarrow -3 < k < 3$

The graph of $f(x) = a(x - p)(x - q)$

If a quadratic function is written in the form $f(x) = a(x - p)(x - q)$ then we can easily identify the x-intercepts of the graph of f. Consider that $f(p) = a(p - p)(p - q) = a(0)(p - q) = 0$ and that $f(q) = a(q - p)(q - q) = a(q - p)(0) = 0$. Therefore, the quadratic function $f(x) = a(x - p)(x - q)$ will intersect the x-axis at the points $(p, 0)$ and $(q, 0)$. We need to factorise in order to rewrite a quadratic function in the form $f(x) = ax^2 + bx + c$ to the form $f(x) = a(x - p)(x - q)$. Hence, $f(x) = a(x - p)(x - q)$ can be referred to as the **factorised form** of a quadratic function. As a parabola is symmetric, the x-intercepts $(p, 0)$ and $(q, 0)$ will be equidistant from the axis of symmetry (see Figure 2.6). As a result, the equation of the axis of symmetry and the x-coordinate of the vertex of the parabola can be found by finding the average of p and q.

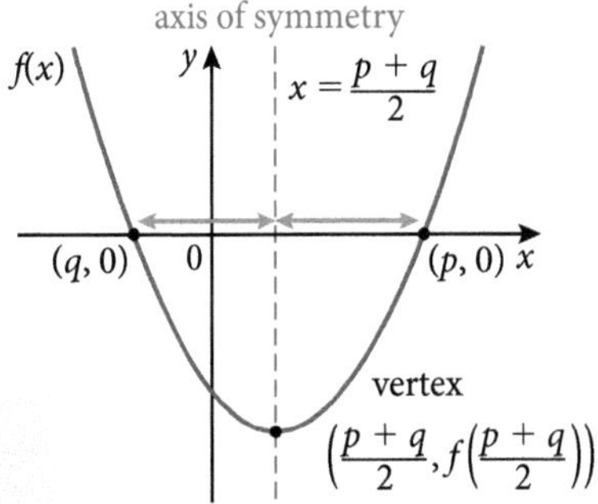

Figure 2.6 Features of the graph of a quadratic function

If a quadratic function is written in the form $f(x) = a(x - p)(x - q)$, with $a \neq 0$, then the graph of f has x-intercepts at $(p, 0)$ and $(q, 0)$, an axis of symmetry with equation $x = \frac{p+q}{2}$, and a vertex at $\left(\frac{p+q}{2}, f\left(\frac{p+q}{2}\right)\right)$.

Example 2.6

Find the equation of each quadratic function from the graph in the form $f(x) = a(x - p)(x - q)$ and also in the form $f(x) = ax^2 + bx + c$.

(a)

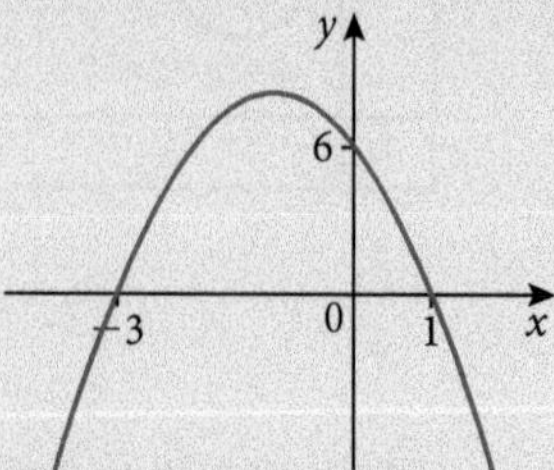

(b)

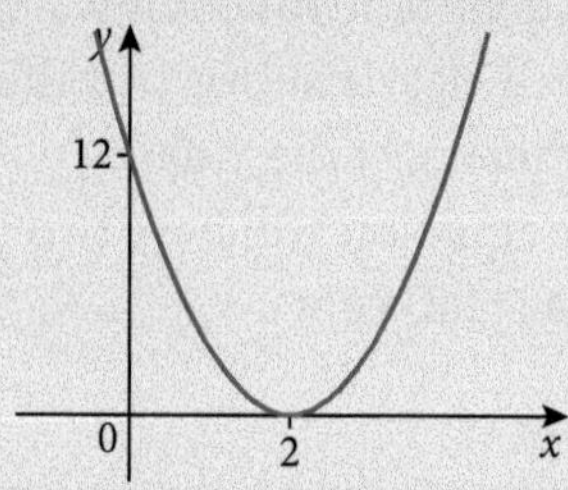

Solution

(a) Since the x-intercepts are -3 and 1 then $y = a(x + 3)(x - 1)$.

The y-intercept is 6, so when $x = 0, y = 6$. Hence, $6 = a(0 + 3)(0 - 1) = -3a \Rightarrow a = -2$ ($a < 0$ agrees with the fact that the parabola is opening downward). The function is $f(x) = -2(x + 3)(x - 1)$.

Expanding the brackets: $f(x) = -2x^2 - 4x + 6$

(b) The function has one x-intercept at 2 (double root), so $p = q = 2$ and $y = a(x - 2)(x - 2) = a(x - 2)^2$. The y-intercept is 12, so when $x = 0, y = 12$. Hence, $12 = a(0 - 2)^2 = 4a \Rightarrow a = 3$ ($a > 0$ agrees with the parabola opening up). The function is $f(x) = 3(x - 2)^2$. Expanding the brackets: $f(x) = 3x^2 - 12x + 12$

Example 2.7

The graph of a quadratic function intersects the x-axis at the points $(-6, 0)$ and $(-2, 0)$ and also passes through the point $(2, 16)$.

(a) Write the function in the form $f(x) = a(x - p)(x - q)$.

(b) Find the coordinates of the vertex of the parabola.

(c) Write the function in the form $f(x) = a(x - h)^2 + k$.

Solution

(a) The x-intercepts of $(-6, 0)$ and $(-2, 0)$ gives $f(x) = a(x + 6)(x + 2)$. Since f passes through $(2, 16)$, then

$f(2) = 16 \Rightarrow f(2) = a(2 + 6)(2 + 2) \Rightarrow 32a = 16 \Rightarrow a = \frac{1}{2}$

Therefore, $f(x) = \frac{1}{2}(x + 6)(x + 2)$

(b) The x-coordinate of the vertex is the average of the x-coordinates of the intercepts. $x = \frac{-6-2}{2} = -4$

The y-coordinate of the vertex is $y = f(-4) = \frac{1}{2}(-4+6)(-4+2) = -2$

Hence, the coordinates of the vertex are $(-4, -2)$.

(c) In vertex form, the quadratic function is $f(x) = \frac{1}{2}(x+4)^2 - 2$

Quadratic function, $a \neq 0$	**Graph of function**	**Results**
General form $f(x) = ax^2 + bx + c$ $\Delta = b^2 - 4ac$ (discriminant)	Parabola opens up if $a > 0$ Parabola opens down if $a < 0$ If $\Delta \geqslant 0$, f has x-intercept(s) $x = -\frac{b}{2a}$ $\left(\frac{-b-\sqrt{\Delta}}{2a}, 0\right)$ $\left(\frac{-b+\sqrt{\Delta}}{2a}, 0\right)$ $\left(-\frac{b}{2a}, c - \frac{b^2}{4a}\right)$ If $\Delta < 0$, f has no x-intercepts	Axis of symmetry is $x = -\frac{b}{2a}$ If $\Delta \geqslant 0$, f has x-intercept(s) $\left(\frac{-b \pm \sqrt{\Delta}}{2a}, 0\right)$ Vertex is $\left(-\frac{b}{2a}, c - \frac{b^2}{4a}\right)$
Vertex form $f(x) = a(x-h)^2 + k$	$x = h$ (h, k)	Axis of symmetry is $x = h$ Vertex is (h, k)
Factorised form (two distinct rational zeros) $f(x) = a(x-p)(x-q)$	$x = \frac{p+q}{2}$ $(q, 0)$ $(p, 0)$ $\left(\frac{p+q}{2}, f\left(\frac{p+q}{2}\right)\right)$	Axis of symmetry is $x = \frac{p+q}{2}$ x-intercepts are: $(p, 0)$ and $(q, 0)$
Factorised form (one rational zero) $f(x) = a(x-p)^2$	$x = p$ $(p, 0)$	Axis of symmetry is $x = p$ Vertex and x-intercept is $(p, 0)$

Table 2.2 Properties of quadratics

Exercise 2.1

1. For each of the quadratic functions f, find:
   (i) the equation for the axis of symmetry and the coordinates of the vertex by algebraic methods
   (ii) the transformation(s) that can be applied to $y = x^2$ to obtain the graph of $y = f(x)$
   (iii) the minimum or maximum value of f.

   Check your results by using your GDC.

   (a) $f:x \mapsto x^2 - 10x + 32$
   (b) $f:x \mapsto x^2 + 6x + 8$
   (c) $f:x \mapsto -2x^2 - 4x + 10$
   (d) $f:x \mapsto 4x^2 - 4x + 9$
   (e) $f:x \mapsto \frac{1}{2}x^2 + 7x + 26$

2. Solve each quadratic equation by using factorisation.
   (a) $x^2 + 2x - 8 = 0$
   (b) $x^2 = 3x + 10$
   (c) $6x^2 - 9x = 0$
   (d) $6 + 5x = x^2$
   (e) $x^2 = 6x - 9$
   (f) $3x^2 + 11x - 4 = 0$
   (g) $3x^2 + 18 = 15x$
   (h) $9x - 2 = 4x^2$

3. Use the method of completing the square to solve each quadratic equation.
   (a) $x^2 + 4x - 3 = 0$
   (b) $x^2 - 4x - 5 = 0$
   (c) $x^2 - 2x + 3 = 0$
   (d) $2x^2 + 16x + 6 = 0$
   (e) $x^2 + 2x - 8 = 0$
   (f) $-2x^2 + 4x + 9 = 0$

4. Let $f(x) = x^2 - 4x - 1$.
   (a) Use the quadratic formula to find the zeros of the function.
   (b) Use the zeros to find the equation for the axis of symmetry of the parabola.
   (c) Find the minimum value of f.

5. Determine the number of real solutions to each equation.
   (a) $x^2 + 3x + 2 = 0$
   (b) $2x^2 - 3x + 2 = 0$
   (c) $x^2 - 1 = 0$
   (d) $2x^2 - \frac{9}{4}x + 1 = 0$

6. Find the value(s) of p for which the equation $2x^2 + px + 1 = 0$ has one real solution.

7. Find the value(s) of k for which the equation $x^2 + 4x + k = 0$ has two distinct real solutions.

8. The equation $x^2 - 4kx + 4 = 0$ has two distinct real solutions. Find the set of all possible values of k.

9. Find all possible values of m so that the graph of the function $g:x \mapsto mx^2 + 6x + m$ does not touch the x-axis.

10. Find the range of values of k such that $3x^2 - 12x + k > 0$ for all real values of x.

For question 10, consider what must be true about the zeros of the quadratic equation $y = 3x^2 - 12x + k$

11. Prove that the expression $x - 2 - x^2$ is negative for all real values of x.

12. Find a quadratic function in the form $y = ax^2 + bx + c$ that satisfies the given conditions.
    (a) The function has zeros of $x = -1$ and $x = 4$ and its graph intersects the y-axis at $(0, 8)$.
    (b) The function has zeros of $x = \frac{1}{2}$ and $x = 3$ and its graph passes through the point $(-1, 4)$.

13. Find the range of values for k in order for the equation $2x^2 + (3 - k)x + k + 3 = 0$ to have two imaginary (non-real) solutions.

14. Find the values of m such that the function $f(x) = 5x^2 - mx + 2$ has two distinct real zeros.

15. Do the following for each function.
    (i) Write it in vertex form: $y = a(x - h)^2 + k$.
    (ii) State the coordinates of the vertex.
    (iii) Indicate whether the vertex is a maximum or minimum point.
    (a) $y = x^2 + 4x + 1$
    (b) $y = -2x^2 + 4x + 3$
    (c) $y = 3x^2 + 12x + 12$
    (d) $y = 3 + 6x - x^2$

16. A quadratic function $h(x)$ passes through the points $(2,0)$ and $(6,0)$. The graph of h is a parabola.
    (a) Write down the equation for the axis of symmetry of the parabola.
    (b) Given that the graph of h also passes through the point $(8,6)$, find an expression for h and write it in the form $h(x) = ax^2 + bx + c$.

17. Show that there is no real value t for which the equation $2x^2 + (2 - t)x + t^2 + 3 = 0$ has real roots.

18. Show that the two roots of $ax^2 + bx + a = 0$ are reciprocals of each other.

2.2 Rational functions

Another important category of functions is **rational functions**, which are functions in the form $R(x) = \frac{f(x)}{g(x)}$, where f and g are polynomials and the domain of the function R is the set of all real numbers not including the real zeros of polynomial g in the denominator. In this course, we will consider rational functions where the numerator and denominator are linear functions. Some examples of such rational functions are

$p(x) = \frac{3}{x - 5}$ and $q(x) = \frac{4x - 2}{2x + 5}$

The domain of function p is all real numbers x not including $x = 5$ (i.e. $x \in \mathbb{R}, x \neq 5$) and the domain of function q is $x \in \mathbb{R}, x \neq -\frac{5}{2}$

Example 2.8

Find the domain and range of $h(x) = \frac{1}{x-2}$. Sketch the graph of h.

Solution

A fraction is only zero if its numerator is zero.

Because the denominator is zero when $x = 2$, the domain of h is all real numbers except $x = 2$, $x \in \mathbb{R}$, $x \neq 2$.

Determining the range of the function is a little less straightforward. It is clear that the function could never take on a value of zero because that will only occur if the numerator is zero. And since the denominator can have any value except zero, it seems that the function values of h could be any real number except zero. To confirm this, and to determine the behaviour of the function (and shape of the graph), some values of the domain and range (pairs of coordinates) are displayed in the tables below.

***x* approaches 2 from the left**	
x	$h(x)$
−98	−0.01
−8	−0.1
0	−0.5
1	−1
1.5	−2
1.9	−10
1.99	−100
1.999	−1000

***x* approaches 2 from the right**	
x	$h(x)$
102	0.01
12	0.1
4	0.5
3	1
2.5	2
2.1	10
2.01	100
2.001	1000

The values in the table provide clear evidence that the range of h is all real numbers except zero; that is, $h(x) \in \mathbb{R}$, $h(x) \neq 0$.

The values in the table also show that as $x \to -\infty$, $h(x) \to 0$ from below (sometimes written $h(x) \to 0^-$) and as $x \to +\infty$, $h(x) \to 0$ from above ($h(x) \to 0^+$). It follows that the line with equation $y = 0$ (the x-axis) is a horizontal asymptote for the graph of h. As $x \to 2$ from the left (sometimes written $x \to 2^-$), $h(x)$ appears to decrease without bound, whereas as $x \to 2$ from the right ($x \to 2^+$), $h(x)$ appears to increase without bound. This indicates that the graph of h will have a vertical asymptote at $x = 2$. This behaviour is confirmed by the graph.

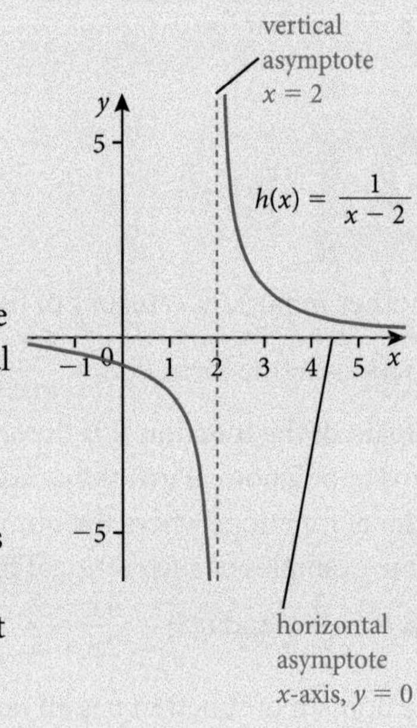

The line $x = c$ is a **horizontal asymptote** of the graph of the function f if at least one of the following statements is true:

- as $x \to +\infty$, then $f(x) \to c^+$
- as $x \to -\infty$, then $f(x) \to c^+$
- as $x \to +\infty$, then $f(x) \to c^-$
- as $x \to -\infty$, then $f(x) \to c^-$

The line $x = d$ is a **vertical asymptote** of the graph of the function f if at least one of the following statements is true:

- as $x \to d^+$, then $f(x) \to +\infty$
- as $x \to d^+$, then $f(x) \to -\infty$
- as $x \to d^-$, then $f(x) \to +\infty$
- as $x \to d^-$, then $f(x) \to -\infty$

Example 2.9

Consider the function $f(x) = \dfrac{3x - 8}{2x + 6}$. Sketch the graph of f and identify any asymptotes and any x- or y-intercepts. Use the sketch to confirm the domain and range of the function.

Solution

Firstly, factorise where possible.

$$f(x) = \frac{3x - 8}{2x + 6} = \frac{3x - 8}{2(x + 3)}$$

An x-intercept occurs when the numerator is zero. Hence, $\left(\frac{8}{3}, 0\right)$ is an x-intercept.

A y-intercept occurs when $x = 0$

$f(0) = \dfrac{3(0) - 8}{2(0) + 6} = -\dfrac{8}{6} = -\dfrac{4}{3}$, so, the y-intercept is $\left(0, -\frac{4}{3}\right)$.

Any vertical asymptote will occur when the denominator is zero – that is, where the function is undefined. From the factorised form of f we see that the line $x = -3$ will be a vertical asymptote for the graph of f.

We need to determine if the graph of f goes down ($f(x) \to -\infty$) or goes up ($f(x) \to \infty$) on either side of the vertical asymptote. It's easiest to do this by analysing what the sign of f will be as x approaches -3 from both the left and right. For example, as $x \to -3^-$ we can use a test value close to and to the left of -3 (e.g. $x = -3.1$) to check whether $f(x)$ is positive or negative to the left of -3.

$$f(-3.1) = \frac{3(-3.1) - 8}{2(-3.1) + 6} \Rightarrow \frac{(-)}{(-)} = (+) \Rightarrow f(x) > 0$$

$\Rightarrow$ as $x \to -3^-$, so $f(x) \to +\infty$ (rises).

As $x \to -3^+$ we use a test value close to and to the right of -3 (e.g. $x = -2.9$) to check whether $f(x)$ is positive or negative to the right of -3.

$$f(2.9) = \frac{3(-2.9) - 8}{2(-2.9) + 6} \Rightarrow \frac{(-)}{(+)} \Rightarrow f(x) < 0$$

$\Rightarrow$ as $x \to -3^+$, so $f(x) \to -\infty$ (falls)

As the number n gets closer to infinity, the number $\frac{1}{n}$ gets closer to 0.

Conversely, as n gets closer to 0, the number $\frac{1}{n}$ gets closer to infinity. These facts can be expressed simply as:

$\frac{1}{\text{BIG}}$ = little and

$\frac{1}{\text{little}}$ = BIG. They can also be expressed more mathematically using the concept of a limit expressed in limit notation as $\lim_{n \to \infty} \frac{1}{n} = 0$ and $\lim_{n \to 0} \frac{1}{n} = \infty$.

A horizontal asymptote (if it exists) is the value that $f(x)$ approaches as $x \to \pm\infty$ (far to the left and right on the graph). To find this value, we divide both the numerator and denominator by x.

$$f(x) = \frac{\frac{3x}{x} - \frac{8}{x}}{\frac{2x}{x} + \frac{6}{x}} = \frac{3-0}{2+0} = \frac{3}{2}$$ Hence, the horizontal asymptote is $y = \frac{3}{2}$

Now we know the behaviour (rising or falling) of the function on either side of the vertical asymptote and that the graph will approach the horizontal asymptote as $x \to \pm\infty$, an accurate sketch of the graph can be made as shown.

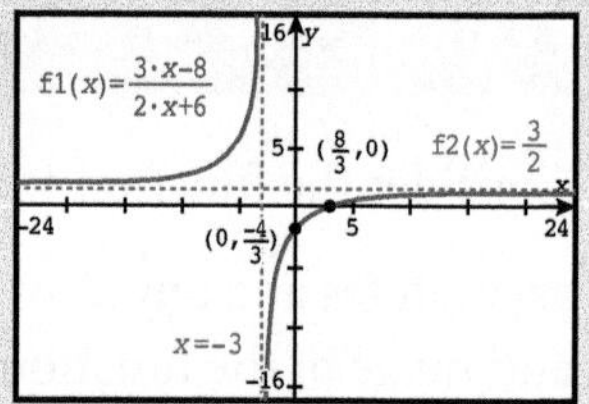

The domain of f is $x \in \mathbb{R}, x \neq -3$, and the range is $f(x) \in \mathbb{R}, f(x) \neq \frac{3}{2}$, which is confirmed in the sketch of the graph of f above.

The steps to analyse a rational function $R(x) = \frac{f(x)}{g(x)}$, given functions f and g have no common factors, are as follows.

1. Factorise: Completely factorise both the numerator and denominator.
2. Intercepts: A zero of f (numerator) will be a zero of R and hence an x-intercept of the graph of R. The y-intercept is found by evaluating $R(0)$.
3. Vertical asymptotes: A zero of g (denominator) will give the location of a vertical asymptote (if any). Then perform a sign analysis to see if $R(x) \to +\infty$ or $R(x) \to -\infty$ on either side of each vertical asymptote.
4. Horizontal asymptotes: Find the horizontal asymptote (if any) by dividing both f (numerator) and g (denominator) by x, and then letting $x \to \pm\infty$.
5. Sketch of graph: Start by drawing dashed lines where the asymptotes are located. Use the information about the intercepts, whether $R(x)$ falls or rises on either side of a vertical asymptote, and additional points as needed to make an accurate sketch.
6. Domain and range: The domain of R will be all real numbers except the zeros of g. You need to study the graph carefully in order to determine the range. The value of the function at the horizontal asymptote will not be included in the range.

Exercise 2.2

1. Sketch the graph of each rational function without the aid of your GDC. On your sketch clearly indicate any x- or y-intercepts and any asymptotes (vertical or horizontal). Use your GDC to verify your sketch.

(a) $f(x) = \frac{1}{x+2}$ **(b)** $g(x) = \frac{3}{x-2}$

(c) $h(x) = \frac{1-4x}{1-x}$ **(d)** $R(x) = \frac{x}{x-5}$

(e) $p(x) = \frac{2x-4}{2x-3}$ **(f)** $M(x) = \frac{3x+8}{x}$

2. Use your GDC to sketch a graph of each function and state the domain and range of the function.

(a) $f(x) = \dfrac{3x + 12}{x + 6}$ (b) $g(x) = \dfrac{4 - x}{3x - 9}$

(c) $h(x) = \dfrac{6}{x + 6}$ (d) $r(x) = \dfrac{2x - 9}{15 - 3x}$

3. Consider the function $g(x) = 3 - \dfrac{x - 4}{x + 3}$.

(a) Graph g on your GDC.

(b) Determine the equations of all asymptotes for the graph of g.

(c) Express g in the form $g(x) = \dfrac{ax - b}{cx + d}$. Graph your result on a GDC to check that it is equivalent to $g(x) = 3 - \dfrac{x - 4}{x + 3}$.

4. If a, b, and c are all positive, sketch the curve $y = \dfrac{x - a}{(x - b)(x - c)}$ for each of the following conditions.

(a) $a < b < c$ (b) $b < a < c$ (c) $b < c < a$

5. A drug is given to a patient and the concentration of the drug in the bloodstream is carefully monitored. At time $0 \leqslant t \leqslant 24$ (in hours after patient receiving the drug), the concentration, in milligrams per litre ($\text{mg}\,\text{l}^{-1}$) is given by the function

$$C(t) = \frac{30 - t}{t + 1}$$

(a) Sketch a graph of the drug concentration ($\text{mg}\,\text{l}^{-1}$) versus time (hours) for the relevant domain.

(b) When does the highest concentration of the drug occur, and what is it?

(c) How long does it take for the concentration to drop below $1.0\,\text{mg}\,\text{l}^{-1}$?

6. Consider the function $f(x) = \dfrac{x + a}{bx + c}$, $x \neq -\dfrac{c}{b}$. The graph of f has asymptotes $x = -6$ and $y = 3$, and the point $\left(6, \dfrac{5}{2}\right)$ lies on the graph. Find the values of a, b, and c.

2.3 Solving equations and inequalities

You are familiar with solving linear equations and, in this chapter, we have studied some approaches to analysing and solving quadratic equations. Some problems lead to other kinds of equations. For example, some equations we encounter may involve radicals, fractions, absolute values, or expressions that are not quadratic but display a quadratic 'structure'. Furthermore, some problems do not involve equations but inequalities. You need to be familiar with effective methods for solving inequalities involving quadratic expressions.

Equations involving a radical

Example 2.10

Solve $\sqrt{3x + 6} = 2x + 1$ for x.

Solution

Squaring both sides gives $3x + 6 = (2x + 1)^2$

$$3x + 6 = 4x^2 + 4x + 1$$

$$4x^2 + x - 5 = 0$$

Factorising $(4x + 5)(x - 1) = 0$

$$x = -\frac{5}{4} \text{ or } x = 1$$

Check both solutions in the original equation.

When

$x = -\frac{5}{4}, \sqrt{3\left(-\frac{5}{4}\right) + 6} = 2\left(-\frac{5}{4}\right) + 1 \Rightarrow \sqrt{\frac{9}{4}} = -\frac{3}{2} \Rightarrow \frac{3}{2} \neq -\frac{3}{2}$

$\Rightarrow \ x = -\frac{5}{4}$ is not a solution.

When $x = 1, \sqrt{3(1) + 6} = 2(1) + 1 \Rightarrow \sqrt{9} = 3 \Rightarrow 3 = 3$

$\Rightarrow \ x = 1$ is the only solution.

If two quantities are equal, for example $a = b$, then it is certainly true that $a^2 = b^2$, and $a^3 = b^3$, and so on. However, the converse is not necessarily true. A simple example can illustrate this.

Consider the trivial equation $x = 3$. There is only one value of x that makes the equation true, and that is 3. Now if we take this original equation and square both sides we transform it to the equation $x^2 = 9$. This transformed equation has two solutions, 3 and -3, so it is not equivalent to the original equation. By squaring both sides, we gained an extra solution, often called an **extraneous solution**, that satisfies the transformed equation but not the original equation, as we saw in Example 2.10. Whenever you raise both sides of an equation by a power, it is imperative that you check all solutions in the original equation.

Equations involving fractions

It is also possible for extraneous solutions to appear when solving equations with fractions.

Example 2.11

Find all real solutions of the equation $\frac{2x}{4 - x^2} + \frac{1}{x + 2} = 3$ and verify solution(s) with a GDC.

Solution

Multiply both sides of the equation by the least common denominator of the fractions.

Factorising $4 - x^2$ gives $(2 - x)(2 + x)$ so $4 - x^2$ is the least common denominator.

$$\frac{4 - x^2}{1} \cdot \frac{2x}{4 - x^2} + \frac{(2 - x)(2 + x)}{1} \cdot \frac{1}{x + 2} = 3(4 - x^2)$$

$$2x + 2 - x = 12 - 3x^2$$

$$3x^2 + x - 10 = 0$$

$$(3x - 5)(x + 2) = 0$$

$$x = \frac{5}{3} \text{ or } x = -2$$

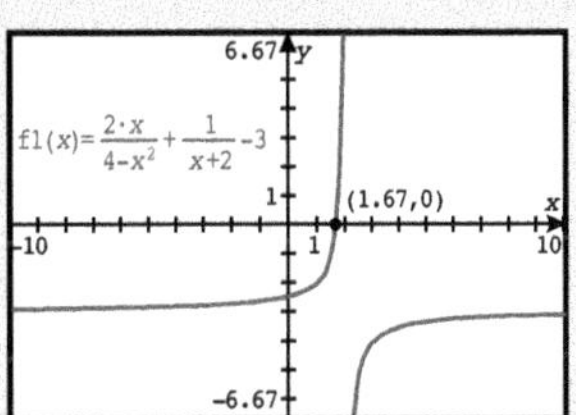

Clearly $x = -2$ cannot be a solution because that would cause division by zero in the original equation. The GDC image shows that the equation $y = \frac{2x}{4 - x^2} + \frac{1}{x + 2} - 3$ has an x-intercept at $\left(\frac{5}{3}, 0\right)$, confirming the solution $x = \frac{5}{3}$.

Not only is it possible to gain an extraneous solution when solving certain equations, it is also possible to lose a correct solution by incorrectly dividing both sides of an equation by a common factor. For example, solve for x in the equation $4(x + 2)^2 = 3x(x + 2)$. Dividing both sides by $(x + 2)$ gives
$4(x + 2) = 3x$
$\Rightarrow \quad 4x + 8 = 3x$
$\Rightarrow \quad x = -8.$
However, there are two solutions, $x = -8$ and $x = -2$. The solution of $x = -2$ was lost because a factor of $x + 2$ was eliminated from both sides of the original equation. This is a common error to be avoided.

Equations in quadratic form

In the first section of this chapter we covered methods of solving quadratic equations. As the two previous examples illustrate, quadratic equations commonly appear in a range of mathematical problems. The methods of solving quadratics can sometimes be applied to other equations. An equation in the form $at^2 + bt + c = 0$, where t is an algebraic expression, is an equation in **quadratic form**. We can solve such equations by substituting for the algebraic expression and then applying an appropriate method for solving a quadratic equation.

Example 2.12

Find all real solutions of the equation $2m^4 - 5m^2 + 2 = 0$.

Solution

The equation can be written as $2(m^2)^2 - 5(m^2) + 2 = 0$, so it is quadratic in terms of m^2.

$$2m^4 - 5m^2 + 2 = 0$$

substitute t for m^2: $\quad 2t^2 - 5t + 2 = 0$

$$(2t - 1)(t - 2) = 0$$

$$t = \frac{1}{2} \text{ or } t = 2$$

substituting m^2 for t: $m^2 = \frac{1}{2}$ or $m^2 = 2$

$m = \pm\sqrt{\frac{1}{2}} = \pm\frac{\sqrt{2}}{2}$ or $m = \pm\sqrt{2}$

Check the solutions (which are in two pairs of opposites) by substituting them directly into the original equation. A value for m will be raised to the 4th and 2nd powers, so we only need to check one value from each pair of opposites.

When $m = \frac{\sqrt{2}}{2}$: $2\left(\frac{\sqrt{2}}{2}\right)^4 - 5\left(\frac{\sqrt{2}}{2}\right)^2 + 2 = 0$

$\Rightarrow\ 2\left(\frac{1}{4}\right) - 5\left(\frac{1}{2}\right) + 2 = 0 \ \Rightarrow\ \frac{1}{2} - \frac{5}{2} + 2 = 0 \ \Rightarrow\ 0 = 0$

When $m = \sqrt{2}$: $2(\sqrt{2})^4 - 5(\sqrt{2})^2 + 2 = 0$

$\Rightarrow\ 2(4) - 5(2) + 2 = 0 \ \Rightarrow\ 8 - 10 + 2 = 0 \ \Rightarrow\ 0 = 0$

Therefore, the solutions to the equation are $m = \frac{\sqrt{2}}{2}, -\frac{\sqrt{2}}{2}, \sqrt{2},$ and $-\sqrt{2}$

Example 2.13

Find all solutions, expressed exactly, to the equation $w^{\frac{1}{2}} = 4w^{\frac{1}{4}} - 2$

Solution

$w^{\frac{1}{2}} - 4w^{\frac{1}{4}} + 2 = 0$	Set the equation to zero
$(w^{\frac{1}{4}})^2 - (4w^{\frac{1}{4}}) + 2 = 0$	Attempt to write in quadratic form $at^2 + bt + c = 0$
$t^2 - 4t + 2 = 0$	Make appropriate substitution: let $w^{\frac{1}{4}} = t$
$t = \frac{-(-4) \pm\sqrt{(-4)^2 - 4(1)(2)}}{2}$	Trinomial does not factorise, so apply quadratic formula
$t = \frac{4 \pm\sqrt{8}}{2} = \frac{4 \pm 2\sqrt{2}}{2} = 2 \pm \sqrt{2}$	
$w^{\frac{1}{4}} = 2 \pm \sqrt{2}$	Substitute $w^{\frac{1}{4}}$ back in for t

$w = (2 + \sqrt{2})^4$ or $w = (2 - \sqrt{2})^4$

$w = ((2 + \sqrt{2})^2)^2$ or $w = ((2 - \sqrt{2})^2)^2$

$w = (6 + 4\sqrt{2})^2$ or $w = (6 - 4\sqrt{2})^2$

$w = 68 + 48\sqrt{2}$ or

$w = 68 - 48\sqrt{2}$

Approximate values found with GDC:

```
68+48√2
          135.882251
68−48√2
        0.1177490061
```

It is difficult to check these two solutions by substituting them directly into the original equation as we did in Example 2.12. It is more efficient to use our GDC.

Most GDC models have an equation solver. A limitation of this GDC feature is that it will usually return only an approximate solution. However, even if exact solutions are required, approximate solutions from a GDC are still very helpful as a check of the exact solutions obtained algebraically. The image shows a GDC solver being used to find the approximate solutions to the equation $4w^{\frac{1}{2}} = w^{\frac{1}{4}} + 2$.

nSolve$\left(w^{\frac{1}{2}}=4\cdot w^{\frac{1}{4}}-2,w\right)$ 0.117749

nSolve$\left(w^{\frac{1}{2}}=4\cdot w^{\frac{1}{4}}-2,w\right)|w>1$ 135.882

If you are solving an equation in an exam for which a GDC is allowed and an exact answer is not required, then usually the most effective and efficient method is to solve the equation with your GDC. Two options include solving the equation with the GDC's solver or graphing and finding relevant x-intercepts. These two approaches, and other GDC methods, will be illustrated throughout this book.

Equations involving absolute value (modulus)

We will encounter equations in quadratic form in later chapters; for example, in equations with logarithms and trigonometric functions.

Equations involving **absolute value** occur in a range of different topics in mathematics. To solve an equation containing one or more absolute value expressions, we apply the definition that states that the absolute value of a real number a, denoted by $|a|$, is given by

$$|a| = \begin{cases} a & \text{if } a \geqslant 0 \\ -a & \text{if } a \leqslant 0 \end{cases}$$

$|a|$ also has the geometric interpretation of being the distance between the coordinate a and the origin on the real number line.

Example 2.14

Use an algebraic approach to solve the equation $|2x + 7| = 13$. Check any solution(s) on a GDC.

Solution

The expression inside the absolute value symbols must be either 13 or -13, so $2x + 7$ equals 13 or -13. Hence, the given equation is satisfied if either

$$2x + 7 = 13 \quad \text{or} \quad 2x + 7 = -13$$
$$2x = 6 \qquad\qquad 2x = -20$$
$$x = 3 \qquad\qquad x = -10$$

The solutions are $x = 3$ and $x = -10$

To check the solutions on a GDC, graph the equation $y = |2x + 7| - 13$ and confirm that $x = 3$ and $x = -10$ are the x-intercepts of the graph.
See Figure 2.7 on the right: the x-intercepts of the graph of $y = |2x + 7| - 13$ agree with the solutions to the equation.

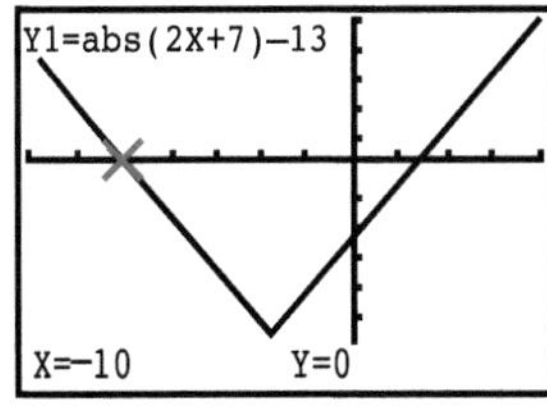

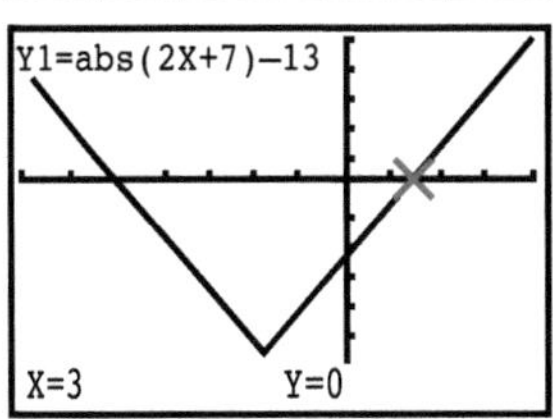

Figure 2.7 GDC screens for checking the solutions to Example 2.14

Example 2.15

Find algebraically the solution(s) to the equation $|2x - 3| = |7 - 3x|$. Check the solution(s) graphically.

Solution

There are four possibilities:

$$2x - 3 = 7 - 3x \quad \text{or} \quad 2x - 3 = -(7 - 3x)$$
$$\text{or} \quad -(2x - 3) = 7 - 3x \quad \text{or} \quad -(2x - 3) = -(7 - 3x)$$

The first and last equations are equivalent, and the second and third equations are also equivalent. So, it is only necessary to solve the first two equations.

$$2x - 3 = 7 - 3x \quad \text{or} \quad 2x - 3 = -(7 - 3x)$$
$$5x = 10 \qquad\qquad 2x - 3 = -7 + 3x$$
$$x = 2 \qquad\qquad 4 = x \Rightarrow x = 4$$

To check, we can graph the equations $y_1 = |2x - 3|$ and $y_2 = |7 - 3x|$, and confirm that the x-coordinates of their points of intersection agree with the solutions to the given equation.

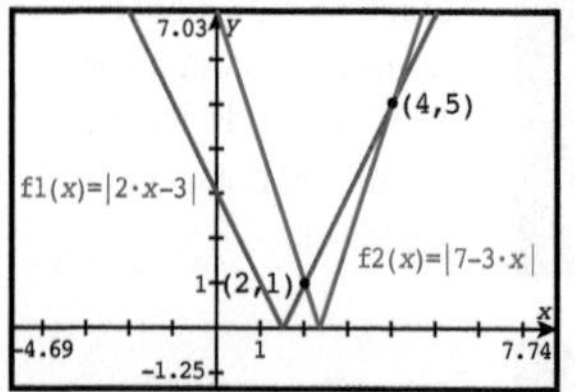

Figure 2.8 GDC screen for the checking solutions to Example 2.15

Quadratic inequalities

In the topics covered in this course, you will need to be as proficient with solving inequalities as with solving equations. Four important properties for inequalities are given below.

For three real numbers a, b, and c:

- If $a > b$ and $b > c$, then $a > c$.
- If $a > b$ and $c > 0$, then $ac > bc$.
- If $a > b$ and $c < 0$, then $ac < bc$.
- If $a > b$, then $a + c > b + c$.

Example 2.16

Find the values of x that solve the inequality $x^2 > x$

Solution

It is possible to determine the solution set to this inequality by a method of trial and error, or simply using a mental process. That may be successful, but generally speaking it is a good idea to attempt to find the solution set by some algebraic method and then check, usually by means of a GDC.

For this example, it is tempting to consider dividing both sides by x, but that cannot be done because it is not known whether x is positive or negative. Recall that when multiplying or dividing both sides of an inequality by a negative number, it is necessary to reverse the inequality sign (see the third property of inequalities listed above). A better approach is to place all terms on one side of the inequality (with zero on the other side) and then try to factorise.

$$x^2 > x$$
$$x^2 - x > 0$$
$$x(x-1) > 0$$

Now analyse the signs of the two different factors in a 'sign chart'.

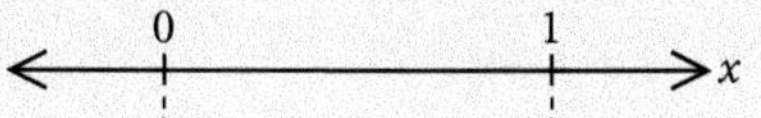

	$x<0$	0	$0<x<1$	1	$x>1$
x	−	0	+		+
$x-1$	−		−	0	+
$x(x-1)$	+	0	−	0	+

The sign chart indicates that the product of the two factors, $x(x-1)$, will be positive when x is less than 0 or greater than 1. Therefore, the solution set is $x < 0$ or $x > 1$.

The solution set, $x < 0$ or $x > 1$, for Example 2.16 comprises two intervals that do not intersect (they are disjoint). It is incorrect to write the solution as $0 > x > 1$, or as $1 < x < 0$. Both of these formats imply that the solution set consists of the values of x **between** 0 and 1, but that is not the case. Only write the 'combined' inequality $a < x < b$ if $x > a$ **and** $x < b$, where the two intervals are intersecting **between** a and b.

Inequalities with quadratic polynomials arise in many different contexts. Problems in which we need to analyse the value of the discriminant of a quadratic equation will usually require us to solve a quadratic inequality, as the next example illustrates.

Example 2.17

Given $f(x) = 3kx^2 - (k+3)x + k - 2 = 0$, find the range of values of k for which f has no real zeros.

Solution

The quadratic function f will have no real zeros when its discriminant is negative. Since f is written in the form $ax^2 + bx + c = 0$ then, in terms of the parameter k, $a = 3k$, $b = -(k+3)$, and $c = k - 2$. Substituting these values into the discriminant, we have the inequality:

$$(-(k+3))^2 - 4(3k)(k-2) < 0$$

$$k^2 + 6k + 9 - 12k^2 + 24k < 0$$

$$-11k^2 + 30k + 9 < 0$$ Easier to factorise if leading coefficient is positive.

$$11k^2 - 30k - 9 > 0$$ Multiply both sides by -1 and reverse inequality sign.

$$k = \frac{-(-30) \pm \sqrt{(-30)^2 - 4(11)(-9)}}{2(11)} = \frac{30 \pm \sqrt{1296}}{22} = \frac{30 \pm 36}{22}$$

$$k = \frac{30+36}{22} = \frac{66}{22} = 3 \text{ or } k = \frac{30-36}{22} = -\frac{6}{22} = -\frac{3}{11}$$

The two rational zeros indicate $11k^2 - 30k - 9$ could have been factorised into $(11k + 3)(k - 3)$.

$(11k + 3)(k - 3) > 0$

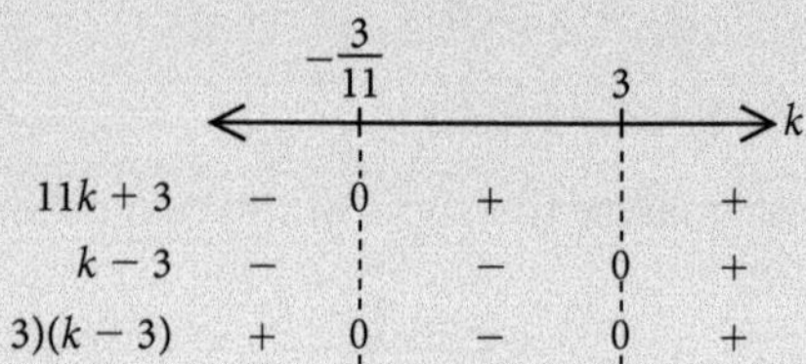

The results of the sign chart indicate that the solution set to the inequality is $k < -\frac{11}{3}$ or $k > 3$.

Therefore, any value of k such that $k < -\frac{11}{3}$ or $k > 3$ will cause the function f to have no real zeros.

Exercise 2.3

1. Solve for x in each equation. If possible, find all real solutions and express them exactly. If this is not possible, then solve using your GDC and approximate any solutions to three significant figures. Be sure to check answers and to recognise any extraneous solutions.

   (a) $\sqrt{x+6} + 2x = 9$ **(b)** $\sqrt{x+7} + 5 = x$

   (c) $\sqrt{7x+14} - 2 = x$ **(d)** $\frac{x-2}{x} = \frac{x+1}{x-2}$

   (e) $\frac{5}{x+4} - \frac{4}{x} = \frac{21}{5x+20}$ **(f)** $\frac{x+1}{2x+3} = \frac{5x-1}{7x+3}$

   (g) $\frac{1}{x} - \frac{1}{x+1} = \frac{1}{x+4}$ **(h)** $x^4 - 2x^2 - 15 = 0$

   (i) $5x^{-2} - x^{-1} - 2 = 0$ **(j)** $x^6 - 35x^3 + 216 = 0$

   (k) $|3x + 4| = 8$ **(l)** $|x - 5| = 3$

   (m) $|5x + 1| = 2x$ **(n)** $x - \sqrt{x+10} = 0$

2. Find the values of x that satisfy each inequality.

   (a) $x^2 - 8 < 2x$ **(b)** $x^2 + 4 < 3x$

   (c) $2x^2 + 5x > 3$ **(d)** $16x \leqslant 3x^2 + 5$

3. Given $f(x) = 3kx^2 - (k + 3)x + k - 2 = 0$, find the range of values of k for which f has no real zeros.

4. Find the values of p for which the equation $px^2 - 3x + 1 = 0$ has:

   (a) one real solution **(b)** two real solutions **(c)** no real solutions.

5. Given $f(x) = x^2 + x(k - 1) + k^2$, find the range of values of k so that $f(x) > 0$ for all real values of x.

6. Show that both of the following inequalities are true for all real numbers m and n such that $m \geqslant n > 0$.

   (a) $m + \frac{1}{n} \geqslant 2$ **(b)** $(m + n)\left(\frac{1}{m} + \frac{1}{n}\right) \geqslant 4$

7. Find all of the exact solutions to the equation $(x^2 + x)^2 = 5x^2 + 5x - 6$.

8. Use your GDC to find the values of x that satisfy the inequality $\frac{3}{x-1} - \frac{2}{x+1} < 1$

Chapter 2 practice questions

1. Solve for x in terms of a and b for the equation $x^2 - (a + 3b)x + 3ab = 0$.

2. Find the value of c such that the vertex of the parabola $y = 3x^2 - 8x + c$ is $\left(\frac{4}{3}, -\frac{1}{3}\right)$.

3. $p(x)$ is a quadratic function that passes through the points $\left(-\frac{5}{2}, 0\right)$ and $\left(\frac{9}{2}, 0\right)$. The graph of p is a parabola.

(a) Write down the equation for the axis of symmetry of the parabola.

(b) Given that the graph of p also passes through the point $(-3, 15)$, find an expression for p and write it in the form $p(x) = ax^2 + bx + c$.

4. The quadratic function $f(x) = ax^2 + bx + c$ has the following characteristics:

- passes through the point $(2, -5)$
- has a maximum value of 4 when $x = -1$
- has a zero of $x = -3$.

Find the values of a, b, and c.

5. Find all values of m such that the equation $mx^2 - 2(m + 2)x + m + 2 = 0$ has:

(a) two real roots

(b) two real roots (one positive and one negative).

6. Solve the inequality $x^2 + 3 \leqslant 4x$.

7. Find the range of values for k in order for the equation $2x^2 + (3 - k)x + k + 3 = 0$ to have no real solutions.

8. Consider the rational function $f(x) = \frac{8x + 4}{x + 2}$. Do not use your GDC for this question.

(a) The graph of f has a vertical asymptote and a horizontal asymptote. Write down the equation for each asymptote.

(b) Find the coordinates of the x-intercept for f.

(c) Find the coordinates of the y-intercept for f.

(d) Sketch a graph of f. Clearly label all asymptotes and axis intercepts.

9. Find the values of k so that the equation $(k - 2)x^2 + 4x - 2k + 1 = 0$ has two distinct real roots.

10. The equation $kx^2 - 3x + (k + 2) = 0$ has two distinct real roots. Find the set of possible values of k.

11. The graph of $f(x) = \frac{5x - 3}{2x - 1}$ has a vertical asymptote and a horizontal asymptote.

(a) The two asymptotes for the graph of f intersect at point P. Find the coordinates of P.

(b) Determine the domain and range of f.

12. Solve the inequality $|x - 2| \geqslant |2x + 1|$.

13. Use your GDC to find the solutions to the equation $\frac{1}{2}|x + 1| = |2x - 10|$.

14. Find the two solutions to the equation $\frac{3x + 1}{x + 1} + \frac{x}{x + 2} = 0$. Express the solutions exactly.

15. Find all solutions to $w^4 - 11w^2 + 28 = 0$ and express them exactly.

3 Sequences and series

3 Sequences and series

Learning objectives

By the end of this chapter, you should be familiar with...

- arithmetic sequences and series
- sum of finite arithmetic sequences
- geometric sequences and series
- sum of finite and infinite geometric series
- sigma notation
- the binomial theorem and the expansion of $(a + b)^n$, $n \in \mathbb{N}$.

The heights of consecutive bounces of a ball, compound interest, population growth, and Fibonacci numbers are only a few of the applications of sequences and series that we have seen in previous courses. In this chapter we will review these concepts, consolidate understanding, and take them one step further.

3.1 Sequences

Look at this pattern:

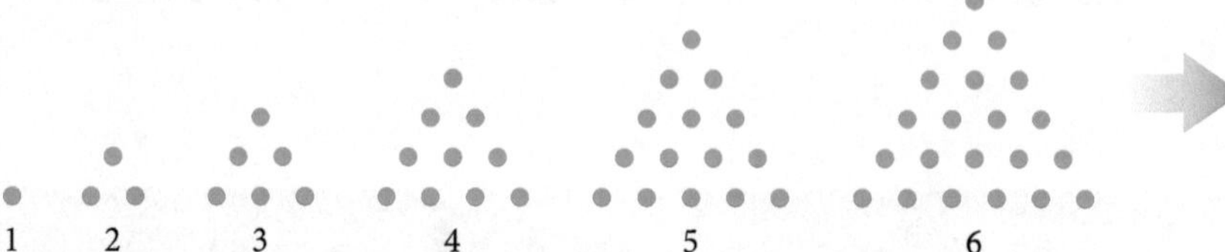

Figure 3.1 Sequence of dots in triangular arrays

The first term represents 1 dot, the second represents 3 dots, etc...
This pattern can be represented as a list of numbers written in a definite order:

$a_1 = 1, a_2 = 3, a_3 = 6, \ldots$

The number a_1 is the first term, a_2 the second term, a_3 the third term, and so on. The nth term is a_n.

If a_n is the nth term of a sequence, then a_{n-1} is the term before it and a_{n+1} is the term after.

While the idea of a sequence of numbers, $a_1, a_2, a_3, \ldots$ is straightforward, it is useful to think of a sequence as a function. The sequence in Figure 3.1 can also be described in function notation as:

$f(1) = 1, f(2) = 3, f(3) = 6$, and so on, where the domain is $\mathbb{Z}^+$

Here are some more examples of sequences:

1 6, 12, 18, 24, 30

2 $3, 9, 27, \ldots, 3^k, \ldots$

3 $\left\{\frac{1}{i^2}; i = 1, 2, 3, \ldots, 10\right\}$

4 $\{b_1, b_2, \ldots, b_n, \ldots\}$, sometimes used with an abbreviation $\{b_n\}$

The first and third sequences are **finite** and the second and fourth are **infinite**. In the second and fourth sequences, we were able to define a rule that yields the nth number in the sequence (called the nth term) as a function of n, the term's number. In this sense, you can think of a sequence, as a **function** that assigns a **unique** number (a_n) to each positive integer n.

Example 3.1

Find the first 5 terms and the 50th term of the sequence $\{b_n\}$ such that $b_n = 2 - \frac{1}{n^2}$

Solution

Since we are given an explicit expression for the nth term as a function of its term number n, we only need to find the value of that function for the required terms:

$b_1 = 2 - \frac{1}{1^2} = 1 \qquad b_2 = 2 - \frac{1}{2^2} = 1\frac{3}{4} \qquad b_3 = 2 - \frac{1}{3^2} = 1\frac{8}{9}$

$b_4 = 2 - \frac{1}{4^2} = 1\frac{15}{16} \qquad b_5 = 2 - \frac{1}{5^2} = 1\frac{24}{25} \qquad b_{50} = 2 - \frac{1}{50^2} = 1\frac{2499}{2500}$

So, informally, a **sequence** is an **ordered set** of **real numbers**. That is, there is a first number, a second, and so on. The notation used for these sets is shown in Example 3.1. The way the function was defined in Example 3.1 is called the **explicit** definition of a sequence. There are other ways to define sequences, one of which is the **recursive** definition (also called the **inductive** definition). The following example will show you how this is used.

Example 3.2

Find the first 5 terms and the 20th term of the sequence $\{b_n\}$ such that $b_1 = 5$ and $b_n = 2(b_{n-1} + 3)$

Solution

The defining formula for this sequence is recursive. It allows us to find the nth term b_n if we know the preceding term b_{n-1}. Thus, we can find the second term from the first, the third from the second, and so on. Since we know the first term $b_1 = 5$, we can calculate the rest:

$b_2 = 2(b_1 + 3) = 2(5 + 3) = 16$

$b_3 = 2(b_2 + 3) = 2(16 + 3) = 38$

$b_4 = 2(b_3 + 3) = 2(38 + 3) = 82$

$b_5 = 2(b_4 + 3) = 2(82 + 3) = 170$

So, the first 5 terms of this sequence are 5, 16, 38, 82, and 170. However, to find the 20th term, we must first find all 19 preceding terms. This is one of the drawbacks of this type of definition, unless we can change the definition into explicit form. This can easily be done using a GDC (Figure 3.2).

```
Plot1  Plot2  Plot3
 nMin=1
 .U(n)=2(u(n−1)+3
 )
 U(nMin)=5
```

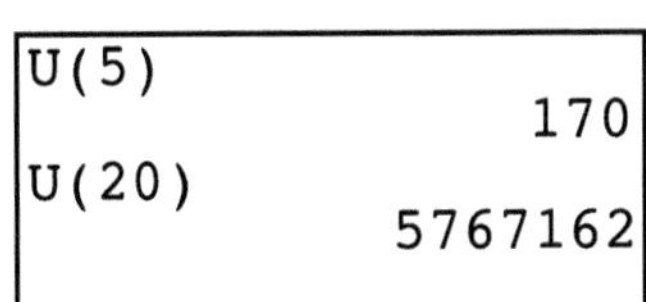

Figure 3.2 GDC screens for Example 3.2

Example 3.3

A Fibonacci sequence is defined recursively as:

$$F_n = \begin{cases} 1 & n = 1 \\ 1 & n = 2 \\ F_{n-1} + F_{n-2} & n > 2 \end{cases}$$

(a) Find the first 10 terms of the sequence.

(b) Find the sum of the first 10 terms of the sequence.

(c) By observing that $F_1 = F_3 - F_2, F_2 = F_4 - F_3$, and so on, derive a formula for the sum of the first n Fibonacci numbers.

Solution

(a) 1, 1, 2, 3, 5, 8, 13, 21, 34, 55

(b) $S_1 = 1, S_2 = 2, S_3 = 4, S_4 = 7, S_5 = 12, S_6 = 20, S_7 = 33, S_8 = 54, S_9 = 88, S_{10} = 143$

(c) Since $F_3 = F_2 + F_1$, then:

$$\begin{aligned} F_1 &= F_3 - F_2 \\ F_2 &= F_4 - F_3 \\ F_3 &= F_5 - F_4 \\ F_4 &= F_6 - F_5 \\ &\vdots \\ F_n &= F_{n+2} - F_{n+1} \end{aligned}$$

$$S_n = F_{n+2} - F_2$$

Notice that $S_5 = 12 = F_7 - F_2 = 13 - 1$ and $S_8 = 54 = F_{10} - F_2 = 55 - 1$

Note: parts (a) and (b) can be made easy by using a spreadsheet:

	A	B	C	D
1	F(n)	S(n)		
2	1	1		
3	1	2		
4	**2**	4		
5	3	7		
6	5	12		
7	8	20		
8	13	33		
9	21	54		
10	34	88		
11	55	**143**		Let this cell be A2 + A3 Then copy it down
12	89	232		
13	144	376		
14	233	609		
15	377	986		Let this cell be B10 + A11 Then copy it down
16	610	1596		
17	987	2583		

Notice that not all sequences have formulae, either recursive or explicit. Some sequences are given only by listing their terms.

Exercise 3.1

1. Find the first 5 terms of each infinite sequence.

(a) $s(n) = 2n - 3$

(b) $g(k) = 2^k - 3$

(c) $f(n) = 3 \times 2^{-n}$

(d) $a_n = (-1)^n(2^n) + 3$

(e) $\begin{cases} a_1 = 5 \\ a_n = a_{n-1} + 3; \quad \text{for } n > 1 \end{cases}$

(f) $\begin{cases} b_1 = 3 \\ b_n = b_{n-1} + 2n; \quad \text{for } n \geqslant 2 \end{cases}$

2. Find the first 5 terms and the 50th term of each infinite sequence.

(a) $a_n = 2n - 3$

(b) $b_n = 2 \times 3^{n-1}$

(c) $u_n = (-1)^{n-1}\left(\dfrac{2n}{n^2 + 2}\right)$

(d) $a_n = n^{n-1}$

(e) $a_n = 2a_{n-1} + 5$ and $a_1 = 3$

(f) $u_{n+1} = \dfrac{3}{2u_n + 1}$ and $u_1 = 0$

(g) $b_n = 3b_{n-1}$ and $b_1 = 2$

(h) $a_n = a_{n-1} + 2$ and $a_1 = -1$

3. Suggest a recursive definition for each sequence.

(a) $\dfrac{1}{3}, \dfrac{1}{12}, \dfrac{1}{48}, \dfrac{1}{192}, \ldots$

(b) $\dfrac{1}{2}a, \dfrac{2}{3}a^3, \dfrac{8}{9}a^5, \dfrac{32}{27}a^7, \ldots$

(c) $a - 5k, 2a - 4k, 3a - 3k, 4a - 2k, 5a - k, \ldots$

4. Write down a possible formula that gives the nth term of each sequence.

(a) 4, 7, 12, 19, …

(b) 2, 5, 8, 11, …

(c) $1, \dfrac{3}{4}, \dfrac{5}{9}, \dfrac{7}{16}, \dfrac{9}{25}, \ldots$

(d) $\dfrac{1}{4}, \dfrac{3}{5}, \dfrac{5}{6}, 1, \dfrac{9}{8}, \ldots$

5. A sequence is defined as $a_n = \dfrac{F_{n+1}}{F_n}, n > 1$, where F_n is a member of the Fibonacci sequence.

(a) Write down the first 10 terms of a_n

(b) Show that $a_n = 1 + \dfrac{1}{a_{n-1}}$

3.2 Arithmetic sequences

There are two types that we will look at here: arithmetic and geometric sequences. We will look at them in the next two sections.

Examine each sequence and the most likely recursive formula for each of them.

7, 14, 21, 28, 35, 42, ...	$a_1 = 7$ and $a_n = a_{n-1} + 7$, for $n > 1$
2, 11, 20, 29, 38, 47, ...	$a_1 = 2$ and $a_n = a_{n-1} + 9$, for $n > 1$
39, 30, 21, 12, 3, −6, ...	$a_1 = 39$ and $a_n = a_{n-1} - 9$, for $n > 1$

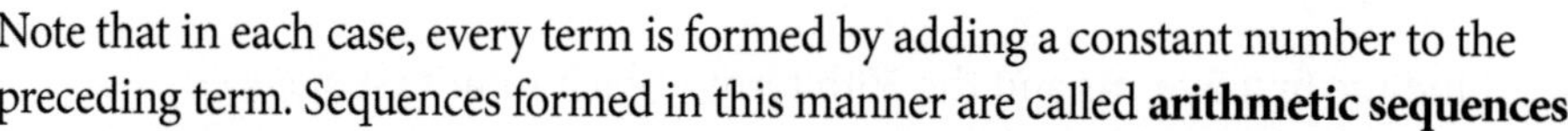

Definition of an arithmetic sequence

A sequence $a_1, a_2, a_3, \ldots$ is an arithmetic sequence if there is a constant d for which $a_n = a_{n-1} + d$ for all integers $n > 1$, where d is called the common difference of the sequence, and $d = a_n - a_{n-1}$ for all integers $n > 1$.

Note that in each case, every term is formed by adding a constant number to the preceding term. Sequences formed in this manner are called **arithmetic sequences**.

So, for the sequences above, 7 is the common difference for the first, 9 is the common difference for the second, and −9 is the common difference for the third.

This description gives us the recursive definition of the arithmetic sequence. It is possible, however, to find the explicit definition of the sequence.

Applying the recursive definition repeatedly will enable us to see the expression we are seeking:

$$a_2 = a_1 + d$$
$$a_3 = a_2 + d = a_1 + d + d = a_1 + 2d$$
$$a_4 = a_3 + d = a_1 + 2d + d = a_1 + 3d$$

The general (nth) term of an arithmetic sequence, a_n with first term a_1 and common difference d may be expressed explicitly as $a_n = a_1 + (n - 1)d$.

So, we get to the nth term by adding d to a_1, $(n - 1)$ times.

This result is useful in finding any term of a sequence without knowing the previous terms.

The arithmetic sequence can be looked at as a linear function as explained in the introduction to this chapter. In other words, for every increase of one unit in n, the value of the sequence will increase by d units. As the first term is a_1, the point $(1, a_1)$ belongs to this function. The constant increase d can be considered to be the gradient (slope) of this linear model, hence the nth term, the dependent variable in this case, can be found by using the point-slope form of the equation of a line:

$$y - y_1 = m(x - x_1)$$
$$a_n - a_1 = d(n - 1) \Leftrightarrow a_n = a_1 + (n - 1)d$$

This agrees with our definition of an arithmetic sequence.

Example 3.4

Find the nth term and the 50th term of the sequence 2, 11, 20, 29, 38, 47, ...

Solution

This is an arithmetic sequence with first term 2 and common difference 9. Therefore:

$$a_n = a_1 + (n - 1)d = 2 + (n - 1) \times 9 = 9n - 7$$
$$\Rightarrow a_{50} = 9 \times 50 - 7 = 443$$

Example 3.5

(a) Find the recursive and the explicit forms of the definition of the sequence:

13, 8, 3, −2, ...

(b) Calculate the value of the 25th term.

Solution

(a) This is clearly an arithmetic sequence, with common difference −5.

Recursive definition: $a_1 = 13$
$a_n = a_{n-1} - 5$

Explicit definition: $a_n = 13 - 5(n - 1) = 18 - 5n$

(b) $a_{25} = 18 - 5 \times 25 = -107$

Example 3.6

Find a definition for the arithmetic sequence whose first term is 5 and fifth term is 11.

Solution

Since the fifth term is given, using the explicit form, we have:

$$a_5 = a_1 + (5 - 1)d \Rightarrow 11 = 5 + 4d \Rightarrow d = \frac{3}{2}$$

This leads to the general term:

$a_n = 5 + \frac{3}{2}(n - 1)$, or equivalently

$$\begin{cases} a_1 = 5 \\ a_n = a_{n-1} + \frac{3}{2}, n > 1 \end{cases}$$

Example 3.7

Insert four arithmetic means between 3 and 7.

Solution

Since there are four means between 3 and 7, the problem can be reduced to a situation similar to Example 3.6, by considering the first term to be 3 and the sixth term to be 7. The rest is left as an exercise for you.

In a finite arithmetic sequence $a_1, a_2, a_3, \ldots, a_k$, the terms $a_2, a_3, \ldots, a_{k-1}$ are called **arithmetic means** between a_1 and a_k.

Exercise 3.2

1. Insert four arithmetic means between 3 and 7.

2. State whether or not each sequence is an arithmetic sequence. If it is, find the common difference and the 50th term. If it is not, say why not.

   (a) $a_n = 2n - 3$ (b) $b_n = n + 2$

   (c) $c_n = c_{n-1} + 2$, and $c_1 = -1$ (d) $u_n = 3u_{n-1} + 2$

   (e) 2, 5, 7, 12, 19, ... (f) 2, −5, −12, −19, ...

3. For each arithmetic sequence find:

   (i) the 8th term

   (ii) an explicit formula for the nth term

   (iii) a recursive formula for the nth term.

   (a) −2, 2, 6, 10, ... (b) 29, 25, 21, 17, ...

   (c) −6, 3, 12, 21, ... (d) 10.07, 9.95, 9.83, 9.71, ...

   (e) 100, 97, 94, 91, ... (f) $2, \frac{3}{4}, -\frac{1}{2}, -\frac{7}{4}, \ldots$

4. Find five arithmetic means between 13 and −23.

5. Find three arithmetic means between 299 and 300.

6. In an arithmetic sequence, $a_5 = 16$ and $a_{14} = 42$.
   Find an explicit formula for the nth term of this sequence.

7. In an arithmetic sequence, $a_3 = -40$ and $a_9 = -18$.
   Find an explicit formula for the nth term of this sequence.

8. The first three terms and the last term are given for each sequence.
   Find the number of terms.

   (a) 3, 9, 15, ..., 525 (b) 9, 3, −3, ..., −201

   (c) $3\frac{1}{8}, 4\frac{1}{4}, 5\frac{3}{8}, \ldots, 14\frac{3}{8}$ (d) $\frac{1}{3}, \frac{1}{2}, \frac{2}{3}, \ldots, 2\frac{5}{6}$

   (e) $1 - k, 1 + k, 1 + 3k, \ldots, 1 + 19k$

9. Find five arithmetic means between 15 and −21.

10. Find three arithmetic means between 99 and 100.

11. In an arithmetic sequence, $a_3 = 11$ and $a_{12} = 47$.
    Find an explicit formula for the nth term of this sequence.

12. In an arithmetic sequence, $a_7 = -48$ and $a_{13} = -10$.
Find an explicit formula for the nth term of this sequence.

13. The 30th term of an arithmetic sequence is 147 and the common difference is 4. Find a formula for the nth term.

14. The first term of an arithmetic sequence is -7 and the common difference is 3. Is 9803 a term of this sequence? If so, which term?

15. The first term of an arithmetic sequence is 9689 and the 100th term is 8996. Show that the 110th term is 8926. Is 1 a term of this sequence? If so, which term?

16. The first term of an arithmetic sequence is 2 and the 30th term is 147. Is 995 a term of this sequence? If so, which term?

3.3 Geometric sequences

Examine the following sequences and the most likely recursive formula for each of them.

7, 14, 28, 56, 112, 224, … $a_1 = 7$ and $a_n = a_{n-1} \times 2$, for $n > 1$

2, 18, 162, 1458, 13 122, … $a_1 = 2$ and $a_n = a_{n-1} \times 9$, for $n > 1$

48, -24, 12, -6, 3, -1.5, … $a_1 = 48$ and $a_n = a_{n-1} \times (-0.5)$, for $n > 1$

Note that in each case, every term is formed by multiplying a constant number with the preceding term. Sequences formed in this manner are called **geometric sequences.**

Definition of a geometric sequence

A sequence a_1, a_2, a_3, is a **geometric sequence** if there is a constant r for which

$a_n = a_{n-1} \times r$

for all integers $n > 1$, where r is the **common ratio** of the sequence, and $r = a_n \div a_{n-1}$ for all integers $n > 1$.

Thus, for the preceding sequences, 2 is the common ratio for the first, 9 is the common ratio for the second and -0.5 is the common ratio for the third.

This description gives us the recursive definition of the geometric sequence. It is possible, however, to find the explicit definition of the sequence.

Applying the recursive definition repeatedly will enable us to see the expression we are seeking:

$$a_2 = a_1 \times r$$

$$a_3 = a_2 \times r = a_1 \times r \times r = a_1 \times r^2$$

$$a_4 = a_3 \times r = a_1 \times r^2 \times r = a_1 \times r^3$$

We can see that we get to the nth term by multiplying r by a_1, $(n - 1)$ times.

This result is useful in finding any term of a sequence without knowing the previous terms.

nth term of a geometric sequence

The general (nth) term of a geometric sequence, a_n with common ratio r and first term a_1 may be expressed explicitly as $a_n = a_1 \times r^{(n-1)}$

Example 3.8

(a) Find the geometric sequence with $a_1 = 2$ and $r = 3$

(b) Describe the sequence $3, -12, 48, -192, 768, \dots$

(c) Describe the sequence $1, \frac{1}{2}, \frac{1}{4}, \frac{1}{8}, \dots$

(d) Graph the sequence $a_n = \frac{1}{4} \cdot 3^{n-1}$

Solution

(a) The geometric sequence is $2, 6, 18, 54, \dots, 2 \times 3^{n-1}$. Notice that the ratio of any two consecutive terms is 3.

(b) This is a geometric sequence with $a_1 = 3$ and $r = -4$. The nth term is $a_n = 3 \times (-4)^{n-1}$. Notice that when the common ratio is negative, the terms of the sequence alternate in sign.

(c) The nth term of this sequence is $a_n = 1 \times \left(\frac{1}{2}\right)^{n-1}$. Notice that the ratio of any two consecutive terms is $\frac{1}{2}$. Also, notice that the terms decrease in value.

(d) Use a GDC to graph the sequence. The terms of the sequence lie on the graph of the exponential function $y = \frac{1}{4} \cdot 3^{x-1}$

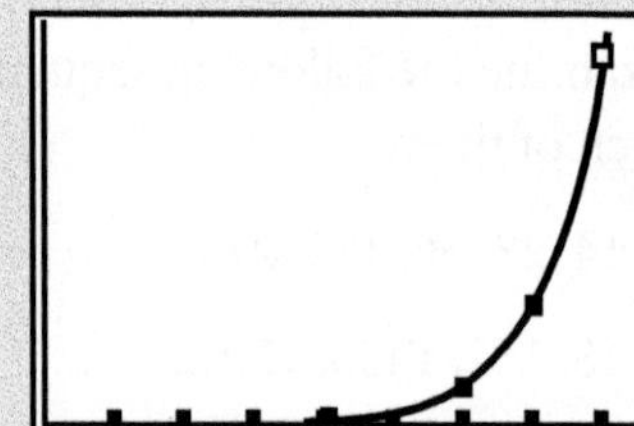

Example 3.9

At 8:00 a.m., 1000 mg of medicine is given to a patient. At the end of each hour, the amount of medicine in the patient's bloodstream is 60% of the amount present at the beginning of the hour.

(a) What portion of the medicine remains in the patient's bloodstream at 12 noon if no additional medication had been given?

(b) If a second dose of 1000 mg is given at 10:00 a.m., what is the total amount of the medication in the patient's bloodstream at 12 noon?

Solution

(a) Use the geometric model, as there is a constant multiple at the end of each hour. Hence, the amount at the end of any hour after giving the medicine is:

$a_n = a_1 \times r^{n-1}$, where n is the number of hours.

So, at 12 noon, $n = 5$ and $a_5 = 1000 \times 0.6^{(5-1)} = 129.6$ mg

(b) For the second dose, the amount of medicine at noon corresponds to $n = 3$:

$$a_3 = 1000 \times 0.6^{(3-1)} = 360$$

So, the amount of medicine is $129.6 + 360 = 489.6$ mg

Compound interest

Compound interest is an example of a geometric sequence.

Interest compounded annually

When we borrow money we pay interest, and when we invest money we receive interest. Suppose an amount of €1000 is put into a savings account that has an annual interest rate of 6%. How much money will we have in the bank at the end of 4 years?

It is important to note that the 6% interest is given annually and is added to the savings account, so that in the following year it will also earn interest, and so on.

Time in years	Amount in the account (€)
0	1000
1	$1000 + 1000 \times 0.06 = 1000(1 + 0.06)$
2	$1000(1 + 0.06) + (1000(1 + 0.06)) \times 0.06 = 1000(1 + 0.06)(1 + 0.06) = 1000(1 + 0.06)^2$
3	$1000(1 + 0.06)^2 + (1000(1 + 0.06)^2) \times 0.06 = 1000(1 + 0.06)^2(1 + 0.06) = 1000(1 + 0.06)^3$
4	$1000(1 + 0.06)^3 + (1000(1 + 0.06)^3) \times 0.06 = 1000(1 + 0.06)^3(1 + 0.06) = \mathbf{1000(1 + 0.06)^4}$

Table 3.1 Compound interest

This appears to be a geometric sequence with five terms. You will notice that the number of terms is five, as both the beginning and the end of the first year are counted. (Initial value, when time = 0, is the first term.)

In general, if a **principal** of P euros is invested in an account that yields an annual interest rate r (expressed as a decimal), and this interest is added at the end of every year to the principal, then we can use the geometric sequence formula to calculate the **future value A,** which is accumulated after t years.

If we repeat the steps above, with

$A_0 = P =$ initial amount

$r =$ annual interest rate

$t =$ number of years

it becomes easier to develop the formula:

Time in years	Amount in the account
0	$A_0 = P$
1	$A_1 = P + Pr = P(1 + r)$
2	$A_2 = A_1(1 + r) = P(1 + r)^2$
⋮	⋮
t	$A_t = P(1 + r)^t$

Table 3.2 Compound interest formula

Notice that since we are counting from 0 to t, we have $t + 1$ terms, so we are using the geometric sequence formula:

$$a_n = a_1 \times r^{n-1} \Rightarrow A_t = A_1 \times (1+r)^{t+1-1}$$

Interest compounded *n* times per year

Suppose that the principal P is invested as before but the interest is paid n times per year. Then $\frac{r}{n}$ is the interest paid every compounding period. Since every year we have n periods, for t years we have nt periods. The amount A in the account after t years is:

$$A = P\left(1 + \frac{r}{n}\right)^{nt}$$

Example 3.10

€1000 is invested in an account paying compound interest at a rate of 6% per annum. Calculate the amount of money in the account after 10 years if the compounding is:

(a) annual (b) quarterly (c) monthly.

Solution

(a) The amount after 10 years is:

$$A = 1000(1 + 0.06)^{10} = €1790.85$$

(b) The amount after 10 years quarterly compounding is:

$$A = 1000\left(1 + \frac{0.06}{4}\right)^{40} = €1814.02$$

(c) The amount after 10 years monthly compounding is:

$$A = 1000\left(1 + \frac{0.06}{12}\right)^{120} = €1819.40$$

Example 3.11

You invest €1000 at 6% per annum, compounded quarterly. How long will it take for this investment to increase to €2000?

Solution

Let $P = 1000$, $r = 0.06$, $n = 4$, and $A = 2000$ in the compound interest formula

$$A = P\left(1 + \frac{r}{n}\right)^{nt}$$

and then solve for t.

$$2000 = 1000\left(1 + \frac{0.06}{4}\right)^{4t} \Rightarrow 2 = 1.015^{4t}$$

Using a GDC, we can graph the functions $y = 2$ and $y = 1.015^{4t}$ and then find the intersection between their graphs.

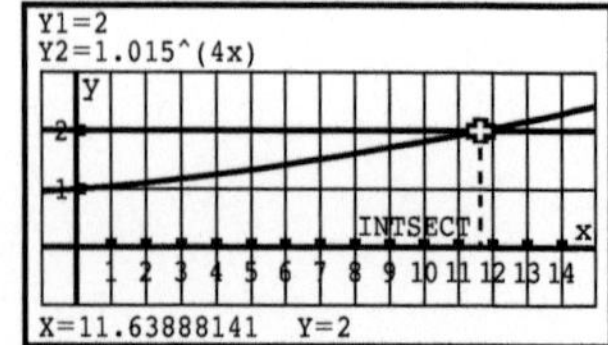

Figure 3.3 GDC screen for the solution to Example 3.11

It will take the €1000 investment 11.64 years to double to €2000. This translates into approximately 47 quarters.

We can check our work to see that this is accurate by using the compound interest formula:

$$A = 1000\left(1 + \frac{0.06}{4}\right)^{47} = €2013.28$$

In the next chapter you will learn how to solve Example 3.11 algebraically, using logarithms.

Example 3.12

You want to invest €1000. What annual interest rate is needed to make this investment grow to €2000 in 10 years if interest is compounded quarterly?

Solution

Let $P = 1000$, $n = 4$, $t = 10$ and $A = 2000$ in the compound interest formula

$$A = P\left(1 + \frac{r}{n}\right)^{nt}$$

and solve for r:

$$2000 = 1000\left(1 + \frac{r}{4}\right)^{40} \Rightarrow 2 = \left(1 + \frac{r}{4}\right)^{40}$$

$$\Rightarrow 1 + \frac{r}{4} = \sqrt[40]{2} \Rightarrow r = 4\left(\sqrt[40]{2} - 1\right) = 0.0699$$

So, at an annual rate of 7% compounded quarterly, the €1000 investment will grow to at least €2000 in 10 years.

We can check to see if our work is accurate by using the compound interest formula:

$$A = 1000\left(1 + \frac{0.07}{4}\right)^{40} = €2001.60$$

Population growth

The same formulae can be applied when dealing with population growth.

Example 3.13

The population of Baden in Austria grows at an annual rate of 0.35%. The population of Baden in 1981 was 23 140. What is the estimate of the population of Baden for 2025?

Solution

This situation can be modelled by a geometric sequence with first term 23 140 and common ratio 1.0035. Since we count the population of 1981 among the terms, the number of terms is 45.

In Chapter 4, more realistic population growth models will be explored and more efficient methods will be developed, including the ability to calculate interest that is continuously compounded.

2025 is equivalent to the 45th term in this sequence. The estimated population for Baden is therefore:

Population (2025) $= a_{45} = 23\,140(1.0035)^{44} = 26\,985$

Exercise 3.3

1. For each sequence:

(i) determine whether the sequence is arithmetic, geometric, or neither.

(ii) find the common difference for the arithmetic ones and the common ratio for the geometric ones.

(iii) find the 10th term for each arithmetic or geometric sequence.

(a) $3, 3^{a+1}, 3^{2a+1}, 3^{3a+1}, \ldots$ **(b)** $a_n = 3n - 3$

(c) $b_n = 2^{n+2}$ **(d)** $c_n = 2c_{n-1} - 2$, and $c_1 = -1$

(e) $u_n = 3u_{n-1}$ and $u_1 = 4$ **(f)** 2, 5, 12.5, 31.25, 78.125, …

(g) 2, −5, 12.5, −31.25, 78.125, … **(h)** 2, 2.75, 3.5, 4.25, 5, …

(i) $18, -12, 8, -\frac{16}{3}, \frac{32}{9}, \ldots$ **(j)** 52, 55, 58, 61, …

(k) −1, 3, −9, 27, −81, … **(l)** 0.1, 0.2, 0.4, 0.8, 1.6, 3.2, …

(m) 3, 6, 12, 18, 21, 27, … **(n)** 6, 14, 20, 28, 34, …

(o) 2.4, 3.7, 5, 6.3, 7.6, …

2. For each arithmetic or geometric sequence, find:

(i) the 8th term

(ii) an explicit formula for the nth term

(iii) a recursive formula for the nth term.

(a) −3, 2, 7, 12, … **(b)** 19, 15, 11, 7, …

(c) −8, 3, 14, 25, … **(d)** 10.05, 9.95, 9.85, 9.75, …

(e) 100, 99, 98, 97, … **(f)** $2, \frac{1}{2}, -1, -\frac{5}{2}, \ldots$

(g) 3, 6, 12, 24, … **(h)** 4, 12, 36, 108, …

(i) 5, −5, 5, −5, … **(j)** 3, −6, 12, −24, …

(k) 972, −324, 108, −36, … **(l)** $-2, 3, -\frac{9}{2}, \frac{27}{4}, \ldots$

(m) $35, 25, \frac{125}{7}, \frac{625}{49}, \ldots$ **(n)** $-6, -3, -\frac{3}{2}, -\frac{3}{4}, \ldots$

(o) 9.5, 19, 38, 76, … **(p)** 100, 95, 90.25, …

(q) $2, \frac{3}{4}, \frac{9}{32}, \frac{27}{256}, \ldots$

3. Find four geometric means between 3 and 96.

4. Find three geometric means between 7 and 4375.

5. Find a geometric mean between 16 and 81.

6. Find four geometric means between 7 and 1701.

7. Find a geometric mean between 9 and 64.

8. The first term of a geometric sequence is 24 and the fourth term is 3. Find the fifth term and an expression for the nth term.

9. The first term of a geometric sequence is 24 and the third term is 6. Find the fourth term and an expression for the nth term.

10. The common ratio in a geometric sequence is $\frac{2}{7}$ and the fourth term is $\frac{14}{3}$. Find the third term.

11. Which term of the geometric sequence 6, 18, 54, … is 118 098?

12. The fourth term and the seventh term of a geometric sequence are 18 and $\frac{729}{8}$. Is $\frac{59\,049}{128}$ a term of this sequence? If so, which term is it?

13. The third term and the sixth term of a geometric sequence are 18 and $\frac{243}{4}$. Is $\frac{19\,683}{64}$ a term of this sequence? If so, which term is it?

14. Vitoria put €1500 into a savings account that pays 4% annual interest compounded semiannually. How much will her account hold 10 years later if she does not make any additional investments in this account?

15. At the birth of her daughter Jane, Charlotte deposited £500 into a savings account. The annual interest rate was 4% compounded quarterly. How much money will Jane have on her 16th birthday?

16. How much money should you invest now if you wish to have an amount of €4000 in your account after 6 years if interest is compounded quarterly at an annual rate of 5%?

17. In 2017, the population of a town in Switzerland was estimated to be 7554. How large would the town's population be in 2022 if it grows at a rate of 0.5% annually?

18. The common ratio in a geometric sequence is $\frac{3}{7}$ and the fourth term is $\frac{14}{3}$. Find the second term.

In a finite geometric sequence $a_1, a_2, a_3, \ldots, a_k$, the terms $a_2, a_3, \ldots a_k - 1$ are called **geometric means** between a_1 and a_k

In questions 5 and 7, this is also called the **mean proportional.**

19. Which term of the geometric sequence 7, 21, 63, … is 137 781?

20. At her son Erik's birth, Astrid deposited £1000 into a savings account. The annual interest rate was 6% compounded quarterly. How much money will Erik have on his 18th birthday?

3.4 Series

In common usage, the word series is the same thing as sequence. But in mathematics, a series is the sum of terms in a sequence. For a sequence of values a_n, the corresponding series is the sequence of S_n with:

$$S_n = a_1 + a_2 + \ldots + a_{n-1} + a_n$$

If the terms are in an arithmetic sequence, then the sum is an **arithmetic series.**

Sigma notation

Most of the series we consider in mathematics are **infinite** series. This is to emphasise that the series contain an infinite number of terms. Any sum in the series S_k will be called a partial sum and is given by:

$$S_k = a_1 + a_2 + \ldots + a_{k-1} + a_k$$

For convenience, this partial sum is written using sigma notation:

$$S_k = \sum_{i=1}^{k} a_i = a_1 + a_2 + \ldots + a_{k-1} + a_k$$

Sigma notation is a concise and convenient way to represent long sums. The symbol $\sum$ is the Greek capital letter *Sigma* that refers to the initial letter of the word 'sum'. So this expression means the sum of all the terms a_i where i takes the values from 1 to k. We can also write $\sum_{i=m}^{n} a_i$ to mean the sum of the terms a_i where i takes the values from m to n. In such a sum, m is called the lower limit and n the upper limit.

This indicates ending with $i = n$

This indicates addition → $\sum_{i=m}^{n} a_i$

This indicates starting with $i = m$

For example, suppose we measure the heights of six children. We will denote their heights by x_1, x_2, x_3, x_4, x_5 and x_6.

The sum of their heights $x_1 + x_2 + x_3 + x_4 + x_5 + x_6$ is written more neatly, by using sigma notation, as $\sum_{i=1}^{6} x_i$.

The symbol $\sum$ means 'add up'. Underneath $\sum$ we see $i = 1$ and on top of it 6. This means that i is replaced by whole numbers starting at the bottom number, 1, until the top number, 6, is reached.

Thus $\sum_{i=3}^{6} x_i = x_3 + x_4 + x_5 + x_6$ and $\sum_{i=3}^{5} x_i = x_3 + x_4 + x_5$

So, the notation $\sum_{i=1}^{n} x_i$ tells us:

- to add the scores x_i
- where to start: x_1
- where to stop: x_n (where n is some integer)

Now take the heights of the children to be $x_1 = 112$ cm, $x_2 = 96$ cm, $x_3 = 120$ cm, $x_4 = 132$ cm, $x_5 = 106$ cm, and $x_6 = 120$ cm.

Then the total height (in cm) is

$$\sum_{k=1}^{6} x_k = x_1 + x_2 + x_3 + x_4 + x_5 + x_6$$
$$= 112 + 96 + 120 + 132 + 106 + 120 = 686 \text{ cm}$$

Notice that we have used k instead of i in the formula above. The i is what we call a dummy variable – any letter can be used.

$$\sum_{k=1}^{n} x_k = \sum_{i=1}^{n} x_i$$

Example 3.14

Write each series in full:

(a) $\sum_{i=1}^{5} i^4$ (b) $\sum_{r=3}^{7} 3^r$ (c) $\sum_{j=1}^{n} x_j p(x_j)$

Solution

(a) $\sum_{i=1}^{5} i^4 = 1^4 + 2^4 + 3^4 + 4^4 + 5^4$

(b) $\sum_{r=3}^{7} 3^r = 3^3 + 3^4 + 3^5 + 3^6 + 3^7$

(c) $\sum_{j=1}^{n} x_j p(x_j) = x_1 p(x_1) + x_2 p(x_2) + \ldots + x_n p(x_n)$

Example 3.15

Evaluate $\sum_{n=0}^{5} 2^n$

Solution

$$\sum_{n=0}^{5} 2^n = 2^0 + 2^1 + 2^2 + 2^3 + 2^4 + 2^5 = 63$$

Example 3.16

Write the sum $\frac{1}{2} - \frac{2}{3} + \frac{3}{4} - \frac{4}{5} + \ldots + \frac{99}{100}$ using sigma notation.

Solution

The numerator and denominator in each term are consecutive integers, so they take on the absolute value of $\frac{k}{k+1}$ or any equivalent form. The signs of the terms alternate and there are 99 terms. To take care of the sign, we use some power of (-1) that will start with a positive value. If we use $(-1)^k$, then the first term will be negative, hence we can use $(-1)^{k+1}$ instead.
We can therefore write the sum as

$$(-1)^{1+1}\left(\frac{1}{2}\right) + (-1)^{2+1}\left(\frac{2}{3}\right) + (-1)^{3+1}\left(\frac{3}{4}\right) + \ldots + (-1)^{99+1}\left(\frac{99}{100}\right)$$

$$= \sum_{k=1}^{99}(-1)^{k+1}\left(\frac{k}{k+1}\right)$$

Properties of sigma notation

There are a number of useful results we can obtain when we use sigma notation.

1. For example, suppose we have a sum of constant terms:

   $$\sum_{i=1}^{5} 2$$

   What does this mean? If we write this out in full, we get:

   $$\sum_{i=1}^{5} 2 = 2 + 2 + 2 + 2 + 2 = 5 \times 2 = 10$$

   In general, if we sum a constant n times then we can write:

   $$\sum_{i=1}^{n} k = k + k + \ldots + k = n \times k = nk$$

2. Suppose we have the sum of a constant multiplied by i. For example:

   $$\sum_{i=1}^{5} 5i = 5 \times 1 + 5 \times 2 + 5 \times 3 + 5 \times 4 + 5 \times 5$$
   $$= 5 \times (1 + 2 + 3 + 4 + 5) = 75$$

   However, this can also be interpreted as:

   $$\sum_{i=1}^{5} 5i = 5 \times 1 + 5 \times 2 + 5 \times 3 + 5 \times 4 + 5 \times 5$$
   $$= 5 \times (1 + 2 + 3 + 4 + 5) = 5\sum_{i=1}^{5} i$$

   which implies that:

   $$\sum_{i=1}^{5} 5i = 5\sum_{i=1}^{5} i$$

   In general, we can say:

   $$\sum_{i=1}^{n} ki = k \times 1 + k \times 2 + \ldots + k \times n$$
   $$= k \times (1 + 2 + \ldots + n)$$
   $$= k\sum_{i=1}^{n} i$$

3. Suppose that we need to consider the summation of two different functions, such as:

$$\sum_{k=1}^{n}(k^2 + k^3) = (1^2 + 1^3) + (2^2 + 2^3) + \dots + (n^2 + n^3)$$
$$= (1^2 + 2^2 + \dots + n^2) + (1^3 + 2^3 + \dots + n^3)$$
$$= \sum_{k=1}^{n} k^2 + \sum_{k=1}^{n} k^3$$

In general

$$\sum_{k=1}^{n}(f(k) + g(k)) = \sum_{k=1}^{n} f(k) + \sum_{k=1}^{n} g(k)$$

4. At times it is convenient to change powers, for example:

$\sum_{i=1}^{k} a_i = a_1 + a_2 + \dots + a_{k-1} + a_k$ is the same as

$$\sum_{i=0}^{k-1} a_{i+1} = a_1 + a_2 + \dots + a_{k-1} + a_k$$

Arithmetic series

In arithmetic series, we are concerned with adding the terms of arithmetic sequences. It is very helpful to be able to find an easy expression for the partial sums of such a series.

Let's start with an example:

Find the partial sum for the first 50 terms of the series

$3 + 8 + 13 + 18 + \dots$

Write the 50 terms in ascending order and then in descending order underneath. Add the terms together as shown.

$$\begin{aligned} S_{50} &= \quad 3 + \quad 8 + \ \ 13 + \dots + 248 \\ S_{50} &= 248 + 243 + 238 + \dots + \quad 3 \\ \hline 2\,S_{50} &= 251 + 251 + 251 + \dots + 251 \end{aligned}$$

There are 50 terms in this sum, and hence

$$2\,S_{50} = 50 \times 251 \Rightarrow S_{50} = 6275$$

This reasoning can be extended to any arithmetic series in order to develop a formula for the nth partial sum S_n.

Let $\{a_n\}$ be an arithmetic sequence with first term a_1 and common difference d. We can construct the series in two ways: Forwards, by adding d to a_1 repeatedly, and backwards by subtracting d from a_n repeatedly. We get the following two expressions for the sum:

$$S_n = a_1 + a_2 + a_3 + \dots + a_n = a_1 + (a_1 + d) + (a_1 + 2d) + \dots + (a_1 + (n-1)\,d)$$

$$S_n = a_n + a_{n-1} + a_{n-2} + \dots + a_1 = a_n + (a_n - d) + (a_n - 2d) + \dots + (a_n - (n-1)\,d)$$

By adding, term by term vertically, we get:

$$S_n = a_1 + \quad (a_1 + d) + (a_1 + 2d) + \ldots + (a_1 + (n-1)d)$$
$$S_n = a_n + \quad (a_n - d) + (a_n - 2d) + \ldots + (a_n - (n-1)d)$$
$$\searrow \qquad \downarrow \qquad \downarrow \qquad \swarrow$$
$$2S_n = (a_1 + a_n) + (a_1 + a_n) + (a_1 + a_n) + \ldots + (a_1 + a_n)$$

Since there are n terms, we can reduce the expression above to:

$$2S_n = n(a_1 + a_n)$$

which can be reduced to:

$$S_n = \frac{n}{2}(a_1 + a_n)$$

which in turn can be changed to give an interesting perspective of the sum:

$$S_n = n\left(\frac{a_1 + a_n}{2}\right)$$

which is n times the average of the first and last terms!

If we substitute $a_1 + (n-1)d$ for a_n then we get an alternative formula for the sum:

$$S_n = \frac{n}{2}(a_1 + a_1 + (n-1)d) = \frac{n}{2}(2a_1 + (n-1)d)$$

Online 

See a geometric derivation for the sum of an arithmetic series formula.

The partial sum S_n of an arithmetic series is given by one of the following:
$S_n = \frac{n}{2}(a_1 + a_n)$, or $S_n = n\left(\frac{a_1 + a_n}{2}\right)$, or $S_n = \frac{n}{2}(2a_1 + (n-1)d)$

Example 3.17

Find the partial sum for the first 50 terms of the series $3 + 8 + 13 + 18 + \ldots$

Solution

Using the second formula for the sum we get:

$$S_{50} = \frac{50}{2}(2 \times 3 + (50 - 1)5) = 25 \times 251 = 6275$$

Using the first formula requires that we know the nth term.
So, $a_{50} = 3 + 49 \times 5 = 248$ which now can be used:

$$S_{50} = 25(3 + 248) = 6275$$

Example 3.18

You are given a sequence of figures as shown in the diagram.

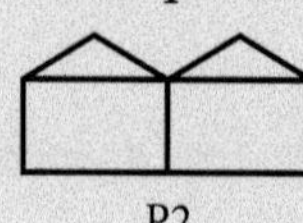

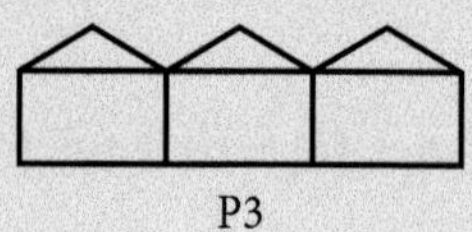

(a) P1 has six line segments. How many line segments are in P20?

(b) Is there a figure with 4401 segments? If so, which one? If not, why not?

(c) Find the total number of line segments in the first 880 figures.

Solution

(a) In each new figure, 5 line segments are added. This is an arithmetic sequence with first term 6, and common difference 5.

Using the nth term form:

$$P20 = 6 + 5(20 - 1) = 101$$

(b) The term whose value is 4401 satisfies the nth term form:

$$4401 = 6 + 5(n - 1), \text{ thus } n = \frac{4401 - 6}{5} + 1 = 880$$

Therefore, 4401 is the 880th figure.

(c) We use one of the formulae for the arithmetic series:

$$S_{880} = \frac{880}{2}(6 + 4401) = 1\,939\,080$$

$$\text{or } S_{880} = \frac{880}{2}(2 \times 6 + 5(880 - 1)) = 1\,939\,080$$

Geometric series

As is the case with arithmetic series, in several cases it is desirable to find a general expression for the nth partial sum of a geometric series.

Let us start with an example:

Find the partial sum for the first 20 terms of the series $3 + 6 + 12 + 24 + \ldots$

We express S_{20} in two different ways and subtract them:

$$\begin{aligned}
S_{20} &= 3 + 6 + 12 + \ldots + 1572864 \\
2S_{20} &= \phantom{3 +{}} 6 + 12 + \ldots + 1572864 + 3\,145\,728 \\
\hline
-S_{20} &= 3 - 3\,145\,728 \\
\Rightarrow S_{20} &= 3\,145\,725
\end{aligned}$$

This reasoning can be extended to any geometric series in order to develop a formula for the nth partial sum S_n.

Let $\{a_n\}$ be a geometric sequence with first term a_1 and common ratio $r \neq 1$. We can construct the series in two ways as before, and, using the definition of the geometric sequence, $a_n = a_{n-1} \times r$, then:

$$\begin{aligned}
S_n &= a_1 + a_2 + a_3 + \ldots + a_{n-1} + a_n \\
rS_n &= ra_1 + ra_2 + ra_3 + \ldots + ra_{n-1} + ra_n \\
&\quad\ \downarrow \qquad \downarrow \qquad\qquad\qquad\quad \downarrow \\
&= a_2 + a_3 + \ldots + a_{n-1} + a_n + ra_n
\end{aligned}$$

Now, we subtract the second expression from the first to get:

$$S_n - rS_n = a_1 - ra_n \Rightarrow S_n(1-r) = a_1 - ra_n \Rightarrow S_n = \frac{a_1 - ra_n}{1-r}, r \neq 1$$

This expression, however, requires that r, a_1, and a_n be known to find the sum. But, using the nth term expression developed earlier, we can simplify this sum formula to:

$$S_n = \frac{a_1 - ra_n}{1-r} = \frac{a_1 - ra_1r^{n-1}}{1-r} = \frac{a_1(1-r^n)}{1-r}, r \neq 1$$

Partial sum of a geometric series

The partial sum, S_n, of n terms of a geometric sequence with common ratio r ($r \neq 1$) and first term a_1 is:

$$S_n = \frac{a_1(1-r^n)}{1-r} \quad \left[\text{equivalent to } S_n = \frac{a_1(r^n-1)}{r-1}\right]$$

Online

An interactive derivation of the geometric sum formula.

Example 3.19

Find the partial sum for the first 20 terms of the series $3 + 6 + 12 + 24 + \ldots$

Solution

$$S_{20} = 3 \times \frac{(1-2^{20})}{1-2} = \frac{3(1-1\,048\,576)}{-1} = 3\,145\,725$$

Infinite geometric series

Consider the series $\sum_{k=1}^{n} 2\left(\frac{1}{2}\right)^{k-1} = 2 + 1 + \frac{1}{2} + \frac{1}{4} + \frac{1}{8} + \ldots$

Consider also finding the partial sums for 10, 20, and 100 terms.

We are looking for the partial sums of a geometric series:

$$\sum_{k=1}^{10} 2\left(\frac{1}{2}\right)^{k-1} = 2 \times \frac{1-\left(\frac{1}{2}\right)^{10}}{1-\frac{1}{2}} \approx 3.996$$

$$\sum_{k=1}^{20} 2\left(\frac{1}{2}\right)^{k-1} = 2 \times \frac{1-\left(\frac{1}{2}\right)^{20}}{1-\frac{1}{2}} \approx 3.999\,996$$

$$\sum_{k=1}^{100} 2\left(\frac{1}{2}\right)^{k-1} = 2 \times \frac{1-\left(\frac{1}{2}\right)^{100}}{1-\frac{1}{2}} \approx 4$$

As the number of terms increases, the partial sum appears to be approaching the number 4. This is no coincidence. In the language of limits,

$$\lim_{n\to\infty} \sum_{k=1}^{n} 2\left(\frac{1}{2}\right)^{k-1} = \lim_{n\to\infty} 2 \times \frac{1-\left(\frac{1}{2}\right)^{k}}{1-\frac{1}{2}} = 2 \times \frac{1-0}{\frac{1}{2}} = 4, \text{ since } \lim_{n\to\infty}\left(\frac{1}{2}\right)^{n} = 0$$

This type of problem allows us to extend the usual concept of a sum of a **finite** number of terms to make sense of sums in which an **infinite** number of terms are involved. Such series are called **infinite series.**

One thing to be made clear about infinite series is that they are not true sums! The associative property of addition of real numbers allows us to extend the definition of the sum of two numbers, such as $a + b$, to three or four or n numbers, but not to an infinite number of numbers. For example, you can add any specific number of 5s together and get a real number, but if you add an infinite number of 5s together, you cannot get a real number. The remarkable thing about infinite series though is that in some cases, such as the example above, the sequence of partial sums (which are true sums) approaches a finite limit L. The limit in our example is 4.

We write this as $\lim_{n\to\infty} \sum_{k=1}^{n} a_k = \lim_{n\to\infty}(a_1 + a_2 + \ldots + a_n) = L$

We say that the series **converges** to L, and it is convenient to define L as the **sum of the infinite series.** We use the notation:

$$\sum_{k=1}^{\infty} a_k = \lim_{n\to\infty} \sum_{k=1}^{n} a_k = L$$

We can therefore write the limit in the previous example:

$$\sum_{k=1}^{\infty} 2\left(\frac{1}{2}\right)^{k-1} = \lim_{n\to\infty} \sum_{k=1}^{n} 2\left(\frac{1}{2}\right)^{k-1} = 4$$

If the series does not have a limit, then it **diverges.**

We are now ready to develop a general rule for **infinite geometric series.** As we know, the sum of a geometric sequence is given by:

$$S_n = \frac{a_1 - ra_n}{1 - r} = \frac{a_1 - ra_1r^{n-1}}{1 - r} = \frac{a_1(1 - r^n)}{1 - r}, r \neq 1$$

If $|r| < 1$, then $\lim_{n\to\infty} r^n = 0$ and hence:

$$\lim_{n\to\infty} S_n = S = \lim_{n\to\infty} \frac{a_1(1 - r^n)}{1 - r} = \frac{a_1}{1 - r}$$

We will call this **the sum of an infinite convergent geometric sequence.** In all other cases the series diverges. The proof is left as an exercise.

In the case of $\sum_{k=1}^{\infty} 2\left(\frac{1}{2}\right)^{k-1} = \frac{2}{1 - \frac{1}{2}} = 4$ as already shown.

Sum of an infinite convergent geometric sequence

The sum, S_∞, of an infinite convergent geometric sequence with first term a_1 such that the common ratio, r, satisfies the condition $|r| < 1$ is given by:

$$S_\infty = \frac{a_1}{1 - r}$$

Example 3.20

A rational number is a number that can be expressed as a quotient of two integers. Show that $0.\overline{6} = 0.666...$ is a rational number.

Solution

$$0.\overline{6} = 0.666... = 0.6 + 0.06 + 0.006 + 0.0006 + ...$$

$$= \frac{6}{10} + \frac{6}{10} \cdot \frac{1}{10} + \frac{6}{10} \cdot \left(\frac{1}{10}\right)^2 + \frac{6}{10} \cdot \left(\frac{1}{10}\right)^3 + ...$$

This is an infinite geometric series with $a_1 = \frac{6}{10}$ and $r = \frac{1}{10}$, therefore:

$$0.\overline{6} = \frac{\frac{6}{10}}{1 - \frac{1}{10}} = \frac{6}{10} \cdot \frac{10}{9} = \frac{2}{3}$$

Example 3.21

A ball has elasticity such that, on each bounce, it bounces to 80% of its previous height. Find the total vertical distance travelled down and up by this ball when it is dropped from a height of 3 m and is allowed to keep bouncing until it comes to rest. Ignore friction and air resistance.

Solution

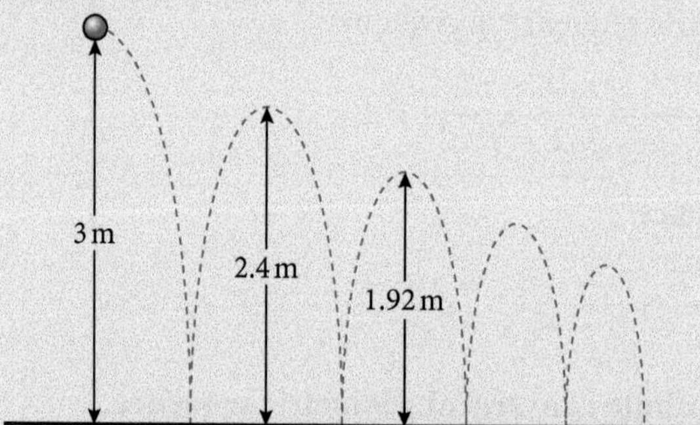

After the ball is dropped the initial 3 m, it bounces up and down a distance of 2.4 m. On each bounce after the first bounce, the ball travels 0.8 times the previous height twice – once upwards and once downwards. So, the total vertical distance is given by

$$h = 3 + 2[2.4 + (2.4 \times 0.8) + (2.4 \times 0.8^2) + ...] = 3 + 2 \times l$$

The terms inside the square brackets form an infinite geometric series with $a_1 = 2.4$ and $r = 0.8$. The value of that quantity is:

$$l = \frac{2.4}{1 - 0.8} = 12$$

Hence the total distance required is $h = 3 + 2(12) = 27$ m

Applications of series to compound interest calculations

Annuities

An **annuity** is a sequence of equal periodic payments. If you are saving money by depositing the same amount at the end of each compounding period, the annuity is called an **ordinary annuity**. Using geometric series, you can calculate the **future value (FV)** of this annuity, which is the amount of money you will have after making the last payment.

You invest €1000 at the end of each year for 10 years at a fixed annual interest rate of 6%, as shown in Table 3.3.

Year	Amount invested (€)	Future value (€)
10	1000	1000
9	1000	$1000(1 + 0.06)$
8	1000	$1000(1 + 0.06)^2$
⋮		
1	1000	$1000(1 + 0.06)^9$

Table 3.3 Calculating the future value

The future value FV of this investment is the sum of the entries in the last column:

$$FV = 1000 + 1000(1 + 0.06) + 1000(1 + 0.06)^2 + \dots + 1000(1 + 0.06)^9$$

This sum is a partial sum of a geometric series with $n = 10$ and $r = 1 + 0.06$
Hence:

$$FV = \frac{1000(1 - (1 + 0.06)^{10})}{1 - (1 + 0.06)} = \frac{1000(1 - (1 + 0.06)^{10})}{-0.06} = €13{,}180.79$$

```
Compound Interest
 I%  =6
 PV  =0
 PMT =-1000
 FV  =0
 P/Y =1
 C/Y =1
```

```
Compound Interest
 FV  =13180.79494
```

Figure 3.4 This result can also be produced with a GDC

We can generalise the previous formula in the same manner. Let the periodic payment be R, and the periodic interest rate be i – that is, $i = \frac{r}{n}$.
Let the number of periodic payments be m.

Period	Amount invested	Future value
m	R	R
$m - 1$	R	$R(1 + i)$
$m - 2$	R	$R(1 + i)^2$
⋮		
1	R	$R(1 + i)^{m-1}$

Table 3.4 Formula for calculating the future value

The future value FV is the sum of the entries in the last column:

$$FV = R + R(1 + i) + R(1 + i)^2 + \dots + R(1 + i)^{m-1}$$

This is a partial sum of a geometric series with m terms and $r = 1 + i$. Hence:

$$FV = \frac{R(1 - (1 + i)^m)}{1 - (1 + i)} = \frac{R(1 - (1 + i)^m)}{-i} = R\left(\frac{(1 + i)^m - 1}{i}\right)$$

If the payment is made at the beginning of the period rather than at the end, then the annuity is called an **annuity due** and the future value after m periods will be slightly different.

Period	Amount invested	Future value
m	R	$R(1 + i)$
$m - 1$	R	$R(1 + i)^2$
$m - 2$	R	$R(1 + i)^3$
1	R	$R(1 + i)^m$

The future value of this investment is the sum of the entries in the last column:

$FV = R(1 + i) + R(1 + i)^2 + \dots + R(1 + i)^{m-1} + R(1 + i)^m$

This is a partial sum of a geometric series with m terms and $r = 1 + i$.

$$FV = \frac{R(1 + i)(1 - (1 + i)^m)}{1 - (1 + i)} = \frac{R(1 + i - (1 + i)^{m+1})}{-i} = R\left(\frac{(1 + i)^{m+1} - 1}{i} - 1\right)$$

If the previous investment is made at the beginning of the year rather than at the end, then in 10 years we have:

$$FV = R\left(\frac{(1 + i)^{m+1} - 1}{i}\right) - 1 = 1000\left(\frac{(1 + 0.06)^{10+1} - 1}{0.06} - 1\right) = 13\,971.64$$

Exercise 3.4

1. Find the sum of the arithmetic sequence $11 + 17 + \dots + 365$

2. Find the sum of this sequence:

   $$2 - 3 + \frac{9}{2} - \frac{27}{4} + \dots - \frac{177\,147}{1024}$$

3. Evaluate $\sum_{k=0}^{13}(2 - 0.3k)$

4. Evaluate $2 - \frac{4}{5} + \frac{8}{25} - \frac{16}{125} + \dots$

5. Evaluate $\frac{1}{3} + \frac{\sqrt{3}}{12} + \frac{1}{16} + \frac{\sqrt{3}}{64} + \frac{3}{256} + \dots$

6. Express each repeating decimal as a fraction:

   (a) $0.\overline{52}$ **(b)** $0.4\overline{53}$ **(c)** $3.01\overline{37}$

7. At the beginning of every month, Maggie invests \$150 in an account that pays a 6% annual rate of interest. How much money will there be in the account after six years?

8. Find the sum of each series.

(a) $9 + 13 + 17 + \ldots + 85$ **(b)** $8 + 14 + 20 + \ldots + 278$

(c) $155 + 158 + 161 + \ldots + 527$

9. The kth term of an arithmetic sequence is $2 + 3k$. Find, in terms of n, the sum of the first n terms of this sequence.

10. For the arithmetic sequence that begins $17 + 20 + 23 + \ldots$, for what value of n will the partial sum S_n of the sequence exceed 678?

11. For the arithmetic sequence that begins $-18 - 11 - 4 \ldots$, for what value of n will the partial sum S_n of the sequence exceed 2335?

12. An arithmetic sequence has a as first term and $2d$ as common difference, i.e., $a, a + 2d, a + 4d, \ldots$. The sum of the first 50 terms is T. Another sequence, with first term $a + d$ and common difference $2d$, is combined with the first one to produce a new arithmetic sequence. Let the sum of the first 100 terms of the new combined sequence be S. If $2T + 200 = S$, find d.

13. Consider the arithmetic sequence $3, 7, 11, \ldots, 999$.

(a) Find the number of terms and the sum of this sequence.

(b) Create a new sequence by removing every third term, i.e., $11, 23, \ldots$. Find the sum of the terms of the remaining sequence.

14. The sum of the first 10 terms of an arithmetic sequence is 235 and the sum of the second 10 terms is 735. Find the first term and the common difference.

15. Use your GDC or a spreadsheet to evaluate each sum.

(a) $\sum_{k=1}^{20}(k^2 + 1)$ **(b)** $\sum_{i=3}^{17}\left(\frac{1}{i^2 + 3}\right)$ **(c)** $\sum_{n=1}^{100}(-1)^n\frac{3}{n}$

16. Find the sum of the arithmetic series $13 + 19 + \ldots + 367$

17. Find the sum of the arithmetic series:

$$2 - \frac{4}{3} + \frac{8}{9} - \frac{16}{27} + \ldots - \frac{4096}{177\,147}$$

18. Evaluate $\sum_{k=0}^{11}(3 + 0.2k)$

19. Evaluate $2 - \frac{4}{3} + \frac{8}{9} - \frac{16}{27} + \ldots$

20. Evaluate $\frac{1}{2} + \frac{\sqrt{2}}{2\sqrt{3}} + \frac{1}{3} + \frac{\sqrt{2}}{3\sqrt{3}} + \frac{2}{9} + \ldots$

21. Find the first four partial sums and the nth partial sum of each sequence.

(a) $u_n = \frac{3}{5^n}$ **(b)** $v_n = \frac{1}{n^2 + 3n + 2}$ **(c)** $u_n = \sqrt{n+1} - \sqrt{n}$

For question 21 part (b), show that $v_n = \frac{1}{n+1} - \frac{1}{n+2}$

22. A ball is dropped from a height of 16 m. Every time it hits the ground it bounces to 81% of its previous height.

(a) Find the maximum height it reaches after the 10th bounce.

(b) Find the total distance travelled by the ball until it comes to rest. (Assume no friction and no loss of elasticity.)

23. The sides of a square are 16 cm in length. A new square is formed by joining the midpoints of the adjacent sides and then two of the resulting triangles are coloured, as shown.

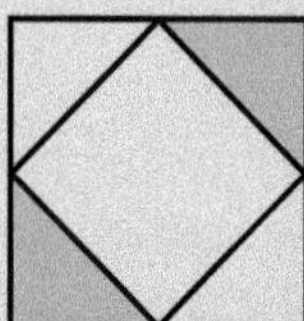 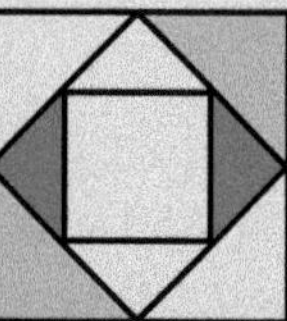 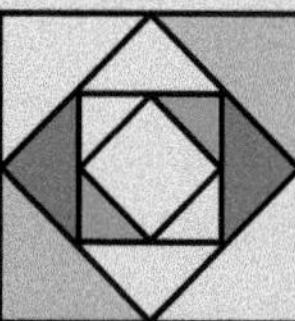 

(a) If the process is repeated six more times, determine the total area of the shaded region.

(b) If the process were to be repeated indefinitely, find the total area of the shaded region.

24. The largest rectangle in the diagram below measures 4 cm by 2 cm. Another rectangle is constructed inside it, measuring 2 cm by 1 cm. The process is repeated. The region surrounding every other inner rectangle is shaded, as shown.

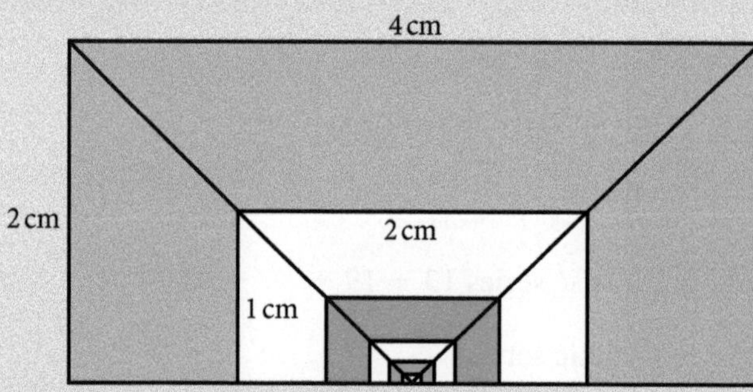

(a) Find the total area for the three regions shaded already.

(b) If the process were to be repeated indefinitely, find the total area of the shaded regions.

25. Find each sum.

(a) $7 + 12 + 17 + 22 + \ldots + 337 + 342$

(b) $9486 + 9479 + 9472 + 9465 + \ldots + 8919 + 8912$

(c) $2 + 6 + 18 + 54 + \ldots + 3\,188\,646 + 9\,565\,938$

(d) $120 + 24 + \frac{24}{5} + \frac{24}{25} + \ldots + \frac{24}{78\,125}$

3.5 The binomial theorem

A binomial is a polynomial with two terms. For example, $x + y$ is a binomial. In principle, it is easy to raise $x + y$ to any power, but raising it to high powers would be tedious. In this chapter, we will find a formula that gives the expansion of $(x + y)^n$ for any positive integer n.

Let's look at some particular cases of the expansion of $(x + y)^n$

$(x+y)^0 = 1$

$(x+y)^1 = x + y$

$(x+y)^2 = x^2 + 2xy + y^2$

$(x+y)^3 = x^3 + 3x^2y + 3xy^2 + y^3$

$(x+y)^4 = x^4 + 4x^3y + 6x^2y^2 + 4xy^3 + y^4$

$(x+y)^5 = x^5 + 5x^4y + 10x^3y^2 + 10x^2y^3 + 5xy^4 + y^5$

$(x+y)^6 = x^6 + 6x^5y + 15x^4y^2 + 20x^3y^3 + 15x^2y^4 + 6xy^5 + y^6$

There are several things that we notice after looking at the expansion:

- There are $n + 1$ terms in the expansion of $(x + y)^n$
- The degree of each term is n.
- The powers on x begin with n and decrease to 0.
- The powers on y begin with 0 and increase to n.
- The coefficients are symmetric.

For instance, notice how the powers of x and y behave in the expansion of $(x + y)^5$

The powers of x decrease:

$(x + y)^5 = x^{\boxed{5}} + 5x^{\boxed{4}}y + 10x^{\boxed{3}}y^2 + 10x^{\boxed{2}}y^3 + 5x^{\boxed{1}}y^4 + y^5$

The powers of y increase:

$(x + y)^5 = x^5 + 5x^4y^{\boxed{1}} + 10x^3y^{\boxed{2}} + 10x^2y^{\boxed{3}} + 5xy^{\boxed{4}} + y^{\boxed{5}}$

With these observations, we can now proceed to expand any binomial raised to power n: $(x + y)^n$. For example, leaving a blank for the missing coefficients, the expansion for $(x + y)^7$ can be written as:

$(x + y)^7 = \square x^7 + \square x^6y + \square x^5y^2 + \square x^4y^3 + \square x^3y^4 + \square x^2y^5 + \square xy^6 + \square y^7$

To finish the expansion, we need to determine these coefficients. In order to see the pattern, look at the coefficients of the expansion at the start of the section.

$(x+y)^0$	1							row 0
$(x+y)^1$	1	1						row 1
$(x+y)^2$	1	2	1					row 2
$(x+y)^3$	1	3	3	1				row 3
$(x+y)^4$	1	4	6	4	1			row 4
$(x+y)^5$	1	5	10	10	5	1		row 5
$(x+y)^6$	1	6	15	20	15	6	1	row 6
	column 0	column 1	column 2	column 3	column 4	column 5	column 6	

Pascal's triangle
Every entry in a row is the sum of the term directly above it and the entry to the left of that. When there is no entry, the value is considered zero.

A triangle like the one above is known as Pascal's triangle. The first and second terms in row 3 give you the second term in row 4, the third and fourth terms in row 3 give you the fourth term of row 4, the second and third terms in row 5 give the third term in row 6, and the fifth and sixth terms in row 5 give you the sixth term in row 6. So now we can state a key property of Pascal's triangle.

Take the last entry in row 5, for example; there is no entry directly above it, so its value is $0 + 1 = 1$.

From this property it is easy to find all the terms in any row of Pascal's triangle from the row above it. So, for the expansion of $(x + y)^7$, the terms are found from row 6 as follows:

$$0 \rightarrow 1 \rightarrow 6 \rightarrow 15 \rightarrow 20 \rightarrow 15 \rightarrow 6 \rightarrow 1 \rightarrow 0$$
$$\downarrow \quad \downarrow \quad \downarrow \quad \downarrow \quad \downarrow \quad \downarrow \quad \downarrow \quad \downarrow$$
$$1 \quad 7 \quad 21 \quad 35 \quad 35 \quad 21 \quad 7 \quad 1$$

So, $(x + y)^7 = x^7 + \boxed{7}x^6y + \boxed{21}x^5y^2 + \boxed{35}x^4y^3 + \boxed{35}x^3y^4 + \boxed{21}x^2y^5 + \boxed{7}xy^6 + y^7$

Several sources use a slightly different arrangement for Pascal's triangle. The common usage considers the triangle as isosceles and uses the principle that every two entries add up to give the entry diagonally below them, as shown in the diagram.

```
                 1
              1     1
           1     2     1
        1     3     3     1
     1     4     6     4     1
  1     5    10    10     5     1
```

Example 3.22

Use Pascal's triangle to expand $(2k - 3)^5$

Solution

We can find the expansion by replacing x by $2k$ and y by -3 in the binomial expansion of $(x + y)^5$.

Using the fifth row of Pascal's triangle for the coefficients will give:

$$1(2k)^5 + 5(2k)^4(-3) + 10(2k)^3(-3)^2 + 10(2k)^2(-3)^3 + 5(2k)(-3)^4 + 1(-3)^5$$
$$= 32k^5 - 240k^4 + 720k^3 - 1080k^2 + 810k - 243$$

Pascal's triangle is a useful tool for finding the coefficients of the binomial expansion for relatively small values of n. It is not very efficient for large values of n. Imagine we want to evaluate $(x + y)^{20}$. Using Pascal's triangle, we would need the terms in the 19th row, and the 18th row and so on. This makes the process tedious and not practical.

Luckily, there is a formula we can use to find the coefficients of any Pascal's triangle row. This formula is the binomial formula, which we will prove in Chapter 5. Every entry in Pascal's triangle is denoted by $\binom{n}{r}$ or ${}_nC_r$ – this is also known as the binomial coefficient. In ${}_nC_r$, n is the row number and r is the column number. To understand the binomial coefficient, we need to understand what factorial notation means.

Factorial notation

The product of the first n positive integers is denoted by $n!$ and is called n **factorial**:

$n! = n \times (n-1) \times (n-2) \times \cdots \times 3 \times 2 \times 1$

We also define $0! = 1$.

This definition of the factorial makes many formulae involving the multiplication of consecutive positive integers shorter and easier to write. That includes the binomial coefficient.

The binomial coefficient

With n and r as non-negative integers such that $n \geqslant r$, the **binomial coefficient** ${}_nC_r$ $\left[\text{or } \binom{n}{r}\right]$ is defined by

$${}_nC_r = \binom{n}{r} = \frac{n!}{r!(n-r)!}$$

${}_nC_r$ is also written as ${}^nC_r$

When simplified, ${}_nC_r$ can be written as ${}_nC_r = \dfrac{n(n-1)(n-2)\cdots(n-r+1)}{r!}$.

For example ${}_nC_3 = \dfrac{n(n-1)(n-2)}{3!}$.

Example 3.23

Find the value of:

(a) ${}_7C_3$ (b) ${}_7C_4$ (c) $\binom{7}{0}$ (d) $\binom{7}{7}$

Solution

(a) $${}_7C_3 = \frac{7!}{3!(7-3)!} = \frac{7!}{3!4!} = \frac{\cancel{1 \cdot 2 \cdot 3 \cdot 4} \cdot 5 \cdot 6 \cdot 7}{(1 \cdot 2 \cdot 3)(\cancel{1 \cdot 2 \cdot 3 \cdot 4})} = \frac{5 \cdot 6 \cdot 7}{1 \cdot 2 \cdot 3} = 35,$$

or using the other form of the expression for the binomial coefficient

$${}_7C_3 = \frac{7 \cdot 6 \cdot 5}{3!} = \frac{210}{6} = 35$$

(b) $${}_7C_4 = \frac{7!}{4!(7-4)!} = \frac{7!}{4!3!} = \frac{\cancel{1 \cdot 2 \cdot 3 \cdot 4} \cdot 5 \cdot 6 \cdot 7}{(\cancel{1 \cdot 2 \cdot 3 \cdot 4})(1 \cdot 2 \cdot 3)} = \frac{5 \cdot 6 \cdot 7}{1 \cdot 2 \cdot 3} = 35$$

(c) $$\binom{7}{0} = \frac{7!}{0!(7-0)!} = \frac{\cancel{7!}}{0!\cancel{7!}} = \frac{1}{1} = 1$$

(d) $$\binom{7}{7} = \frac{7!}{7!(7-7)!} = \frac{\cancel{7!}}{\cancel{7!}0!} = \frac{1}{1} = 1$$

Although the binomial coefficient $\binom{n}{r}$ appears as a fraction, all the results where n and r are non-negative integers are positive integers. Also notice the **symmetry** of the coefficient in Example 3.23.

Example 3.24

Calculate each binomial coefficient:

$\binom{6}{0}, \binom{6}{1}, \binom{6}{2}, \binom{6}{3}, \binom{6}{4}, \binom{6}{5}, \binom{6}{6}$

Solution

$\binom{6}{0} = 1 \quad \binom{6}{1} = 6 \quad \binom{6}{2} = 15 \quad \binom{6}{3} = 20 \quad \binom{6}{4} = 15 \quad \binom{6}{5} = 6 \quad \binom{6}{6} = 1$

The values in Example 3.24 are the entries in the 6th row of Pascal's triangle.

We can write Pascal's triangle in the following manner:

$$\begin{array}{llllllll}
\binom{0}{0} \\
\binom{1}{0} & \binom{1}{1} \\
\binom{2}{0} & \binom{2}{1} & \binom{2}{2} \\
\binom{3}{0} & \binom{3}{1} & \binom{3}{2} & \binom{3}{3} \\
\vdots & \vdots & \vdots & \vdots \\
\binom{n}{0} & \binom{n}{1} & \cdots & \cdots & \cdots & \cdots & \binom{n}{n}
\end{array}$$

Example 3.25

Calculate ${}_nC_{r-1} + {}_nC_r$ (This is called Pascal's rule.)

Solution

You will be able to provide reasons for the steps after you do the exercises.

$$\begin{aligned}
{}_nC_{r-1} + {}_nC_r &= \frac{n!}{(r-1)!(n-r+1)!} + \frac{n!}{r!(n-r)!} \\
&= \frac{n! \cdot r}{r \cdot (r-1)!(n-r+1)!} + \frac{n! \cdot (n-r+1)}{r!(n-r)! \cdot (n-r+1)} \\
&= \frac{n! \cdot r}{r!(n-r+1)!} + \frac{n! \cdot (n-r+1)}{r!(n-r+1)!} \\
&= \frac{n!\,(r+n-r+1)}{r!(n-r+1)!} \\
&= \frac{(n+1)!}{r!(n+1-r)!} \\
&= {}_{n+1}C_r
\end{aligned}$$

If we read the result in Example 3.25 carefully, it says that the sum of the terms in the nth row, $(r - 1)$th and rth columns, is equal to the entry in the $(n + 1)$th row and rth column.

	$(r - 1)$th column		rth column
nth row	${}_nC_{r-1}$	$+$	${}_nC_r$
			$\Vert$
$(n + 1)$th row			${}_{n+1}C_r$

That is, the two entries in blue are adjacent entries in the nth row of Pascal's triangle and the entry in red is the entry in the $(n + 1)$th row directly below the rightmost entry. This is precisely the principle behind Pascal's triangle!

Using the binomial theorem

We are now prepared to state the binomial theorem:

$$(x + y)^n = \binom{n}{0}x^n + \binom{n}{1}x^{n-1}y + \binom{n}{2}x^{n-2}y^2 + \binom{n}{3}x^{n-3}y^3 + \cdots + \binom{n}{n-1}xy^{n-1} + \binom{n}{n}y^n$$

or

$$(x + y)^n = {}_nC_0x^n + {}_nC_1x^{n-1}y + {}_nC_2x^{n-2}y^2 + {}_nC_3x^{n-3}y^3 + \cdots + {}_nC_ny^n$$

In a compact form, we can use sigma notation to express the theorem as follows:

$$(x + y)^n = \sum_{i=0}^{n}\binom{n}{i}x^{n-i}y^i = \sum_{i=0}^{n}{}_nC_ix^{n-i}\,y^i$$

Example 3.26

Use the binomial theorem to expand $(x + y)^7$

Solution

$$(x + y)^7 = \binom{7}{0}x^7 + \binom{7}{1}x^{7-1}y + \binom{7}{2}x^{7-2}y^2 + \binom{7}{3}x^{7-3}y^3 + \binom{7}{4}x^{7-4}y^4$$
$$+ \binom{7}{5}x^{7-5}y^5 + \binom{7}{6}x^{7-6}y^6 + \binom{7}{7}y^7$$
$$= x^7 + 7x^6y + 21x^5y^2 + 35x^4y^3 + 35x^3y^4 + 21x^2y^5 + 7xy^6 + y^7$$

Example 3.27

Find the expansion for $(2k - 3)^5$

Solution

$$(2k - 3)^5 = \binom{5}{0}(2k)^5 + \binom{5}{1}(2k)^4(-3) + \binom{5}{2}(2k)^3(-3)^2 + \binom{5}{3}(2k)^2(-3)^3$$
$$+ \binom{5}{4}(2k)(-3)^4 + \binom{5}{5}(-3)^5$$
$$= 32k^5 - 240k^4 + 720k^3 - 1080k^2 + 810k - 243$$

Example 3.28

Find the term containing a^3 in the expansion $(2a - 3b)^9$

Solution

To find the term, we do not need to expand the whole expression.
Since $(x + y)^n = \sum_{i=0}^{n} {}_nC_i\, x^{n-i}y^i$, the term containing a^3 is the term where $n - i = 3$, (i.e. when $i = 6$). So the required term is

${}_9C_6(2a)^{9-6}(-3b)^6 = 84 \cdot 8a^3 \cdot 729b^6 = 489\,888a^3b^6$

Example 3.29

Find the term independent of x in:

(a) $\left(2x^2 - \frac{3}{x}\right)^9$ (b) $\left(4x^3 - \frac{2}{x^2}\right)^5$

Solution

(a) 'Independent of x' means the term with no x variable – that is, the constant term in the expansion of $\left(2x^2 - \frac{3}{x}\right)^9 = \sum_{k=0}^{9} {}_9C_k(2x^2)^k\left(-\frac{3}{x}\right)^{9-k}$

The constant term contains x^0. Thus it is the term where $(x^2)^k\left(\frac{1}{x}\right)^{9-k} = x^0$,

so $2k = 9 - k \Rightarrow k = 3$, thus the term is ${}_9C_3(2x^2)^3\left(-\frac{3}{x}\right)^6 = 84 \cdot 8x^6 \cdot \frac{729}{x^6}$

$= 489\,888$

(b) Similarly, $(x^3)^k\left(\frac{1}{x^2}\right)^{5-k} = x^0$, so $3k = 10 - 2k \Rightarrow k = 2$, thus the term is

${}_5C_2(4x^3)^2\left(-\frac{2}{x^2}\right)^3 = -1280$

You can also use your GDC.

```
(9C3)×(2³)×(-3)⁶
                489888
(5C2)×4²×(-2)³
                 -1280
```

Example 3.30

Find the coefficient of b^6 in the expansion of $\left(2b^2 - \frac{1}{b}\right)^{12}$

Solution

The general term is $\binom{12}{i}(2b^2)^{12-i}\left(-\frac{1}{b}\right)^i = \binom{12}{i}(2)^{12-i}(b^2)^{12-i}\left(-\frac{1}{b}\right)^i$

$= \binom{12}{i}(2)^{12-i}b^{24-2i}b^{-i}(-1)^i$

$= \binom{12}{i}(2)^{12-i}b^{24-3i}(-1)^i$

$24 - 3i = 6 \Rightarrow i = 6$. So the coefficient in question is $\binom{12}{6}(2)^6(-1)^6 = 59\,136$

Exercise 3.5

1. Use Pascal's triangle to expand each binomial.

   (a) $(x + 2y)^5$ (b) $(a - b)^4$ (c) $(x - 3)^6$

   (d) $(2 - x^3)^4$ (e) $(x - 3b)^7$ (f) $\left(2n + \frac{1}{n^2}\right)^6$

   (g) $\left(\frac{3}{x} - 2\sqrt{x}\right)^4$

2. Evaluate each expression.

   (a) $\binom{8}{3}$ (b) $\binom{18}{5} - \binom{18}{13}$ (c) $\binom{7}{4}\binom{7}{3}$

   (d) $\binom{5}{0} + \binom{5}{1} + \binom{5}{2} + \binom{5}{3} + \binom{5}{4} + \binom{5}{5}$

   (e) $\binom{6}{0} - \binom{6}{1} + \binom{6}{2} - \binom{6}{3} + \binom{6}{4} - \binom{6}{5} + \binom{6}{6}$

3. Use the binomial theorem to expand each expression.

   (a) $(x - 2y)^7$ (b) $(2a - b)^6$ (c) $(x - 4)^5$

   (d) $(2 + x^3)^6$ (e) $(3x - b)^7$ (f) $\left(2n - \frac{1}{n^2}\right)^6$

   (g) $\left(\frac{2}{x} - 3\sqrt{x}\right)^4$ (h) $(1 + \sqrt{5})^4 + (1 - \sqrt{5})^4$

   (i) $(\sqrt{3} + 1)^8 - (\sqrt{3} - 1)^8$

4. Consider the expression $\left(x - \frac{2}{x}\right)^{45}$

   (a) Find the first three terms of this expansion.

   (b) Find the constant term if it exists or justify why it doesn't exist.

   (c) Find the last three terms of the expansion.

   (d) Find the term containing x^3 if it exists or justify why it doesn't exist.

5. Prove that ${}_nC_k = {}_nC_{n-k}$ for all $n, k \in \mathbb{N}$ and $n \geqslant k$

6. Prove that for any positive integer n:

   $\binom{n}{1} + \binom{n}{2} + \ldots + \binom{n}{n-1} + \binom{n}{n} = 2^n - 1.$

For question 6: $2^n = (1 + 1)^n$

7. Consider all $n, k \in \mathbb{N}$ and $n \geqslant k$

   (a) Verify that $k! = k(k - 1)!$

   (b) Verify that $(n - k + 1)! = (n - k + 1)(n - k)!$

   (c) Justify the steps given in the proof of $\binom{n}{r-1} + \binom{n}{r} = \binom{n+1}{r}$ in Example 3.25.

8. Find the value of the expression:

   $$\binom{6}{0}\left(\frac{1}{3}\right)^6 + \binom{6}{1}\left(\frac{1}{3}\right)^5\left(\frac{2}{3}\right) + \binom{6}{2}\left(\frac{1}{3}\right)^4\left(\frac{2}{3}\right)^2 + \ldots + \binom{6}{6}\left(\frac{2}{3}\right)^6$$

9. Find the value of the expression:

$$\binom{8}{0}\left(\frac{2}{5}\right)^8 + \binom{8}{1}\left(\frac{2}{5}\right)^7\left(\frac{3}{5}\right) + \binom{8}{2}\left(\frac{2}{5}\right)^6\left(\frac{3}{5}\right)^2 + \cdots + \binom{8}{8}\left(\frac{3}{5}\right)^8$$

10. Find the value of the expression:

$$\binom{n}{0}\left(\frac{1}{7}\right)^n + \binom{n}{1}\left(\frac{1}{7}\right)^{n-1}\left(\frac{6}{7}\right) + \binom{n}{2}\left(\frac{1}{7}\right)^{n-2}\left(\frac{6}{7}\right)^2 + \cdots + \binom{n}{n}\left(\frac{6}{7}\right)^n$$

11. Find the term independent of x in the expansion of $\left(x^2 - \frac{1}{x}\right)^6$

12. Find the term independent of x in the expansion of $\left(3x - \frac{2}{x}\right)^8$

13. Find the term independent of x in the expansion of $\left(2x - \frac{3}{x^3}\right)^8$

14. Find the first three terms in the expansion of $(1 + x)^{10}$ and use them to find an approximation to:

 (a) 1.01^{10} **(b)** 0.99^{10}

15. Show that $\binom{n}{r-1} + 2\binom{n}{r} + \binom{n}{r+1} = \binom{n+2}{r+1}$ and interpret your result on the entries in Pascal's triangle.

16. Express each repeating decimal as a fraction:

 (a) $0.\overline{7}$ **(b)** $0.3\overline{45}$ **(c)** $3.21\overline{29}$

17. Find the coefficient of x^6 in the expansion of $(2x - 3)^9$

18. Find the coefficient of x^3b^4 in the expansion of $(ax + b)^7$

19. Find the constant term in the expansion of $\left(\frac{2}{z^2} - z\right)^{15}$

20. Expand $(3n - 2m)^5$

21. Find the coefficient of r^{10} in the expansion of $(4 + 3r^2)^9$

22. In the expansion of $(2 - kx)^5$, the coefficient of x^3 is -1080. Find the constant k.

Chapter 3 practice questions

1. In an arithmetic sequence, the first term is 4, the 4th term is 19 and the nth term is 99. Find the common difference and the number of terms, n.

2. How much money should you invest now if you wish to have \$3000 in your account after 6 years, if interest is compounded quarterly at an annual rate of 6%?

3. Two students, Nick and Maxine, decide to start preparing for their IB exams 15 weeks ahead of the exams. Nick starts by studying for 12 hours in the first week and plans to increase the amount by 2 hours per week. Maxine starts with 12 hours in the first week and decides to increase her time by 10% every week.

   (a) How many hours will each student study in week 5?

   (b) How many hours in total will each student study for the 15 weeks?

   (c) In which week will Maxine exceed 40 hours per week?

   (d) In which week will Maxine catch up with Nick in the number of hours spent studying per week?

4. Two diet schemes are available for people to lose weight. Plan A promises the patient an initial weight loss of 1000 grams the first month with a steady loss of an additional 80 grams every month after the first, for a maximum duration of 12 months.

   Plan B starts with a weight loss of 1000 grams the first month and an increase in weight loss by 6% more every subsequent month.

   (a) Write down the number of grams lost under Plan B in the second and third months.

   (b) Find the weight lost in the 12th month for each plan.

   (c) Find the total weight loss during a 12-month period under

   (i) Plan A **(ii)** Plan B.

5. You start a savings plan to buy a car, where you invest €500 at the beginning of each year for 10 years. Your bank offers a fixed rate of 6% per year, compounded annually.

   Calculate, giving your answers to the nearest euro(€):

   (a) how much the first €500 is worth at the end of 10 years

   (b) the total value of your investment at the end of the 10 years.

6. $\{a_n\}$ is defined as follows:

   $$a_n = \sqrt[3]{8 - a_{n-1}^3}$$

   (a) Given that $a_1 = 1$, evaluate a_2, a_3, and a_4. Describe $\{a_n\}$.

   (b) Given that $a_1 = 2$, evaluate a_2, a_3, and a_4. Describe $\{a_n\}$.

7. A marathon runner plans her training program for a 20 km race. On the first day she plans to run 2 km, then she wants to increase her distance by 500 m on each subsequent training day.

   (a) On which day of her training does she first run a distance of 20 km?

   (b) By the time she manages to run the 20 km distance, what is the total distance she would have run for the whole training program?

8. In a certain country, smartphones were first introduced in the year 2010. During the first year, 1600 people bought a smartphone. In 2011, the number of new participants was 2400, and in 2012 the new participants numbered 3600.

(a) You notice that the trend follows a geometric sequence. Find the common ratio.

(b) Assuming that the trend continues:

(i) how many participants will join in 2022?

(ii) in what year would the number of new participants first exceed 50 000?

Between 2010 and 2012, the total number of participants reached 7600.

(c) What is the total number of participants between 2010 and 2022? Assume that all users continue to have a smartphone.

During this period, the total adult population remains approximately 800 000.

(d) Use this information to suggest a reason why this trend in growth would not continue.

9. The midpoints M, N, P, and Q of the sides of a square of side 1 cm are joined to form a new square.

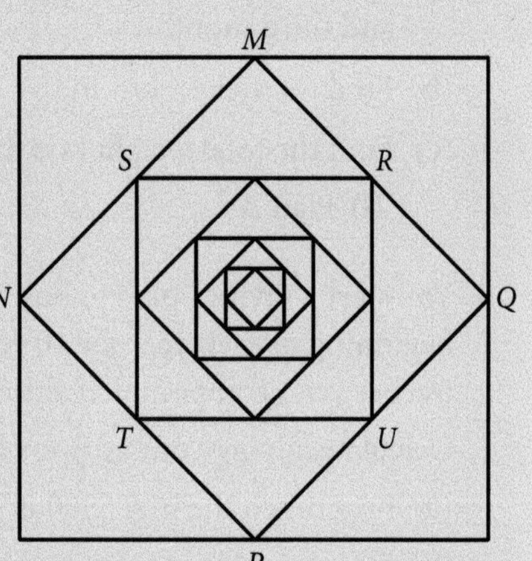

(a) Show that the side length of the square $MNPQ$ is $\frac{\sqrt{2}}{2}$

(b) Find the area of square $MNPQ$.

A new third square $RSTU$ is constructed in the same manner.

(c) **(i)** Find the area of the square $RSTU$.

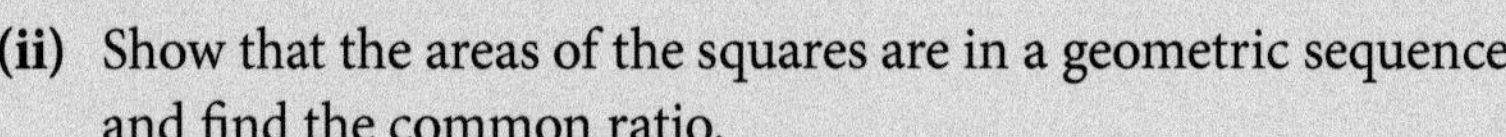

(ii) Show that the areas of the squares are in a geometric sequence and find the common ratio.

The procedure continues indefinitely.

(d) **(i)** Find the area of the 10th square.

(ii) Find the sum of the areas of all the squares.

10. Aristede is a dedicated swimmer. He goes swimming once every week. He starts the first week of the year by swimming 200 metres. Each week after that he swims 20 metres more than the previous week. He does this for the whole year (52 weeks).

(a) How far does he swim in the final week?

(b) How far does he swim altogether?

11. The diagram shows three iterations of constructing squares in the following manner: A square of side 3 units is drawn, then it is divided into nine smaller squares and the middle square is shaded (below, left). Each of the unshaded squares is in turn divided into nine squares and the process is repeated. The area of the first shaded square is 1 unit.

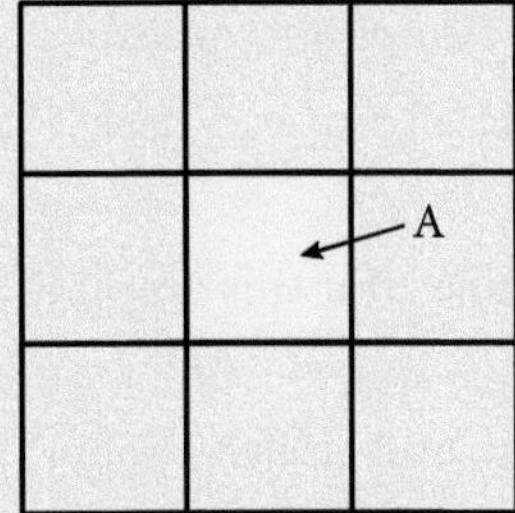

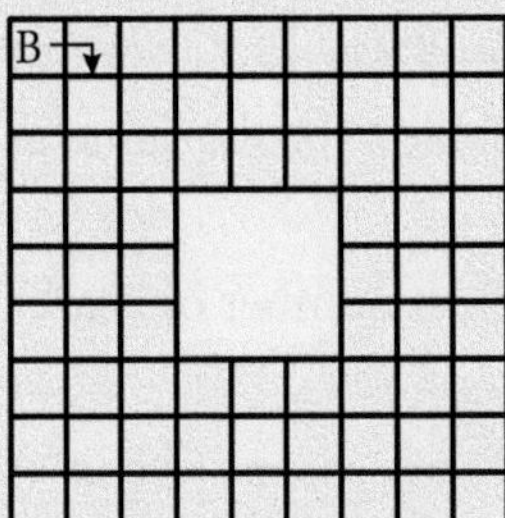

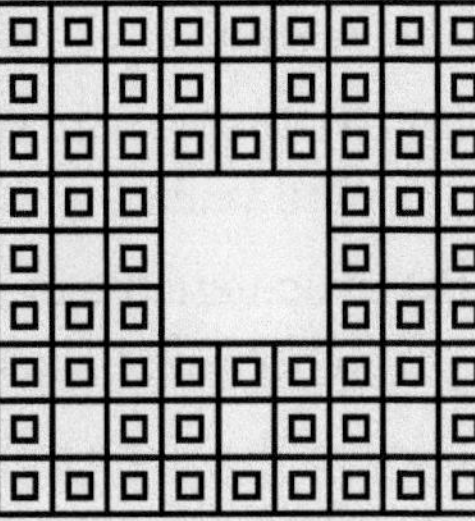

(a) Find the area of each of the squares A and B.

(b) Find the area of any small square in the third diagram.

(c) Find the area of the shaded regions in the second and third iterations.

(d) If the process was continued indefinitely, find the area left unshaded.

12. The table shows four series of numbers. One is an arithmetic series, one is a converging geometric series, one is a diverging geometric series, and the fourth is neither geometric nor arithmetic.

	Series	Type of series
(i)	$2 + 22 + 222 + 2222 + \dots$	
(ii)	$2 + \frac{4}{3} + \frac{8}{9} + \frac{16}{27} + \dots$	
(iii)	$0.8 + 0.78 + 0.76 + 0.74 + \dots$	
(iv)	$2 + \frac{8}{3} + \frac{32}{9} + \frac{128}{27} + \dots$	

(a) Copy and complete the table by stating the type of each series.

(b) Find the sum of the infinite geometric series.

13. Two IT companies offer apparently similar salary schemes for their new appointees. Kell offers a starting salary of €18,000 per year and an annual increase of €400 each year after the first. YBO offers a starting salary of €17,000 per year and an annual increase of 7% each year after the first year.

(a) **(i)** Write down the salary paid in the 2nd and 3rd years for each company.

(ii) Calculate the total amount that an employee working for 10 years will accumulate over 10 years in each company.

(iii) Calculate the salary paid in the tenth year in each company.

(b) Tim works at Kell and Merijayne works at YBO.

(i) When would Merijayne start earning more than Tim?

(ii) What is the minimum number of years that Merijayne requires so that her total earnings exceed Tim's total earnings?

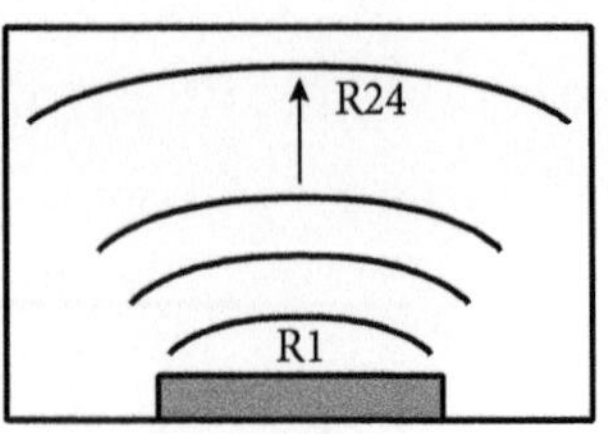

Figure 3.5 Diagram for question 14

14. A theatre has 24 rows of seats. There are 16 seats in the first row, and each successive row increases by 2 seats.

(a) Calculate the number of seats in the 24th row.

(b) Calculate the number of seats in the whole theatre.

15. The amount of €7000 is invested at 5.25% annual compound interest.

(a) Write down an expression for the value of this investment after t full years.

(b) Calculate the minimum number of years required for this amount to become €10,000.

(c) For the same number of years as in part **(b)**, would an investment of the same amount be better if it were invested at a 5% rate compounded quarterly?

16. With S_n denoting the sum of the first n terms of an arithmetic sequence, we are given that $S_1 = 9$ and $S_2 = 20$.

(a) Find the second term.

(b) Calculate the common difference of the sequence.

(c) Find the fourth term.

17. Consider an arithmetic sequence whose second term is 7. The sum of the first four terms of this sequence is 12. Find the first term and the common difference of the sequence.

18. Given that

$$(1 + x)^5(1 + ax)^6 \equiv 1 + bx + 10x^2 + \ldots + a^6x^{11},$$

find the values of $a, b \in \mathbb{Z}$.

19. In an arithmetic sequence of positive terms, a_n represents the nth term. Given that $\frac{a_5}{a_{12}} = \frac{6}{13}$ and $a_1 \times a_3 = 32$, find $\sum_{i=1}^{100} a_i$

20. In an arithmetic sequence, $a_1 = 5$ and $a_2 = 13$.

(a) Write down, in terms of n, an expression for the nth term, a_n.

(b) Find n such that $a_n < 400$.

21. Consider the arithmetic sequence 85, 78, 71, ...
Find the sum of its positive terms.

22. When we expand $\left(x + \frac{1}{kx^2}\right)^7$, the coefficient of x is $\frac{7}{3}$.
Find all possible values of k.

23. The sum to infinity of a geometric sequence is $\frac{27}{2}$, and the sum of its first three terms is 13. Find the first term.

24. A geometric sequence is defined by $u_n = 3(4)^{n+1}, n \in \mathbb{Z}^+$, where u_n is the nth term.

(a) Find the common ratio r.

(b) Hence, find S_n, the sum of the first n terms of this sequence.

25. Consider the infinite geometric series:

$$1 + \left(\frac{3x}{5}\right) + \left(\frac{3x}{5}\right)^2 + \left(\frac{3x}{5}\right)^3 + \dots$$

(a) For what values of x does the series converge?

(b) Find the sum of the series if $x = 1.5$

26. Consider the arithmetic series $S_n = 2 + 5 + 8 + \dots$

(a) Find an expression for the partial sum S_n, in terms of n.

(b) For what value of n is $S_n = 1365$?

27. Find $\sum_{r=1}^{50} \ln(2^r)$, giving the answer in the form $a \ln 2$, where $a \in \mathbb{Q}$

28. Consider the sequence $\{a_n\}$ defined recursively by:

$$a_{n+1} = 3a_n - 2a_{n-1}, n \in \mathbb{Z}^+, \text{ with } a_0 = 1, a_1 = 2$$

(a) Find a_2, a_3, and a_4.

(b) (i) Find the explicit form for a_n in terms of n.

(ii) Verify that your answer to part (i) satisfies the given recursive definition.

29. The sum to infinity of a geometric sequence with all positive terms is 27, and the sum of the first two terms is 15. Find the value of:

(a) the common ratio (b) the first term.

30. The first four terms of an arithmetic sequence are 2, $a - b$, $2a + b + 7$, and $a - 3b$, where a and b are constants. Find a and b.

31. Three consecutive terms of an arithmetic sequence are: a, 1, and b. The terms 1, a, and b are consecutive terms of a geometric sequence. If $a \neq b$, find the value of a and of b.

32. The diagram in Figure 3.6 shows a sector AOB of a circle of radius 1 and centre O, where $A\hat{O}B = \theta$.

The lines (AB_1), (A_1B_2), and (A_2B_3) are perpendicular to OB. A_1B_1 and A_2B_2 are arcs of circles with centre O.

Calculate the sum to infinity of the arc lengths:

$AB + A_1B_1 + A_2B_2 + A_3B_3 + \dots$

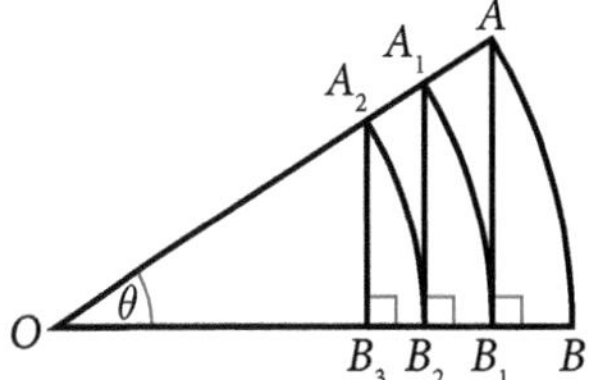

Figure 3.6 Diagram for question 32

33. (a) Expand $(2 + x)^5$, giving your answer in ascending powers of x.

(b) Hence, find the exact value of 2.01^5

34. You invest \$5000 at an annual compound interest rate of 6.3%.

(a) Write an expression for the value of this investment after t full years.

(b) Find the value of this investment at the end of five years.

(c) After how many full years will the value of the investment exceed \$10,000?

35. The sum of the first n terms of an arithmetic sequence $\{u_n\}$ is given by the formula $S_n = 4n^2 - 2n$. Three terms of this sequence, u_2, u_m and u_{32}, are consecutive terms in a geometric sequence. Find m.

36. The sum of the first 16 terms of an arithmetic sequence $\{u_n\}$ is 12. Find the first term and the common difference if the ninth term is zero.

37. **(a)** Write down the first four terms of the expansion of $(1 - x)^n$, with $n > 2$, in ascending powers of x.

(b) The absolute values of the coefficients of 2nd, 3rd, and 4th terms of the expansion in **(a)** are consecutive terms in an arithmetic sequence.

(i) Show that $n^3 - 9n^2 + 14n = 0$ **(ii)** Find the value of n.

38. **(a)** Write down the full expansion of $(3 + x)^4$ in ascending powers of x.

(b) Find the exact value of 3.1^4

39. **(a)** Write down how many integers between 10 and 300 are divisible by 7.

(b) Express the sum of these integers in sigma notation.

(c) Find the sum in (b) above.

(d) Given an arithmetic sequence with first term 1000 and common difference -7, find the smallest n so that the sum of the first n terms of this sequence is negative.

40. Let $\{u_n\}$, $n \in \mathbb{Z}^+$, be an arithmetic sequence with first term a and common difference d, where $d \neq 0$. Let another sequence $\{v_n\}$, $n \in \mathbb{Z}^+$, be defined by $v_n = 2^{u_n}$.

(a) **(i)** Show that $\frac{v_{n+1}}{v_n}$ is a constant.

(ii) Write down the first term of the sequence $\{v_n\}$.

(iii) Write down a formula for v_n in terms of a, d, and n.

Let S_n be the sum of the first n terms of the sequence $\{v_n\}$.

(b) **(i)** Find S_n in terms of a, d, and n.

(ii) Find the values of d for which $\sum_1^{\infty} v_i$ exists.

You are now told that $\sum_1^{\infty} v_i$ does exist and is denoted by S_∞.

(iii) Write down S_∞ in terms of a and d.

(iv) Given that $S_\infty = 2^{a+1}$, find the value of d.

4 Exponential and logarithmic functions

4 Exponential and logarithmic functions

Learning objectives

By the end of this chapter, you should be familiar with...

- exponential functions and their graphs
- concepts of exponential growth and decay, and applications
- the nature and significance of the number e
- logarithmic functions and their graphs
- properties of logarithms
- solving equations involving exponential expressions
- solving equations involving logarithmic expressions.

Exponential functions help us model a wide variety of physical phenomena. The natural exponential function (or simply, *the* exponential function), $f(x) = e^x$, is one of the most important functions in calculus. Exponential functions and their applications – especially to situations involving growth and decay – will be covered at length.

Logarithms, which were originally invented as a computational tool, lead to logarithmic functions. These functions are closely related to exponential functions and play an equally important part in calculus and a range of applications. We will learn that certain exponential and logarithmic functions are inverses of each other.

4.1 Exponential functions

Characteristics of exponential functions

x	$y = x^2$	$y = 2^x$
0	0	1
1	1	2
2	4	4
3	9	8
4	16	16
5	25	32
6	36	64
7	49	128
8	64	256
9	81	512
10	100	1024

Table 4.1 Contrast between a power function, $y = x^2$, and an exponential function, $y = 2^x$

The two equations, $y = x^2$ and $y = 2^x$, are similar in that they both contain a **base** and an **exponent** (or **power**). In $y = x^2$, the base is the variable x and the exponent is the constant 2. In $y = 2^x$, the base is the constant 2 and the exponent is the variable x.

The quadratic function $y = x^2$ is in the form 'variable base$^{\text{constant power}}$' where the base is a variable and the exponent is an integer greater than or equal to zero (non-negative integer). Any function in this form is called a **power function.**

The function $y = 2^x$ is in the form 'constant base$^{\text{variable power}}$' where the base is a positive real number (not equal to one) and the exponent is a variable. Any function in this form is called an **exponential function.**

To illustrate a fundamental difference between exponential functions and power functions, consider the function values for $y = x^2$ and $y = 2^x$ when x is an integer from 0 to 10. Table 4.1 shows clearly how the values for the exponential function increase at a significantly faster rate than the power function.

Power functions can easily be defined, and computed, for any real number. For any power function $y = x^n$, where n is any positive integer, y is found by taking x and repeatedly multiplying it n times. Hence, the domain of all power functions is all real numbers. For example, for the power function $y = x^3$, if $x = \pi$, then $y = \pi^3 = \pi \cdot \pi \cdot \pi \approx 31.006\,198\,11$. Since a power function like $y = x^3$ has a domain of all real numbers, its graph is a continuous curve (no gaps) where every real number is the x-coordinate of some point on the curve. Is the same true for exponential functions? In other words, can we compute a value for y for any real number x? For example, considering the exponential function $y = 2^x$, is it possible to compute 2^π? Table 4.2 shows a list of exponential expressions getting closer and closer to the value of 2^π. Clearly, there will be a limiting value to the list that will give us the value of 2^π. With our GDC we can evaluate 2^π to at least 10 significant figures and we could carry out a similar computation for any value of x that is a real number. Although we have not formally proved it, it seems reasonable to accept that the domain of any exponential function (with the base being a positive real number) is all real numbers.

To demonstrate just how quickly $y = 2^x$ increases, consider what would happen if you were able to repeatedly fold a piece of paper in half 50 times. A typical piece of paper is about five thousandths of a centimetre thick. Each time you fold the paper, the thickness of the paper doubles, so after 50 folds the thickness of the folded paper is the height of a stack of 2^{50} pieces of paper. The thickness of the paper after being folded 50 times would be $2^{50} \times 0.005$ cm which is more than 56 million kilometres. Compare that with the height of a stack of 50^2 pieces of paper that would be a meagre 12.5 centimetres – only 0.000 125 kilometres.

x	2^x (12 s.f.)
3	8.000 000 000 00
3.1	8.574 187 700 29
3.14	8.815 240 927 01
3.141	8.821 353 304 55
3.1415	8.824 411 082 48
3.141 59	8.824 961 595 06
3.141 592	8.824 973 829 06
3.141 5926	8.824 977 499 27
3.141 592 65	8.824 977 805 12

Table 4.2 Limiting value for 2^π

Graphs of exponential functions

Thus, the graph of any exponential function $f(x) = b^x$, $b > 0$, $b \neq 1$, is a continuous curve.

Example 4.1

Graph each exponential function by plotting several points and then drawing a smooth curve through the points.

(a) $f(x) = 3^x$ (b) $g(x) = \left(\frac{1}{3}\right)^x$

Solution

Calculate values for each function for integral values of x from -3 to 3.

x	$f(x) = 3^x$	$g(x) = \left(\frac{1}{3}\right)^x$
-3	$\frac{1}{27}$	27
-2	$\frac{1}{9}$	9
-1	$\frac{1}{3}$	3
0	1	1
1	3	$\frac{1}{3}$
2	9	$\frac{1}{9}$
3	27	$\frac{1}{27}$

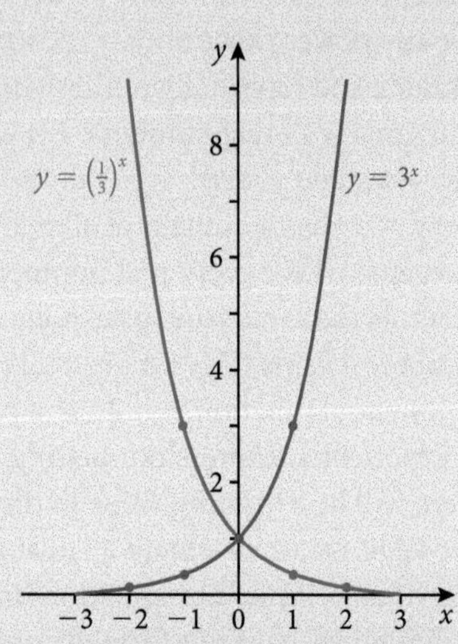

Knowing that exponential functions are defined for all real numbers, not just integers, we can sketch a smooth curve filling in between the ordered pairs shown in the table.

Remember that in Section 1.5 we established that the graph of $y = f(-x)$ is obtained by reflecting the graph of $y = f(x)$ in the y-axis. It is clear that the graph of function g is a reflection of function f about the y-axis. We can use some laws of powers to show that $g(x) = f(-x)$.

$$g(x) = \left(\frac{1}{3}\right)^x = \frac{1^x}{3^x} = \frac{1}{3^x} = 3^{-x} = f(-x)$$

It is useful to point out that both graphs, $y = 3^x$ and $y = \left(\frac{1}{3}\right)^x$, pass through the point (0, 1) and have a horizontal asymptote of $y = 0$ (x-axis). The same is true for the graph of all exponential functions in the form $y = b^x$ given that $b \in \mathbb{R}$, $b \neq 1$. If $b = 1$, then $y = 1^x = 1$, and the graph is a horizontal line rather than a constantly increasing or decreasing curve.

If $b > 0$ and $b \neq 1$, an **exponential function** with base b is the function defined by $f(x) = b^x$
The **domain** of f is the set of real numbers ($x \in \mathbb{R}$) and the **range** of f is the set of positive real numbers ($y > 0$). The graph of f passes through (0, 1), has the x-axis as a **horizontal asymptote**, and, depending on the value of the base of the exponential function b, will either be a continually increasing **exponential growth curve** or a continually decreasing **exponential decay curve**.

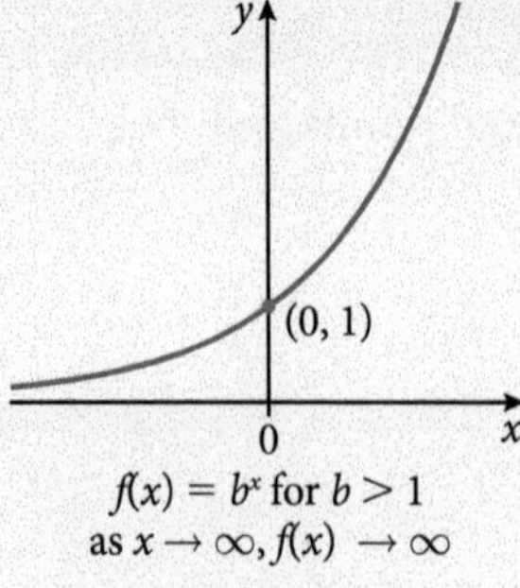

$f(x) = b^x$ for $b > 1$
as $x \to \infty$, $f(x) \to \infty$

Exponential growth curve: f is an increasing function. As $x \to \infty$, $f(x) \to \infty$.

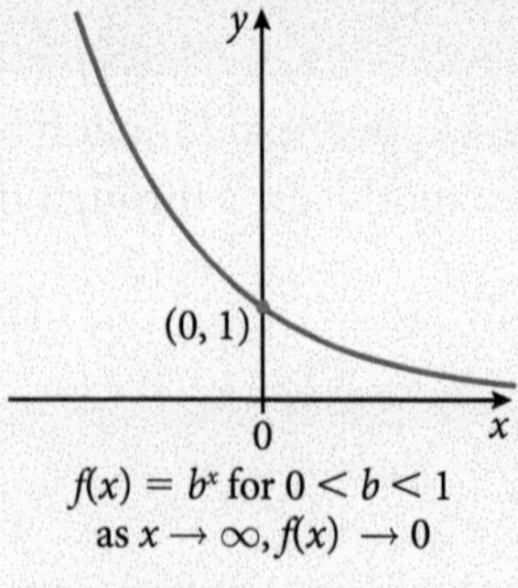

$f(x) = b^x$ for $0 < b < 1$
as $x \to \infty$, $f(x) \to 0$

Exponential decay curve: f is a decreasing function. As $x \to \infty$, $f(x) \to 0$.

The graphs of all exponential functions will display a characteristic growth or decay curve. As we shall see, many natural phenomena exhibit exponential growth or decay. Also, the graphs of exponential functions behave asymptotically for either very large positive values of x (decay curve) or very large negative values of x (growth curve). This means that there will exist a horizontal line that the graph will approach but not intersect as either $x \to \infty$ or as $x \to -\infty$.

Transformations of exponential functions

Recalling from Section 1.5 how the graphs of functions are translated and reflected, we can efficiently sketch the graph of many exponential functions.

Example 4.2

Using the graph of $f(x) = 2^x$, sketch the graph of each function. State the domain and range for each function and the equation of its horizontal asymptote.

(a) $g(x) = 2^x + 3$ (b) $h(x) = 2^{-x}$ (c) $p(x) = -2^x$

(d) $r(x) = 2^{x-4}$ (e) $v(x) = 3(2^x)$

Solution

(a) The graph of $g(x) = 2^x + 3$ can be obtained by translating the graph of $f(x) = 2^x$ vertically three units up.

The domain of g is all real numbers ($x \in \mathbb{R}$) and the range is $y > 3$.

The horizontal asymptote is $y = 3$.

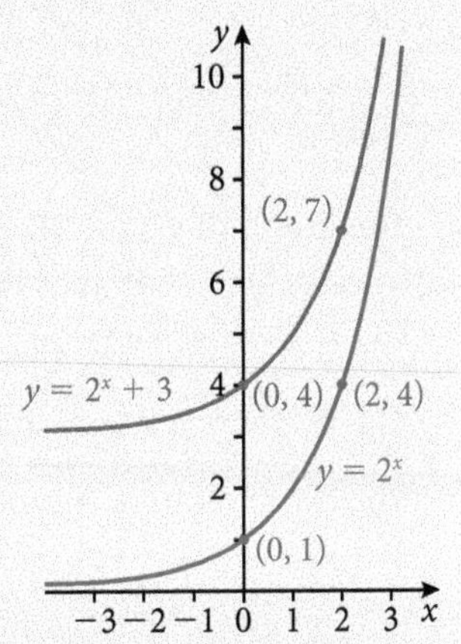

(b) The graph of $h(x) = 2^{-x}$ can be obtained by reflecting the graph of $f(x) = 2^x$ in the y-axis.

The domain is $x \in \mathbb{R}$ and the range is $y > 0$.

The horizontal asymptote is $y = 0$ (x-axis).

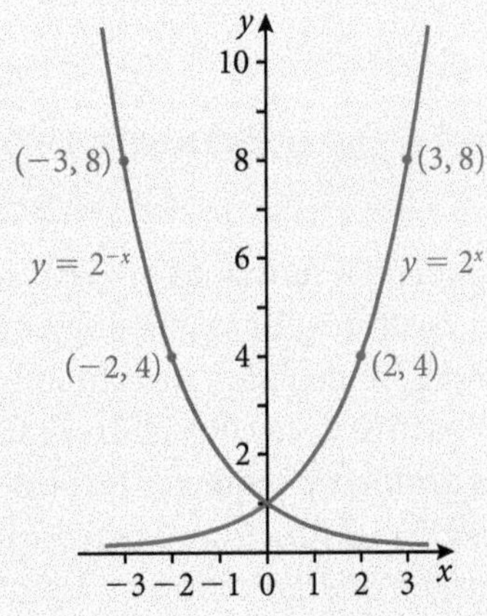

Note that for function $p(x) = -2^x$ in part (c) of Example 4.2, the horizontal asymptote is an upper bound (i.e. no function value is equal to or greater than $y = 0$). In parts (a), (b), (d), and (e) the horizontal asymptote for each function is a lower bound (i.e. no function value is equal to or less than the y value of the asymptote).

(c) The graph of $p(x) = -2^x$ can be obtained by reflecting the graph of $f(x) = 2^x$ in the x-axis.

The domain is $x \in \mathbb{R}$ and the range is $y < 0$.

The horizontal asymptote is $y = 0$ (x-axis).

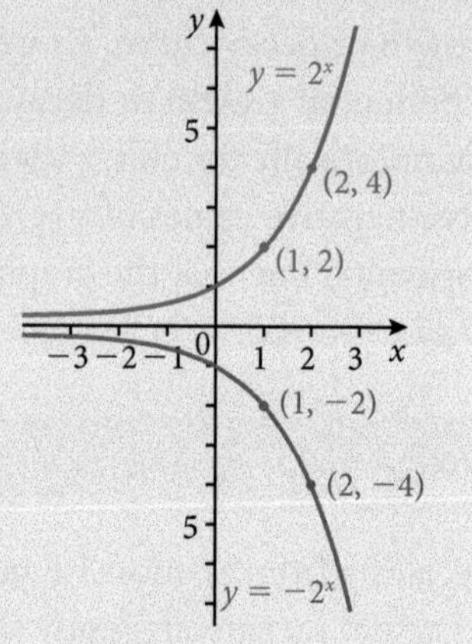

(d) The graph of $r(x) = 2^{x-4}$ can be obtained by translating the graph of $f(x) = 2^x$ four units to the right.

The domain is $x \in \mathbb{R}$ and the range is $y > 0$.

The horizontal asymptote is $y = 0$ (x-axis).

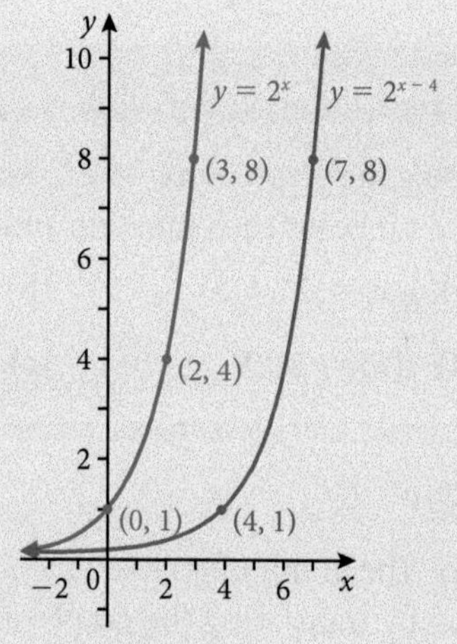

Online

Explore changes to the graphs of exponential functions.

(e) The graph of $v(x) = 3(2^x)$ can be obtained by a vertical stretch of the graph of $f(x) = 2^x$ by scale factor 3.

The domain is $x \in \mathbb{R}$ and the range is $y > 0$.

The horizontal asymptote is $y = 0$ (x-axis).

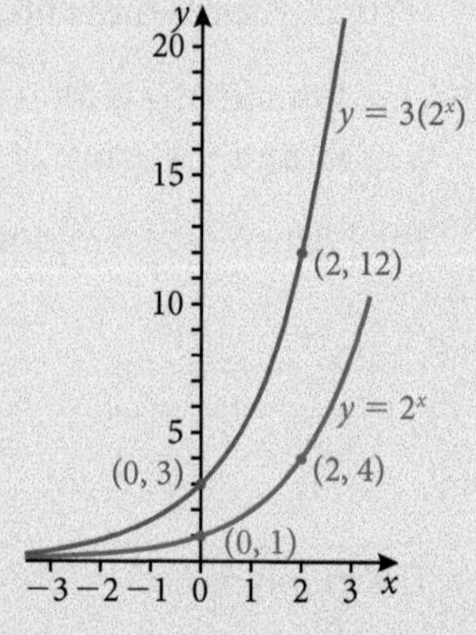

Exponential models are equations of the form $A(t) = A_0 b^t$, where $A_0 \neq 0$, $b > 0$, and $b \neq 1$. $A(t)$ is the amount or value of a variable after time t. $A(0) = A_0 b^0 = A_0(1) = A_0$, so A_0 is called the **initial amount** (often the value at time $t = 0$). If $b > 1$, then $A(t)$ is an **exponential growth model**, and if $0 < b < 1$, then $A(t)$ is an **exponential decay model**. The value of b, the base of the exponential function, is often called the **growth** or **decay factor**.

Mathematical models of growth and decay

Exponential functions are well suited as a mathematical model for a wide variety of steadily increasing or decreasing phenomena, including population growth (or decline), investment of money with compound interest, and radioactive decay. Recall from Chapter 3 that the formula for finding terms in a geometric sequence (repeated multiplication by common ratio r) is an exponential function. Many instances of growth or decay occur where there is repeated multiplication by a growth or decay factor that can be modelled with an exponential function.

Example 4.3

A sample count of bacteria in a culture indicates that the number of bacteria is doubling every hour. Given that the estimated count at 15:00 was 12 000 bacteria:

(a) find the estimated count three hours earlier, at 12:00.

(b) write an exponential growth function for the number of bacteria at any hour t.

Solution

(a) Consider the time at 12:00 to be the starting, or initial, time and label it $t = 0$ hours. Then the time at 15:00 is $t = 3$. The amount at any time t (in hours) will double after an hour, so the growth factor, b, is 2. Therefore, $A(t) = A_0(2)^t$. Knowing that $A(3) = 12\,000$, compute A_0.

$12\,000 = A_0(2)^3$

$12\,000 = 8A_0$

$A_0 = 1500$

Therefore, the estimated count at 12:00 was 1500.

(b) The growth function for the number of bacteria at time t is $A(t) = 1500(2)^t$.

Recall from Chapter 3 how the formula for the nth term in a geometric sequence was used to develop a formula (an exponential function) for an investment of money with interest added to the account (compounded) a certain number of times per year. This exponential function for calculating the future value, A, after t years, where P is the initial amount (principal), the annual interest rate is r, and n is the number of times interest is compounded per year, is given by $A(t) = P\left(1 + \frac{r}{n}\right)^{nt}$.

Example 4.4

An initial amount of 1000 euros is deposited into an account earning $5\frac{1}{4}\%$ interest per year. Find the amounts in the account after 8 years if interest is compounded:

(a) annually (b) semi-annually (c) quarterly

(d) monthly (e) daily.

Solution

We use the exponential function associated with compound interest with values of $P = 1000$, $r = 0.0525$ and $t = 8$ to compute the results below.

Compounding	n	Amount (€) after 8 years
(a) Annually	1	$1000\left(1 + \frac{0.0525}{1}\right)^{8} = 1505.83$
(b) Semi-annually	2	$1000\left(1 + \frac{0.0525}{2}\right)^{2(8)} = 1513.74$
(c) Quarterly	4	$1000\left(1 + \frac{0.0525}{4}\right)^{4(8)} = 1517.81$
(d) Monthly	12	$1000\left(1 + \frac{0.0525}{12}\right)^{12(8)} = 1520.57$
(e) Daily	365	$1000\left(1 + \frac{0.0525}{365}\right)^{365(8)} = 1521.92$

Example 4.5

A new car is purchased for \$22,000. If the value of the car decreases (depreciates) at a rate of approximately 15% per year, what will be the approximate value of the car to the nearest whole dollar in $4\frac{1}{2}$ years?

Solution

The decay factor for the exponential function is $1 - r = 1 - 0.15 = 0.85$. In other words, after each year the car's value is 85% of what it was one year before. We use the exponential decay model $A(t) = A_0\, b^t$ with values $A_0 = 22\,000$, $b = 0.85$ and $t = 4.5$.

$$A(4.5) = 22\,000(0.85)^{4.5} \approx 10\,588$$

The value of the car will be approximately \$10,588.

Exercise 4.1

1. **(a)** Write the equation for an exponential equation with base $b > 0$.

(b) Given $b \neq 1$, state the domain and range of this function.

(c) Sketch the general shape of the graph of this exponential function for each of two cases:

(i) $b > 1$ **(ii)** $0 < b < 1$

2. Sketch a graph of each function and state:

(i) the coordinates of any x-intercept(s) and y-intercept

(ii) the equation of any asymptote(s)

(iii) the domain and range.

(a) $f(x) = 3^{x+4}$ **(b)** $g(x) = 2^{-x} + 3$ **(c)** $h(x) = 2^{x-2}$

(d) $p(x) = \dfrac{1}{2^x}$ **(e)** $q(x) = 2(3^x) - 1$

3. A general exponential function is written in the form $f(x) = a(b)^{x-c} + d$. State the domain, range, y-intercept and the equation of the horizontal asymptote in terms of the parameters a, b, c, and d.

4. Using your GDC and a graph viewing window with Xmin $= -2$, Xmax $= 2$, Ymin $= 0$ and Ymax $= 4$, sketch a graph for each exponential equation on the same set of axes.

(a) $y = 2^x$ **(b)** $y = 4^x$ **(c)** $y = 8^x$

(d) $y = 2^{-x}$ **(e)** $y = 4^{-x}$ **(f)** $y = 8^{-x}$

5. Write equations that are equivalent to those in question **4(d)**, **(e)**, and **(f)** but with an exponent of positive x rather than negative x.

6. Given that $1 < a < b$, state which is steeper: the graph of $y = a^x$ or the graph of $y = b^x$. Give reasons for your answer.

7. The population of a city triples every 25 years. At time $t = 0$, the population is 100 000.

   (a) Write a function for the population $P(t)$ as a function of t.

   (b) Calculate the population after:

   (i) 50 years (ii) 70 years (iii) 100 years.

8. An experiment involves a colony of bacteria in a solution. It is determined that the number of bacteria doubles approximately every 3 minutes and the initial number of bacteria at the start of the experiment is 10^4.

   (a) Write a function for the number of bacteria $N(t)$ as a function of t (in minutes).

   (b) Calculate approximately how many bacteria there are after:

   (i) 3 minutes (ii) 9 minutes (iii) 27 minutes (iv) one hour.

9. A bank offers an investment account that will double your money in 10 years.

   (a) Write a function for the amount of money in the account $A(t)$ after t years for an initial investment of A_0.

   (b) If interest was added into the account only at the end of each year, find the annual interest rate for the account (to 3 significant figures).

10. \$10 000 is invested at an annual interest rate of 11%, compounded quarterly. Find the value of the investment after:

    (a) 5 years (b) 10 years (c) 15 years.

11. A sum of \$5000 is deposited into an investment account that earns interest at a rate of 9% per year compounded monthly.

    (a) Write the function $A(t)$ that computes the value of the investment after t years.

    (b) Use your GDC to sketch a graph of $A(t)$ with values of t on the horizontal axis ranging from $t = 0$ years to $t = 25$ years.

    (c) Use the graph on your GDC to determine the minimum number of years (to the nearest whole year) that it will take for this investment to have a value greater than \$20,000.

12. \$10,000 is invested at an annual interest rate of 11% for a period of 5 years. Find the value of the investment if the interest is compounded:

    (a) annually (b) monthly (c) daily (d) hourly

13. The population of deer in a national park increases at a steady rate of 3.2% per year. The present population is approximately 248 000.

(a) Calculate the approximate number of deer one year before.

(b) Calculate the approximate number of deer six years in the future.

14. An open can is filled with 1000 ml of fluid that evaporates at a rate of 30% per week.

(a) Write a function, $A(w)$, that gives the amount of fluid after w weeks.

(b) Use your GDC to find how many weeks, to the nearest week, it will take for the volume of fluid to be less than 1 ml.

15. You are offered a very highly paid job that lasts for one month (exactly 30 days). You may choose from one of the following two payment plans.

(I) One dollar on the first day of the month, two dollars on the second day, three dollars on the third day, and so on (getting paid one dollar more each day), until the end of the 30 days. (You would have a total of $55 after 10 days.)

(II) One cent ($0.01) on the first day of the month, two cents ($0.02) on the second day, four cents on the third day, eight cents on the fourth day and so on (each day getting paid double from the previous day), until the end of the 30 days. (You would have a total of $10.23 after 10 days.)

(a) State which of the payment plans, (I) or (II), would give you the greater salary.

(b) Calculate how much you would get paid with this plan?

16. Each exponential function graphed below can be written in the form $f(x) = k(a)^x$. Find the value of a and k for each function.

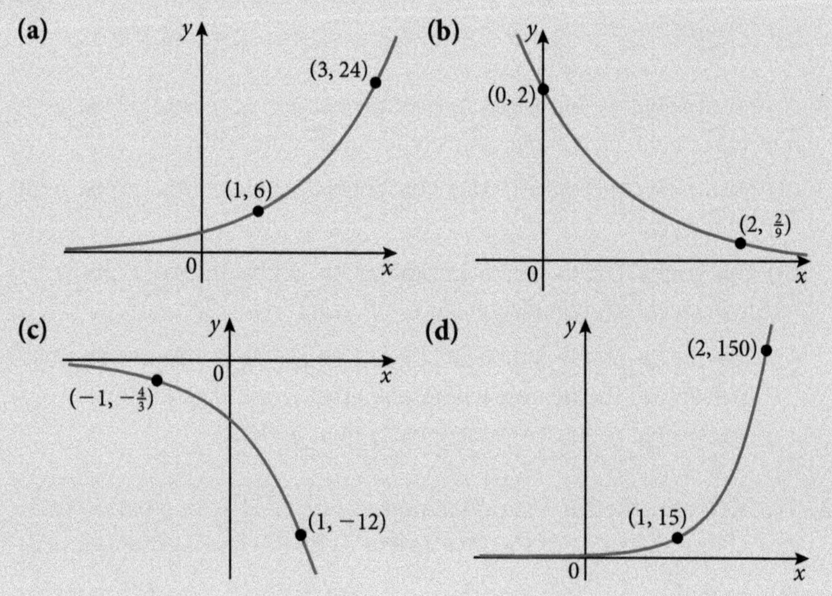

4.2 The number e and continuous growth or decay

In an exponential function, $f(x) = b^x$, b is any positive constant and x is any real number. Graphs of $y = b^x$ for a few values where $b \geqslant 1$ are shown in Figure 4.1. As noted in Section 4.1, all the graphs pass through the point (0, 1).

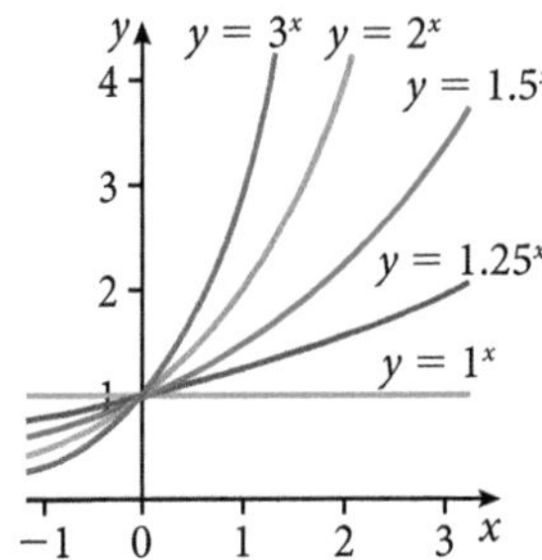

Figure 4.1 Graphs of $y = b^x$ for some values when $b \geqslant 1$

The question arises: what is the **best** number to choose for the base b when we are modelling a particular phenomenon? There is a good argument for $b = 10$, since we most commonly use a base 10 number system. Your GDC will have the expression 10^x as a built-in command. The base $b = 2$ is also plausible because a binary number system (base 2) is used in many processes, especially in computer systems. However, the most important base is an irrational number that is denoted with the letter e. As we will see, the value of e approximated to five significant figures is 2.718 28. The importance of e will be clearer when we get to calculus topics. You already know about the number π, another very useful irrational number, which has a natural geometric significance as the ratio of circumference to diameter for any circle. The number e also occurs in a 'natural' manner. We will illustrate this by taking another look at compound interest and considering **continuous change** rather than **incremental change.**

Continuously compounded interest

In the previous section of this chapter, and in Chapter 3, we computed amounts of money resulting from an initial amount (principal) with interest being compounded (added in) at discrete intervals (e.g. yearly, monthly, or daily). In the formula that we used, $A(t) = P\left(1 + \frac{r}{n}\right)^{nt}$, n is the number of times that interest is compounded per year. Instead of adding interest only at discrete intervals, let's investigate what happens if we try to add interest continuously. That is, let the value of n increase without bound ($n \to \infty$).

Consider investing \$1 at a very generous annual interest rate of 100%. How much will be in the account at the end of one year? It depends on how often the interest is compounded; in other words, it depends on the value of n in the compound interest formula $A(t) = P\left(1 + \frac{r}{n}\right)^{nt}$. Results for different values of n are shown in Table 4.3 on the next page.

As the number of times, n, per year that interest is compounded increases, the amount at the end of the year appears to approach a limiting value. We realise that as $n \to \infty$ the quantity of $\left(1 + \frac{1}{n}\right)^n$ approaches the number e. To twelve decimal places, e is approximately 2.718 281 828 459.

Compounding	n	$A(1) = \left(1 + \frac{1}{n}\right)^n$
Annual	1	2
Semi-annual	2	2.25
Quarterly	4	2.441 406 25...
Monthly	12	2.613 035 290 22...
Daily	365	2.714 567 482 02...
Hourly	8760	2.718 126 691 62...
Every minute	525 600	2.718 279 242 57...
Every second	31 536 000	2.718 281 785 36...

Table 4.3 Values for $\left(1 + \frac{1}{n}\right)^n$ as $n \to \infty$

Leonhard Euler (1701–1783) was the dominant mathematical figure of the 18th century.

He proved mathematically that the limit of $\left(1 + \frac{1}{n}\right)^n$ as n tends to infinity is precisely equal to an irrational constant, which he labelled e.

$$\mathrm{e} = \lim_{n \to \infty}\left(1 + \frac{1}{n}\right)^n$$

This definition for the number e is read 'e equals the limit of $\left(1 + \frac{1}{n}\right)^n$ as n tends to infinity'.

As the number of compoundings, n, increases without bound, we approach continuous compounding, where interest is being added continuously. In the formula for calculating amounts resulting from compound interest, if we let $m = \frac{n}{r}$, we get

$$A(t) = P\left(1 + \frac{r}{n}\right)^{nt} = P\left(1 + \frac{1}{m}\right)^{mrt} = P\left[\left(1 + \frac{1}{m}\right)^m\right]^{rt}$$

Now if $n \to \infty$ and the interest rate r is constant then $\frac{n}{r} = m \to \infty$. From the limit definition of e, we know that if $m \to \infty$, then $\left(1 + \frac{1}{m}\right)^m \to \mathrm{e}$. Therefore, for continuous compounding, it follows that $A(t) = P\left[\left(1 + \frac{1}{m}\right)^m\right]^{rt} = P[\mathrm{e}]^{rt}$.

Examples of **continuous growth** or **decay**: area covered by a growing oil spill, radioactive decay, height of a plant.

Examples of **discrete growth** or **decay**: population of rabbits, number of people remaining in a tennis tournament, number of books being added to a library.

This result is part of the reason that e is the best choice for the base of an exponential function modelling change that occurs **continuously** (changing every instant) rather than change that occurs **discretely** (changing at specific intervals).

Continuous compound interest formula

An exponential function for calculating the amount of money after t years, $A(t)$, for interest compounded continuously, where P is the initial amount or principal and r is the annual interest rate, is given by $A(t) = P\mathrm{e}^{rt}$

Example 4.6

The starting balance of an investment account is €1000. The account earns interest at an annual rate of $5\frac{1}{4}$%. Assuming there are no further withdrawals or deposits (other than interest), find the total amount in the account after 10 years if the interest is added to the account:

(a) annually (b) quarterly (c) continuously

Solution

(a) $A(t) = P(1 + r)^t = 1000(1 + 0.0525)^{10} =$ €1669.10

(b) $A(t) = P\left(1 + \frac{r}{n}\right)^{nt} = 1000\left(1 + \frac{0.0525}{4}\right)^{4(10)} =$ €1684.70

(c) $A(t) = Pe^{rt} = 1000e^{0.0525(10)} =$ €1690.46

The natural exponential function and continuous change

For many applications involving continuous change, the most suitable choice for a mathematical model is an exponential function with a base having the value of e.

The **natural exponential function** is the function defined as

$$f(x) = e^x$$

As with other exponential functions, the domain of the natural exponential function is the set of all real numbers ($x \in \mathbb{R}$), and its range is the set of positive numbers ($y > 0$). The natural exponential function is often referred to as **the** exponential function.

The formula developed for continuously compounded interest, $A(t) = Pe^{rt}$, does not only apply to applications involving adding interest to financial accounts. It can be used to model growth or decay of a quantity that is changing exponentially (i.e. repeated multiplication by a constant ratio, or growth/decay factor) when the change is continuous or approaching continuous.

Continuous exponential growth/decay
If an initial quantity P grows or decays continuously at a rate r over a certain time period, then the amount $A(t)$ after t time periods is given by the function $A(t) = Pe^{rt}$. If $r > 0$, then the quantity is increasing (growing). If $r < 0$, then the quantity is decreasing (decaying).

Example 4.7

A particular new commercial jet costs $150 million. The aeroplane will lose value at a continuous rate. This is modelled by the continuous decay function $A(t) = 150e^{-0.053t}$ where $A(t)$ is the value of the aeroplane (in millions) after t years.

(a) Find how much (to the nearest million dollars) the aeroplane would be worth precisely five years after being purchased.

(b) If the aeroplane is purchased in 2020, determine the first year that the jet is worth less than half of its original cost.

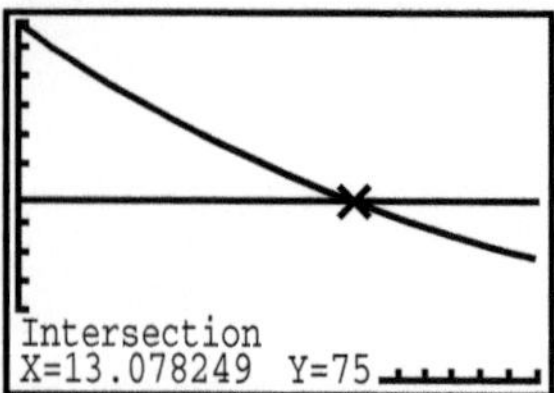

Figure 4.2 GDC screens for the solution to Example 4.7(b)

Solution

(a) $C(5) = 150e^{-0.053(5)} \approx 115$

The value is approximately \$115 million after five years.

(b) Using a GDC, we graph the decay equation $y = 150e^{-0.053x}$ and the horizontal line $y = 75$ and determine the intersection point.
The x-coordinate of the intersection point is approximately 13.08. At the start of 2033, the jet's value is not yet half of its original value. Therefore, the first year that the jet is worth less than half of its original cost is 2034.

Exercise 4.2

1. Sketch a graph of each function and state:

(i) the coordinates of any x-intercept(s) and y-intercept

(ii) the equation of any asymptote(s)

(iii) the domain and range.

(a) $f(x) = e^{x-1}$ **(b)** $g(x) = e^{-x}$ **(c)** $h(x) = 2e^x$

(d) $p(x) = e^x + 3$ **(e)** $h(x) = \frac{1}{e^x}$

2. Two different banks, Bank A and Bank B, offer accounts with exactly the same annual interest rate of 6.85%. The account from Bank A has the interest compounded monthly whereas the account from Bank B compounds the interest weekly (52 weeks in a year). To decide which bank to open an account with, you calculate the amount of interest you would earn from each bank after three years, given an initial deposit of €500. It is assumed that you make no further deposits and no withdrawals during the three years.

(a) Calculate how much interest you would earn from each account.

(b) State which account earns more, and by how much.

3. Dina wishes to deposit \$1000 into an investment account and then withdraw the total in the account in five years. She has the choice of two different accounts.

Blue Star account: Interest is earned at an annual interest rate of 6.1% compounded weekly (52 weeks in a year).

Red Star account: Interest is earned at an annual interest rate of 6.2% compounded monthly.

(a) Determine which investment account will result in the greatest total at the end of five years.

(b) Calculate the total after five years for this account.

4. Strontium-90 is a radioactive isotope of strontium. Strontium-90 decays according to the function $A(t) = Ce^{-0.0239t}$, where t is time in years, and C is the initial amount of strontium-90 when $t = 0$. Find the percentage remaining of a sample of strontium-90 after:

(a) 1 year **(b)** 10 years **(c)** 100 years **(d)** 250 years

5. A radioactive substance decays in such a way that the mass (in kilograms) remaining after t days is given by the function $A(t) = 5e^{-0.0347t}$.

(a) Find the mass at time $t = 0$ (i.e. the initial mass).

(b) Calculate how much of the initial mass remains after 10 days.

6. **(a)** Consider that £1000 is invested at 4.5% interest compounded continuously. Calculate the amount of money in the account after:

(i) 10 years **(ii)** 20 years.

(b) Use your GDC to determine how many years (to the nearest tenth of a year) it takes for the initial investment to double to £2000.

7. In certain conditions the bacteria that causes cholera, *Vibrio cholerae*, can grow rapidly. In a laboratory experiment a culture of *Vibrio cholerae* is started with 20 bacteria. The bacteria's growth is modelled with the continuous growth model $A(t) = 20e^{0.068t}$, where $A(t)$ is the number of bacteria after t minutes.

(a) Find the number of bacteria after 10 minutes.

(b) Find how many minutes it takes, to the nearest minute, for the number of bacteria to be at least twice the number at the start of the experiment

4.3 Logarithmic functions

In Example 4.6 we used the equation $A(t) = 1000e^{0.0525t}$ to compute the amount of money in an account after t years. Now suppose we wish to determine how much time, t, it takes for an initial investment of €1000 to double. To find this, we need to solve the following equation for t:

$$2000 = 1000e^{0.0525t} \Rightarrow 2 = e^{0.0525t}$$

The unknown, t, is in the exponent. So far, we have not seen an algebraic method to solve such an equation. Developing the concept of a **logarithm** will provide us with the means to do so.

For $b > 0$ and $b \neq 1$, the **logarithmic function** $y = \log_b x$ (read 'logarithm with base b of x') is the inverse of the exponential function with base b.

$y = \log_b x$ if and only if $x = b^y$

The domain of the logarithmic function $y = \log_b x$ is the set of positive numbers ($x > 0$) and its range is all real numbers ($y \in \mathbb{R}$).

The inverse of an exponential function

For $b > 1$, an exponential function with base b is increasing for all x, and for $0 < b < 1$ an exponential function is decreasing for all x. It follows from this that all exponential functions must be one to one. In Section 1.4 we demonstrated that an inverse function exists for any one to one function. Thus, an exponential function with base b such that $b > 0$ and $b \neq 1$ will have an inverse function. Also recall from Section 1.6 that the domain of a function $f(x)$ is the range of its inverse function $f^{-1}(x)$, and, similarly, the range of $f(x)$ is the domain of $f^{-1}(x)$. The domain and range are switched around for a function and its inverse.

Logarithmic expressions and equations

When evaluating logarithms, note that a logarithm is essentially an exponent. This means that the value of $\log_b x$ is the exponent to which b must be raised to obtain x. For example, $\log_2 8 = 3$ because 2 must be raised to the power of 3 to obtain 8. That is, $\log_2 8 = 3$ if and only if $2^3 = 8$.

We can use the definition of a logarithmic function to translate a logarithmic equation into an exponential equation and vice versa. When doing this, it is helpful to remember, as the definition stated, that in either form, logarithmic or exponential, the base is the same.

Logarithmic equation | **Exponential equation**

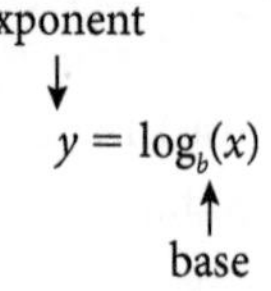

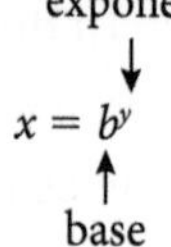

Example 4.8

Find the value of each logarithm without using a calculator.

(a) $\log_7 49$ (b) $\log_5\left(\frac{1}{5}\right)$ (c) $\log_6\sqrt{6}$ (d) $\log_4 64$ (e) $\log_{10} 0.001$

Solution

We set each logarithmic expression equal to y and use the definition of a logarithmic function to obtain an equivalent equation in exponential form. We then solve for y by applying the logical fact that if $b > 0$, $b \neq 1$ and $b^y = b^k$, then $y = k$.

(a) Let $y = \log_7 49$, which is equivalent to the exponential equation $7^y = 49$. Since $49 = 7^2$, then $7^y = 7^2$. Therefore, $y = 2 \Rightarrow \log_7 49 = 2$

(b) Let $y = \log_5\left(\frac{1}{5}\right)$, which is equivalent to the exponential equation $5^y = \frac{1}{5}$. Since $\frac{1}{5} = 5^{-1}$, then $5^y = 5^{-1}$. Therefore, $y = -1 \Rightarrow \log_5\left(\frac{1}{5}\right) = -1$

(c) Let $y = \log_6 \sqrt{6}$, which is equivalent to the exponential equation $6^y = \sqrt{6}$.

Since $\sqrt{6} = 6^{\frac{1}{2}}$, then $6^y = 6^{\frac{1}{2}}$. Therefore, $y = \frac{1}{2} \Rightarrow \log_6 \sqrt{6} = \frac{1}{2}$

(d) Let $y = \log_4 64$, which is equivalent to the exponential equation $4^y = 64$.

Since $64 = 4^3$, then $4^y = 4^3$. Therefore, $y = 3 \Rightarrow \log_4 64 = 3$

(e) Let $y = \log_{10} 0.001$, which is equivalent to the exponential equation

$10^y = 0.001$. Since $0.001 = \frac{1}{1000} = \frac{1}{10^3} = 10^{-3}$, then $10^y = 10^{-3}$.

Therefore, $y = -3 \Rightarrow \log_{10} 0.001 = -3$

Example 4.9

Find the domain of the function $f(x) = \log_2(4 - x^2)$.

Solution

From the definition of a logarithmic function, the domain of $y = \log_b x$ is $x > 0$, thus for $f(x)$ it follows that $4 - x^2 > 0 \Rightarrow (2 + x)(2 - x) > 0 \Rightarrow -2 < x < 2$. Hence, the domain is $-2 < x < 2$.

Properties of logarithms

For any function with an inverse, the graph of the inverse is the graph of the function reflected in the line $y = x$. Figure 4.3 illustrates this relationship for exponential and logarithmic functions, and also confirms the domain and range for the logarithmic function.

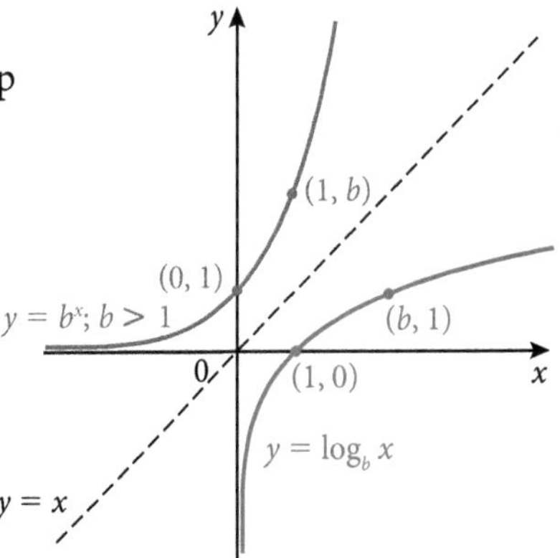

Figure 4.3 Reflection of $y = \log_b x$ in the line $y = x$

Notice that the points (0, 1) and (1, 0) are mirror images of each other in the line $y = x$. This corresponds to the fact that since $b^0 = 1$, then $\log_b 1 = 0$. Another pair of mirror image points, $(1, b)$ and $(b, 1)$, highlights the fact that $\log_b b = 1$.

Notice also that since the x-axis is a horizontal asymptote of $y = b^x$, then the y-axis is a vertical asymptote of $y = \log_b x$.

In Section 1.4 we established that a function f and its inverse function f^{-1} satisfy the equations

$f^{-1}(f(x)) = x$ for x in the domain of f

$f(f^{-1}(x)) = x$ for x in the domain of f^{-1}

When applied to $f(x) = b^x$ and $f^{-1}(x) = \log_b x$ these equations become

$\log_b(b^x) = x$ $\quad x \in \mathbb{R}$

$b^{\log_b x} = x$ $\quad x > 0$

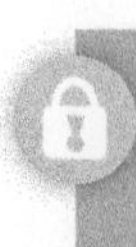

Properties of logarithms I

For $b > 0$ and $b \neq 1$, the following statements are true:

1. $\log_b 1 = 0$ because $b^0 = 1$
2. $\log_b b = 1$ because $b^1 = b$
3. $\log_b(b^x) = x$ because $b^x = b^x$
4. $b^{\log_b x} = x$ because $\log_b x$ is the power to which b must be raised to get x

Common logarithm:
$\log_{10} x = \log x$

The logarithmic function with base 10 is called the **common logarithmic function**. On calculators this function is denoted by **log**. The value of the expression $\log_{10} 1000$ is 3 because 10^3 is 1000. Generally, for common logarithms (i.e. base 10) we omit writing the base of 10. Hence, if **log** is written with no base indicated, it is assumed to have a base of 10.

For example, $\log 0.01 = -2$ because $10^{-2} = \frac{1}{10^2} = \frac{1}{100} = 0.01$

Natural logarithm:
$\log_e x = \ln x$

The other logarithmic function supplied on calculators is the logarithmic function with the base of e. This function is known as the **natural logarithmic function** and it is the inverse of the natural exponential function $y = e^x$. The natural logarithmic function is denoted by the symbol **ln,** and the expression $\ln x$ is read as 'the natural logarithm of x'.

Example 4.10

Evaluate each expression. Find an exact value without using a GDC if possible. Otherwise, use your GDC to approximate the value to 4 significant figures.

(a) $\log\left(\frac{1}{10}\right)$ (b) $\log(\sqrt{10})$

(c) $\log 1$ (d) $10^{\log 47}$

(e) $\log 50$ (f) $\ln e$

(g) $\ln\left(\frac{1}{e^3}\right)$ (h) $\ln 1$

(i) $e^{\ln 5}$ (j) $\ln 50$

Solution

(a) $\log\left(\frac{1}{10}\right) = \log(10^{-1}) = -1$ (b) $\log(\sqrt{10}) = \log(10^{\frac{1}{2}}) = \frac{1}{2}$

(c) $\log 1 = \log(10^0) = 0$ (d) $10^{\log 47} = 47$

(e) $\log 50 \approx 1.699$ [using GDC] (f) $\ln e = 1$

(g) $\ln\left(\frac{1}{e^3}\right) = \ln(e^{-3}) = -3$ (h) $\ln 1 = \ln(e^0) = 0$

(i) $e^{\ln 5} = 5$ (j) $\ln 50 \approx 3.912$ [using GDC]

Example 4.11

The diagram shows the graph of the line $y = x$ and of two curves. Curve A is the graph of the equation $y = \log x$. Curve B is the reflection of curve A in the line $y = x$.

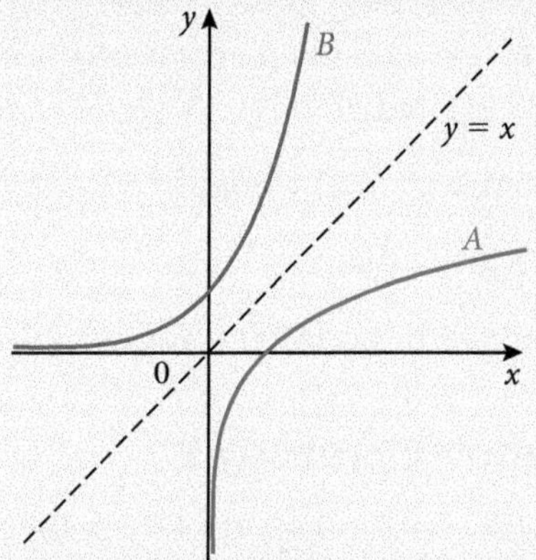

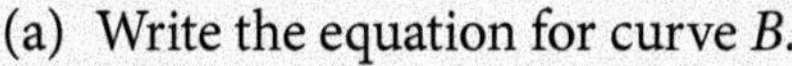

(a) Write the equation for curve B.

(b) Write the coordinates of the y-intercept of curve B.

Solution

(a) Curve A is the graph of $y = \log x$, the common logarithm with base 10, which could also be written as $y = \log_{10} x$. Curve B is the inverse of $y = \log_{10} x$ since it is the reflection of curve A in the line $y = x$. Hence, the equation for curve B is the exponential equation $y = 10^x$.

(b) The y-intercept occurs when $x = 0$. For curve B, $y = 10^0 = 1$. Therefore, the y-intercept for curve B is $(0, 1)$.

The logarithmic function with base b is the inverse of the exponential function with base b. Therefore, it makes sense that the laws of exponents (Section 1.1) should have corresponding properties involving logarithms. For example, the exponential property $b^0 = 1$ corresponds to the logarithmic property $\log_b 1 = 0$. We will state and prove three further important logarithmic properties that correspond to the following three exponential properties.

- $b^m \cdot b^n = b^{m+n}$
- $\dfrac{b^m}{b^n} = b^{m-n}$
- $(b^m)^n = b^{mn}$

Properties of logarithms II

Given $M > 0$, $N > 0$ and any real number k, the following properties are true for logarithms with $b > 0$ and $b \neq 1$.

	Property	Description
1	$\log_b (MN) = \log_b M + \log_b N$	the log of a product is the sum of the logs of its factors
2	$\log_b \left(\frac{M}{N}\right) = \log_b M - \log_b N$	the log of a quotient is the log of the numerator minus the log of the denominator
3	$\log_b (M^k) = k \log_b M$	the log of a number raised to a power is the power times the log of the number

Any of these properties can be applied in either direction.

Proof of property 1

Let $x = \log_b M$ and $y = \log_b N$.

The corresponding exponential forms of these two equations are $b^x = M$ and $b^y = N$.

Then, $\log_b(MN) = \log_b(b^x b^y) = \log_b(b^{x+y}) = x + y$.

It is given that $x = \log_b M$ and $y = \log_b N$, hence $x + y = \log_b M + \log_b N$.

Therefore, $\log_b(MN) = \log_b M + \log_b N$.

Proof of property 2

Again, let $x = \log_b M$ and $y = \log_b N \Rightarrow b^x = M$ and $b^y = N$.

Then, $\log_b\left(\frac{M}{N}\right) = \log_b\left(\frac{b^x}{b^y}\right) = \log_b(b^{x-y}) = x - y$.

With $x = \log_b M$ and $y = \log_b N$, then $x - y = \log_b M - \log_b N$.

Therefore, $\log_b\left(\frac{M}{N}\right) = \log_b M - \log_b N$.

The notation $f(x)$ uses brackets **not** to indicate multiplication, but to indicate the argument of the function f. The symbol f is the name of a function, not a variable; it is not multiplying the variable x. Therefore, in general, $f(x + y)$ is not equal to $f(x) + f(y)$. Likewise, the symbol **log** is also the name of a function. Therefore, $\log_b(x + y)$ is not equal to $\log_b(x) + \log_b(y)$. Other mistakes to avoid include incorrectly simplifying quotients or powers of logarithms. Specifically, $\frac{\log_b x}{\log_b y} \neq \log\left(\frac{x}{y}\right)$ and $(\log_b x)^k \neq k(\log_b x)$.

Proof of property 3

Let $x = \log_b M \Rightarrow b^x = M$.

Now, take the logarithm of M^k and substitute b^x for M.

$$\log_b(M^k) = \log_b[(b^x)^k] = \log_b(b^{kx}) = kx$$

It is given that $x = \log_b M$, hence $kx = k\log_b M$.

Therefore, $\log_b(M^k) = k\log_b M$.

Example 4.12

Use the properties of logarithms to write each logarithmic expression as a sum, difference, and/or constant multiple of simple logarithms (i.e. logarithms without sums, products, quotients, or exponents).

(a) $\log_2(8x)$ (b) $\ln\left(\frac{3}{y}\right)$ (c) $\log(\sqrt{7})$

(d) $\log_b\left(\frac{x^3}{y^2}\right)$ (e) $\ln(5e^2)$ (f) $\log\left(\frac{m+n}{n}\right)$

Solution

(a) $\log_2(8x) = \log_2 8 + \log_2 x = 3 + \log_2 x$

(b) $\ln\left(\frac{3}{y}\right) = \ln 3 - \ln y$

(c) $\log(\sqrt{7}) = \log(7^{\frac{1}{2}}) = \frac{1}{2}\log 7$

(d) $\log_b\left(\frac{x^3}{y^2}\right) = \log_b(x^3) - \log_b(y^2) = 3\log_b x - 2\log_b y$

(d) $\log_b\left(\frac{x^3}{y^2}\right) = \log_b(x^3) - \log_b(y^2) = 3\log_b x - 2\log_b y$

(e) $\ln(5e^2) = \ln 5 + \ln(e^2) = \ln 5 + 2\ln e = \ln 5 + 2(1) = 2 + \ln 5$
[$2 + \ln 5 \approx 3.609$ using GDC.]

(f) $\log\left(\frac{m+n}{m}\right) = \log(m+n) - \log m$
(Remember $\log(m+n) \neq \log m + \log n$.)

Example 4.13

Write each expression as the logarithm of a single quantity.

(a) $\log 6 + \log x$

(b) $\log_2 5 + 2\log_2 3$

(c) $\ln y - \ln 4$

(d) $\log_b 12 - \frac{1}{2}\log_b 9$

(e) $\log_3 M + \log_3 N - 2\log_3 P$

(f) $\log_2 80 - \log_2 5$

Solution

(a) $\log 6 + \log x = \log(6x)$

(b) $\log_2 5 + 2\log_2 3 = \log_2 5 + \log_2(3^2) = \log_2 5 + \log_2 9$
$= \log_2(5 \times 9) = \log_2 45$

(c) $\ln y - \ln 4 = \ln\left(\frac{y}{4}\right)$

(d) $\log_b 12 - \frac{1}{2}\log_b 9 = \log_b 12 - \log_b(9^{\frac{1}{2}}) = \log_b 12 - \log_b(\sqrt{9})$
$= \log_b 12 - \log_b 3 = \log_b\left(\frac{12}{3}\right) = \log_b 4$

(e) $\log_3 M + \log_3 N - 2\log_3 P = \log_3(MN) - \log_3(P^2) = \log_3\left(\frac{MN}{P^2}\right)$

(f) $\log_2 80 - \log_2 5 = \log_2\left(\frac{80}{5}\right) = \log_2 16 = 4$ [because $2^4 = 16$]

Change of base

The answer to Example 4.13, part (f), was $\log_2 16$, which we can compute to be exactly 4 because we know that $2^4 = 16$. The answer to Example 4.12, part (e), was $2 + \ln 5$, which we approximated to 3.609 using the natural logarithm function key (**ln**) on our GDC. But, what if we wanted to compute an approximate value for $\log_2 45$, the answer to Example 4.13, part (b)? Some calculators can only evaluate common logarithms (base 10) and natural logarithms (base e). To evaluate logarithmic expressions and graph logarithmic functions with other bases, we need to apply a change of base formula.

Change of base formula
Let a, b, and x be positive real numbers such that $a \neq 1$ and $b \neq 1$. Then $\log_b x$ can be expressed in terms of logarithms to any other base a, as:

$$\log_b x = \frac{\log_a x}{\log_a b}$$

Proof of change of base formula

$y = \log_b x \Rightarrow b^y = x$ Convert from logarithmic form to exponential form.

$\log_a x = \log_a(b^y)$ If $b^y = x$, then log of each with same bases must be equal.

$\log_a x = y\log_a b$ Applying the property $\log_b(M^k) = k\log_b M$.

$y = \dfrac{\log_a x}{\log_a b}$ Divide both sides by $\log_a b$.

To apply the change of base formula, let $a = 10$ or $a = \mathrm{e}$. Then the logarithm of any base b can be expressed in terms of either common logarithms or natural logarithms. For example:

$$\log_2 x = \frac{\log x}{\log 2} \text{ or } \frac{\ln x}{\ln 2} \qquad \log_5 x = \frac{\log x}{\log 5} \text{ or } \frac{\ln x}{\ln 5}$$

$$\log_2 45 = \frac{\log 45}{\log 2} = \frac{\ln 45}{\ln 2} \approx 5.492 \quad \text{[using GDC]}$$

Example 4.14

Use the change of base formula and common or natural logarithms to evaluate each logarithmic expression. Start by making a rough mental estimate. Write your answer to four significant figures.

(a) $\log_3 30$ (b) $\log_9 6$

Solution

(a) The value of $\log_3 30$ is the power to which 3 is raised to obtain 30. Because $3^3 = 27$ and $3^4 = 81$, the value of $\log_3 30$ is between 3 and 4, and will be much closer to 3 than 4, perhaps around 3.1.

Using the change of base formula and common logarithms, we obtain $\log_3 30 = \dfrac{\log 30}{\log 3} \approx 3.096$. This agrees well with the mental estimate.

After computing the answer on your GDC, use your GDC to also check it by raising 3 to the answer and confirming that it gives a result of 30.

```
log(30)/log(3)
              3.095903274
3^Ans
                       30
█
```

(b) The value of $\log_9 6$ is the power to which 9 is raised to obtain 6. Because $9^{\frac{1}{2}} = \sqrt{9} = 3$ and $9^1 = 9$, the value of $\log_9 6$ is between $\dfrac{1}{2}$ and 1, perhaps around 0.75.

Using the change of base formula and natural logarithms, we obtain $\log_9 6 = \dfrac{\ln 6}{\ln 9} \approx 0.8155$. This agrees well with the mental estimate.

```
ln(6)/ln(9)
             .8154648768
9^Ans
                       6
█
```

Exercise 4.3

1. Express each logarithmic equation as an exponential equation.

(a) $\log_2 16 = 4$ (b) $\ln 1 = 0$ (c) $\log 100 = 2$

(d) $\log 0.01 = -2$ (e) $\log_7 343 = 3$ (f) $\ln\left(\frac{1}{e}\right) = -1$

(g) $\log 50 = y$ (h) $\ln x = 12$ (i) $\ln(x + 2) = 3$

2. Express each exponential equation as a logarithmic equation.

(a) $2^{10} = 1024$ (b) $10^{-4} = 0.0001$ (c) $4^{-\frac{1}{2}} = \frac{1}{2}$

(d) $3^4 = 81$ (e) $10^0 = 1$ (f) $e^x = 5$

(g) $2^{-3} = 0.125$ (h) $e^4 = y$ (i) $10^{x+1} = y$

3. Find the exact value of each expression without using your GDC.

(a) $\log_2 64$ (b) $\log_4 64$ (c) $\log_2\left(\frac{1}{8}\right)$

(d) $\log_3(3^5)$ (e) $\log_{16} 8$ (f) $\log_{27} 3$

(g) $\log_{10}(0.001)$ (h) $\ln e^{13}$ (i) $\log_8 1$

(j) $10^{\log 6}$ (k) $\log_3\left(\frac{1}{27}\right)$ (l) $e^{\ln\sqrt{2}}$

(m) $\log 1000$ (n) $\ln(\sqrt{e})$ (o) $\ln\left(\frac{1}{e^2}\right)$

(p) $\log_3(81^{22})$ (q) $\log_4 2$ (r) $3^{\log_3 18}$

(s) $\log_5(\sqrt[3]{5})$ (t) $10^{\log \pi}$

4. Use a GDC to evaluate each expression, correct to 3 significant figures.

(a) $\log 50$ (b) $\log\sqrt{3}$ (c) $\ln 50$

(d) $\ln\sqrt{3}$ (e) $\log 25$ (f) $\log\left(\frac{1+\sqrt{5}}{2}\right)$

(g) $\ln 100$ (h) $\ln(100^3)$

5. Find the domain of each function exactly.

(a) $y = \log(x - 2)$ (b) $y = \ln(x^2)$

(c) $y = \log(x) - 2$ (d) $y = \log_7(8 - 5x)$

(e) $y = \sqrt{x+2} - \log_3(9 - 3x)$ (f) $y = \sqrt{\ln(1 - x)}$

6. Find the domain and range of each function exactly.

(a) $y = \frac{1}{\ln x}$ (b) $y = |\ln(x - 1)|$ (c) $y = \frac{x}{\log x}$

7. Find the equation of the function that is graphed in the form $f(x) = \log_b x$.

(a)

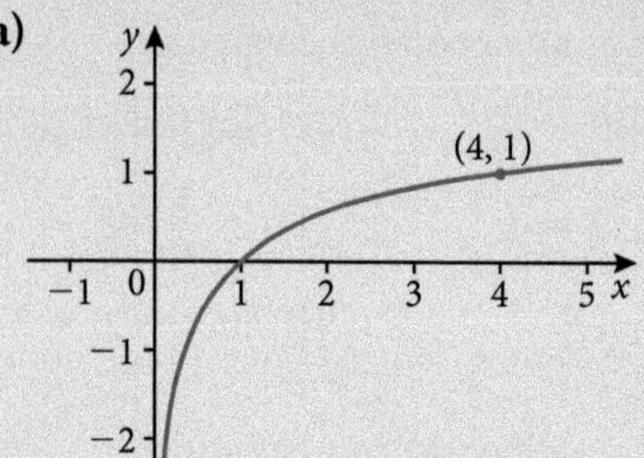

(b)

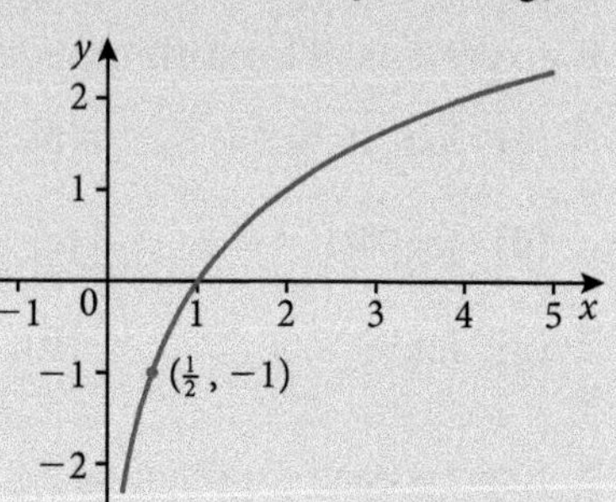

(c)

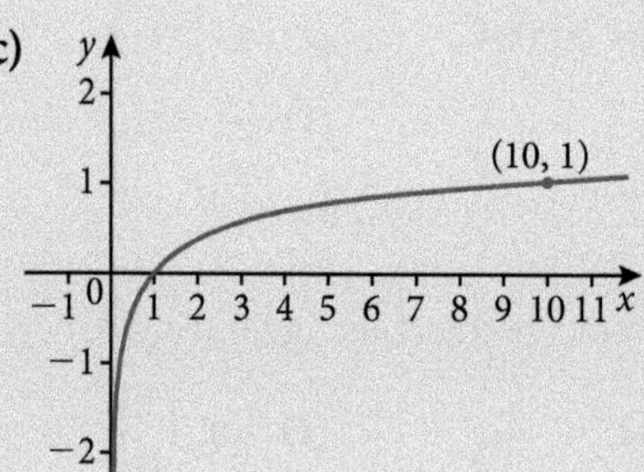

(d)

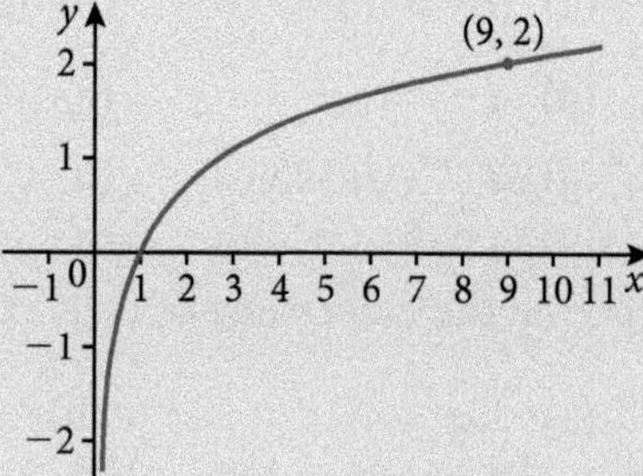

8. Use properties of logarithms to write each logarithmic expression as a sum, difference, and/or constant multiple of simple logarithms (i.e. logarithms without sums, products, quotients, or exponents).

(a) $\log_2(2m)$ (b) $\log\left(\frac{9}{x}\right)$ (c) $\ln(\sqrt[5]{x})$

(d) $\log_3(ab^3)$ (e) $\log[10x(1 + r)^t]$ (f) $\ln\left(\frac{m^3}{n}\right)$

9. Write each expression in terms of $\log_b p$, $\log_b q$ and $\log_b r$.

(a) $\log_b pqr$ (b) $\log_b\left(\frac{p^2q^3}{r}\right)$ (c) $\log_b(\sqrt[4]{pq})$

(d) $\log_b\sqrt{\frac{qr}{p}}$ (e) $\log_b\frac{p\sqrt{q}}{r}$ (f) $\log_b\frac{(pq)^3}{\sqrt{r}}$

10. Write each expression as the logarithm of a single quantity.

(a) $\log(x^2) + \log\left(\frac{1}{x}\right)$ (b) $\log_3 9 + 3\log_3 2$ (c) $4\ln y - \ln 4$

(d) $\log_b 12 - \frac{1}{2}\log_b 9$ (e) $\log x - \log y - \log z$ (f) $2\ln 6 - 1$

$\ln(?) = 1$

11. Use the change of base formula and common or natural logarithms to evaluate each logarithmic expression. Approximate your answer to three significant figures.

(a) $\log_2 1000$ (b) $\log_{\frac{1}{2}} 40$ (c) $\log_6 40$ (d) $\log_5(0.75)$

12. Use the change of base formula to rewrite each function in terms of natural logarithms. Then graph the function on a GDC and use the graph to approximate to three significant figures the value of the function when $x = 20$.

(a) $f(x) = \log_2 x$ **(b)** $f(x) = \log_5 x$

13. Use the change of base formula to prove the statement

$$\log_b a = \frac{1}{\log_a b}$$

14. Show that $\log e = \frac{1}{\ln 10}$.

15. The relationship between the number of decibels dB (one variable) and the intensity I of a sound in watts per square metre is given by the formula

$$dB = 10 \log\left(\frac{I}{10^{-16}}\right)$$

(a) Use properties of logarithms to write the formula in simpler form.

(b) Find the number of decibels of a sound with an intensity of 10^{-4} watts per square metre.

4.4 Exponential and logarithmic equations

Solving exponential equations

At the start of Section 4.3, we wanted to find a way to determine how much time t (in years) it would take for an investment of €1000 to double if the investment earned interest at an annual rate of 5.25%. Since the interest is compounded continuously, we need to solve the equation

$$2000 = 1000e^{0.0525t} \Rightarrow 2 = e^{0.0525t}$$

This equation has the variable, t, in the exponent. With the properties of logarithms established in Section 4.3, we now have a way to algebraically solve such equations. Along with these properties, we need to apply the logic that if two expressions are equal then their logarithms must also be equal. That is, if $m = n$, then $\log_b m = \log_b n$.

Example 4.15

Solve the equation for the variable t. Give your answer accurate to three significant figures.

$$2 = e^{0.0525t}$$

Solution

$2 = e^{0.0525t}$

$\ln 2 = \ln(e^{0.0525t})$ Take natural logarithm of both sides.

$\ln 2 = 0.0525t$ Apply the property $\log_b(b^x) = x$.

$t = \dfrac{\ln 2}{0.0525} \approx 13.2$

Example 4.15 shows a general strategy for solving exponential equations. To solve an exponential equation, first isolate the exponential expression, then take the logarithm of both sides. Then apply a property of logarithms so that the variable is no longer in the exponent and can be isolated on one side of the equation.

By taking the logarithm of both sides of an exponential equation, we are making use of the inverse relationship between exponential and logarithmic functions. Symbolically, this method can be represented as follows, solving for x:

$$y = b^x \Rightarrow \log_b y = \log_b b^x \Rightarrow \log_b y = x$$

Although most calculators can evaluate a logarithm with a base other than 10 or e, some calculators cannot. If your calculator is not able to evaluate a logarithm with a base of b, you will need to use the change of base formula to convert into a base that your calculator can handle (e.g. base 10, or base e).

$$y = b^x \Rightarrow \log_a y = \log_a b^x \Rightarrow \log_a y = x\log_a b \Rightarrow x = \frac{\log_a y}{\log_a b}$$

Example 4.16

Solve for x in the equation $3^{x-4} = 24$. Approximate the answer to three significant figures.

Solution

$3^{x-4} = 24$

$\log(3^{x-4}) = \log 24$ Take common logarithm of both sides.

$(x - 4)\log 3 = \log 24$ Apply the property $\log_b(M^k) = k\log_b M$.

$x - 4 = \dfrac{\log 24}{\log 3}$ Divide both sides by $\log 3$.

$x = \dfrac{\log 24}{\log 3} + 4$ $\left[\text{note: } \dfrac{\log 24}{\log 3} \neq \log 8\right]$

$x \approx 6.89$ Using GDC.

We could have used natural logarithms instead of common logarithms to solve the equation in Example 4.16. Using the same method but with natural logarithms will give you the same answer: $x = \dfrac{\ln 24}{\ln 3} + 4 \approx 6.89$.

Recall Example 3.11, in which we solved an exponential equation graphically because we did not yet have the tools to solve it algebraically. Let's solve it now using logarithms.

Example 4.17

You invested €1000 at 6% compounded quarterly. How long will it take for this investment to increase to €2000?

Solution

Using the compound interest formula from Section 4.2, $A(t) = P\left(1 + \frac{r}{n}\right)^{nt}$, with $P = €1000$, $r = 0.06$ and $n = 4$, we need to solve for t when $A(t) = 2P$.

$2P = P\left(1 + \frac{0.06}{4}\right)^{4t}$ Substitute $2P$ for $A(t)$.

$2 = 1.015^{4t}$ Divide both sides by P.

$\ln 2 = \ln(1.015^{4t})$ Take natural logarithm of both sides.

$\ln 2 = 4t \ln 1.015$ Apply the property $\log_b(M^k) = k \log_b M$.

$t = \frac{\ln 2}{4 \ln 1.015}$

$t \approx 11.6389$ Evaluate on GDC.

The investment will double in about 11.64 years, or about 11 years and 8 months.

Be sure to use brackets appropriately when entering the expression $\frac{\ln 2}{4 \ln 1.015}$ on your GDC. Following the rules for order of operations, your GDC will give an incorrect result (as shown in the GDC image here) if entered without the necessary brackets. Some GDC models will enter the necessary brackets automatically.

```
ln(2)/4ln(1.015)
     .0025799999
```

missing brackets

Example 4.18

The number of the bacteria that causes strep throat will grow at a rate of about 2.3% per minute. Find, to the nearest whole minute, how long it will take for these bacteria to double in number.

Solution

Let t represent time in minutes and let A_0 represent the number of bacteria at $t = 0$.

Using the exponential growth model from Section 4.1, $A(t) = A_0 b^t$, the growth factor, b, is $1 + 0.023 = 1.023$ giving $A(t) = A_0(1.023)^t$.

Our mathematical model assumes that the number of bacteria increases incrementally by 2.3% at the end of each minute. To find the doubling time, find the value of t so that $A(t) = 2A_0$.

$2A_0 = A_0(1.023)^t$ Substitute $2A_0$ for $A(t)$.

$2 = 1.023^t$ Divide both sides by A_0.

$\ln 2 = \ln(1.023^t)$ Take natural logarithm of both sides.

The same equation would apply to money earning 2.3% annual interest with the money being added (compounded) once per year rather than once per minute.

$\ln 2 = t \ln 1.023$ Apply the property $\log_b(M^k) = k\log_b M$.

$$t = \frac{\ln 2}{\ln 1.023} \approx 30.482$$

The number of bacteria will double in about 30 minutes.

Example 4.19

\$1000 is invested in an account that earns interest at an annual rate of 10%, compounded monthly. Calculate the minimum number of years needed for the amount in the account to exceed \$4000.

Solution

We use the exponential function associated with compound interest, $A(t) = P\left(1 + \frac{r}{n}\right)^{nt}$, with $P = 1000$, $r = 0.1$ and $n = 12$.

$$4000 = 1000\left(1 + \frac{0.1}{12}\right)^{12t} \Rightarrow 4 = (1.008\dot{3})^{12t} \Rightarrow \log 4 = \log[(1.008\dot{3})^{12t}]$$

$$\Rightarrow \log 4 = 12t\log(1.008\dot{3}) \Rightarrow t = \frac{\log 4}{12\log(1.008\dot{3})} \approx 13.92 \text{ years}$$

The minimum time needed for the account to exceed \$4000 is 14 years.

Example 4.20

A 20 gram sample of radioactive iodine decays so that the mass remaining after t days is given by the equation $A(t) = 20e^{-0.087t}$, where $A(t)$ is measured in grams. Find how many days (to the nearest whole day) it will take until there are only 5 grams remaining.

Solution

$$5 = 20e^{-0.087t} \Rightarrow \frac{5}{20} = e^{-0.087t} \Rightarrow \ln 0.25 = \ln(e^{-0.087t}) \Rightarrow \ln 0.25 = -0.087t$$

$$\Rightarrow t = \frac{\ln 0.25}{-0.087} \approx 15.93$$

After about 16 days there are only 5 grams remaining.

Example 4.21

Solve for x in the equation $3^{2x} - 18 = 3^{x+1}$. Express any answers in exact form.

Solution

The key to solving this equation is recognising that it can be written in quadratic form.

We need to apply some laws of exponents to show that the equation is quadratic for the expression 3^x.

$$3^{2x} - 18 = 3^{x+1}$$

$(3^x)^2 - (3^1)(3^x) - 18 = 0$ Applying rules $b^{mn} = (b^m)^n$ and $b^{m+n} = b^m b^n$.

Substituting a single variable, for example y, for the expression 3^x clearly makes the equation quadratic in terms of 3^x. We solve first for y and then solve for x after substituting 3^x back for y.

$y^2 - 3y - 18 = 0$

$(y + 3)(y - 6) = 0$

$y = -3$ or $y = 6$

$3^x = -3$ or $3^x = 6$

$3^x = -3$ has no solution. Raising a positive number to a power cannot produce a negative number.

$3^x = 6$

$\ln(3^x) = \ln 6$ Take logarithms of both sides.

$x \ln 3 - \ln 6$

Therefore, the one solution to the equation is exactly $x = \dfrac{\ln 6}{\ln 3}$.

There are a couple of common errors to avoid in Example 4.21.

- If $3^x = -3$, then it does **not** follow that $x = -1$. An exponent of -1 indicates a reciprocal.
- If $x = \dfrac{\ln 6}{\ln 3}$, it does **not** follow that $x = \ln 2$. The rule $\log m - \log n = \log\left(\frac{m}{n}\right)$ does not apply to the expression $\dfrac{\ln 6}{\ln 3}$

Solving logarithmic equations

A logarithmic equation is an equation where the variable appears within the argument of a logarithm. For example, $\log x = \frac{1}{2}$ or $\ln x = 4$. We can solve both of these logarithmic equations directly by applying the definition of a logarithmic function from Section 4.3:

$$y = \log_b x \text{ if and only if } x = b^y$$

The logarithmic equation $\log x = \frac{1}{2}$ is equivalent to the exponential equation $x = 10^{\frac{1}{2}} = \sqrt{10}$, which leads directly to the solution. Likewise, the equation $\ln x = 4$ is equivalent to $x = e^4 \approx 54.598$. Both of these equations could have been solved by means of another method that makes use of the following two facts:

$$\text{if } a = b \text{ then } n^a = n^b$$

$$b^{\log_b x} = x$$

To understand the second fact, remember that a logarithm is an exponent. The value of $\log_b x$ is the exponent to which b is raised to give x. And b is being raised to this value, hence the expression $b^{\log_b x}$ is equivalent to x.

In the context of solving equations, to **exponentiate** means to use two equal expressions as powers to construct another equality involving equal bases. For example: If $a = b$, then $m^a = m^b$.

Therefore, another method for solving the logarithmic equation $\ln x = 4$ is to **exponentiate** both sides; that is, use the expressions on either side of the equal sign as exponents for exponential expressions with equal bases. The base needs to be the base of the logarithm.

$$\ln x = 4 \Rightarrow e^{\ln x} = e^4 \Rightarrow x = e^4$$

Example 4.22

Solve for x in the equation $\log_3(2x - 5) = 2$.

Solution

$\log_3(2x - 5) = 2$

Method 1

$2x - 5 = 3^2$ $\quad y = \log_b x \Leftrightarrow x = b^y$

$2x = 9 + 5$

$x = 7$

Method 2

$3^{\log_3(2x-5)} = 3^2$ $\quad$ Exponentiate with base = 3

$2x - 5 = 9$

$x = 7$

Example 4.23

Solve for x (in terms of k) in the equation $\log_2(5x) = 3 + k$.

Solution

$\log_2(5x) = 3 + k \Rightarrow 2^{\log_2(5x)} = 2^{3+k}$ $\quad$ Exponentiate both sides with base = 2

$5x = (2^3)(2^k)$ $\quad$ Law of powers $(b^m)(b^n) = b^{m+n}$ used 'in reverse'

$x = \frac{8}{5}(2^k)$

For some logarithmic equations it is necessary to simplify by applying a property, or properties, of logarithms before solving.

Example 4.24

Solve for x in the equation $\log_2 x + \log_2(10 - x) = 4$.

Solution

$\log_2 x + \log_2(10 - x) = 4$

$\log_2[x(10 - x)] = 4$ $\quad$ Property of logarithms: $\log_b M + \log_b N = \log_b(MN)$

$10x - x^2 = 2^4$ $\quad$ Changing from logarithmic form to exponential form

$x^2 - 10x + 16 = 0$

$(x - 2)(x - 8) = 0$

$x = 2$ or $x = 8$

When solving a logarithmic equation, we must be careful to always check if the original equation is a true statement when any solutions are substituted in for the variable. For Example 4.24, both of the solutions $x = 2$ and $x = 8$ produce true statements when substituted into the original equations. Sometimes 'extra' (extraneous) invalid solutions are produced, as illustrated in Example 4.25.

Example 4.25

Solve for x in the equation $\ln(x - 2) + \ln(2x - 3) = 2\ln x$.

Solution

$\ln(x - 2) + \ln(2x - 3) = 2\ln x$

$\ln[(x - 2)(2x - 3)] = \ln x^2$ Properties of logarithms

$\ln(2x^2 - 7x + 6) = \ln x^2$

$e^{\ln(2x^2-7x+6)} = e^{\ln x^2}$ Exponentiate both sides

$2x^2 - 7x + 6 = x^2$

$x^2 - 7x + 6 = 0$

$(x - 6)(x - 1) = 0$ Factorise

$x = 6$ or $x = 1$

Substituting these two possible solutions indicates that $x = 1$ is not a valid solution. If we try to substitute 1 in for x into the original equation, we get the expression $\ln(2x - 3) = \ln(-1)$. This cannot be evaluated because we can only take the logarithm of a positive number. Therefore, $x = 6$ is the only solution. $x = 1$ is an extraneous solution that is not valid.

Solving, or checking the solutions to, a logarithmic equation on your GDC will help you avoid, or determine, extraneous solutions. To solve Example 4.25 on your GDC, a useful approach is to first rearrange the equation so that the right hand side is equal to zero. Then graph the expression (after setting it equal to y) and observe where the graph intersects the x-axis (i.e. $y = 0$).

For Example 4.25:

$\ln(x - 2) + \ln(2x - 3) = 2\ln x \Rightarrow \ln(x - 2) + \ln(2x - 3) - 2\ln x = 0$

Graph the equation $y = \ln(x - 2) + \ln(2x - 3) - 2\ln x$ on your GDC and find the x-intercepts.

The graph only intersects the x-axis at $x = 6$ and not at $x = 1$. Hence, $x = 6$ is the only valid solution and $x = 1$ is an extraneous solution.

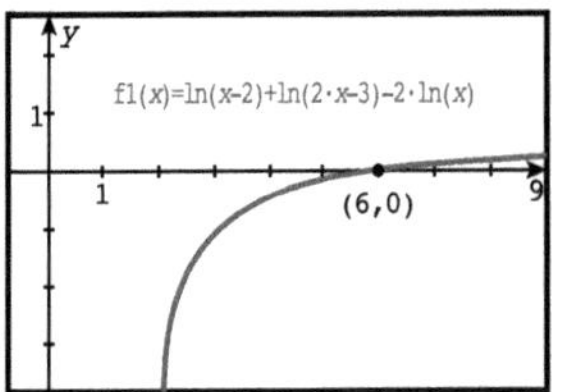

Figure 4.4 Graphical solution for Example 4.25

Exercise 4.4

1. Solve each equation for x. Give x accurate to three significant figures.
   (a) $10^x = 5$ (b) $4^x = 32$ (c) $8^{x-6} = 60$
   (d) $2^{x+3} = 100$ (e) $\left(\frac{1}{5}\right)^x = 22$ (f) $e^x = 15$
   (g) $10^x = e$ (h) $3^{2x-1} = 35$ (i) $2^{x+1} = 3^{x-1}$
   (j) $2e^{10x} = 19$ (k) $6^{\frac{x}{2}} = 5^{1-x}$ (l) $\left(1 + \frac{0.05}{12}\right)^{12x} = 3$

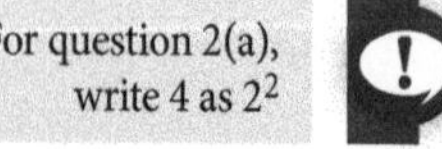
For question 2(a), write 4 as 2^2

2. Solve each equation for x. Give exact answers.
   (a) $4^x - 2^{x+1} = 48$ (b) $2^{2x+1} - 2^{x+1} + 1 = 2^x$

3. $5000 is invested in an account that pays 7.5% interest per year, compounded quarterly.
   (a) Find the amount in the account after 3 years.
   (b) Find how long it will take for the money in the account to double. Give your answer to the nearest quarter of a year.

4. Find how long it will take for an investment of €500 to triple in value when the interest is 8.5% per year, compounded:
   (a) monthly (b) continuously.
   Give your answers in number of years accurate to four significant figures.

5. A single bacterium begins a colony in a laboratory dish. If the colony doubles every hour, find how many hours it takes for the colony to first contain more than one million bacteria.

6. Find the least number of years for an investment to double if interest is compounded annually with the following interest rates.
   (a) 3% (b) 6% (c) 9%

7. A new car purchased in 2005 decreases in value by 11% per year. Find the first year that the car is worth less than half of its original value.

8. Uranium-234 is a radioactive substance that has a half-life of 2.46×10^5 years.
   (a) Find the amount remaining from a one-gram sample after a thousand years.
   (b) Find how long it will take a one-gram sample to decompose until its mass is 700 milligrams (i.e. 0.7 grams). Give your answer in years accurate to three significant figures.

9. The stray dog population in a town is growing exponentially, with about 18% more stray dogs each year. In 2008, there are 16 stray dogs.
   (a) Find the projected population of stray dogs after 5 years.
   (b) Find the first year that the number of stray dogs is greater than 70.

10. A water tank initially contains one thousand litres of water. At time $t = 0$ (in minutes) a tap is opened and water flows out of the tank. The volume, V litres, remaining in the tank after t minutes is given by the exponential function $V(t) = 1000(0.925)^t$.

(a) Find the value of V after ten minutes.

(b) Find how long, to the nearest second, it takes for half of the initial amount of water to flow out of the tank.

(c) The tank is considered empty when only 5% of the water remains. From when the tap is first opened, find how many whole minutes have passed before the tank can first be considered empty.

11. The mass, m kilograms, of a radioactive substance at time t days is given by $m = 5e^{-0.13t}$.

(a) Find the initial mass.

(b) Find how long it takes for the substance to decay to 0.5 kilograms. Give your answer in days accurate to three significant figures.

12. Solve for x in the logarithmic equation. Give exact answers and be sure to check for extraneous solutions.

(a) $\log_2(3x - 4) = 4$ **(b)** $\log(x - 4) = 2$

(c) $\ln x = -3$ **(d)** $\log_{16} x = \frac{1}{2}$

(e) $\log\sqrt{x + 2} = 1$ **(f)** $\ln(x^2) = 16$

(g) $\log_2(x^2 + 8) = \log_2 x + \log_2 6$ **(h)** $\log_3(x - 8) + \log_3 x = 2$

(i) $\log 7 - \log(4x + 5) + \log(2x - 3) = 0$

(j) $\log_3 x + \log_3(x - 2) = 1$

13. Solve each inequality.

(a) $5\log x + 2 > 0$

(b) $2\log x^2 - 3\log x < \log 8x - \log 4x$

(c) $(e^x - 2)(e^x - 3) < 2e^x$

(d) $3 + \ln x > e^x$

Chapter 4 practice questions

1. Solve for x in each equation.

(a) $\log_x 16 = 4$ **(b)** $\log_3 27 = x$

(c) $\log_8 x = -\frac{1}{3}$ **(d)** $\log_6(x - 1) + \log_6 x = 1$

2. Solve for x in each equation.

(a) $4^x = 36$ **(b)** $5 \times 3^x = 18$

(c) $8^{-x} = \left(\frac{1}{4}\right)^3$ **(d)** $6^x = 0.25^{2x-1}$

3. Write each expression as the logarithm of a single quantity.

(a) $\log_2 x^2 - \log_2 x + 2\log_2 3$ **(b)** $\ln 3 + \frac{1}{2}\ln(x - 4) - \ln x$

4. If $\log_b M = 5.42$ and $\log_b N^2 = 3.78$, find the following:

(a) $\log_b N$ **(b)** $\log_b\left(\frac{N^4}{\sqrt{M}}\right)$

5. Pablo invested €2000 at an annual rate of 6.75%, compounded annually.

(a) Find the value of Pablo's investment after four years. Give your answer to the nearest whole euro.

(b) Find how many years it will take for Pablo's investment to double in value.

(c) Determine what the interest rate should be for Pablo's initial investment to double in value in ten years?

6. \$1000 is deposited into a bank account that earns interest at an annual rate of 4% compounded annually. After three years, the annual interest rate is increased to 7% for a further four years.

(a) Find how much money is in the account after the seven years.

(b) Find what constant rate of annual interest compounded annually would have given the same amount of money in the seven years. Give your answer as a percentage to one decimal place.

7. Express each of the following expressions as simply as possible.

(a) $\log_2 5 \times \log_5 2$ **(b)** $\log_4 8$ **(c)** $4^{\log_2 6}$

8. At the start of the year 2000 there were 500 elephants in a game reserve. After t years, the number of elephants E is given by $500(1.032)^t$.

(a) Find the number of elephants at the start of 2006.

(b) Determine how many full years it will take for the number of elephants to first become greater than 750.

9. A certain car, when purchased new, had an initial value of \$25,000. After one year the car had decreased in value to \$22,000.

(a) Find the value of the car after one year as a percentage of the initial value.

(b) If the car continues to decrease in value at the same annual rate, find the car's value after six years. Give your answer to the nearest dollar.

(c) If the car was purchased in 2002, determine in what year the car is first worth less than \$8000.

10. Consider the function $f: x \mapsto e^{x-2}$.

(a) Write down the domain and range of f.

(b) Write down the coordinates of any y-intercept, and the equation of any asymptotes for the graph of f.

(c) Find f^{-1}.

(d) Write down the domain and range of f^{-1}.

11. An insect population grows at a rate of 6% per month. Initially there are 500 insects.

(a) Find the size of the population after four months.

(b) Find the size of the population after a further year.

(c) Let the size of the population after t months be given by the function $f(t) = A_0 b^t$. Write down:

(i) the value A_0 (ii) the value of b.

An alternative way of modelling the size of the insect population is given by the function $g(t) = 500e^{kt}$.

(d) By equating $f(t)$ and $g(t)$, find the value of k. Give your answer correct to five decimal places.

12. (a) State the domain for each of the following two functions.

(i) $f(x) = \log\left(\frac{x}{x-2}\right)$ (ii) $g(x) = \log x - \log(x-2)$

(b) Solve each of the following equations.

(i) $\log\left(\frac{x}{x-2}\right) = -2$ (ii) $\log x - \log(x-2) = -2$

13. An experiment is designed to study a certain type of bacteria. The number of bacteria after t minutes is given by an exponential function of the form $A(t) = Ce^{kt}$ where C and k are constants. At the start of the experiment (when $t = 0$) there are 5000 bacteria. After 22 minutes the number of bacteria has increased to 17 000.

(a) Find the exact value of C and an approximate value of k (to three significant figures).

(b) Find how many bacteria the exponential function predicts there will be after one hour.

14. Part of the graph $y = 2 - \log_3(x+1)$ is shown. It intersects the x-axis at point P, the y-axis at point Q, and the line $y = 3$ at point R.

Find :

(a) the x-coordinate of point P

(b) the y-coordinate of point Q

(c) the coordinates of point R.

15. Solve for x in the equation $\log_2(5x^2 - x - 2) = 2 + 2\log_2 x$.

16. If $\log_2 4\sqrt{2} = x$, $\log_z y = 4$, and $y = 4x^2 - 2x - 6 + z$, find y.

17. Find the exact value of x for each equation.

(a) $\log_3 x - 4\log_x 3 + 3 = 0$ **(b)** $\log_2(x - 5) + \log_2(x + 2) = 3$

18. Express each expression as a single logarithm.

(a) $2\log a + 3\log b - \log c$ **(b)** $3\ln x - \frac{1}{2}\ln y + 1$

19. A piece of wood is recovered from an ancient building during an archaeological excavation. The formula $A(t) = A_0 e^{-0.000124t}$ is used to determine the age of the wood, where A_0 is the activity of carbon-14 in any living tree, $A(t)$ is the activity of carbon in the wood being dated and t is the age of the wood in years. For the ancient piece of wood, it is found that $A(t)$ is 79% of the activity of the carbon in a living tree. Determine how old the piece of wood is, to the nearest 100 years.

20. The graph of the equation $y = \log_3(2x - 3) - 4$ intersects the x-axis at the point $(c, 0)$. Without using your GDC, find the exact value of c.

21. Solve $2(\ln x)^2 = 3\ln x - 1$ for x. Give your answers in exact form.

22. A sum of \$100 is invested.

(a) If the interest is compounded annually at a rate of 5% per year, find the total value V of the investment after 20 years.

(b) If the interest is compounded monthly at a rate of $\frac{5}{12}$% per month, find the minimum number of months for the value of the investment to exceed V.

23. Solve the equation $9\log_5 x = 25\log_x 5$, expressing your answer in the form $5^{\frac{p}{q}}$, where $p, q \in \mathbb{Z}$.

24. An experiment is carried out in which the number, n, of bacteria in a liquid is given by the formula $n = 650e^{kt}$, where t is the time in minutes after the beginning of the experiment and k is a constant. The number of bacteria doubles every 20 minutes. Find the exact value of k.

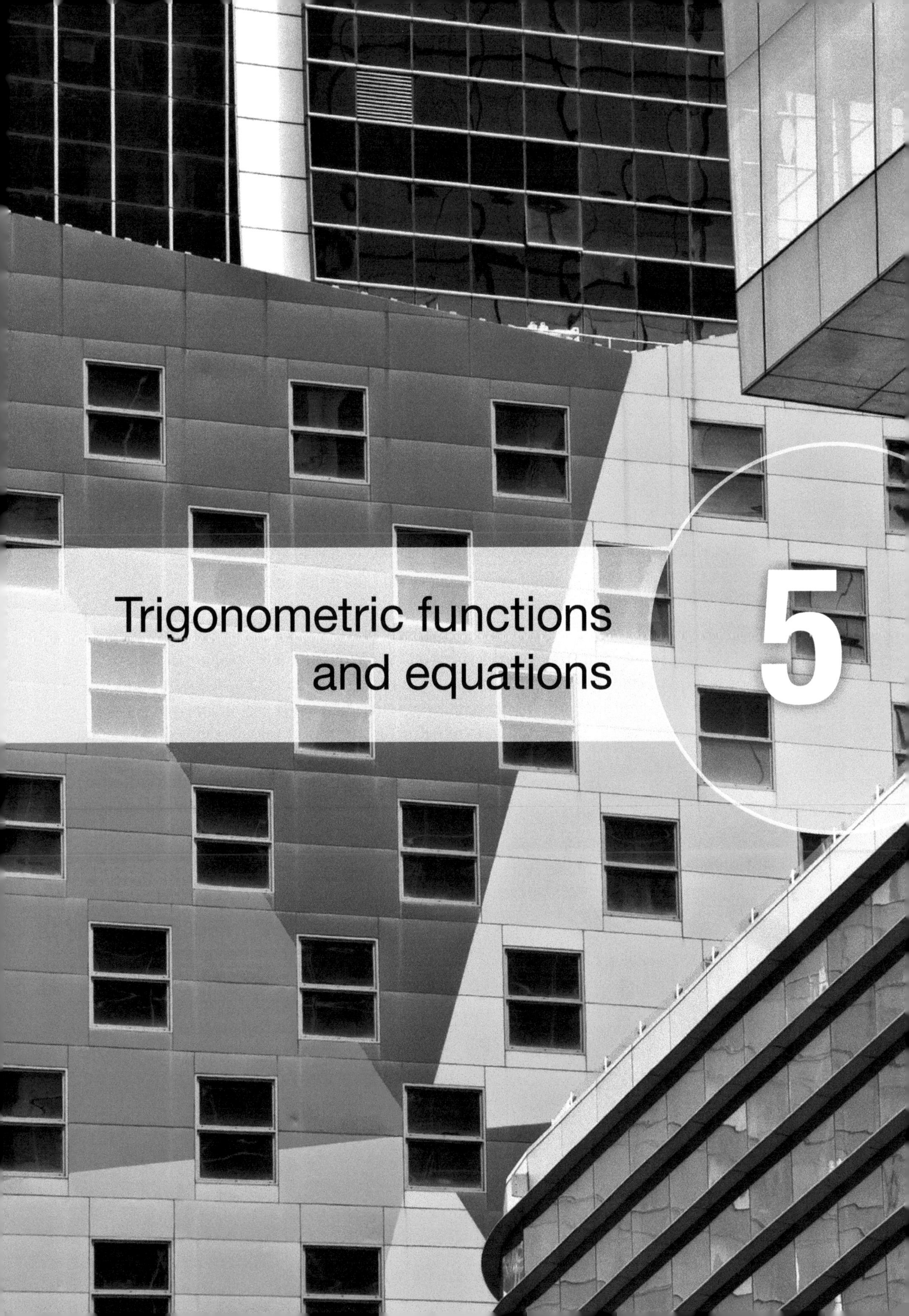

5 Trigonometric functions and equations

5 Trigonometric functions and equations

Learning objectives

By the end of this chapter, you should be familiar with...

- the radian measure of angles
- finding the length of an arc and area of a sector
- the unit circle and definitions of $\sin\theta$, $\cos\theta$, and $\tan\theta$
- the exact values of trigonometric ratios of $0, \frac{\pi}{6}, \frac{\pi}{4}, \frac{\pi}{3}, \frac{\pi}{2}$ and their multiples
- the Pythagorean identity $\sin^2\theta + \cos^2\theta = 1$
- double angle identities for sine and cosine
- the graphs of $\sin\theta$, $\cos\theta$, and $\tan\theta$
- the amplitude and period for the graphs of $\sin\theta$, $\cos\theta$, and $\tan\theta$
- composite functions of the form $a\sin(b(x + c)) + d$ and $a\cos(b(x + c)) + d$ and their graphs
- transformations of the graphs of trigonometric functions and their applications
- applying trigonometric functions to real-life problems
- solving trigonometric equations in a finite interval.

The word *trigonometry* comes from two Greek words, *trigonon* and *metron*, meaning 'triangle measurement'. Trigonometry developed out of the use and study of triangles in surveying, navigation, architecture, and astronomy to find relationships between lengths of sides of triangles and measurement of angles. As a result, trigonometric functions were initially defined as functions of angles – that is, functions with angle measurements as their domains. With the development of calculus in the 17th century and the growth of knowledge in the sciences, the application of trigonometric functions grew to include a wide variety of periodic (repetitive) phenomena such as wave motion, vibrating strings, oscillating pendulums, alternating electrical current, and biological cycles. These applications of trigonometric functions require their domains to be sets of real numbers without reference to angles or triangles. Hence, trigonometry can be approached from two different perspectives: functions of angles or functions of real numbers. This chapter focuses on the latter; that is, viewing trigonometric functions as defined in terms of a real number that is the length of an arc along the unit circle. Although this chapter will not refer much to triangles (this is covered in Chapter 6), it is appropriate to begin by looking at angles and arc lengths – geometric objects that are essential to the two different ways of approaching trigonometry.

5.1 Angles, circles, arcs, and sectors

An **angle** in a plane is made by rotating a ray about its endpoint, called the **vertex** of the angle. The starting position of the ray is called the **initial side** and the position of the ray after rotation is called the **terminal side** of the angle (Figure 5.1). An angle with its vertex at the origin and its initial side on the positive x-axis is in **standard position** (Figure 5.2). A **positive angle** is produced when a ray is rotated in an anti-clockwise direction, and a **negative angle** when a ray is rotated in a clockwise direction. Two angles in standard position that have the same terminal sides, regardless of the direction or number of rotations, are called **coterminal angles**. Greek letters are often used to represent angles, and the direction of rotation is indicated by an arc with an arrow at its endpoint.

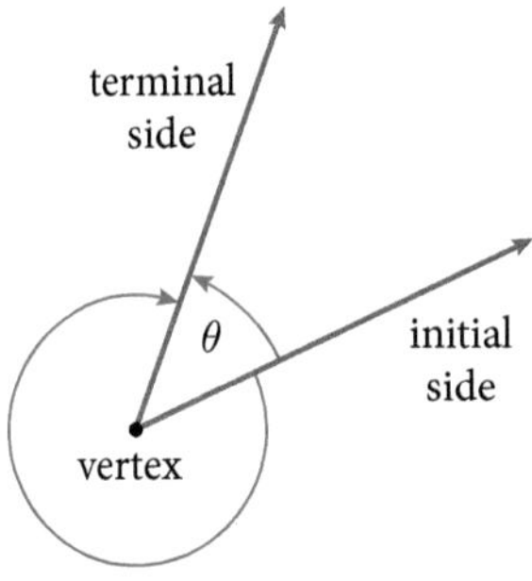

Figure 5.1 Components of an angle

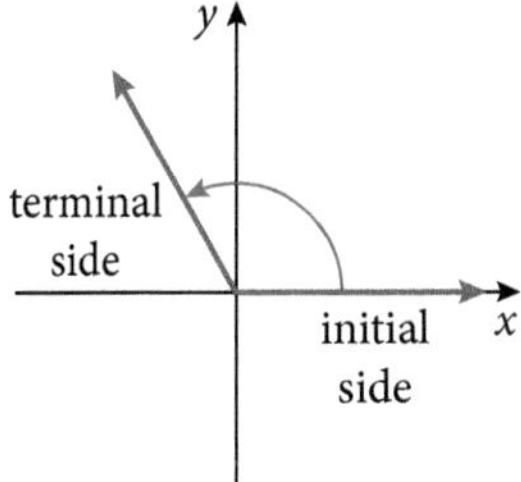

Figure 5.2 Standard position of an angle

The x- and y-axes divide the coordinate plane into four quadrants (numbered with Roman numerals). Figure 5.3 shows a positive angle α and a negative angle β that are coterminal in quadrant III.

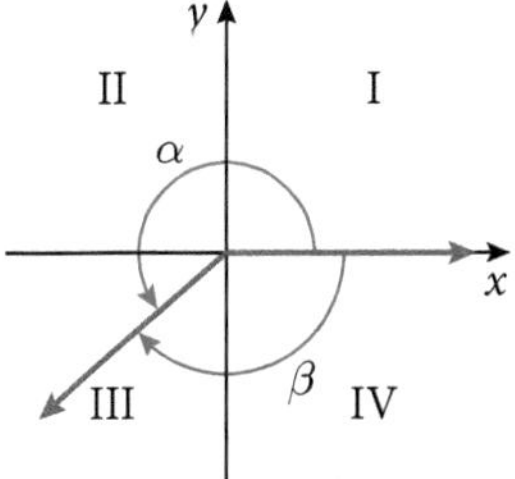

Figure 5.3 Coterminal angles

Radian measure

Instead of dividing a full revolution into 360 degrees, consider an angle that has its vertex at the centre of a circle (a **central angle**) and subtends (or intercepts) an **arc of the circle**. Figure 5.4 shows three circles with radii of different lengths ($r_1 < r_2 < r_3$) and the same central angle θ (theta) subtending the arc lengths s_1, s_2, and s_3. Regardless of the size of the circle, the ratio of arc length, s, to radius, r, for a given circle will be constant. For the angle θ in Figure 5.4, $\frac{s_1}{r_1} = \frac{s_2}{r_2} = \frac{s_3}{r_3}$. Because this ratio is an arc length divided by another length (radius), it is just an ordinary real number and has no units.

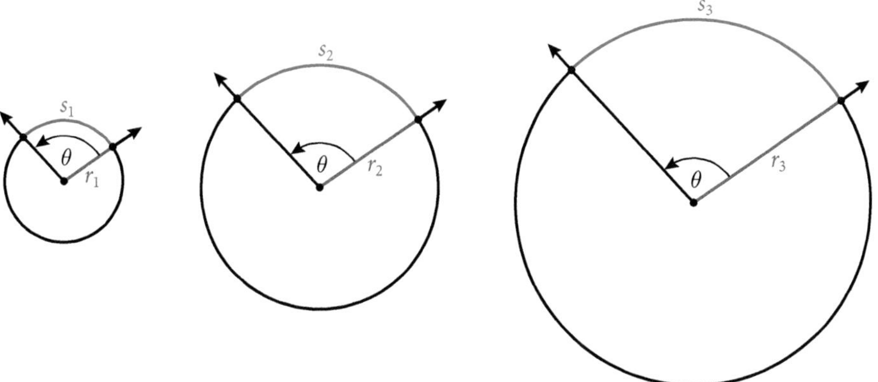

Figure 5.4 Different circles with the same central angle θ subtending different arcs, but the ratio of arc length to radius remains constant

Major and minor arcs

If a central angle is less than 180°, then the subtended arc is referred to as a minor arc. If a central angle is greater than 180°, then the subtended arc is referred to as a major arc.

The ratio $\frac{s}{r}$ indicates how many radius lengths, r, fit into the length of the arc s. For example, if $\frac{s}{r} = 2$, then the length of s is equal to two radius lengths. This accounts for the name **radian** and leads to the following definition.

One radian is the measure of a central angle θ of a circle that subtends an arc s of the circle that is exactly the same length as the radius r of the circle. That is, when $\theta = 1$ radian, arc length = radius.

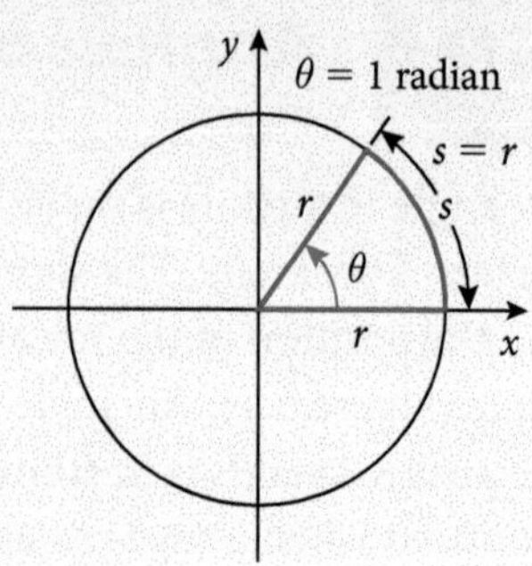

The unit circle

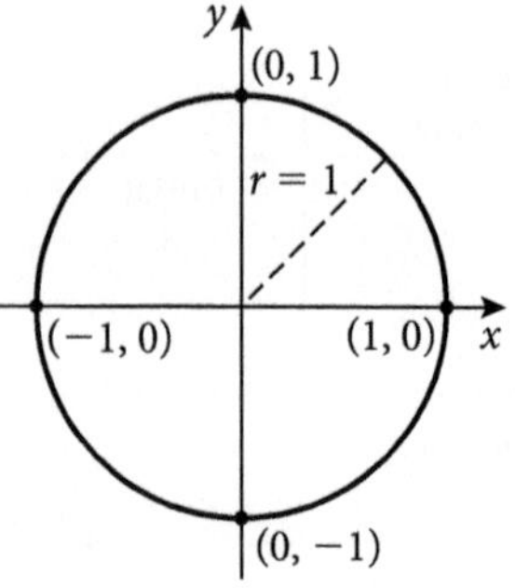

Figure 5.5 The unit circle

Conversion between degrees and radians

Because $180° = \pi$ radians, then $1° = \frac{\pi}{180}$ radians, and 1 radian $= \frac{180°}{\pi}$.

An angle with a radian measure of 1 has a degree measure of approximately 57.3° (to three significant figures).

When an angle is measured in radians it makes sense to draw it so that it is in standard position. It follows that the angle will be a central angle of a circle whose centre is at the origin. As Figure 5.4 illustrated, it makes no difference what size circle is used. The most practical circle to use is the circle with a radius of one unit so the radian measure of an angle will simply be equal to the length of the subtended arc.

$$\text{radian measure: } \theta = \frac{s}{r}$$

$$\text{If } r = 1, \text{ then } \theta = \frac{s}{1} = s$$

The circle with a radius of one unit and centre at the origin (0, 0) is called the **unit circle** (Figure 5.5). The equation for the unit circle is $x^2 + y^2 = 1$. Because the circumference of a circle with radius r is $2\pi r$, then a central angle of one full anti-clockwise revolution (360°) subtends an arc on the unit circle equal to 2π units. Hence, if an angle has a degree measure of 360°, then its radian measure is exactly 2π. It follows that an angle of 180° has a radian measure of exactly π. This fact can be used to convert between degree measure and radian measure and vice versa.

Example 5.1

(a) Convert 30° and 45° to radian measure and sketch the corresponding arc on the unit circle.

(b) Use these results to convert 60° and 90° to radian measure.

Solution

(a) Note that the 'degree' units cancel.

$$30° = 30°\left(\frac{\pi}{180°}\right) = \frac{30°}{180°}\pi = \frac{\pi}{6}$$

$$45° = 45°\left(\frac{\pi}{180°}\right) = \frac{45°}{180°}\pi = \frac{\pi}{4}$$

(b) Since $60° = 2(30°)$ and $30° = \frac{\pi}{6}$, then $60° = 2\left(\frac{\pi}{6}\right) = \frac{\pi}{3}$.

Similarly, $90° = 2(45°)$ and $45° = \frac{\pi}{4}$, then $90° = 2\left(\frac{\pi}{4}\right) = \frac{\pi}{2}$

The results given in Example 5.1 are very useful and often appear in problems and applications. Knowing these four facts can help you to quickly convert mentally between degrees and radians for many common angles.
For example, to convert 225° to radians, apply the fact that $225° = 5(45°)$. Since $45° = \frac{\pi}{4}$, then

$$225° = 5(45°) = 5\left(\frac{\pi}{4}\right) = \frac{5\pi}{4}.$$

This is also helpful for conversions in the opposite direction. For example, to convert $\frac{11\pi}{6}$ to degrees:

$$\frac{11\pi}{6} = 11\left(\frac{\pi}{6}\right) = 11(30°) = 330°.$$

Example 5.2

(a) Convert the following radian measures to degrees. Express exactly, if possible. Otherwise express accurate to three significant figures.

(i) $\frac{4\pi}{3}$ (ii) $-\frac{3\pi}{2}$ (iii) 5 (iv) 1.38

(b) Convert the following degree measures to radians. Express exactly.

(i) 135° (ii) −150° (iii) 175° (iv) 10°

Solution

(a) (i) $\frac{4\pi}{3} = 4\left(\frac{\pi}{3}\right) = 4(60°) = 240°$

(ii) $-\frac{3\pi}{2} = -\frac{3}{2}(\pi) = -\frac{3}{2}(180°) = -270°$

(iii) $5\left(\frac{180°}{\pi}\right) \approx 286.479° \approx 286°$

(iv) $1.38\left(\frac{180°}{\pi}\right) \approx 79.068° \approx 79.1°$

(b) (i) $135° = 3(45°) = 3\left(\frac{\pi}{4}\right) = \frac{3\pi}{4}$

(ii) $-150° = -5(30°) = -5\left(\frac{\pi}{6}\right) = -\frac{5\pi}{6}$

(iii) $175°\left(\frac{\pi}{180°}\right) = \frac{29\pi}{30}$ (iv) $10°\left(\frac{\pi}{180°}\right) = \frac{\pi}{18}$

Because 2π is 6.28 (to three significant figures), there are a little more than six radius lengths in one revolution, as shown in Figure 5.6.

Figure 5.7 shows all of the angles between 0° and 360°, inclusive, that are multiples of 30° or 45°, and their equivalent radian measure. You will benefit by being able to quickly convert between degree measure and radian measure for these common angles.

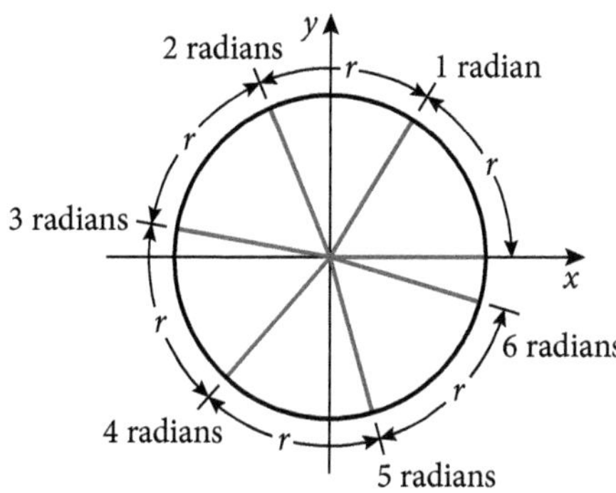

Figure 5.6 Arcs with lengths equal to the radius placed along the circumference of a circle

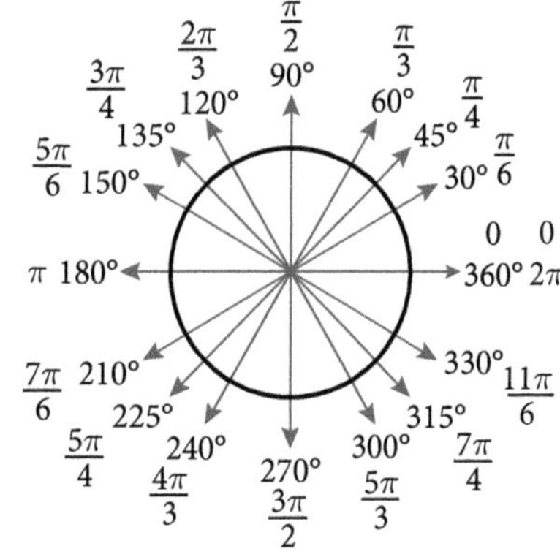

Figure 5.7 Degree measure and radian measure for common angles

All GDCs will have a degree mode and a radian mode. Before doing any calculations with angles on your GDC, be certain that the mode setting for angle measurement is set correctly. Although you may be more familiar with degree measure, as you progress further in mathematics, and especially in calculus, radian measure is far more useful.

Arc length

For any angle θ, its radian measure is given by $\theta = \frac{s}{r}$. Simple rearrangement of this formula leads to another formula for computing arc length.

For a circle of radius r, a central angle θ subtends an arc of the circle of length s given by

$$s = r\theta$$

where θ is measured in radians.

Example 5.3

A circle has a radius of 10 centimetres. Find the length of the arc of the circle subtended by a central angle of 150°.

Solution

To use the formula $s = r\theta$, we must first convert 150° to radian measure.

$$150° = 150°\left(\frac{\pi}{180°}\right) = \frac{150\pi}{180} = \frac{5\pi}{6}$$

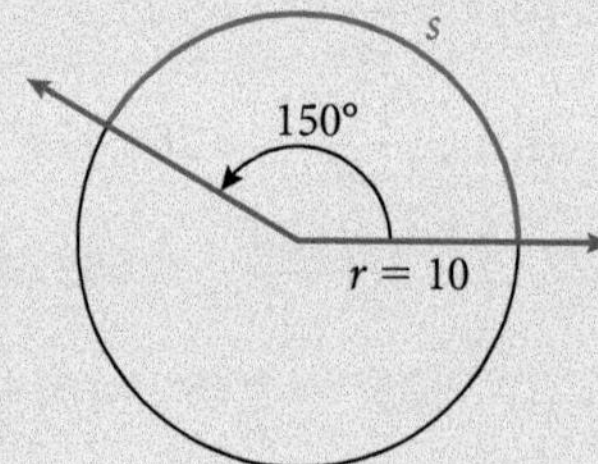

Substituting $r = 10$ cm into the formula:

$$s = r\theta \Rightarrow s = 10\left(\frac{5\pi}{6}\right) = \frac{25\pi}{3} \approx 26.17994$$

The length of the arc is 26.18 cm (4 significant figures).

The units of the product $r\theta$ are equal to the units of r because θ has no units in radian measure.

Example 5.4

The diagram shows a circle of centre O with radius $r = 6$ cm. Angle AOB subtends the minor arc AB such that the length of the arc is 10 cm. Find the size of angle AOB in degrees to three significant figures.

Solution

Rearrange the arc length formula $s = r\theta$, to make θ the subject:

$$\theta = \frac{s}{r}$$

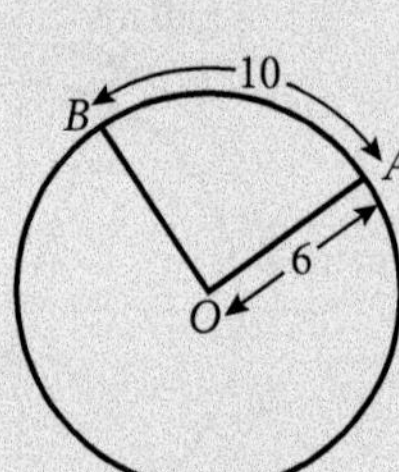

Remember that θ will be in radians. Therefore,

angle $AOB = \frac{10}{6} = \frac{5}{3}$ or $1.\dot{6}$ radians.

Now, we convert to degrees: $\frac{5}{3}\left(\frac{180°}{\pi}\right) \approx 95.49297°$.

The degree measure of angle AOB is approximately 95.5°.

Geometry of a circle

inscribed circle of a polygon – the radius is perpendicular to the side of the polygon at the point of tangency

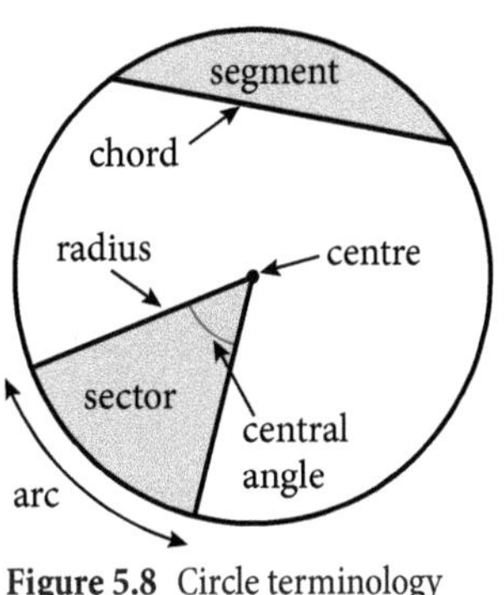

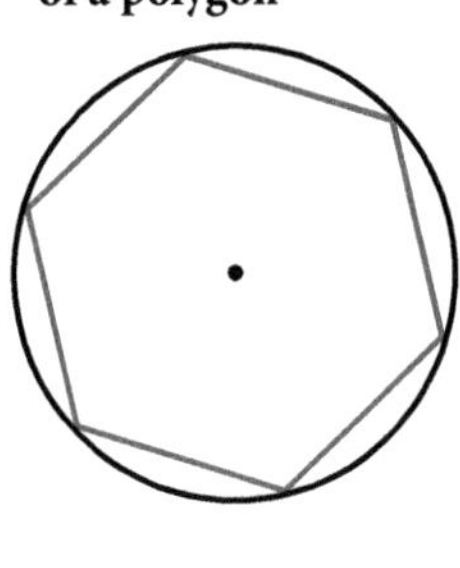

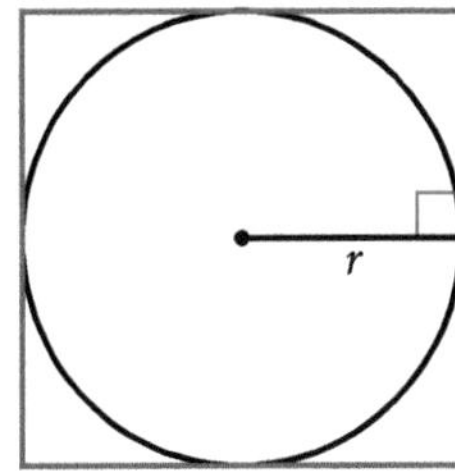

Figure 5.8 Circle terminology

Sector of a circle

A sector of a circle is the region bounded by an arc of the circle and the two sides of a central angle (Figure 5.8). The ratio of the area of a sector to the area of the circle, πr^2, is equal to the ratio of the length of the subtended arc to the circumference of the circle, $2\pi r$. If s is the arc length and A is the area of the sector, then we can write $\frac{A}{\pi r^2} = \frac{s}{2\pi r}$.

Solving for A gives $A = \frac{\pi r^2 s}{2\pi r} = \frac{1}{2}rs$.

We already know that for arc length, $s = r\theta$, where θ is the central angle in radians.

Substituting $r\theta$ for s, gives the area of a sector to be

$$A = \frac{1}{2}rs = \frac{1}{2}r(r\theta) = \frac{1}{2}r^2\theta$$

If the sector is the entire circle then $\theta = 2\pi$, and the area is

$$A = \frac{1}{2}r^2\theta = \frac{1}{2}r^2(2\pi) = \pi r^2,$$

which is the formula for the area of a circle.

Area of a sector

In a circle of radius r, the area of a sector with a central angle θ measured in radians is

$$A = \frac{1}{2}r^2\theta$$

Example 5.5

A circle of radius 9 cm has a sector whose central angle is $\frac{2\pi}{3}$. Find the exact values of:

(a) the length of the arc subtended by the central angle

(b) the area of the sector.

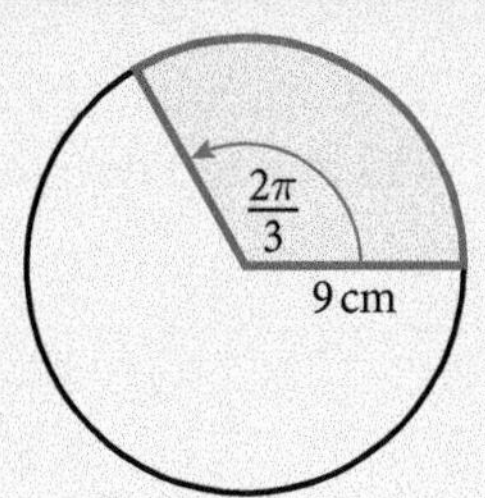

Solution

(a) $s = r\theta \Rightarrow s = 9\left(\frac{2\pi}{3}\right) = 6\pi$

The length of the arc is exactly 6π cm.

(b) $A = \frac{1}{2}r^2\theta \Rightarrow A = \frac{1}{2}(9)^2\left(\frac{2\pi}{3}\right) = 27\pi$

The area of the sector is exactly $27\pi\,\text{cm}^2$.

The formula for arc length $s = r\theta$, and the formula for area of a sector $A = \frac{1}{2}r^2\theta$, are true only when θ is in radians.

Exercise 5.1

1. Convert each angle into radians. Give your answers in exact form.

(a) 60° **(b)** 150° **(c)** −270°
(d) 36° **(e)** 135° **(f)** 50°
(g) −45° **(h)** 400° **(i)** −480°

2. Convert each angle to degrees. If possible, express exactly, otherwise express accurate to three significant figures.

(a) $\frac{3\pi}{4}$ **(b)** $-\frac{7\pi}{2}$ **(c)** 2
(d) $\frac{7\pi}{6}$ **(e)** −2.5 **(f)** $\frac{5\pi}{3}$
(g) $\frac{\pi}{12}$ **(h)** 1.57 **(i)** $\frac{8\pi}{3}$

3. The size of an angle in standard position is given. Find two angles, one positive and one negative, that are coterminal with each given angle. If no units are given, assume the angle is in radian measure.

(a) 30° **(b)** $\frac{3\pi}{2}$ **(c)** 175°
(d) $-\frac{\pi}{6}$ **(e)** $\frac{5\pi}{3}$ **(f)** 3.25

4. Find the length of the arc s in each diagram.

(a)

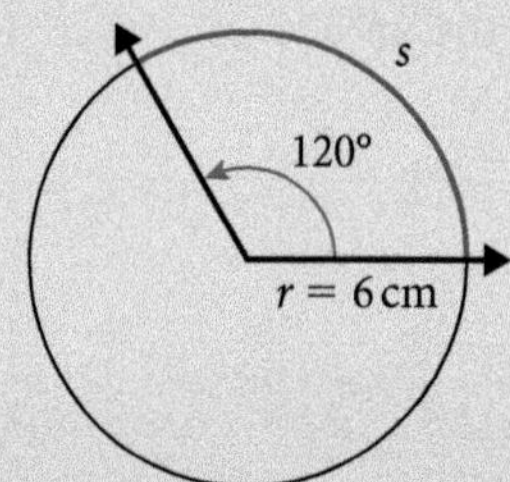

(b)

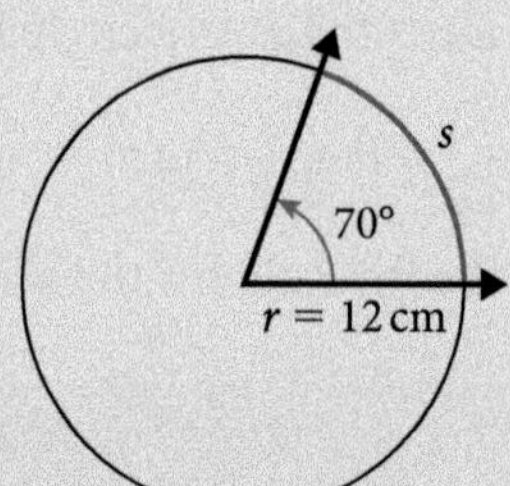

5. Find the angle θ in both radians and degrees.

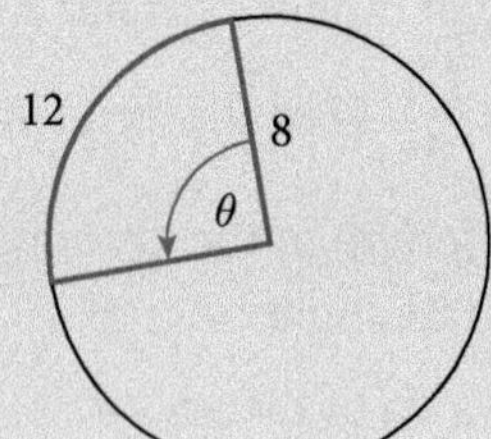

6. Find the radius, r, of the circle in the diagram.

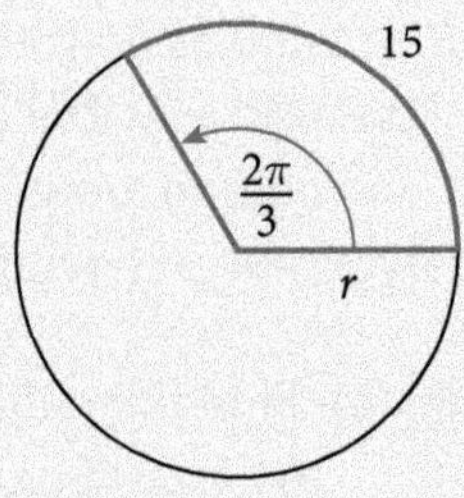

7. Find the area of the sector in each diagram.

(a)

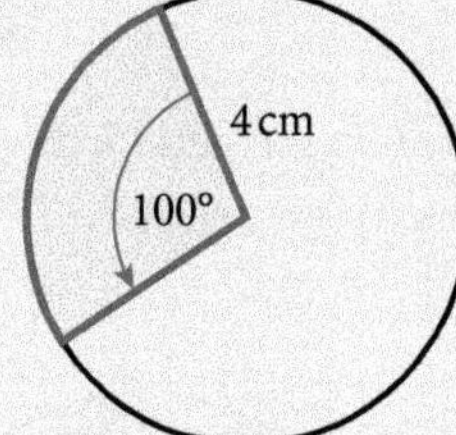

(b)

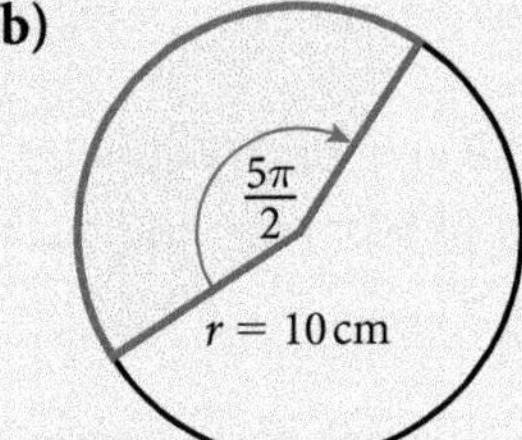

8. An arc of length 60 cm subtends a central angle α in a circle of radius 20 cm. Find the size of α in both degrees and radians. Give your answer to three significant figures.

9. Find the length of an arc that subtends a central angle of 2 radians in a circle of radius 16 cm.

10. The area of a sector of a circle with a central angle of 60° is 24 cm^2. Find the radius of the circle.

11. A bicycle with tyres 70 cm in diameter is travelling such that its tyres complete one and a half revolutions every second. That is, the angular velocity of the wheel is 1.5 revolutions per second.
    (a) What is the angular velocity of a wheel in radians per second?
    (b) At what speed (in km per hour) is the bicycle travelling along the ground? (This is the linear velocity of any point on the tyre that touches the ground.)

12. A bicycle with tyres 70 cm in diameter is travelling along a road at 25 km h^{-1}. What is the angular velocity of a wheel of the bicycle in radians per second?

13. Given that ω is the angular velocity, in radians per second, of a point on a circle with radius r cm, express the linear velocity, v, in cm s^{-1}, of the point as a function in terms of ω and r.

14. A chord of 26 cm is in a circle of radius 20 cm. Find the length of the arc that the chord subtends.

15. A circular irrigation system consists of a 400 m pipe that is rotated around a central pivot point. If the irrigation pipe makes one full revolution around the pivot point in a day, what area, in m², does it irrigate each hour?

16. (a) Find the radius of a circle circumscribed about a regular polygon of 64 sides if one side is 3 cm.

(b) Find the difference between the circumference of the circle and the perimeter of the polygon.

17. Find the area of an equilateral triangle that has an inscribed circle with an area of $50\pi\,\text{cm}^2$, and a circumscribed circle with an area of $200\pi\,\text{cm}^2$.

18. In the diagram, the sector of a circle is subtended by two perpendicular radii. The area of the shaded segment is A square units. Find an expression for the area of the circle in terms of A.

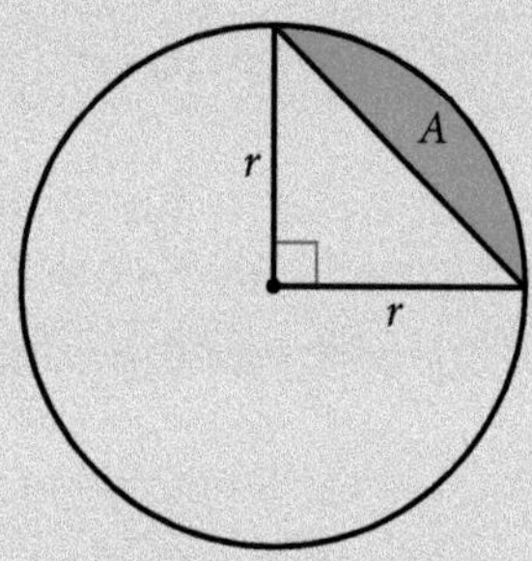

5.2 The unit circle and trigonometric functions

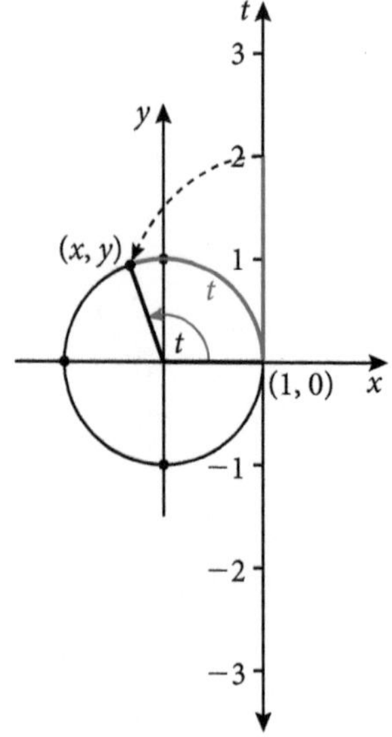

Figure 5.9 The wrapping function

Several important functions can be described by mapping the coordinates of points on the real number line onto points on the unit circle. Recall from Section 5.1 that the unit circle has its centre at (0, 0), a radius of one unit, and equation $x^2 + y^2 = 1$.

Suppose that the real number line is tangent to the unit circle at the point (1, 0), and that zero on the number line matches with (1, 0) on the circle (Figure 5.9). Because of the properties of circles, the real number line in this position will be perpendicular to the x-axis. The scales on the number line, the x-axis, and the y-axis need to be the same. Imagine that the real number line is flexible like a string and can wrap around the circle, with zero on the number line remaining fixed to the point (1, 0) on the unit circle. When the top portion of the string moves along the circle, the wrapping is anti-clockwise ($t > 0$), and when the bottom portion of the string moves along the circle, the wrapping is clockwise ($t < 0$).

As the string wraps around the unit circle, each real number t on the string is mapped onto a point (x, y) on the circle. Hence the real number line from 0 to t makes an arc of length t starting on the circle at $(1, 0)$ and ending at the point (x, y) on the circle. For example, since the circumference of the unit circle is 2π the number $t = 2\pi$ will be wrapped anti-clockwise around the circle to the point $(1, 0)$. Similarly, the number $t = \pi$ will be wrapped anti-clockwise halfway around the circle to the point $(-1, 0)$ on the circle. And the number $t = -\frac{\pi}{2}$ will be wrapped clockwise one-quarter of the way around the circle to the point $(0, -1)$ on the circle. Note that each number t on the real number line is mapped (corresponds) to *exactly one* point on the unit circle, thereby satisfying the definition of a function. Consequently, this mapping is called a **wrapping function**.

A function f such that $f(x) = f(x + p)$ is a **periodic function**. If p is the least positive constant for which $f(x) = f(x + p)$ is true, then p is called the **period** of the function.

We are surrounded by periodic functions. A few examples include: the average daily temperature during the year, sunrise and the day of the year, animal populations over many years, the height of tides and the position of the Moon, and electrocardiograms, which give a visual image of the heart's electrical activity.

Before we leave our mental picture of the string (representing the real number line) wrapping around the unit circle, consider any pair of points on the string that are exactly 2π units from each other. Let these two points represent the real numbers t_1 and $t_1 + 2\pi$. Because the circumference of the unit circle is 2π, these two numbers will be mapped to the same point on the unit circle. Furthermore, consider the infinite number of points whose distance from t_1 is any integer multiple of 2π – that is, $t_1 + k \cdot 2\pi, k \in \mathbb{Z}$. Again, all of these numbers will be mapped to the same point on the unit circle. Consequently, the wrapping function is not a one-to-one function. Output for the function (points on the unit circle) are unchanged by the addition of any integer multiple of 2π to any input value (a real number). Functions that behave in such a repetitive (or cyclic) manner are called **periodic**.

Trigonometric functions

The x- and y-coordinates of the points on the unit circle can be used to define the trigonometric functions **sine, cosine**, and **tangent**. The names of these functions are usually abbreviated as **sin, cos**, and **tan**, respectively.

When the real number t is wrapped to a point (x, y) on the unit circle, the value of the y-coordinate is assigned to the sine function; the x-coordinate is assigned to the cosine function; and the ratio of the two coordinates, $\frac{y}{x}$, is assigned to the tangent function. Sine, cosine, and tangent are defined by means of the length of an arc on the unit circle as follows.

Let t be any real number and (x, y) a point on the unit circle to which t is mapped. Then the definitions of the trigonometric functions are:

$\sin t = y \quad \cos t = x \quad \tan t = \frac{y}{x}, x \neq 0$

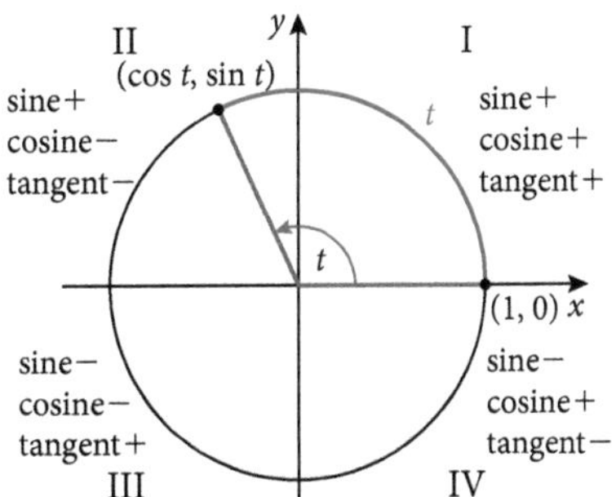

On the unit circle: $x = \cos t$, $y = \sin t$.

Figure 5.10 Signs of the trigonometric functions depend on the quadrant where the arc t terminates

When trigonometric functions are defined as circular functions based on the unit circle, radian measure is used. The values for the domain of the sine and cosine functions are real numbers that are arc lengths on the unit circle. As we know from the previous section, the arc length on the unit circle subtends an angle in standard position whose radian measure is equivalent to the arc length (see Figure 5.10).

Evaluating the trigonometric functions for any value of t involves finding the coordinates of the point on the unit circle where the arc of length t will 'wrap to' (or terminate) starting at the point (1, 0). It is useful to remember that an arc of length π is equal to one-half of the circumference of the unit circle. All of the values for t in Example 5.6 are positive, so the arc length will wrap along the unit circle in an anti-clockwise direction.

We can use the definitions for the three trigonometric, or circular, functions to evaluate them for some 'easy' values of t.

Example 5.6

Evaluate the three trigonometric functions for each value of t.

(a) $t = 0$ (b) $t = \frac{\pi}{2}$ (c) $t = \pi$ (d) $t = \frac{3\pi}{2}$ (e) $t = 2\pi$

Solution

(a) An arc of length $t = 0$ has no length so it terminates at the point (1, 0). By definition:

$\sin 0 = y = 0$ $\cos 0 = x = 1$ $\tan 0 = \frac{y}{x} = \frac{0}{1} = 0$

(b) An arc of length $t = \frac{\pi}{2}$ is equivalent to one-quarter of the circumference of the unit circle, so it terminates at the point (0, 1). By definition:

$\sin\frac{\pi}{2} = y = 1$

$\cos\frac{\pi}{2} = x = 0$

$\tan\frac{\pi}{2} = \frac{y}{x} = \frac{1}{0}$ is undefined

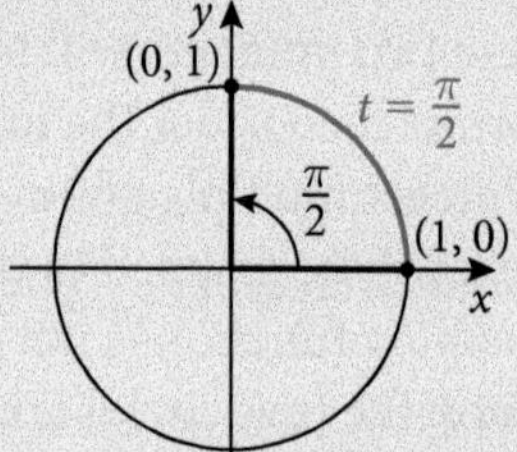

Arc length of $\frac{\pi}{2}$ or one-quarter of an anti-clockwise revolution

(c) An arc of length $t = \pi$ is equivalent to one-half of the circumference of the unit circle, so it terminates at the point $(-1, 0)$. By definition:

$\sin\pi = y = 0$

$\cos\pi = x = -1$

$\tan\pi = \frac{y}{x} = \frac{0}{-1} = 0$

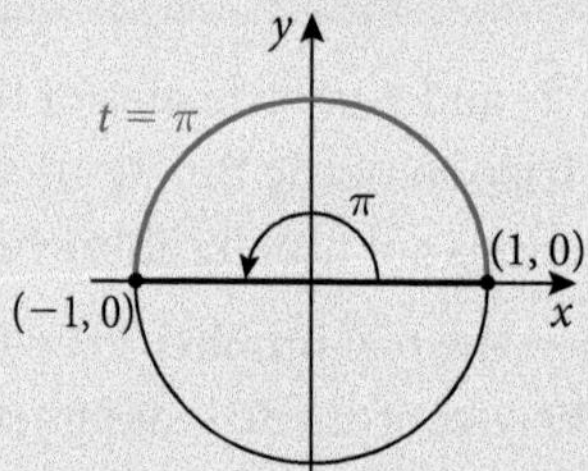

Arc length of π or one-half of an anti-clockwise revolution

(d) An arc of length $t = \frac{3\pi}{2}$ is equivalent to three-quarters of the circumference of the unit circle, so it terminates at the point $(0, -1)$. By definition:

$\sin\frac{3\pi}{2} = y = -1$

$\cos\frac{3\pi}{2} = x = 0$

$\tan\frac{3\pi}{2} = \frac{y}{x} = \frac{-1}{0}$ is undefined

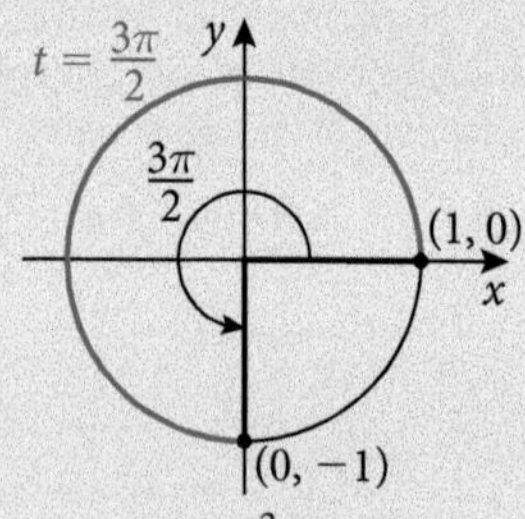

Arc length of $\frac{3\pi}{2}$ or three-quarters of an anti-clockwise revolution

(e) An arc of length $t = 2\pi$ terminates at the same point as an arc of length $t = 0$, so the values of the trigonometric function are the same as those found in part (a):

$\sin 0 = y = 0 \qquad \cos 0 = x = 1$

$\tan 0 = \frac{y}{x} = \frac{0}{1} = 0$

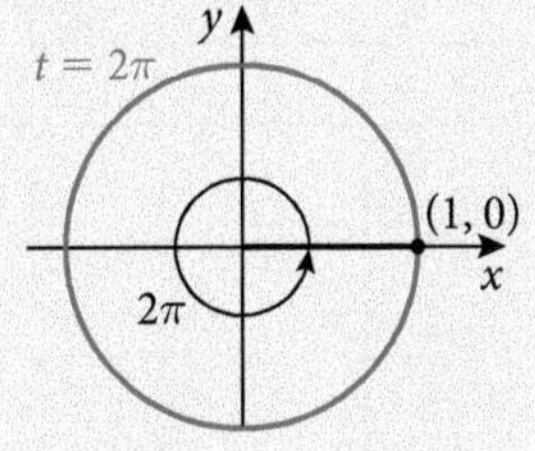

Arc length of 2π or one full anti-clockwise revolution

Domains of the three trigonometric functions

$f(t) = \sin t$ and $f(t) = \cos t$ domain: $\{t : t \in \mathbb{R}\}$

$f(t) = \tan t$ domain: $\left\{t : t \in \mathbb{R}, t \neq \frac{\pi}{2} + k\pi, k \in \mathbb{Z}\right\}$

From our previous discussion of periodic functions, we can conclude that all of the trigonometric functions are periodic. Given that the sine and cosine functions are generated directly from the wrapping function, the period of each of these functions is 2π. That is,

$$\sin t = \sin(t + k \cdot 2\pi), k \in \mathbb{Z} \text{ and } \cos t = \cos(t + k \cdot 2\pi), k \in \mathbb{Z}$$

Initial evidence from Example 5.6 indicates that the period of the tangent function is π. That is, $\tan t = \tan(t + k \cdot \pi), k \in \mathbb{Z}$.

Evaluating trigonometric functions

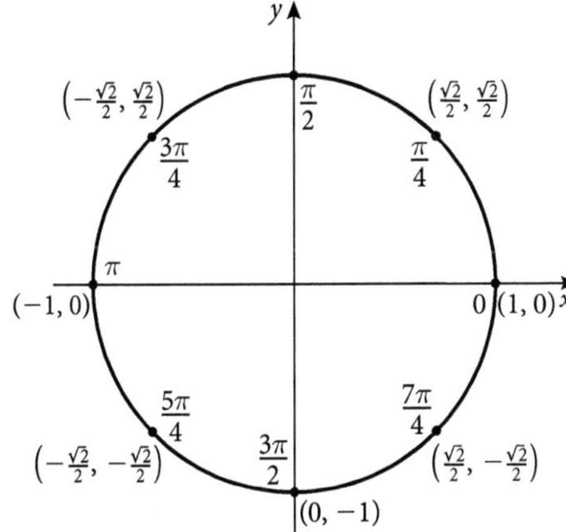

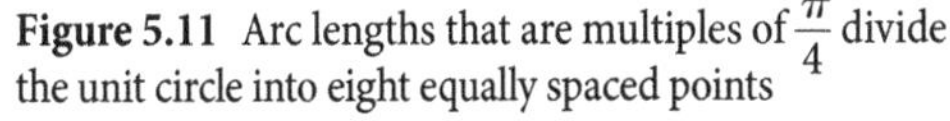

Figure 5.11 Arc lengths that are multiples of $\frac{\pi}{4}$ divide the unit circle into eight equally spaced points

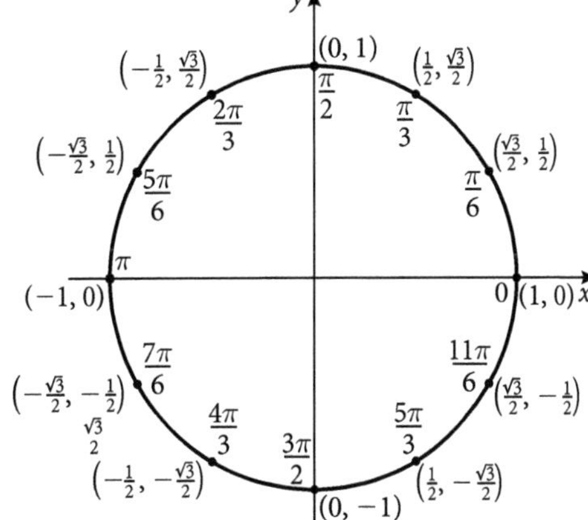

Figure 5.12 Arc lengths that are multiples of $\frac{\pi}{6}$ divide the unit circle into twelve equally spaced points

You will find it very helpful to know the exact values of sine and cosine for numbers that are multiples of $\frac{\pi}{6}$ and $\frac{\pi}{4}$ from memory. Use the unit circle diagrams shown in Figures 5.11 and 5.12 as a guide to help you do this and to visualise the location of the terminal points of different arc lengths. With the symmetry of the unit circle and a point's location in the coordinate plane telling us the sign of x and y (see Figure 5.10), we need to remember only the sine and cosine of common values of t in the first quadrant and on the positive x- and y-axes. These are organised in Table 5.1.

t	$\sin t$	$\cos t$	$\tan t$
0	0	1	0
$\frac{\pi}{6}$	$\frac{1}{2}$	$\frac{\sqrt{3}}{2}$	$\frac{\sqrt{3}}{3}$
$\frac{\pi}{4}$	$\frac{\sqrt{2}}{2}$	$\frac{\sqrt{2}}{2}$	1
$\frac{\pi}{3}$	$\frac{\sqrt{3}}{2}$	$\frac{1}{2}$	$\sqrt{3}$
$\frac{\pi}{2}$	1	0	undefined

Table 5.1 The sine, cosine, and tangent functions evaluated for special values of t

If t is not a multiple of $\frac{\pi}{6}, \frac{\pi}{4}, \frac{\pi}{3}$, or $\frac{\pi}{2}$ then the approximate values of the trigonometric functions can be found using your GDC.

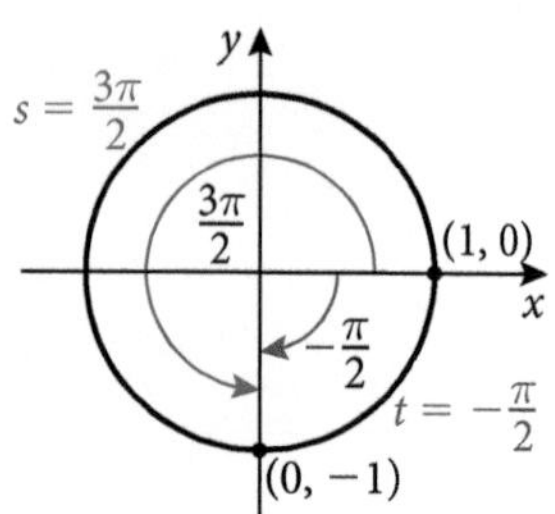

Figure 5.13 Coterminal arcs $\frac{3\pi}{2}$ and $-\frac{\pi}{2}$

If s and t are coterminal arcs (i.e. terminate at the same point) then the trigonometric functions of s are equal to those of t. That is, $\sin s = \sin t$, $\cos s = \cos t$, and so on. For example, the arcs $s = \frac{3\pi}{2}$ and $t = -\frac{\pi}{2}$ are coterminal (Figure 5.13). Thus, $\sin\frac{3\pi}{2} = \sin\left(-\frac{\pi}{2}\right)$, $\tan\frac{3\pi}{2} = \tan\left(-\frac{\pi}{2}\right)$, and so on.

Exercise 5.2

1. By knowing the ratios of sides in any triangle with angles measuring 30°, 60° and 90° (see diagram), find the coordinates of the points on the unit circle where an arc of length $t = \frac{\pi}{6}$ and $t = \frac{\pi}{3}$ terminate in the first quadrant.

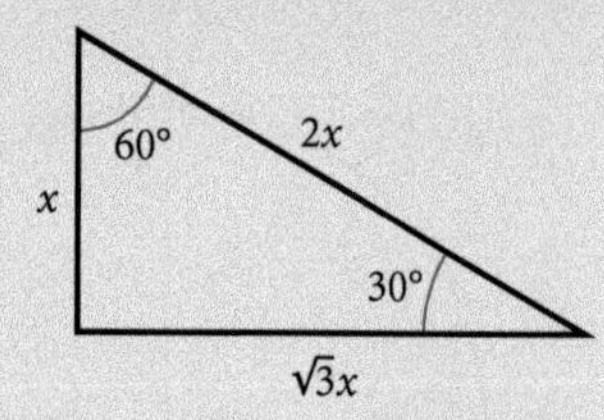

2. The diagram of quadrant I of the unit circle shown indicates angles in intervals of 10 degrees and also indicates angles in radian measure of 0.5, 1, and 1.5. Use the diagram and the definitions of the sine and cosine functions to approximate each function value to one decimal place. Check your answers with your GDC (be sure to be in the correct angle measure mode).

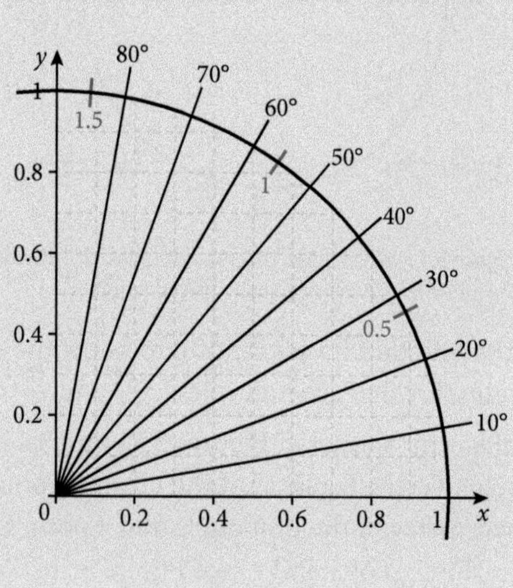

(a) $\cos 50°$ **(b)** $\sin 80°$ **(c)** $\cos 1$

(d) $\sin 0.5$ (e) $\tan 70°$ (f) $\cos 1.5$

(g) $\sin 20°$ (h) $\tan 1$

3. t is the length of an arc on the unit circle starting from (1, 0). For each value of t:

(i) State the quadrant in which the terminal point of the arc lies.

(ii) Find the coordinates of the terminal point (x, y) on the unit circle. Give exact values for x and y if possible. Otherwise approximate to three significant figures.

(a) $t = \frac{\pi}{6}$ (b) $t = \frac{5\pi}{3}$ (c) $t = \frac{7\pi}{4}$

(d) $t = \frac{3\pi}{2}$ (e) $t = 2$ (f) $t = -\frac{\pi}{4}$

(g) $t = -1$ (h) $t = -\frac{5\pi}{4}$ (i) $t = 3.52$

4. State the exact value (if possible) of the sine, cosine, and tangent of each real number.

(a) $\frac{\pi}{3}$ (b) $\frac{5\pi}{6}$ (c) $-\frac{3\pi}{4}$

(d) $\frac{\pi}{2}$ (e) $-\frac{4\pi}{3}$ (f) 3π

(g) $\frac{3\pi}{2}$ (h) $-\frac{7\pi}{6}$ (i) $t = 1.25\pi$

5. Use the periodic properties of the sine and cosine functions to find the exact value of $\sin x$ and $\cos x$.

(a) $x = \frac{13\pi}{6}$ (b) $x = \frac{10\pi}{3}$

(c) $x = \frac{15\pi}{4}$ (d) $x = \frac{17\pi}{6}$

6. Find the exact function values, if possible. Do not use your GDC.

(a) $\cos\frac{5\pi}{6}$ (b) $\sin 315°$ (c) $\tan\frac{3\pi}{2}$

7. Find the exact function values, if possible. Otherwise, find the approximate value accurate to three significant figures.

(a) $\sin 2.5$ (b) $\cos\frac{5\pi}{4}$ (c) $\tan \pi$

8. Specify in which quadrant(s) an angle θ in standard position could be for each set of conditions.

(a) $\sin\theta > 0$ (b) $\sin\theta > 0$ and $\cos\theta < 0$

(c) $\sin\theta < 0$ and $\tan\theta > 0$ (d) $\cos\theta < 0$ and $\tan\theta < 0$

(e) $\cos\theta > 0$

5.3 Graphs of trigonometric functions

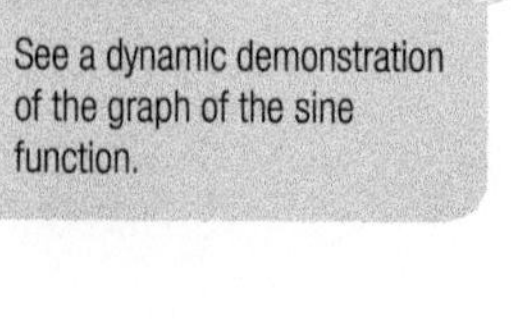

Figure 5.14 The period of $y = \sin x$ is 2π

From the previous section, we know that trigonometric functions are periodic; that is, their values repeat in a regular manner. The graphs of the trigonometric functions should provide a picture of this periodic behaviour. In this section, we will graph the sine, cosine, and tangent functions and transformations of the sine and cosine functions.

Graphs of the sine and cosine functions

Online

See a dynamic demonstration of the graph of the sine function.

Since the period of the sine function is 2π, we know that two values of t (domain) that differ by 2π will produce the same value for y (range). This means that any portion of the graph of $y = \sin t$ with a t-interval of length 2π (called one **period** or **cycle** of the graph) will repeat. Remember that the domain of the sine function is all real numbers, so one period of the graph of $y = \sin t$ will repeat indefinitely in the positive and negative direction. Therefore, in order to construct a complete graph of $y = \sin t$, we need to graph just one period of the function; for example, from $t = 0$ to $t = 2\pi$, and then repeat the pattern in both directions.

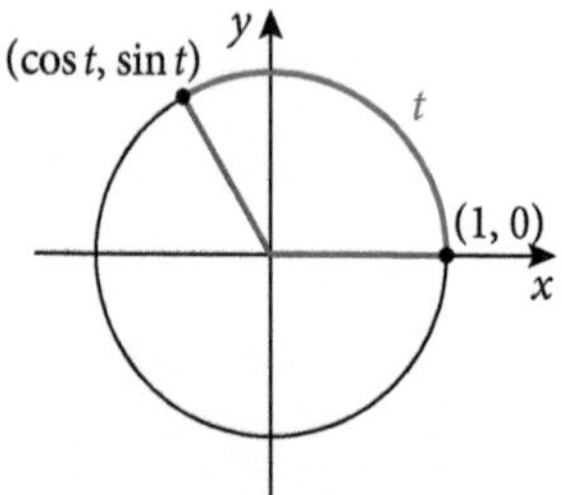

Figure 5.15 The coordinates of the terminal point of arc t give the values of $\cos t$ and $\sin t$

We know from the previous section that $\sin t$ is the y-coordinate of the terminal point on the unit circle corresponding to the real number t (Figure 5.10). In order to generate one period of the graph of $y = \sin t$, we need to record the y-coordinates of a point on the unit circle and the corresponding value of t as the point travels anti-clockwise one revolution starting from the point (1, 0). These values are then plotted on a graph with t on the horizontal axis and y ($\sin t$) on the vertical axis. Figure 5.16 illustrates this process in a sequence of diagrams.

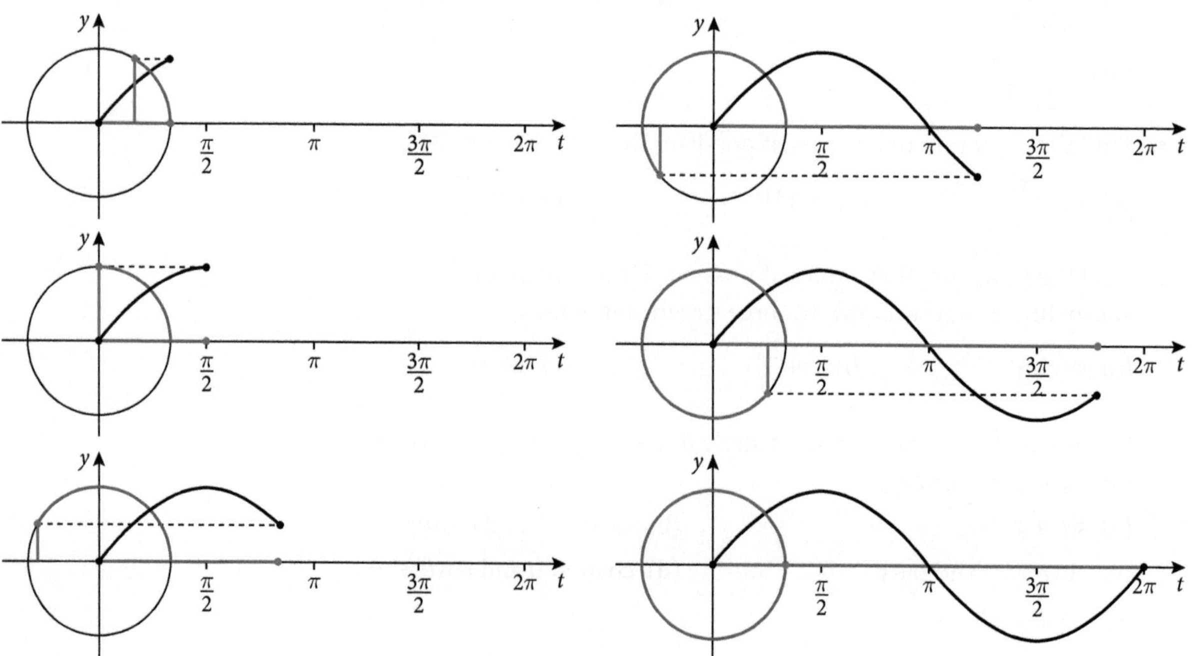

Figure 5.16 Graph of the sine function for $0 \leqslant t \leqslant 2\pi$ generated from a point travelling along the unit circle

As the point $(\cos t, \sin t)$ travels along the unit circle, the x-coordinate $(\cos t)$ goes through the same cycle of values as the y-coordinate $(\sin t)$ does. The only difference is that the x-coordinate begins at a different place in the cycle; when $t = 0$, $y = 0$, but $x = 1$. The result is that the graph of $y = \cos t$ is the exact same shape as $y = \sin t$ but has been shifted to the left by $\frac{\pi}{2}$ units. The graph of $y = \cos t$ for $0 \leqslant t \leqslant 2\pi$ is shown in Figure 5.17.

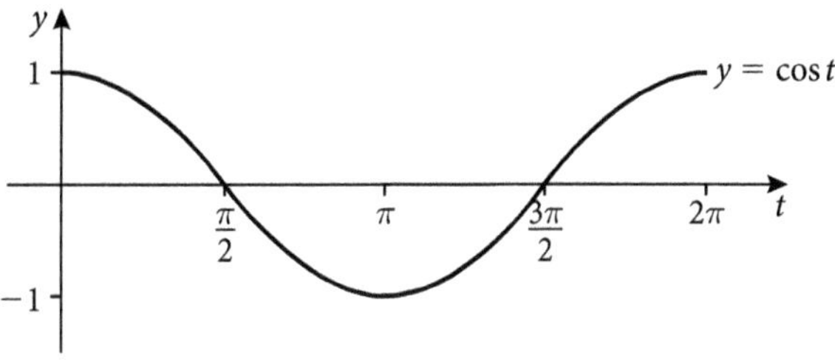

Figure 5.17 Graph of $y = \cos t$ for $0 \leqslant t \leqslant 2\pi$

The convention is to use the letter x to denote the variable in the domain of the function. Hence, from here on we will use the letter x rather than t and write the trigonometric functions as $y = \sin x$, $y = \cos x$, and $y = \tan x$.

Figure 5.18 shows the graphs of $y = \sin x$ and $y = \cos x$ for $-4\pi \leqslant x \leqslant 4\pi$

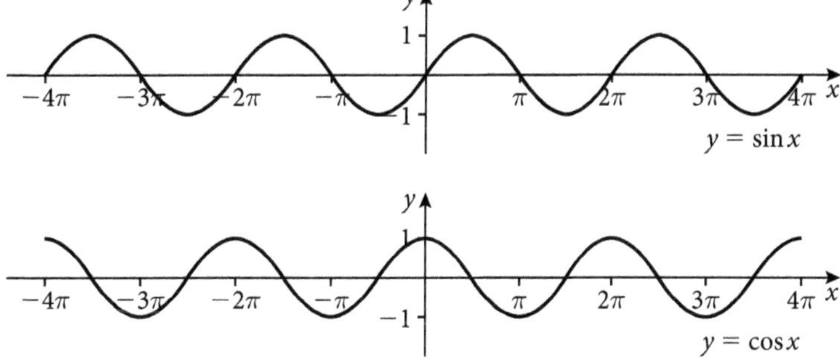

Figure 5.18 Graphs of $y = \sin x$ and $y = \cos x$ for $-4\pi \leqslant t \leqslant 4\pi$

Aside from their periodic behaviour, these graphs reveal further properties of the graphs of $y = \sin x$ and $y = \cos x$. Note that the sine function has a maximum value of $y = 1$ for all $x = \frac{\pi}{2} + k \cdot 2\pi, k \in \mathbb{Z}$, and has a minimum value of $y = -1$ for all $x = k \cdot 2\pi, k \in \mathbb{Z}$. The cosine function has a maximum value of $y = 1$ for all $x = k \cdot 2\pi, k \in \mathbb{Z}$, and has a minimum value of $y = -1$ for all $x = \pi + k \cdot 2\pi, k \in \mathbb{Z}$. This also confirms that both functions have a domain of all real numbers and a range of $-1 \leqslant y \leqslant 1$.

Online

See a dynamic demonstration of the graph of the cosine function.

Graphs of transformations of the sine and cosine functions

In Section 1.5, we learned how to transform the graph of a function by horizontal and vertical translations, by reflections in the coordinate axes, and by stretching and shrinking, both horizontally and vertically.

Review of transformations of graphs of functions
Assume that a, b, c, and d are real numbers.

To obtain the graph of	From the graph of $y = f(x)$
$y = f(x) + d$	Translate d units up for $d > 0$, d units down for $d < 0$
$y = f(x + c)$	Translate c units left for $c > 0$, c units right for $c < 0$
$y = -f(x)$	Reflect in the x-axis
$y = af(x)$	Vertical stretch $(a > 1)$ or shrink $(0 < a < 1)$ of factor a
$y = f(-x)$	Reflect in the y-axis
$y = f(bx)$	Horizontal stretch $(0 < b < 1)$ or shrink $(b > 1)$ of factor $\frac{1}{b}$

In this section we will look at the composition of sine and cosine functions of the form $f(x) = a\sin[b(x + c)] + d$ and $f(x) = a\cos[b(x + c)] + d$.

Example 5.7

Sketch the graph of each function on the interval $-\pi \leqslant x \leqslant 3\pi$.

(a) $f(x) = 2\cos x$

(b) $g(x) = \cos x + 3$

(c) $h(x) = 2\cos x + 3$

(d) $p(x) = \frac{1}{2}\sin x - 2$

Solution

(a) Since $a = 2$, the graph of $y = 2\cos x$ is obtained by stretching the graph of $y = \cos x$ vertically by a factor of 2.

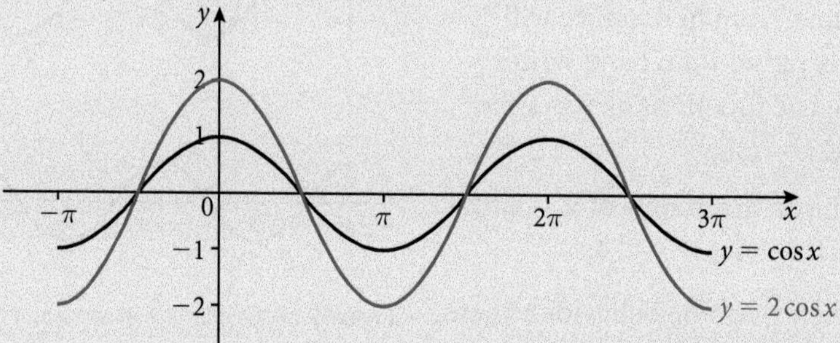

(b) Since $d = 3$, the graph of $y = \cos x + 3$ is obtained by translating the graph of $y = \cos x$ up by 3 units.

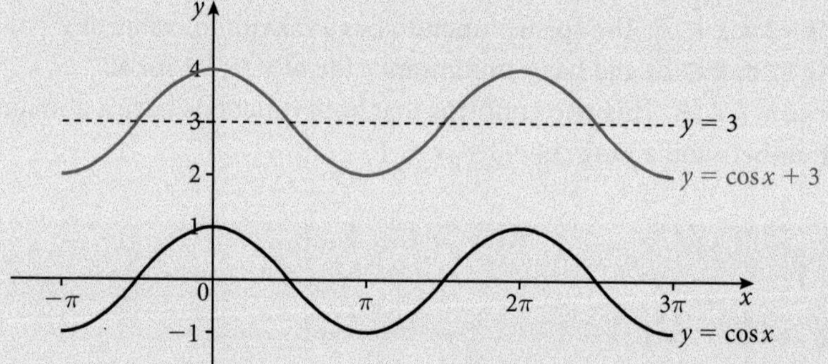

(c) We can obtain the graph of $y = 2\cos x + 3$ by combining both of the transformations to the graph of $y = \cos x$ performed in parts (a) and (b), namely, a vertical stretch of factor 2 and a translation 3 units up.

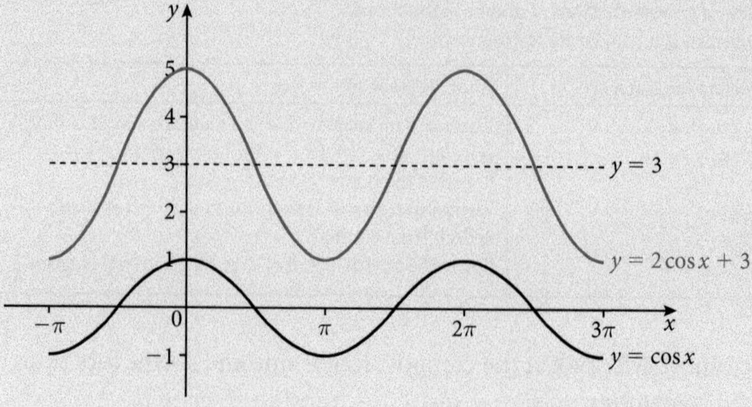

(d) The graph of $y = \frac{1}{2}\sin x - 2$ can be obtained by shrinking the graph of $y = \sin x$ vertically by a factor of $\frac{1}{2}$ and then translating it down 2 units.

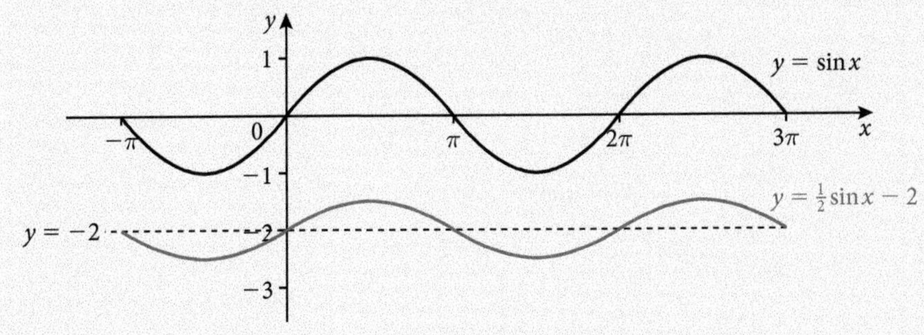

In Example 5.7 (a), the graph of $y = 2\cos x$ has many of the same properties as the graph of $y = \cos x$: it has the same period, and the maximum and minimum values occur at the same x values. However, the graph ranges between -2 and 2 instead of -1 and 1. This difference is best described by referring to the graph's **amplitude**. The amplitude of $y = \cos x$ is 1 and the amplitude of $y = 2\cos x$ is 2. The amplitude (always positive) is not always equal to the maximum value. In (b), the amplitude of $y = \cos x + 3$ is 1, in (c), the amplitude of $y = 2\cos x + 3$ is 2, and in (d) the amplitude of $y = \frac{1}{2}\sin x - 2$ is $\frac{1}{2}$. For all three of these, the graphs oscillate about the horizontal midline $y = d$. To find the graph's amplitude, we look at how high and low it oscillates with respect to the midline. With respect to the general form $y = af(x)$, changing the amplitude is equivalent to a vertical stretching or shrinking. Thus, we can give a more precise definition of amplitude in terms of the parameter a.

Amplitude of the graph of sine and cosine functions
The graphs of $f(x) = a\sin[b(x + c)] + d$ and $f(x) = a\cos[b(x + c)] + d$ have an amplitude equal to $|a|$.

Example 5.8

Waves are produced in a long tank of water. The depth of the water, d metres, at t seconds at a fixed location in the tank is modelled by the function $d(t) = M\cos\left(\frac{\pi}{2}t\right) + K$, where M and K are positive constants. The diagram on the next page shows the graph of $d(t)$ for $0 \leqslant t \leqslant 12$ indicating that the point (2, 5.1) is a minimum and the point (8, 9.7) is a maximum.

(a) Find the value of K and the value of M.

(b) After $t = 0$, find the first time when the depth of the water is 9.7 metres.

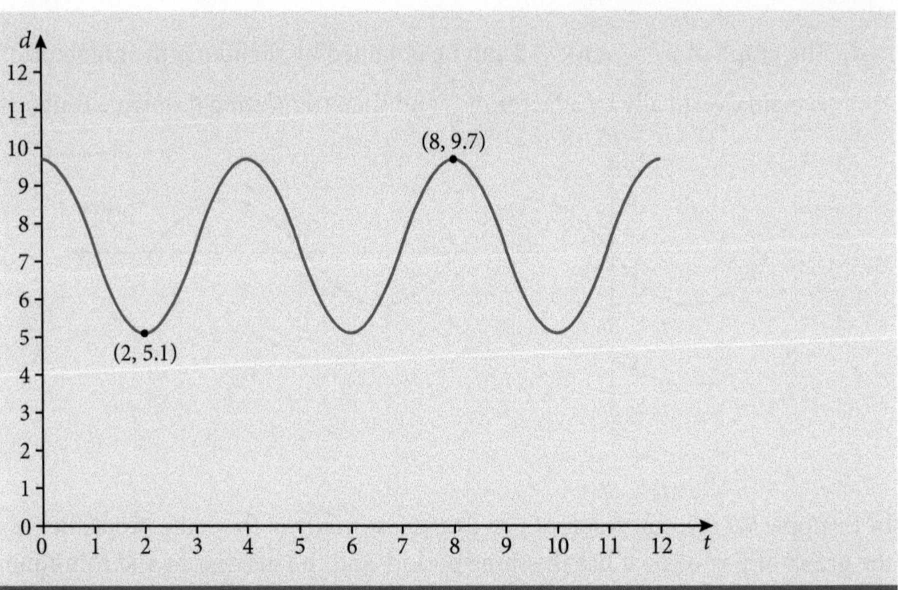

Solution

(a) The constant K is equivalent to the constant d in the general form $f(x) = a\cos[b(x + c)] + d$. To find the value of K and the equation of the horizontal midline, $y = K$, find the average of the function's maximum and minimum values: $K = \frac{9.7 + 5.1}{2} = 7.4$

The constant M is equivalent to the constant a in the general form – its absolute value is the amplitude. The amplitude is the difference between the function's maximum value and the midline:
$|M| = 9.7 - 7.4 = 2.3$. So, $M = 2.3$ or $M = -2.3$

Try $M = 2.3$ by evaluating the function at one of the known values.
$d(2) = 2.3\cos\left(\frac{\pi}{2}(2)\right) + 7.4 = 2.3\cos\pi + 7.4 = 2.3(-1) + 7.4 = 5.1$

This agrees with the point (2, 5.1) on the graph. Therefore, $M = 2.3$

(b) Maximum values of the function ($d = 9.7$) occur at values of t that differ by a value equal to the period. The graph shows that the difference in t-values from the minimum (2, 5.1) to the maximum (8, 9.7) is equivalent to one and a half periods. Therefore, the period is 4 and the first time after $t = 0$ at which $d = 9.7$ is $t = 4$

All four of the functions in Example 5.7 had the same period of 2π, but the function in Example 5.8 had a period of 4. Because $y = \sin x$ completes one period from $x = 0$ to $x = 2\pi$, it follows that $y = \sin bx$ completes one period from $bx = 0$ to $bx = 2\pi$. This implies that $y = \sin bx$ completes one period from $x = 0$ to $x = \frac{2\pi}{b}$. This agrees with the period for the function $d(t) = 2.3\cos\left(\frac{\pi}{2}t\right) + 7.4$ in Example 5.8: period $= \frac{2\pi}{b} = \frac{2\pi}{\frac{\pi}{2}} = \frac{2\pi}{1} \cdot \frac{2}{\pi} = 4$

Note that the change in amplitude and vertical translation had no effect on the period. We should also expect that a horizontal translation of a sine or cosine curve should not affect the period. Example 5.9 looks at a function that is horizontally translated (shifted) and has a period different from 2π.

Example 5.9

Sketch the function $f(x) = \sin\left(2x + \frac{2\pi}{3}\right)$

Solution

To determine how to transform the graph of $y = \sin x$ to the graph of $y = \sin\left(2x + \frac{2\pi}{3}\right)$ so that we can sketch the function, we need to make sure the function is written in the form $f(x) = a\sin[b(x + c)] + d$. Clearly, $a = 1$ and $d = 0$, but we need to take out a common factor of 2 from $2x + \frac{2\pi}{3}$ to get $f(x) = \sin\left[2\left(x + \frac{\pi}{3}\right)\right]$. From the transformations studied in Chapter 1, the graph of f is obtained by first translating the graph of $y = \sin x$ to the left by $\frac{\pi}{3}$ units and then applying a horizontal shrink of factor $\frac{1}{2}$. The following graphs show the two steps of transforming $y = \sin x$ to $y = \sin\left[2\left(x + \frac{\pi}{3}\right)\right]$

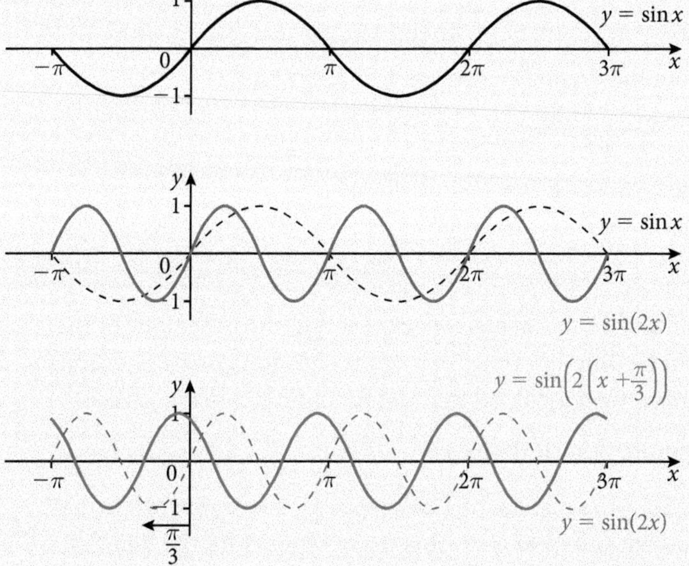

A horizontal translation of a sine or cosine curve is often referred to as a **phase shift**. The equations $y = \sin\left(x + \frac{\pi}{3}\right)$ and $y = \sin\left[2\left(x + \frac{\pi}{3}\right)\right]$ both underwent a phase shift of $-\frac{\pi}{3}$.

Period and horizontal translation (phase shift) of sine and cosine functions
Given that b is a positive real number, $y = a\sin[b(x + c)] + d$ and $y = a\cos[b(x + c)] + d$ have a period of $\frac{2\pi}{b}$ and a horizontal translation (phase shift) of $-c$.

Example 5.10

The graph of a function in the form $y = a\cos bx$ is given below.

(a) Write down the value of a.

(b) Calculate the value of b.

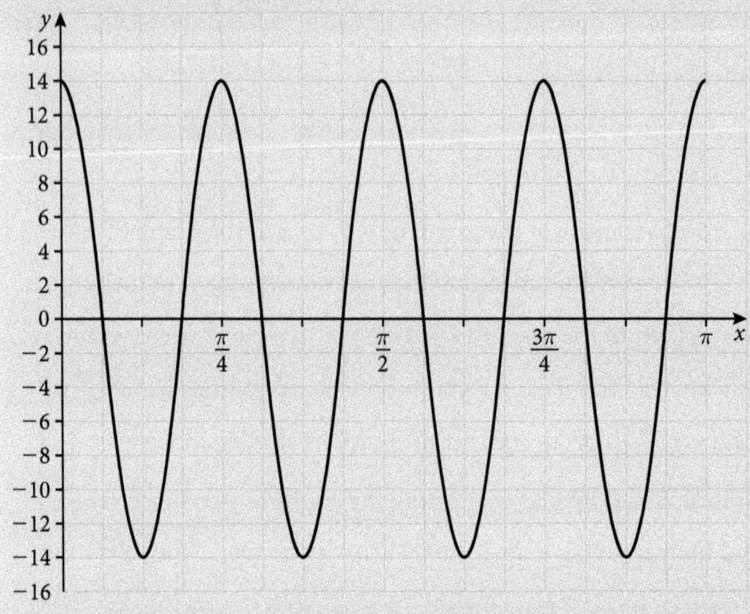

Solution

(a) The amplitude of the graph is 14. Therefore, $a = 14$.

(b) From inspecting the graph, we can see that the period is $\frac{\pi}{4}$.

Period $= \frac{2\pi}{b} = \frac{\pi}{4} \Rightarrow b\pi = 8\pi \Rightarrow b = 8$

Graph of the tangent function

From work done earlier in this chapter, we expect that the behaviour of the tangent function will be different from that of the sine and cosine functions. In the previous section, we concluded that the function $f(x) = \tan x$ has a domain of all real numbers such that $x \neq \frac{\pi}{2} + k\pi, k \in \mathbb{Z}$, and that its range is all real numbers. Also, the results for Example 5.6 led us to speculate that the period of the tangent function is π. This make sense since the identity $\tan x = \frac{\sin x}{\cos x}$ informs us that $\tan x$ will be zero whenever $\sin x = 0$, which occurs at values of x that differ by π. The values of x for which $\cos x = 0$, causing $\tan x$ to be undefined (gaps in the domain), also differ by π.

As x approaches the values where $\cos x = 0$, the value of $\tan x$ will become very large – either very large negative or very large positive. Thus, the graph of $y = \tan x$ has vertical asymptotes at $x = \frac{\pi}{2} + k\pi, k \in \mathbb{Z}$. Consequently, the graphical behaviour of the tangent function will not be a wave pattern such as those produced by the sine and cosine functions, but rather a series of separate curves that repeat every π units (Figure 5.19).

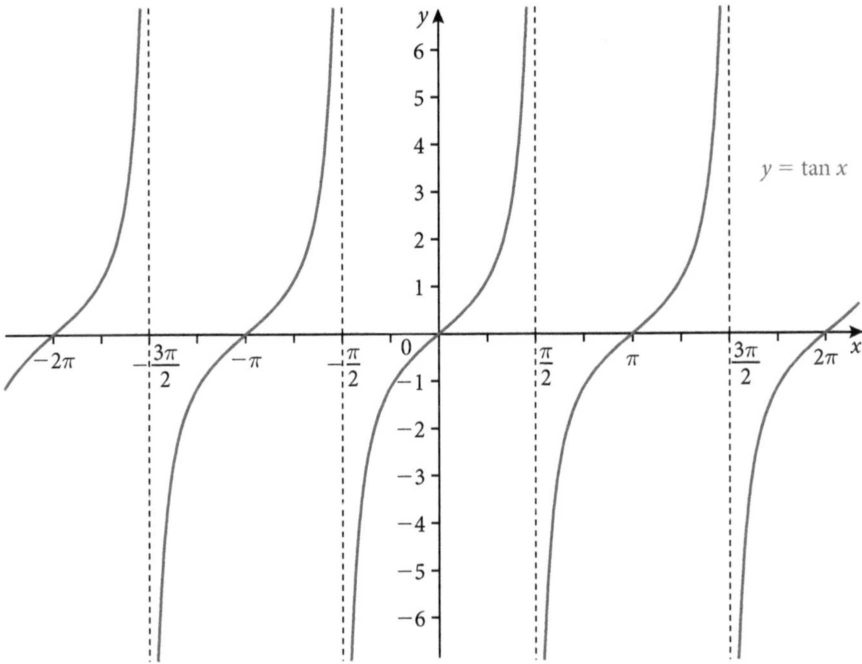

Figure 5.19 Graph of $y = \tan x$ for $-2\pi \leqslant x \leqslant 2\pi$

The graph gives clear confirmation that the period of the tangent function is π.

Although the graph of $y = \tan x$ can undergo a vertical stretch or shrink, it does not have an amplitude since the tangent function has no maximum or minimum values. However, other transformations can affect the period of the tangent function.

Example 5.11

Sketch each function.

(a) $f(x) = \tan 2x$ (b) $g(x) = \tan\left[2\left(x - \frac{\pi}{4}\right)\right]$

Solution

(a) An equation in the form $y = f(bx)$ indicates a horizontal shrink of $f(x)$ by a factor of $\frac{1}{b}$. Hence, the period of $y = \tan 2x$ is $\frac{1}{2} \cdot \pi = \frac{\pi}{2}$.

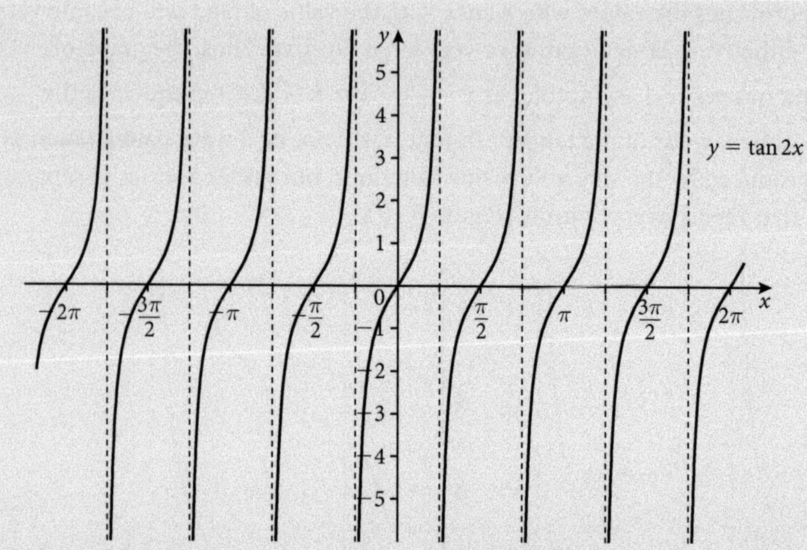

(b) The graph of $y = \tan\left[2\left(x - \frac{\pi}{4}\right)\right]$ is obtained by first translating the graph of $y = \tan x$ to the right by $\frac{\pi}{4}$ units, and then applying a horizontal shrink by a factor of $\frac{1}{2}$. As for $f(x) = \tan 2x$ in part (a), the period of $g(x) = \tan\left[2\left(x - \frac{\pi}{4}\right)\right]$ is $\frac{\pi}{2}$.

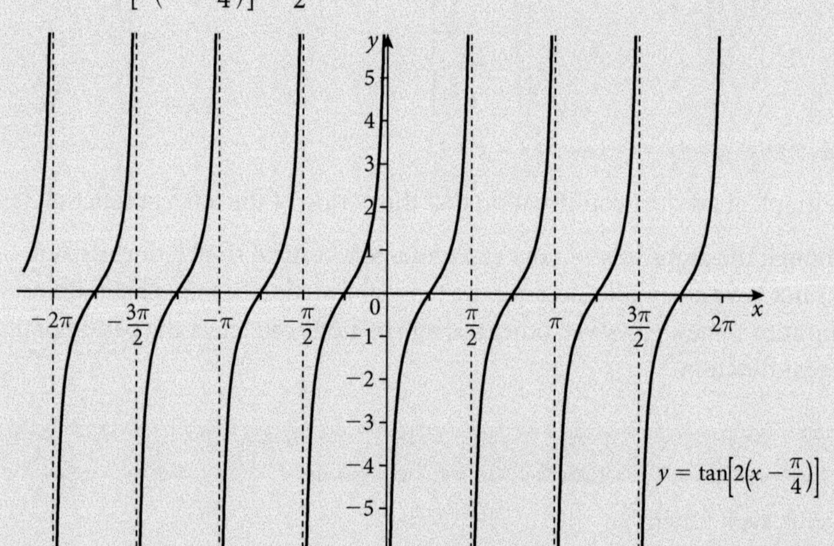

Exercise 5.3

1. Without using your GDC, sketch a graph of each equation on the interval $-\pi \leqslant x \leqslant 3\pi$.

(a) $y = 2\sin x$ **(b)** $y = \cos x - 2$ **(c)** $y = \frac{1}{2}\cos x$

(d) $y = \sin\left(x - \frac{\pi}{2}\right)$ **(e)** $y = \cos(2x)$ **(f)** $y = 1 + \tan x$

(g) $y = \sin\left(\frac{x}{2}\right)$ **(h)** $y = \tan\left(x + \frac{\pi}{2}\right)$ **(i)** $y = \cos\left(2x - \frac{\pi}{4}\right)$

2. For each function:

(i) Sketch the function for the interval $-\pi \leqslant x \leqslant 5\pi$. Write down its amplitude and period.

(ii) Determine the domain and range.

(a) $f(x) = \frac{1}{2}\cos x - 3$ **(b)** $g(x) = 3\sin(3x) - \frac{1}{2}$

(c) $g(x) = 1.2\sin\left(\frac{x}{2}\right) + 4.3$

3. Each graph shows a trigonometric equation on the interval $0 \leqslant x \leqslant 12$ that can be written in the form $y = A\sin\left(\frac{\pi}{4}x\right) + B$. Two points – one a minimum and the other a maximum – are indicated on each graph. Find the values of A and of B for each function.

(a)

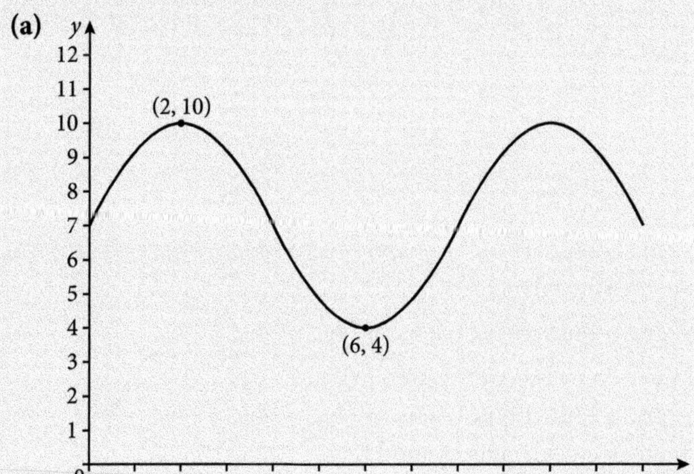

(b)

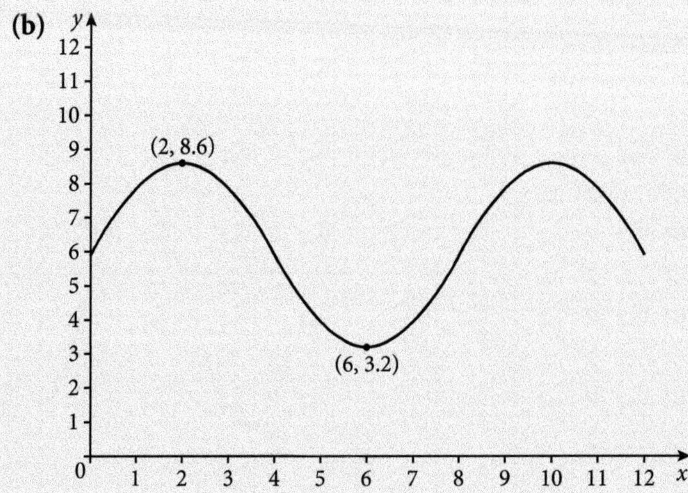

4. The graph of a function in the form $y = p\cos qx$ is given in the diagram.
   (a) Write down the value of p.
   (b) Calculate the value of q.

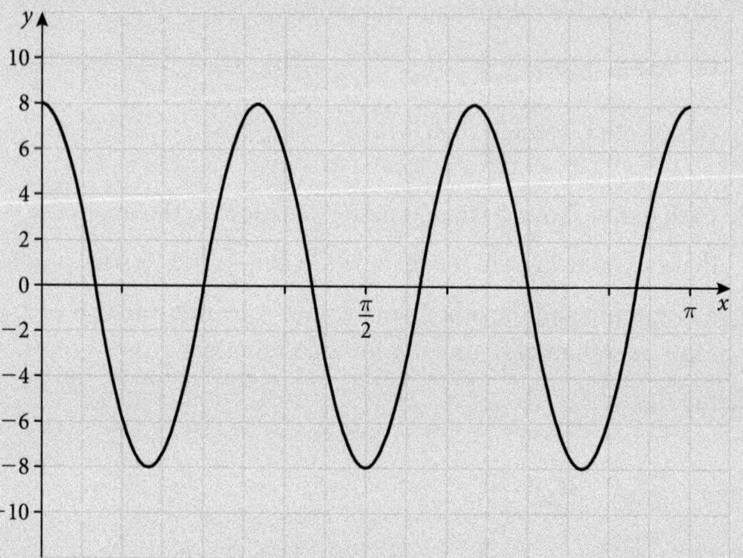

5. The diagram shows part of the graph of a function whose equation is in the form $y = a\sin(bx) + c$.
   (a) Write down the values of a, b, and c.
   (b) Find the exact value of the x-coordinate of the point P, the point where the graph crosses the x-axis as shown in the diagram.

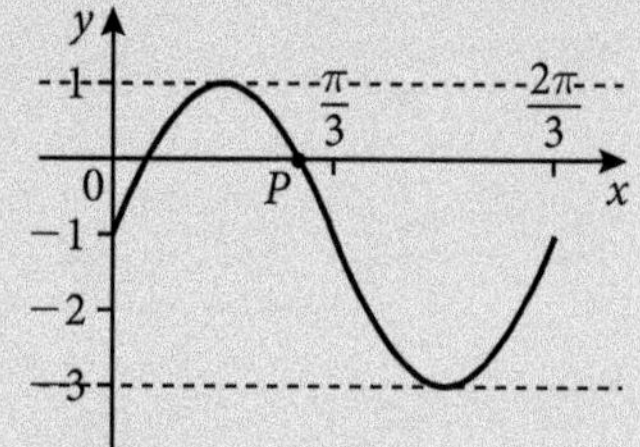

6. The graph represents $y = a\sin(x + b) + c$, where a, b, and c are constants. Find values for a, b, and c.

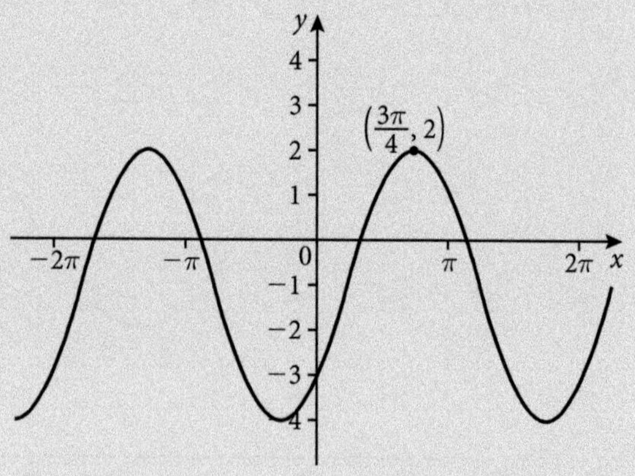

5.4 Trigonometric equations and identities

The primary focus of this section is to examine methods for solving equations that contain the sine, cosine, and tangent functions. For example, the following are **trigonometric equations**:

$\sin x = \frac{1}{2}$ $\quad 3\cos x = 5\sin x$ $\quad \tan x = \frac{\sin x}{\cos x}$

$1 + \sin x = 3\cos^2 x$ $\quad \sin^2 x + \cos^2 x = 1$

The equations $\tan x = \frac{\sin x}{\cos x}$ and $\sin^2 x + \cos^2 x = 1$ are examples of special equations called **identities**. As we learned in Section 1.1, an identity is an equation that is true for all possible values of the variable. The other equations are true for only certain values of x. Identities can be helpful in solving trigonometric equations by allowing us to simplify some trigonometric expressions. Equations that contain trigonometric functions can often be solved using the same graphical and algebraic methods used to solve other equations.

The unit circle and exact solutions to trigonometric equations

When we are asked to solve a trigonometric equation, there are two important questions we need to consider:

1. Is it possible, or required, to express any solution(s) exactly?
2. For what interval of the variable (usually x) are all solutions to be found?

With regard to the first question, exact solutions are only attainable, in most cases, if they are an integer multiple of $\frac{\pi}{6}$ or $\frac{\pi}{4}$. The variable for which we are trying to solve in trigonometric equations is a real number that can be interpreted as the length of an arc on the unit circle. As explained in Section 5.2, arc lengths that are multiples of $\frac{\pi}{6}$ or $\frac{\pi}{4}$ commonly occur and it is important to be familiar with the sine, cosine, and tangent of these numbers.

Concerning the second question, for most trigonometric equations there are infinitely many values of the variable that satisfy the equation. In order to restrict the number of solutions, we are asked for the solutions to be contained within a suitable interval. For example, we may search for all the values of x that solve an equation such that $0 \leqslant x < 2\pi$. Although it is certainly possible to write a general expression using a parameter that specifies the infinite values that solve a trigonometric equation (this is called the general solution), it is not required for this course. A solution interval will always be given, as in the next example.

Example 5.12

Find the exact solution(s) to the equation $\sin x = \frac{1}{2}$ for $0 \leqslant x < 2\pi$.

Solution

Recalling the definition of the sine function, this equation can be interpreted as asking for the length, x, of arcs along the unit circle that have a terminal point with a y-coordinate equal to $\frac{1}{2}$.

We know from Section 5.2 that arc lengths of $\frac{\pi}{6}$ and $\frac{5\pi}{6}$ have terminal points with y-coordinates of $\frac{1}{2}$ (see diagram).

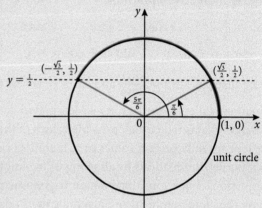

There are clearly an infinite number of arcs, both positive and negative, that will terminate at the same points.

This can be written as $x = \frac{\pi}{6} + k \cdot 2\pi$ and $x = \frac{5\pi}{6} + k \cdot 2\pi, k \in \mathbb{Z}$.

However, we are asked for solutions in the interval $0 \leqslant x < 2\pi$.

Thus, $x = \frac{\pi}{6}$ or $x = \frac{5\pi}{6}$.

Another way to see that the equation $\sin x = \frac{1}{2}$ has infinitely many solutions is to graph the equations $y = \sin x$ and $y = \frac{1}{2}$, as in Figure 5.20, and search for intersection points, i.e. points where the two equations are equal.

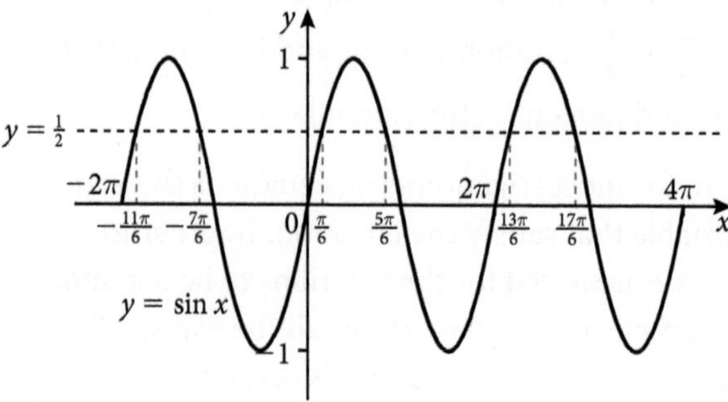

Figure 5.20 Graphs of $y = \sin x$ and $y = \frac{1}{2}$

The graphs of the two equations will intersect repeatedly as they extend indefinitely in both directions.

Your GDC can be a very effective tool to search for solutions graphically. However, it can be limited when exact solutions are required. The graph in Figure 5.21 shows a graphical solution for the equation in Example 5.12.

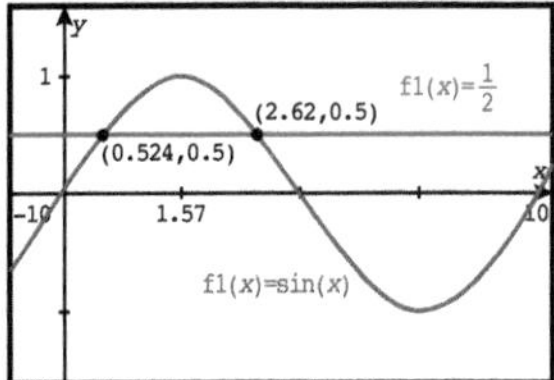

Figure 5.21 Finding intersection points of $y = \sin x$ and $y = \frac{1}{2}$ in the interval $0 \leqslant x < 2\pi$

The GDC gives the two solutions in the interval $0 \leqslant x < 2\pi$ as $x = 0.524$ and $x = 2.62$. These are approximations (to 3 significant figures) of $\frac{\pi}{6}$ and $\frac{5\pi}{6}$. Therefore, if exact solutions are required, you will need to remember the trigonometric function values for the multiples of $\frac{\pi}{6}$ and $\frac{\pi}{4}$ (see Table 5.1 in Section 5.2).

Example 5.13

Find the exact solution(s) to the equation $1 + \tan x = 0$ for $-\pi \leqslant x < \pi$.

Solution

It is important to note that the solution interval is different from Example 5.12. With respect to the unit circle, the solutions will correspond to points in any of the four quadrants (as in Example 5.12) but points in quadrants III and IV will correspond to arcs rotating clockwise (negative direction). Solutions to this equation are values of x such that $\tan x = -1$. Since

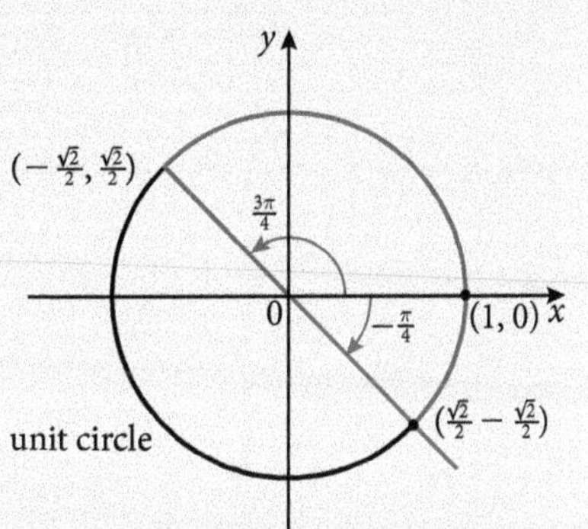

$\tan x = \frac{\sin x}{\cos x} = \frac{y}{x}$, any solutions will be in quadrants II and IV, where the x- and y-coordinates have opposite signs. The arcs terminating midway in the quadrants will terminate at points having opposite values for x and y. Therefore, as shown in the diagram, the solutions are exactly $x = -\frac{\pi}{4}$ or $x = \frac{3\pi}{4}$.

It is possible to arrive at exact answers that are not multiples of $\frac{\pi}{6}$ or $\frac{\pi}{4}$, as the next example illustrates.

When writing the square of a trigonometric function, the square symbol is often written next to the function name.

$\sin^2(x)$ is the same as $[\sin(x)]^2$

$\cos^2(x)$ is the same as $[\cos(x)]^2$

$\tan^2(x)$ is the same as $[\tan(x)]^2$

Example 5.14

Find the exact solution(s) to the equation $\cos^2\left(x - \frac{\pi}{3}\right) = \frac{1}{2}$ for $0 \leqslant x \leqslant 2\pi$.

Solution

The expression $\cos^2\left(x - \frac{\pi}{3}\right)$ can also be written as $\left[\cos\left(x - \frac{\pi}{3}\right)\right]^2$

The first step is to take the square root of both sides, remembering that every positive number has two square roots, which gives

$\cos\left(x - \frac{\pi}{3}\right) = \pm\sqrt{\frac{1}{2}} = \pm\frac{1}{\sqrt{2}} = \pm\frac{\sqrt{2}}{2}$

All of the odd integer multiples of $\frac{\pi}{4}$ $\left(\ldots, -\frac{3\pi}{4}, -\frac{\pi}{4}, 0, \frac{\pi}{4}, \frac{3\pi}{4}, \ldots\right)$ have a cosine equal to either $\frac{\sqrt{2}}{2}$ or $-\frac{\sqrt{2}}{2}$. That is, $x - \frac{\pi}{3} = \frac{\pi}{4} + k \cdot \frac{\pi}{2}$

Now, solve for x.

$x = \frac{\pi}{4} + \frac{\pi}{3} + k \cdot \frac{\pi}{2} = \frac{7\pi}{12} + k \cdot \frac{6\pi}{12}$

The last step is to substitute in different integer values for k to generate all the possible values for x so that $0 \leqslant x \leqslant 2\pi$.

When $k = 0$: $x = \frac{7\pi}{12}$

When $k = 1$: $x = \frac{7\pi}{12} + \frac{6\pi}{12} = \frac{13\pi}{12}$

When $k = 2$: $x = \frac{7\pi}{12} + \frac{12\pi}{12} = \frac{19\pi}{12}$

When $k = 3$: $x = \frac{7\pi}{12} + \frac{18\pi}{12} = \frac{25\pi}{12}$, which is outside the required interval as $\frac{25\pi}{12} > 2\pi$

When $k = -1$: $x = \frac{7\pi}{12} - \frac{6\pi}{12} = \frac{\pi}{12}$

There are four exact solutions in the interval $0 \leqslant x \leqslant 2\pi$. They are:

$x = \frac{\pi}{12}, \frac{7\pi}{12}, \frac{13\pi}{12}$ or $\frac{19\pi}{12}$

Check the solutions to trigonometric equations with your GDC. The GDC images below verify the four exact solutions to the equation.

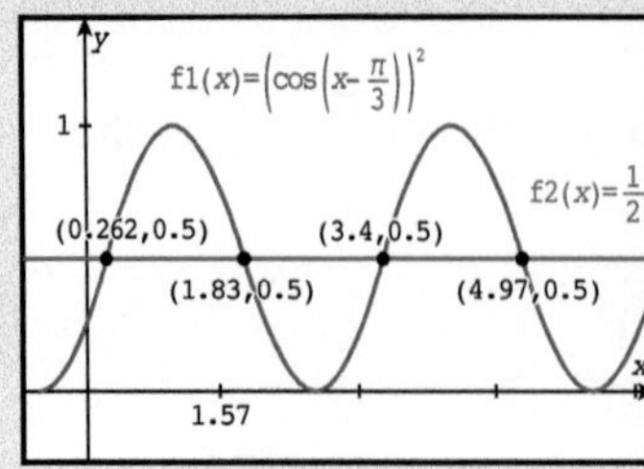

$\frac{\pi}{12}$	0.261799
$\frac{7\cdot\pi}{12}$	1.8326
$\frac{13\cdot\pi}{12}$	3.40339
$\frac{19\cdot\pi}{12}$	4.97419

Graphical solutions to trigonometric equations

If exact solutions are not required, then a graphical solution using your GDC is a very effective way to find approximate solutions to trigonometric equations. Unless instructed to do otherwise, you should give approximate solutions to an accuracy of three significant figures.

Let's solve the equation in Example 5.13 again. If the instructions do not explicitly ask for exact solutions, then approximate solutions are acceptable.

Example 5.15

Find the solution(s) to the equation $1 + \tan x = 0$ for $-\pi \leqslant x < \pi$

Solution

Graph the equation $y = 1 + \tan x$ and find all of its zeros (x-intercepts) in the interval $-\pi \leqslant x < \pi$.

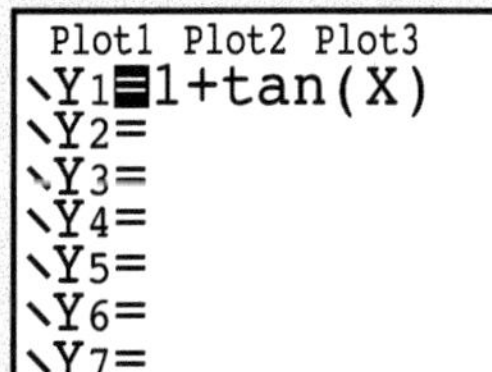

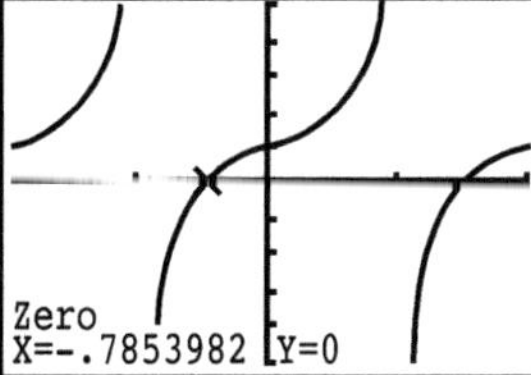

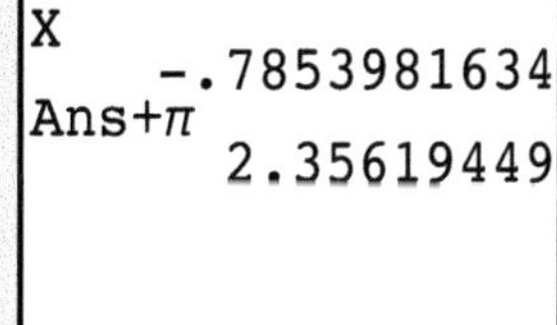

This GDC images above indicate an approximate solution of $x \approx -0.785$ between $-\pi$ and 0. Since we know that the period of $y = 1 + \tan x$ is π (same as $y = \tan x$), we can simply add π to this first solution to find the one between 0 and π, as shown in the final GDC image. Therefore, the two solutions for x in the interval $-\pi \leqslant x < \pi$ are:

$x \approx -0.785$ and $x \approx 2.36$ (three significant figures).

Of course, a graphical approach is most effective when it is not possible, or very difficult, to find exact solutions.

Example 5.16

The peak height, h metres, of ocean waves during a storm is given by the equation $h = 9 + 4\sin\left(\frac{t}{2}\right)$, where t is the number of hours after midnight. A tsunami alarm is triggered when the peak height goes above 12.5 metres. Find the value of t when the alarm first sounds.

Solution

Graph the equations $y = 9 + 4\sin\left(\frac{x}{2}\right)$ and $y = 12.5$ and find the first point of intersection for $x > 0$.

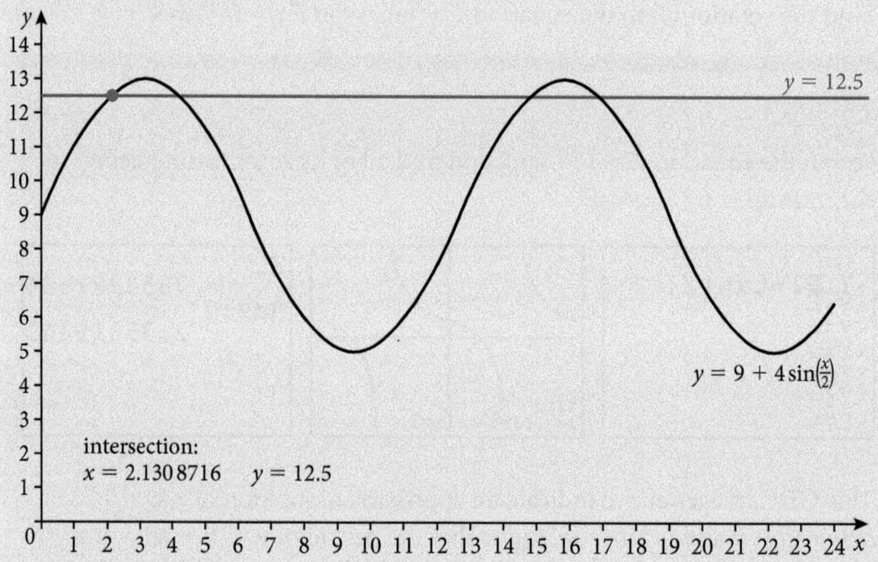

Analysing the graphs on a GDC shows that the first point of intersection has an x-coordinate of approximately 2.13. Therefore, the alarm will first sound when $t \approx 2.13$ hours.

Analytic solutions to trigonometric equations

Now we will see how combining algebraic techniques with trigonometric identities can be used to solve trigonometric equations. An analytical approach requires us to devise a solution strategy using algebraic methods that we have applied to other types of equations, such as quadratic equations. Often trigonometric equations that demand an analytic approach will result in exact solutions, but not always. Although our approach for equations in this section focuses on algebraic techniques, it is important to use graphical methods to support or confirm our analytic solutions.

Example 5.17

Solve $2\sin^2 x - \sin x = 0$ for $-\pi \leqslant x \leqslant \pi$

Solution

We can factorise and apply the rule that if $a \cdot b = 0$ then either $a = 0$ or $b = 0$.

$2\sin^2 x - \sin x = 0 \Rightarrow \sin x(2\sin x - 1) = 0 \Rightarrow \sin x = 0 \quad \text{or} \quad \sin x = \frac{1}{2}$

For $\sin x = 0$: $x = -\pi, 0, \pi$

For $\sin x = \frac{1}{2}$: $x = \frac{\pi}{6}, \frac{5\pi}{6}$

Therefore, $x = -\pi, 0, \frac{\pi}{6}, \frac{5\pi}{6}, \pi$

Although exact answers were not required in Example 5.17, given our knowledge of the unit circle and familiarity with the sine of common values (i.e. multiples of $\frac{\pi}{6}$ and $\frac{\pi}{4}$), we are able to give exact answers without any difficulty. It would have been acceptable to give approximate solutions, but it is worth recognising that this would have required considerably more effort than providing exact solutions. Entering and graphing the equation $y = 2\sin^2 x - \sin x$ on your GDC would not be the most efficient or appropriate solution method, but if enough time is available, it is an effective way to confirm your exact solutions.

The next example illustrates how the application of a trigonometric identity can help us to rewrite an equation in a way that allows us to solve it algebraically.

Example 5.18

Solve $3\sin x + \tan x = 0$ for $0 \leqslant x \leqslant 2\pi$

Solution

Since the structure of this equation is such that an expression is set equal to zero, it would be nice to be able to use the same algebraic technique as the previous example – that is, factorise and solve for when each factor is zero. However, it is not possible to factorise the expression $3\sin x + \tan x$, and rewriting the equation as $3\sin x = -\tan x$ does not help.

Are there any expressions in the equation for which we can substitute an equivalent expression that will make the equation accessible to an algebraic solution? We do not have any equivalent expressions for $\sin x$, but we do have an identity for $\tan x$. From the definition of $\tan x$ we know that $\tan x = \frac{\sin x}{\cos x}$.

Let's see what happens when we substitute $\frac{\sin x}{\cos x}$ for $\tan x$.

$3\sin x + \tan x = 0 \Rightarrow 3\sin x + \frac{\sin x}{\cos x} = 0$

Now, multiply both sides by $\cos x$ while recognising that $\cos x \neq 0$ $(x \neq \frac{\pi}{2} + k \cdot \pi, k \in \mathbb{Z})$

$3\sin x + \frac{\sin x}{\cos x} = 0 \Rightarrow 3\sin x\cos x + \sin x = 0 \Rightarrow \sin x(3\cos x + 1) = 0$

$\Rightarrow \sin x = 0$ or $\cos x = -\frac{1}{3}$

For $\sin x = 0$: $x = 0, \pi, 2\pi$

We know that (1, 0) and (−1, 0) are the points on the unit circle that correspond to $\sin x = 0$, giving the three exact solutions above. Although we know that the points on the unit circle that correspond to $\cos x = -\frac{1}{3}$ are in quadrants II and III, we do not know their exact coordinates. So, we need a GDC to find approximate solutions to $\cos x = -\frac{1}{3}$ for $0 \leqslant x \leqslant 2\pi$.

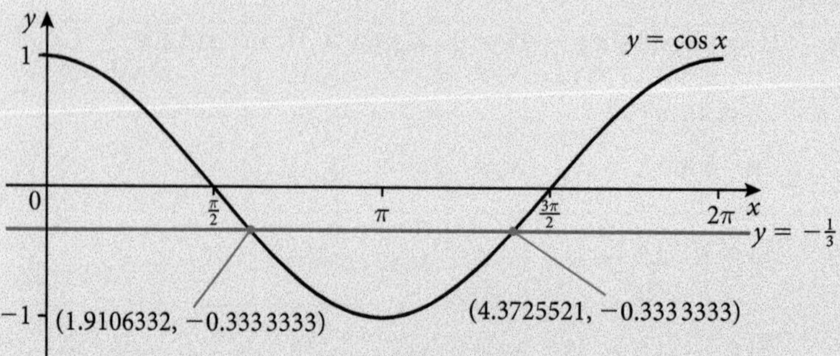

Thus, for $\cos x = -\frac{1}{3}$: $x \approx 1.91$ or $x \approx 4.37$ (three significant figures).

Therefore, the full solution set for the equation is
$x = 0, \pi, 2\pi; x \approx 1.91, 4.37$

Trigonometric identities

A strategy that often proves fruitful is to try to rewrite a trigonometric equation in terms of just one trigonometric function. If that is not possible, then try to rewrite it in terms of only the sine and cosine functions. This strategy was used in Example 5.18.

As Example 5.18 illustrated, sometimes an analytical method for solving a trigonometric equation relies on a trigonometric identity providing a useful substitution. There are a few trigonometric identities other than $\tan x = \frac{\sin x}{\cos x}$ required for this course. They can be used to help simplify trigonometric expressions and solve equations.

At the start of this section, it was stated that the equation $\sin^2 x + \cos^2 x = 1$ is an identity; that is, it is true for all possible values of x. Let's prove that this is the case.

Recall from Section 5.1 that the equation for the unit circle is $x^2 + y^2 = 1$. That is, the coordinates (x, y) of any point on the circle will satisfy the equation $x^2 + y^2 = 1$. Also, in Section 5.2, we learned that the sine and cosine functions are defined in terms of the coordinates of the terminal point of an arc on the unit circle starting at (1, 0), as shown in Figure 5.22.

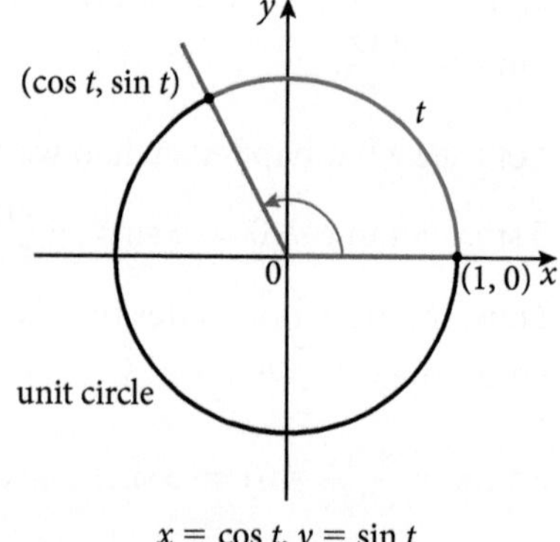

Figure 5.22 sine and cosine functions defined in terms of the coordinates of the terminal point of an arc on the unit circle

If t is any real number that is the length of an arc on the unit circle that terminates at (x, y), then $x = \cos t$ and $y = \sin t$. Substituting directly into the equation for the circle gives $\sin^2 t + \cos^2 t = 1$.

As mentioned in the previous section, the convention is to use x to denote the domain variable rather than t. Therefore, the equation $\sin^2 x + \cos^2 x = 1$ is true for any real number x.

The Pythagorean identities for sine and cosine

The following equations are true for all real numbers x:

$\sin^2 x + \cos^2 x = 1 \qquad \sin^2 x = 1 - \cos^2 x \qquad \cos^2 x = 1 - \sin^2 x$

The identity $\sin^2 x + \cos^2 x = 1$ is often referred to as a Pythagorean identity because, as we will see in the next chapter, $\sin x$ and $\cos x$ can represent the legs of a right triangle with a hypotenuse equal to one. Substituting into the Pythagorean theorem gives $\sin^2 x + \cos^2 x = 1$.

Another useful set of trigonometric identities is referred to as the double angle identities because they are equations involving $\sin 2x$ and $\cos 2x$. As discussed back in Section 5.1, the argument of a trigonometric function (x in $\sin x$, θ in $\cos\theta$) can be interpreted as an angle (in degrees or radians), or as just a real number. Even though these identities are called double angle identities, they apply for either interpretation.

Double angle identities for sine and cosine

The following equations are true for all real numbers x:

$\sin 2x = 2\sin x\cos x$

$$\cos 2x = \begin{cases} \cos^2 x - \sin^2 x \\ 2\cos^2 x - 1 \\ 1 - 2\sin^2 x \end{cases}$$

It is quite easy to verify the double angle identities by means of graphical analysis on your GDC. The GDC screen images shown in Figure 5.23 illustrate that $\sin 2x$ is equivalent to $2\sin x\cos x$. Use your GDC to verify that $\cos 2x$ is equivalent to $\cos^2 x - \sin^2 x$. Once the identity $\cos 2x = \cos^2 x - \sin^2 x$ is established, we can use one of the Pythagorean identities to rewrite it in terms of only sine or only cosine, thus establishing the other two double angle identities for cosine.

$\cos 2x = \cos^2 x - \sin^2 x$

$\cos 2x = \cos^2 x - (1 - \cos^2 x)$ Substitute $1 - \cos^2 x$ for $\sin^2 x$

$\cos 2x = 2\cos^2 x - 1$

Similar steps can be performed to show that $\cos 2x = 1 - 2\sin^2 x$

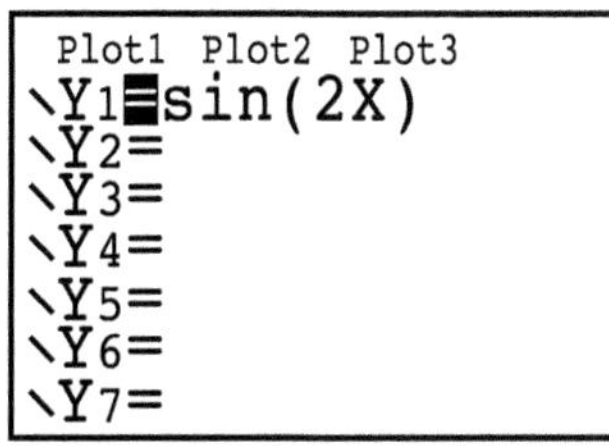

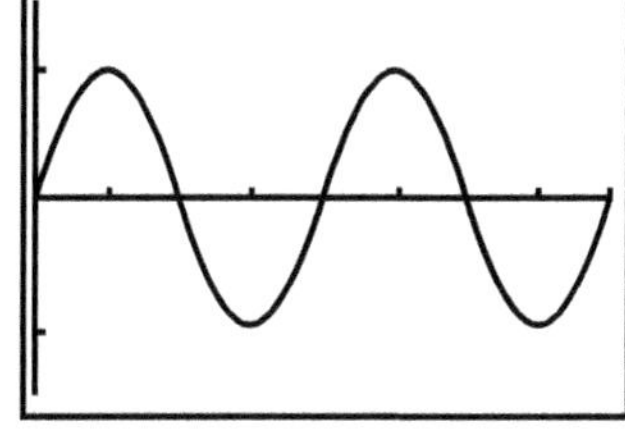

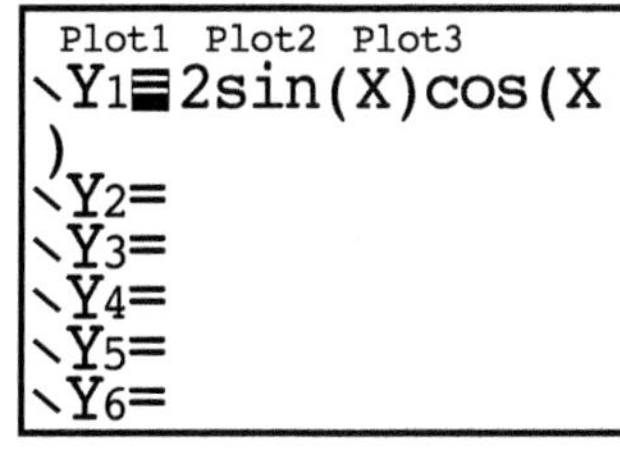

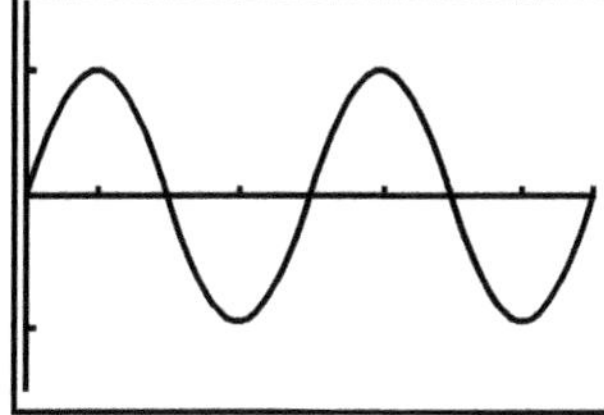

Figure 5.23: GDC screens showing that $\sin 2x$ is equivalent to $2\sin x\cos x$

Now let's see how these identities can help us with algebraic solutions of trigonometric equations.

Example 5.19

Solve the equation $\cos 2x + \cos x = 0$ for $0 \leqslant x \leqslant 2\pi$

Solution

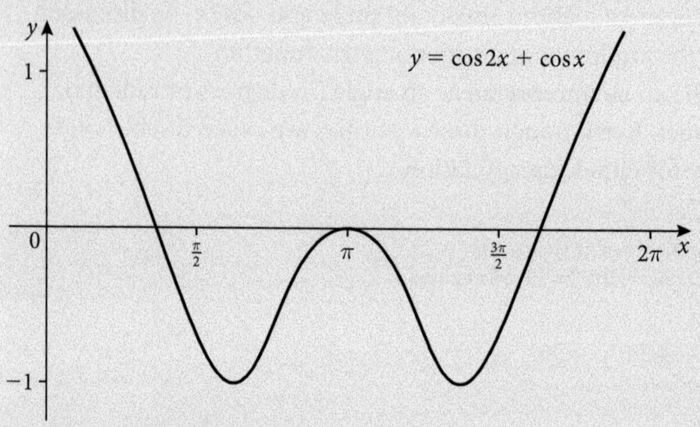

Taking an initial look at the graph of $y = \cos 2x + \cos x$ suggests that there are possibly three solutions in the interval $x \in [0, 2\pi]$. Although the expression $\cos 2x + \cos x$ contains terms with only the cosine function, it is not possible to perform any algebraic operations on them because they have different arguments. In order to solve algebraically, we need both cosine functions to have arguments of x (rather than $2x$). There are three different double angle identities for $\cos 2x$. It is best to have the equation in terms of one trigonometric function, so we choose to substitute $2\cos^2 x - 1$ for $\cos 2x$.

$\cos 2x + \cos x = 0 \Rightarrow 2\cos^2 x - 1 + \cos x = 0 \Rightarrow 2\cos^2 x + \cos x - 1 = 0$

$(2\cos x - 1)(\cos x + 1) = 0 \Rightarrow \cos x = \frac{1}{2}$ or $\cos x = -1$

For $\cos x = \frac{1}{2}, x = \frac{\pi}{3}, \frac{5\pi}{3}$

For $\cos x = -1, x = \pi$

Therefore, the solutions in the interval $0 \leqslant x \leqslant 2\pi$ are: $x = \frac{\pi}{3}, \pi, \frac{5\pi}{3}$

Example 5.20

(a) Express $2\cos^2 x + \sin x$ in terms of $\sin x$ only.

(b) Solve the equation $2\cos^2 x + \sin x = -1$ for x in the interval $0 \leqslant x \leqslant 2\pi$, expressing your answer(s) exactly.

Solution

(a) $2\cos^2 x + \sin x = 2(1 - \sin^2 x) + \sin x$ — Use the Pythagorean identity $\cos^2 x = 1 - \sin^2 x$

$2\cos^2 x + \sin x = 2 - 2\sin^2 x + \sin x$

(b)

$$2\cos^2 x + \sin x = -1$$

$$2 - 2\sin^2 x + \sin x = -1$$ Substitute result from (a)

$$2\sin^2 x - \sin x - 3 = 0$$

$$(2\sin x - 3)(\sin x + 1) = 0$$

$$\sin x = \frac{3}{2} \text{ or } \sin x = -1$$

For $\sin x = \frac{3}{2}$, no solution because $\frac{3}{2}$ is not in the range of the sine function.

For $\sin x = -1$, $x = \frac{3\pi}{2}$

Therefore, there is only one solution in $0 \leqslant x \leqslant 2\pi$, which is $x = \frac{3\pi}{2}$

Use your GDC to check this result by rewriting $2\cos^2 x + \sin x = -1$ as $2\cos^2 x + \sin x + 1 = 0$ and then graph $y = 2\cos^2 x + \sin x + 1$, confirming a single zero at $x = \frac{3\pi}{2}$ in the interval $x \in [0, 2\pi]$

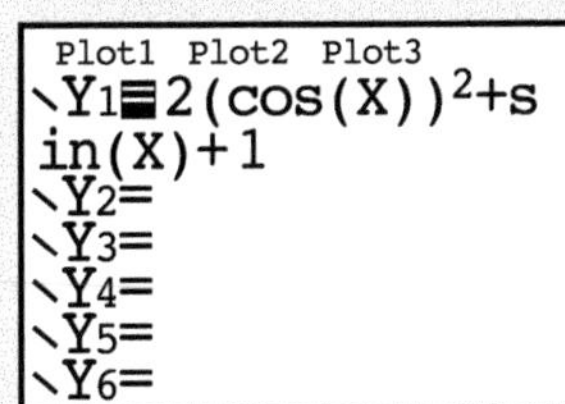

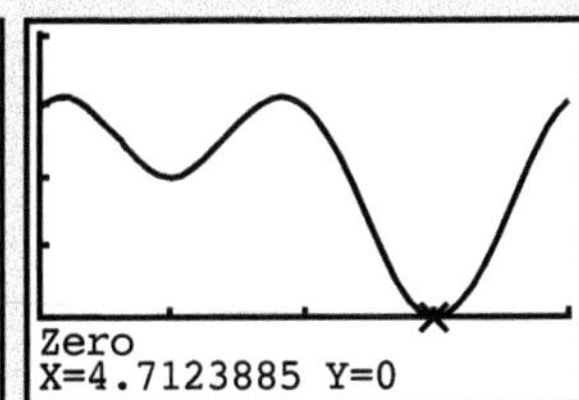

X
4.712388457
3π/2
4.71238898

Example 5.21

Solve the equation $2\sin 2x = 3\cos x$ for $0 \leqslant x \leqslant \pi$

Solution

$$2\sin 2x = 3\cos x$$

$$2(2\sin x\cos x) = 3\cos x$$ Use double angle identity for sine

$$4\sin x\cos x = 3\cos x$$ Do not divide by $\cos x$; solution(s) may be eliminated

$4\sin x\cos x - 3\cos x = 0$	Set equal to zero to prepare for solving by factorisation
$\cos x(4\sin x - 3) = 0$	Factorise

$\cos x = 0$ or $\sin x = \frac{3}{4}$

For $\cos x = 0$: $x = \frac{\pi}{2}$

For $\sin x = \frac{3}{4}$: $x \approx 0.848$ or 2.29

Approximate solutions can be found using on a GDC. All solutions in interval $0 \leqslant x \leqslant \pi$ are: $x = \frac{\pi}{2}$; $x \approx 0.848, 2.29$

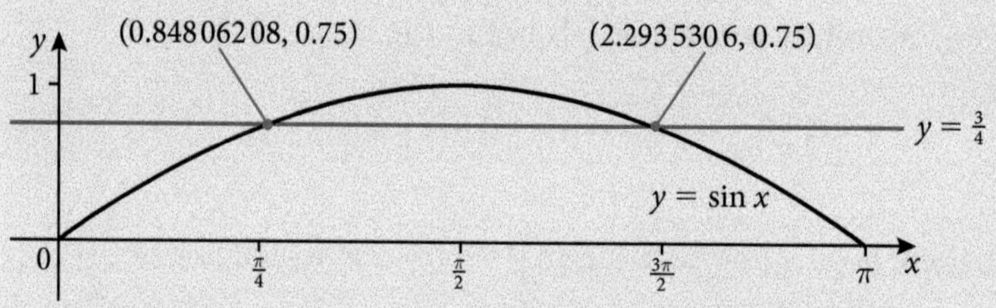

The final example illustrates how trigonometric identities can be applied to find exact values for trigonometric expressions.

Example 5.22

Given that $\cos x = \frac{1}{4}$ and that $0 < x < \frac{\pi}{2}$, find the exact values of:

(a) $\sin x$ (b) $\sin 2x$

Solution

(a) Given $0 \leqslant x \leqslant \frac{\pi}{2}$, it follows that $\sin x > 0$, because the arc with length x will terminate in the first quadrant. The Pythagorean identity is useful when relating $\sin x$ and $\cos x$.

$\sin^2 x = 1 - \cos^2 x \Rightarrow \sin x = \sqrt{1 - \cos^2 x}$

$$\Rightarrow \sin x = \sqrt{1 - \left(\frac{1}{4}\right)^2} = \sqrt{\frac{15}{16}} = \frac{\sqrt{15}}{4}$$

(b) $\sin 2x = 2\sin x\cos x = 2\left(\frac{\sqrt{15}}{4}\right)\left(\frac{1}{4}\right) = \frac{\sqrt{15}}{8}$

Summary of trigonometric identities

Definition of tangent function: $\tan x = \dfrac{\sin x}{\cos x}$

Co-function identities: $\sin\left(\dfrac{\pi}{2} - x\right) = \cos x$

$\cos\left(\dfrac{\pi}{2} - x\right) = \sin x$

Pythagorean identities: $\sin^2 x + \cos^2 x = 1$

$\sin^2 x = 1 - \cos^2 x$

$\cos^2 x = 1 - \sin^2 x$

Double angle identities: $\sin 2x = 2\sin x \cos x$

$\cos 2x = \cos^2 x - \sin^2 x$

$\cos 2x = 2\cos^2 x - 1$

$\cos 2x = 1 - 2\sin^2 x$

Exercise 5.4

1. Find the exact solution(s) for each equation for $0 \leqslant x \leqslant 2\pi$. Verify your solution(s) with your GDC.

(a) $\cos x = \dfrac{1}{2}$ **(b)** $2\sin x + 1 = 0$

(c) $1 - \tan x = 0$ **(d)** $\sqrt{3} = 2\sin x$

(e) $2\sin^2 x = 1$ **(f)** $4\cos^2 x = 3$

(g) $\tan^2 x - 1 = 0$ **(h)** $4\cos^2 x = 1$

(i) $\tan x(\tan x + 1) = 0$ **(j)** $\sin x \cos x = 0$

2. Use your GDC to find approximate solution(s) for $0 \leqslant x \leqslant 2\pi$ for each equation. Express solutions accurate to three significant figures.

(a) $\sin x = 0.4$ **(b)** $3\cos x + 1 = 0$

(c) $\tan x = 2$ **(d)** $\sin 2x = 0.85$

(e) $\cos(x - 1) = -0.38$ **(f)** $3\tan x = 10$

3. Given that k is any integer, list all of the possible values for x that are in the specified interval for each expression.

(a) $\dfrac{\pi}{2} + k \cdot \pi, -3\pi \leqslant x \leqslant 3\pi$ **(b)** $\dfrac{\pi}{6} + k \cdot 2\pi, -2\pi \leqslant x \leqslant 2\pi$

(c) $\dfrac{7\pi}{12} + k \cdot \pi, 0 \leqslant x \leqslant 2\pi$ **(d)** $\dfrac{\pi}{4} + k \cdot \dfrac{\pi}{4}, 0 \leqslant x \leqslant 4\pi$

4. Find the exact solutions for $0 \leqslant x \leqslant 2\pi$ for each equation.

(a) $\cos\left(x - \dfrac{\pi}{6}\right) = -\dfrac{1}{2}$ **(b)** $\tan(x + \pi) = 1$

(c) $\sin 2x = \dfrac{\sqrt{3}}{2}$ **(d)** $\sin^2\left(x + \dfrac{\pi}{2}\right) = \dfrac{3}{4}$

5. The number, N, of empty birds' nests in a park is approximated by the function $N = 74 + 42\sin\left(\frac{\pi}{12}t\right)$, where t is the number of hours after midnight. Find the value of t when the number of empty nests first equals 90. Approximate the answer to one decimal place.

6. In Edinburgh, the number of hours of daylight on day D is modelled by the function

$$H = 12 + 7.26\sin\left[\frac{2\pi}{365}(D - 80)\right]$$

where D is the number of days after December 31 (e.g. January 1 is $D = 1$, January 2 is $D = 2$, and so on). Do not use your GDC on part (a).

(a) Determine which days of the year have 12 hours of daylight.

(b) Determine which days of the year have about 15 hours of daylight.

(c) Find how many days of the year have more than 17 hours of daylight.

7. Solve each equation for the stated solution interval. Find exact solutions when possible, otherwise give solutions to three significant figures. Verify solutions with your GDC.

(a) $2\cos^2 x + \cos x = 0;\ 0 \leqslant x \leqslant 2\pi$

(b) $2\sin^2 x - \sin x - 1 = 0;\ 0 \leqslant x \leqslant 2\pi$

(c) $2\cos x + \sin 2x = 0;\ -\pi \leqslant x \leqslant \pi$

(d) $2\sin x = \cos 2x;\ -\pi \leqslant x \leqslant \pi$

(e) $\tan^2 x - \tan x = 2;\ -\frac{\pi}{2} \leqslant x \leqslant \frac{\pi}{2}$

(f) $\sin^2 x = \cos^2 x;\ 0 \leqslant x \leqslant \pi$

(g) $2\sin^2 x + 3\cos x - 3 = 0;\ 0 \leqslant x \leqslant 2\pi$

(h) $2\sin x = 3\cos x;\ 0 \leqslant x \leqslant 2\pi$

8. Given that $\sin x = \frac{3}{5}$ and that $0 < x < \frac{\pi}{2}$, find the exact values of

(a) $\cos x$ **(b)** $\cos 2x$ **(c)** $\sin 2x$

9. Given that $\cos x = -\frac{2}{3}$ and that $\frac{\pi}{2} < x < \pi$, find the exact values of

(a) $\sin x$ **(b)** $\sin 2x$ **(c)** $\cos 2x$

Chapter 5 practice questions

1. A toy on an elastic string is attached to the top of a doorway. It is pulled down and released, allowing it to bounce up and down. The length of the elastic string, L cm, is modelled by the function $L = 110 + 25\cos(2\pi t)$, where t is time in seconds after release.

(a) Find the length of the elastic string after 2 seconds.

(b) Find the minimum length of the string.

(c) Find the first time after release that the string is 85 cm.

(d) What is the period of the motion?

2. Find the exact solution(s) to the equation $2\sin^2 x - \cos x - 1 = 0$

3. The diagram shows a circle of radius 6 cm.

The perimeter of the shaded sector is 36 cm. Find the radian measure of the angle θ.

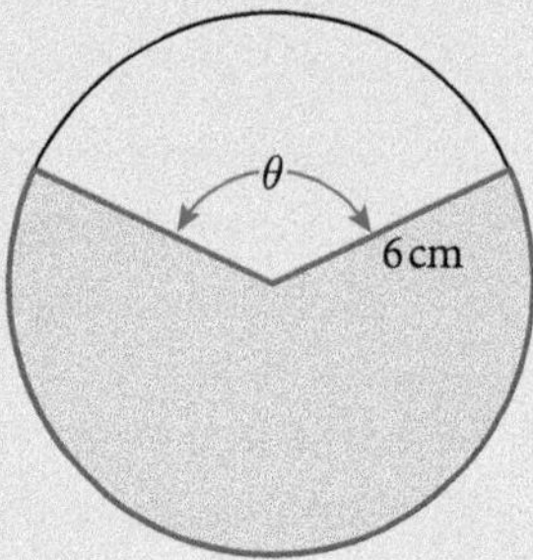

4. Consider the two functions $f(x) = \cos 4x$ and $g(x) = \cos\left(\frac{x}{2}\right)$.

(a) Write down:

(i) the minimum value of the function f

(ii) the period of g.

(b) For the equation $f(x) = g(x)$, find the number of solutions in the interval $0 \leqslant x \leqslant \pi$.

5. A reflector is attached to the spoke of a bicycle wheel. As the wheel rolls along the ground, the distance, d cm, that the reflector is above the ground after t seconds is modelled by the function

$$d = p + q\sin\left(\frac{2\pi}{m}t\right)$$

where p, q, and m are constants

The distance d is at a maximum of 64 cm at $t = 0$ seconds and at $t = 0.5$ seconds, and is at a minimum of 6 cm at $t = 0.25$ seconds and at $t = 0.75$ seconds. Write down the value of:

(a) p **(b)** q **(c)** m

6. Find all solutions to $1 + \sin 3x = \cos(0.25x)$ such that $x \in [0, \pi]$.

7. Find all solutions to each trigonometric equation in the interval $x \in [0, 2\pi]$. Express the solutions exactly.

(a) $2\cos^2 x + 5\cos x + 2 = 0$ **(b)** $\sin 2x - \cos x = 0$

8. The value of x is in the interval $\frac{\pi}{2} < x < \pi$ and $\cos^2 x = \frac{8}{9}$.

Without using your GDC, find the exact values for the following:

(a) $\sin x$ **(b)** $\cos 2x$ **(c)** $\sin 2x$

9. The depth, d m, of water in a harbour varies with the tides during each day. The first high (maximum) tide after midnight occurs at 5:00 a.m. with a depth of 5.8 metres. The first low (minimum) tide occurs at 10:30 a.m. with a depth of 2.6 metres.

(a) Find a trigonometric function that models the depth, d, of the water t hours after midnight.

(b) Find the depth of the water at 12 noon.

(c) A large boat needs at least 3.5 metres of water to dock in the harbour. Determine the time interval after 12 noon during which the boat can dock safely?

10. Solve the equation $\tan^2 x + 2\tan x - 3 = 0$ for $0 \leqslant x \leqslant \pi$. Give solutions exactly if possible, otherwise give to three significant figures.

11. The diagram shows a circle of centre O and radius 10 cm. The arc ABC subtends an angle of $\frac{3}{2}$ radians at the centre O.

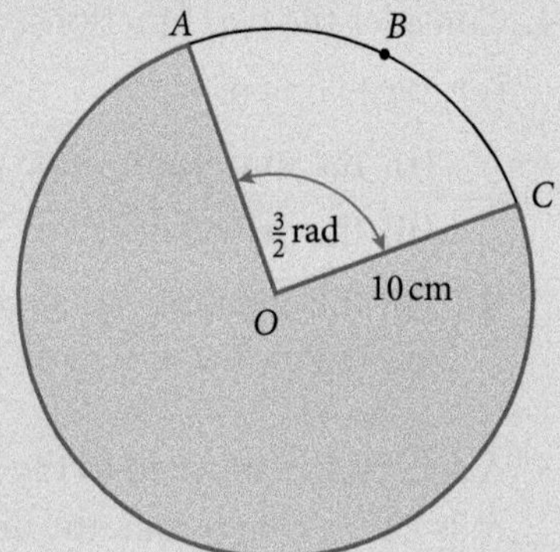

(a) Find the length of the arc ABC.

(b) Find the area of the shaded region.

12. Consider the function $f(x) = \frac{5}{2}\cos\left(2x - \frac{\pi}{2}\right)$. Find the values of k for which the equation $f(x) = k$ will have no solutions.

13. The diagram shows a portion of the graph of $y = k + a\sin x$. The graph passes through the points $(0, 1)$ and $\left(\frac{3\pi}{2}, 3\right)$. Find the value of k and the value of a.

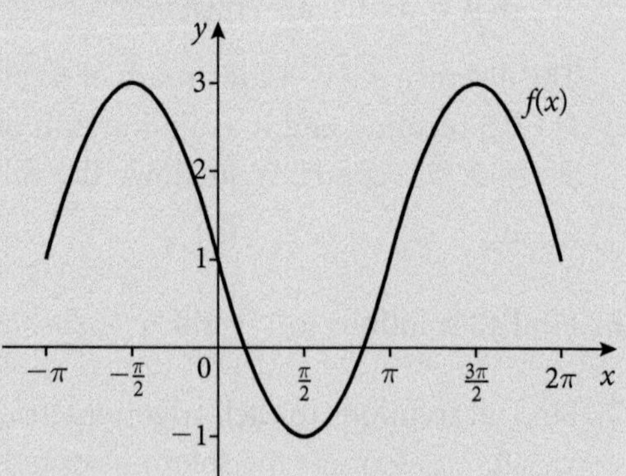

14. The obtuse angle B is such that $\tan B = -\frac{5}{12}$. Find the exact values of

(a) $\sin B$ **(b)** $\cos B$ **(c)** $\sin 2B$ **(d)** $\cos 2B$

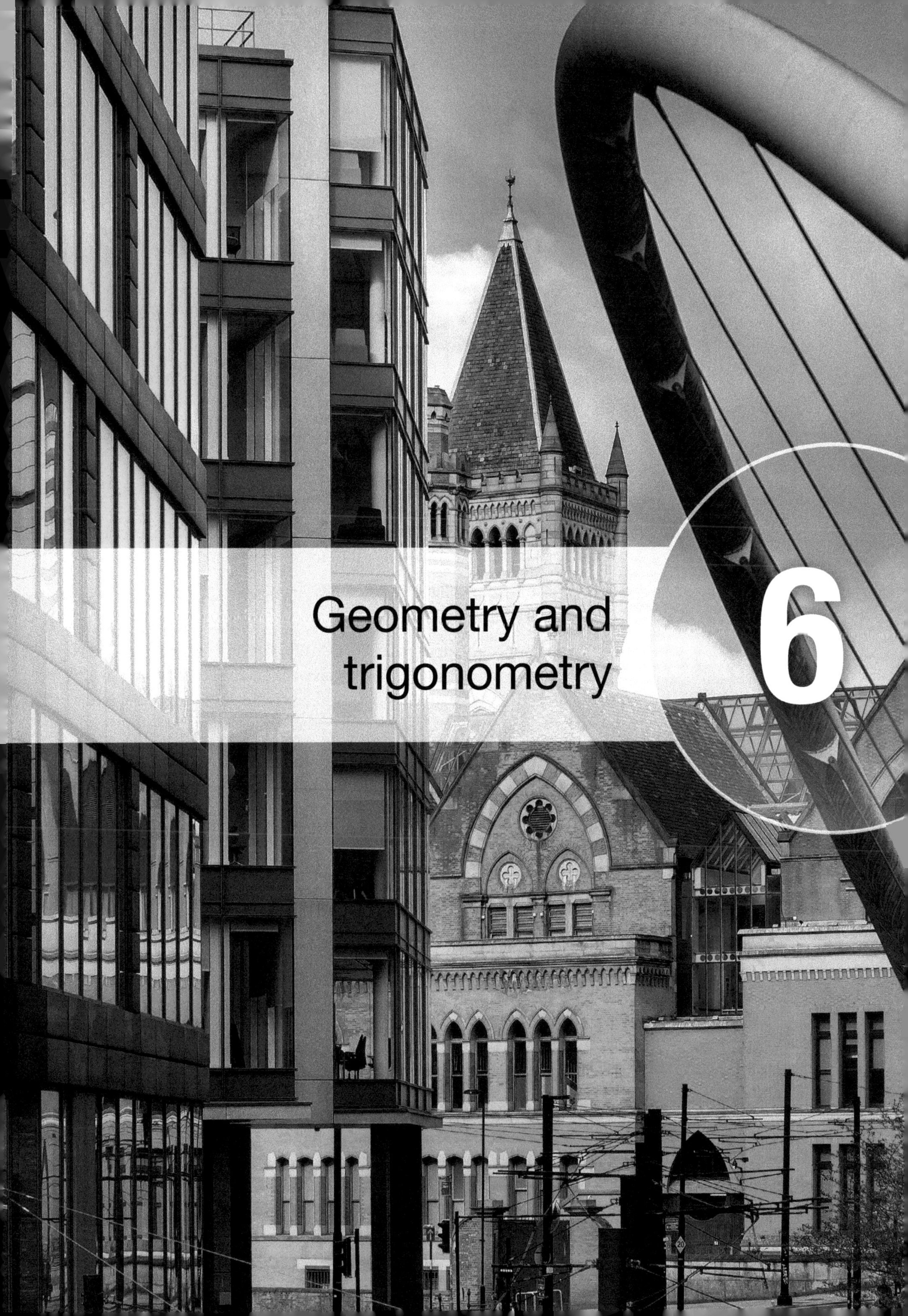

6 Geometry and trigonometry

6 Geometry and trigonometry

Learning objectives

By the end of this chapter, you should be familiar with...

- finding the distance between two points in 3-dimensional space
- finding the midpoint of a line segment in 3-dimensional space
- computing the volume and surface area of a solid such as a pyramid, cone, sphere, hemisphere or a solid made from a combination of these
- determining the size of an angle between two lines
- finding the sides and angles of a right-angled triangle using the sine, cosine and tangent ratios
- applying the sine rule and the cosine rule to find an unknown length or an angle
- computing the area of a triangle using the formula $\frac{1}{2}ab \sin C$
- solving problems involving 2-dimensional or 3-dimensional figures by means of right-angled and non-right-angled trigonometry
- solving problems involving compass bearings.

In this chapter, we cover some basic 3-dimensional geometry and trigonometry using right-angled triangles. This chapter may contain topics that you have studied before. Trigonometric functions will be defined in terms of the ratios of the sides of a right-angled triangle rather than in terms of an arc on the unit circle.

6.1 Measurements in three dimensions

Given the coordinates of two points $A(x_1, y_1)$ and $B(x_2, y_2)$ in a 2-dimensional plane, recall that the length AB of the segment $[AB]$ is the hypotenuse of a right-angled triangle where the length of one side of the right-angled triangle is the difference in the x-coordinates and the length of the other perpendicular side is the difference in the y-coordinates (Figure 6.1).

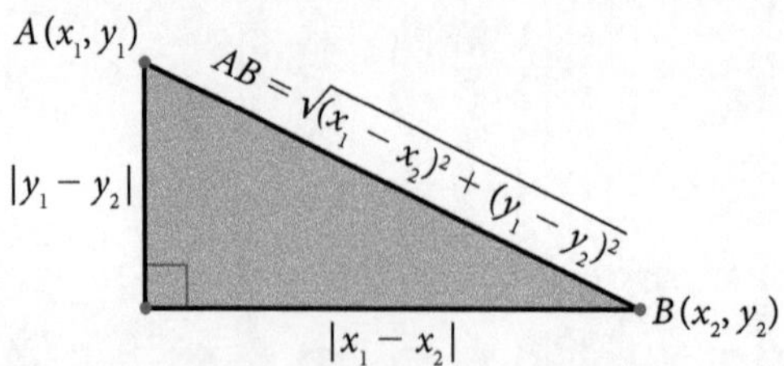

Figure 6.1 Distance between two points in a plane derived from the Pythagorean theorem

Thus, from the Pythagorean theorem, the distance, d, between the points with coordinates (x_1, y_1) and (x_2, y_2) is $d = \sqrt{(x_1 - x_2)^2 + (y_1 - y_2)^2}$.

The **midpoint** M of $[AB]$ is the point whose coordinates are the average of the x-coordinates and the average of the y-coordinates, respectively (Figure 6.2). Thus, the midpoint, M, of the line segment with endpoints (x_1, y_1) and (x_2, y_2) has coordinates $M\left(\frac{x_1 + x_2}{2}, \frac{y_1 + y_2}{2}\right)$.

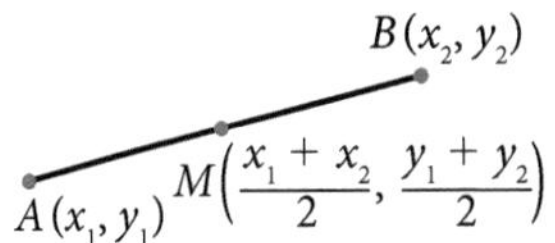

Figure 6.2 Midpoint of a line segment in a plane

The rule for drawing the x, y, and z axes of a 3D coordinate system is that the direction of the positive z axis points in the direction of the thumb on a right hand when the fingers curl in the direction from the positive x axis to the positive y axis, as illustrated in Figure 6.3. The three axes in Figure 6.4 also conform to the right-hand rule.

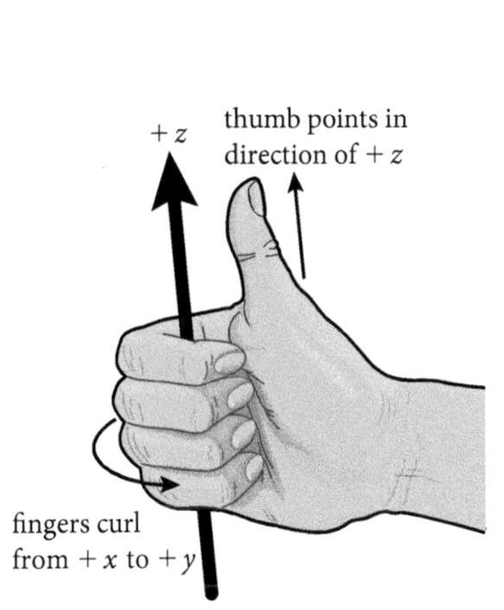

Figure 6.3 The orientation of the $+x$, $+y$ and $+z$ axes must follow the right hand rule

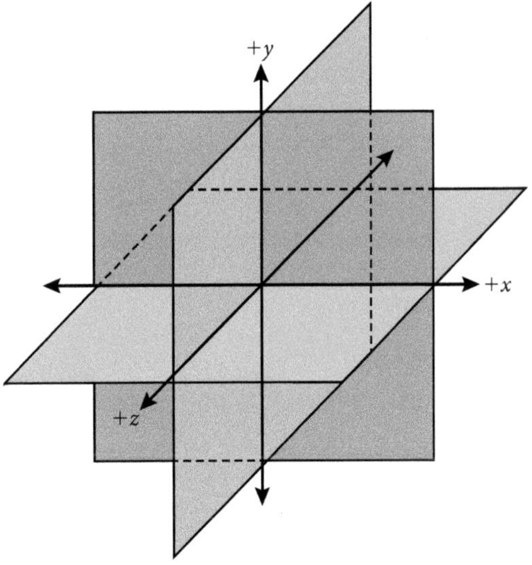

Figure 6.4 3D coordinate system showing axes and the x-y plane, x-z plane and y-z plane

It is straightforward to extend the formulas for the distance between two points and the midpoint of a line segment from points in a 2-dimensional coordinate system to points in a 3-dimensional coordinate system (Figure 6.4). Consider a right rectangular prism (all six faces are rectangles: a cuboid) with a width of a units, a depth of b units, and a height of c units (Figure 6.5). Consider a diagonal $[PQ]$ of the right rectangular prism. Applying the Pythagorean theorem twice – first to find the length of the diagonal of the bottom face and then to find the length, PQ, of the diagonal of the prism – gives $PQ = \sqrt{a^2 + b^2 + c^2}$

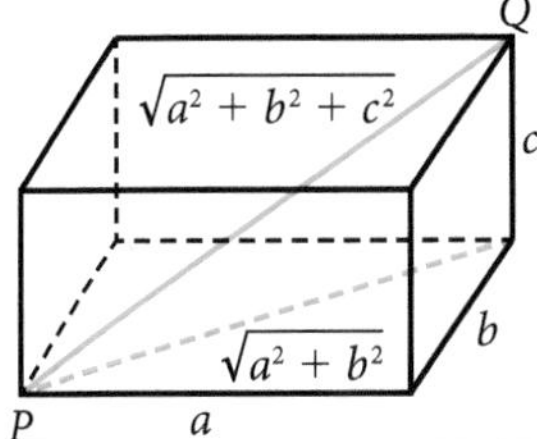

Figure 6.5 Pythagoras' theorem for three-dimensional space

A line segment in space with endpoints $P(x_1, y_1, z_1)$ and $Q(x_2, y_2, z_2)$ is the diagonal of a right rectangular prism as shown in Figure 6.6 and we can use Pythagoras' theorem, as we did for the prism in Figure 6.5, to find PQ.

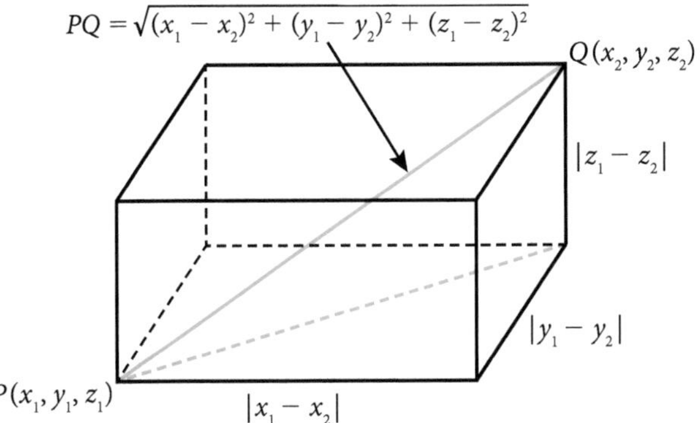

Figure 6.6 The distance between two points in space

The **distance**, d, between the points (x_1, y_1, z_1) and (x_2, y_2, z_2) is
$d = \sqrt{(x_1 - x_2)^2 + (y_1 - y_2)^2 + (z_1 - z_2)^2}$

The **midpoint** M of the line segment with endpoints (x_1, y_1, z_1) and (x_2, y_2, z_2) has coordinates
$M\left(\frac{x_1 + x_2}{2}, \frac{y_1 + y_2}{2}, \frac{z_1 + z_2}{2}\right)$

Example 6.1

Use the distance formula to show that the three points $F(1, -1, 3)$, $G(2, -4, 5)$, and $H(5, -13, 11)$ are collinear.

Solution

Find the exact distance between each of the three pairs of points. The three points are collinear (lie on the same line) if the sum of any two distances is equal to the third distance.

$FG = \sqrt{(1-2)^2 + (-1-(-4))^2 + (3-5)^2} = \sqrt{1+9+4} = \sqrt{14}$

$FH = \sqrt{(1-5)^2 + (-1-(-13))^2 + (3-11)^2} = \sqrt{16+144+64} = 4\sqrt{14}$

$GH = \sqrt{(2-5)^2 + (-4-(-13))^2 + (5-11)^2} = \sqrt{9+81+36} = 3\sqrt{14}$

It is true that $FG + GH = FH$. Therefore, points F, G, and H are collinear.

Example 6.2

Given that three of the vertices of parallelogram $ABCD$ are $A(3, -1, 2)$, $B(1, 2, -4)$, and $C(-1, 1, 2)$, determine the coordinates of vertex D.

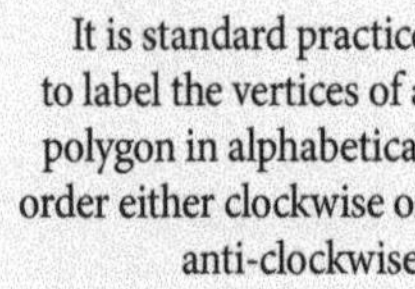
It is standard practice to label the vertices of a polygon in alphabetical order either clockwise or anti-clockwise.

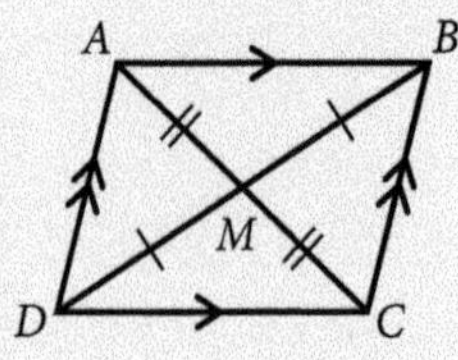

Solution

The diagonals of a parallelogram bisect each other. Thus, the diagonals AC and BD must have the same midpoint M.

$M\left(\frac{3-1}{2}, \frac{-1+1}{2}, \frac{2+2}{2}\right) = M(1, 0, 2)$

Let (x, y, z) be the coordinates of vertex D.

Then $\frac{x+1}{2} = 1, \frac{y+2}{2} = 0, \frac{z-4}{2} = 2.$

Therefore, the coordinates of D are $(1, -2, 8)$.

3-dimensional solids: volumes and surface areas

Table 6.1 lists the solids and their respective formulae that you need to know for this course.

Solid	Volume	Surface area	Parameters
cuboid	$V = l \cdot w \cdot h$	$S = 2(l \cdot w + l \cdot h + w \cdot h)$	l = length; w = width; h = height
sphere	$V = \frac{4}{3}\pi r^3$	$S = 4\pi r^2$	r = radius
prism	$V = A_{base} \cdot h$	$S = 2 \cdot A_{base} + \sum_{i=1}^{n} A_{lateral face}$	A_{base} = area of polygonal base $\sum_{i=1}^{n} A_{lateral face}$ = sum of the n lateral faces, each of which is a rectangle
pyramid	$V = \frac{1}{3} A_{base} \cdot h$	$S = A_{base} + \sum_{i=1}^{n} A_{lateral face}$	A_{base} = area of polygonal base $\sum_{i=1}^{n} A_{lateral face}$ = sum of the n lateral faces, each of which is a triangle
cylinder	$V = \pi r^2 h$	$S = 2\pi r^2 + 2\pi rh$ or $= 2\pi r(r + h)$	r = radius; h = height
cone	$V = \frac{1}{3}\pi r^2 h$	$S = \pi r^2 + \pi rl$ where $l = \sqrt{r^2 + h^2}$	r = radius; h = height l = lateral height (or slant height)

Table 6.1 Volume and surface area formulae for different solids

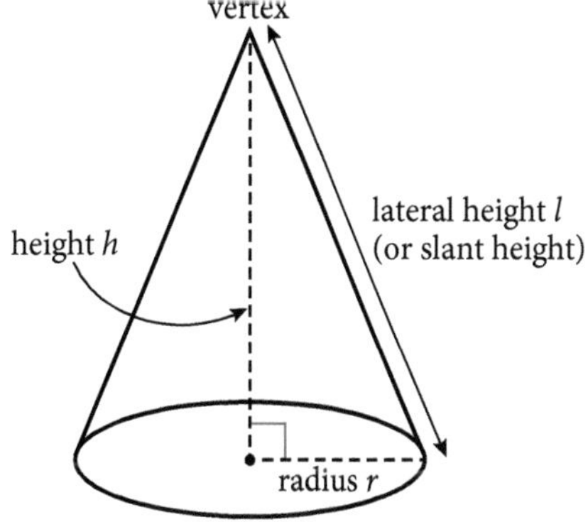

Figure 6.7 Features of a cone

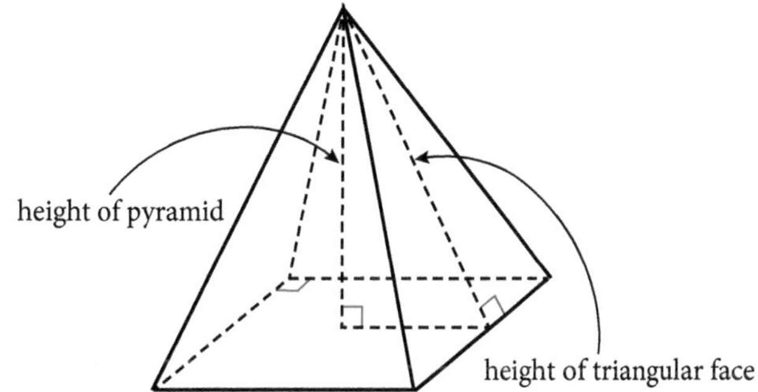

Figure 6.8 Features of a pyramid

Example 6.3

A triangular pyramid sits on top of a triangular right prism. The prism has a height of h. Find the height of the pyramid – in terms of h – so that the prism and the pyramid have the same volume.

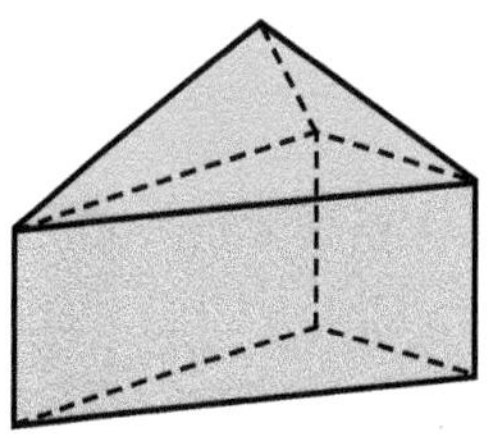

Figure 6.9 Diagram for Example 6.3

Solution

The volume of the prism is $V = A_{base} \cdot h$ and the volume of the pyramid is $V = \frac{1}{3} A_{base} \cdot h$. The two solids have the same base. Therefore, for the two solids to have the same volume, the height of the pyramid needs to be $3h$.

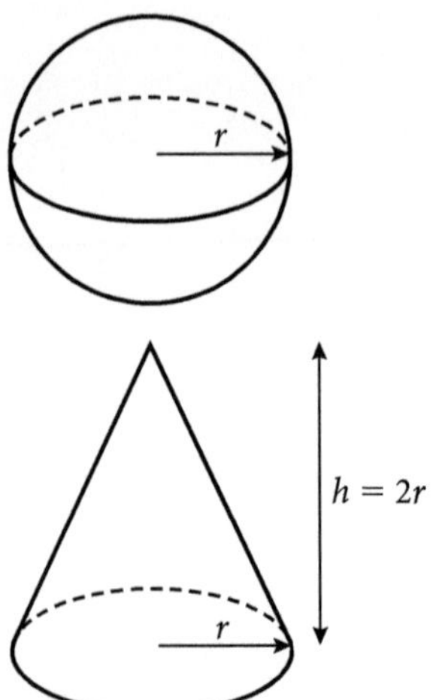

Figure 6.10 Diagrams for Example 6.4

Example 6.4

Compare a sphere and a cone. The radius of the sphere and the radius of cone are both r. The height of the cone is $2r$. Find the exact value of each ratio:

(a) The ratio of the volume of the sphere to the volume of the cone.

(b) The ratio of the surface area of the cone to the surface area of the sphere.

Solution

(a) The volume of the sphere is $V = \frac{4}{3}\pi r^3$

The volume of the cone is $V = \frac{1}{3}\pi r^2 h$. Since $h = 2r$, then

$V = \frac{1}{3}\pi r^2(2r) = \frac{2}{3}\pi r^3$

Thus, $\frac{V_{\text{sphere}}}{V_{\text{cone}}} = \frac{\frac{4}{3}\pi r^3}{\frac{2}{3}\pi r^3} = 2$

(b) The surface area of the cone is $S = \pi r^2 + \pi rl = \pi r^2 + \pi r\left(\sqrt{r^2 + h^2}\right)$

Substituting $h = 2r$, gives $S = \pi r^2 + \pi r\left(\sqrt{r^2 + (2r)^2}\right) = \pi r^2 + \pi r\left(\sqrt{5}\, r^2\right)$
$= \pi r^2(1 + \sqrt{5})$

The surface area of the sphere is $S = 4\pi r^2$

Thus, $\frac{S_{\text{cone}}}{S_{\text{sphere}}} = \frac{\pi r^2(1 + \sqrt{5})}{4\pi r^2} = \frac{1 + \sqrt{5}}{4}$

For three line segments to form a triangle, their lengths must satisfy the **triangle inequality theorem** which states that the sum of the lengths of any two sides of a triangle must be greater than the length of the third side.

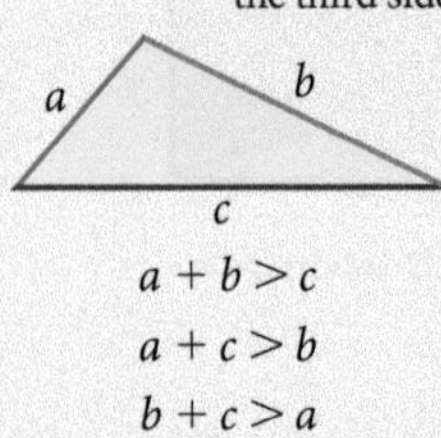

$a + b > c$
$a + c > b$
$b + c > a$

Exercise 6.1

1. For each set of three points, determine whether they are the vertices of a scalene, isosceles, or equilateral triangle.

(a) $(3, -1, 5), (-4, 0, 2), (2, 2, -1)$

(b) $(-2, 4, -3), (4, -3, -2), (-3, -2, 4)$

(c) $(4, 5, 0), (2, 6, 2), (2, 3, -1)$

(d) $(a, b, c), (b, c, a), (c, a, b)$

2. Find the point on the y-axis that is a distance of $\sqrt{10}$ from the point $(1, 2, 3)$
.

3. For each set of three points, determine whether or not they could be the vertices of a triangle.

(a) $A(-1, 2, 3), B(1, 4, 5), C(5, 4, 0)$

(b) $P(2, -3, 3), Q(1, 2, 4), R(3, -8, 2)$

4. Show that the points $(0, 7, 10), (-1, 6, 6)$, and $(-4, 9, 6)$ are the vertices of a right-angled isosceles triangle.

5. For each set of three points, determine whether they are collinear.
   (a) $(0, -1, -7), (2, 1, -9), (6, 5, -13)$
   (b) $(-2, 0, 4), (5, -1, 1), (4, -6, 3)$
   (c) $(1, 8, -4), (-3, 5, -1), (2, 7, 2)$
   (d) $(2, 3, 4), (-1, 2, -3), (-4, 1, -10)$

6. Find the distance of each of these points from:
   (i) the origin (ii) the x-axis (iii) the y-axis (iv) the z-axis.
   (a) $(2, 6, -3)$
   (b) $(2, -\sqrt{3}, 3)$

7. $PQRS$ is a parallelogram. Given that three of the vertices are $P(6, -2, 4)$, $Q(2, 4, -8)$, and $R(-2, 2, 4)$, determine the coordinates of vertex S.

8. Consider the triangle with vertices $X(2, 2, 3)$, $Y(3, 7, 5)$, and $Z(1, 4, -2)$.
   (a) Show that triangle XYZ is isosceles.
   (b) Find the exact area of triangle XYZ.

9. A line segment connecting two antipodal (diametrically opposite) points on a sphere will pass through the centre of the sphere.
   Points $A(2, -7, -4)$ and $B(6, 1, 2)$ are a pair of antipodal points on a sphere. Find the exact surface area and exact volume of this sphere.

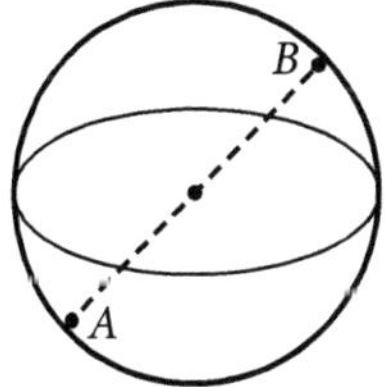

Figure 6.11 Diagram for question 9

10. A rectangular wooden box has dimensions: 62 cm × 44 cm × 20 cm. Find the length (to the nearest whole cm) of the longest piece of straight wire that can be placed completely inside the box.

11. A solid consists of a cone, cylinder and hemisphere joined together as shown in the diagram. Given the dimensions indicated in the diagram, find the volume and surface area of the solid, accurate to 3 significant figures.

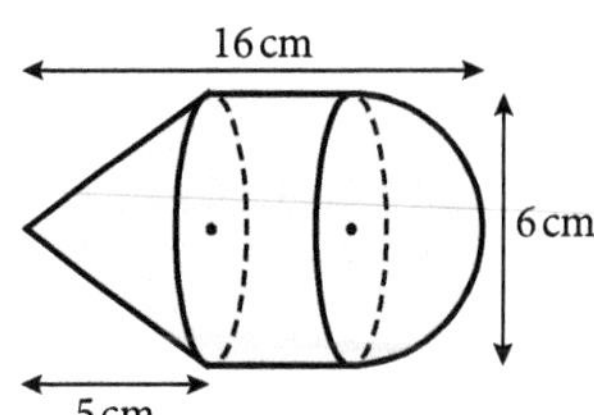

Figure 6.12 Diagram for question 11

12. The midpoints of the sides of a triangle are $(1, 5, -1)$, $(0, 4, -2)$, and $(2, 3, 4)$. Find the coordinates of the three vertices of the triangle.

13. A sphere with radius r is inscribed in a cylinder such that the sphere is tangent to the bases of the cylinder and touches the inside of the cylinder along a circle. Find:
    (a) the ratio of the sphere's volume to the cylinder's volume.
    (b) the ratio of the sphere's surface area to the cylinder's surface area.

14. A large building for storing grain consists of a cylinder with a cone on top (Figure 6.13). Given the dimensions indicated in the diagram, find the surface area and the volume of the building accurate to 3 significant figures. (The circular base is not included in the building's surface.)

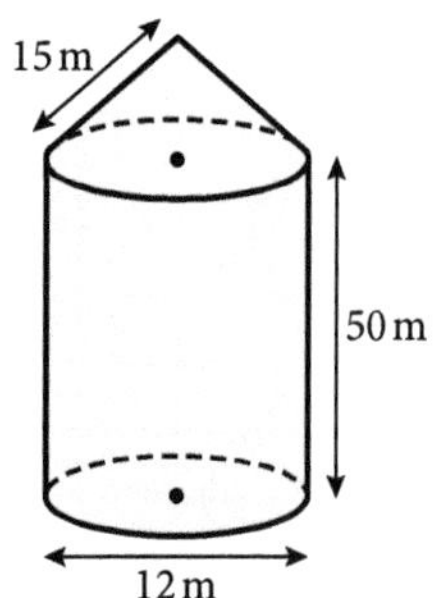

Figure 6.13 Diagram for question 14

15. A metal spike is made from a cube with a pyramid attached to one of the cube's faces, as shown in the diagram. Given the dimensions indicated in the diagram, find the exact volume and exact surface area of the spike.

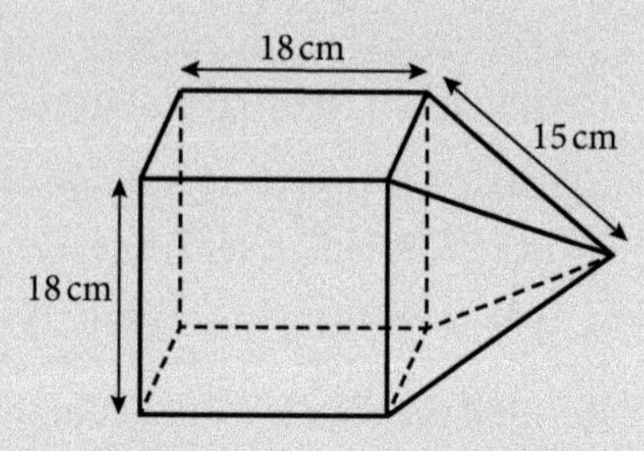

6.2 Right-angled triangles and trigonometric functions of acute angles

In IB notation, the symbol $A\hat{B}C$ denotes the angle with its vertex at point B, with one side of the angle containing point A and the other side containing point C.

Right-angled triangles

The conventional notation for triangles is to label the three vertices with capital letters, for example A, B, and C. The same capital letters can be used to represent the angles at these vertices, but we will often use a Greek letter, such as α (alpha), β (beta), or θ (theta) instead – as we did in Chapter 5. The corresponding lower-case letters, a, b, and c represent the lengths of the sides opposite the vertices. For example, b represents the length of the side opposite angle B – that is, the line segment AC (Figure 6.14).

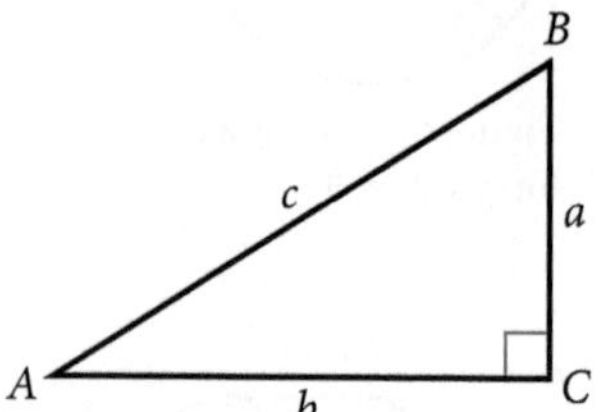

Figure 6.14 Conventional triangle notation

Trigonometric functions of an acute angle

We can use properties of similar triangles and the definitions of the sine, cosine, and tangent functions from Chapter 5 to define these functions in terms of the sides of a right-angled triangle.

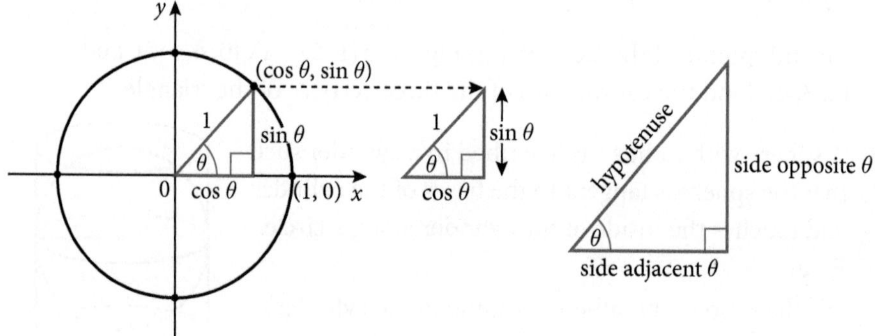

Figure 6.15 Trigonometric functions defined in terms of the sides of a right-angled triangle

The right-angled triangles shown in Figure 6.15 are **similar triangles** because corresponding angles are the same size: each has a right angle and an acute angle of θ. It follows that the ratios of corresponding sides are equal, allowing us to write these three proportions involving the sine, cosine, and tangent of the acute angle θ.

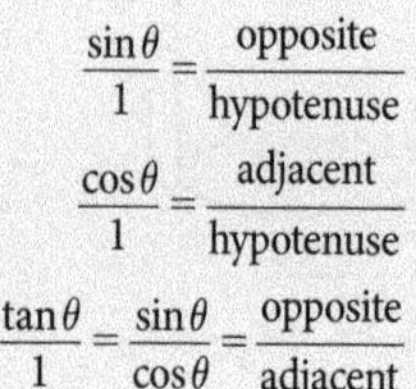

$$\frac{\sin\theta}{1} = \frac{\text{opposite}}{\text{hypotenuse}}$$

$$\frac{\cos\theta}{1} = \frac{\text{adjacent}}{\text{hypotenuse}}$$

$$\frac{\tan\theta}{1} = \frac{\sin\theta}{\cos\theta} = \frac{\text{opposite}}{\text{adjacent}}$$

The definitions of the trigonometric functions in terms of the sides of a right-angled triangle follow directly from these three equations.

Let θ be an **acute angle** of a right-angled triangle. Then the sine, cosine and tangent functions of the angle θ are defined as:

$$\sin\theta = \frac{\text{side opposite angle } \theta}{\text{hypotenuse}}$$

$$\cos\theta = \frac{\text{side adjacent angle } \theta}{\text{hypotenuse}}$$

$$\tan\theta = \frac{\text{side opposite angle } \theta}{\text{side adjacent angle } \theta}$$

It follows that the trigonometric functions of an acute angle are positive.

It is important to understand that properties of similar triangles are the foundation of right-angled triangle trigonometry. Regardless of the size (i.e. lengths of the sides) of a right-angled triangle, so long as the angles don't change, the ratio of any two sides in the right-angled triangle will remain constant. All the right-angled triangles in Figure 6.16 have an acute angle with a measure of 30°. For each triangle, the ratio of the side opposite the 30° angle to the hypotenuse is exactly $\frac{1}{2}$.

In other words, the sine of 30° is always $\frac{1}{2}$. This agrees with results from the previous chapter that an angle of 30° is equivalent to $\frac{\pi}{6}$ in radian measure.

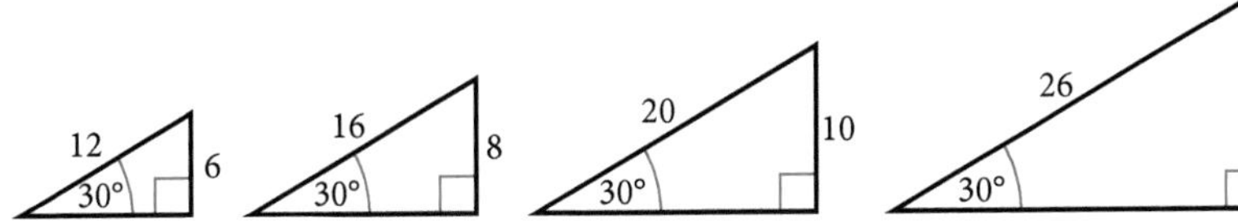

Figure 6.16 Corresponding ratios of a pair of sides for similar triangles are equal

It is important that you are able to recall – without a GDC – the exact trigonometric values for these common angles.

Values of sine, cosine and tangent for common acute angles:

$\sin 30° = \sin\frac{\pi}{6} = \frac{1}{2}$ $\quad\cos 30° = \cos\frac{\pi}{6} = \frac{\sqrt{3}}{2}$ $\quad\tan 30° = \tan\frac{\pi}{6} = \frac{\sqrt{3}}{3}$

$\sin 45° = \sin\frac{\pi}{4} = \frac{\sqrt{2}}{2}$ $\quad\cos 45° = \cos\frac{\pi}{4} = \frac{\sqrt{2}}{2}$ $\quad\tan 45° = \tan\frac{\pi}{4} = 1$

$\sin 60° = \sin\frac{\pi}{3} = \frac{\sqrt{3}}{2}$ $\quad\cos 60° = \cos\frac{\pi}{3} = \frac{1}{2}$ $\quad\tan 60° = \tan\frac{\pi}{3} = \sqrt{3}$

Observe that $\sin 30° = \cos 60° = \frac{1}{2}$, $\sin 60° = \cos 30° = \frac{\sqrt{3}}{2}$ and $\sin 45° = \cos 45° = \frac{\sqrt{2}}{2}$.

Complementary angles (sum of 90°) have equal function values for sine and cosine. That is, for all angles x measured in degrees:

$\sin x = \cos(90° - x)$ or $\sin(90° - x) = \cos x$.

As noted in Chapter 5, it is for this reason that sine and cosine are called co-functions.

Finding unknowns of right-angled triangles

We can use Pythagoras' theorem and trigonometric functions to find the size of any unknown side or angle. We will use trigonometric functions in two different ways – to find the length of a side, and to find the measure of an angle. Finding unknowns in right-angled triangles using the sine, cosine and tangent functions is essential to finding solutions to problems in fields such as astronomy, navigation, engineering, and architecture. In Section 6.4, we will see how trigonometry can also be used to find missing parts in triangles that are not right-angled triangles.

Angles of depression and elevation

An imaginary line segment from an observation point O to a point P (representing the location of an object) is called the **line of sight** of P. If P is above O, the acute angle between the line of sight of P and a horizontal line passing through O is called the **angle of elevation** of P. If P is below O, the angle between the line of sight and the horizontal is called the **angle of depression** of P. This is illustrated in Figure 6.17.

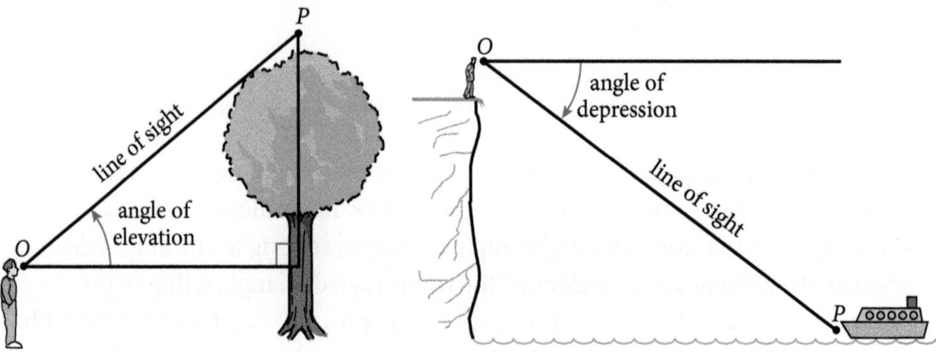

Figure 6.17 An angle of elevation or depression is always measured from the horizontal. Also, note that the angle of elevation from O to P is equal to the angle of depression from P to O

Example 6.5

Determine the lengths of the missing sides in triangle ABC given $c = 8.76$ cm, angle $A = 30°$, and the right angle is at C. Give exact answers when possible, otherwise give to an accuracy of 3 significant figures.

Solution

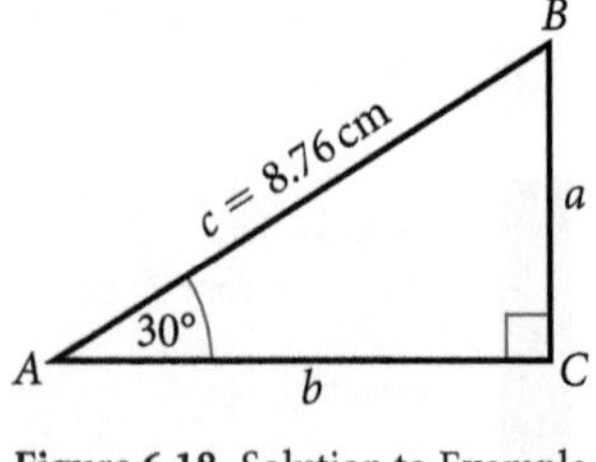

Figure 6.18 Solution to Example 6.5

Sketch triangle ABC indicating the known measurements.

From the definition of sine and cosine functions, we have:

$$\sin 30° = \frac{\text{opposite}}{\text{hypotenuse}} = \frac{a}{8.76} \quad a = 8.76 \sin 30°$$

$$a = 8.76\left(\frac{1}{2}\right) = 4.38$$

$$\cos 30° = \frac{\text{adjacent}}{\text{hypotenuse}} = \frac{b}{8.76}$$

$$b = 8.76 \cos 30°$$

$$b = 8.76\left(\frac{\sqrt{3}}{2}\right) \approx 7.586382537 \approx 7.59$$

Therefore $a = 4.38$ cm, $b \approx 7.59$ cm and, it's clear that angle $B = 60°$.

We can use Pythagoras' theorem to check our results for a and b.

$a^2 + b^2 = c^2 \Rightarrow \sqrt{a^2 + b^2} = 8.76$

Be aware that the result for a is exactly 4.38 cm (assuming measurements given for angle A and side c are exact), but the result for b can only be approximated. To reduce error when performing the check, we should use the most accurate value (i.e. most significant figures) for b possible. The most effective way to do this on your GDC is to use results that are stored to several significant figures.

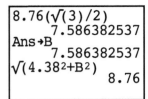

```
8.76(√(3)/2)
                7.586382537
Ans→B
                7.586382537
√(4.38²+B²)
                       8.76
```

Figure 6.19 Using stored results on your GDC

Example 6.6

A man who is 183 cm tall casts a shadow 72 cm long on horizontal ground. What is the angle of elevation of the sun to the nearest tenth of a degree?

Solution

In Figure 6.20, the angle of elevation of the sun is labelled θ.

GDC computation in degree mode.

$$\tan\theta = \frac{183}{72}$$

$$\theta = \tan^{-1}\left(\frac{183}{72}\right)$$

$$\theta \approx 68.5°$$

The angle of elevation of the sun is approximately 68.5°

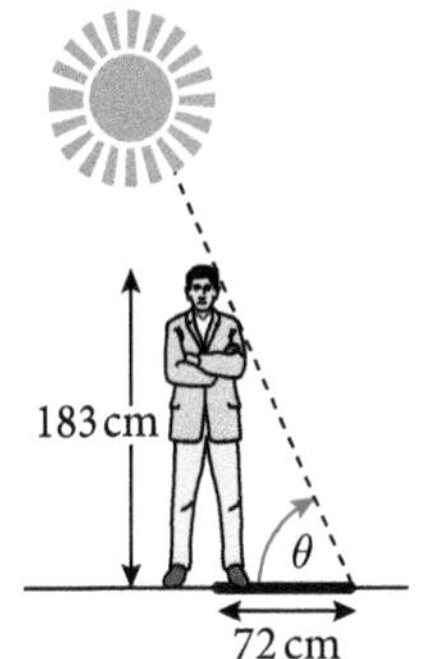

Figure 6.20 Diagram for Example 6.6

```
tan-1(183/72)
        68.52320902
```

Figure 6.21 GDC screen for the solution to Example 6.6

The notation for indicating the inverse of a function is a superscript of negative one. For example, the inverse of the tangent function is denoted as $\tan^{-1}$ on your GDC. The negative one is not an exponent, so it does not denote a reciprocal.

$$\tan^{-1} x \neq \frac{1}{\tan x}$$

Example 6.7

During a training exercise, an Air Force pilot is flying his jet at a constant altitude of 1200 metres. His task is to fire a missile at a target on the ground. At the moment he fires his missile, he is able to see the target at an angle of depression of 18.5°. If the missile travels in a straight line, what distance will the missile cover (to the nearest metre) from the jet to the target?

Solution

Draw a diagram to represent the information and let x be the distance that the missile travels from the jet to the target. A right-angled triangle can be 'extracted' from the diagram with one side 1200 metres, the angle opposite that side is 18.5°, and the hypotenuse is x.

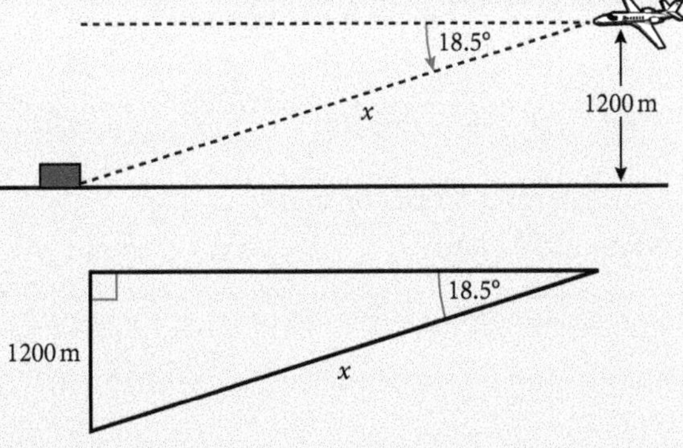

Applying the sine ratio, we can write the equation $\sin 18.5° = \frac{1200}{x}$

Then $x = \frac{1200}{\sin 18.5°} \approx 3781.85$

Hence, the missile travels approximately 3782 metres.

Example 6.8

A boat is sailing directly towards a cliff. The angle of elevation of a point on the top of the cliff and straight ahead of the boat increases from 10° to 15° as the ship sails a distance of 50 metres. Find the height of the cliff.

Solution

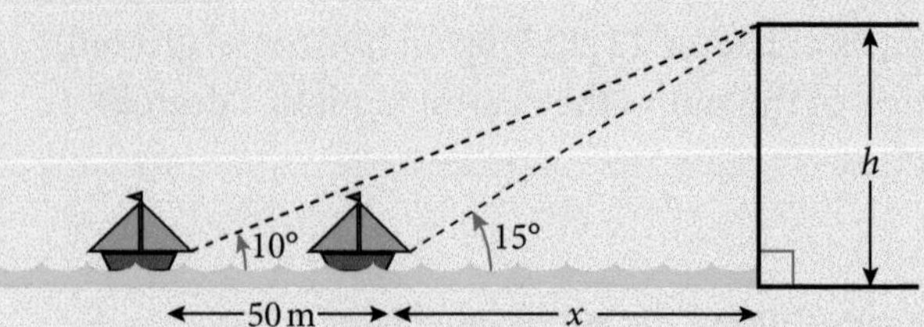

Draw a diagram that accurately represents the information, with the height of the cliff labelled h and the distance from the base of the cliff to the later position of the boat labelled x. There are two right-angled triangles that can be extracted from the diagram. From the smaller triangle, we have:

$$\tan 15° = \frac{h}{x} \Rightarrow h = x\tan 15°$$

From the larger triangle, we have:

$$\tan 10° = \frac{h}{x + 50} \Rightarrow h = (x + 50)\tan 10°$$

We can solve for x by setting the two expressions for h equal to each other. Then we can solve for h by substitution.

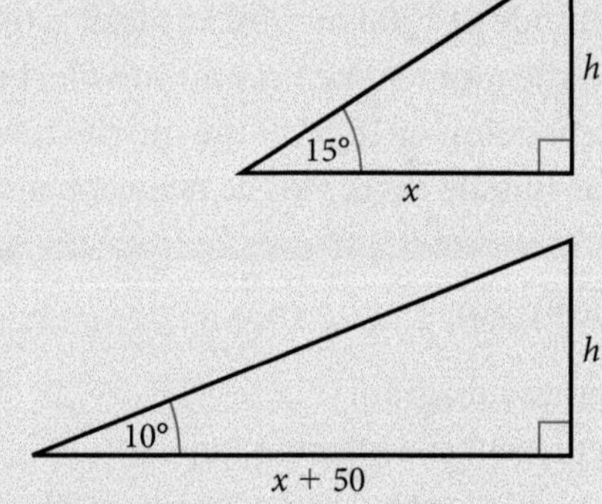

$$x\tan 15° = (x + 50)\tan 10°$$

$$x\tan 15° = x\tan 10° + 50\tan 10°$$

$$x(\tan 15° - \tan 10°) = 50\tan 10°$$

$$x = \frac{50\tan 10°}{\tan 15° - \tan 10°} \approx 96.225$$

Substituting this value for x into $h = x\tan 15°$, gives:

$$h \approx 96.225\tan 15° \approx 25.783$$

Therefore, the height of the cliff is approximately 25.8 metres.

Example 6.9

Using a suitable right-angled triangle, find the exact minimum distance from the point (8, 3) to the line with the equation $2x - y + 2 = 0$.

Solution

Graph the line with equation $2x - y + 2 = 0$. The minimum distance from the point (8, 3) to the line is the length of the line segment drawn from the point perpendicular to the line. This minimum distance is labelled d in the diagram. d is also the height of the large yellow triangle formed by drawing vertical and horizontal line segments from (8, 3) to the line.

The area of the right-angled triangle is $A = \frac{1}{2}\left(\frac{15}{2}\right)(15) = \frac{225}{4}$

The area of the triangle can also be found by using the hypotenuse as the base and the distance d as the height.
By Pythagoras' theorem, we have

$$\text{hypotenuse} = \sqrt{\left(\frac{15}{2}\right)^2 + 15^2} = \sqrt{\frac{1125}{4}} = \frac{\sqrt{225}\sqrt{5}}{\sqrt{4}} = \frac{15\sqrt{5}}{2}$$

Thus, the area can also be expressed as $A = \frac{1}{2}\left(\frac{15\sqrt{5}}{2}\right)d$. We can solve for d by equating the two results for the area of the triangle.

$$\frac{1}{2}\left(\frac{15\sqrt{5}}{2}\right)d = \frac{225}{4}$$

$$\frac{15\sqrt{5}}{4}d = \frac{225}{4}$$

$$d = \frac{225}{4} \cdot \frac{4}{15\sqrt{5}}$$

$$d = \frac{15}{\sqrt{5}} = \frac{15}{\sqrt{5}} \cdot \frac{\sqrt{5}}{\sqrt{5}} = \frac{15\sqrt{5}}{5} = 3\sqrt{5}$$

Therefore, the minimum distance from the point (8, 3) to the line with equation $2x - y + 2 = 0$ is $3\sqrt{5}$ units.

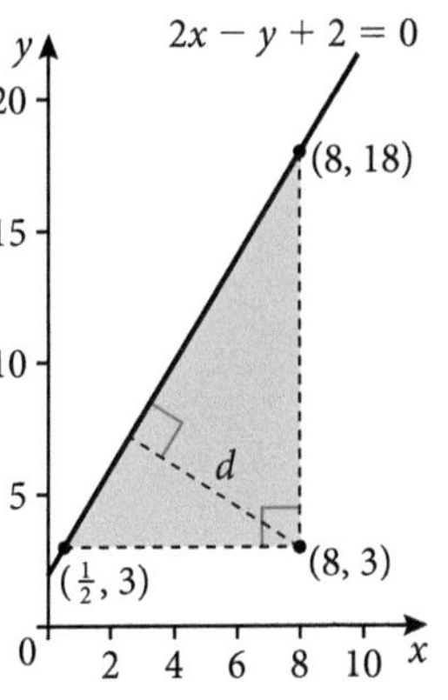

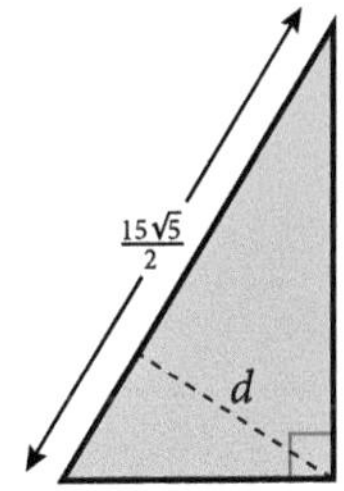

Figure 6.22 Solution to Example 6.9

Exercise 6.2

1. For each **(a)** to **(f)**
   (i) sketch a right-angled triangle corresponding to the acute angle θ
   (ii) find the exact value of the other five trigonometric functions associated with the angle
   (iii) use your GDC to find the degree measure of θ and the other acute angle (approximate to 3 significant figures).

   (a) $\sin\theta = \frac{3}{5}$ **(b)** $\cos\theta = \frac{5}{8}$ **(c)** $\tan\theta = 2$

2. Find the exact value of θ in degrees $(0 < \theta < 90°)$ and in radians $\left(0 < \theta < \frac{\pi}{2}\right)$ without using your GDC.

(a) $\cos\theta = \frac{1}{2}$ (b) $\sin\theta = \frac{\sqrt{2}}{2}$ (c) $\tan\theta = \sqrt{3}$

3. Find the values of x and y in each triangle. If possible, give an exact answer; otherwise, give the answer correct to 3 significant figures.

(a)

(b)

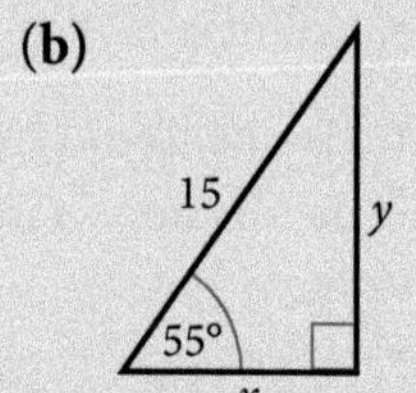

(c)

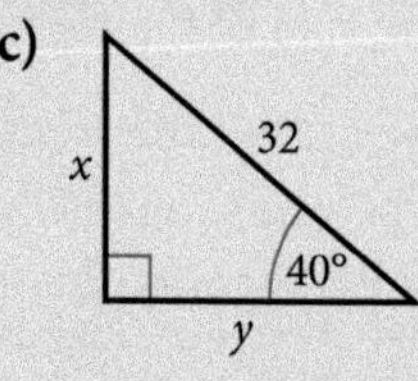

(d)

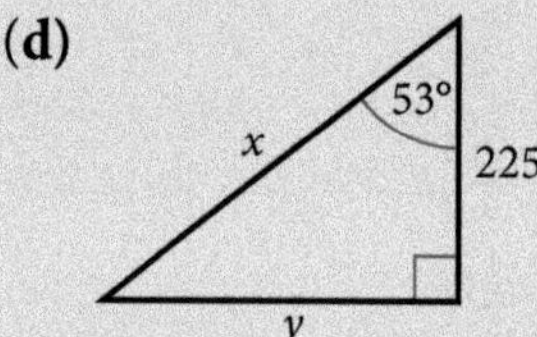

(e)

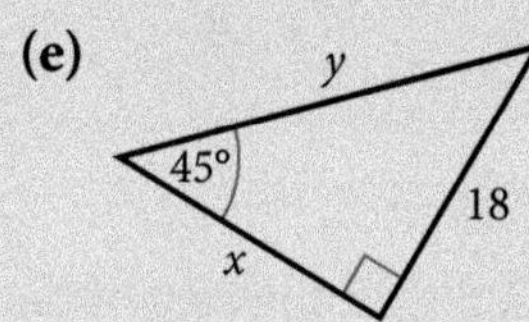

(f)

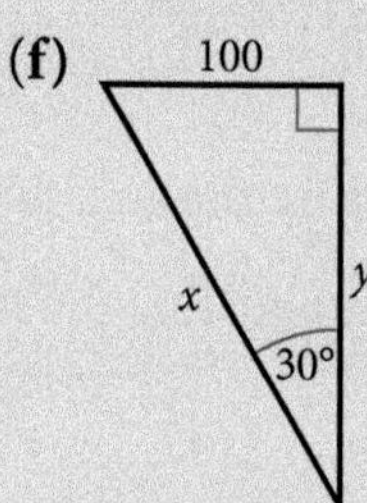

4. Find the size of the angles α and β in degrees. If possible, give an exact answer; otherwise, give your answer correct to 3 significant figures.

(a)

(b)

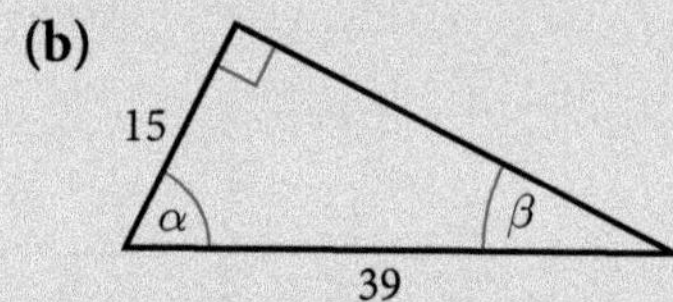

(c)

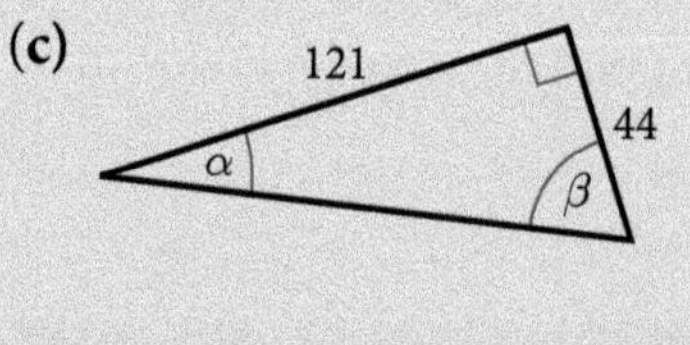

(d)

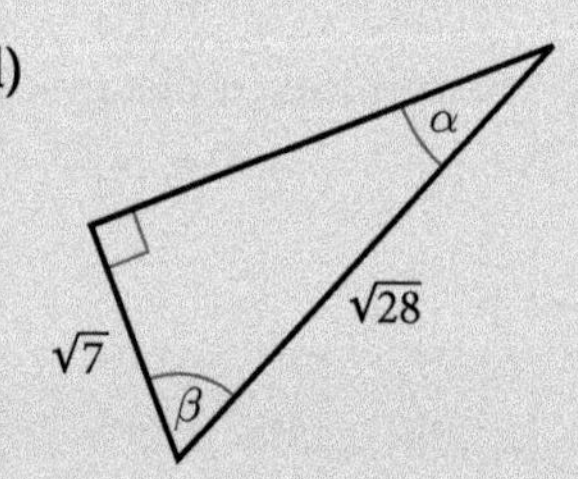

5. The tallest tree in the world is reputed to be a giant redwood named *Hyperion* located in Redwood National Park in California, USA. At a point 41.5 metres from the centre of its base and on the same elevation, the angle of elevation of the top of the tree is 70°. How tall is the tree? Give your answer to 3 significant figures.

6. The Eiffel Tower in Paris is 300 metres high (not including the antenna on top). What is the angle of elevation of the top of the tower from a point on the ground (assumed level) that is 125 metres from the centre of the tower's base?

7. A 1.62 m tall woman, standing 3 metres from a streetlight, casts a 2 m shadow. What is the height of the streetlight?

8. A pilot measures the angles of depression to two ships to be 40° and 52°. The pilot is flying at an elevation of 10 000 metres. Find the distance between the two ships.

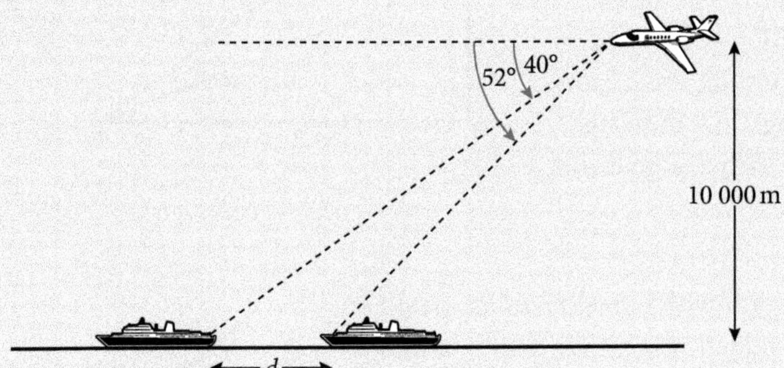

9. Find the size of all three angles in a triangle with sides of length 8 cm, 8 cm, and 6 cm.

10. A boat is sighted from a 50-metre observation tower on the shoreline at an angle of depression of 4° moving directly towards the shore at a constant speed. Five minutes later the angle of depression of the boat is 12°. What is the speed of the boat in kilometres per hour?

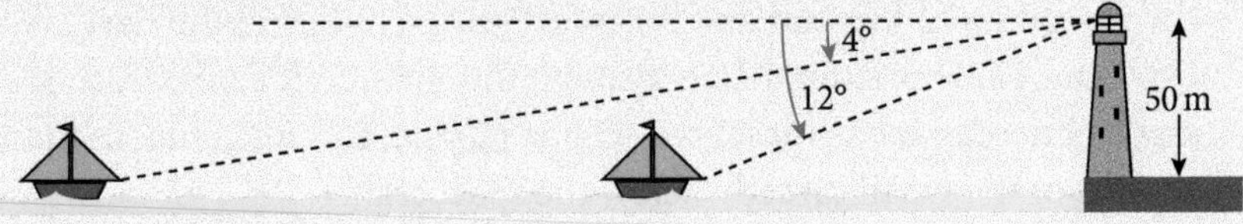

11. Find the length of x indicated in Figure 6.23. Give your answer to 3 significant figures.

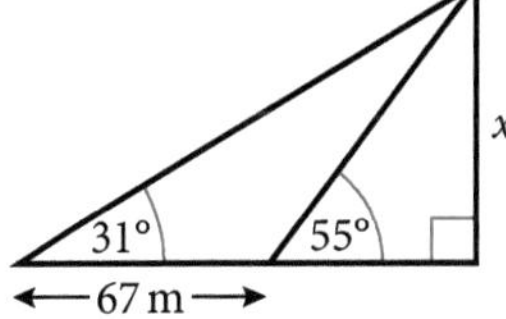

Figure 6.23 Diagram for question 11

12. A support wire for a tower is connected from an anchor point on level ground to the top of the tower. The straight wire makes a 65° angle with the ground at the anchor point. At a point 25 metres farther from the tower than the wire's anchor point and on the same side of the tower, the angle of elevation to the top of the tower is 35°. Find the wire length to the nearest tenth of a metre.

13. A 30-metre high building sits on top of a hill. The angles of elevation of the top and bottom of the building from the same spot at the base of the hill are measured to be 55° and 50° respectively. How high is the hill to the nearest metre?

14. The angle of elevation of the top of a vertical pole as seen from a point 10 metres away from the pole is double its angle of elevation as seen from a point 70 metres from the pole. Find the height (to the nearest tenth of a metre) of the pole above the level of the observer's eyes.

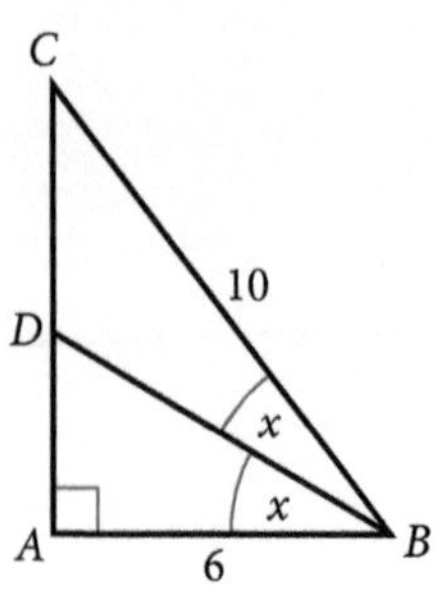

Figure 6.24 Diagram for question 15

15. Angle ABC of a right-angled triangle is bisected by segment BD. The lengths of sides AB and BC are given in Figure 6.24. Find the exact length of BD, expressing your answer in its simplest form.

16. In the diagram, $D\hat{E}C = C\hat{E}B = x°$ and $C\hat{D}E = B\hat{E}A = 90°$, $CD = 1$ unit, $DE = 3$ units. By writing $D\hat{E}A$ in terms of x, find the exact value of $\cos(D\hat{E}A)$.

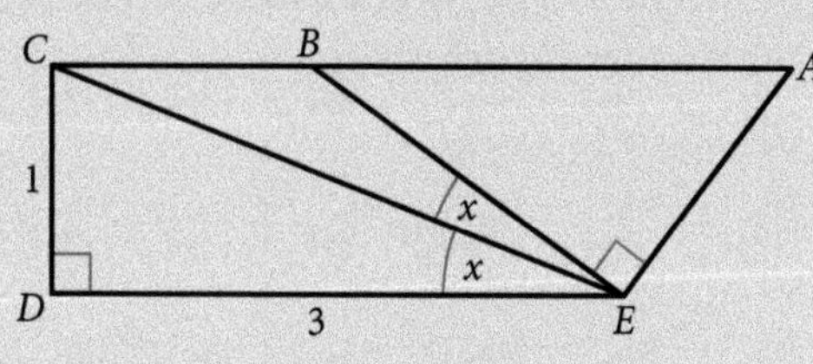

17. For any point with coordinates (p, q) and any line with equation $ax + by + c = 0$, find a formula in terms of a, b, c, p, and q that gives the minimum (perpendicular) distance, d, from the point to the line.

$$x = \frac{d}{\cot\alpha - \cot\beta}$$

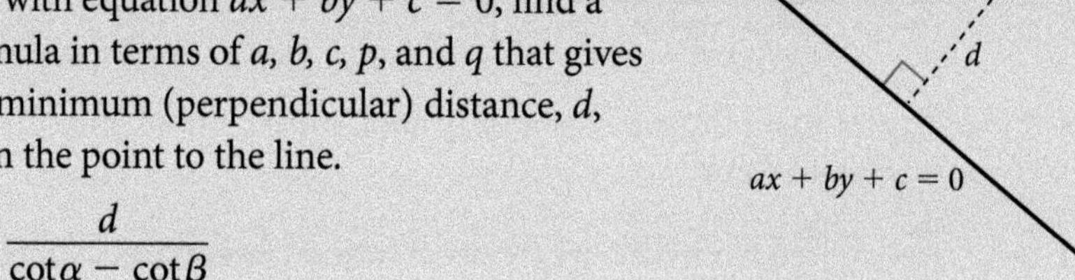

18. A spacecraft is travelling in a circular orbit 200 km above the surface of the Earth. Find the angle of depression (to the nearest degree) from the spacecraft to the horizon. Assume that the radius of the Earth is 6400 km. The 'horizontal' line through the spacecraft from which the angle of depression is measured will be parallel to a line tangent to the surface of the Earth directly below the spacecraft.

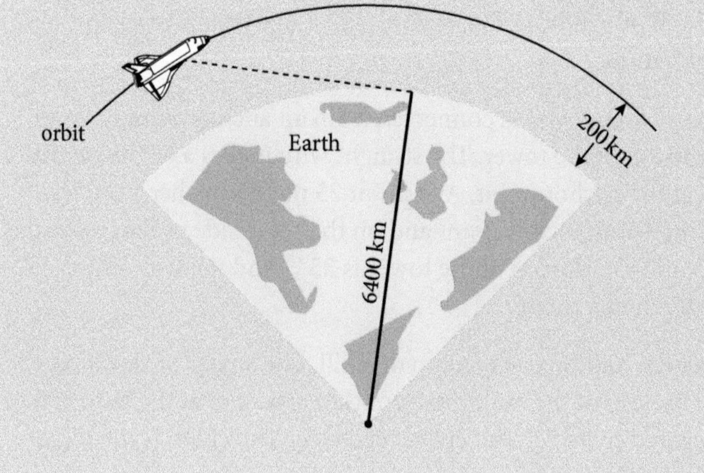

6.3 Trigonometric functions of any angle

In this section we will extend the trigonometric ratios to all angles, allowing us to solve problems involving any size angle.

Defining trigonometric functions for any angle in standard position

Consider the point $P(x, y)$ on the terminal side of an angle θ in standard position (Figure 6.25) such that r is the distance from the origin O to P. If θ is an acute angle, then we can construct a right-angled triangle POQ (Figure 6.26) by dropping a perpendicular from P to a point Q on the x-axis. It follows that:

$$\sin\theta = \frac{y}{r} \qquad \cos\theta = \frac{x}{r} \qquad \tan\theta = \frac{y}{x}\ (x \neq 0)$$

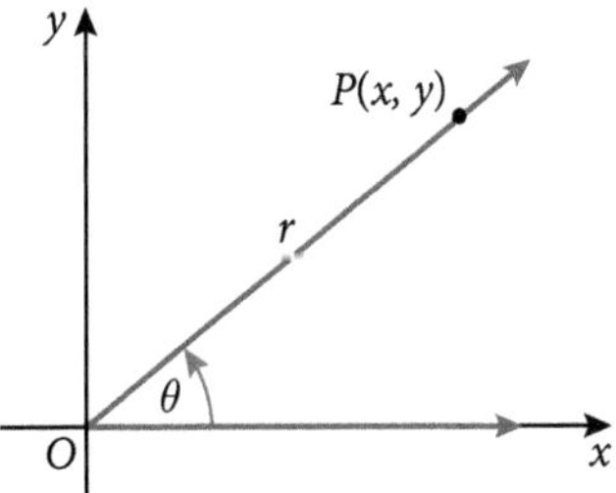

Figure 6.25 Angle θ in standard position

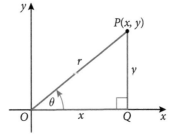

Figure 6.26 θ is an acute angle in ΔPOQ

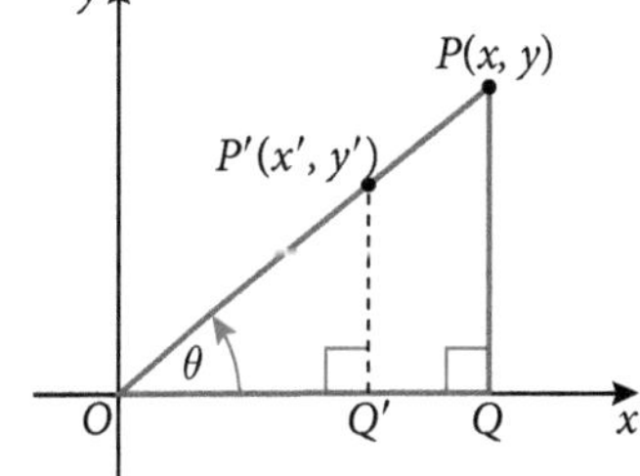

Figure 6.27 Similar right-angled triangles POQ and $P'OQ'$

Extending this to angles other than acute angles allows us to define the trigonometric functions for any angle – positive or negative. It is important to note that the values of the trigonometric ratios do not depend on the choice of the point $P(x, y)$. If $P'(x',y')$ is any other point on the terminal side of angle θ, as in Figure 6.27, then triangles POQ and $P'OQ'$ are similar and the trigonometric ratios for corresponding angles are equal.

Example 6.10

Find the sine, cosine and tangent of an angle α that contains the point $(-3, 4)$ on its terminal side when in standard position.

Solution

$$r = \sqrt{x^2 + y^2} = \sqrt{(-3)^2 + 4^2} = \sqrt{25} = 5$$

Then, $\sin\alpha = \frac{y}{r} = \frac{4}{5}$

$$\cos\alpha = \frac{x}{r} = \frac{-3}{5} = -\frac{3}{5}$$

$$\tan\alpha = \frac{y}{x} = \frac{4}{-3} = -\frac{4}{3}$$

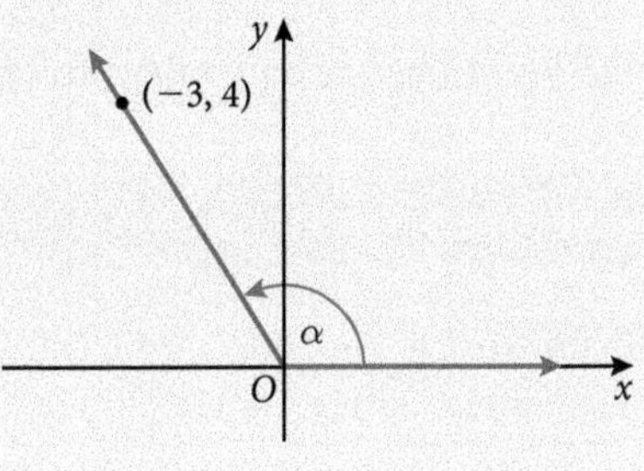

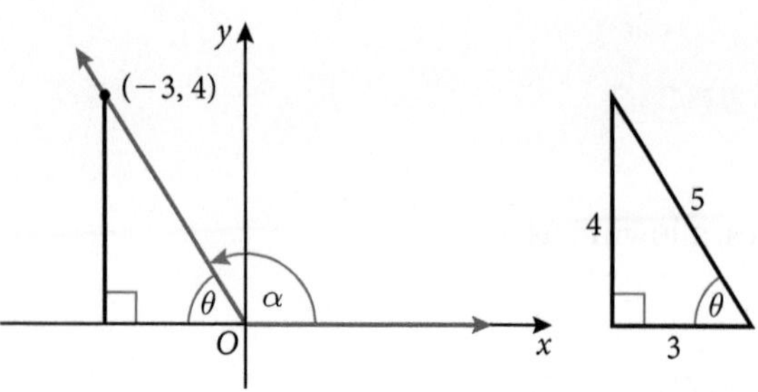

Figure 6.28 Reference triangle for computing trigonometric values for angle a

Note that for the angle α in Example 6.10, we can form a right-angled triangle by constructing a line segment from the point $(-3, 4)$ perpendicular to the x-axis, as shown in Figure 6.28. Clearly, $\theta = 180° - \alpha$. Furthermore, the values of the sine, cosine, and tangent of the angle θ are the same as those for the angle α, except that the sign may be different.

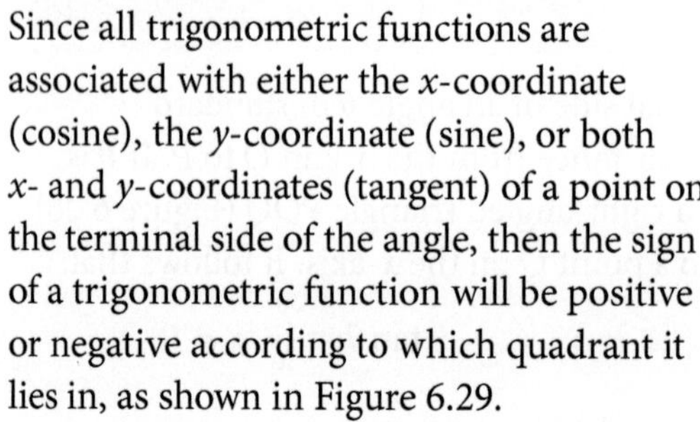

Since all trigonometric functions are associated with either the x-coordinate (cosine), the y-coordinate (sine), or both x- and y-coordinates (tangent) of a point on the terminal side of the angle, then the sign of a trigonometric function will be positive or negative according to which quadrant it lies in, as shown in Figure 6.29.

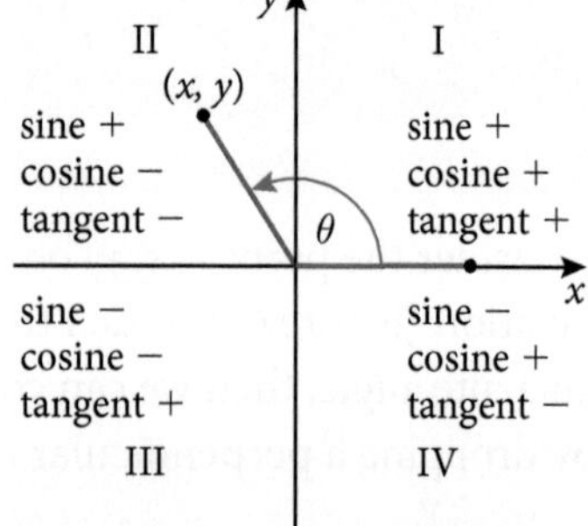

Figure 6.29 Signs of trigonometric function values in each quadrant

Example 6.11

Find the sine, cosine and tangent of the obtuse angle that measures 150°.

Solution

The terminal side of the angle forms a 30° angle with the x-axis. The sine values for 150° and 30° will be exactly the same, and the cosine and tangent values will be the same but of opposite sign.

We know that $\sin 30° = \frac{1}{2}$, $\cos 30° = \frac{\sqrt{3}}{2}$

and $\tan 30° = \frac{\sqrt{3}}{3}$

Therefore, $\sin 150° = \frac{1}{2}$, $\cos 150° = -\frac{\sqrt{3}}{2}$ and $\tan 150° = -\frac{\sqrt{3}}{3}$

Figure 6.30 Solution to Example 6.11

Example 6.11 illustrates three trigonometric identities for angles whose sum is 180° (i.e. a pair of supplementary angles). The following identities are true for any acute angle θ:

$\sin(180° - \theta) = \sin\theta \qquad \cos(180° - \theta) = -\cos\theta \qquad \tan(180° - \theta) = -\tan\theta$

These identities are equivalent to ones in radian measure ($180° = \pi$).

Example 6.12

Given $\sin\theta = \frac{5}{13}$, $90° < \theta < 180°$, find the exact values of $\cos\theta$ and $\tan\theta$.

Solution

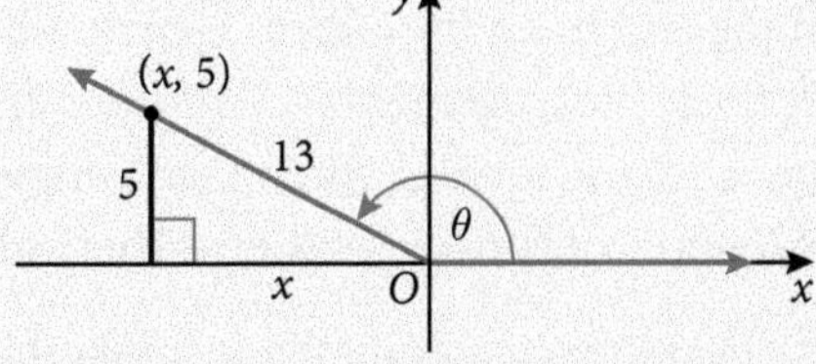

θ is an angle in the 2nd quadrant. It follows from the definition $\sin\theta = \frac{y}{r}$ that with θ in standard position, there must a be a point on the terminal side of the angle that is 13 units from the origin ($r = 13$) and has a y-coordinate of 5, as shown in the diagram.

Thus, $x = \sqrt{13^2 - 5^2} = \sqrt{144} = 12$. Because θ is in the 2nd quadrant, the x-coordinate of the point is negative; thus, $x = -12$.

Therefore, $\cos\theta = \frac{-12}{13} = -\frac{12}{13}$, and $\tan\theta = \frac{5}{-12} = -\frac{5}{12}$

Example 6.13

(a) Find the acute angle with the same sine ratio as (i) 135°, and (ii) 117°

(b) Find the acute angle with the same cosine ratio as (i) 300° and (ii) 342°

Solution

(a) (i) Angles in the 1st and 2nd quadrants have the same sine ratio. Hence, the identity $\sin(180° - \theta) = \sin\theta$.
Since $180° - 135° = 45°$, then $\sin 135° = \sin 45°$

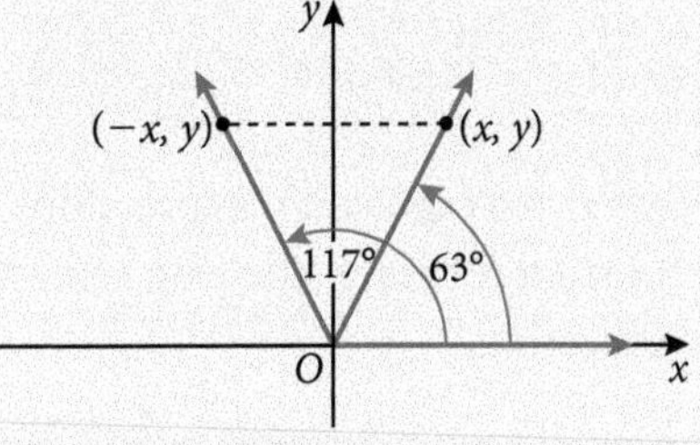

(ii) Since $180° - 117° = 63°$, then $\sin 117° = \sin 63°$

(b) (i) Angles in the 1st and 4th quadrants have the same cosine ratio. Hence, the identity $\cos(360° - \theta) = \cos\theta$.
Since $360° - 300° = 60°$, then $\cos 300° = \cos 60°$

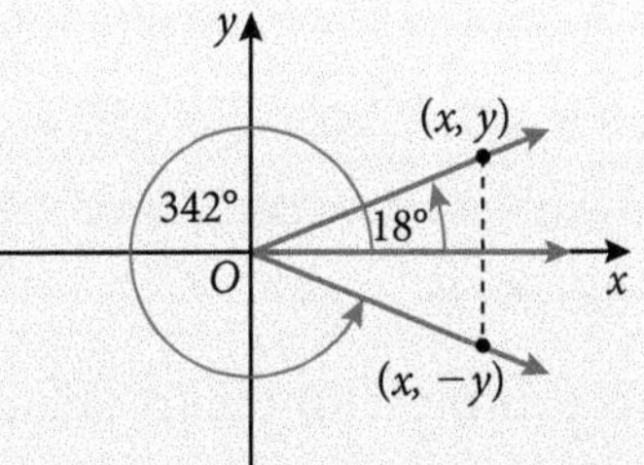

(ii) Since $360° - 342° = 18°$, then $\cos 342° = \cos 18°$

Areas of triangles

We are familiar with the standard formula for the area of a triangle, area $= \frac{1}{2} \times$ base $\times$ height $\left(A = \frac{1}{2}bh\right)$, where the base, b, is a side of the triangle and the height, h, (or altitude) is a line segment perpendicular to the base (or the line containing it) and drawn to the vertex opposite the base, as shown in Figure 6.31.

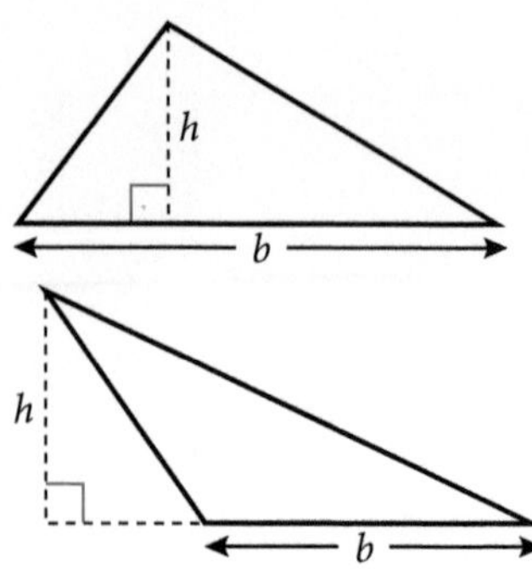

Figure 6.31 Area of a triangle
$A = \frac{1}{2}bh$

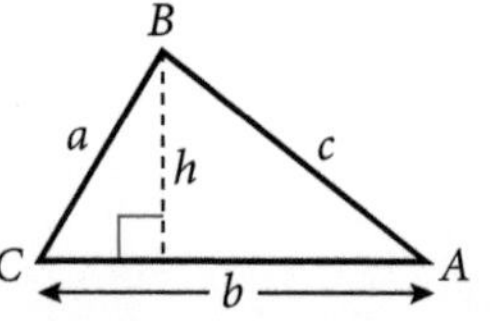

Figure 6.32 An acute triangle

For a triangle with sides of lengths a and b and included angle C: area of $\Delta = \frac{1}{2}ab\sin C$.

For any triangle labelled in the manner of the triangles in Figures 6.32 and 6.33, its area is given by any of the expressions.

If the lengths of two sides of a triangle and the measure of the angle between these sides (often called the included angle) are known, then the triangle is unique and has a fixed area. Hence, we should be able to calculate the area from just these measurements – two sides and the included angle. This calculation is quite straightforward when the triangle is a right-angled triangle and we know the lengths of the two legs on either side of the right angle.

Let's develop a general area formula that will apply to any triangle – right-angled, acute, or obtuse. For triangle ABC shown in Figure 6.32, suppose we know the lengths of the two sides a and b and the included angle C. If the height is h, then the area of ABC is $\frac{1}{2}bh$. From trigonometry, we know that $\sin C = \frac{h}{a}$, or $h = a\sin C$. Substituting $a\sin C$ for h gives area $= \frac{1}{2}bh = \frac{1}{2}b(a\sin C) = \frac{1}{2}ab\sin C$.

If the angle C is obtuse, then from Figure 6.33 we see that $\sin(180^\circ - C) = \frac{h}{a}$.

So, the height is $h = a\sin(180^\circ - C)$. However, $\sin(180^\circ - C) = \sin C$. Thus, $h = a\sin C$ and, again, area$= \frac{1}{2}ab\sin C$.

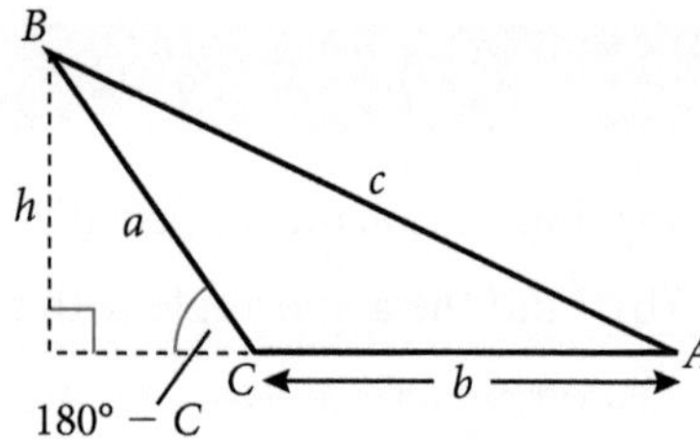

Figure 6.33 An obtuse triangle

$$\text{area of } \Delta = \frac{1}{2}ab\sin C = \frac{1}{2}ac\sin B = \frac{1}{2}bc\sin A$$

These three equivalent expressions will prove to be helpful for developing an important formula (the sine rule) for solving non-right-angled triangles in the next section.

Example 6.14

The circle shown has a radius of 1 cm and the central angle θ subtends an arc of length $\frac{2\pi}{3}$ cm. Find the area of the shaded region (a segment of the circle).

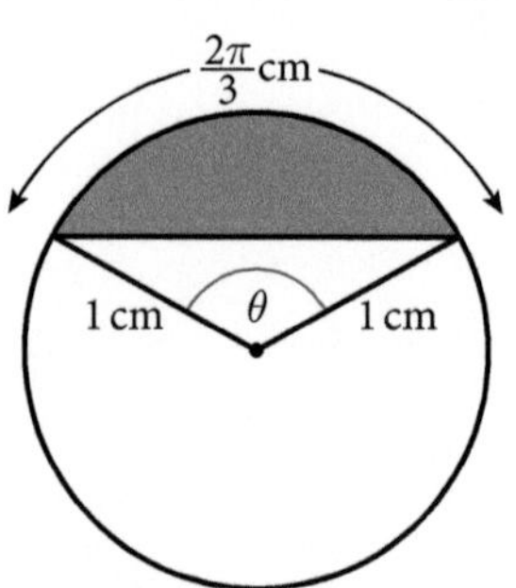

Figure 6.34 Diagram for Example 6.14

Solution

The formula for the area of a sector is $A = \frac{1}{2}r^2\theta$ where θ is the central angle in radians. Since the radius of the circle is 1, the length of the arc subtended by θ is the same as the size of θ. Thus, area of sector $= \frac{1}{2}(1)^2\left(\frac{2\pi}{3}\right) = \frac{\pi}{3}$ cm². The area of the triangle formed by the two radii and the chord is equal to $\frac{1}{2}(1)(1)\sin\left(\frac{2\pi}{3}\right) = \frac{1}{2}\left(\frac{\sqrt{3}}{2}\right) = \frac{\sqrt{3}}{4}$ cm²

We find the area of the shaded region (segment) by subtracting the area of the triangle from the area of the sector. Area$= \frac{\pi}{3} - \frac{\sqrt{3}}{4}$ or $\frac{4\pi - 3\sqrt{3}}{12}$ or approximately 0.614 cm² (3 s.f.)

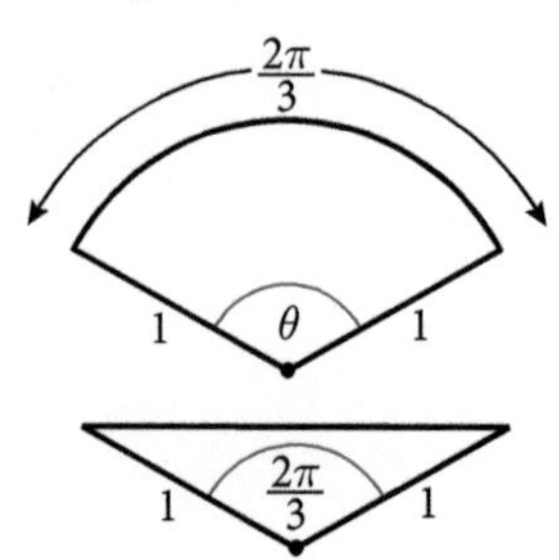

Figure 6.35 Solution for Example 6.14

Example 6.15

Show that it is possible to construct two different triangles with an area of 35 cm^2 that have sides measuring 8 cm and 13 cm. For each triangle, find the size of the (included) angle between the sides of 8 cm and 13 cm to the nearest tenth of a degree.

Solution

We can visualise the two different triangles with equal areas: one with an acute included angle (α) and the other with an obtuse included angle (β).

$$\text{area} = \frac{1}{2}(\text{side})(\text{side})(\text{sine of included angle}) = 35\text{ cm}^2$$

$$\Rightarrow \frac{1}{2}(8)(13)(\sin\alpha) = 35$$

$$52\sin\alpha = 35$$

$$\sin\alpha = \frac{35}{52}$$

$$\alpha = \sin^{-1}\left(\frac{35}{52}\right)$$

$$= 42.3° \text{ (to the nearest tenth of a degree)}$$

Knowing that $\sin(180° - \alpha) = \sin\alpha$, the obtuse angle β is equal to $180° - 42.3° = 137.7°$

Therefore, there are two different triangles with sides 8 cm and 13 cm and an area of 35 cm^2, one with an included angle of 42.3° and the other with an included angle of 137.7°

Your GDC will give only the acute angle value for the $\sin^{-1}$ function.

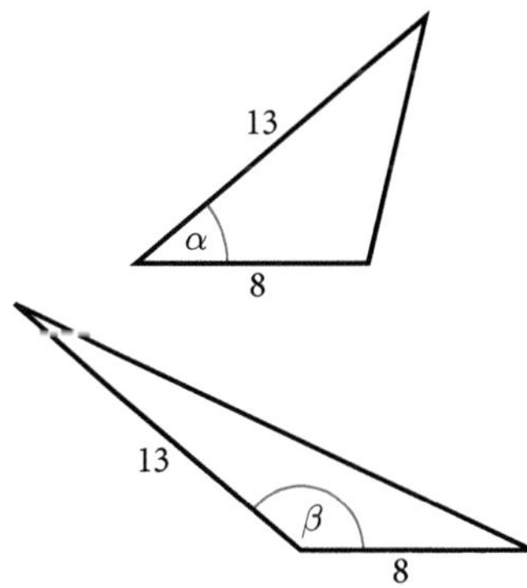

Figure 6.36 Solution for Example 6.15

Equations of lines and angles between two lines

Recall that the gradient m of a non-vertical line is defined as

$$m = \frac{y_2 - y_1}{x_2 - x_1} = \frac{\text{vertical change}}{\text{horizontal change}}$$

The equation of the line shown in Figure 6.37 has a gradient $m = \frac{1}{2}$ and a y-intercept of $(0, -1)$. So, the equation of the line is $y = \frac{1}{2}x - 1$. We can find the size of the acute angle θ between the line and the x-axis by using the tangent function (Figure 6.38).

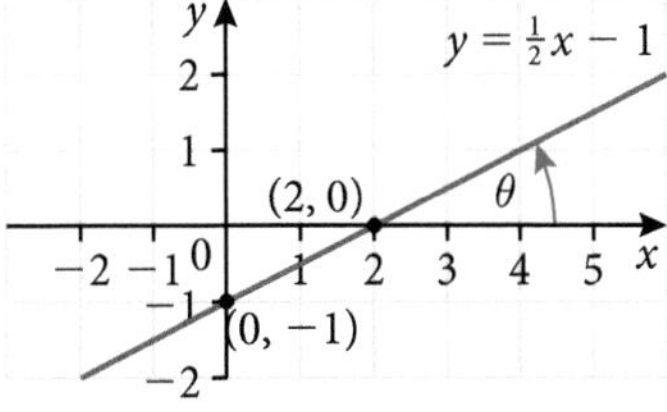

Figure 6.37 The line $y = \frac{1}{2}x - 1$

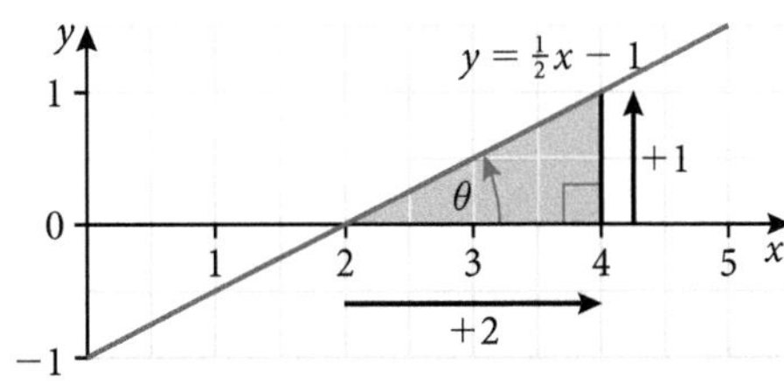

Figure 6.38 Angle between $y = \frac{1}{2}x - 1$ and x-axis

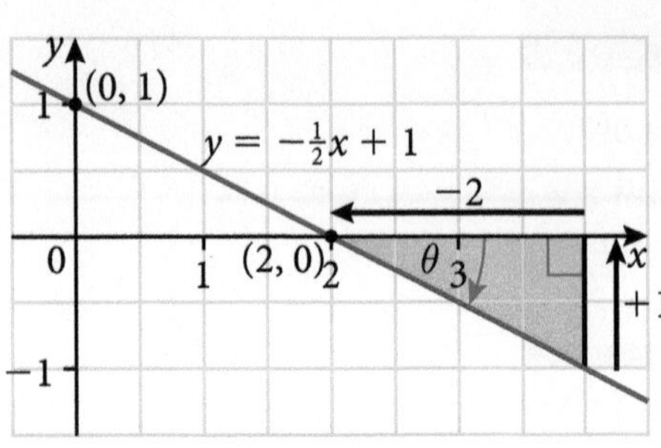

Figure 6.39 The line $y = -\frac{1}{2}x + 1$

$\theta = \tan^{-1}(m) = \tan^{-1}\left(\frac{1}{2}\right) \approx 26.6°$

Clearly, the gradient, m, of this line is equal to $\tan\theta$. If we know the angle between the line and the x-axis, and the y-intercept $(0, c)$, we can write the equation of the line in gradient-intercept form as $y = (\tan\theta)x + c$

Before we can generalise for any non-horizontal line, let's look at a line with a negative gradient.

The gradient of the line is $-\frac{1}{2}$. In order for $\tan\theta$ to be equal to the gradient of the line, the angle θ must be the angle that the line makes with the x-axis in the positive direction, as shown in Figure 6.39. In this example, $\theta = \tan^{-1}(m) = \tan^{-1}\left(-\frac{1}{2}\right) \approx -26.6°$. Remember, a negative angle indicates a clockwise rotation from the initial side to the terminal side of the angle.

If a line has a y-intercept of $(0, c)$ and makes an angle of θ with the positive direction of the x-axis, such that $-90° < \theta < 90°$, then the gradient of the line is $m = \tan\theta$ and the equation of the line is $y = (\tan\theta)x + c$.

The angle this line makes with any horizontal line will be θ.

We will now use triangle trigonometry to find the angle between any two intersecting lines – not just for a line intersecting the x-axis. Any pair of intersecting lines that are not perpendicular will have both an acute angle and an obtuse angle between them. When asked for an angle between two lines, the convention is to give the acute angle.

Example 6.16

Find the acute angle between the lines $y = 3x$ and $y = -x$

Solution

The angle between the line $y = 3x$ and the positive x-axis is α, and the angle between the line $y = -x$ and the positive x-axis is β.

$$\alpha = \tan^{-1}(3) \approx 71.565°$$

$$\beta = \tan^{-1}(-1) = -45°$$

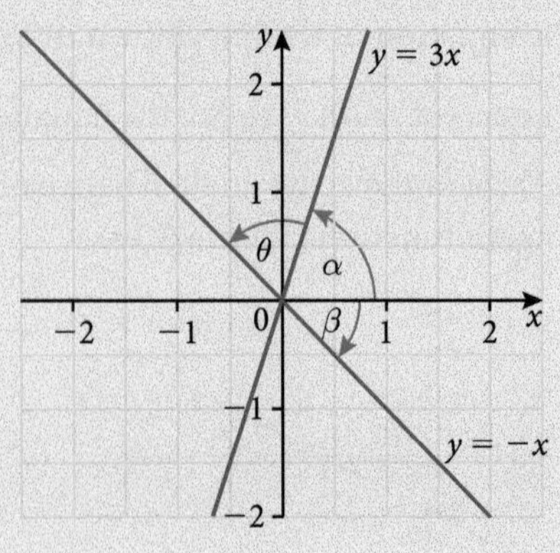

The obtuse angle between the two lines is $\alpha - \beta \approx 71.565° - (-45°) \approx 116.565°$

Therefore, the acute angle θ between the two lines is $\theta \approx 180° - 116.565 \approx 63.4°$

Example 6.17

Find the acute angle between the lines $y = 5x - 2$ and $y = \frac{1}{3}x - 1$

Solution

A horizontal line is drawn through the point of intersection.

The angle between $y = 5x - 2$ and this horizontal line is α, and the angle between $y = \frac{1}{3}x - 1$ and this horizontal line is β.

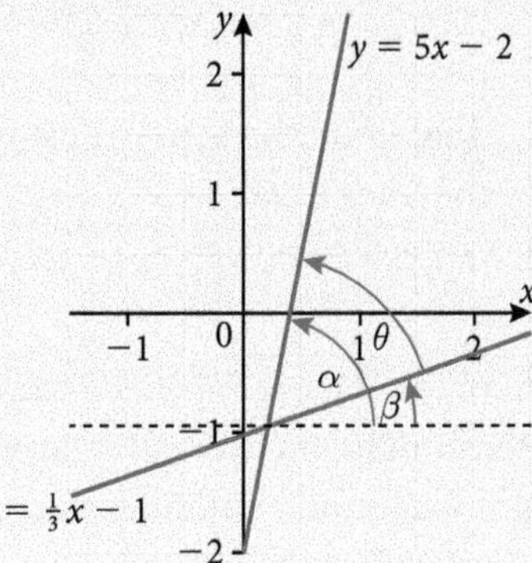

$\alpha = \tan^{-1}(5) \approx 78.690°$ and

$\beta = \tan^{-1}\left(\frac{1}{3}\right) \approx 18.435°$

The acute angle θ between the two lines is $\theta = \alpha - \beta \approx 78.690° - 18.435° \approx 60.3°$

We can generalise the procedure for finding the angle between two lines as follows.

Given two non-vertical lines with equations of $y_1 = m_1x + c_1$ and $y_2 - m_2x + c_2$, the angle between the two lines is $|\tan^{-1}(m_1) - \tan^{-1}(m_2)|$. This angle may be acute or obtuse.

Example 6.18

(a) Find the exact equation of line L_1 that passes through the origin and makes an angle of $-60°$ (or $120°$) with the positive direction of the x-axis.

(b) The equation of line L_2 is $7x + y + 1 = 0$. Find the acute angle between L_1 and L_2.

Solution

(a) The equation of the line is given by
$y = (\tan\theta)x$

$$\Rightarrow [\tan(-60°)]x = \left[\frac{\sin(-60°)}{\cos(-60°)}\right]x$$

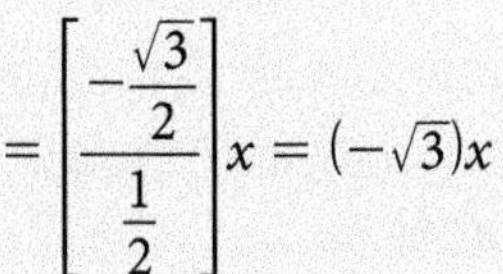

$$= \left[\frac{-\frac{\sqrt{3}}{2}}{\frac{1}{2}}\right]x = (-\sqrt{3})x$$

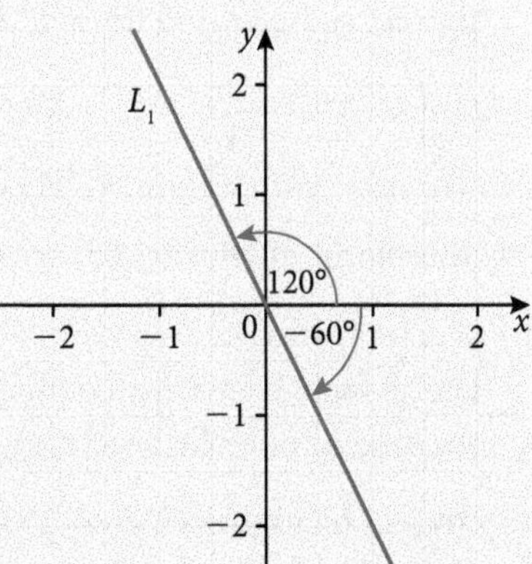

Therefore, the equation of L_1 is
$y = (-\sqrt{3})x$ or $y = -x\sqrt{3}$

Remember: $\tan(-60°) = \tan 120° = -\sqrt{3}$

(b) $L_2: 7x + y + 1 = 0 \Rightarrow y = -7x - 1$

θ is the acute angle between L_1 and L_2.

$\theta = |\tan^{-1}(m_1) - \tan^{-1}(m_2)|$

$= |\tan^{-1}(-\sqrt{3}) - \tan^{-1}(-7)|$

$\Rightarrow \theta \approx -60° - (-81.870°) \approx -21.87°$

Therefore, the acute angle between the lines is 21.9° (3 s.f.)

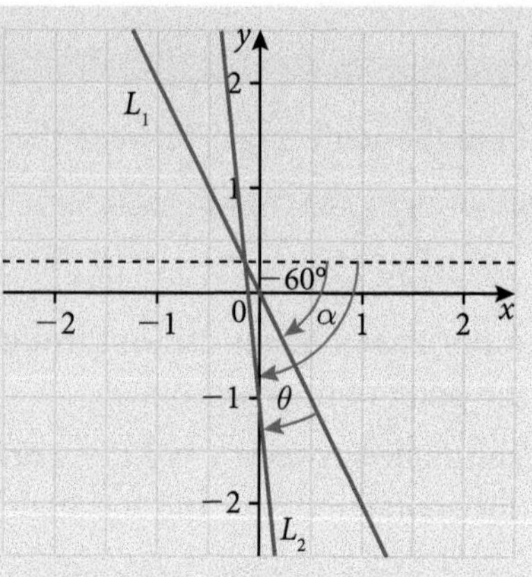

In many problems, it is necessary to calculate lengths and angles in three-dimensional structures. It is very important to analyse the three-dimensional diagram carefully and to extract any relevant triangles to find the missing angle(s) or length(s).

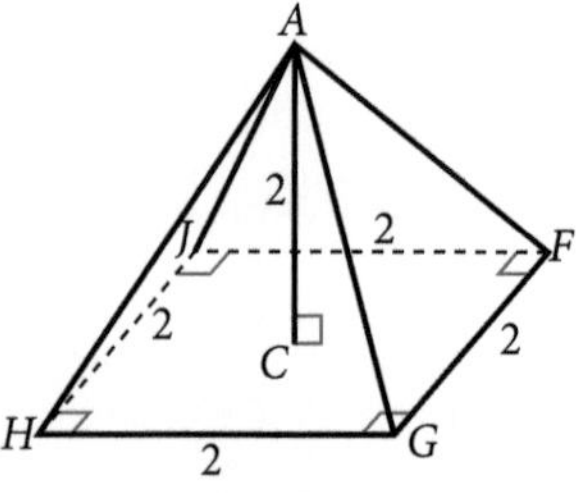

Figure 6.40 Diagram for Example 6.19

Example 6.19

The diagram shows a pyramid with a square base. It is a right pyramid, so the line segment (the height) drawn from the top vertex A perpendicular to the base will intersect the square base at its centre C. Each side of the square base has a length of 2 cm and the height of the pyramid is also 2 cm. Find:

(a) the size of $A\hat{G}F$

(b) the total surface area of the pyramid.

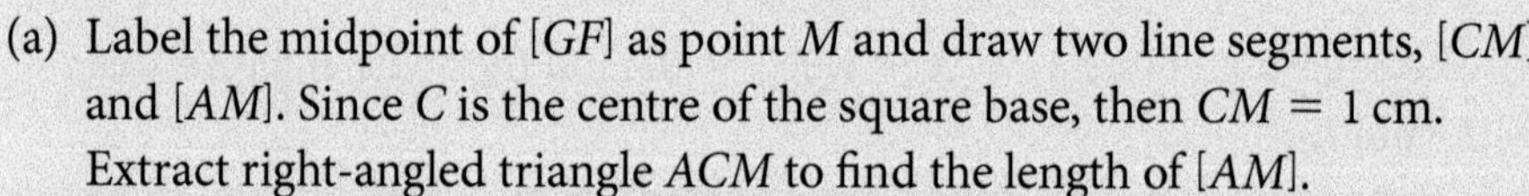

Solution

(a) Label the midpoint of $[GF]$ as point M and draw two line segments, $[CM]$ and $[AM]$. Since C is the centre of the square base, then $CM = 1$ cm. Extract right-angled triangle ACM to find the length of $[AM]$.

$AM = \sqrt{1^2 + 2^2} = \sqrt{5}$ $\quad [AM]$ is perpendicular to $[GF]$

Extract right-angled triangle AMG and use the tangent ratio to find $A\hat{G}M$.

$\tan(A\hat{G}M) = \dfrac{\sqrt{5}}{1}$

$A\hat{G}M = A\hat{G}F = \tan^{-1}(\sqrt{5}) \approx 65.905°$

Therefore, $A\hat{G}F \approx 65.9°$

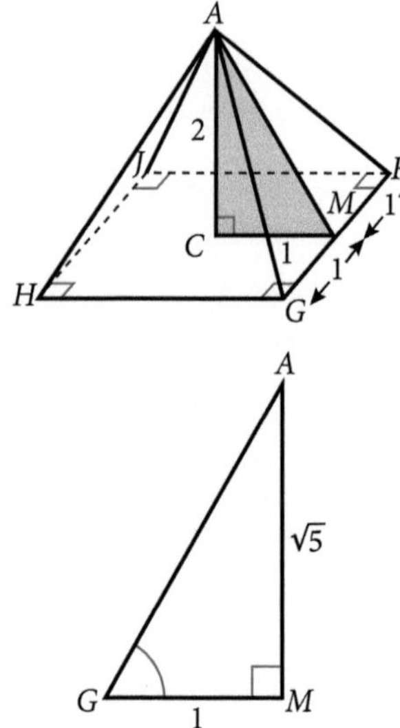

Figure 6.41 Solution to Example 6.19

(b) The total surface area comprises the square base plus four identical lateral faces that are all equilateral triangles. Triangle AGM is one-half the area of one of these triangular faces.

Area of triangle $AGM = \frac{1}{2}(1)(\sqrt{5}) = \dfrac{\sqrt{5}}{2} \Rightarrow$

Area of triangle $= AGF = 2\left(\dfrac{\sqrt{5}}{2}\right) = \sqrt{5}$

Surface area = area of square base + area of 4 lateral faces $= 2^2 + 4\sqrt{5}$

$= 4 + 4\sqrt{5} \approx 12.94 \text{ cm}^2$

Exercise 6.3

1. In each diagram, find the exact value of the three trigonometric functions of the angle θ. Simplify your answers.

(a)

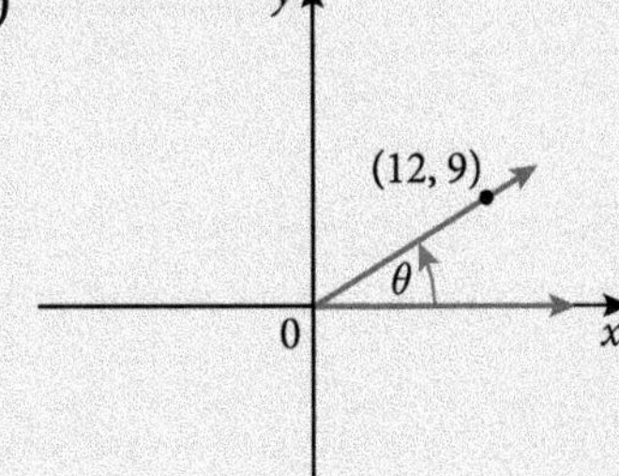

(b)

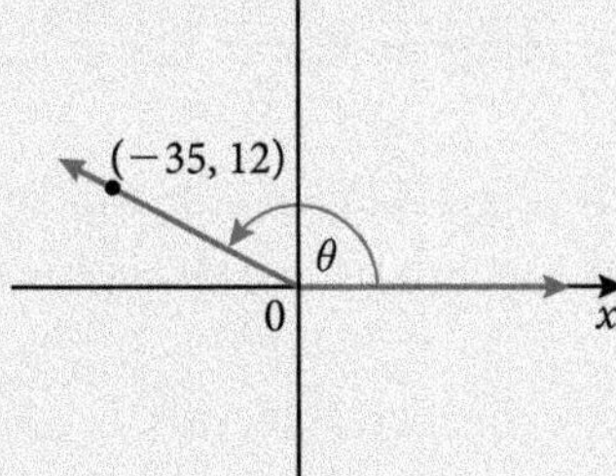

(c)

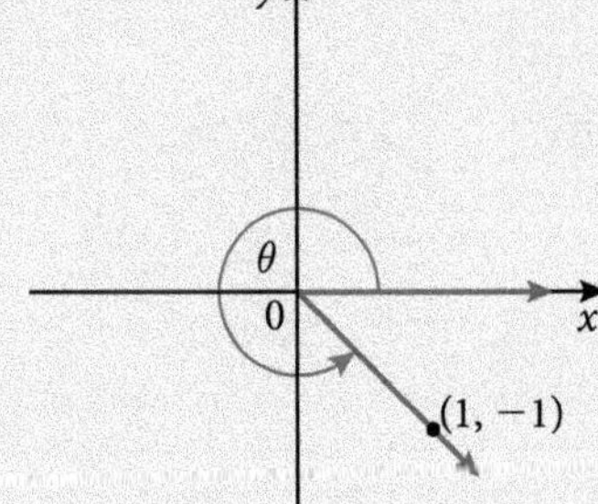

(d)

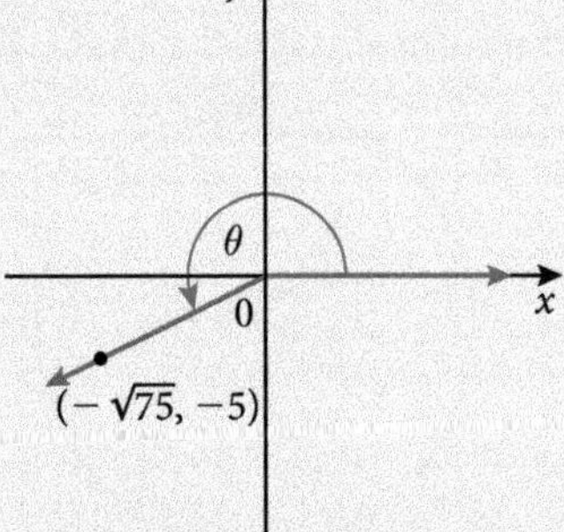

2. Without using your GDC, determine the exact values of all three trigonometric functions for each angle.

(a) $120°$ (b) $135°$ (c) $330°$ (d) $270°$ (e) $240°$

(f) $\frac{5\pi}{4}$ (g) $\frac{\pi}{6}$ (h) $\frac{7\pi}{6}$ (i) $-60°$ (j) $-\frac{3\pi}{2}$

(k) $\frac{5\pi}{3}$ (l) $-210°$ (m) $-\frac{\pi}{4}$ (n) π (o) 4.25π

3. Given that $\cos\theta = \frac{8}{17}$, $0° < \theta < 90°$, find the exact values of the other two trigonometric functions.

4. Given that $\tan\theta = -\frac{6}{5}$, $\sin\theta < 0$, find the exact values of $\sin\theta$ and $\cos\theta$.

5. Given that $\sin\theta = 0$, $\cos\theta < 0$, find the exact values of the other two trigonometric functions.

6. (a) Find the acute angle with the same sine ratio as **(i)** 150°, and **(ii)** 95°

(b) Find the acute angle with the same cosine ratio as **(i)** 315°, and **(ii)** 353°

(c) Find the acute angle with the same tangent ratio as **(i)** 240°, and **(ii)** 200°

7. Find the area of each triangle. Express the area exactly, or, if not possible, express it correct to 3 significant figures.

(a) 4 60° 6

(b)
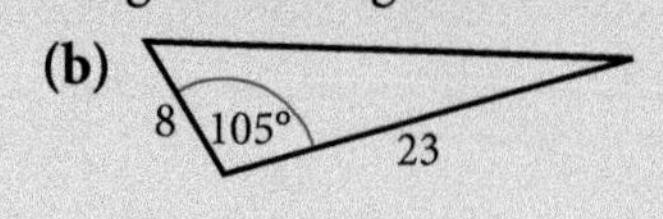

(c)
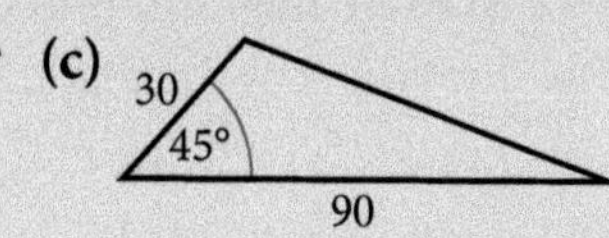

8. Triangle ABC has an area of 43 cm^2. The length of side AB is 12 cm and the length of side AC is 15 cm. Find the degree measure of angle A.

9. A chord AB subtends an angle of 120° at O, the centre of a circle with radius 15 centimetres. Find the area of **(a)** the sector AOB, and **(b)** the triangle AOB.

10. Find the area of the shaded region in each circle.

(a)
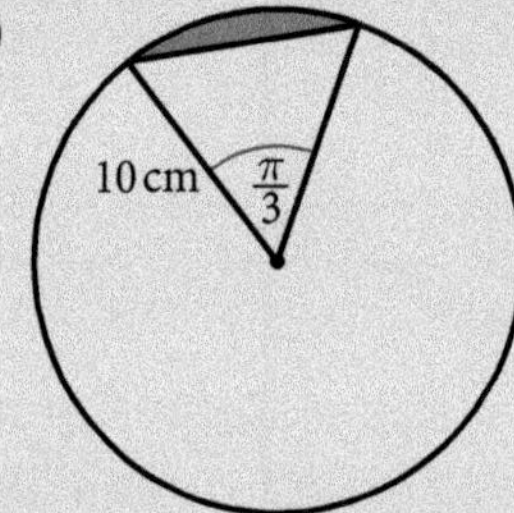

(b)
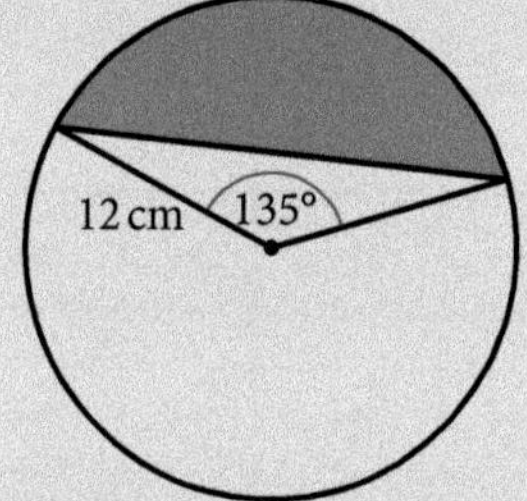

11. Two adjacent sides of a parallelogram have lengths a and b and the angle between these two sides is θ. Express the area of the parallelogram in terms of a, b, and θ.

12. For the triangle shown, express y in terms of x.

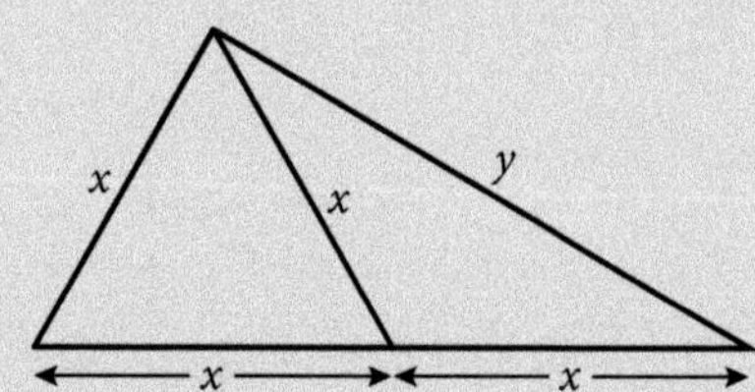

13. GJ bisects $F\hat{G}H$ such that $F\hat{G}J = H\hat{G}J = \theta$. Express x in terms of h, f, and $\cos\theta$.

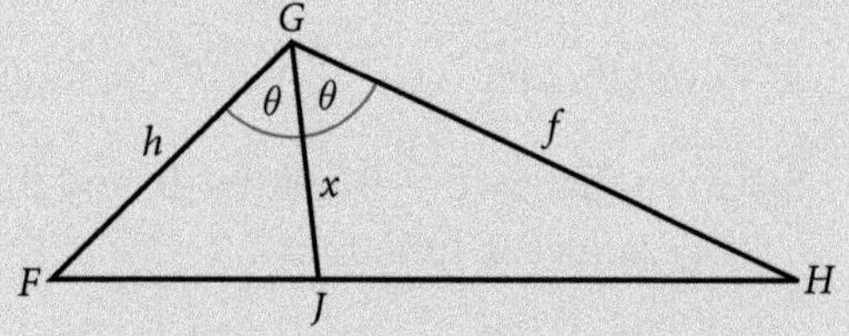

14. s is the length of each side of a regular polygon with n sides and r is the radius of the circumscribed circle. Show that $s = 2r\sin\left(\frac{180°}{n}\right)$

15. A triangle has two sides of lengths 6 cm and 8 cm and an included angle x.

   (a) Express the area of the triangle as a function of x.

   (b) State the domain and range of the function and sketch its graph for a suitable interval of x.

   (c) Find the exact coordinates of the maximum point of the function. What type of triangle corresponds to this maximum? Explain why this triangle gives a maximum area.

16. Find the angle that the line through the given pair of points makes with the positive direction of the x-axis.

   (a) $(1, 4)$ and $(-1, 2)$ **(b)** $(-3, 1)$ and $(6, -5)$

   (c) $\left(2, \frac{1}{2}\right)$ and $(-4, -10)$

17 Find the acute angle between the two given lines.

   (a) $y = -2x$ and $y = x$ **(b)** $y = -3x + 5$ and $y = 2x$

6.4 The sine rule and the cosine rule

So far, we have used trigonometry to find an unknown angle or side of a right-angled triangle. We will now study methods for finding unknown lengths and angles in triangles that are not right-angled triangles. These general methods are effective for solving problems involving any kind of triangle – right-angled, acute or obtuse.

Possible triangles constructed from three given parts

We need to know at least three parts of a triangle to solve for other unknown parts. Different arrangements of the three known parts can be given. Before solving for unknown parts, it is helpful to know whether the three known parts determine a unique triangle, or more than one possible triangle. Table 6.2 summarises the five different arrangements of three parts and the number of possible triangles for each.

Known parts	Number of possible triangles
Three angles (AAA)	Infinite triangles (not possible to solve)
Three sides (SSS) (sum of any two must be greater than the third)	One unique triangle
Two sides and their included angle (SAS)	One unique triangle
Two angles and any side (ASA or AAS)	One unique triangle
Two sides and a non-included angle (SSA)	No triangle, one triangle or two triangles

Table 6.2 Possible triangles formed with three known parts

Arrangements ASA, AAS, and SSA can be solved using the **sine rule**, whereas arrangements SSS and SAS can be solved using the **cosine rule**.

Online 

Apply the sine rule to solve triangles.

The sine rule

In Section 6.3, we showed that we can write three equivalent expressions for the area of any triangle for which we know two sides and the included angle.

Area of $\Delta = \frac{1}{2}ab\sin C = \frac{1}{2}ac\sin B = \frac{1}{2}bc\sin A$

Divide each expression by $\frac{1}{2}abc$:

$$\frac{\frac{1}{2}ab\sin C}{\frac{1}{2}abc} = \frac{\frac{1}{2}ac\sin B}{\frac{1}{2}abc} = \frac{\frac{1}{2}bc\sin A}{\frac{1}{2}abc}$$

We obtain the following three equivalent ratios – each containing the sine of an angle divided by the length of the side opposite the angle.

If A, B, and C are the angle measures of any triangle and a, b, and c are, respectively, the lengths of the sides opposite these angles, then, according to the **sine rule**:

$$\frac{\sin A}{a} = \frac{\sin B}{b} = \frac{\sin C}{c}$$

Alternatively, the sine rule can also be written as $\frac{a}{\sin A} = \frac{b}{\sin B} = \frac{c}{\sin C}$

Online 

See a demonstration of the sine rule for any triangle.

Finding unknowns given two angles and any side (ASA or AAS)

When we know two angles and any side of a triangle, we can use the sine rule to find any of the other angles or sides of the triangle.

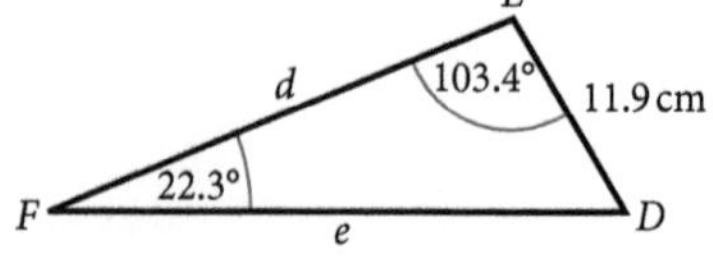

Figure 6.42 Diagram for Example 6.20

Example 6.20

Find all the unknown angles and sides of triangle DEF shown in the diagram. Give all measurements correct to 1 decimal place.

Solution

The third angle of the triangle is

$$D = 180° - E - F = 180° - 103.4° - 22.3° = 54.3°$$

We can write an equation using the sine rule to solve for e

$$\frac{\sin 22.3°}{11.9} = \frac{\sin 103.4°}{e}$$

$$e = \frac{11.9\sin 103.4°}{\sin 22.3°} \approx 30.507 \text{ cm}$$

We can write another equation using the sine rule to solve for d

$$\frac{\sin 22.3°}{11.9} = \frac{\sin 54.3°}{d}$$

$$d = \frac{11.9\sin 54.3°}{\sin 22.3°} \approx 25.467 \text{ cm}$$

Therefore, the other parts of the triangle are:

$D = 54.3°$, $e \approx 30.5$ cm and $d \approx 25.5$ cm

When using your GDC to find angles and lengths with the sine rule (or the cosine rule), remember to store intermediate answers on the GDC for greater accuracy. By not rounding until the final answer, you reduce the amount of rounding error.

Example 6.21

A tree on a sloping hill casts a shadow 45 metres down the slope of the hill. The gradient of the hill is $\frac{60}{12}$ and the angle of elevation of the sun is 35°. How tall is the tree, to the nearest tenth of a metre?

Solution

α is the angle that the hill makes with the horizontal. We can work out its size using the inverse tangent of $\frac{12}{60} = \frac{1}{5}$.

$\alpha = \tan^{-1}\left(\frac{1}{5}\right) \approx 11.3099°$

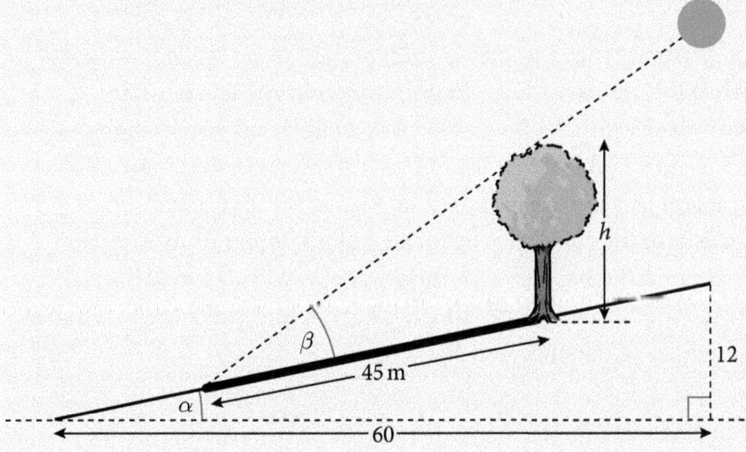

The height of the tree is h. The angle of elevation of the sun is the angle between the sun's rays and the horizontal. In the diagram, this angle of elevation is the sum of α and β. Thus, $\beta \approx 35° - 11.3099° \approx 23.6901°$

If a line segment is dropped from the base of the tree perpendicular to the length of 60 m, then we can sketch a right-angled triangle with $\alpha + \beta = 35°$ as one of its acute angles; the other acute angle – and the angle in the obtuse triangle opposite the side of 45 metres – must be 55°. We can apply the sine rule for the obtuse triangle to solve for h.

$$\frac{\sin 23.7°}{h} = \frac{\sin 55°}{45} \Rightarrow h = \frac{45 \sin 23.7°}{\sin 55°} \approx 22.0809$$

Therefore, the tree is approximately 22.1 metres tall.

Two sides and a non-included angle (SSA) – the ambiguous case

The SSA arrangement – two sides of a triangle and the size of an angle not between these two sides – can produce three different results: no triangle, one unique triangle, or two different triangles.

Example 6.22

Find all of the unknown angles and sides of triangle ABC where $a = 35$ cm, $b = 50$ cm, and $A = 30°$. Give all measurements correct to 1 decimal place.

Solution

Here are the three parts we have in our attempt to construct triangle ABC:

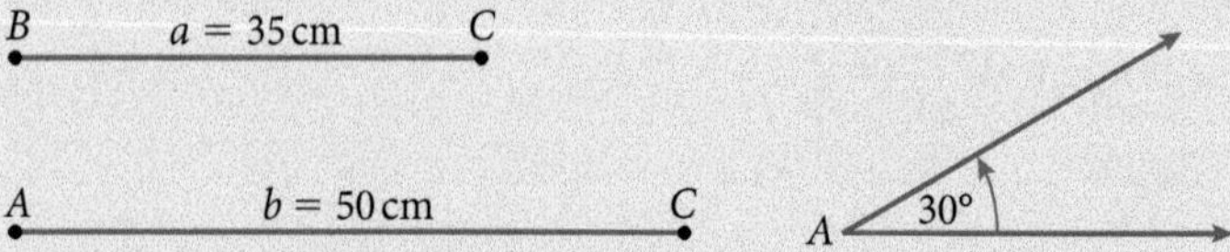

We attempt to construct the triangle, as shown. First draw angle A with its initial and terminal sides extended. Then measure off the known side $b = AC = 50$ on the terminal side. To construct side a (opposite angle A), we take point C as the centre and with radius $a = 35$ we draw an arc of a circle. The points on this arc are all possible positions for vertex B. Point B must be on the base line, so B can be located at any point of intersection of the circular arc and the base line. In this instance, with these particular measurements for the two sides and non-included angle, there are two points of intersection, which we label B_1 and B_2.

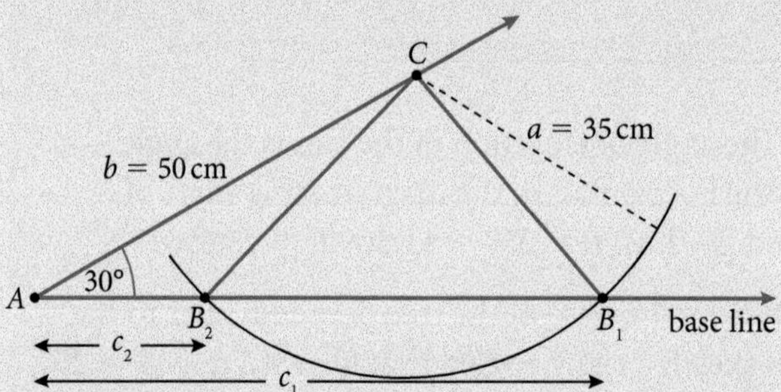

Therefore, we can construct two different triangles, triangle AB_1C and triangle AB_2C. Angle B_1 is acute and angle B_2 is obtuse. To complete the solution of this problem, we need to solve each of these triangles.

Solve triangle AB_1C:

We can solve for acute angle B_1 using the sine rule.

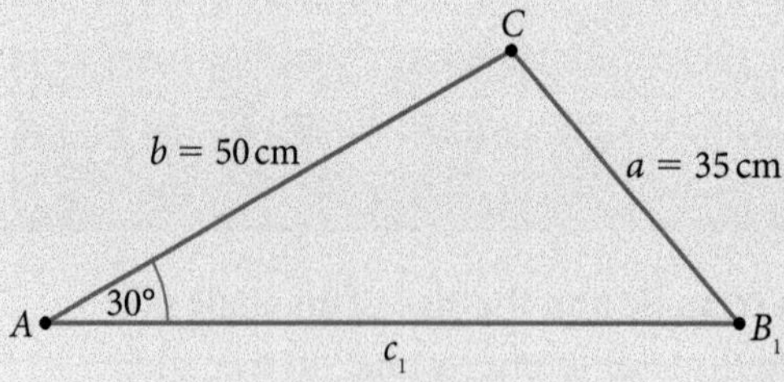

$$\frac{\sin 30°}{35} = \frac{\sin B_1}{50}$$

$$\sin B_1 = \frac{50 \sin 30°}{35} = \frac{50(0.5)}{35}$$

$$B_1 = \sin^{-1}\left(\frac{5}{7}\right) \approx 45.5847°$$

Then, $C \approx 180° - 30° - 45.5847° \approx 104.4153°$

With another application of the sine rule, we can solve for side c_1.

$$\frac{\sin 30°}{35} = \frac{\sin 104.4153°}{c_1}$$

$$c_1 \approx \frac{35 \sin 104.4153°}{\sin 30°} \approx \frac{35(0.96852)}{0.5} \approx 67.7964\text{ cm}$$

Therefore, for triangle AB_1C: $B_1 \approx 45.6°$, $C \approx 104.4°$ and $c_1 \approx 67.8$ cm

Solve triangle AB_2C:

Solving for obtuse angle B_2 using the sine rule gives the same result as above, except we know that $90° < B_2 < 180°$. We also know that $\sin(180° - \theta) = \sin\theta$.

Thus, $B_2 = 180° - B_1 \approx 180° - 45.5847° \approx 134.4153°$

Then, $C \approx 180° - 30° - 134.4153° \approx 15.5847°$

Applying the sine rule again, we can solve for side c_2.

$$\frac{\sin 30°}{35} = \frac{\sin 15.5847°}{c_2}$$

$$c_2 \approx \frac{35 \sin 15.5847°}{\sin 30°} \approx \frac{35(0.26866)}{0.5}$$

$$\approx 18.8062\text{ cm}$$

Therefore, for triangle AB_2C: $B_2 \approx 134.4°$, $C \approx 15.6°$ and $c_2 \approx 18.8$ cm

We will now take a more general look and examine all the possible conditions and outcomes for the SSA arrangement. In general, we are given the lengths of two sides – call them a and b – and a non-included angle; for example, angle A that is opposite side a. From these measurements, we can determine the number of different triangles. Figure 6.43 shows the four different possibilities (or cases) when angle A is acute. The number of triangles depends on the length of side a.

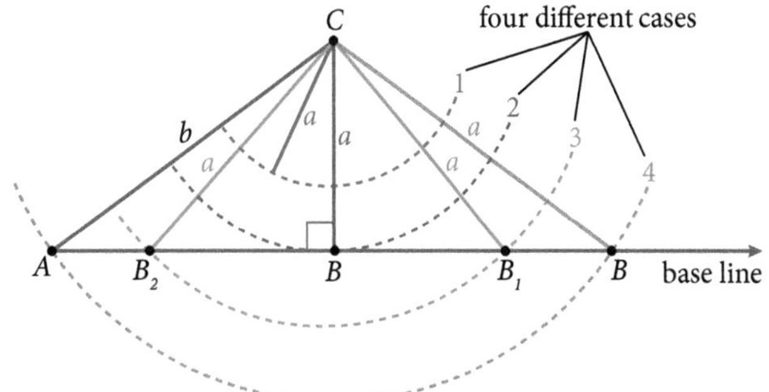

Figure 6.43 Four distinct cases for SSA when angle A is acute

In case 2, side a is perpendicular to the base line resulting in a single right-angled triangle, shown in Figure 6.44. In this case, $\sin A = \frac{a}{b}$ and $a = b \sin A$.

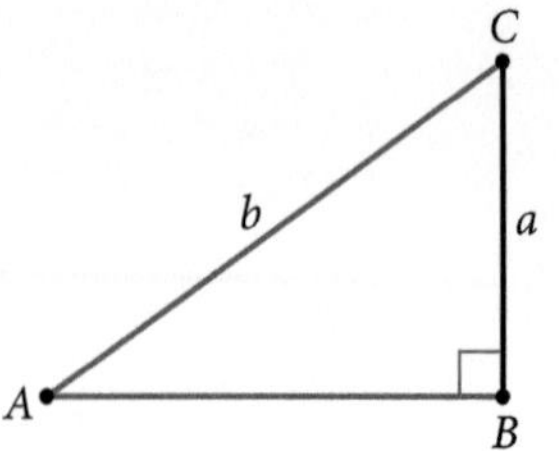

Figure 6.44 Case 2 for SSA: $a = b \sin A$, one right angle

Given the length of sides a and b and the non-included angle A is acute, the four cases and resulting triangles shown in Table 6.3 can occur.

In case 1, a is shorter than it is in case 2, i.e. $b \sin A$. In case 3, which occurred in Example 6.22, a is longer than $b \sin A$, but less than b. And, in case 4, a is longer than b. These results are summarised in Table 6.3. Because the number of triangles may be none, one, or two, depending on the length of a (the side opposite the given angle), the SSA arrangement is called the **ambiguous case**.

Length of a	Number of triangles	Case number in Figure 6.43
$a < b \sin A$	No triangle	1
$a = b \sin A$	One right-angled triangle	2
$b \sin A < a < b$	Two triangles	3
$a > b$	One triangle	4

Table 6.3 Possible triangles formed with two known sides and an acute non-included angle

The situation is considerably simpler if angle A is obtuse rather than acute. Figure 6.46 shows that if $a > b$ then there is only one possible triangle, and if $a \leqslant b$ then no triangle is possible that contains angle A.

Example 6.23 uses the same SSA information given in Example 6.22 with the exception that side a is not fixed at 35 cm, but is allowed to vary.

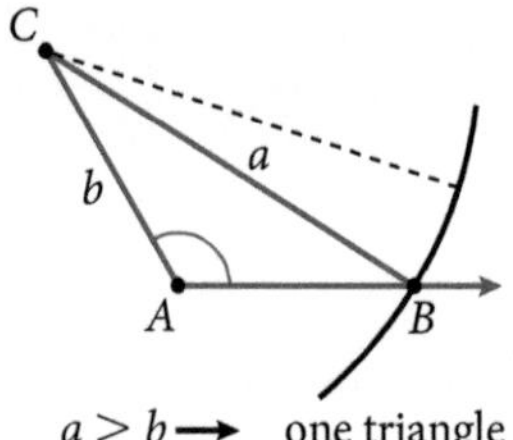

$a > b \longrightarrow$ one triangle

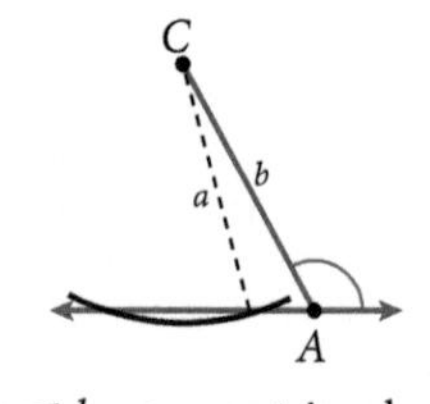

$a \leqslant b \longrightarrow$ no triangle

Figure 6.45 Angle A is obtuse

Example 6.23

For triangle ABC, side $b = 50$ cm and angle $A = 30°$. Find the values for the length of side a that will produce: (i) no triangle, (ii) one triangle, (iii) two triangles.

Solution

Because this is an SSA arrangement (ambiguous case) and given A is an acute angle, then the number of different triangles that can be constructed depends on the length of a.

First calculate the value of $b \sin A$: $b \sin A = 50 \sin 30° = 50(0.5) = 25$ cm

Thus, if a is exactly 25 cm, triangle ABC is a right-angled triangle.

(i) When $a < 25$ cm, there is no triangle.

(ii) When $a = 25$ cm, or $a > 50$ cm, there is one unique triangle.

(iii) When 25 cm $< a <$ 50 cm, there are two different possible triangles.

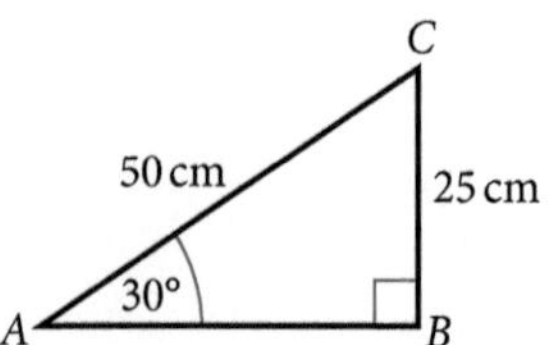

Figure 6.46 Diagram for solution to Example 6.23

Example 6.24

The diagrams show two different triangles both satisfying the conditions:

$HK = 18$ cm, $JK = 15$ cm, $J\hat{H}K = 53°$

(a) Calculate the size of $H\hat{J}K$ in triangle 2.

(b) Calculate the area of triangle 1.

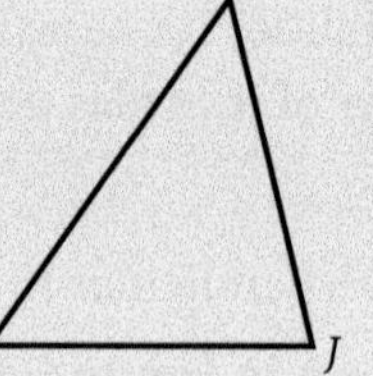

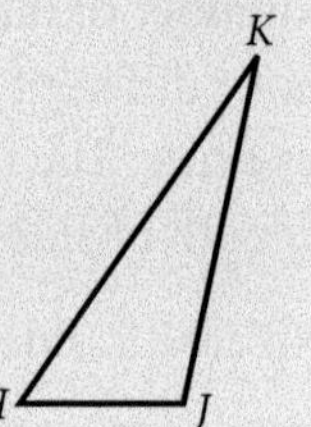

Solution

From the sine rule, $\sin\left(\frac{H\hat{J}K}{18}\right) = \frac{\sin 53°}{15} \Rightarrow \sin(H\hat{J}K) = \frac{18 \sin 53°}{15} \approx 0.95836$

$\Rightarrow \sin^{-1}(0.95836) \approx 73.408°$

However, $H\hat{J}K > 90° \Rightarrow H\hat{J}K \approx 180° - 73.408° \approx 106.592°$

Therefore in triangle 2, $H\hat{J}K \approx 107°$ (3 s.f.)

In triangle 1, $H\hat{J}K < 90° \Rightarrow H\hat{J}K \approx 73.408°$

$\Rightarrow H\hat{K}J \approx 180° - (73.408° + 53°) \approx 53.592°$

Area $= \frac{1}{2}(18)(15)\sin(53.592°) \approx 108.649 \text{ cm}^2$

Therefore, the area of triangle 1 is approximately 109 cm^2 (3 s.f.)

The cosine rule

Two arrangements remain in our list (Table 6.2) of different ways to arrange three known parts of a triangle. If three sides of a triangle are known (SSS arrangement), or two sides of a triangle and the angle between them are known (SAS arrangement), then a unique triangle is determined. However, in both of these cases the sine rule cannot be used to work out the unknowns in the triangle.

For example, it is not possible to set up an equation using the sine rule to solve triangle *PQR* or triangle *STU* in Figure 6.47

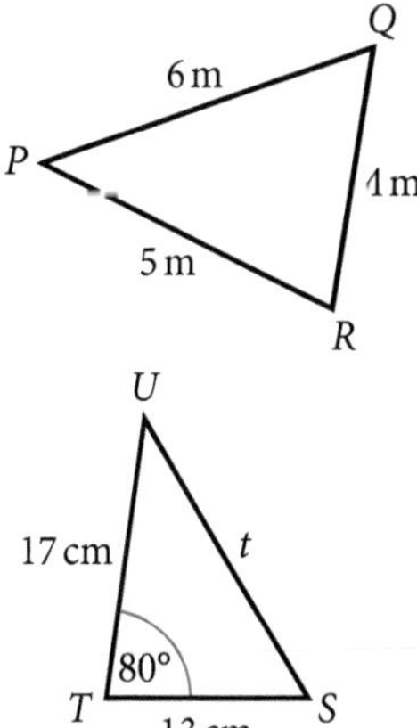

Figure 6.47 Two triangles that cannot be solved with the sine rule

Trying to solve Δ*PQR*:

$\frac{\sin P}{4} = \frac{\sin R}{6} \Rightarrow$ two unknowns; cannot solve for angle *P* or angle *R*

Trying to solve Δ*STU*:

$\frac{\sin 80°}{t} = \frac{\sin U}{13} \Rightarrow$ two unknowns; cannot solve for angle *U* or side *t*

We will need the **cosine rule** to solve triangles with SSS and SAS arrangements. To derive this law, we need to place a general triangle *ABC* in the coordinate plane so that one of the vertices is at the origin and one of the sides is on the positive *x*-axis. Figure 6.48 (top of next page) shows both an acute triangle *ABC* and an obtuse triangle *ABC*. In both cases, the coordinates of vertex *A* are $x = b\cos C$ and $y = b\sin C$. Because *c* is the distance from *A* to *B*, we can use the distance formula to write the following:

distance between $(b\cos C, b\sin C)$ and $(a, 0)$	$c = \sqrt{(b\cos C - a)^2 + (b\sin C - 0)^2}$
squaring both sides	$c^2 = (b\cos C - a)^2 + (b\sin C - 0)^2$
expand brackets	$c^2 = b^2\cos^2 C - 2ab\cos C + a^2 + b^2\sin^2 C$
take out common factor of b^2 from two terms	$c^2 = b^2(\cos^2 C + \sin^2 C) - 2ab\cos C + a^2$

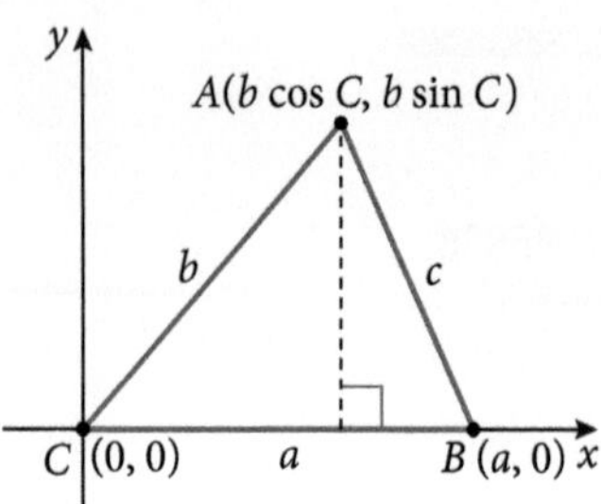

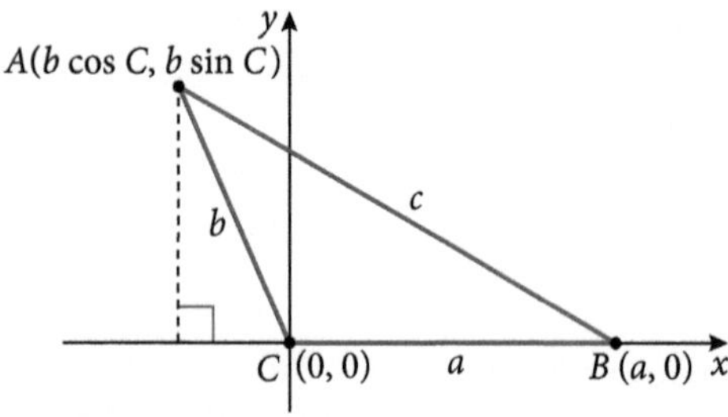

Figure 6.48 Deriving the cosine rule

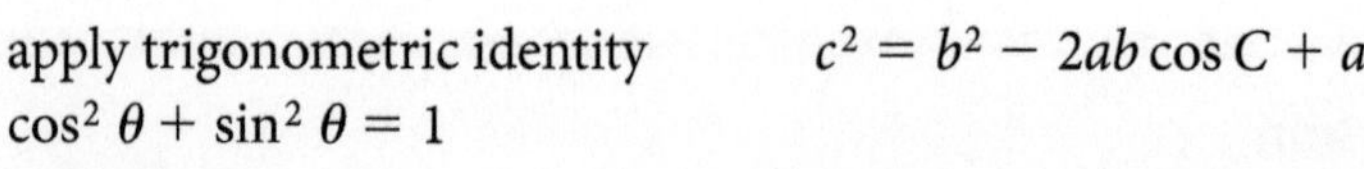

apply trigonometric identity $\cos^2\theta + \sin^2\theta = 1$ $\quad c^2 = b^2 - 2ab\cos C + a^2$

rearrange terms $\quad c^2 = a^2 + b^2 - 2ab\cos C$

This equation gives one form of the cosine rule. Two other forms are obtained in a similar manner by having either vertex A or vertex B, rather than C, located at the origin.

It is helpful to understand the underlying pattern of the cosine rule when applying it to solve for parts of triangles. The pattern relies on choosing one particular angle of the triangle and then identifying the two sides that are adjacent to the angle and the one side that is opposite to it (Figure 6.49). The cosine rule can be used to solve for the chosen angle or the side opposite the chosen angle.

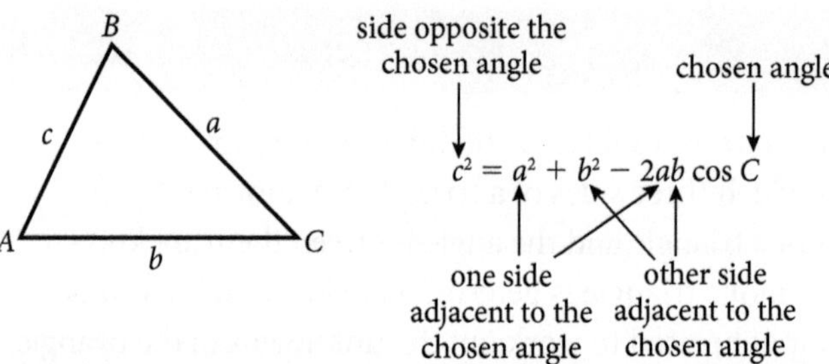

Figure 6.49 Applying the cosine rule

Finding unknowns given two sides and the included angle (SAS)

If we know two sides and the included angle, we can use the cosine rule to solve for the side opposite the given angle. Then it is best to solve for one of the two remaining angles using the sine rule.

Example 6.25

Find all of the unknown angles and sides of triangle STU. Give all measurements to 1 decimal place.

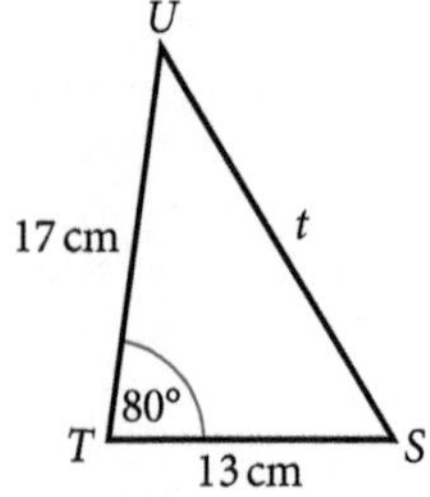

Figure 6.50 Diagram for Example 6.25

Solution

We first solve for side t opposite known angle $S\hat{T}U$ using the cosine rule.

$$t^2 = 13^2 + 17^2 - 2(13)(17)\cos 80°$$

$$t = \sqrt{13^2 + 17^2 - 2(13)(17)\cos 80°}$$

$$t \approx 19.5256$$

Now use the sine rule to solve for one of the other angles, say $T\hat{S}U$

$$\frac{\sin T\hat{S}U}{17} = \frac{\sin 80°}{19.5256}$$

$$\sin T\hat{S}U = \frac{17 \sin 80°}{19.5256}$$

$$T\hat{S}U = \sin^{-1}\left(\frac{17 \sin 80°}{19.5256}\right)$$

$$T\hat{S}U \approx 59.0288°$$

Then, $S\hat{U}T \approx 180° - (80° + 59.0288°) \approx 40.9712$

Therefore, the other parts of the triangle are $t \approx 19.5$ cm, $T\hat{S}U \approx 59.0°$ and $S\hat{U}T \approx 41.0°$

You may have noticed that the cosine rule looks similar to Pythagoras' theorem. In fact, Pythagoras' theorem can be considered a special case of the cosine rule. When the chosen angle in the cosine rule is 90°, then since $\cos 90° = 0$, the cosine rule becomes Pythagoras' theorem.

If angle $C = 90°$, then $c^2 = a^2 + b^2 - 2ab\cos C \Rightarrow c^2 = a^2 + b^2 - 2ab\cos 90°$

$\Rightarrow c^2 = a^2 + b^2 - 2ab(0) \Rightarrow c^2 = a^2 + b^2$ or $a^2 + b^2 = c^2$

In any triangle ABC with corresponding sides a, b, and c, the cosine rule states:

$c^2 = a^2 + b^2 - 2ab\cos C$
$b^2 = a^2 + c^2 - 2ac\cos B$
$a^2 = b^2 + c^2 - 2bc\cos A$

Example 6.26

A sailboat travels 50 kilometres due west, then changes its course 18° northward, as shown in the diagram. After travelling 75 kilometres in that direction, how far is the sailboat from its point of departure? Give your answer to the nearest tenth of a kilometre.

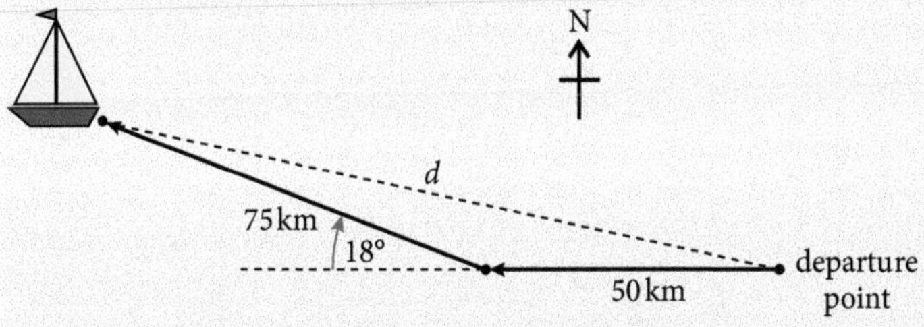

Solution

Let d be the distance from the departure point to the position of the sailboat. A large obtuse triangle is formed by the three distances of 50 km, 75 km, and d km. The angle opposite side d is $180° - 18° = 162°$. Using the cosine rule, we can write this equation to solve for d:

$$d^2 = 50^2 + 75^2 - 2(50)(75)\cos 162°$$

$$d = \sqrt{50^2 + 75^2 - 2(50)(75)\cos 162°} \approx 123.523$$

Therefore, the sailboat is approximately 123.5 km from its departure point.

Finding unknowns given three sides (SSS)

The triangle formed by three line segments where the sum of the lengths of any two is greater than the length of the third will be unique. Therefore, if we know three sides of a triangle we can solve for the three angle measures.
To use the cosine rule to solve for an unknown angle, it is best to first rearrange the formula so that the chosen angle is the subject of the formula.

Solve for angle C in:

$$c^2 = a^2 + b^2 - 2ab\cos C \Rightarrow 2ab\cos C = a^2 + b^2 - c^2 \Rightarrow \cos C = \frac{a^2 + b^2 - c^2}{2ab}$$

Then, $C = \cos^{-1}\left(\frac{a^2 + b^2 - c^2}{2ab}\right)$

Example 6.27

Find all of the unknown angles of triangle PQR. Give all measurements to 1 decimal place.

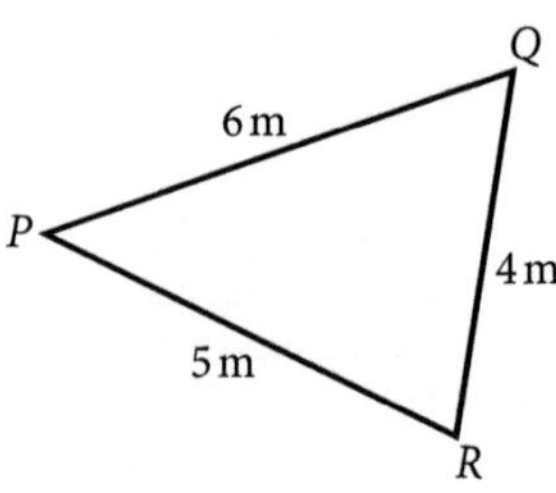

Figure 6.51 Diagram for Example 6.27

Solution

Note that the smallest angle will be opposite the shortest side. First find the smallest angle; writing the cosine rule with chosen angle P:

$$P = \cos^{-1}\left(\frac{5^2 + 6^2 - 4^2}{2(5)(6)}\right) \approx 41.4096°$$

Now that we know the size of angle P, we have two sides and a non-included angle (SSA) and the sine rule can be used to find the other non-included angle. Consider the sides $QR = 4$, $RP = 5$ and the angle $P \approx 41.4096°$. Substituting into the sine rule, we can solve for angle Q that is opposite RP.

$$\frac{\sin Q}{5} = \frac{\sin 41.4096°}{4}$$

$$\sin Q = \frac{5\sin 41.4096°}{4}$$

$$Q = \sin^{-1}\left(\frac{5\sin 41.4096°}{4}\right) \approx 55.7711°$$

Then, $R \approx 180° - (41.4096° + 55.7711°) \approx 82.8192°$
Therefore, the three angles of are $P \approx 41.4°$, $Q \approx 55.8°$ and $R \approx 82.8°$

Example 6.28

A ladder that is 8 metres long is leaning against a non-vertical wall that slopes away from the ladder. The foot of the ladder is 3.5 metres from the base of the wall, and the distance from the top of the ladder down the wall to the ground is 5.75 metres. To the nearest tenth of a degree, work out the acute angle at which the ladder is inclined to the horizontal.

Solution

Start by drawing a diagram that represents the given information accurately.
θ marks the acute angle of inclination of the ladder. Its supplement is $F\hat{B}T$.
From the cosine rule:

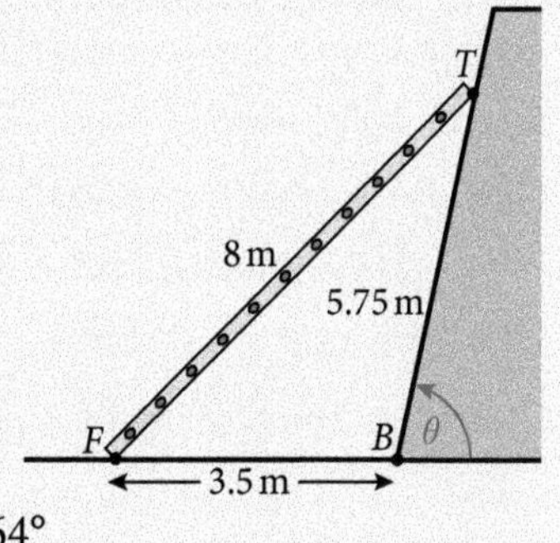

$$\cos F\hat{B}T = \frac{3.5^2 + 5.75^2 - 8^2}{2(3.5)(5.75)}$$

$$F\hat{B}T = \cos^{-1}\left(\frac{3.5^2 + 5.75^2 - 8^2}{2(3.5)(5.75)}\right) \approx 117.664°$$

$$\theta \approx 180° - 117.664° \approx 62.336°$$

Therefore, the angle of inclination of the ladder is approximately 62.3°

Example 6.29

The diagram shows a point P that is 10 kilometres due south of a point D. A straight road PQ is such that the (compass) bearing of Q from P is 45°. A and B are two points on this road that are both 8 km from D. Find the bearing of B from D, to 3 significant figures.

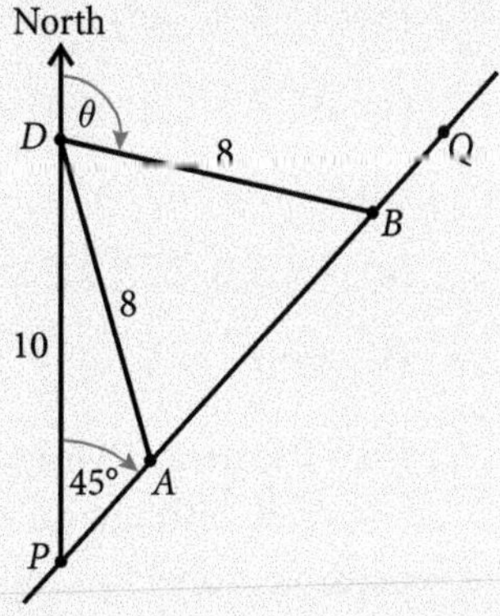

Compass bearings are measured **clockwise** from north. For example, the diagram shows a bearing of 225° from point A to point B.

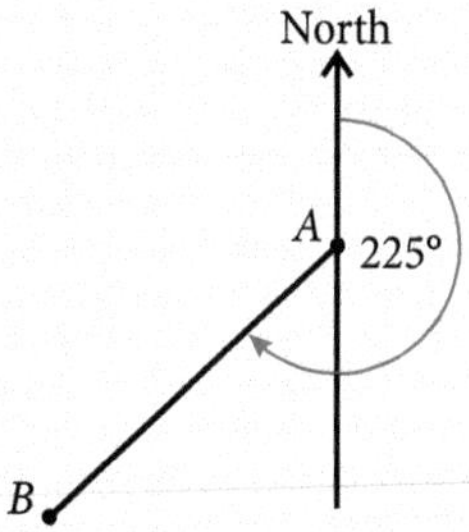

Solution

The angle θ is the bearing of B from D.

First extract triangle PDB and use the sine rule to solve for $D\hat{B}P$

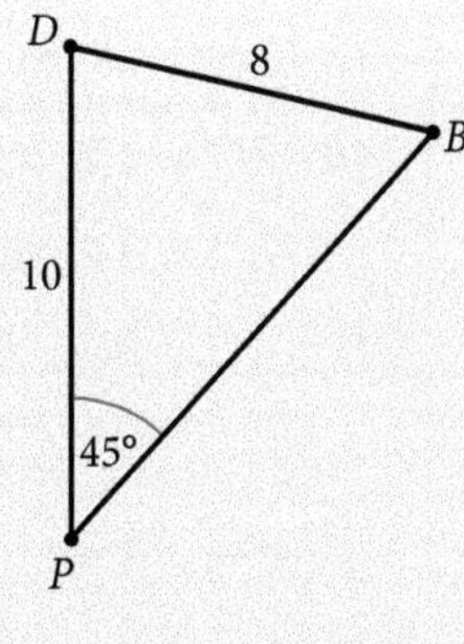

$$\frac{\sin D\hat{B}P}{10} = \frac{\sin 45°}{8}$$

$$\sin D\hat{B}P = \frac{10 \sin 45°}{8}$$

$$D\hat{B}P = \sin^{-1}\left(\frac{10 \sin 45°}{8}\right) \approx 62.11°$$

Triangle ADB is isosceles (two sides equal), so $D\hat{A}B = D\hat{B}P$, and since the sum of angles in triangle ADB is 180°, we can solve for $A\hat{D}B$.

$$D\hat{A}B = D\hat{B}P \approx 62.11°$$

$$A\hat{D}B \approx 180° - 2(62.11°) \approx 55.78°$$

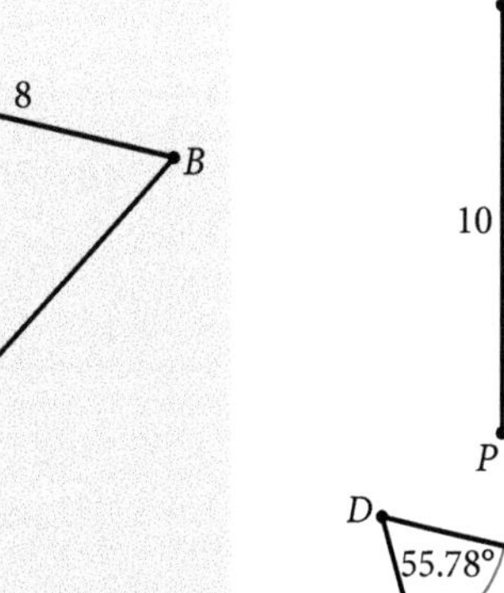

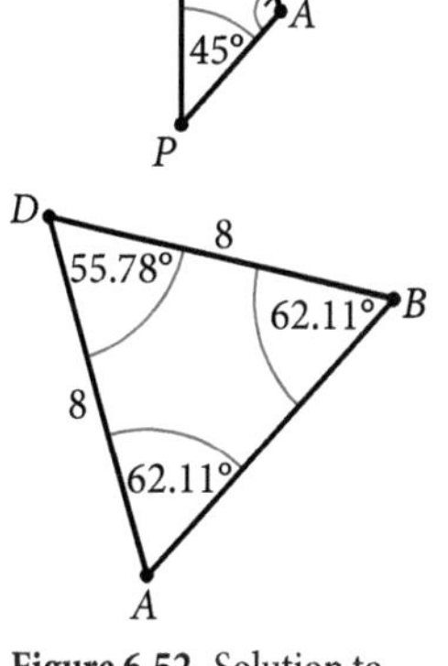

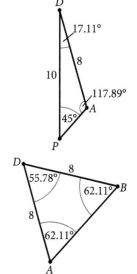

Figure 6.52 Solution to Example 6.29

Online

Apply cosine rule to solve for missing sides or angles in a randomly generated triangle.

Solve for $D\hat{A}P$ because it is supplementary to $D\hat{A}B$. Then we can find the third angle in triangle APD. Since $\theta + A\hat{D}B + A\hat{D}P = 180°$, we can solve for θ.

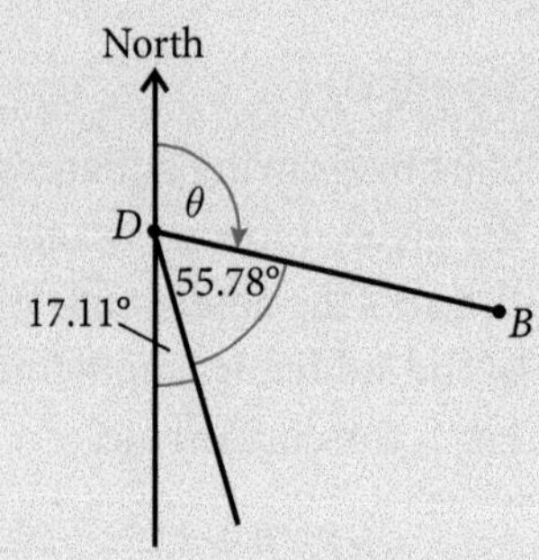

$$P\hat{A}D \approx 180° - 62.11° \approx 117.89°$$

$$A\hat{D}P \approx 180° - (45° + 117.89°) \approx 17.11°$$

$$\theta \approx 180° - (17.11° + 55.78°) \approx 107.11°$$

Therefore, the bearing of B from D is 107° (3 s.f.)

Exercise 6.4

1. State the number of distinct triangles (none, one, two, or infinite) that can be constructed with the given measurements. If the answer is one or two triangles, provide a sketch of each triangle.

(a) $A\hat{C}B = 30°$, $A\hat{B}C = 50°$, and $B\hat{A}C = 100°$

(b) $A\hat{C}B = 30°$, $AC = 12$ cm, and $BC = 17$ cm

(c) $A\hat{C}B = 30°$, $AB = 7$ cm, and $AC = 14$ cm

(d) $A\hat{C}B = 47°$, $BC = 20$ cm, and $A\hat{B}C = 55°$

(e) $B\hat{A}C = 25°$, $AB = 12$ cm, and $BC = 7$ cm

(f) $AB = 23$ cm, $AC = 19$, and $BC = 11$ cm

2. Find the measurements of all unknown sides and angles in the triangles with the following measurements. If two triangles are possible, solve for both.

(a) $B\hat{A}C = 37°$, $A\hat{B}C = 28°$, and $AC = 14$

(b) $A\hat{B}C = 68°$, $A\hat{C}B = 47°$, and $AC = 23$

(c) $BC = 68$, $A\hat{C}B = 71°$, and $AC = 59$

(d) $BC = 42$, $AC = 37$, and $AB = 26$

(e) $BC = 34$, $A\hat{B}C = 43°$, and $AC = 28$

(f) $AC = 0.55$, $B\hat{A}C = 62°$, and $BC = 0.51$

3. Find the lengths of the diagonals of a parallelogram whose sides measure 14 cm and 18 cm and which has one angle with measure 37°.

4. Find the measures of the angles of an isosceles triangle whose sides are 10 cm, 8 cm, and 8 cm.

5. Given that for triangle DEF, $E\hat{D}F = 43°$, $DF = 24$ and $FE = 18$, find the two possible measures of $D\hat{F}E$.

6. Find the measure of the smallest angle in the triangle shown.

4 6 9

Figure 6.53 Diagram for question 6

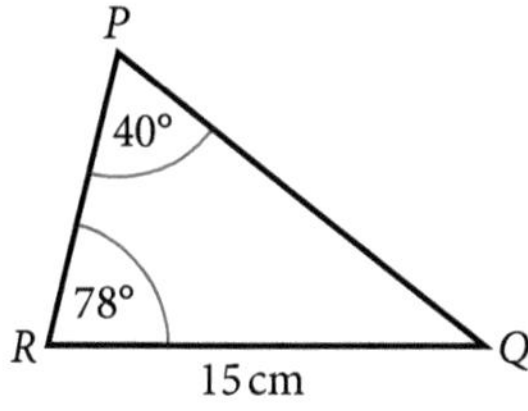

Figure 6.54 Diagram for question 7

7. Find the area of triangle PQR in Figure 6.54.

8. Find a value for the length of AC so that the number of possible triangles is: **(i)** one **(ii)** two **(iii)** none.

   (a) $B\hat{A}C = 36°, AB = 5$ **(b)** $B\hat{A}C = 60°, AB = 10$

9. A 50-metre vertical pole is to be erected on the side of a sloping hill that makes an 8° angle with the horizontal (see diagram). Find the length of each of the two supporting wires (x and y) that will be anchored 35 metres uphill and downhill from the base of the pole.

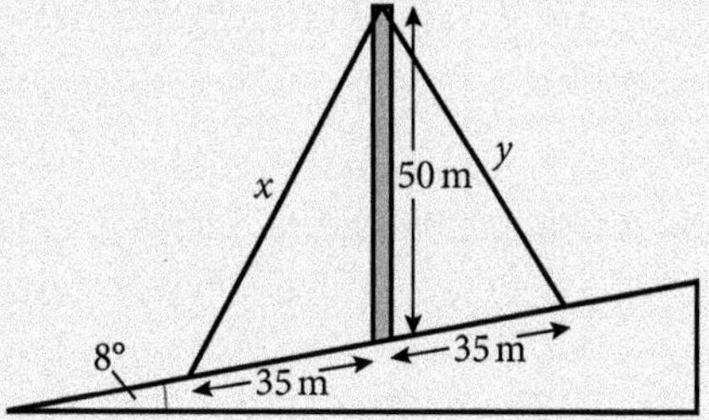

10. The lengths of the sides of a triangle ABC are $x - 2$, x and $x + 2$. The largest angle is 120°.

    (a) Find the value of x.

    (b) Show that the area of the triangle is $\frac{15\sqrt{3}}{4}$

    (c) Find $\sin A + \sin B + \sin C$, giving your answer in the form $\frac{p\sqrt{q}}{r}$ where $p, q, r \in \mathbb{Z}$.

11. Find the area of a triangle that has sides of lengths 6, 7, and 8 cm.

12. Let a, b, and c be the sides of a triangle where c is the longest side.

    (a) If $c^2 > a^2 + b^2$, what is true about triangle ABC?

    (b) If $c^2 < a^2 + b^2$, what is true about triangle ABC?

    (c) Use the cosine rule to prove each of your conclusions for **(a)** and **(b)**.

13. Consider triangle DEF with $E\hat{D}F = 43.6°$, $DE = 19.3$, and $EF = 15.1$ Find DF.

14. In the diagram, $WX = x$ cm, $XY = 3x$ cm, $YZ = 20$ cm, $\sin\theta = \frac{4}{5}$ and $W\hat{X}Y = 120°$.

    (a) If the area of triangle WZY is 112 cm^2, find the length of $[WZ]$.

    (b) Given that θ is an acute angle, state the value of $\cos\theta$ and hence find the length of $[WY]$.

    (c) Find the exact value of x.

    (d) Find the degree measure of $X\hat{Y}Z$ to 3 significant figures.

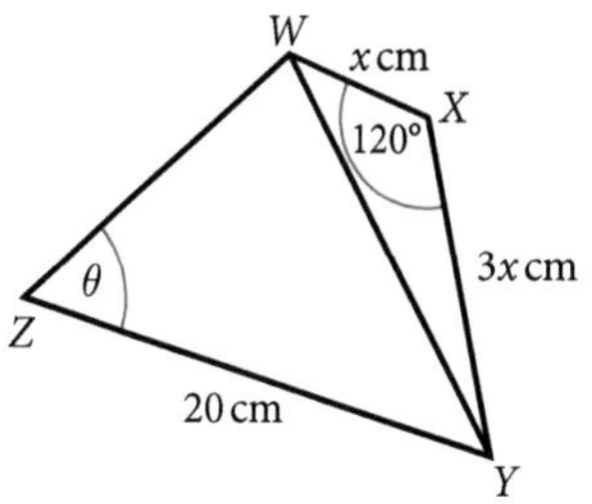

Figure 6.55 Diagram for question 14

15. In triangle FGH, $FG = 12$ cm, $FH = 15$ cm, and $\hat{G}$ is twice the size of $\hat{H}$. Find the approximate degree measure of $\hat{H}$ to 3 significant figures.

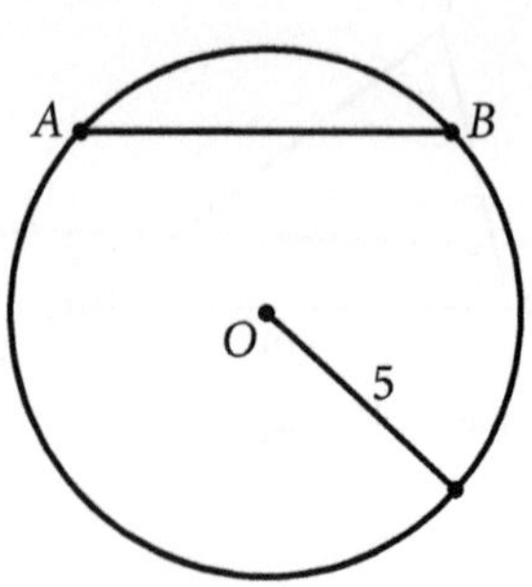

Figure 6.56 Diagram for question 1

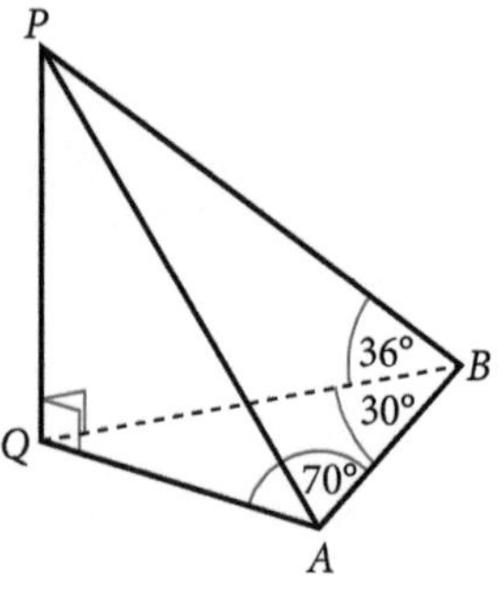

Figure 6.57 Diagram for question 5

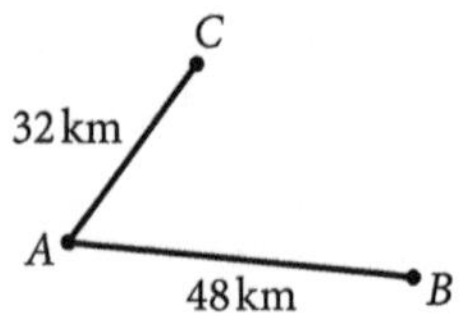

Figure 6.58 Diagram for question 6

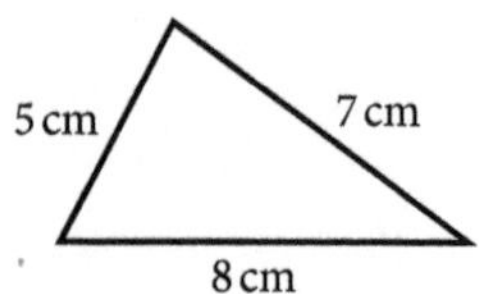

Figure 6.59 Diagram for question 7

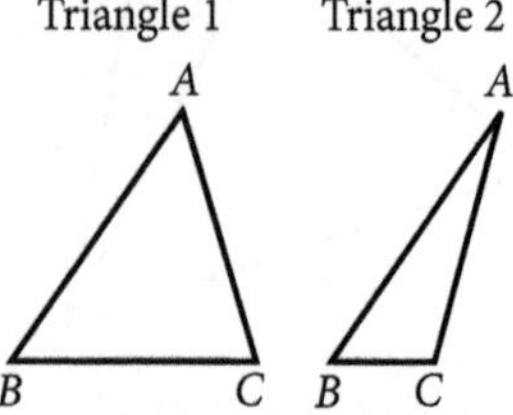

Figure 6.60 Diagram for question 8

Chapter 6 practice questions

1. The shortest distance from a chord [*AB*] to the centre *O* of a circle is 3 units. The radius of the circle is 5 units. Find the exact value of $\sin A\hat{O}B$.

2. In a right-angled triangle, $\tan\theta = \frac{3}{7}$. Find the exact value of $\sin 2\theta$ and $\cos 2\theta$.

3. A triangle has sides of length 4, 5 and 7 units. Find, to the nearest tenth of a degree, the size of the largest angle.

4. *A* is an obtuse angle in a triangle and $\sin A = \frac{5}{13}$. Calculate the exact value of $\sin 2A$.

5. The diagram shows a vertical pole *PQ*, which is supported by two wires fixed to the horizontal ground at *A* and *B*.

$BQ = 40$ m, $P\hat{B}Q = 36°$, $B\hat{A}Q = 70°$, $A\hat{B}Q = 30°$

Find:

(a) the height of the pole, *PQ*

(b) the distance between *A* and *B*.

6. Town *A* is 48 km from town *B* and 32 km from town *C* as shown in the diagram.

Given that town *B* is 56 km from town *C*, find the size of the angle $C\hat{A}B$ to the nearest tenth of a degree.

7. The diagram shows a triangle with sides 5 cm, 7 cm, and 8 cm.

Find:

(a) the size of the smallest angle, in degrees

(b) the area of the triangle.

8. The diagrams show two different triangles both satisfying the conditions $AB = 20$ cm, $AC = 17$ cm, $A\hat{B}C = 50°$

(a) Calculate the size of $A\hat{C}B$ in Triangle 2.

(b) Calculate the area of Triangle 1.

9. Two boats, *A* and *B*, start moving from the same point *P*. Boat *A* moves in a straight line at 20 km per hour and boat *B* moves in a straight line at 32 km per hour. The angle between their paths is 70°. Find the distance between the two boats after 2.5 hours.

10. In triangle JKL, $JL = 25$, $KL = 38$ and $K\hat{J}L = 51°$, as shown in the diagram.

Find $J\hat{K}L$, giving your answer correct to the nearest degree.

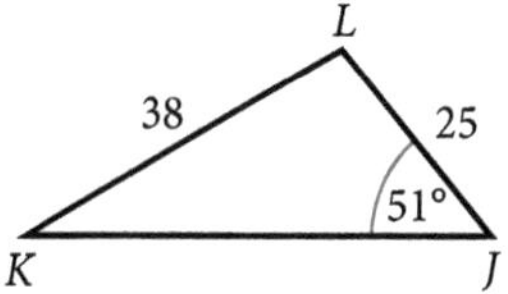

Figure 6.61 Diagram for question 10

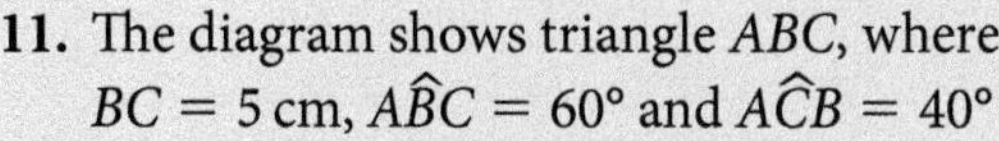

11. The diagram shows triangle ABC, where $BC = 5$ cm, $A\hat{B}C = 60°$ and $A\hat{C}B = 40°$

(a) Calculate AB.

(b) Find the area of the triangle.

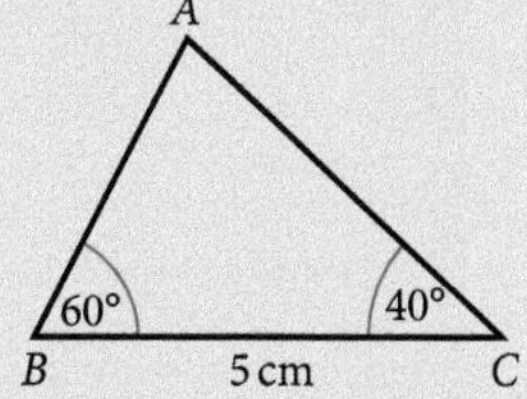

12. Find the size of the acute angle between a pair of diagonals of a cube.

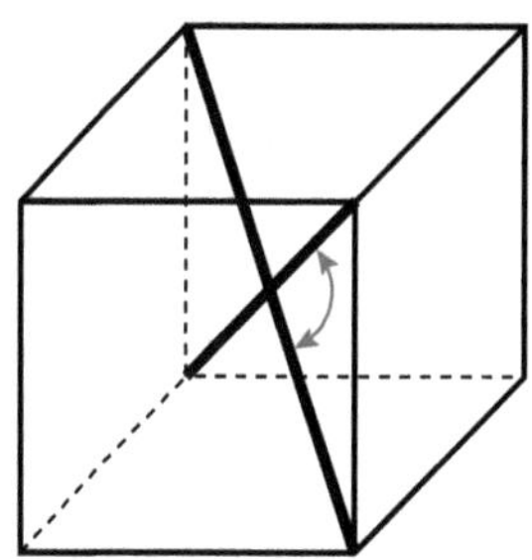

Figure 6.62 Diagram for question 12

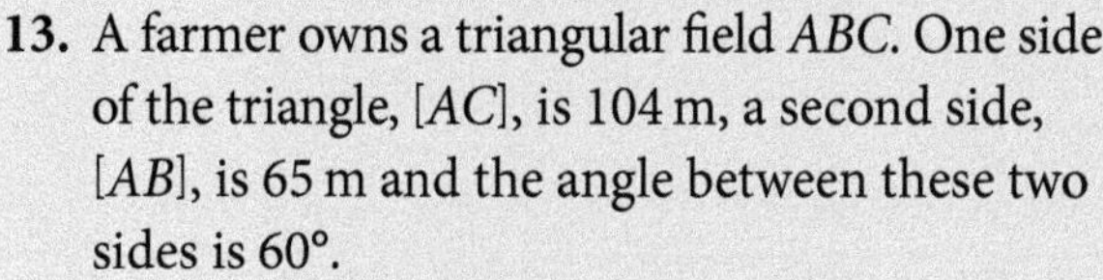

13. A farmer owns a triangular field ABC. One side of the triangle, $[AC]$, is 104 m, a second side, $[AB]$, is 65 m and the angle between these two sides is 60°.

(a) Use the cosine rule to calculate the length of the third side, $[BC]$, of the field.

(b) Given that $\sin 60° = \frac{\sqrt{3}}{2}$, find the area of the field in the form $p\sqrt{3}$ where p is an integer.

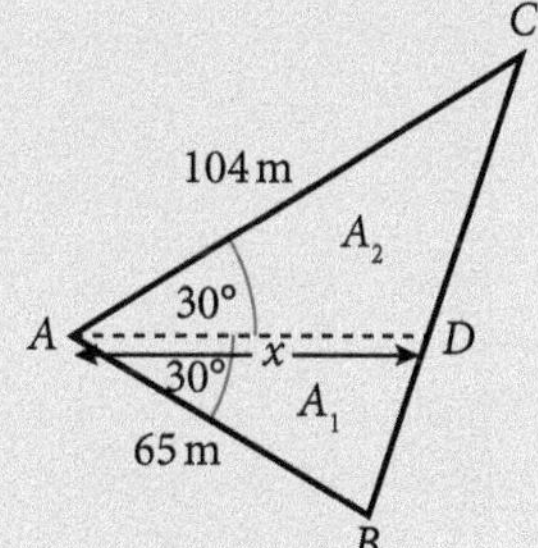

Let D be a point on $[BC]$ such that $[AD]$ bisects the 60° angle.
The farmer divides the field into two parts, A_1 and A_2, by constructing a straight fence $[AD]$ of length x metres, as shown in the diagram.

(c) **(i)** Show that the area of A_1 is $\frac{65x}{4}$

(ii) Find a similar expression for the area of A_2

(iii) Hence, find the value of x in the form $q\sqrt{3}$, where $q \in \mathbb{Z}$

(d) **(i)** Explain why $\sin A\hat{D}C = \sin A\hat{D}B$

(ii) Use the result of part **(i)** and the sine rule to show that $\frac{BD}{DC} = \frac{5}{8}$

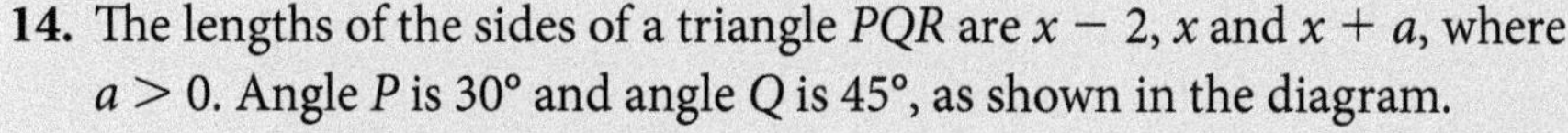

14. The lengths of the sides of a triangle PQR are $x - 2$, x and $x + a$, where $a > 0$. Angle P is 30° and angle Q is 45°, as shown in the diagram.

(a) Find the exact value of x.

(b) Find the exact area of triangle PQR.

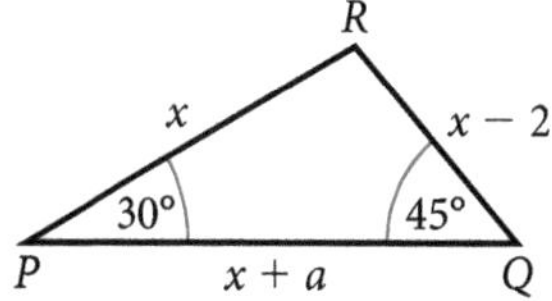

Figure 6.63 Diagram for question 14

15. In the diagram, the radius of the circle with centre C is 7 cm and the radius of the circle with centre D is 5 cm. If the length of the chord AB is 9 cm, find the area of the shaded region enclosed by the two minor arcs AB.

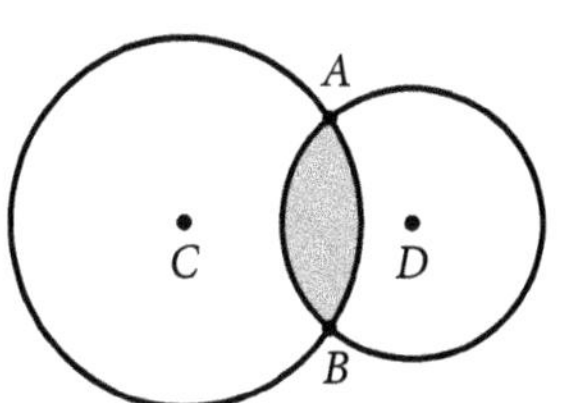

Figure 6.64 Diagram for question 15

16. One corner, K, of a field consists of two stone walls, $[KJ]$ and $[KL]$, at an angle of 60° to each other. A wooden fence $[JL]$ is to be built to create a triangular enclosure JKL, as shown in the diagram.

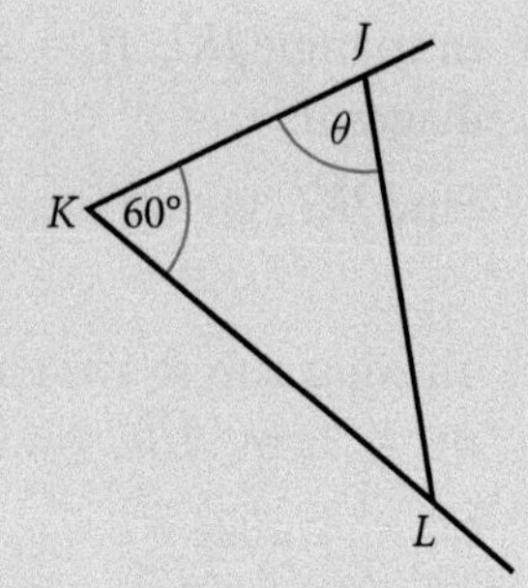

(a) If $K\hat{J}L$ is denoted by θ, state the range of possible values for θ.

(b) Show that the area of triangle JKL is $300\sqrt{3}\sin\theta\sin(\theta + 60°)$

(c) Use your GDC to determine the value of θ that gives the maximum area for the enclosure.

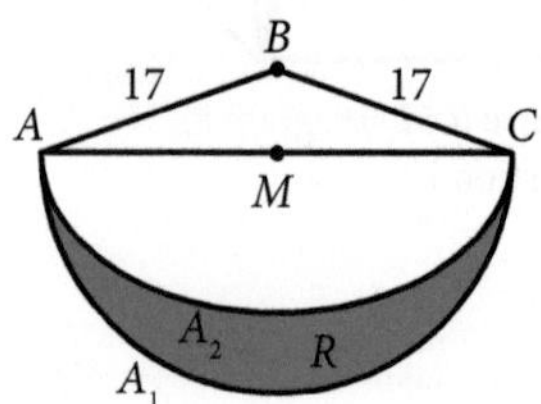

Figure 6.65 Diagram for question 17

17. The diagram shows the triangle ABC with $AB = BC = 17$ cm and $AC = 30$ cm. The midpoint of AC is M. The circular arc A_1 is half the circle (semicircle) with centre M. Another circular arc A_2 is drawn with centre B. The shaded region R is bounded by the arcs A_1 and A_2.
Find:

(a) the area of triangle ABC

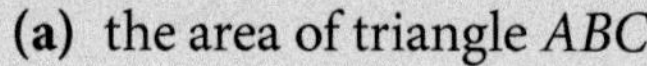

(b) the size of $A\hat{B}C$ in radians

(c) the area of the shaded region R.

18. (a) In the diagram, radii drawn to endpoints of a chord of the unit circle determine a central angle α. Show that the length of the chord is
$L = \sqrt{2 - 2\cos\theta}$

(b) By using the substitution $\theta = \frac{\alpha}{2}$ in the double-angle formula $\cos 2\theta = 1 - 2\sin^2\theta$, derive a formula for $\sin\frac{\alpha}{2}$, that is a half-angle formula for the sine function.

(c) Use the result in **(a)** and your result in **(b)** to show that the length of the chord is $L = 2\sin\left(\frac{\theta}{2}\right)$

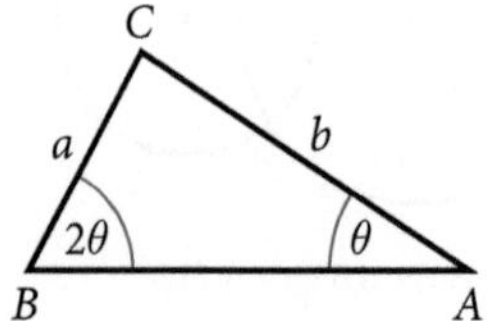

Figure 6.66 Diagram for question 19

19. In triangle ABC, $A\hat{B}C = 2\theta$ and $B\hat{A}C = \theta$.
Determine an expression for $\cos\theta$ in terms of a and b.

20. The traditional bicycle frame consists of tubes connected together in the shape of a triangle and a quadrilateral (four-sided polygon). In the diagram, AB, BC, CD, and AD represent the four tubes of the quadrilateral section of the frame. A frame maker has prepared three tubes such that $AD = 53$ cm, $AB = 55$ cm, and $BC = 11$ cm. $D\hat{A}B = 76°$ and $A\hat{B}C = 97°$. What must be the length of tube CD? Give your answer to the nearest tenth of a centimetre.

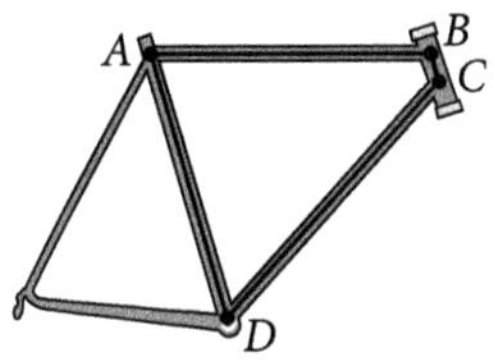

Figure 6.67 Diagram for question 20

21. The tetrahedron shown in the figure has the following measurements: $AB = 12$ cm, $DC = 10$ cm, $A\hat{C}B = 45°$, and $A\hat{D}B = 60°$.

AB is perpendicular to the triangle BCD. Find the area of each of the four triangular faces: ABC, ABD, BCD, and ACD.

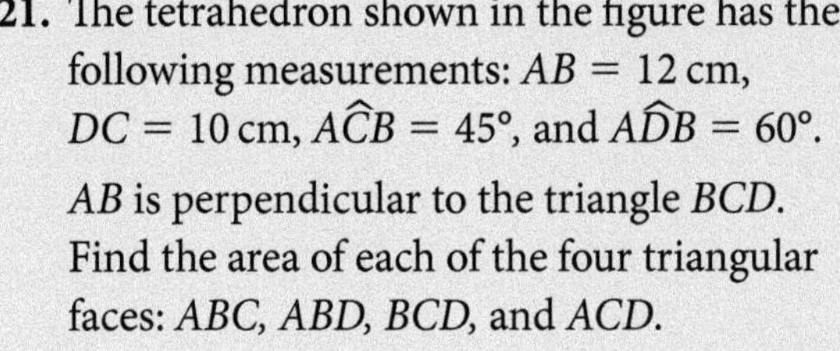

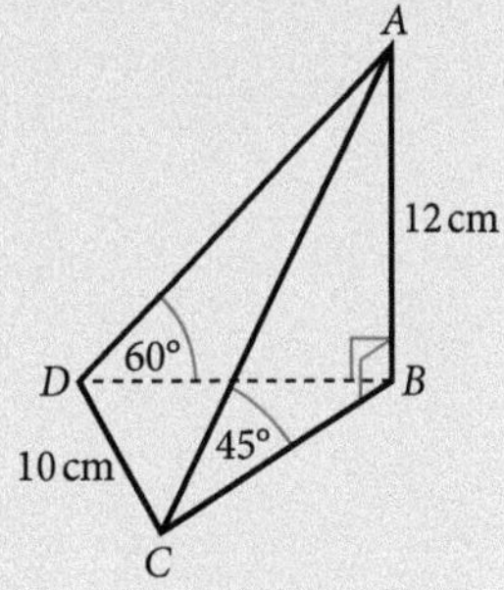

22. Find the size of angle DEF in the rectangular box.

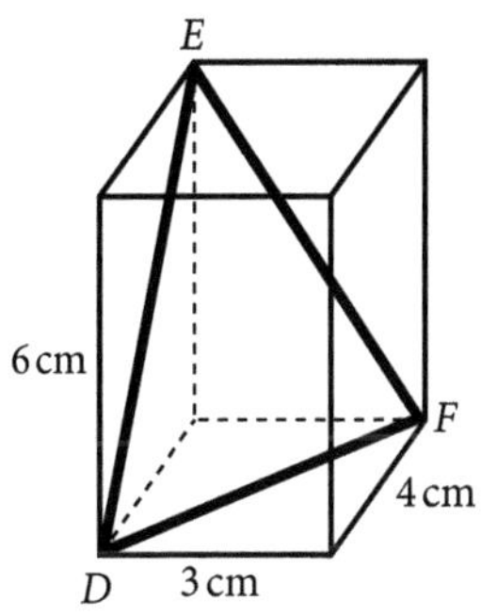

Figure 6.68 Diagram for question 22

23. At a point A due south of a building, the angle of elevation from the ground to the top of a building is 58°. At a point B (on level ground with A), 80 metres due west of A, the angle of elevation to the top of the building is 27°. Find the height of the building.

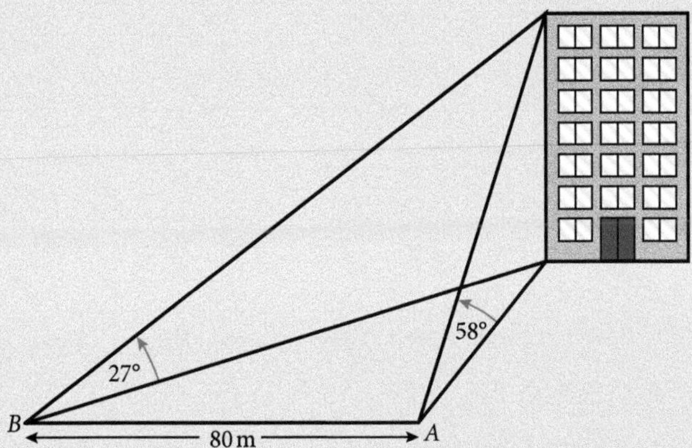

24. A right pyramid has a square base with sides of length 8 cm. The height of the pyramid is 10 cm. There are four lateral faces that are isosceles triangles, and one square base. Two adjacent lateral faces are shaded in the diagram. Calculate the angle between two adjacent lateral faces.

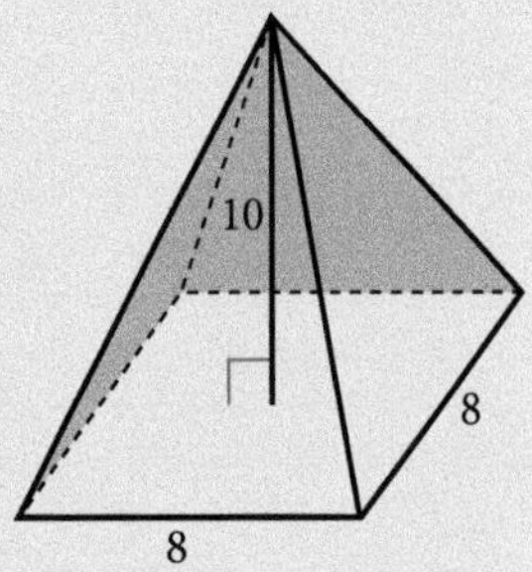

25. (a) Find the exact equation of line L_1 that passes through the origin and makes an angle of 30° with the positive x-axis.
(b) The equation of line L_2 is $x + 2y = 6$. Find the acute angle between L_1 and L_2.

26. A helicopter leaves from point P and flies in a straight line on a bearing of 125° for 150 km to point Q. It then flies in a straight line for 275 km from point Q to point R on a bearing of 230°. From point R, the helicopter flies directly back to point P. Calculate the length and bearing of the flight from R to P.

7
Statistics

7 Statistics

Learning objectives

By the end of this chapter, you should be familiar with...

- concepts of population, sample, random sample, and frequency distribution of discrete and continuous data
- reliability of data sources and bias in sampling
- sampling techniques and their effectiveness
- interpretation of outliers
- presentation of data using frequency tables and diagrams and box-and-whisker plots
- working with grouped data: mid-interval values, interval width, upper and lower interval boundaries, and frequency histograms
- calculating and interpreting the mean, median, mode, quartiles, and percentiles
- calculating and interpreting the range, interquartile range, variance, and standard deviation
- calculating and interpreting cumulative frequency graphs and using them to find the median, quartiles, and percentiles
- understanding and interpreting linear correlation of bivariate data
- working with linear regression.

Statistics are a part of everyday life, and you are likely to encounter them in one form or another on a daily basis.

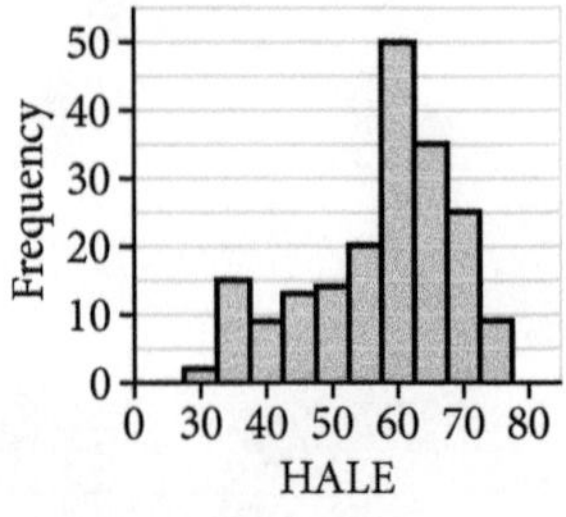

Figure 7.1 HALE data

For example, the World Health Organization (WHO) collects and reports data about worldwide population health on all 192 UN-member countries. Among the indicators reported is the health-adjusted life expectancy (HALE). This is based on life expectancy at birth, but includes an adjustment for time spent in poor health. It is most easily understood as the equivalent number of years in full health that a newborn can expect to live, based on current rates of ill-health and mortality. According to WHO rankings, lost years due to disability are substantially higher in poorer countries. Several factors contribute to this trend, including injury, blindness, paralysis, and the debilitating effects of tropical disease.

Of the 192 countries ranked by WHO, Japan has the highest healthy life expectancy (75 years) and Sierra Leone has the lowest (29 years).

Reports like this are commonplace in business publications, newspapers, magazines, and on the internet. There are some questions that come to mind as we read such a report. How did the researchers collect the data? How can we be sure that these results are reliable? What conclusions should be drawn from this report? The increased frequency with which statistical techniques are used in all fields, from business to agriculture to social and natural sciences, leads to the need for statistical literacy – familiarity with the goals and methods of these techniques – to be a part of any well-rounded educational programme.

Since statistical methods for summary and analysis provide us with powerful tools for making sense out of the data we collect, in this chapter we will first start by introducing two basic components of most statistical problems – population and sample – and then delve into the methods of presenting and making sense of data. This will include some basic techniques in **descriptive statistics** – the branch of statistics concerned with describing sets of measurements, both samples and populations.

7.1 Graphical tools

Once you have collected a set of measurements, how can you display this set in a clear, understandable, and readable form? First, you must be able to define what is meant by measurement or 'data' and to categorise the types of data you are likely to encounter. We begin by introducing some definitions of the new terms in the statistical language that you need to know.

In the language of statistics, one of the most basic concepts is **sampling**. In most statistical problems, we draw a specified number of measurements or data – a **sample** – from a much larger body of measurements, called the **population**. On the basis of our observation of the data in the well-chosen sample, we try to describe or predict the behaviour of the population.

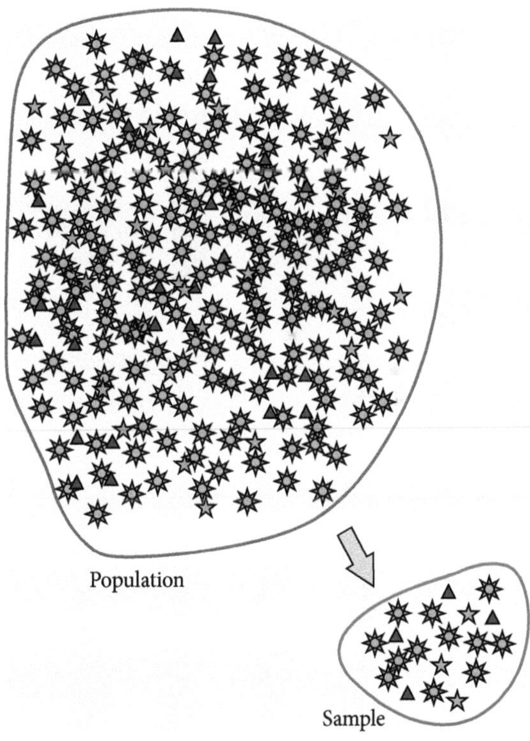

Figure 7.2 A sample is drawn from a population

A population is any entire collection of people, animals, plants, or things from which we may collect data. It is the entire group we are interested in, which we might wish to describe or draw conclusions about.

In order to make generalisations about a population, a sample is often studied. The sample should be representative of the population. For each population there are many possible samples.

For example, a study about the usage of resources in the households of an EU country stated that:

'… in the sample of 1674 households surveyed, the amount of water used by each washing cycle is given in the following…, The average time for each cycle was reported to be 42 minutes…. It was also discovered that the amount of laundry done by a household every year is related in some way to the household's income…'

In this example, the population is households' usage of water for washing, the average time spent on laundry, income, and so on. The sample is the set of measurements of 1674 households that took part in the study. Notice that the population and sample are the measurements and not the people. The households are 'experimental units' or subjects in this study.

7 Statistics

Height is a variable that changes with time for an individual and from person to person. If you gather the heights of the students at your school, the set of measurements you get is called a **data set**.

In everyday life, the terms 'reliability' and 'validity' are often used interchangeably. In statistics, however, these terms have specific meanings relating to different properties of the statistical or experimental method.

A **variable** is a characteristic that might vary over time or for different objects under consideration. When a variable is measured, the set of measurements obtained is called the **data** about that variable.

When a large amount of data is collected, it becomes difficult to see what it means. The statistician's job is to summarise the data succinctly, bringing out the important characteristics of the numbers so that a clear and accurate picture emerges. There are several ways of summarising and describing data, including tables, graphs, and numerical measures.

When looking at statistical results, we must be aware of how the data has been collected by assessing its **reliability** and **validity**.

Reliability, or **reproducibility**, is another word for consistency. It refers to the capacity of a test or method to produce the same result for two identical states or, more operationally, the closeness of the initial estimated values to the subsequent estimated values. For example, if one person takes the same personality test several times and always receives the same results, the test is reliable.

A test is valid if it measures what it is supposed to measure. If the results of the personality test claimed that a very shy person was in fact outgoing, the test would be invalid.

Reliability and validity are independent of each other. A measurement may be valid but not reliable, or reliable but not valid. Suppose your bathroom scale was reset to read 5 kg lighter than the actual weight. The reading it gave would be reliable, as it would be the same every time, but it would not be valid, since it would be lower than your correct weight.

Classification of variables

Numerical or categorical

Data can be classified into two main types: **numerical** (or **quantitative**) and **categorical** (or **qualitative**) data.

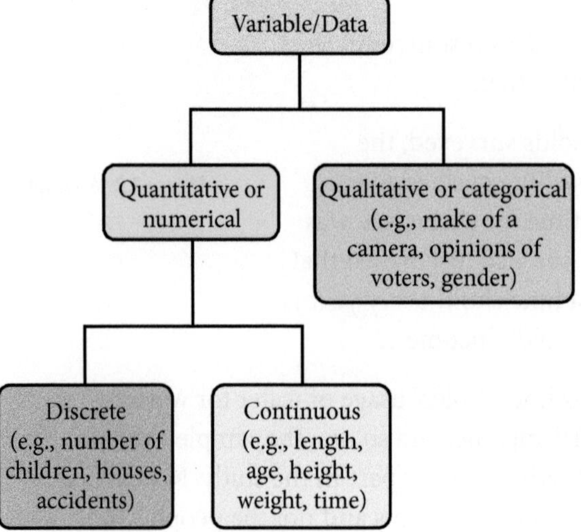

Figure 7.3 Data classifications

Numerical, or quantitative, variables measure a numerical quantity or amount on each experimental unit. This type of data always yields a numerical response.

There are two types of numerical data.

- **Discrete** data can take only particular values. For example, if you are counting the number of students that take a particular class, the values will all be integers. It makes no sense to have 0.5 students.
- **Continuous** data can take any value, subject to the accuracy with which you can measure it. For example, the time it takes a student to travel from home to school could potentially be measured to the nearest second, although it might not be appropriate to measure to this level of accuracy.

Examples of numerical data include yearly income of company presidents, heights of students at school, time taken for students to finish their lunch at school, and total score received on exams.

There are two types of continuous variable.

- **Interval** variables can be measured along a continuum, and the difference between two values on the continuum is meaningful.
- **Ratio** variables are interval variables with the additional condition that a value of 0 indicates that there is none of that variable. The name ratio reflects the fact that you can use the ratio of the measurements. So, for example, a distance of 20 m is twice as large as a distance of 10 m.

Temperature measured in degrees Celsius or Fahrenheit is an example of an interval variable. The difference between 20°C and 30°C is the same as the difference between 30°C and 40°C, but a temperature of 40°C is not twice as hot as a temperature of 20°C, because 0°C does not mean there is no temperature.
However, temperature measured in kelvin is a ratio variable, because 0 kelvin (often called absolute zero) indicates that there is no temperature whatsoever. A temperature of 100 kelvin is twice as hot as 50 kelvin. Other examples of ratio variables include height, mass, distance, and many more.

Categorical, or qualitative, variables measure a quality or characteristic of the experimental unit. Categorical data yields a qualitative response, such as colour.

We often use **pie charts** to summarise categorical data or to display the different values of a given variable (for example, percentage distribution). This type of chart is a circle divided into a series of segments. Each segment represents a particular category. The ratio of the area of each segment to the area of the circle is the same as the ratio of the corresponding category to the total data set.

Pie charts usually show the component parts of a whole. Often you will see a segment of the drawing separated from the rest of the pie in order to emphasise an important piece of information.

This pie chart shows how students in a large school are categorised into the IB Mathematics class they are taking. This is an example of qualitative data. There are 230 students in the Math AI/SL class, 180 students in the Math AA/SL class and 90 students in the Math AA/HL class. The pie chart shows what percentage of students take each class.

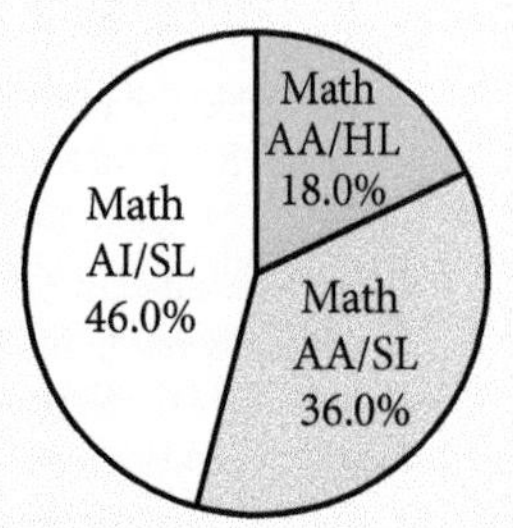

Frequency distributions

After collecting data, you should try to organise it so that it can be read easily. Methods for organising data include **ordered arrays** and **stem-and-leaf diagrams** – not required.

In its raw form, your data may be listed in the order you collected it:

24, 26, 24, 21, 27, 27, 30, 41, 32, 38

Ordering the data in a ordered array, in either ascending or descending order, makes it easier to spot patterns and to start to understand the data:

21, 24, 24, 26, 27, 27, 30, 32, 38, 41

Suppose a consumer organisation is interested in studying weekly food and living expenses of college students. A survey of 80 students yielded the data shown in Table 7.1. Expenses are given to the nearest euro.

38	50	55	60	46	51	58	64	50	49	48	65	58	61	65	53
39	51	56	61	48	53	59	65	54	54	54	59	65	66	47	49
40	51	56	62	47	55	60	63	60	59	59	50	46	45	54	47
41	52	57	64	50	53	58	67	67	66	65	58	54	52	55	52
44	52	57	64	51	55	61	68	67	54	55	48	57	57	66	66

Table 7.1 Weekly food and living expenses of college students

In its raw form, it is difficult to find any patterns or draw conclusions from this data. The first step in analysing data is to create a summary. This should show the following information:

- What values of the variable have been measured?
- How often has each value occurred?

Such summaries can be done in many ways. The most useful are **frequency distributions** and **histograms**. There are other methods of presenting data, some of which we will discuss later.

Consider a frequency distribution for the living expenses of 80 college students. If the frequency distribution contained the intervals '35–40' and '40–45', to which of these two classes would a person spending €40 belong? More appropriate intervals would be '35 or more but less than 40' and '40 or more but less than 45'.

If classes are described with discrete limits such as '30–34', '35–39', then the boundaries are midway between the neighbouring endpoints. That is, the classes will be considered as '29.5 or more but less than 34.5', '34.5 or more but less than 39.5', Here the boundaries are 29.5, 34.5, 39.5, and each class width is 5.

A **frequency distribution** is a table used to organise data. The left column, called **classes** or **groups**, includes numerical intervals on the variable being studied. The right column is a list of the **frequencies**, or **number of observations**, for each class. Intervals are normally of equal size. They must cover the range of the sample observations and they must not overlap.

Construction of a frequency distribution

There are some general rules for preparing frequency distributions that make it easier to summarise data and to communicate results.

Rule 1: Classes must be **inclusive** and **non-overlapping**. Each observation must belong to one, and only one, class interval. The **boundaries**, or **endpoints**, of each class must be clearly defined.

Rule 2: Determine k, the number of classes. Practice and experience are the best guidelines for deciding on the number of classes. In general, it is reasonable to have between 5 and 10 classes, but this is not an absolute rule. Practitioners use their judgement in these issues. If there are too few classes, some characteristics of the distribution will be hidden. If there are too many, some characteristics will be lost with the detail.

Rule 3: Intervals should be the same width. The width is determined by the formula

$$\text{interval width} = \frac{\text{largest number} - \text{smallest number}}{\text{number of intervals}}$$

Both the number of intervals and the interval width should be rounded up, possibly to the next integer. The above formula can be used when there are no natural ways of grouping the data. If this formula is used, the interval width is generally rounded to a convenient integer for easy interpretation.

Example 7.1

Organise the data from Table 7.1 into a frequency distribution, using appropriate class intervals.

Solution

Start by putting the data in ascending order.

38	39	40	41	44	45	46	46	47	47	47	48	48	48	49	49
50	50	50	50	51	51	51	51	52	52	52	52	53	53	53	54
54	54	54	54	54	55	55	55	55	55	56	56	57	57	57	57
58	58	58	58	59	59	59	59	60	60	60	61	61	61	62	63
64	64	64	65	65	65	65	65	66	66	66	66	67	67	67	68

With the data in order, we can immediately see that the smallest value is €38 and the largest value is €68. A reasonable grouping with nice round numbers is '35 or more but less than 40' and '40 or more but less than 45', and so on. This gives a class width of 5.

Living expenses (l)	Number of students	Percentage of students
$35 \leq l < 40$	2	2.50
$40 \leq l < 45$	3	3.75
⋮	⋮	⋮
$65 \leq l < 70$	13	16.25
Total	80	100.00

Frequency and percentage frequency distributions of weekly expenses

Grouping the data in a table, as in Example 7.1, allows us to see some of its characteristics. For example, we can observe that there are few students who spend as little as €35 to €45, while the majority of the students spend more than €45. Grouping the data also causes some loss of detail, as we cannot see from the table what the real values in each class are.

In the table in Example 7.1, the **class midpoint**, also known as the **mid-interval value**, can be used to represent the data in that interval. For example, 37.5 can represent the data in the first class, while 42.5 can represent the data in the 40 to 45 class. This will be discussed in more detail later in the chapter. The values at the ends of each class, such as 35 and 40, are known as the **interval boundaries**.

Histograms

We can visualise a frequency distribution graphically using a **histogram**.

A histogram is a graph that consists of vertical bars constructed on a horizontal line that is marked off with intervals for the variable being displayed. The intervals correspond to the class intervals in a frequency distribution table. The height of each bar is proportional to the number of observations in that interval. The number of observations can also be displayed above the bars.

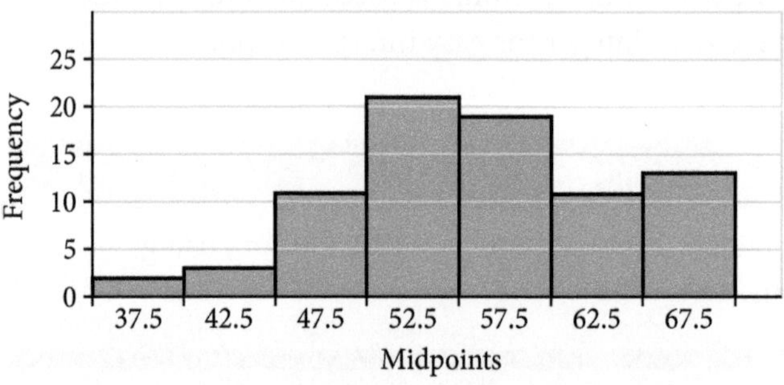

Figure 7.4 Histogram for the data in Example 7.1

Your GDC allows you to draw histograms. Different models will have different procedures. Here is a sample.

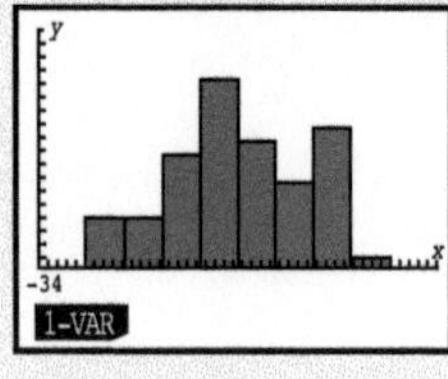

Figure 7.4 shows a histogram for the data in Example 7.1. By looking at the histogram, it becomes visually clear that our previous observation is true. From the histogram we can also see that the distribution is not **symmetric**. You will find out more about the shape of frequency distributions later in this chapter.

Cumulative and relative cumulative frequency distributions

A **cumulative frequency distribution** contains the total number of observations whose values are less than the upper limit for each interval. It is constructed by adding the frequencies of all the intervals up to and including the present interval. A **relative cumulative frequency distribution** converts all cumulative frequencies to cumulative percentages.

Table 7.2 shows a cumulative distribution and a relative cumulative distribution for the data in Example 7.1.

Living expenses (l)	No. of students	Cumulative number of students	Percentage of students	Cumulative Percentage of students
$35 \leqslant l < 40$	2	2	2.50	2.50
$40 \leqslant l < 45$	3	5	3.75	6.25
$45 \leqslant l < 50$	11	16	13.75	20.00
$50 \leqslant l < 55$	21	37	26.25	46.25
$55 \leqslant l < 60$	19	56	23.75	70.00
$60 \leqslant l < 65$	11	67	13.75	83.75
$65 \leqslant l < 70$	13	80	16.25	100.00
Total	80		100.00	

Table 7.2 Cumulative frequency and cumulative relative frequency distributions of weekly expenses

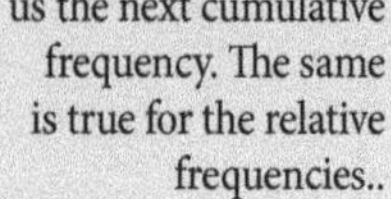

Notice how every cumulative frequency is added to the frequency in the next interval to give us the next cumulative frequency. The same is true for the relative frequencies..

As we will see later, cumulative frequencies and their graphs help in analysing data given in group form.

Cumulative frequency graphs

A **cumulative frequency graph**, sometimes called a **cumulative line graph** or an **ogive**, is a line that connects points that are the cumulative percentage of observations below the upper limit of each class in a cumulative frequency distribution. Figure 7.5 shows a cumulative frequency graph for the data in Example 7.1.

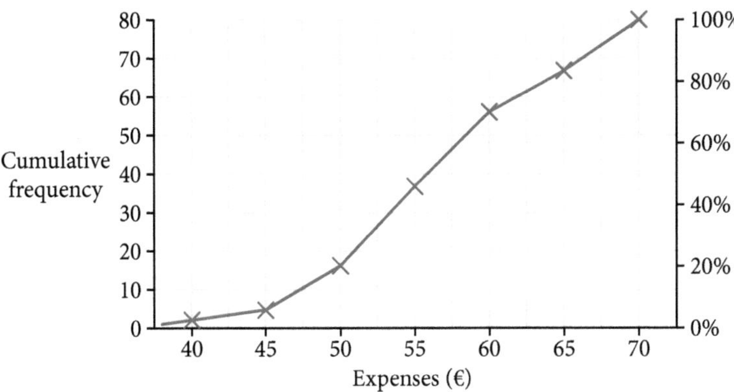

Figure 7.5 Cumulative frequency graph for the data in Example 7.1

Notice how the height of each line at the upper boundary represents the cumulative frequency for that interval. For example, at 50 the height is 16 and at 60 it is 56.

Example 7.2

The WHO data discussed in the introduction is given here in raw form.

(a) Prepare a frequency table, starting with a lower class boundary of 20 and a class interval of 5.

(b) Draw a histogram to represent the data.

(c) Draw a cumulative frequency graph to represent the data.

29	36	40	44	48	52	54	56	59	60	61	61	62	63	64	66	68	71	72	73	63	64	66	68
31	36	41	44	49	52	54	57	59	60	61	62	62	64	64	66	68	71	72	75	63	64	66	68
33	36	41	44	49	52	55	57	59	60	61	62	62	64	65	66	69	71	72	35	38	43	47	71
34	37	41	45	49	53	55	58	59	60	61	62	63	64	65	66	69	71	73	36	40	44	48	71
34	37	42	45	50	53	55	58	59	60	61	62	63	64	65	67	70	71	73	50	54	56	59	72
35	37	42	45	50	53	55	58	59	60	61	62	63	64	65	67	70	71	73	51	54	56	59	72
35	37	43	46	50	54	55	58	59	60	61	62	63	64	65	67	70	71	73	60	60	61	62	73
35	38	43	46	50	54	55	58	59	60	61	62	63	64	65	67	70	72	73	60	61	61	62	73

$25 \leqslant l < 30$ contains all observations larger than or equal to 25 but less than 30.

Solution

(a) First sort the data, then count every number in each class.

Life expectancy, l	Number of countries	Life expectancy	Number of countries
$25 \leqslant l < 30$	1	$55 \leqslant l < 60$	26
$30 \leqslant l < 35$	4	$60 \leqslant l < 65$	54
$35 \leqslant l < 40$	14	$65 \leqslant l < 70$	22
$40 \leqslant l < 45$	14	$70 \leqslant l < 75$	27
$45 \leqslant l < 50$	11	$75 \leqslant l < 80$	1
$50 \leqslant l < 55$	18		

(b) The histogram is shown on the right. Since all classes have equal width, the height and the area give the same impression about the frequency of the class interval. For example, the 60–65 class contains almost twice as many countries the 55–60 class, and the heights of the bars in the histogram reflect this, as do the areas. Similarly, the height of the 65–70 class is double that of the 45–50 class.

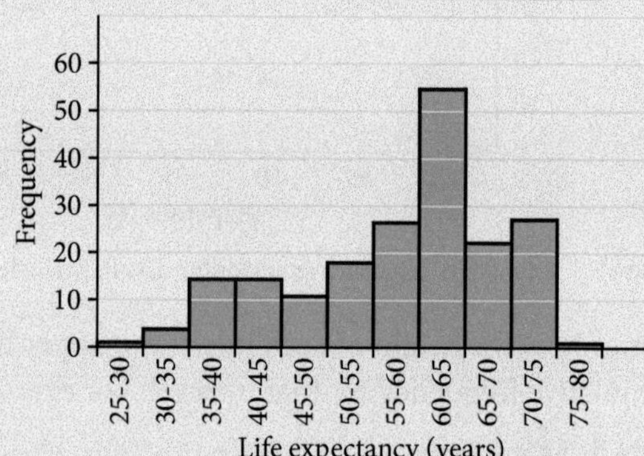

(c) In order to construct a cumulative frequency graph, we must first construct a cumulative frequency table.

Life expectancy	Number of countries	Cumulative number of countries	Life expectancy	Number of countries	Cumulative number of countries
$25 \leqslant l < 30$	1	1	$55 \leqslant l < 60$	26	88
$30 \leqslant l < 35$	4	5	$60 \leqslant l < 65$	54	142
$35 \leqslant l < 40$	14	19	$65 \leqslant l < 70$	22	164
$40 \leqslant l < 45$	14	33	$70 \leqslant l < 75$	27	191
$45 \leqslant l < 50$	11	44	$75 \leqslant l < 80$	1	192
$50 \leqslant l < 55$	18	62			

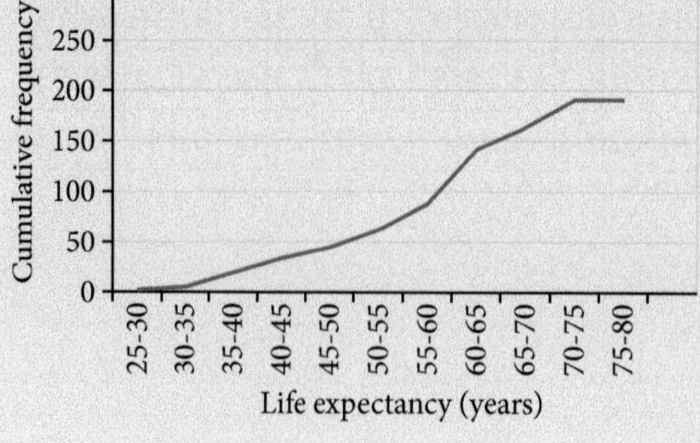

Sampling

Any study concerning populations needs data to be collected. Usually we do not collect data from the entire population. For statistical studies, data from samples is used. The method used to conduct a study is usually something like this:

1. Specify the population of interest.
2. Choose an appropriate sampling method.
3. Collect the sample data.
4. Analyse the pertinent information in the sample.
5. Use the results of the sample analysis to make an inference about the population.
6. Provide a measure of the inference's reliability.

Reasons for sampling

Taking a sample instead of conducting a census offers several advantages.

A **census** is a survey of the entire population.

A sample can save money and time. If an eight-minute interview is being undertaken, conducting the interviews with a sample of 100 people rather than with a population of 100 000 is obviously less expensive. In addition to the cost savings, the significantly smaller number of interviews usually requires less total time.

For given resources, the sample can broaden the scope of the study. With fixed resources, more detailed information can be gathered by taking a sample than by gathering information from the whole population. By concentrating on fewer individuals or items, the study can be broadened in scope to allow for more specialised questions.

Some research processes are destructive to the product or item being studied. For example, if light bulbs are being tested to determine how long they burn or if candy bars are being taste tested to determine whether the taste is acceptable, the product is destroyed.

If accessing the entire population is impossible, using a sample is the only option.

A **representative sample** is a sample that represents the characteristics of the population as closely as possible.

If sampling is deemed to be appropriate, it must be decided how to select a sample. Since the sample will be employed to draw conclusions about the entire population, it is crucial that the sample is **representative** of that population. It should reflect the relevant parameter of the population under consideration as closely as possible.

Random and non-random sampling

The two main types of sampling are **random** and **non-random**. In random sampling, every unit of the population has the same probability of being selected into the sample. Random sampling implies that chance enters into the process of selection.

In non-random sampling, not every unit of the population has the same probability of being selected into the sample.

Non-random sampling methods are not appropriate techniques for gathering data to be analysed by most of the statistical methods presented in this book.

Random sampling is also called **probability sampling** and non-random sampling is called **non-probability sampling**. Because every unit of the population is not equally likely to be selected in non-random sampling, assigning a probability of occurrence is impossible. The statistical methods presented and discussed in the IB syllabus assume that the data comes from random samples.

However, several non-random sampling techniques are described in this section, primarily to alert you to their characteristics and limitations.

Random sampling

Sampling error occurs when, by chance, the sample does not represent the population.

We will discuss three basic random sampling techniques: **simple random sampling**, **stratified random sampling**, and **systematic random sampling**. Each technique offers advantages and disadvantages. Some techniques are simpler to use, some are less costly, and others show greater potential for reducing **sampling error**.

Generally, all samples selected from the same population will give different results because they contain different elements of the population. Additionally, the results obtained from any one sample will not be exactly the same as those obtained from a census. The difference between a sample result and the result we would have obtained by conducting a census is called the **sampling error**, assuming that the sample is random and no non-sampling error has been made.

The sampling error is the difference between the result obtained from a sample survey and the result that would have been obtained if the whole population had been included in the survey.

Non-sampling errors can occur in both a sample survey and a census. Such errors occur because of human mistakes and not through chance.

Simple random sampling

A **sampling frame** is a list of all the elements of a population from which a sample can be taken, such as a register of all the students in a college.

The most elementary random sampling technique is **simple random sampling**. Simple random sampling can be viewed as the basis for the other random sampling techniques. With simple random sampling, each unit of the sampling frame is numbered from 1 to N (where N is the size of the population). Next, a **random number generator** (or a **table of random numbers**) is used to select n items into the sample.

Example 7.3

The screenshot shows the first five from a list of random numbers generated by a GDC. Computer programs may be more efficient.

```
RanInt#(1,659,20)
{217,100,191,518,252▸
```

Suppose it has been decided to interview 20 students from a school of 659 to form an understanding of their views of a new block-scheduling the school wants to adopt.

To find a simple random sample, number the students from 001 (or simply 1) to 659 and have a random generator choose 20 numbers. The students allocated to the chosen numbers form the sample.

Stratified random sampling

In **stratified random sampling**, the population is divided into non-overlapping subpopulations called **strata**. The researcher then carries out simple random sampling on each of the subpopulations. The main reason for using stratified random sampling is that it has the potential for reducing sampling error.

With stratified random sampling, the potential to match the sample closely to the population is greater than it is with simple random sampling because portions of the total sample are taken from different population subgroups. However, stratified random sampling is generally more costly than simple random sampling because each unit of the population must be assigned to a stratum before the random selection process begins.

Strata selection is usually based on available information. Such information may have been gleaned from previous censuses or surveys. The more different the strata are, the greater the benefits of using stratification. Internally, a stratum should be relatively homogeneous; externally, strata should contrast with each other. The process is demonstrated in Example 7.4.

Example 7.4

In FM radio markets, 'age of listener' is an important determinant of the type of programming used by a station.

The figure shows a stratification by age with three strata, based on the assumption that age makes a difference in preference of programming. This stratification assumes that listeners of 20 to 30 years of age tend to prefer the same type of programming, which is different from that preferred by listeners of 30 to 40 and 40 to 50 years of age. Within each age subgroup (stratum), **homogeneity** or alikeness is present; between each pair of subgroups, **heterogeneity** or difference is present. A simple random sample is taken from each stratum. Together, the samples constitute a representative sample of the whole population.

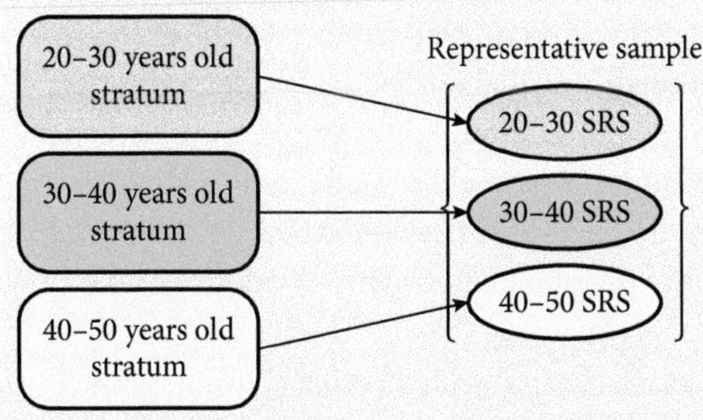

An advantage of stratified random sampling is that, in addition to collecting information about the entire population, we can also compare different strata. In Example 7.4, the information we get will also help us compare the different age groups.

Systematic random sampling

With **systematic random sampling**, every kth item is selected to produce a sample of size n from a population of size N. The value of k, sometimes called the sampling cycle, can be determined by the formula

$$k = \frac{N}{n}$$

If k is not an integer value, it should be rounded to the nearest integer.

Unlike stratified random sampling, systematic sampling is not done in an attempt to reduce sampling error. Rather, it is used because of its convenience and relative ease of administration.

Example 7.5

Given the data in Example 7.3, suppose we need to take a sample of 20 students using systematic sampling. First find k.

$$k = \frac{659}{20} \approx 32$$

From the list of 659 students, we randomly choose a starting number between 1 and 32. This might be 11, for example. After that we choose every 32nd number: 43, 75, 107, ….

Systematic sampling has other advantages besides convenience. Because systematic sampling is evenly distributed across the population, a knowledgeable person can easily determine whether a sampling plan has been followed in a study.

Non-random sampling

Sampling techniques used to select elements from the population by any mechanism that does not involve a random selection process are called **non-random sampling techniques**. Because chance is not used to select items from the samples, these techniques are **non-probability** techniques and are not desirable for use in gathering data to be analysed by standard methods of inferential statistics. Sampling error cannot be determined objectively for these sampling techniques. Two non-random sampling techniques are presented here: convenience sampling and quota sampling.

Convenience sampling

In **convenience sampling**, elements for the sample are selected for the convenience of the researcher. The researcher typically chooses elements that are readily available, nearby, or willing to participate. The sample tends to be less variable than the population because in many environments the extreme elements of the population are not readily available. The researcher will select more elements from the middle of the population. For example, a convenience sample of homes for door-to-door interviews might include houses where people are at home, houses with no dogs, houses near the street, first-floor apartments, and houses with friendly people. In contrast, a random sample would require the researcher to gather data only from houses and apartments that have been selected randomly, no matter how inconvenient or unfriendly the location. If a research firm is located in a mall, a convenience sample might be selected by interviewing only shoppers who pass the shop and look friendly.

Quota sampling

Quota sampling appears to be similar to stratified random sampling at first glance. However, instead of selecting a simple random sample from each stratum, a non-random sampling method is used to gather data from one stratum until the desired quota of samples is filled. Quotas are described by setting the sizes of the samples to be obtained from the subgroups. Generally, a quota is based on the proportions of the subclasses in the population.

For example, a company is test marketing a new soft drink and is interested in how age groups react to it. An interviewer goes to a shopping mall and interviews shoppers of age group 16–20, for example, until enough responses are obtained to fill the quota.

In quota sampling, an interviewer starts by asking a few filter questions. If the respondent represents a subclass whose quota has been filled, the interviewer stops the interview.

Quota sampling can be useful if no previous information is available for the population. For example, suppose we want to stratify the population into cars using different types of winter tyres but we do not have lists of users of the 'Continental' brand of tyres. Through quota sampling, we would proceed by interviewing all car owners and casting out non-Continental users until the quota of Continental users is filled.

Quota sampling is less expensive than most random sampling techniques because it is a technique of convenience. Another advantage of quota sampling is the speed of data gathering. We do not have to call back or send out a second questionnaire if we do not receive a response; we just move on to the next element.

The problem with quota sampling is that it is a non-random sampling technique. Some researchers believe that a solution to this issue can be achieved if the quota is filled by randomly selecting elements and discarding those not from a stratum. This way, quota sampling is essentially a version of stratified random sampling. The object is to gain the benefits of stratification without the high costs. However, it remains a non-probability sampling method.

Exercise 7.1

1. Identify the experimental units, sensible population, and sample on which each of the following variables is measured. Then indicate whether the variable is quantitative or qualitative.

(a) Gender of a student.

(b) Number of errors on a final exam for 10th grade students.

(c) Height of a newborn child.

(d) Eye colour for children aged less than 14.

(e) Amount of time it takes to travel to work.

(f) Rating of a country's leader: excellent, good, fair, poor.

(g) Country of origin of students at international schools.

2. State what you expect the shapes of the distributions of the following variables to be: uniform, unimodal, bimodal, symmetric, etc. Explain why.

(a) Number of goals shot by football players during the last season.

(b) Weights of newborn babies in a major hospital during the course of 10 years.

(c) Number of countries visited by a student at an international school.

(d) Number of emails received by a high school student at your school per week.

3. Identify each variable as quantitative or qualitative.
   (a) Amount of time to finish your extended essay.
   (b) Number of students in each section of IB Maths SL.
   (c) Rating of your textbook as excellent, good, satisfactory, terrible.
   (d) Country of origin of each student in Maths SL courses.

4. Identify each variable as discrete or continuous.
   (a) Population of each country represented by SL students in your session of the exam.
   (b) Weight of IB Maths SL exams printed every May since 1976.
   (c) Time it takes to mark an exam paper by an examiner.
   (d) Number of customers served at a bank counter.
   (e) Time it takes to finish a transaction at a bank counter.
   (f) Amount of sugar used in preparing your favourite cake.

5. Grade point averages (GPA) in several colleges are on a scale of 0–4. Here are the GPAs of 45 students at a certain college.

1.8	1.9	1.9	2.0	2.1	2.1	2.1	2.2	2.2	2.3	2.3	2.4	2.4	2.4	2.5
2.5	2.5	2.5	2.5	2.5	2.6	2.6	2.6	2.6	2.6	2.7	2.7	2.7	2.7	2.7
2.8	2.8	2.8	2.9	2.9	2.9	3.0	3.0	3.0	3.1	3.1	3.1	3.2	3.2	3.4

Draw a histogram, a relative frequency histogram, and a cumulative frequency graph. Describe the data in two to three sentences.

6. The following are the grades of an IB course with 40 students on a 100-point test. Use the graphical methods you have learned so far to describe the grades.

61	62	93	94	91	92	86	87	55	56
63	64	86	87	82	83	76	77	57	58
94	95	89	90	67	68	62	63	72	73
87	88	68	69	65	66	75	76	84	85

7. The lengths of time (in months) between repeated speeding violations of 50 young drivers are given in the table below.

2.1	1.3	9.9	0.3	32.3	8.3	2.7	0.2	4.4	7.4
9	18	1.6	2.4	3.9	2.4	6.6	1	2	14.1
14.7	5.8	8.2	8.2	7.4	1.4	16.7	24	9.6	8.7
19.2	26.7	1.2	18	3.3	11.4	4.3	3.5	6.9	1.6
4.1	0.4	13.5	5.6	6.1	23.1	0.2	12.6	18.4	3.7

   (a) Construct a histogram for the data.
   (b) Would you describe the shape as symmetric?
   (c) The law in this country requires that the driving licence be taken away if the driver repeats the violation within a period of 10 months. Estimate the proportion of drivers who may lose their licence.

8. To decide on the number of counters needed to be open during busy times in a supermarket, the management collected data from 60 customers for the time they spent waiting to be served. The times in minutes are given in the following table.

3.6	0.7	5.2	0.6	1.3	0.3	1.8	2.2	1.1	0.4
1	1.2	0.7	1.3	0.7	1.6	2.5	0.3	1.7	0.8
0.3	1.2	0.2	0.9	1.9	1.2	0.8	2.1	2.3	1.1
0.8	1.7	1.8	0.4	0.6	0.2	0.9	1.8	2.8	1.8
0.4	0.5	1.1	1.1	0.8	4.5	1.6	0.5	1.3	1.9
0.6	0.6	3.1	3.1	1.1	1.1	1.1	1.4	1	1.4

(a) Construct a relative frequency histogram for the times.

(b) Construct a cumulative frequency graph and estimate the number of customers who have to wait 2 minutes or more.

9. The histogram below shows the number of days spent in hospital by heart patients in a certain country's hospitals in the 2015–2017 period.

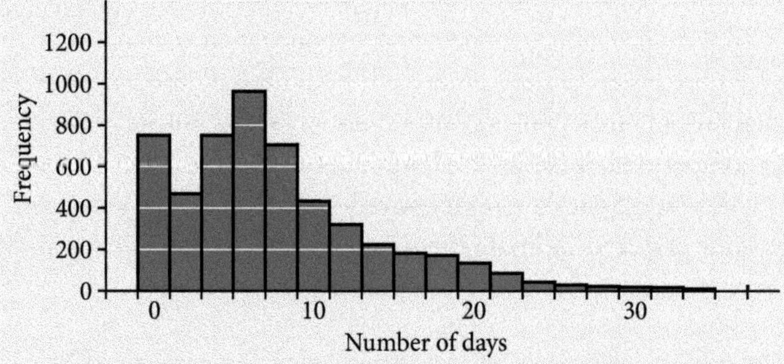

(a) Describe the data in a few sentences.

(b) Draw a cumulative frequency graph for the data.

(c) What percentage of the patients stayed less than 6 days?

10. One of the authors exercises on almost a daily basis. He records the length of time of his exercise on most of the days. Here is what he recorded for 2017.

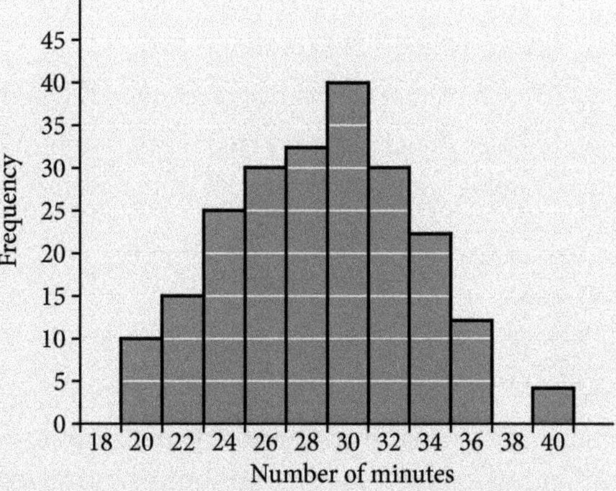

(a) What is the longest time he has spent doing his exercises?

(b) What percentage of the days did he exercise more than 30 minutes?

(c) Draw a cumulative frequency graph for his exercise time.

11. Radar devices are installed at several locations on a main highway. Speeds, s, in km h^{-1} of 400 cars travelling on that highway are measured and summarised in the following table.

Speed	$60 \leqslant s < 75$	$75 \leqslant s < 90$	$90 \leqslant s < 105$	$105 \leqslant s < 120$	$120 \leqslant s < 135$	Over 135
Frequency	20	70	110	150	40	10

(a) Construct a frequency table for the data.

(b) Draw a histogram to illustrate the data.

(c) Draw a cumulative frequency graph for the data.

(d) The speed limit in this country is 130 km h^{-1}. Use your graph in **(c)** to estimate the percentage of the drivers driving faster than this limit?

12. Electronic components used in the production of computers are manufactured in a factory and their measures must be very accurate. Here are the lengths of a sample of 400 such components.

Length, l (mm)	< 5.00	$5.00 \leqslant l < 5.05$	$5.05 \leqslant l < 5.10$	$5.10 \leqslant l < 5.15$	$5.15 \leqslant l < 5.20$	More than 5.20
Frequency	16	100	123	104	48	9

(a) Construct a cumulative relative frequency graph for the data.

(b) The components must have a length between 5.01 and 5.18 mm, and any component with a length above 5.18 mm has to be scrapped. Use your graph to estimate the percentage of components that must be scrapped from this production facility.

13. The time, t, in seconds, that 300 customers wait at a supermarket checkout are recorded in the table below.

Time	$t < 60$	$60 \leqslant t < 120$	$120 \leqslant t < 180$	$180 \leqslant t < 240$	$240 \leqslant t < 300$	$300 \leqslant t < 360$	$t > 360$
Frequency	12	15	42	105	66	45	15

(a) Draw a histogram of the data.

(b) Construct a cumulative frequency graph of the data.

(c) Use the cumulative frequency graph to estimate the waiting time that is exceeded for 25% of the customers.

7.2 Measures of central tendency

When a data set is large, **summary measures** can help us to understand it. This section presents several ways to summarise quantitative data by calculating a **measure of central tendency**, also called a **measure of location** (a value that is representative of a typical data item), and a **measure of spread** (a value that indicates how well the typical value represents the data). These measures can be used in addition to or instead of tables and graphs.

The farthest we can reduce a set of data, and still retain any information at all, is to summarise the data with a single value. Measures of location do just that: they try to capture with a single number what is typical of the data. What single number is most representative of an entire list of numbers? We cannot say without defining 'representative' more precisely.

We will study three common measures of location: the **mean**, the **median**, and the **mode**. The mean, median, and mode are all 'most representative', but for different, related notions of representativeness.

- The arithmetic mean is commonly called the average. It is the sum of the data, divided by the number of items of data:

$$\text{mean} = \frac{\text{sum of data}}{\text{number of data}} = \frac{\text{total}}{\text{number of data}}$$

- The median of a set of measurements is the value that falls in the middle position when the data are sorted in ascending order. In a histogram, the median is the value that divides the histogram into two equal areas.
- The mode of a set of data is the most common value among the data. It is rare that several data coincide exactly, unless the variable is discrete, or the measurements are reported with low precision.

When these measures are computed for a population they are called **parameters**. When they are computed from a sample they are called **statistics**.

A **statistic** is a descriptive measure computed from a sample of data.
A **parameter** is a descriptive measure computed from an entire population of data.

Measures of central tendency provide information about a typical observation in the data or locate the data set.

Mean, median, mode

The mean

The most common measure of central tendency is the arithmetic mean, usually referred to simply as the 'mean' or the 'average.'

Example 7.6

The five closing prices of the NASDAQ Index for the first business week in November 2007 are given below. This is a sample of size $n = 5$ for the closing prices from the entire population.

2794.83   2810.38   2795.18   2825.18   2748.76

Find the average closing price.

Solution

$$\text{Average} = \frac{2794.83 + 2810.38 + 2795.18 + 2825.18 + 2748.76}{5} = 2794.87$$

Because this average was calculated from a sample, it is called the **sample mean**.

A second measure of central tendency is the **median**, which is the value in the middle position when the measurements are ordered from smallest to largest. The median of this data can only be calculated if we first sort them in ascending order.

2748.76   2794.83   **2795.18**   2810.38   2825.18

↑

The **arithmetic mean** or **average** of a set of n measurements is equal to the sum of the measurements divided by n.

The sample mean: $\bar{x} = \frac{\sum_{i=1}^{n} x_i}{n} = \frac{x_1 + x_2 + x_3 + \cdots + x_n}{n}$, where n is the sample size.

This is a **statistic**.

The population mean: $\mu = \frac{\sum_{i=1}^{N} x_i}{N} = \frac{x_1 + x_2 + x_3 + \cdots + x_N}{N}$, where N is the population size.

This is a **parameter**.

It is important to observe that you normally do not know the population mean, μ. It is usually estimated using the sample mean, $\bar{x}$.

The median

The **median** of a set of n measurements is the value that falls in the middle position when the data are sorted in ascending order.

In Example 7.6, we calculated the sample median by finding the third measurement to be in the middle position. If the number of measurements is even, the process is slightly different.

Let us assume that you took six tests last term and that your marks were, in ascending order,

52, 63, 74, 78, 80, 89.

When the data are arranged in order, there are two 'middle' observations, 74 and 78.

52   63   (74)   ↑   (78)   80   89

To find the median, choose a value halfway between the two middle observations. This is done by calculating the mean of the two middle vales:

$$m = \frac{74 + 78}{2} = 76$$

The position of the median can be given by $\frac{n+1}{2}$. If this number ends with a decimal, you need to find the mean of the adjacent values.

In the NASDAQ Index case, we have five observations. The position of the median is then at $\frac{5+1}{2} = 3$.

In the marks example, the position of the median mark is at $\frac{6+1}{2} = 3.5$, hence we find the mean of the numbers at positions 3 and 4.

Although both the mean and median are good measures for the centre of a distribution, the median is less sensitive to **extreme values** or **outliers**. For example, the value 52 in the previous example is lower than all the other test scores and is the only failing score. The median, 76, would not be affected by this outlier even if it were much lower than 52. Assume, for example, that the lowest score is 12 rather than 52.

12 63 (74) (78) 80 89

The median still gives the same answer of 76.

If we were to calculate the mean of the original set, we would get

$$\bar{x} = \frac{\sum x}{6} = \frac{436}{6} = 72.\dot{6}$$

The new mean, with 12 as the lowest score, is

$$\bar{x} = \frac{\sum x}{6} = \frac{396}{6} = 66$$

Clearly, the low outlier 'pulled' the mean towards it while leaving the median untouched. However, because the mean depends on every observation and uses all the information in the data, it is generally, wherever possible, the preferred measure of central tendency.

Online

Use an interactive dot plot to explore the mean and median values of a data set.

The mode

A third way to locate the centre of a distribution is to look for the value that occurs with the highest frequency. This measure of the centre is called the **mode**. When the data are given as a frequency distribution, we call the most frequent class the **modal class**.

Example 7.7

The table gives the frequency distribution of 25 families in Lower Austria that were polled in a marketing survey to find the number of litres of milk consumed during a particular week.

Construct a frequency histogram and find the modal class, median and mean.

Number of litres	Frequency	Relative frequency
0	2	0.08
1	5	0.20
2	9	0.36
3	5	0.20
4	3	0.12
5	1	0.04

Solution

The histogram shows a relatively symmetric shape with a modal class at $x = 2$. Apparently, the mean and median are not far from each other. The median is the 13th observation, which is 2, and the mean is calculated to be 2.2.

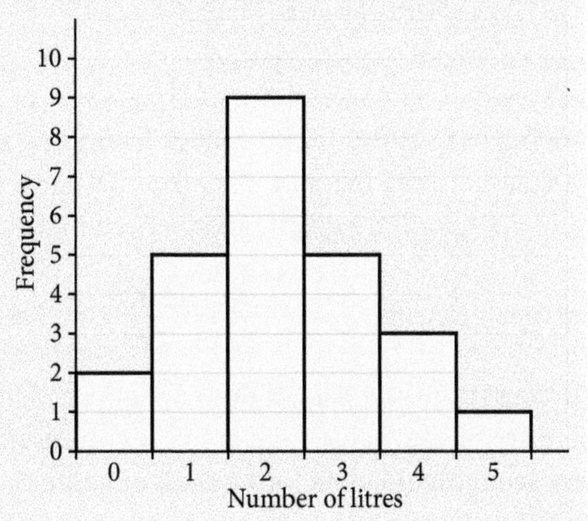

The symmetric shape of the histogram shows that the median, mean and mode are all close together. This will be discussed further in the next section.

For lists, the mode is a most common (frequent) value. A list can have more than one mode. For histograms, a mode is a relative maximum.

Shape of the distribution

The shape of a distribution indicates how the distribution is centred around the mean. Distributions are either **symmetric** or they are **not symmetric.** If they are not symmetric, the shape of the distribution is described as **asymmetric** or **skewed.**

Symmetry

The shape of a distribution is said to be symmetric if the observations are balanced, or evenly distributed, about the mean. In a symmetric distribution, the mean, the median, and the mode are equal, as seen in the next section.

Skewness

A distribution is **skewed** if the observations are not symmetrically distributed above and below the mean.

A **positively skewed** (or skewed to the right) distribution has a tail that extends to the right in the direction of positive values. A **negatively skewed** (or skewed to the left) distribution has a tail that extends to the left in the direction of negative values.

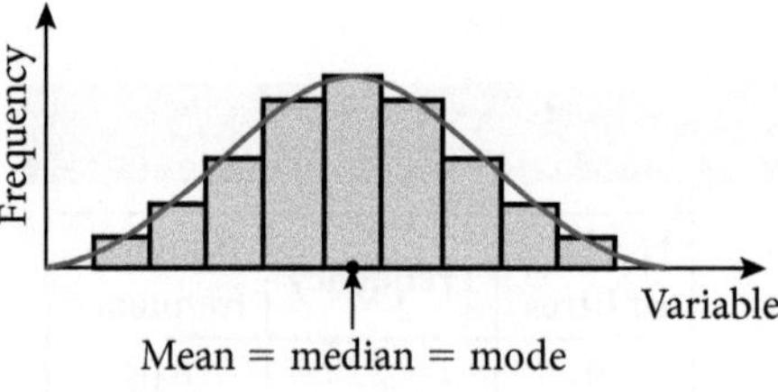

Figure 7.6 Symmetric distribution

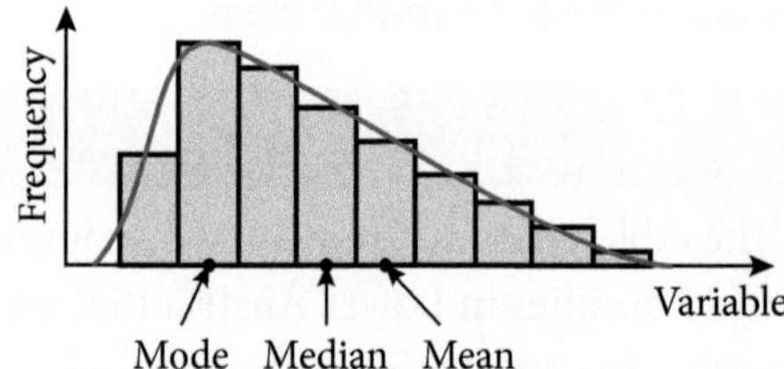

Figure 7.7 Positively skewed distribution

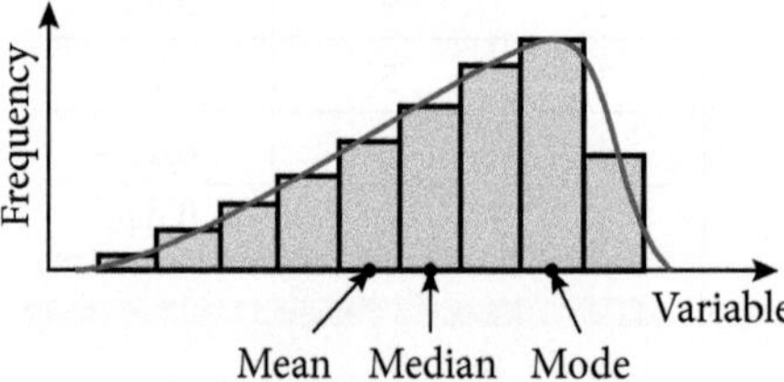

Figure 7.8 Negatively skewed distribution

Looking back at the WHO data in Example 7.2, we can clearly see that the data are skewed to the left. Few countries have low life expectancies. The bulk of the countries have life expectancies between 50 and 65.

The mean HALE is $\mu = \frac{\sum x}{n} = \frac{11\,028}{192} = 57.44$.

Looking at the raw data, it does not appear sensible to search for the mode, as there are several values that are very common (59, 60, 61, and 62). However, after grouping the data into classes of width 5, we can see that the modal class is 60–65.

As there are 192 observations, the median is the $\frac{n+1}{2} = \frac{192+1}{2} = 96.5$th value. We therefore take the mean of the 96th and 97th observations, which are both 60. So the median is 60.

Knowing the median, we could say that a typical life expectancy is 60 years. How much does this really tell us? How well does this median describe the real situation? After all, not all countries have a life expectancy of 60 years. Whenever we find the centre of a data set, the next step is always to ask how well it summarises the data. When we describe a distribution numerically, we always report a measure of its **spread** along with its centre. This will be discussed further in Section 7.3.

Exercise 7.2

1. You are given eight measurements: 5, 4, 7, 8, 6, 6, 5, 7.

(a) Find $\bar{x}$.

(b) Find the median.

(c) Based on the previous results, are the data symmetric or skewed? Explain and support your conclusion with an appropriate graph.

2. You are given ten measurements: 5, 7, 8, 6, 12, 7, 8, 11, 4, 10.

(a) Find $\bar{x}$.

(b) Find the median.

(c) Find the mode.

3. The following table gives the number of DVD players owned by a sample of 50 typical families in a large city in Germany.

Number of DVD players	0	1	2	3
Number of households	12	24	8	6

Find the average and the median number of DVD players. What is more appropriate here? Explain.

4. Ten businesses are listed below along with their 2017 revenue in millions of US dollars.

Company	Revenue ($ millions)	Company	Revenue ($ millions)
A	500 343	F	265 172
B	348 903	G	260 028
C	326 953	H	244 582
D	326 008	I	244 363
E	311 870	J	242 137

Calculate the mean and median of the revenues. Which measure is more appropriate in this case. Explain.

5. Even on a crucial examination, students tend to lose focus while writing their tests. In a psychology experiment, 20 students were given a 10-minute quiz and were observed for the number of seconds they spent 'on task'. Here are the results:

350	380	500	460	480	400	370	380	450	530
520	460	390	360	410	470	470	490	390	340

Find the mean and median of the time spent on task.
If you were writing a report to describe these times, which measure of central tendency would you use and why?

6. At 5:30 p.m. during the holiday season, a toy shop counted the number of items sold and the revenue collected for that day; the result was $n = 90$ toys with a total revenue of $\sum x = €4460$.

   (a) Find the average amount spent on each toy that day.

   Shortly before the shop closed at 6 p.m., two new purchases of €74 and €60 were made.

   (b) Calculate the new mean amount spent on each toy that day.

7. A farmer has 144 bags of new potatoes weighing 2.15 kg each. He also has 56 bags of potatoes from last year with an average weight of 1.80 kg. Find the mean weight of a bag of potatoes available from this farmer.

8. The following are the marks earned by 25 students on a 50-mark test in statistics.

   26, 27, 36, 38, 23, 26, 20, 35, 19, 24, 25, 27,34,
   27, 26, 42, 46, 18, 22, 23, 24, 42, 46, 33, 40

   (a) Calculate the mean of the marks.

   (b) Draw a stem-and-leaf plot of the marks. Use the plot to estimate where the median is.

   (c) Draw a histogram of the marks.

   (d) Develop a cumulative frequency graph of the marks. Use your graph to estimate the mean.

9. The following are data concerning the injuries in road accidents in a certain country classified by severity.

Year	Fatal	Serious	Slight
1970	758	7860	13515
1975	699	6912	13041
1980	644	7218	13926
1985	550	6507	13587
1990	491	5237	14443
1995	361	4071	12102
2000	297	3007	11825
2005	264	2250	10922

(a) Draw bar graphs for the total number of injuries and describe any patterns you observe.

(b) Draw pie charts for the different types of injuries for the years 1970, 1990, and 2005.

10. The data on the right report the car driver casualties in a certain district for 2017.

(a) Draw a histogram of the data.

(b) Estimate the mean of the data.

(c) Develop a cumulative frequency graph and use it to estimate the median of the data.

Age	Number
15–19	103
20–24	125
25–29	103
30–34	80
35–39	88
40–44	96
45–49	78
50–54	60
55–59	45
60–64	33
65–69	17
70–74	13
75–79	26

11. Use the data in question 9 of Exercise 7.1 to estimate the median and the mean of the number of days spent in hospital by heart patients.

12. Use the data in question 10 of Exercise 7.1 to estimate the median and the mean of the exercise time of the author for 2006.

13. Use the data in question 11 of Exercise 7.1 to estimate the median and the mean speed of cars on the highway.

14. Use the data in question 12 of Exercise 7.1 to estimate the median and the mean length of components at this facility.

15. Use the data in question 13 of Exercise 7.1 to estimate the median and the mean of the waiting time for customers at this supermarket.

16. (a) Given that $\sum_{i=1}^{40} x_i = 1664$, find $\bar{x}$.

(b) Given that $\sum_{i=1}^{40} (x_i - 20) = 1664$, find $\bar{x}$.

17. For 60 students in a large class, 12 marks are added to each score to boost the students' scores on a relatively difficult test.

(a) Knowing that $\sum (x + 12) = 4404$, find the mean score of this group of 60 students.

(b) Another section of the class has 40 students and their average score is 67.4. Find the average of the whole class of 100 students.

7.3 Measures of variability

Measures of location summarise what is typical of elements of a data set, but not every element is typical. Are all the elements close to each other? Are most of the elements close to each other? What is the biggest difference between elements? On average, how far are the elements from each other? The answers lie in the **measures of spread** or **variability**.

It is possible that two data sets have the same mean, but the individual observations in one set could vary more from the mean than the observations in the second set. It takes more than the mean alone to describe data. Measures of variability (also called measures of dispersion or spread), which include the **range**, the **variance**, the **standard deviation**, the **interquartile range**, and the **coefficient of variation**, help to summarise the data.

Range

The **range** in a data set is the difference between the largest and smallest observations.

Consider the expenses data given in Table 7.1. Also consider the same data with the largest value of 68 replaced by 120. What is the range for these two sets of data?

Notice that the range is a single number, not an interval of values.

	Expenses data	Expenses data with outlier
Minimum	38	38
Maximum	68	120
Range	30	82

Table 7.3 Expenses data with outlier

The maximum of the WHO data in Example 7.2 is 79 and the minimum is 29, so the range is 50.

Range doesn't take into account how the data are distributed. It is affected by extreme values (outliers), as we see above.

Variance and standard deviation

The most comprehensive measures of variability are those given in terms of the average deviation from some location parameter.

Variance

The **sample variance** is denoted s_n^2 and is evaluated as

$$s_n^2 = \frac{\sum_{i=1}^{n}(x_i - \bar{x})^2}{n}$$

where $\bar{x}$ is the sample mean, n is the size of the sample and the x_i are the values of the items in the sample.

This sample variance is the sum of the squared differences between each observation and the sample mean, divided by the sample size.

In most statistics references, the sample variance, s^2, is called the **unbiased estimate** of the population variance, σ^2, and is denoted as s^2_{n-1}. It is calculated slightly differently from the sample variance required by the IB syllabus. Its value is $s^2_{n-1} = \frac{\sum_{i=1}^{n}(x_i - \bar{x})^2}{n-1}$.

The reason for defining the sample variance in this manner is beyond the scope of this book. The use of $n - 1$ in the denominator has to do with the use of the sample variance as an estimate of the population variance. Such an estimate has to be unbiased, and this sample variance is the most unbiased estimate of the population variance. However, the IB syllabus uses a different definition of the sample variance as already discussed.

Be careful when using a calculator, because the s_x function on GDCs corresponds to s^2_{n-1}. When you use your GDC to calculate the sample variance, make sure you use the σ_x function.

The population variance, σ^2, is the sum of the squared differences between each observation and the population mean, divided by the population size, N.

$$\sigma^2 = \frac{\sum_{i=1}^{N}(x_i - \mu)^2}{N}$$

where μ is the population mean and x_i is the value of each item in the population.

The variance is a measure of the variation about the mean, squared. This means that the unit used for the variance is the square of the unit used for the measurements. In order for the measure to have the same unit as the data measurements, the square root is taken. This gives a new measure, the **standard deviation**.

If the unit of the data measurements is kg, the unit of the variance is kg^2.

Standard deviation

The standard deviation measures the **standard amount of deviation** or **spread** around the mean.

The sample standard deviation, s_n, is the (positive) square root of the variance, and is defined as:

$$s_n = \sqrt{s_n^2} = \sqrt{\frac{\sum_{i=1}^{n}(x_i - \bar{x})^2}{n}}$$

where $\bar{x}$ is the sample mean, n is the number of items in the sample and x_i is the value of each item in the sample.

The population standard deviation is

$$\sigma = \sqrt{\sigma^2} = \sqrt{\frac{\sum_{i=1}^{N}(x_i - \mu)^2}{N}}$$

where μ is the sample mean and x_i is the value of each item in the population.

These are measures of variation about the mean.

- When is $\sigma = 0$? When all the data takes on the same value and there is no variability about the mean.
- When is σ large? When there is a large amount of variability about the mean.

Stock A	Stock B
4	1
4.25	3
5	2.5
4.75	5
5.75	7
5.25	6.5
6	10
$\bar{x}_A = 5$	$\bar{x}_B = 5$
Median (A) = 5	Median (B) = 5

Table 7.4 Stock closing prices

Consider the following example.

In business, investors invest their money in stocks whose prices fluctuate with market conditions. Stocks are considered risky if they have high fluctuations. Table 7.4 gives the closing prices of two stocks traded on Vienna's stock market for the first seven business days in September 2017.

Even though the two stocks have similar central values, they behave very differently. It is obvious that stock B is more variable and it becomes more obvious when we calculate the standard deviations.

We will calculate the standard deviation manually in this example to demonstrate the process. You do not have to do this manually all the time!

$$s_A^2 = \frac{\sum_{i=1}^{7}(x_i - 5)^2}{7} = \frac{(4-5)^2 + (4.25-5)^2 + (5-5)^2 + (4.75-5)^2 + (5.75-5)^2 + (5.25-5)^2 + (6-5)^2}{7}$$
$$= 0.464$$

$$s_B^2 = \frac{\sum_{i=1}^{7}(x_i - 5)^2}{7} = \frac{(1-5)^2 + (3-5)^2 + (2.5-5)^2 + (5-5)^2 + (7-5)^2 + (6.5-5)^2 + (10-5)^2}{7}$$
$$= 8.21$$

This means that the standard deviations are:

$$s_A = \sqrt{0.464} = 0.681$$

$$s_B = \sqrt{8.21} = 2.865$$

Stock B is 4.2 times as variable as stock A.

When computing s_n^2 manually, you might find it easier to use the following shortcut formula:

$$s_n^2 = \frac{\sum_{i=1}^{n} x_i^2}{n} - \bar{x}^2$$

The shortcut formula is found by manipulating the original formula:

$$s_n^2 = \frac{\sum_{i=1}^{n} x_i - \bar{x}^2}{n} = \frac{\sum_{i=1}^{n} x_i^2 - 2x_i\bar{x} + \bar{x}^2)}{n} = \frac{\sum_{i=1}^{n} x_i^2 - 2\sum_{i=1}^{n} x_i\bar{x} + \sum_{i=1}^{n}\bar{x}^2}{n}$$
$$= \frac{\sum_{i=1}^{n} x_i^2}{n} - \frac{2\bar{x}\sum_{i=1}^{n} x_i}{n} + \frac{\sum_{i=1}^{n}\bar{x}^2}{n} = \frac{\sum_{i=1}^{n} x_i^2}{n} - 2\bar{x}\sum_{i=1}^{n}\frac{x_i}{n} + \frac{n\bar{x}^2}{n} = \frac{\sum_{i=1}^{n} x_i^2}{n} - \bar{x}^2$$

However, remember that once you have a good understanding of the standard deviation, you will rely on a GDC or software to do most of the calculation for you.

The output from a GDC is shown below.

```
1—Var Stats
 x̄=5
 Σx=35
 Σx²=178.25
 Sx=0.68138514
 σx=0.73598007
↓n=7
```

```
1—Var Stats
↑minX=4
 Q1=4.25
 Med=5
 Q3=5.75
 maxX=6
↓mod=4
```

Get to know the specifics of how your GDC works with statistical calculations.

The s_x shown on the GDC uses $s_{n-1} = \sqrt{\frac{\sum_{i=1}^{n}(x_i - \bar{x})^2}{n-1}}$. For IB exams, you need to use the σ_x value, which uses the calculation $\sigma = \sqrt{\frac{\sum_{i=1}^{n}(x_i - \bar{x})^2}{n}}$.

The screenshots also show you that the GDC gives you $\sum x^2$. You can use this to find the variance by hand.

$$s_n^2 = \frac{\sum_{i=1}^{n} x_i^2}{n} - \bar{x}^2 = \frac{178.25}{7} - 5^2 = 0.464 \Rightarrow s_n = 0.681$$

The interquartile range and measures of non-central tendency

Another measure of spread is the **interquartile range**. To understand this measure, we must first define **percentiles** and **quartiles**.

Percentiles and quartiles

Percentiles separate large ordered data sets into 100ths. The pth percentile is a number such that p per cent of the observations are at or below that number.

Quartiles are descriptive measures that separate large ordered data sets into four quarters.

A test score in the 90th percentile means that 90% of the test scores were less than or equal to your score. An excellent performance! The score is in the upper 10% of all test scores.

The **first quartile, Q_1**, is another name for the 25th percentile. The first quartile divides the ordered data such that 25% of the observations are at or below this value. Q_1 is located in position $0.25(n + 1)$ when the data are in ascending order. That is,

$$Q_1 = \frac{n+1}{4}\text{th ordered observation}$$

The **third quartile, Q_3**, is another name for the 75th percentile. The third quartile divides the ordered data such that 75% of the observations are at or below this value. Q_3 is located in position $0.75(n + 1)$ when the data are in ascending order. That is,

$$Q_3 = \frac{3(n+1)}{4}\text{th ordered observation}$$

The median is the 50th percentile, or the second quartile, Q_2.

To find percentiles and quartiles, data must first be in ascending order.

A practical method to calculate the quartiles is to split the data into two halves at the median. (When n is odd, include the median in both halves.) The lower quartile is the median of the first half and the upper quartile is the median of the second half. For example, with the stocks data, {4, 4.25, 4.75, 5, 5.25, 5.75, 6}, $n = 7$ and the median is the 4th observation, 5.

The first quartile is then the median of {4, 4.25, 4.75, 5}, which is 4.5, and the third quartile is the median of {5, 5.25, 5.75, 6}, which is 5.5.

Interquartile range

A measure which helps to measure variability and is not affected by extreme values is the **interquartile range** (IQR). It avoids the problem of extreme values by just looking at the range of the middle 50% of the data.

The interquartile range measures the spread in the middle 50% of the data. It is the difference between the observations at the 25th and the 75th percentiles:

$$\text{IQR} = Q_3 - Q_1$$

If we consider the student expenses data in Table 7.1, both with and without the outlier 120 replacing the largest value 68, we have the results shown in Table 7.5.

Range doesn't take into account how the data are distributed and is affected by extreme values. We can see in Table 7.5 that the IQR does not have this problem.

	Expenses data	Expenses data with outlier
Minimum	38	38
Q_1	50	50
Median	55	55
Q_3	61	61
Maximum	68	120
Range	30	82
IQR	11	11

Table 7.5 The IQR is not affected by extreme values

Box-and-whisker plots

minimum $< Q_1$ $<$ median $< Q_3$ $<$ maximum

The five descriptive measures, minimum, first quartile, median, third quartile, and maximum, give us a **five-number summary** of a data set.

Whenever we have a five-number summary, we can put the information together in one graphical display called a **box-and-whisker plot**, also known as **box plot**.

Let us make a box plot with the student expenses data.

- Draw an axis spanning the range of the data. Mark the numbers corresponding to the median, minimum, maximum, and the first and third quartiles.
- Draw a rectangle with the lower end at Q_1 and the upper end at Q_3, as shown in Figure 7.9.
- To help us consider outliers, we calculate **lower** and **upper fences**. Any point outside these fences is considered an **outlier**. Mark the fences with a dotted line since they are not part of the box. The fences are constructed at the following positions:

  Lower fence: $Q_1 - 1.5 \times \text{IQR}$ (in this case: $50 - 1.5(11) = 33.5$)

  Upper fence: $Q_3 + 1.5 \times \text{IQR}$ (in this case: $61 + 1.5(11) = 77.5$)

An outlier is an unusual observation. It lies at an abnormal distance from the rest of the data. There is no unique way of describing what an outlier is. A common practice is to consider any observation that is further than $1.5 \times \text{IQR}$ from the first quartile or the third quartile an outlier.

- Mark any outlier with an asterisk (*) on the graph.
- Extend horizontal lines, called **whiskers**, from the ends of the box to the smallest and largest observations that are not outliers. In the first case these are 38 and 68, while in the second they are 38 and 67.
- Outliers are important in statistical analysis. They may contain important information not shared with the rest of the data. Statisticians look very carefully at outliers because of their influence on the shapes of distributions

and their effect on the values of the other statistics such as the mean and standard deviation.

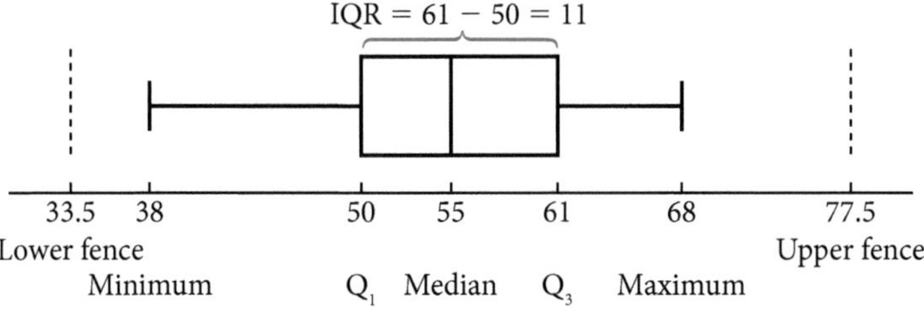

Figure 7.9 Box plot for student expenses data

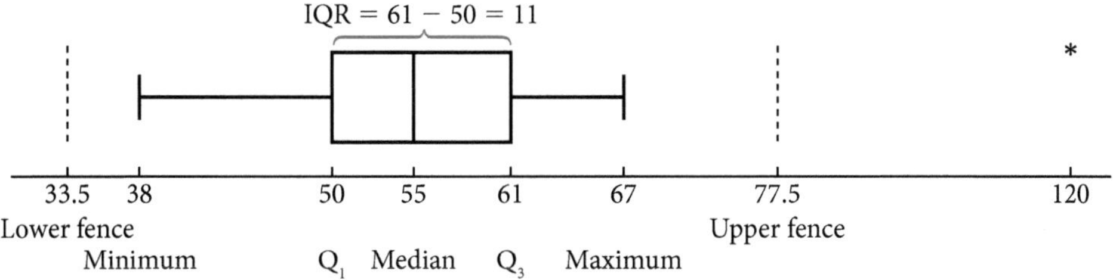

Figure 7.10 Box plot for student expenses data with outlier

Using the box plots, we can immediately see that the IQR is €11, the difference between 50 and 61.

We can also construct box plots using software packages, as in Figure 7.11. Again, this shows that the box contains the middle 50% of the data. The width of the box is the IQR.

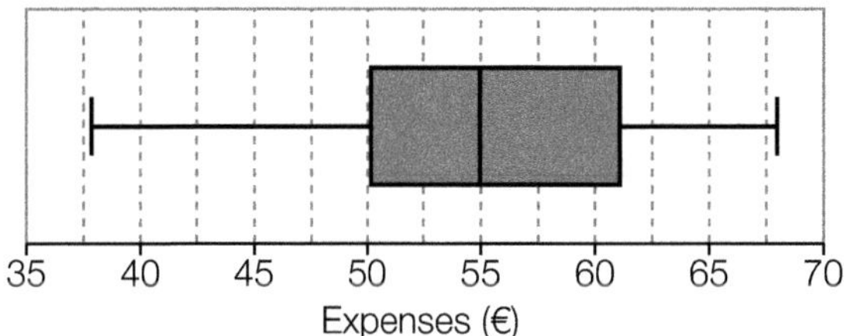

Figure 7.11 Box plot for student expenses data

You can also use your GDC to draw box plots.

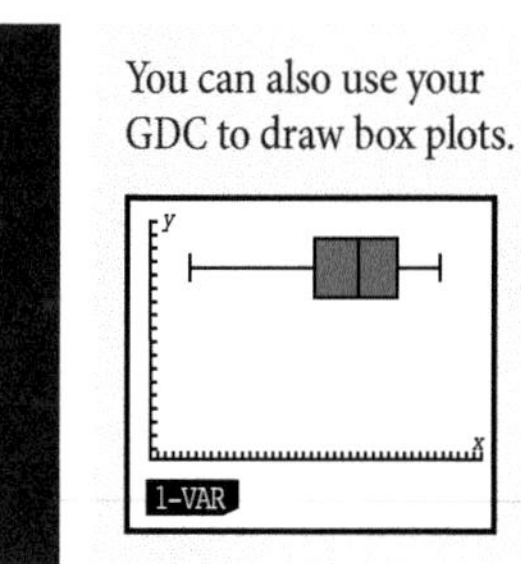

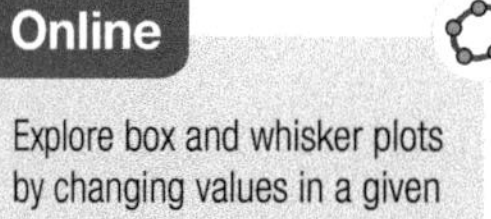

Explore box and whisker plots by changing values in a given data set.

This is a reasonable summary of the spread of the distribution, as you can see by comparing it to the histogram in Figure 7.12. Locating the IQR on the histogram gives another visual indication of the spread of the data.

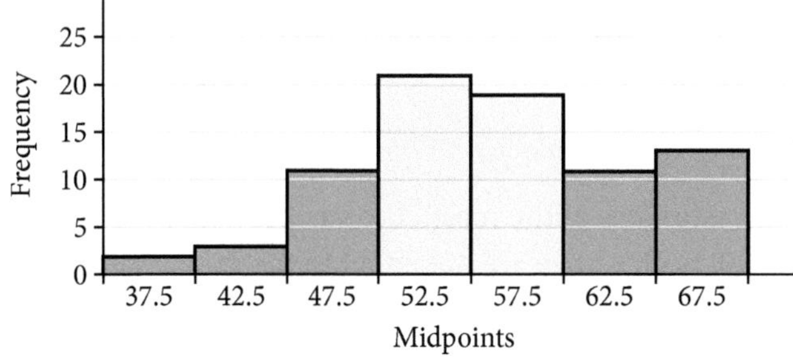

Figure 7.12 Histogram showing student expenses data

Grouped data

The calculation of the mean and variance for grouped data is similar to the calculation for raw data. The difference lies in the use of frequencies instead of individual data points. A comparison is given in Table 7.6.

Statistic	Raw data	Grouped data	Grouped data with intervals
$\bar{x}$	$\bar{x} = \dfrac{\sum_{i=1}^{n} x_i}{n}$	$\bar{x} = \dfrac{\sum_{i=1}^{n} x_i \cdot f(x_i)}{n} = \dfrac{\sum_{i=1}^{n} x_i \cdot f(x_i)}{\sum_{i=1}^{n} f(x_i)}$	$\bar{x} = \dfrac{\sum_{i=1}^{n} m_i \cdot f(m_i)}{n} = \dfrac{\sum_{i=1}^{n} m_i \cdot f(m_i)}{\sum_{i=1}^{n} f(m_i)}$
s_n^2	$s_n^2 = \dfrac{\sum_{i=1}^{n} (x_i - \bar{x})^2}{n}$	$s_n^2 = \dfrac{\sum_{i=1}^{n} (x_i - \bar{x})^2 \cdot f(x_i)}{n} = \dfrac{\sum_{i=1}^{n} (x_i - \bar{x})^2 \cdot f(x_i)}{\sum_{i=1}^{n} f(x_i)}$	$s_n^2 = \dfrac{\sum_{i=1}^{n} (m_i - \bar{x})^2 \cdot f(m_i)}{n} = \dfrac{\sum_{i=1}^{n} (m_i - x)^2 \cdot f(m_i)}{\sum_{i=1}^{n} f(m_i)}$

x_i: data point
m_i: interval midpoint (mid-mark or mid-value)
$\sum f(x_i), \sum f(m_i)$: total number of data points
$f(x_i)$: frequency of x_i
$f(m_i)$: frequency of interval i

Table 7.6 Formulae for calculating the mean and variance of raw and grouped data

Table 7.7 shows how we estimate the mean and variance for the grouped data from Example 7.1.

Living expenses (x)	Midpoint m	Number of students $f(m)$	$m_i \times f(m_i)$	$(m_i - \bar{x})^2$	$(m_i - \bar{x})^2 \times f(m_i)$
$35 \leqslant x < 40$	37.5	2	75	344.5	688.9
$40 \leqslant x < 45$	42.5	3	127.5	183.9	551.6
$45 \leqslant x < 50$	47.5	11	522.5	73.3	806.0
$50 \leqslant x < 55$	52.5	21	1102.5	12.7	266.1
$55 \leqslant x < 60$	57.5	19	1092.5	2.1	39.4
$60 \leqslant x < 65$	62.5	11	687.5	41.5	456.2
$65 \leqslant x < 70$	67.5	13	877.5	130.9	1701.4
Totals		$\sum f(m_i) = 80$	$\sum_{\text{all } m} m_i \cdot f(m_i) = 4485$	$\sum_{\text{all } m} (m_i - \bar{x})^2 \cdot f(m_i) = 4509.6$	
			Mean $= \dfrac{4485}{80} = 56.06$	Variance $= \dfrac{4509.6}{80} = 56.37$	
				Standard deviation $= 7.51$	

Table 7.7 Estimating mean and variance for grouped data

The numbers in Table 7.7 are estimates of the mean, the variance, and the standard deviation. As you will notice, they are not equal to the values we calculated earlier, but they are close. The reason for the difference is that, with grouping, we lose the detail in each interval. For example, the interval between 45 and 50 is represented by the mid-interval value 47.5. In essence, we are assuming that every number in the interval is equal to 47.5.

An ogive can also be produced from this data, as shown in Figure 7.13.

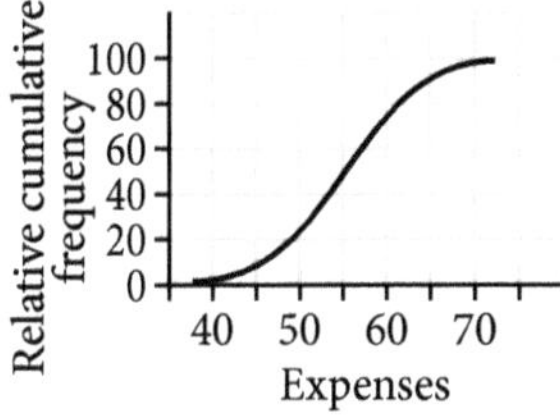

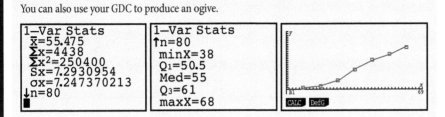

Figure 7.13 Cumulative distribution for expenses data

You can also use your GDC to produce an ogive.

```
1-Var Stats
 x̄=55.475
 Σx=4438
 Σx²=250400
 Sx=7.2930954
 σx=7.247370213
↓n=80
```

```
1-Var Stats
↑n=80
 minX=38
 Q1=50.5
 Med=55
 Q3=61
 maxX=68
```

```
CALC  DefG
```

Figure 7.14 is a realistic ogive.

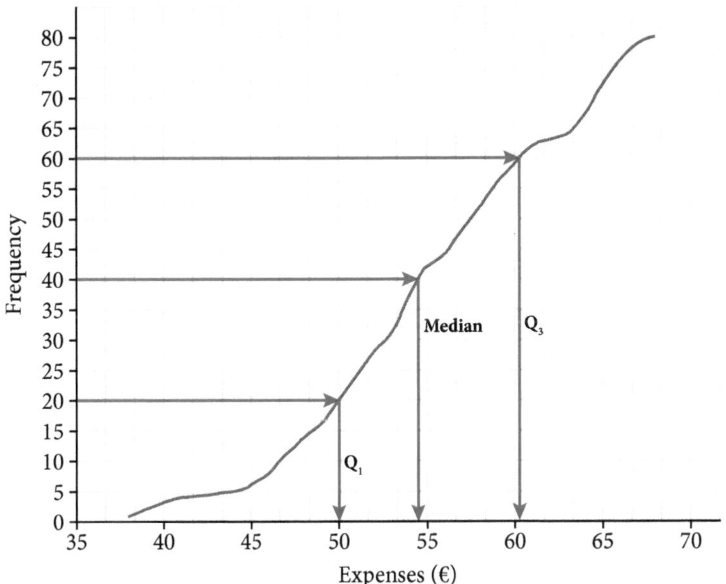

Figure 7.14 Ogive constructed from real data

Notice how we locate the first quartile. Since there are 80 observations, the first quartile is at approximately the $\frac{n+1}{4} = \frac{81}{4} \approx$ 20th position. Read from 20 on the y-axis across to the curve, then down to the x-axis to find that it is around 50.

The median is at the $\frac{n+1}{2} = \frac{81}{2} =$ 40.5th position, i.e. approximately 55.

Similarly, the third quartile is at $\frac{3(n+1)}{4} = \frac{243}{4} \approx$ 61st position, which happens to be approximately 61.

Example 7.8

Speed limits in some European cities are set to 50 km h^{-1}. Drivers in various cities react to such limits differently. A study was undertaken to compare drivers' behaviour in Brussels, Vienna, and Stockholm. The table shows the recorded speeds of different drivers. Use box plots to compare the results.

Brussels	64	61	63	57	49	49	46	58	45	60	51	36	65	45	47	46				
Vienna	62	60	59	50	61	63	53	46	58	49	51	37	47	51	63	52	44	50	45	44
Stockholm	43	44	34	35	31	34	29	33	36	38	45	47	29	48	51	49	48			

Solution

Parallel box plots are an appropriate tool for comparing the three data sets. After arranging each data set in ascending order and finding the minimum, first quartile, median, third quartile, and maximum for each data set, we can draw three box plots on the same set of axes.

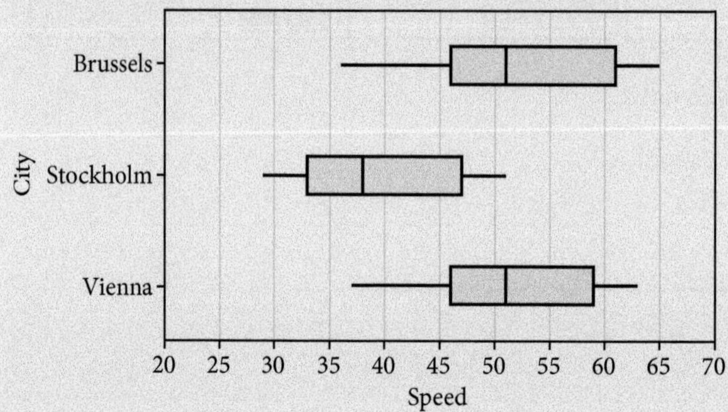

The box plots show that, on average, drivers in Brussels and Vienna tend to drive faster. The median in both cities is higher than 50, which means that more than 50% of the drivers in the two cities do not respect the speed limit. The variation in these two cities is similar, with Brussels having a slightly wider range than Vienna.

Almost all drivers in Stockholm appear to adhere to the 50 km h^{-1} limit. The median is around 40 km h^{-1} and the third quartile about 47 km h^{-1}, which means that more than 75% of the drivers in this city drive at a speed less than the 50 km h^{-1} limit.

Shape, centre, and spread

Statistics is about variation, so spread is an important fundamental concept. Measures of spread help us to precisely analyse what we do not know. If the values we are looking at are scattered very far from the centre, then the IQR and the standard deviation will be large. If these are large, then our central values will not represent data well. That is why we always report spread with any central value.

A practical way of seeing the significance of the standard deviation can be demonstrated with the following (optional) observations.

Empirical rule

If the data are close to being symmetrical, as in Figure 7.15, then the following is true:

- The interval $\mu \pm \sigma$ contains approximately 68% of the measurements.
- The interval $\mu \pm 2\sigma$ contains approximately 95% of the measurements.
- The interval $\mu \pm 3\sigma$ contains approximately 99.7% of the measurements.

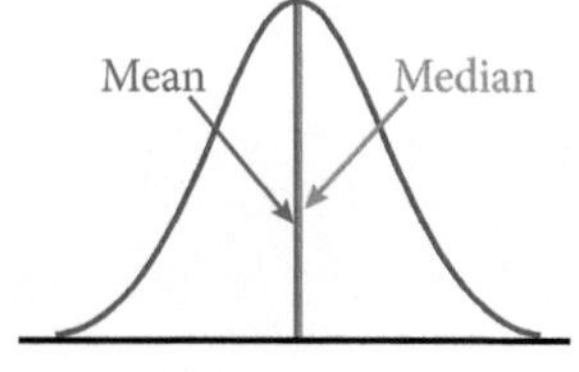

Figure 7.15 Close to symmetrical data

The empirical rule usually indicates whether or not an observation is very far from the expected value. Take the following example.

A car's fuel efficiency is recorded each time it is refuelled. 98 efficiency measurements are taken. The data are given in Table 7.8.

Fuel effiency (km l^{-1})	Frequency	Fuel effiency (km l^{-1})	Frequency
6.0	1	10.0	14
7.0	1	10.5	7
7.5	4	11.0	9
8.0	8	11.5	5
8.5	14	12.0	1
9.0	21	12.5	2
9.5	11		

Table 7.8 Data about a car's fuel efficiency

The summary measures are given in Table 7.9.

Mean	9.454
σ	1.223
Median	9.25
Q_1	8.5
Q_3	10.125
IQR	1.625

Table 7.9 Summary data about a car's fuel efficiency

We can draw a histogram and a box plot from the data.

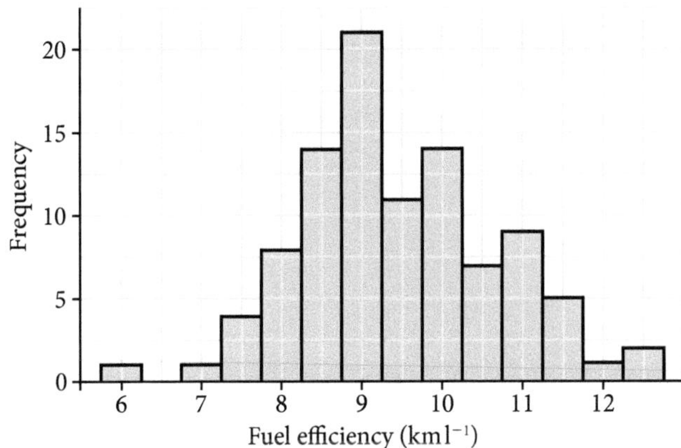

Figure 7.16 Fuel efficiency histogram

Figure 7.17 Fuel efficiency box plot

The histogram shows that the distribution is almost symmetric. The possible outlier has little effect on the mean and standard deviation. That is why the mean and median are almost the same.

Looking at the box plot, we can see that there is one outlier. This is confirmed by calculation.

Lower fence $= 8.5 - 1.5 \times 1.625 = 6.1$. As 6 is smaller than this, it is considered an outlier.

Upper fence = $10.125 + 1.5 \times 1.625 = 12.6$. Hence there are no outliers on this side.

If we use the empirical rule, we can expect about 99.7% of the data to lie within three standard deviations of the mean.

Three standard deviations below the mean $= 9.454 - 3 \times 1.223 = 5.8$.

Three standard deviations above the mean $= 9.454 + 3 \times 1.223 = 13.1$.

In fact, all the data are within the specified interval, including the potential outlier.

If you are asked to describe a quantitative variable, you should report the shape of its distribution, and include a measure of centre and a measure of spread.

- If the shape is skewed, report the median and IQR. You may want to include the mean and standard deviation, but you should point out that the mean and median differ because the data are skewed. A histogram can help.
- If the shape is symmetrical, report the mean and standard deviation. You may report the median and IQR as well.
- If there are clear outliers, report the data with and without the outliers. The differences may be revealing.

Example 7.9

The records of a large high school show the heights of their students for the year 2018.

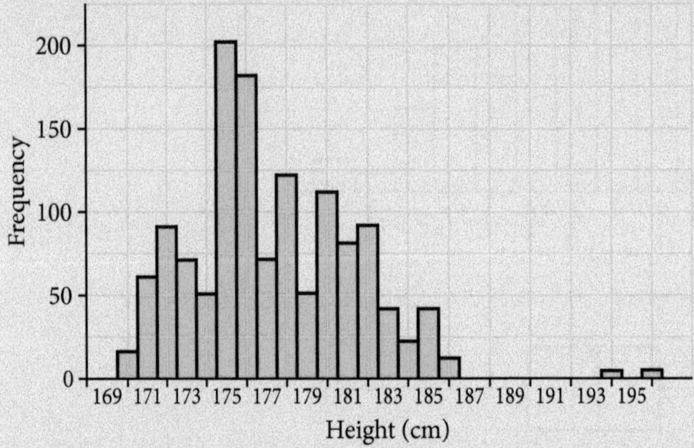

(a) State which statistics would best represent the data. Give reasons for your answer.

(b) Calculate the mean and standard deviation.

(c) Construct a cumulative frequency table and a cumulative frequency graph of the data.

(d) Use your cumulative frequency table or graph to estimate the median, Q_1, Q_3, and IQR.

(e) Are there any outliers in the data? Give reasons for your answer.

(f) Write a few sentences describing the distribution.

Solution

(a) The data appear to have outliers and are slightly skewed to the right. The most appropriate measure is the median since the mean is influenced by the extreme values.

(b) Set up a table to calculate the mean and standard deviation. Read the values from the histogram.

Height x_i	Number of students $f(x)$	$x_i \times f(x_i)$	$(x_i - \bar{x})^2$	$(x_i - \bar{x})^2 \times f(x_i)$
170	15	2550	51.84	777.6
171	60	10 260	38.44	2306.4
172	90	15 480	27.04	2433.6
⋮	⋮	⋮	⋮	⋮
194	2	388	282.24	564.5
196	3	588	353.44	1060.3
Totals	$\sum f(x_i) = 1300$	$\sum_{all\,x} x_i \cdot f(x_i) = 230\,376$	$\sum_{all\,x} (x_i - \bar{x})^2 \cdot f(x_i) = 19\,927.6$	
		Mean $= \frac{230\,376}{1300} = 177.2$	Variance $= \frac{19\,927.4}{1300} = 15.33$	
			Standard deviation $= 3.92$	

Using the shortcut formula for the variance will give the same result. (Answers may differ slightly due to rounding.)

$$s_n^2 = \frac{\sum_{i=1}^{n} x_i^2 \times f(x_i)}{n} - \bar{x}^2 = \frac{40\,845\,390}{1300} - 177.2123^2 = 15.3315$$

(c) To construct a cumulative frequency graph, we first need a cumulative frequency table. This is constructed by accumulating the frequencies as shown.

x	$f(x)$	Cum $f(x)$	x	$f(x)$	Cum $f(x)$
170	15	15	181	80	1095
171	60	75	182	90	1185
172	90	165	183	40	1225
173	70	235	184	20	1245
174	50	285	185	40	1285
175	200	485	186	10	1295
176	180	665	187	0	1295
177	70	735	⋮	⋮	⋮
178	120	855	194	2	1297
179	50	905	195	0	1297
180	110	1015	196	3	1300

The cumulative frequency table is constructed such that the cumulative frequency corresponding to any measurement is the number of observations that are less than or equal to its value. So, for example, the cumulative frequency corresponding to a height of 174 cm is 285, which consists of the 50 observations with height 174 cm and the 235 observations with heights less than 174 cm.

The cumulative frequency graph plots the observations on the horizontal axis against their cumulative frequencies on the vertical axis as shown below.

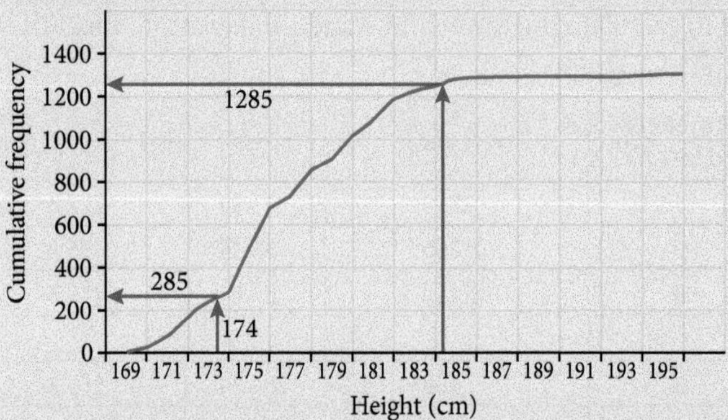

If a data set has two modes, it is said to be **bimodal**.

(d) As the number of observations is even and $\frac{1301}{2} = 650.5$, the median is the observation between the 650th and 651st observations. From the cumulative table, we can see that the median is in the 176 cm interval. So the median is 176 cm.

Q_1 is at the $\frac{1301}{4} \approx 325$th observation. From the table, as 174 cm has a cumulative frequency of 285, and 175 cm has 485, then Q_1 has to be 175 cm.

Q_3 is at the $\frac{3 \times 1301}{4} \approx 976$th observation. So it is 180 cm.

IQR = 180 − 175 = 5

The same calculations can be made using a GDC.

```
1-Var Stats
 x̄=177.2123077
 Σx=230376
 Σx²=40845390
 Sx=3.916704232
 σx=3.915197517
↓n=1300
```

```
1-Var Stats
↑n=1300
 minX=170
 Q1=175
 Med=176
 Q3=180
 maxX=196
```

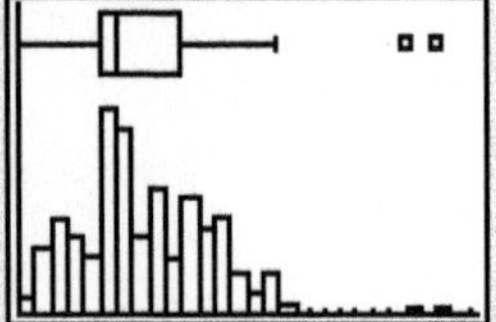

(e) To check for outliers, we can calculate the lower and upper fences.

Lower fence = 175 − 1.5 × 5 = 167.5, which is lower than the minimum value, so there are no outliers on the left.

Upper fence = 180 + 1.5 × 5 = 187.5. There are five outliers: 194 cm occurs twice and 196 cm occurs three times.

(f) The distribution appears to have two modes, at 175 cm and 176 cm. It is slightly skewed to the right with a few extreme values at 194 cm and 196 cm. This is further confirmed by the fact that the mean of 177.2 is higher than the median of 176.

Exercise 7.3

1. The pulse rates of 15 patients chosen at random from visitors at a local clinic are given below.

72, 80, 67, 68, 80, 68, 80, 56, 76, 68, 71, 76, 60, 79, 71

(a) Calculate the mean and standard deviation of the pulse rate of the patients at the clinic.

(b) Draw a box plot of the data and indicate the values of the different parts of the box.

(c) Check if there are any outliers.

2. The numbers of passengers on 50 flights from Washington to London on a commercial airline are given below.

165	173	158	171	177	156	178	210	160	164
141	127	119	146	147	155	187	162	185	125
163	179	187	174	166	174	139	138	153	142
153	163	185	149	154	154	180	117	168	182
130	182	209	126	159	150	143	198	189	218

(a) Calculate the mean and standard deviation of the number of passengers on this airline between the two cities.

(b) Set up a stem-and-leaf diagram for the data and use it to find the median of the number of passengers.

(c) Develop a cumulative frequency graph. Estimate the median, first, and third quartiles. Draw a box plot.

(d) Find the IQR and use it to check whether there are any outliers.

(e) Use the Empirical rule to check for outliers.

3. At a school, 100 students took a practice IB exam using Paper 3. The paper was marked out of 60 marks. Here are the results

Marks	0–9	10–19	20–29	30–39	40–49	50–60
No. of students	5	9	16	24	27	19

(a) Draw a cumulative frequency curve.

(b) Estimate the median and quartiles.

4. 130 first year IB students were given a placement test to decide whether they go for SL or HL. The times, in minutes, for these students to finish the test are given in the table below.

Time, t	30–40	40–50	50–60	60–70	70–80	80–90	90–100	100–110	110–120
No. of students	8	12	24	29	19	16	12	8	2

(a) Develop a cumulative frequency curve.

(b) Estimate the median and the IQR.

(c) 20 students did not manage to finish the test after 120 minutes and had to hand it in uncompleted. Estimate the median finishing time for all 150 students.

5. The mean score of 26 students on a 40-point paper is 22. The mean for another group of 84 other students is 32. Find the mean of the combined group of 110 students.

6. The scores on a 100-mark test of a sample of 80 students in a large school are given in the table.

Score	59–63	63–67	67–71	71–75	75–79	79–83	83–87
No. of students	6	10	18	24	10	8	4

(a) Find the mean and standard deviation of the scores of all students.

(b) A bonus of 13 points is to be added to these scores. What is the new value of the mean and standard deviation?

7. In a large theatre in London (1744 capacity), during a period of 10 years, there are 1000 performances of a particular production. The manager of the group kept a record of the empty seats on the days it played. Here is the table.

No. of empty seats	1–10	11–20	21–30	31–40	41–50
Days	15	50	100	170	260

No. of empty seats	51–60	61–70	71–80	81–90	91–100
Days	220	90	45	30	20

(a) Copy and complete the cumulative frequency table below for the above information

No. of empty seats	$x \leqslant 10$	$x \leqslant 20$	$x \leqslant 30$	$x \leqslant 40$	$x \leqslant 50$
Days	15		165		

No. of empty seats	$x \leqslant 60$	$x \leqslant 70$	$x \leqslant 80$	$x \leqslant 80$	$x \leqslant 100$
Days	815				1000

(b) Draw a cumulative frequency graph of this distribution. Use 1 unit on the vertical axis to represent every 100 days and 1 unit on the horizontal axis to represent every 10 seats.

(c) Use the graph from **(b)** to answer the following questions.

(i) Find an estimate of the median number of empty seats.

(ii) Find an estimate for the lower and upper quartiles, and the IQR.

(iii) The days the number of empty seats was less than 35 seats were considered bumper days (lots of profit). How many days were considered bumper days?

(iv) The highest 15% of the days with empty seats were categorised as loss days. What is the number of empty seats above which a day is claimed as a loss?

8. Aptitude tests sometimes use jigsaw puzzles to test the ability of job applicants to perform precision assembly work in electronic instruments. One company, which produces the computerised parts of video and CD players, gave the results on the right.

Time to finish the puzzle, t (nearest second)	No. of applicants
$10 \leqslant t < 30$	16
$30 \leqslant t < 40$	24
$40 \leqslant t < 45$	22
$45 \leqslant t < 50$	26
$50 \leqslant t < 55$	38
$55 \leqslant t < 60$	36
$60 \leqslant t < 65$	32
$65 \leqslant t < 70$	18

(a) Draw a histogram of the data.

(b) Draw a cumulative frequency curve and estimate the median and IQR.

(c) Calculate estimates of the mean and standard deviation of all the applicants.

9. The heights (to the nearest cm) of football players at a given school are given in the table on the right.

(a) Find the five-number summary for this data.

(b) Display the data with a box plot and a histogram.

(c) Find the mean and standard deviation of the data.

(d) Describe the data in a few sentences.

(e) Draw a cumulative frequency graph and estimate the height of a player that is in the 90th percentile.

(f) 10 players' data were missing when the data was collected. The average height of the 10 players is 182. Find the average height of all the players, including the last 10.

Height	Frequency
152	2
155	6
157	9
160	7
163	5
165	20
168	18
170	7
173	12
175	5
178	11
180	8
183	9
185	4
188	2
191	4
193	1

Table 7.10 Data for question 9

10. Consider ten data measures.

(a) If the mean of the first nine measures is 12, and the tenth measure is 12, what is the mean of the ten measures?

(b) If the mean of the first nine measures is 11, and the tenth measure is 21, what is the mean of the ten measures?

(c) If the mean of the first nine measures is 11, and the mean of the ten measures is 21, what is the value of the tenth measure?

11. Suppose that the mean of a set of 10 data points is 30.

(a) It is discovered that a data point having a value of 25 was incorrectly entered as 15. What should be the revised value of the mean?

(b) Suppose an additional point of value 32 was added. Will this increase or decrease the value of the mean?

12. Half the values of a sample are equal to 20, one-sixth are equal to 40, and one-third are equal to 60. What is the sample mean?

13. The seven numbers 7, 10, 12, 17, 21, x, and y have a mean $\mu = 12$ and a variance $\sigma^2 = \frac{172}{7}$. Find x and y given that $x < y$.

14. A sample of 25 observations was taken out of a large population of measurements. If it is given that $\sum_{i=1}^{25} x_i = 278$ and $\sum_{i=1}^{n} x_i^2 = 3682$, estimate the mean and the variance of the population of measurements.

15. Use the data in question 9 of Exercise 7.1 to estimate the IQR and the standard deviation of the number of days spent in hospital by heart patients.

16. Use the data in question 10 of Exercise 7.1 to estimate the IQR and the standard deviation of the exercise time of the author for 2006.

17. Use the data in question 11 of Exercise 7.1 to estimate the IQR and the standard deviation of the speed of cars on the highway.

18. Use the data in question 12 of Exercise 7.1 to estimate the IQR and the standard deviation of the length of components at this facility.

19. Use the data in question 13 of Exercise 7.1 to estimate the IQR and the standard deviation of the waiting time for customers at this supermarket.

7.4 Linear regression

Correlation and covariance

Scatter plots

Consider the following statements.

- The time you spend getting ready for an exam affects the score you obtain in that exam.
- In general, the foot size of an adult is related to the height of that adult.
- Smoking increases the chances of a heart attack.

Such statements concern the relationship between two variables. So far we have considered how to describe the characteristics of one variable. In this section, we will look at relationships between two variables. This study is called **bivariate statistics.**

To study the relationship between two variables, we measure both variables on the same subjects. For example, if we are interested in the relationship between height and foot size, then for a group of individuals we record each person's height and foot size. This way we know which foot size goes with which height. Similarly, we record the grades of each individual in the study along with the time that person spent preparing for the exam. So our data are sets of ordered pairs. These data allow us to study the link between height and foot size or time and grade. In fact, taller people tend to have larger foot sizes, and the more you prepare for an exam, the higher your grade is. We say that pairs of variables like these are **associated**.

Table 7.11 shows the grades of 10 students in an IB Economics SL class. The table also gives the time they spent preparing for a test and the score they achieved, out of a maximum score of 100.

Student	Tim	Joon	S-youn	Kevin	Steve	Niki	Henry	Anton	Cindy	Lukas
Hours	4	4.5	6	3.5	3	5	5.5	6.5	7	6.5
Grade	65	80	83	61	55	79	85	89	92	95

Table 7.11 IB Economics student grades

Figure 7.18 shows a scatter plot of the data given in the table. The horizontal axis shows the number of hours spent studying and the vertical axis shows the grades received. As you will notice, it appears that the more hours spent studying, the higher the grade. We say that the grades on tests and the time spent preparing for them are **associated**. We call the time the **explanatory variable** (or **independent variable**) and the grade the **response variable** (or **dependent variable**). The students whose times and grades are recorded are the **subjects** of the **experiment/study**.

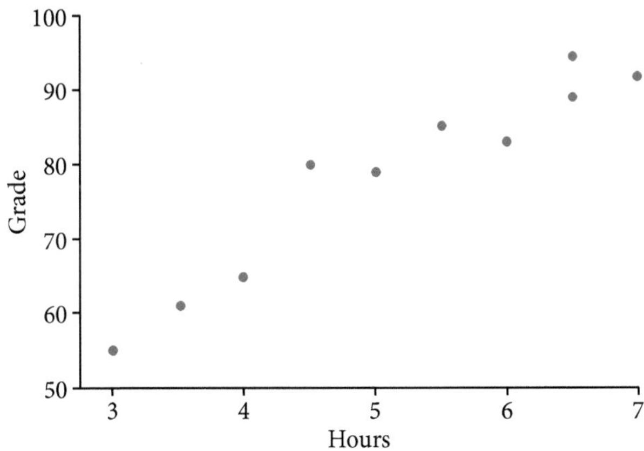

Figure 7.18 Scatter plot of IB Economics student data

A **response variable** measures an outcome of a study. An **explanatory variable** explains the changes in the response variable. If the study is to determine the relationship between weight and blood pressure, then weight is the explanatory variable and blood pressure is the response variable. If the study is to investigate the relationship between the level of fertiliser and the crop volume during an agricultural season, then the level of fertiliser is the explanatory variable and the crop is the response variable.

To study the nature of the relationship between two variables, we look into how changes in the values of one variable help explain the variation in the other

We have already seen how to use histograms and box plots to display data with one variable. In bivariate statistics, we use a **scatter plot**, or **scatter diagram**. In a scatter plot, each observation is represented by a point on a grid. The horizontal component represents the explanatory variable and the vertical component represents the response variable.

Two variables measured on the same subjects are **associated** if specific values of one variable tend to occur in connection with particular values of the other variable.

Larger values for the foot size of an individual tend to occur in connection with taller individuals. A higher rate of serious road accidents happens in connection with drivers that have a high level of alcohol in their blood. We claim that height and foot size are **positively associated**, and we can also claim that alcohol level and involvement in serious road accidents are positively associated. There may be a **negative association** between time spent watching TV and scores on weekly tests for teenagers.

variable. For instance, we look at how the increase in a person's height can explain the increase in foot size.

The principles that guide this work are:

- Start with a graphical display, and then explore numerical summaries.
- Look for overall patterns and deviations from those patterns.
- When the overall pattern is quite regular, use a mathematical model to describe it.

Example 7.10

The data presented below is for 80 adults in a dieting programme. The researchers are testing the hypothesis that the metabolic rate (calories burnt per 24 hours) is influenced by the lean body mass (in kg without fat).

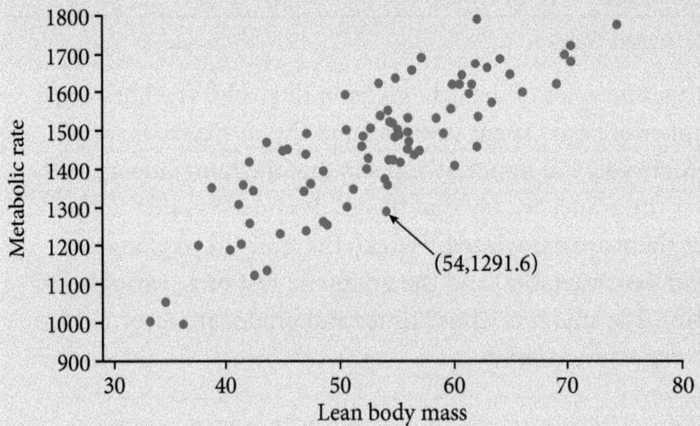

The scatter plot shows that there is an association between the metabolic rate and lean body mass. There is a positive association between these two variables: the greater the lean body mass, the higher the metabolic rate.

How to analyse a scatter plot

As a rule of thumb, when we examine a scatter plot, we may look at the following characteristics:

- Overall pattern (form, direction, and strength)
- Striking deviations from the pattern (outliers)

In Example 7.10, the form is roughly linear. That is, the points appear to cluster around a straight line. The direction, as mentioned earlier, appears to be a positive association. The strength is determined by how closely the points follow the form. We will discuss this idea in more detail later in the chapter. Even though some points stray away from the line, in this case it does not appear that there are any outliers.

In bivariate statistics, an outlier is an observation whose values fall outside the overall pattern of the relationship.

Example 7.11

The table lists the fuel consumption of 34 small cars in km l^{-1} during city driving and highway driving. Make a scatter plot of the data and comment on any patterns you observe.

City	7.3	7.7	7.3	5.1	8.5	5.1	6.8	10.7	6.0
Highway	10.2	10.7	9.8	7.3	11.1	8.1	9.8	13.7	9.4

City	8.5	5.1	3.8	9.4	8.5	9.0	7.7	9.8	25.6
Highway	11.9	8.5	6.4	11.9	12.4	12.4	11.1	13.2	28.2

City	8.5	4.7	3.8	6.8	6.4	8.1	6.8	8.5
Highway	11.9	6.8	5.5	9.8	9.8	11.5	9.8	12.4

City	7.3	4.3	6.4	5.5	11.1	8.1	7.7	7.7
Highway	10.7	6.8	9.4	8.1	13.7	11.9	9.8	11.1

Solution

A scatter plot of the data is given, with highway fuel consumption plotted on the x-axis and city fuel consumption plotted on the y-axis.

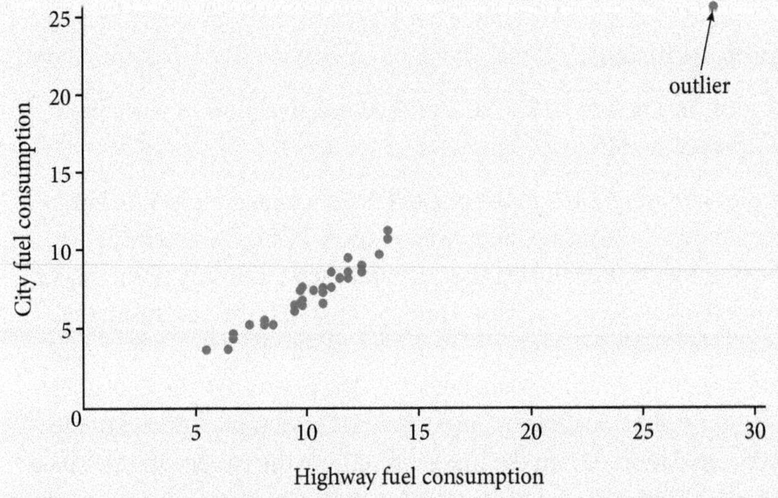

The data are tightly clustered around a positively sloped line. This indicates that the fuel consumption in highway driving and city driving are, as expected, positively associated, and the relationship is strong.

However, we can see that there is one observation that is positioned quite far from the rest of the data. This observation is an outlier.

Outliers in statistics are important. Sometimes they indicate a problem in the data being observed and sometimes they may have a special significance. In this case, the data corresponds to a 'hybrid' car, which uses battery power in addition to fuel and hence can travel much further on a litre of fuel. This observation is not typical of the study and should be removed in order to get

You can use either a spreadsheet or your GDC to produce a scatter plot. Here are two samples. Note that you need to enter your data into lists to perform any calculations.

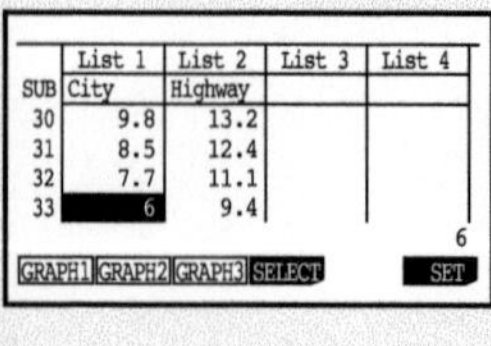

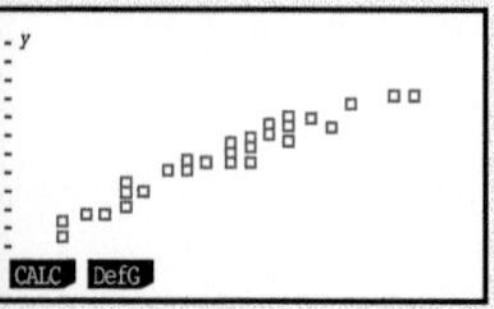

a clear indication of the nature of the relationship between the two variables. Here is an adjusted scatter plot without the hybrid car.

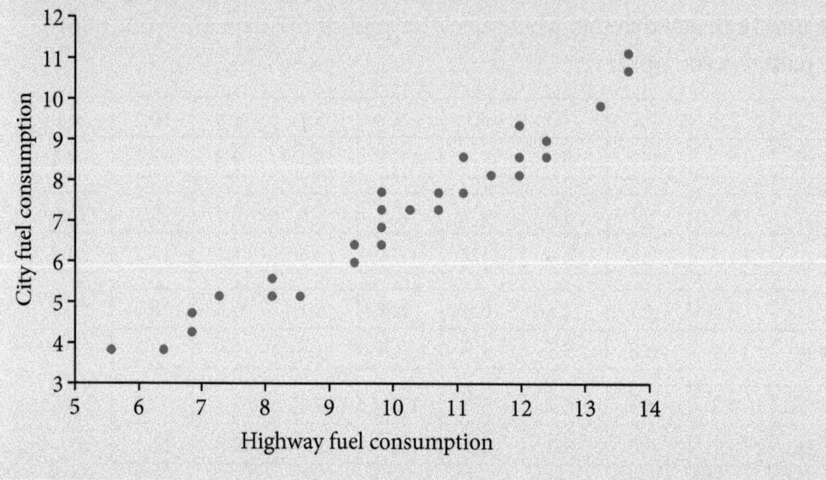

Covariance

Intuitively, we think of the dependence of two variables X and Y as implying that one variable, Y for example, either increases or decreases as the other variable, X, changes. In this book, we will confine our discussion to two measures of dependence: the **covariance** between two random variables and their **correlation coefficient**.

In the scatter plots in Figures 7.19, 7.20, and 7.21 we give plots of variables X and Y, for samples of size 15.

E(X) is the **expected value** of X. This is the average value that you would expect after a large number of observations and is usually the same as the population mean.

In the scatter plot in Figure 7.19, all the points fall on a straight line. Obviously X and Y are dependent in this case. Suppose we know $E(X) = \mu_X$ and $E(Y) = \mu_Y$. Locate the point with coordinates (μ_X, μ_Y) and then locate any point, say (x_1, y_1), and measure the deviations $(x_1 - \mu_X)$ and $(y_1 - \mu_Y)$. If the point is in the upper right quadrant, both deviations are positive. Similarly, if the point is in the lower left quadrant, both deviations are negative. The product of the deviations $(x_1 - \mu_X)(y_1 - \mu_Y)$ will be positive in both cases. This is a typical and extreme case of positive association. When the line representing the pattern in the data is positively sloped, the product of deviations from the mean is on average positive; that is, $E((X - \mu_X)(Y - \mu_Y)) > 0$.

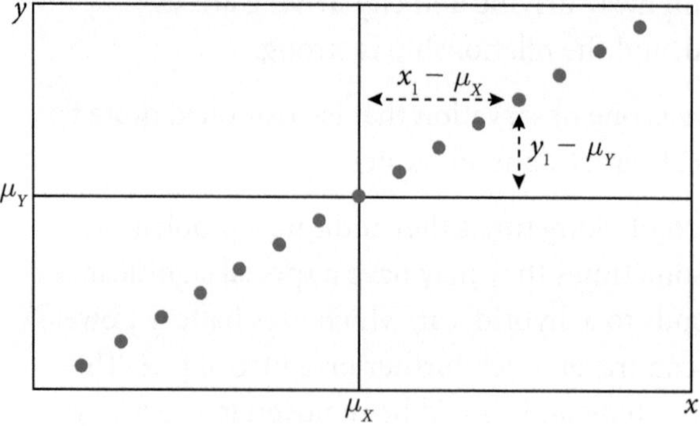

Figure 7.19 Positively sloped pattern

In the scatter plot in Figure 7.20, the data follow a negatively sloped pattern. If the point is in the upper left quadrant, the X-deviations are negative while the Y-deviations are positive. Similarly, if the point is in the lower right quadrant, the X-deviations are positive while the Y-deviations are negative. The product of the deviations $(x_1 - \mu_X)(y_1 - \mu_Y)$ is negative in both cases.

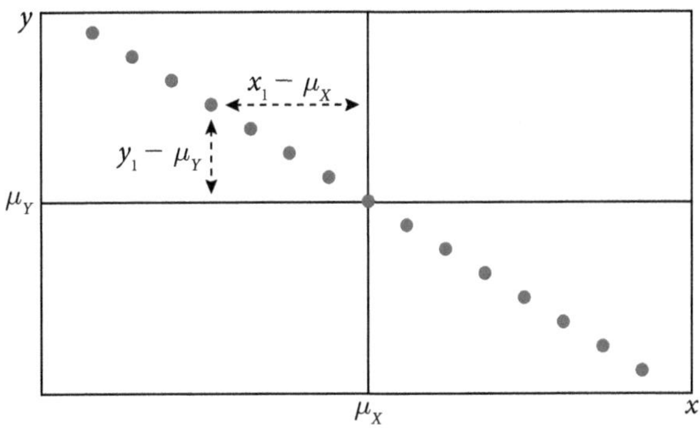

Figure 7.20 Negatively sloped pattern

Figure 7.21 shows a scatter plot where little dependence (if any) exists between the variables.

In this case, the deviations $(x_1 - \mu_X)(y_1 - \mu_Y)$ sometimes assume the same algebraic sign and sometimes opposite signs. Thus the product $(x_1 - \mu_X)(y_1 - \mu_Y)$ will be positive sometimes and negative other times, and the average may be close to zero.

The average $E((X - \mu_X)(Y - \mu_Y))$ provides a measure of the linear dependence between X and Y. This quantity is called the **covariance** of X and Y.

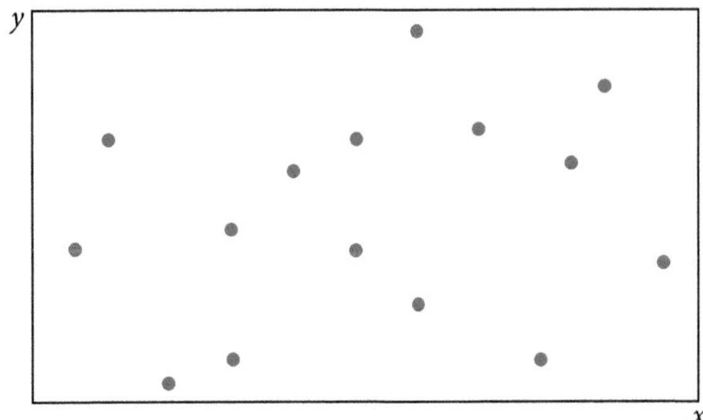

Figure 7.21 Scatter plot with little (if any) dependence between variables

If X and Y are random variables with means μ_X and μ_Y, the **covariance** of X and Y is

$$\text{cov}(X, Y) = E[(X - \mu_X)(Y - \mu_Y)]$$

The larger the absolute value of the covariance of X and Y, the greater the linear dependence between X and Y. Positive values indicate that Y increases as X increases and negative values indicate that Y decreases as X increases. A zero value of the covariance indicates that the variables are linearly **uncorrelated** and that there is no linear association between X and Y.

You will usually use your GDC or software to calculate the covariance, but there is a shortcut calculation formula that can be helpful if you need to do the calculations manually.

$$\begin{aligned}\text{cov}(X, Y) &= E((X - \mu_X)(Y - \mu_Y)) \\ &= E(XY - X\mu_Y - \mu_X Y + \mu_X\mu_Y) \\ &= E(XY) - E(X\mu_Y) - E(\mu_X Y) + E(\mu_X\mu_Y) \\ &= E(XY) - \mu_Y E(X) - \mu_X E(Y) + \mu_X\mu_Y \\ &= E(XY) - \mu_Y\mu_X - \mu_X\mu_Y + \mu_X\mu_Y \\ &= E(XY) - \mu_X\mu_Y\end{aligned}$$

The above result leads to

$$\text{cov}(X, X) = E(XX) - \mu_X\mu_X = E(X^2) - \mu_X^2 = V(X)$$

If X and Y are **not independent**, then

$$V(X + Y) = V(X) + 2\,\text{cov}(X,Y) + V(Y)$$

V(X) is the variance of X.

If X and Y are **independent,** then

$$\text{cov}(X, Y) = E(XY) - \mu_X\mu_Y = E(X)E(Y) - \mu_X\mu_Y = 0$$

and consequently

$$V(X + Y) = V(X) + V(Y)$$

Note that the converse of the theorem above is not true: if $\text{cov}(X, Y) = 0$, then X and Y are not necessarily independent.

Unfortunately, it is difficult to employ the covariance of X and Y as an absolute measure of association between variables because its value depends on the scales used. In Example 7.11, the covariance of the data expressed as $\text{km}\,\text{l}^{-1}$ is 3.8. However, if we change the scale from $\text{km}\,\text{l}^{-1}$ to $\text{mile}\,\text{l}^{-1}$, then the covariance will be 1.49, even though the scatter plot does not indicate any change in the form or the strength of association between the two variables.

This problem with covariance can be eliminated by 'standardising' its value and using the **correlation coefficient**, ρ, instead.

$$\rho_{XY} = \frac{\text{cov}(X, Y)}{\sigma_X \sigma_Y}$$

Since σ_X and σ_Y are both positive, the sign of the correlation coefficient is the same as that of the covariance.

Correlation

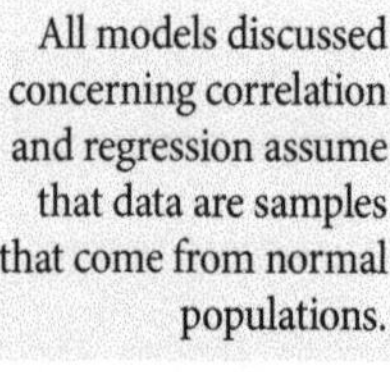
All models discussed concerning correlation and regression assume that data are samples that come from normal populations.

A scatter plot is a good device that reveals the form, trend, and strength of the association between two quantitative variables. At this level, we are only interested in linear relations. As mentioned earlier, we say that a linear relationship is strong if the data are tightly packed around the line, and weak if they are widely dispersed around the line. Our judgement using our eyes only may be misleading though. Look at the scatter plots in Figure 7.22.

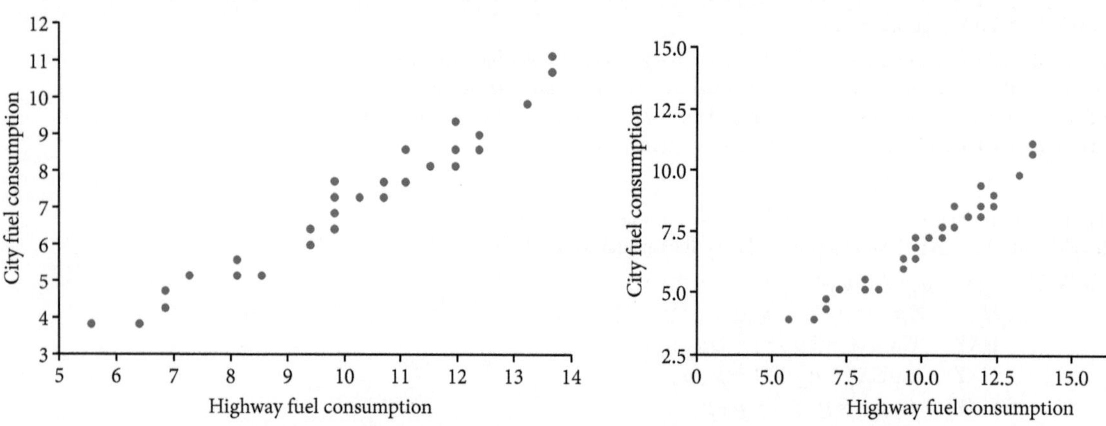

Figure 7.22 The second graph gives the impression that the association is stronger than it is in the first graph.

The graph on the left is a copy of the scatter plot (excluding the outlier) in Example 7.11. The second graph gives the impression that the association is stronger than it is in the first graph. This is due to the change in scale on the vertical axis. However, both scatter plots represent the same situation. We need a more robust measure to support our initial graphical impressions.

This measure is the **correlation coefficient**.

Let us consider the height and weight data collected from 130 19-year-olds in Figure 7.23.

The measurements were made in metric units. Not surprisingly, the association between the two variables is strong. To measure the strength of this association, we use the correlation coefficient.

The correlation coefficient measures the strength and direction of the linear relationship between two quantitative variables, when such a relationship exists.

For a set of data (x_i, y_i) of size n, the correlation coefficient is

$$R = \frac{1}{n-1}\sum\left(\frac{x_i - \bar{x}}{s_x}\right)\left(\frac{y_i - \bar{y}}{s_y}\right)$$

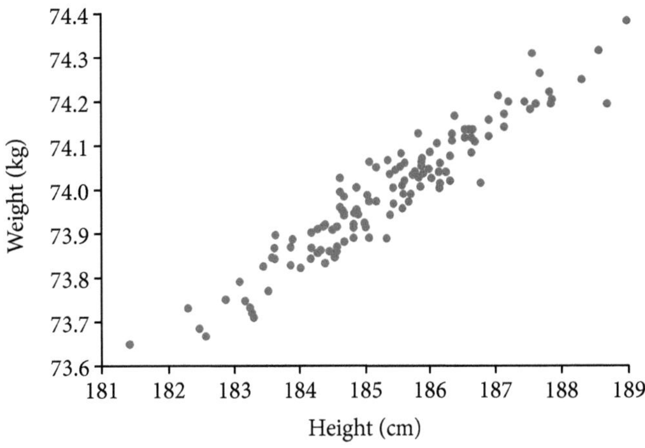

Figure 7.23 Height and weight data for 130 19-year-olds

where $\bar{x}$ and $\bar{y}$ are the means of the variables and s_x and s_y are the standard deviations. Specific observed values of R are denoted by r.

R is also called the **Pearson product-moment correlation coefficient**. In fact, R is an **unbiased estimate of the population coefficient**, which is given by

$$\rho = \frac{\text{cov}(X, Y)}{\sigma_X \sigma_Y} = \frac{1}{n}\sum\left(\frac{x_i - \mu_x}{\sigma_x}\right)\left(\frac{y_i - \mu_y}{\sigma_y}\right)$$

The GDCs use r.
In exams, you will not be asked to calculate the coefficient by hand but to interpret the GDC result. There are several equivalent forms for the equation but it is not necessary at this stage to calculate any of them.

This formula is somewhat complex to calculate. However, it helps us to see what correlation is. In practice, you will read the result from your calculator or computer output.

If we look at the formula we see that the first component, $\frac{x_i - \bar{x}}{s_x}$, is the standardised value for x_i. Similarly, the second component $\frac{y_i - \bar{y}}{s_y}$ is the standardised value for y_i. So, the correlation coefficient can be written as $R = \frac{\sum z_x z_y}{n-1}$. That is, the correlation coefficient is an average of the products of the standardised values of the two variables.

In fact, if we take the height and weight data and express it in inches and pounds instead of cm and kg, we get the scatter plot in Figure 7.24.

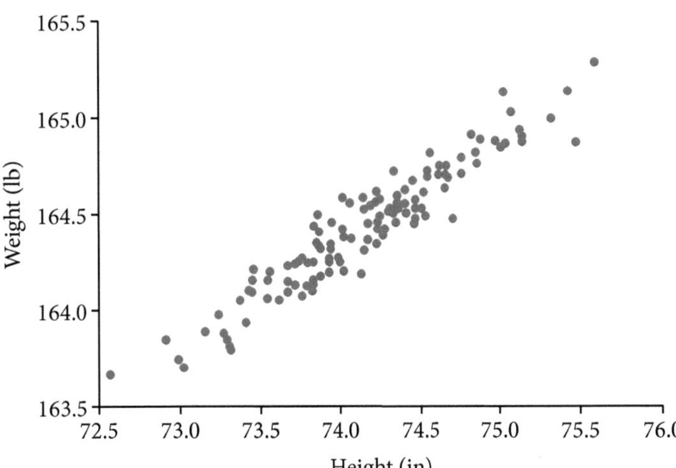

Figure 7.24 Height and weight data for 130 19-year-olds (inches and pounds)

As you notice, other than the scale on the axes being inches and pounds, the plot has the same form, direction, and strength as the original one. Similarly, when you standardise the variables, you are subtracting a constant from each value and dividing by another constant. If you plot the standardised variables, you get the scatter plot shown in Figure 7.25.

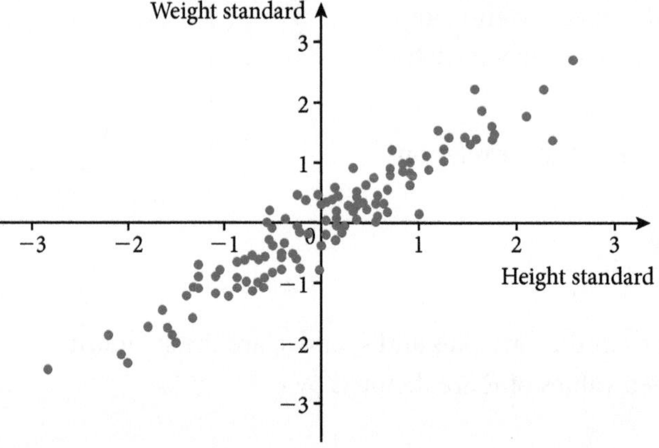

Figure 7.25 Standardised height and weight data for 130 19-year-olds

As you will notice, other than the centre of the data being at the origin, the form, direction, and strength appear to be the same.

This fact is verified by calculating the correlation coefficient for all three forms of the data. The result is always the same, 0.95.

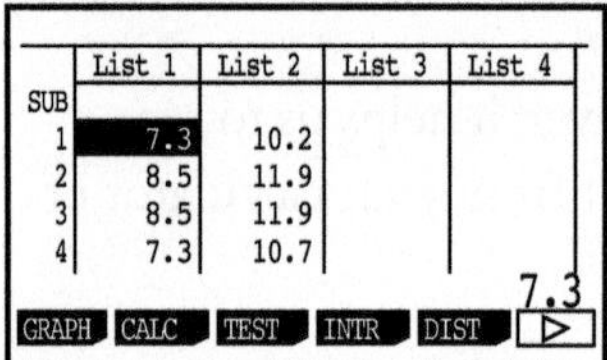

LinearReg(ax+b)
a =1.10816325
b =2.14112562
r =0.93772097
r² =0.87932062
MSe=0.60072376
y=ax+b
COPY

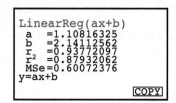

Figure 7.26 Your GDC can produce the correlation coefficient as part of the linear regression output.

You may have observed in the technology output that r^2 is also reported. This measure is not required by your exam. However, it is an extremely useful and powerful tool, known as the **coefficient of determination**. It reports the portion of variation in the response variable that can be explained by the variation in the explanatory variable. As such, r^2 can be expressed as a percentage. For the data in Example 7.11, $r^2 = 0.879$, which can be interpreted as 'all else being equal, 88% of the variation in city consumption can be explained by variation in the highway fuel consumption.' That is, on average for cars with the same characteristics, if there is a 1 km l^{-1} change in city fuel consumption, we expect that 88% of this change can be explained by changes in the highway fuel consumption.

For the data in Example 7.10, $r = 0.84$ and $r^2 = 0.7056$, which means that approximately 70.6% of the changes in the metabolic rate can be explained by changes in the lean body mass. Finally, in the height–weight example, $r^2 = 0.9025$, which means that, all else being equal, approximately 90% of the variation in weight could be explained by the variation in the height of these teenagers.

Properties of the correlation coefficient

The correlation coefficient is a measure of the strength of the linear association between two quantitative variables. Do not apply correlation to non-quantitative data!

The coefficient makes sense only if there is a linear relationship. It does not prove a linear relationship. If there is a linear association, the coefficient will describe its strength.

Outliers can distort the correlation. Special attention must be paid to such outliers.

The correlation is always a number between -1 and $+1$. Values of R near 0 indicate a weak relationship. Values close to $+1$ or -1 indicate strong association.

R does not change as we change the units of measurement. R has no units and is not a percentage. Don't express a correlation of 0.85 as 85%.

Correlation between two variables means there is some association between them. It does not mean that one of them causes the other. So, correlation does not mean causation. That is, two variables can have a strong correlation without one of them being the cause of the changes in the other. For example, there may be a strong correlation between the amount of crude oil imported by country X and the birth rate in country Y. That should not mean that the increase of oil imports causes an increase in birth rate. However, in some cases, there may be a causal relationship. For example, the increase in level of income in a certain country and the decrease of unemployment can have a strong negative correlation. This association is also causal. However, the task of proving a causal relationship has many subtle difficulties. This is something that has been studied in depth by economists.

There are statistical tests to report the strength of correlation between variables. They are beyond the scope of this course. However, some guidelines are suggested below.

If $|r| < 0.1$, then there is a very weak to no association indicated.

If $0.1 < |r| < 0.3$, then there is a small association indicated.

If $0.3 < |r| < 0.5$, then there is a medium association indicated.

If $|r| > 0.5$, then there is relatively strong association (depending on how close we are to 1).

Remember that these values are guidelines.

When there is no association,

$\text{cov}(X, Y) = 0$ and hence

$$\rho = \frac{\text{cov}(X, Y)}{\sigma_X \sigma_Y} = \frac{0}{\sigma_X \sigma_Y} = 0.$$

A proof for the values ± 1 is beyond the scope of this book.

Example 7.12

The table below gives data from a lab experiment involving the length (in mm) of a metal alloy bar used in electronic equipment when it is exposed to heat (temperature in °C).

Temperature (°C)	40	45	50	55	60	65	70	75	80
Length (mm)	20	20.12	20.20	20.21	20.25	20.25	20.34	20.47	20.61

Draw a scatter plot to represent this data. Comment on the strength of the relationship, using both r and r^2.

Solution

The scatter plot is shown on the right, with temperature on the x-axis and length on the y-axis.

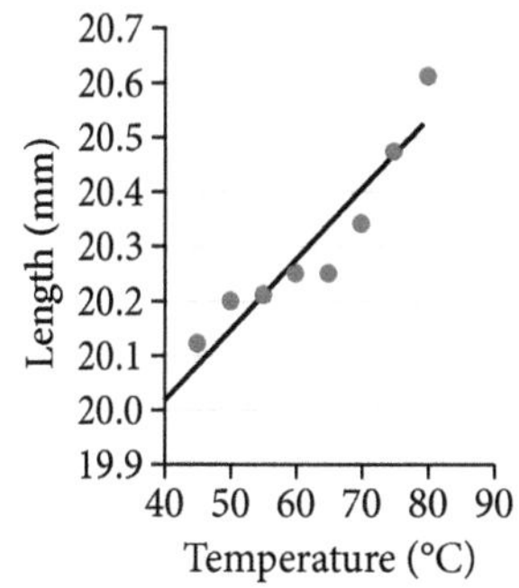

Figure 7.27 Scatter plot

This shows a relatively strong relationship, where the points are tightly spread around the trend line.

This is confirmed by calculating the correlation coefficient. In this case, regardless of which formula we use (r or ρ), the correlation is approximately 0.955 21. Using $r^2 = 0.912$ implies that 91.2% of the variation in the length can be explained by variation in the temperature.

You may now attempt the first four questions in Exercise 7.4.

Least-squares regression

We have seen above that correlation measures the strength and direction of a linear relationship between two quantitative variables. If we suspect from a scatter plot that the relationship is linear, then we need to summarise this linear behaviour. We need to find an equation of a straight line that best fits the trend in the data. In this section, we will discuss how to find a **line of best fit** that describes the linear relationship between an explanatory and response variable, when such a relationship exists.

Finding a line of best fit means finding a line that comes as close as possible to the points in the data set. Usually, there is no straight line that contains all the points in the set.

Regression line

A **regression line** is a straight line that describes how a response variable changes with changes in an explanatory variable.

Let Y be the response variable and X be the explanatory variable. We can expect several values of Y for the same value of X. Our linear model enables us to predict, on average, the value of Y given a value of $X = x$, and hence we write the equation of the linear regression line in the form

The regression model can be stated formally as $E(Y|X = x) = \alpha + \beta x$.

$$E(Y) = \alpha + \beta x$$

That is to say, given a specific value of x, the expected value of Y is equal to $\alpha + \beta x$, where α is the value corresponding to $x = 0$, and β is the slope representing the rate with which the response variable changes with every change of one unit in the explanatory variable (in other words, the gradient).

The interpretation of a in this case is peculiar. As you know from algebra, a stands for the value of y (which is weight in this case) corresponding to a zero value of x (which is height in this case). However, for this problem, the interpretation is not ideal, as it corresponds to a height of zero. The general rule is that if 0 is not included in the domain of the explanatory variable, then trying to interpret the intercept is pointless.

This issue has to do with **extrapolation**. Extrapolation is the use of the regression line for predicting values far off the range of values of the explanatory variable X used to find the equation of that line. Such predictions are often inaccurate.

When we are working with data from a sample of a population, we can only estimate the regression equation. We write our estimate as

$$y = a + bx = bx + a$$

where b, the slope of the line, is an estimate of β and reflects how the response variable, Y, changes according to changes in the explanatory variable, X. The constant a is an estimate of α and is the value of the response variable corresponding to a zero value in X.

In the height–weight example, the equation is

$$w = 56.1 + 0.0966h$$

where w is the weight in kg and h is the height in cm. That is, $b = 0.0966$ and $a = 56.1$.

This means that on average, for every increase (or decrease) of 1 cm in height, we predict an increase (or decrease) of 0.0966 kg in weight.

Why the least-squares regression line?

The graph in Figure 7.28 represents a few points in a data set. The diagonal line is the line of best fit. Take, for example, the point (x_1, y_1). The point on the line

$(x_1, \hat{y}_1)$ is the point whose y-coordinate, $\hat{y}_1$, predicts the real y-coordinate using the line of best fit. The distance $y_1 - \hat{y}_1$ is the error in this prediction and similarly for $y_2 - \hat{y}_2$ and all other $y_i - \hat{y}_i$. The line of best fit is the line that minimises the sum of all these errors. However, like the variance, some of these errors are positive and some are negative and may eventually cancel each other. To avoid this, as we did in the variance, we try to minimise the squares of these errors. That is, the line of best fit is the line that minimises the sum $\sum(y_i - \hat{y}_i)^2$. Hence it is called the **least-squares line of regression** $\hat{y} = bx + a$.

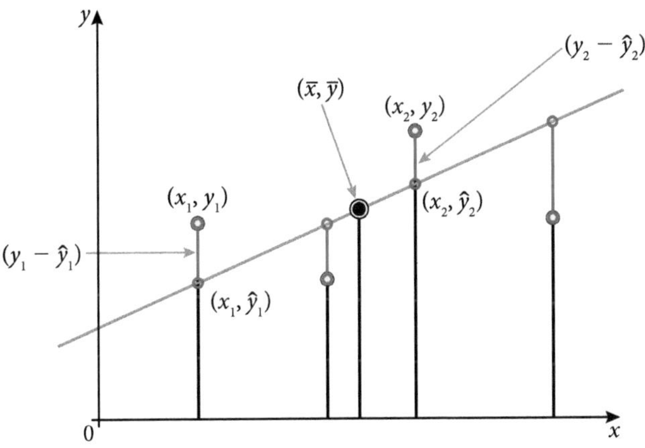

Figure 7.28 Least-squares line of regression

The process of finding the slope of such a line is beyond the scope of this book. Here are some of the many forms of the resulting formulae for the slope and intercept:

$$b = \frac{\text{cov}(X, Y)}{V(X)} = \frac{\sum(x_i - \bar{x})(y_i - \bar{y})}{\sum(x_i - \bar{x})^2} = \frac{\sum x_i y_i - n\bar{x}\bar{y}}{\sum x_i^2 - n\bar{x}^2} = r\frac{s_y}{s_x}$$

Here r is the correlation coefficient, and $\bar{x}$, $\bar{y}$, s_x, and s_y are the means and standard deviations of the explanatory and response variables. The last form demonstrates the close relationship between the slope of the regression line and the correlation coefficient. One conclusion we can draw from this formula is that along a line of regression with slope b, a change of 1 standard deviation in the x direction will result in a change of r standard deviations in the y direction.

After estimating the slope, and using the fact that the line has to contain the point with coordinates $(\bar{x}, \bar{y})$, the intercept, a, can be found using

$$a = \bar{y} - b\bar{x}$$

As you may notice from the equations, every regression line should contain the point $(\bar{x}, \bar{y})$ with the means of the variables as coordinates.

Example 7.13

The following scatter plot represents a random sample of IB students who went on to study at university. It gives a comparison of their scores on the IB exams they took with their grade point averages (GPA) in their university studies (scale 1–4).

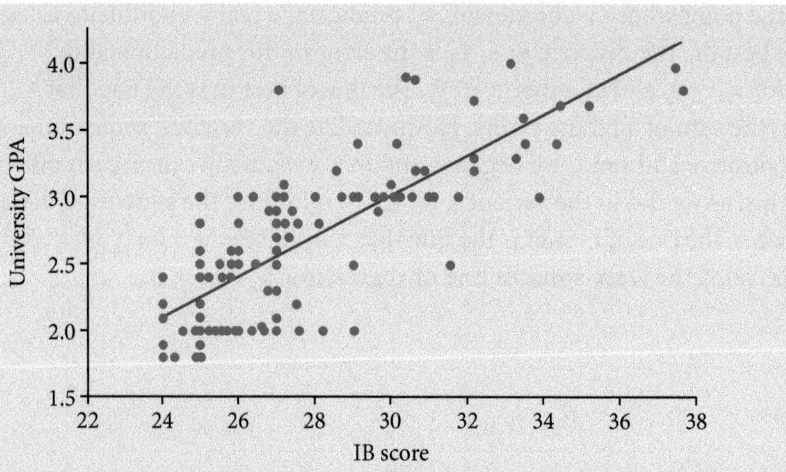

There appears to be a linear relationship between them. When we run a linear regression, the equation is

$$y = -1.51 + 0.151x$$

This means that on average, for every increase of 1 point in the total IB score, we expect an increase of 0.151 points in university GPA. If we want to predict the GPA of students who scored 30 on an IB exam, the model predicts, on average, a grade of

$$y = -1.51 + 0.151(30) = 3.02$$

The correlation coefficient of this relationship is $r = 0.758$, which is a relatively strong correlation. Add to this a value of $r^2 = 0.575$, which means that changes in the IB score may help us explain 57.5% of the variation in the university GPA.

Does that mean high IB scores cause high university averages? The answer is no. They only help predict the future university averages.

Features of the regression line:

- The regression line equation can be used to predict the response variable for given values of the explanatory variable.
- The regression line must pass through the point $(\bar{x}, \bar{y})$.

Exceptional cases of the regression line:
- If $r = 0$, the regression line is horizontal: its slope is zero.
- If $r = 1$, all the points fall on a line with positive slope.
- If $r = -1$, all the points fall on a line with negative slope.

When the regression line is used for prediction, the predicted value $\hat{y}$ of the response variable is an average value. For example, when we use the height–weight example equation $w = 56.1 + 0.0966h$ to predict the weight corresponding to a height of 182 cm, the value of 73.68 kg that we get is an average weight of 19-year-old students with height 182 cm.

Estimating the value of Y for a value of X that is within the range of the observed values of X but is not equal to any of the observed values is called **interpolation.**

Estimating the value of Y for a value of X that is larger or smaller than any of those observed is called **extrapolation.**

Extrapolation is extremely suspect. Without data in the range in which the estimate is wanted, there is no reason to believe that the relationship between X and Y is the same as it is in the region in which there are data.

Interpolation is sometimes reasonable when the scatter plot shows a strong relationship, especially if there are many points near the value of X or Y at which the estimate is sought.

Example 7.14

The table on the right shows the data for two variables. Draw the line of regression and indicate the distances, the sum of whose squares is minimised by the choice of the line of regression.

x	y
11	21
12	43
13	31
14	34
15	29
16	55
17	33

Solution

The scatter plot below shows the data and line of regression. The solid vertical lines show the distances required.

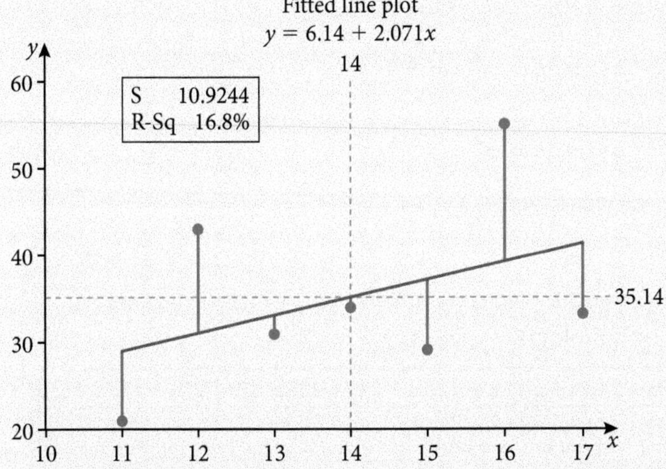

The line has the equation $\hat{y} = 6.14 + 2.071x$.

In the table below, we introduce the value of each predicted y (in the Fit column) and calculate the directed distances whose squares have been minimised.

x	y	Fit	Distance	Distance squared
11	21	28.928 57	−7.928 57	62.862 244 9
12	43	31	12	144
13	31	33.071 43	−2.071 43	4.290 816 327
14	34	35.142 86	−1.142 86	1.306 122 449
15	29	37.214 29	−8.214 29	67.474 489 8
16	55	39.285 71	15.714 29	246.938 775 5
17	33	41.357 14	28.357 14	69.841 836 73

The minimum sum is 596.71. You can try any other line and perform the same calculations. You will notice that 596.71 is the minimum sum of the squares of distances.

Moreover, since $\bar{x} = 14$ and $\bar{y} = 35.14$ then

$$35.14 = 6.14 + 2.071 \times 14$$

This indicates that the line contains the point $(\bar{x}, \bar{y})$.

In cases where the explanatory variable is not controlled, we can regress x on y. The equation of regression will be $\hat{x} = dy + c$.

The resulting formulae for the slope and intercept are $d = \frac{\text{cov}(X, Y)}{\text{V}(Y)} = \frac{\sum(x_i - \bar{x})(y_i - \bar{y})}{\sum(y_i - \bar{y})^2} = \frac{\sum x_i y_i - n\bar{x}\bar{y}}{\sum y_i^2 - n\bar{y}^2} = r\frac{s_x}{s_y}$ and $c = \bar{x} - d\bar{y}$.

Thus, taking into consideration our discussion of regressing y on x, a remarkable relationship appears between the gradients of the two regression lines and r.

When $b = r\frac{S_y}{S_x}$ and $d = r\frac{S_x}{S_y}$, then $bd = r\frac{S_y}{S_x} \cdot r\frac{S_x}{S_y} = r^2$

The product of the gradients of the two regression lines is equal to the square of the correlation coefficient (or coefficient of determination).

Example 7.15

The following data represent the volume in mm³ and weight in grams of a certain fruit studied by a biologist.

Volume (x)	223	236	242	226	223	221	233	222	222	218	232	223
Weight (y)	165	171	173	170	168	172	168	167	162	166	164	164

Obtain the least-squares regression line of y on x and the regression line of x on y. Use the model to predict the weight of a 230 mm³ fruit. Predict the volume of a 168 g fruit.

Solution

Use software or a GDC to find the regression equations.

The least-squares regression of y on x is

$$Y = 115 + 0.233x$$

The predicted weight is $y = 115 + 0.233(230) = 168.22$ g

The least-squares regression of x on y is

$$X = 56.1 + 1.02y$$

The predicted volume is $x = 56.1 + 1.02(168) = 227.26\text{ mm}^3$

Here are two samples of GDC output. Notice that in order to get the regression of x on y you switch the choice of List 1 and List 2, but the GDC still reports it as $y = a + bx$.

```
LinearReg(a+bx)
 a   =114.731215
 b   =0.2327179
 r   =0.48692124
 r²  =0.2370923
 MSe =10.1466723
y=a+bx          COPY
```

```
LinearReg(a+bx)
 a   =56.1015037
 b   =1.01879699
 r   =0.48692124
 r²  =0.2370923
 MSe =44.4203007
y=a+bx          COPY
```

```
LinReg
y=ax+b
a=.2327179047
b=114.7312151
r²=.2370923014
r=.4869212476
```

```
LinReg
y=ax+b
a=1.018796992
b=56.10150376
r²=.2370923014
r=.4869212476
```

Notice too that regardless of which regression we have (y on x or x on y), the values of r and r^2 do not change.

The product of the gradients (0.233) and (1.02) is 0.237, which is the same as the value of r^2 given by the software.

Online

Linear regression illustrated for random sets of six data points.

Segmented regression, also known as **piecewise regression** or **broken-stick regression**, is a method in regression analysis in which the independent variable is partitioned into intervals and a separate line segment is fitted to each interval. For example, if we are analysing data concerning road accidents before and after a certain law has been enforced, we collect accident data for the period before the law and for the period after the law. For our purposes at this level, a regression line can be fitted to each interval separately. If we judge that a scatter plot shows that the data implies a change in pattern at some point, called a **break point**, then we split our data into two or more segments as shown.

Exercise 7.4

1. The following table lists the values of a response variable y against an explanatory variable x. Draw a scatter plot and comment on the strength of the relationship.

x	12	6	12	11	16	13	11	12	11	12	12	12	15	16	14	13	13	8	10	11
y	8	10	9	6	14	10	10	9	15	14	10	6	12	8	13	11	11	9	9	6

Develop a regression model for this situation and interpret the gradient.

2. The data below represents the outcome of an experiment on a small car, relating fuel consumption to speed.

Speed ($km\,h^{-1}$)	60	65	70	75	80	85	90	95
Fuel consumption ($km\,L^{-1}$)	16.9	16.8	15.9	15.9	14.4	14.3	13.2	14.3

Speed ($km\,h^{-1}$)	100	105	110	120	130	140	150
Fuel consumption ($km\,L^{-1}$)	12.1	12.0	10.2	9.8	9.0	8.0	7.1

(a) Make a scatter plot to show the data.

(b) Describe the relationship and justify your choice of which variable is the explanatory variable and which is the response variable.

(c) Is the relationship strong? Explain your answer.

(d) Develop a regression model for this situation and interpret the gradient.

3. The following data are from the World Bank statistics relating the gross national income per capita (GNI/Cap) to purchasing power parity (PPP) for a few developed countries. The exchange rate adjusts so that an identical good in two different countries has the same price when expressed in the same currency. For example, a chocolate bar that sells for C$1.50 in a Canadian city should cost US$1.00 in a US city when the exchange rate between Canada and the USA is 1.50 USD/CDN.

Country	GNI/Cap	PPP
NOR	85 380	57 130.0
CH	70 350	49 180.0
DK	58 980	40 140.0
SWE	49 930	39 600.0
NL	49 720	42 590.0
FIN	47 170	37 180.0
USA	47 140	47 020.0
AUT	46 710	39 410.0
BEL	45 420	37 840.0
D	43 330	38 170.0
F	42 390	34 440.0
JPN	42 150	34 790.0
SGP	40 920	54 700.0

(a) Make a scatter plot.

(b) Describe the relationship and justify your choice of which variable is the explanatory variable and which is the response variable.

(c) Is the relationship strong? Explain your answer.

(d) Develop a regression model for this situation and interpret the gradient.

4. In hotel management, there is a need to estimate the electricity consumption in relation to the number of guests. The data for a large hotel is given in the table.

Guests	232	311	321	334	352	375	412	447	456	472	480	495	512
Consumption	237	278	270	303	298	328	387	390	376	402	431	430	432

(a) Make a scatter plot.

(b) Describe the relationship and justify your choice of which variable is the explanatory variable and which is the response variable.

(c) Is the relationship strong? Explain your answer.

(d) Develop a regression model for this situation and interpret the gradient.

5. To test the benefit of using an online tutoring course for exam preparation, 20 students were given a test before they took the course and then afterwards. The tests were similar and the scores before and after the course were recorded. The intention was to find how improved the scores were due to taking the online course.

Analyse the data. For a student whose score before the course was 60, what do you expect, on average, the student's new score to be?

Student	1	2	3	4	5	6	7	8	9	10
Before	98	24	6	8	56	54	40	40	68	30
After	122	46	16	28	84	68	64	62	82	50

Student	11	12	13	14	15	16	17	18	19	20
Before	32	80	102	30	12	16	60	58	50	48
After	40	100	129	56	32	56	90	73	74	70

6. A large electronics company produces LCD monitors to be used in the computer industry. The monthly total cost of production over the period of one year is given in the table below. Number of units produced is in thousands and the cost is in €1000.

Number of units produced	16	31	57	76	13	25
Cost	1875	2586	3716	4712	1690	2191

Number of units produced	49	71	20	38	63	81
Cost	3319	4362	2005	2775	4116	4860

(a) Draw a scatter plot of the data.

(b) Write down the equation of the regression line representing the association between units of production and cost. Draw the line on your scatter plot.

(c) Interpret the slope of the line and comment on the strength of this association.

(d) If the selling price of each unit during this year is €105, what is the production level where the sales are equal to the cost?

7. The table shows the marks of 12 students sitting for both IB Economics SL and IB Physics SL.

Economics	7	6	5	5	6	3	7	7	5	4	5	7
Physics	6	6	6	4	7	4	6	5	6	4	6	5

(a) Find the correlation coefficient and comment on your result.

(b) Find the regression equation that enables us to predict the economics scores from the physics scores.

(c) What mark in economics would you expect for a candidate with a mark of 4 in physics?

8. Diamonds are usually priced according to weight. The carat is the usual measure and it is the weight of the diamond. 1 carat is equivalent to 200 mg. Some experts use points as the measure instead. 1 point is equivalent to 2 mg. Therefore, every carat is equivalent to 100 points. So, a 0.5 carat diamond is equivalent to 50 points. Here is the data for 20 diamonds and their prices.

Points	73	103	106	21	31	100
Price (€)	5909	15 260	13 640	1287	2177	12 837

Points	26	82	101	100	63	66
Price (€)	1911	6927	16 143	10 945	9117	6020

(a) Construct a scatter plot of the data. What type of trend do you observe?

(b) Write down the equation of a straight-line model relating the price to the number of points.

(c) Give a practical interpretation of the coefficients. If a practical interpretation is not possible, explain why.

(d) How well does the line fit the given data?

(e) Use the line you found to predict the price of a diamond with 63 points.

(f) Find the residual corresponding to your estimate in part **(e)**.

9. The amount of blood that is pumped out of the left ventricle to the body with each heartbeat is called the stroke volume. Researchers studying, among other factors, the effect of age (in years) on the amount of blood for every heartbeat (in ml) collected the following data from a random sample of patients.

Age	24	30	35	39	44	50	56	60	64	68	73
Blood volume	73	75	75	72	70	71	68	69	66	62	61

(a) Draw a scatter plot of the data.

(b) Find the product-moment correlation coefficient and comment on the result.

(c) Find an equation for the regression line and interpret the coefficients.

(d) For a 45-year-old person, what stroke volume can we predict? What stroke volume can we predict for a 90-year-old?

10. The table below shows the speed of a car, in km h^{-1} against time in seconds for the first 7 seconds as it tries to accelerate to maximum speed.

Time	0	0.5	0.6	1	1.1	1.3	1.5	1.8	2	2.5	2.6	2.8	3	3.3	3.5
Speed	0	18	19	42	38	58	56	77	79	92	89	101	100	114	106

Time	3.9	4	4.2	4.5	4.8	5	5.4	5.5	5.7	6	6.3	6.5	6.8	7
Speed	116	118	125	124	136	135	139	140	153	155	154	158	157	161

(a) Draw a scatter plot of the data and comment on the pattern you observe.

(b) Considering the whole data as one set, find the equation of the regression line. Plot the regression line along with the data and comment on its appearance.

(c) Write down the product-moment correlation coefficient and explain your answer.

(d) Split the data into two intervals, [0, 3.0] and [3.1, 7.0], and find the equations of the regression lines for each interval.

(e) Predict the speed of a car at 4 seconds using your equation in **(b)** and that in **(d)**. Comment on the result.

Chapter 7 practice questions

1. Given that μ is the mean of a data set $y_1, y_2, \frac{1}{4}, y_{30}$, and that $\sum_{i=1}^{30} y_i = 360$ and $\sum_{i=1}^{30}(y_i - \mu)^2 = 925$, find

(a) the value of μ

(b) the standard deviation of the set.

2. Laura made a survey of some students at school, asking them about the time it takes each of them to come to school every morning. She scribbled the numbers on a piece of paper and, unfortunately, could not read the number of students who spend 40 minutes on their trip to school. The average number of minutes she had originally found was 34 minutes. Find out how many students spend 40 minutes on their trip.

Time in minutes	10	20	30	40	50
Number of students with this time	1	2	5	??	3

3. The following table gives 50 measurements of the time it took a certain reaction to complete in a laboratory experiment.

3.1	5.1	4.9	1.8	2.8	5.6	3.6	2.2	2.5	3.4
4.5	2.5	3.5	3.6	3.7	5.1	4.1	4.8	4.9	1.6
2.9	3.6	2.1	6.1	3.5	4.7	4	3.9	3.7	3.9
2.7	4.3	4	5.7	4.4	3.7	3.7	4.6	4.2	4
3.8	5.6	6.2	4.9	2.5	4.2	2.9	3.1	2.8	3.9

(a) Construct a frequency table and histogram starting at 1.6 with interval length of 0.5

(b) What fraction of the measurements is less than 5.1?

(c) Estimate, from your histogram, the median of this data set.

(d) Estimate the mean and standard deviation using your frequency table.

(e) Construct a cumulative frequency table.

(f) From your cumulative frequency graph, estimate each of the five numbers in the five-number summary.

4. In large cities around the world, governments offer parking facilities for public use. The histogram below gives a picture of the number of parking sites available with the capacity of each in a number of cities chosen at random.

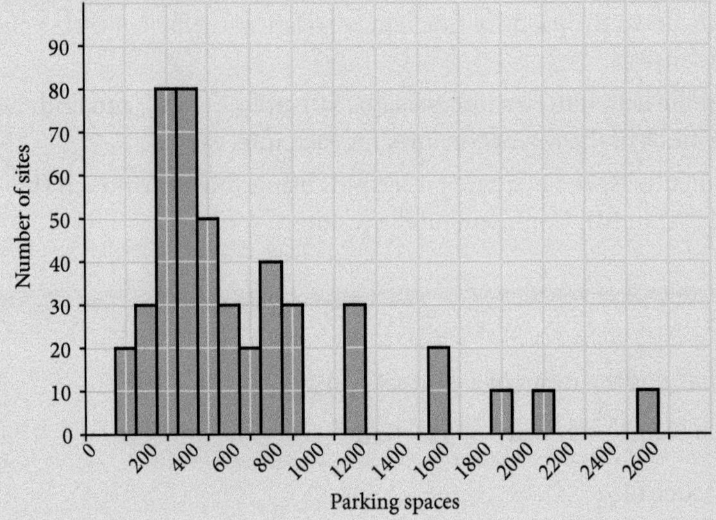

(a) Which statistics would best represent the data here? Why?

(b) Calculate the mean and standard deviation.

(c) Develop a cumulative frequency graph of the data.

(d) Use the result of (c) above to estimate the median, Q_1, Q_3, and IQR.

(e) Are there any outliers in the data? Why?

(f) Write a few sentences describing the distribution.

5. The box plots display the case prices (in €) of red wines produced in France, Italy, and Spain.

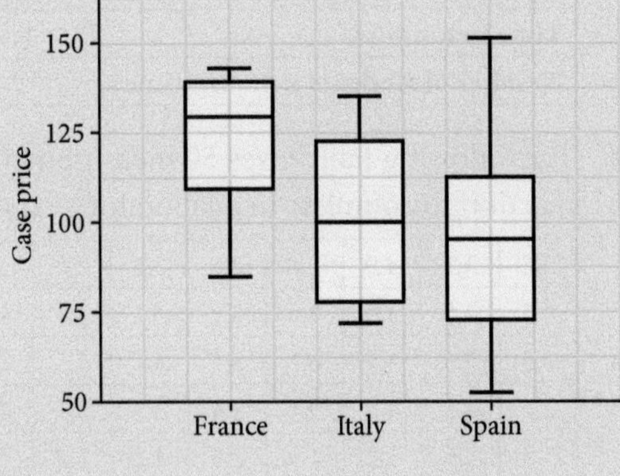

(a) Which country produces the most expensive red wine? The cheapest?

(b) In which country are the red wines generally more expensive?

(c) Write a few sentences comparing the pricing of red wines in the three countries.

6. The table shows the record for the times (in seconds) of the 71 male swimmers competing in the 100-metre swim on the first day of the 2000 Summer Olympics in Sydney.

112.72	53.55	54.12	54.33	58.79	59.26	60.39	62.45	52.22	52.52	52.58	52.85
54.06	51.34	51.93	52.09	52.14	52.24	52.24	52.53	53.5	51.82	51.93	52
52.78	52.82	50.28	50.49	51.28	51.28	51.52	51.62	52.4	52.43	49.83	50.46
50.95	51.07	51.11	49.45	49.45	49.73	49.76	49.93	50.19	50.32	50.63	48.64
49.79	50.19	50.62	50.96	49.09	49.16	49.29	49.74	49.74	49.75	49.84	49.76
52.9	52.91	53.4	52.18	52.57	52.72	50.56	50.87	50.9	49.32	49.7	

(a) Calculate the mean time and the standard deviation.

(b) Calculate the median and IQR.

(c) Explain the differences between these two sets of measures.

7. In a survey of universities in major cities in the world, the percentage of first-year students who graduate on time (some require 4 years and some 5 years) was reported. The summary statistics are given below.

Number of universities surveyed	120
Mean percentage	69
Median percentage	70
Standard deviation	9.8
Minimum	42
Maximum	86
Range	44
Q_1	60.25
Q_3	75.75

(a) Is this distribution symmetric? Explain.

(b) Check for outliers.

(c) Create a box plot of the data.

(d) Describe the data in a short paragraph.

8. The International Heart Association studies, among other factors, the influence of cholesterol level (in mg dl^{-1}) on the conditions of heart patients. In a study of 2000 subjects, the cumulative relative frequency graph on the right was recorded.

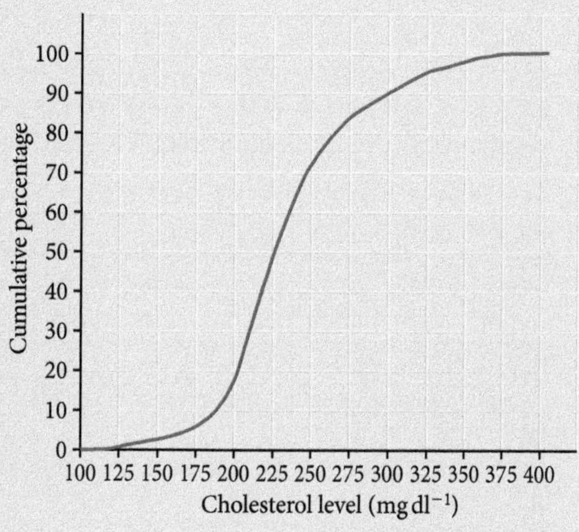

(a) Estimate the median cholesterol level of heart patients in the study.

(b) Estimate the first and third quartiles, and the 90th and the 10th percentiles.

(c) Estimate the IQR. Also estimate the number of patients in the middle 50% of this distribution.

(d) Create a box plot of the data.

(e) Give a short description of the distribution.

9. Many of the streets in Vienna, Austria, have a speed limit of 30 km h^{-1}. On one Sunday evening, the police registered the speed of cars passing an important intersection in order to give speeding tickets when drivers exceeded the limit. Here is a random sample of 100 cars recorded that afternoon.

26	46	39	41	44	37	38	35	34	31
27	47	39	41	44	37	38	35	34	32
27	47	39	41	44	37	38	35	34	32
27	48	39	41	44	37	38	35	34	32
29	48	40	41	45	37	38	36	34	33
30	48	40	41	45	37	38	36	35	33
30	48	40	42	45	38	39	36	35	33
30	49	40	42	46	38	39	36	35	33
30	50	41	42	46	38	39	36	35	33
31	54	41	43	46	38	39	36	35	33

(a) Prepare a frequency table for the data.

(b) Draw a histogram of the data and describe the shape.

(c) Calculate, showing all working, the mean and standard deviation of the data.

(d) Prepare a cumulative frequency table of the data.

(e) Find the median, Q_1, Q_3, and IQR.

(f) Are there any outliers in the data? Explain using an appropriate diagram.

10. The following are the data collected from 50 industrial countries chosen at random in 2001. The data represent the per capita gasoline consumption in these countries. The Netherlands' consumption was 1123 litres per capita, while Italy's was 2220 litres per capita.

2062	2076	1795	1732	2101	2211	1748	1239	1936	1658
1639	1924	2086	1970	2220	1919	1632	1894	1934	1903
1714	1689	1123	1671	1950	1705	1822	1539	1976	1999
2017	2055	1943	1553	1888	1749	2053	1963	2053	2117
1600	1795	2176	1445	1727	1751	1714	2024	1714	2133

(a) Calculate the mean, median, standard deviation, Q_1, Q_3, and IQR.

(b) Are there any outliers?

(c) Draw a box plot.

(d) What consumption levels are within 1 standard deviation from the mean?

(e) Germany, with a consumption level of 2758 litres per capita, was not included in the sample. What effect on the different statistics calculated would adding Germany have? Do not recalculate the statistics.

11. 90 students on a statistics course were given an experiment where each reported the time, x, it took them to commute to school on a specific day to the nearest minute. The teachers then reported back that the total travelling time for the course participants was $\sum x = 4460$ minutes.

(a) Find the mean number of minutes the students spent travelling to school that day.

Four students who were absent when the data was first collected reported that they spent 35, 39, 28, and 32 minutes, respectively.

(b) Calculate the new mean including these two students.

12. Two thousand students at a large university take the final statistics examination, which is marked out of 100, and the distribution of marks received is given in the table below:

Marks	1−10	11−20	21−30	31−40	41−50
Number of candidates	30	100	200	340	520

Marks	51−60	61−70	71−80	81−90	91−100
Number of candidates	440	180	90	60	40

(a) Complete the table below so that it represents the cumulative frequency for each interval.

Marks	⩽10	⩽20	⩽30	⩽40	⩽50
Number of candidates	30	130			

Marks	⩽60	⩽70	⩽80	⩽90	⩽100
Number of candidates	1630				

(b) Draw a cumulative frequency graph of the distribution, using a scale of 1 cm for 100 students on the vertical axis and 1 cm for 10 marks on the horizontal axis.

(c) Use your graph to answer parts **(i)**–**(iii)** below,

(i) Find an estimate for the median score.

(ii) Candidates who scored less than 35 were required to retake the examination. How many candidates had to retake the exam?

(iii) The highest scoring 15% of candidates were awarded a distinction. Find the mark above which a distinction was awarded.

13. 100 physicians are taking part in a symposium. 72 of the physicians are male and 28 are female. The mean height of the men is 179 cm and that for the women is 162 cm. Find the mean height of the 100 physicians.

14. Consider a population $x_1, x_2, ..., x_{25}$ such that $\sum_{i=1}^{25} x_i = 300$ and $\sum_{i=1}^{25} (x_i - \mu)^2 = 625$, where μ is the mean.

(a) Find the value of μ.

(b) Find the standard deviation of the population.

15. The table below lists the scores of students in a small class on a 50-mark test.

Score	10	20	30	40	50
Number of students with this score	1	2	5	k	3

The mean score is 34. Find the value of k.

16. Waiting times for 100 customers at a supermarket's cash counter are recorded in the table on the right.

Waiting time, t (seconds)	**Number of customers**
$0 \leqslant t < 30$	5
$30 \leqslant t < 60$	15
$60 \leqslant t < 90$	33
$90 \leqslant t < 120$	21
$120 \leqslant t < 150$	11
$150 \leqslant t < 180$	7
$180 \leqslant t < 210$	5
$210 \leqslant t < 240$	3

(a) Estimate the mean waiting time for a customer.

(b) Set up a cumulative frequency table for these data.

(c) Use the table in **(b)** to draw a cumulative frequency graph.

(d) Use the graph in **(c)** to find estimates for the median and the first (lower) and third (upper) quartiles.

17. The diagram on the right represents the lengths, in cm, of 80 plants grown in a laboratory.

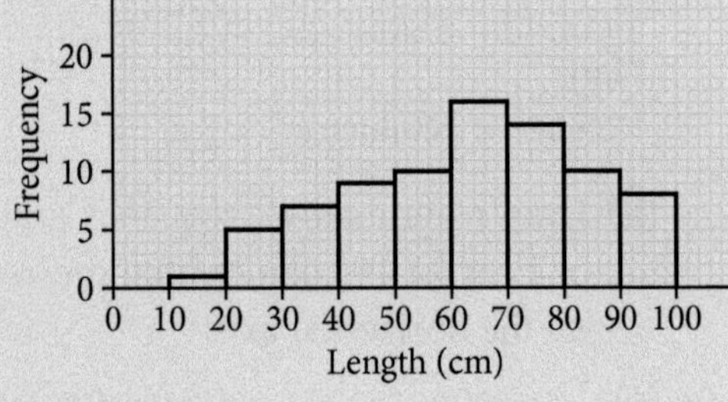

(a) How many plants have lengths in cm between

(i) 50 and 60?

(ii) 70 and 90?

(b) Calculate estimates for the mean and the standard deviation of the lengths of the plants.

(c) Explain what feature of the diagram suggests that the median is different from the mean.

(d) The following is an extract from the cumulative frequency table.

Use the information in the table to estimate the median.

Length in cm	Cumulative frequency
< 50	22
< 60	32
< 70	48
< 80	62

18. The table below represents the weights, W, in grams, of 80 packets of roasted peanuts.

Weight (W)	$80 < W \leqslant 85$	$85 < W \leqslant 90$	$90 < W \leqslant 95$	$95 < W \leqslant 100$
Number of packets	5	10	15	26

Weight (W)	$100 < W \leqslant 105$	$105 < W \leqslant 110$	$110 < W \leqslant 115$	
Number of packets	13	7	4	

(a) Use the midpoint of each interval to find an estimate for the standard deviation of the weights.

(b) Copy and complete the following cumulative frequency table for the above data.

Weight (W)	$W \leqslant 85$	$W \leqslant 90$	$W \leqslant 95$	$W \leqslant 100$
Number of packets	5	15		

Weight (W)	$W \leqslant 105$	$W \leqslant 110$	$W \leqslant 115$	
Number of packets			80	

(c) A cumulative frequency graph of the distribution is shown on the right.

Use the graph to estimate

(i) the median;

(ii) the upper quartile (that is, the third quartile).

Give your answers to the nearest gram.

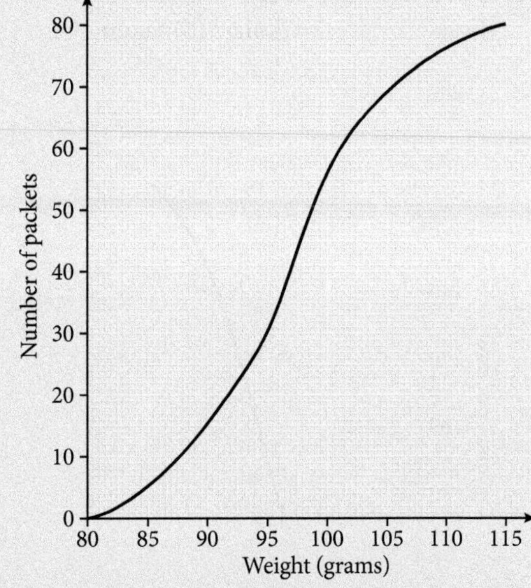

(d) Let $W_1, W_2, \ldots, W_{80}$ be the individual weights of the packets, and let $\overline{W}$ be their mean. What is the value of the sum $(W_1 - \overline{W}) + (W_2 - \overline{W}) + (W_3 - \overline{W}) + \ldots + (W_{79} - \overline{W}) + (W_{80} - \overline{W})$?

(e) One of the 80 packets is selected at random. Given that its weight satisfies $80 < W \leqslant 110$, find the probability that its weight is greater than 100 grams.

19. A TV camera is mounted at a critical stretch of a highway and records the speeds in km h^{-1} of cars passing a certain point. The table shows the record of a 5-minute interval.

Speed s	$s \leqslant 60$	$60 < s \leqslant 70$	$70 < s \leqslant 80$	$80 < s \leqslant 90$	$90 < s \leqslant 100$
No. of cars	0	7	25	63	70

Speed s	$100 < s \leqslant 110$	$110 < s \leqslant 120$	$120 < s \leqslant 130$	$130 < s \leqslant 140$	$s > 140$
No. of cars	71	39	20	5	0

(a) Estimate of the mean speed of cars passing this point.

(b) The table lists cumulative frequencies for the speeds above.

(i) Write down the values of m and n.

(ii) Construct a cumulative frequency curve to represent information in **(b)(i)**.

Speed s	**Cumulative frequency**
$s \leqslant 60$	0
$s \leqslant 70$	7
$s \leqslant 80$	32
$s \leqslant 90$	95
$s \leqslant 100$	m
$s \leqslant 110$	236
$s \leqslant 120$	n
$s \leqslant 130$	295
$s \leqslant 140$	300

(c) Use the graph in **(b)** to determine

(i) the percentage of cars travelling at a speed in excess of 105 km h^{-1}

(ii) the speed that is exceeded by 15% of the cars.

20. A taxi company has 200 taxi cabs. The cumulative frequency curve below shows the fares in dollars ($) taken by the cabs on a particular morning.

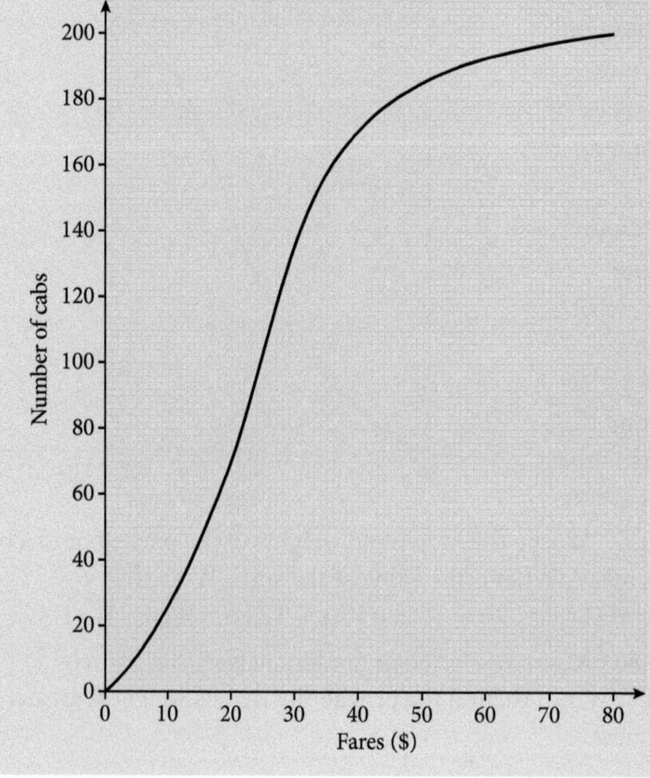

(a) Use the curve to estimate

(i) the median fare

(ii) the number of cabs in which the fare taken is \$35 or less.

The company charges 55 cents per kilometre for distance travelled. There are no other charges. Use the curve to answer the following.

(b) On that morning, 40% of the cabs travel less than a km. Find the value of a.

(c) What percentage of the cabs travel more than 90 km on that morning?

21. Three positive integers a, b, and c, where $a < b < c$, are such that their median is 11, their mean is 9, and their range is 10. Find the value of a.

22. A real estate agent keeps records of small houses sold in a suburb of Vienna, Austria. The table below is a cumulative frequency table showing 100 houses that were sold in the second half of 2017. Prices are in thousands of euros.

Selling price P (€1000)	$P \leqslant 100$	$P \leqslant 200$	$P \leqslant 300$	$P \leqslant 400$	$P \leqslant 500$
Total number of houses	12	58	87	94	100

(a) Draw a cumulative frequency curve for the information in the table.

(b) Use your curve to find the lower and upper quartiles as well as the interquartile range.

Below is the frequency distribution of the information above.

Selling price P (€1000)	$0 < P \leqslant 100$	$100 < P \leqslant 200$	$200 < P \leqslant 300$	$300 < P \leqslant 400$	$400 < P \leqslant 500$
Number of houses	12	46	29	m	n

(c) Find the values of m and n.

(d) Use mid-interval values to calculate an estimate for the mean selling price.

(e) Houses that sell for more than €350,000 are described as luxury houses.

(i) Use your graph to estimate the number of luxury houses sold.

(ii) Two luxury houses are selected at random. Find the probability that both have a selling price of more than €400,000.

23. A student measured the diameters of 80 snail shells. His results are shown in the following cumulative frequency graph. The lower quartile (LQ) is 14 mm and is marked clearly on the graph.

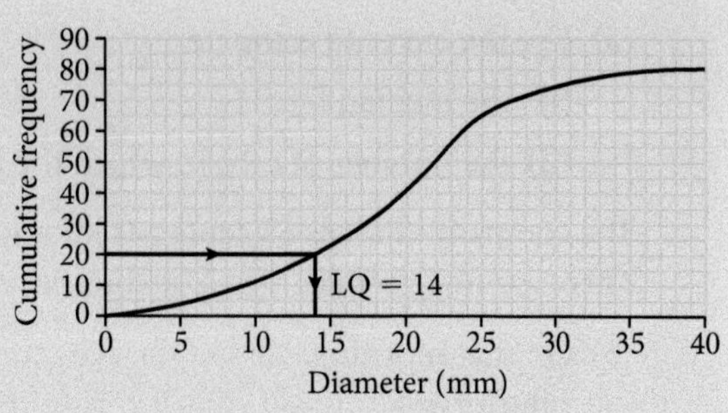

(a) On the graph, mark clearly in the same way, and write down the value of:

(i) the median

(ii) the upper quartile.

(b) Write down the interquartile range.

24. The cumulative frequency curve on the right shows the marks obtained in an examination by a group of 200 students.

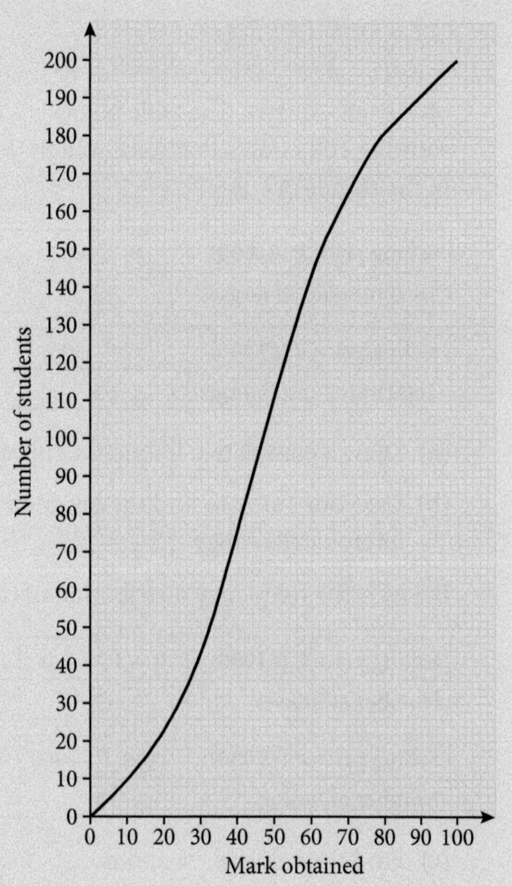

(a) Use the cumulative frequency curve to complete the frequency table below.

Mark (x)	Number of students
$0 \leqslant x < 20$	22
$20 \leqslant x < 40$	
$40 \leqslant x < 60$	
$60 \leqslant x < 80$	
$80 \leqslant x < 100$	

(b) Forty percent of the students fail. Find the pass mark.

25. The cumulative frequency curve on the right shows the heights of 120 basketball players in centimetres.

Use the curve to estimate

(a) the median height

(b) the interquartile range.

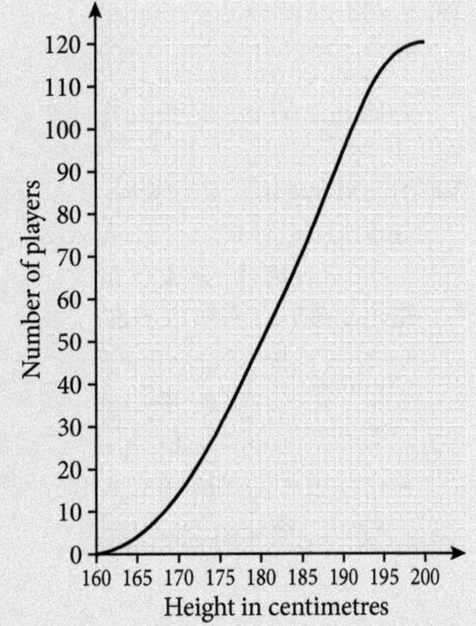

26. Let a, b, c, and d be integers such that $a < b$, $b < c$, and $c = d$.

The mode of these four numbers is 11.

The range of these four numbers is 8.

The mean of these four numbers is 8.

Calculate the value of each of the integers a, b, c, d.

27. A test marked out of 100 is written by 800 students. The cumulative frequency graph for the marks is given below.

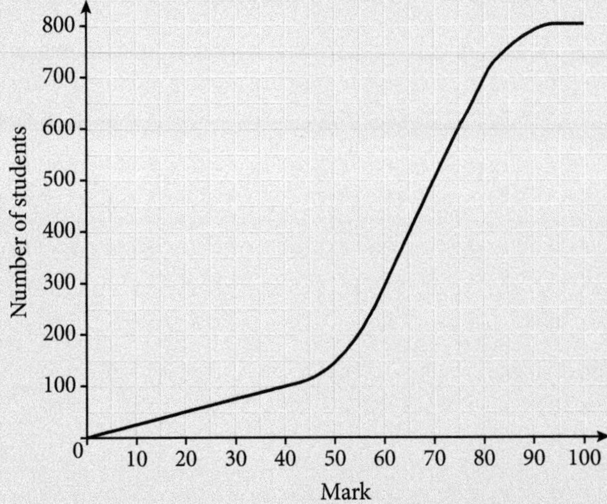

(a) Write down the number of students who scored 40 marks or less on the test.

(b) The middle 50% of test results lie between marks a and b, where $a < b$. Find a and b

28. x and y are integers with $x < y$.

The set of numbers $\{x, y, 10, 12, 16, 16, 18, 18\}$ have a mean of 13 and a variance σ^2 of 21. Find x and y.

29. The following table gives the average yield of olives per tree, in kg, and the rainfall, in cm, for nine separate regions of Italy. You may assume that these data are a random sample from a bivariate normal distribution, with correlation coefficient ρ.

Rainfall (x)	11	10	15	13	7	18	22	20	28
Yield (y)	56	53	67	61	54	78	86	88	78

A scientist wishes to use these data to determine whether there is a positive correlation between rainfall and yield.

(a) Draw a scatter plot of the data and comment on its shape.

(b) Determine the product-moment correlation coefficient for these data and comment on the strength of the relationship.

(c) Find the equation of the regression line of y on x and interpret its parameters.

(d) Hence, estimate the yield per tree in a tenth region where the rainfall was 19 cm.

(e) Determine the angle between the regression line of y on x and that of x on y. Give your answer to the nearest degree.

8 Probability

8 Probability

Learning objectives

By the end of this chapter, you should be familiar with...

- the concepts of trial, outcome, equally likely outcomes, sample space (U) and event
- the probability of an event A as $\mathrm{P}(A) = \frac{n(A)}{n(U)}$
- complementary events A and A' (not A), and the identity $\mathrm{P}(A) + \mathrm{P}(A') = 1$
- combined events and use of the formula $\mathrm{P}(A \cup B) = \mathrm{P}(A) + \mathrm{P}(B) - \mathrm{P}(A \cap B)$
- mutually exclusive events and the fact that $\mathrm{P}(A \cap B) = 0$
- conditional probability and the formula $\mathrm{P}(A|B) = \frac{\mathrm{P}(A \cap B)}{\mathrm{P}(B)}$
- probabilities of independent events: $\mathrm{P}(A|B) = \mathrm{P}(A) = \mathrm{P}(A|B')$
- the use of Venn and tree diagrams and tables of outcomes to solve problems.

Now that we have learned to describe a data set in Chapter 7, how can we use sample data to draw conclusions about the populations from which we drew our samples? The techniques we use in drawing conclusions are part of **inferential** statistics. Inferential statistics uses **probability** as one of its tools. To use this tool properly, we must first understand how it works. This chapter will introduce you to the language and basic tools of probability.

The variables we discussed in Chapter 7 can now be redefined as **random variables**, whose values depend on the chance selection of the elements in the sample. Using probability as a tool, later in Chapter 12, you will be able to create **probability distributions** that serve as models for random variables. You can then describe these using a mean and a standard deviation.

8.1 Randomness

Probability is the study of randomness and uncertainty.

The reasoning in statistics rests on asking, 'How often would this method give a correct answer if I used it very many times?' When we produce data by random sampling or by experiments, the laws of probability enable us to answer the question 'What would happen if we did this many times?'

It is important to distinguish between 'random' and 'haphazard' (or chaos). At first glance they might seem to be the same because neither of their outcomes can be anticipated with certainty. However, random events have a long-term predictability, where haphazard events do not.

What does 'random' mean? In ordinary speech, we use 'random' to denote things that are unpredictable. Events that are **random** are not perfectly predictable, but they have long-term regularities that we can describe and quantify using probability. In contrast, **haphazard** events do not necessarily have long-term regularities.

Throwing an unbiased coin and observing the number of heads that appear gives an example of random behaviour. When you throw the coin, there are only two outcomes: heads or tails. Figure 8.1 shows the results of the first

50 throws in an experiment that involved throwing the coin 5000 times. Two trials are shown. The red graph shows the result of first trial, in which the first toss was a head, followed by a tail, making the proportion of heads after two throws 0.5. The next two throws were also tails, so the proportion of heads after three throws was 0.33, and after four throws it was 0.25.

The second trial, shown in blue, starts with a series of tails. The fifth throw was a head, which raised the proportion of heads to 0.2.

The proportion of heads is quite variable at first. However, in the long run, and as the number of throws increases, the proportion of heads stabilises at around 0.5. We say that 0.5 is the **probability** of a head.

It is important to realise that the proportion of heads in a small number of tosses can be far from the probability. Probability describes only what happens in the long run. How a fair coin lands when it is thrown is an example of a random event. One cannot predict perfectly whether the coin will land on heads or tails. However, in repeated throws, the percentage of times the coin lands on heads will tend to settle down to a limit of 50%. The outcome of an individual throw is not predictable, but the long-term average behaviour is predictable. Thus, it is reasonable to consider the outcome of tossing a fair coin to be random.

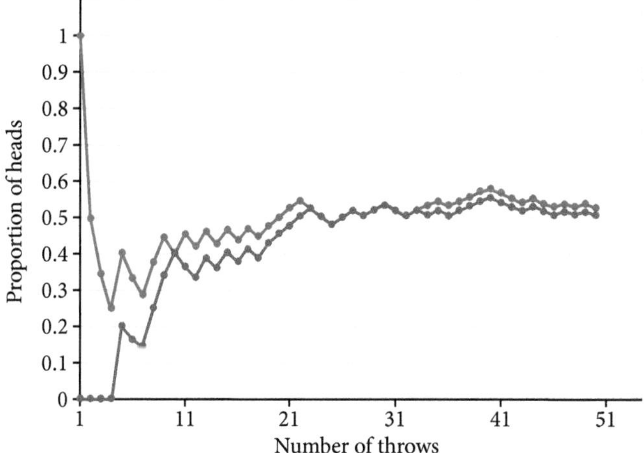

Figure 8.1 Probability of throwing a head

Imagine the following scenario. I drive to school every day. Shortly before school, there is a traffic light. It always seems to be red when I get there. I collected data over the course of one year (180 school days) and considered the green light to be a 'success'. Table 8.1 shows some of the collected data.

The first day it was red, so the proportion of success is 0% (0 out of 1). The second day it was green, so the proportion is now 50% (1 out of 2). The third day it was red again, so the proportion is 33.3% (1 out of 3), and so on. As more data is collected, the new measurement becomes a smaller and smaller fraction of the accumulated frequency, so, in the long run, the graph settles to the real chance of finding a green light, which in this case is about 30%. This is shown in Figure 8.2.

Day	Light	% green
1	red	0
2	green	50
3	red	33.3
4	green	50
5	red	40
6	red	33.3
7	red	28.6
⋮	⋮	⋮

Table 8.1 Traffic light data

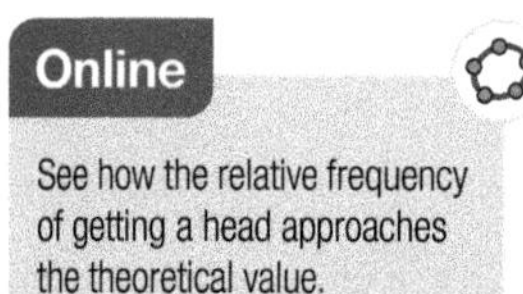

See how the relative frequency of getting a head approaches the theoretical value.

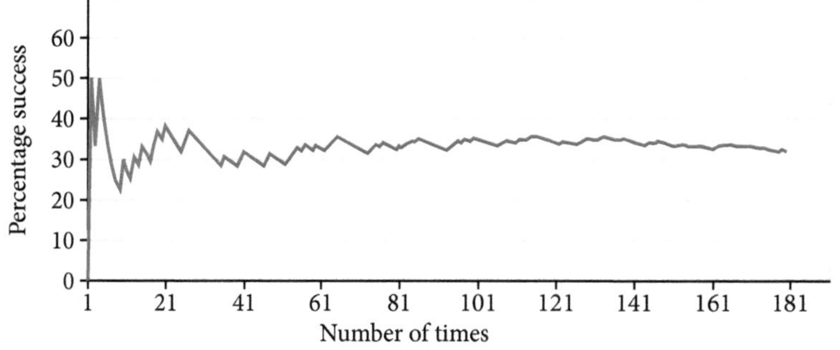

Figure 8.2 The chance of finding the traffic light green is about 30% over time

If you run a simulation for a longer period, as shown in Figure 8.3, you can see that it really stabilises around 30%.

Note that the randomness in the experiment is not in the traffic light itself, as this is controlled by a timer. In fact, if the system works well, then it may turn green at the same time every day. The randomness of the event is the time I arrive at the traffic light.

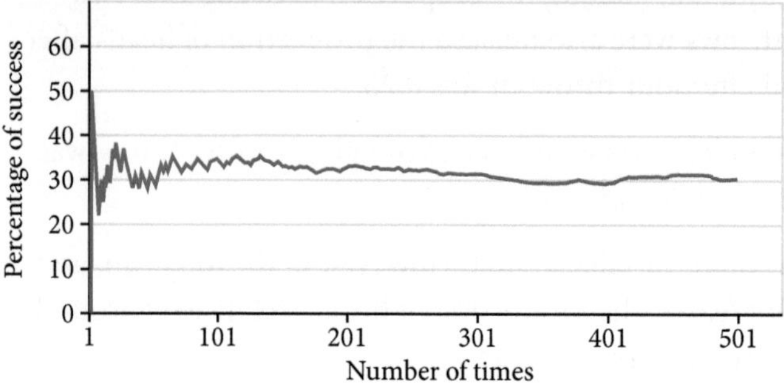

Figure 8.3 Further stabilisation at 30% over time

If we ask for the probability of finding the traffic light green in the above example, our answer will be about 30%. We base our answer on knowing that, in the long run, the fraction of time that the traffic light was green is 30%. We could also say that the **long-run relative frequency** of a green light settles down to about 30%.

Basic definitions

An **experiment** is the process by which an observation (or measurement) is obtained. A **random experiment** (or **chance experiment**) is an experiment where there is uncertainty concerning which of two or more possible outcomes will result.

Data are obtained by observing either uncontrolled events in nature or controlled situations in a laboratory. We use the term **experiment** to describe either method of data collection.

Throwing a coin, rolling a dice and observing the number on the top surface, counting cars at a traffic light when it turns green, measuring daily rainfall in a certain area, and so on, are a few experiments in this sense of the word.

A description of a random phenomenon in the language of mathematics is called a **probability model**. For example, when we throw a coin, we cannot know the outcome in advance. What do we know? We can say that the outcome will be either heads or tails. Because the coin appears to be balanced, we believe that each of these outcomes has probability 0.50. This description of coin throwing has two parts:

- a list of possible outcomes
- a probability for each outcome.

This two-part description is the starting point for a probability model. We will begin by describing the outcomes of a random phenomenon and learning how to assign probabilities to the outcomes by using one of the definitions of probability.

The **sample space**, U, of a random experiment (or phenomenon) is the set of all possible outcomes.

For example, for one toss of a coin, the sample space is

$$U = \{\text{heads, tails}\}, \text{ or simply } \{\text{h, t}\}$$

The notation for sample space could also be **S** or any other letter.

Example 8.1

Throw a coin twice (or two coins once each) and record the results. What is the sample space?

Solution

$U = \{hh, ht, th, tt\}$

Use 'h' to represent throwing a head, and 't' to represent throwing a tail.

Example 8.2

Throw a coin twice (or two coins once) and count the number of heads thrown. What is the sample space?

Solution

$U = \{0, 1, 2\}$

There could be zero heads, one head, or two heads.

A **simple event** is the outcome we observe in a single repetition (trial) of the experiment.

For example, an experiment is throwing a dice and observing the number that appears on the top face. The simple events in this experiment are {1}, {2}, {3}, {4}, {5}, and {6}. The set of all these simple events is the sample space of the experiment.

We are now ready to define an **event**. There are several ways of looking at it, which in essence are all the same.

Probability and set theory
Set theory provides a foundation for all of mathematics. The language of probability is much the same as the language of set theory. Logical statements can be interpreted as statements about sets. Later, this will enable us to introduce a method for setting up probability problems.

An **event** is an **outcome** or a **set of outcomes** of a random experiment.

With this understanding, we can also look at the event as a **subset of the sample space** or as a **collection of simple events.**

Example 8.3

When rolling a standard six-sided dice, the event A is 'observe an odd number', and the event B is 'observe a number less than 5'. Write a set to represent each of these events.

Solution

Event A is the set $\{1, 3, 5\}$.

Event B is the set $\{1, 2, 3, 4\}$.

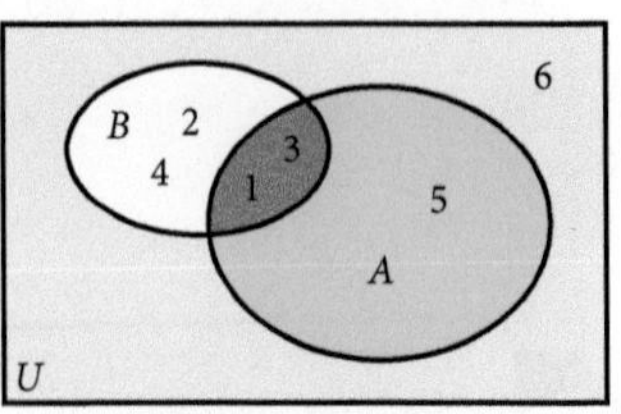

Figure 8.4 Events A and B from Example 8.3

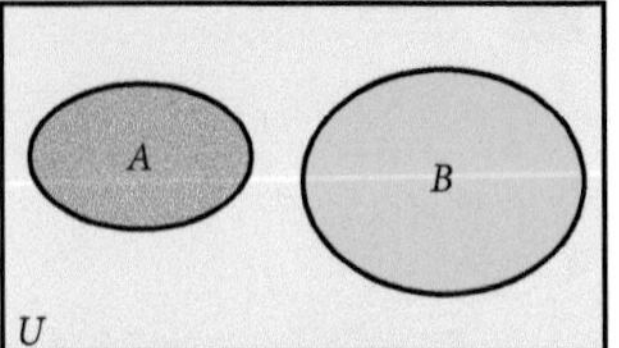

Figure 8.5 The rectangle represents the sample space

Sometimes it helps to visualise an experiment using some tools of set theory. There are several similarities between the ideas of set theory and probability and it is very helpful when we see the connection. A simple, but powerful, diagram is the **Venn diagram.** Figure 8.4 shows the events A and B from Example 8.3.

In general, in this book, we will use a rectangle to represent the sample space and closed curves to represent events, as shown in Figure 8.5.

To understand these definitions more clearly, let's look at the following example.

Example 8.4

Suppose we choose one card at random from a deck of 52 playing cards. What is the sample space U?

Solution

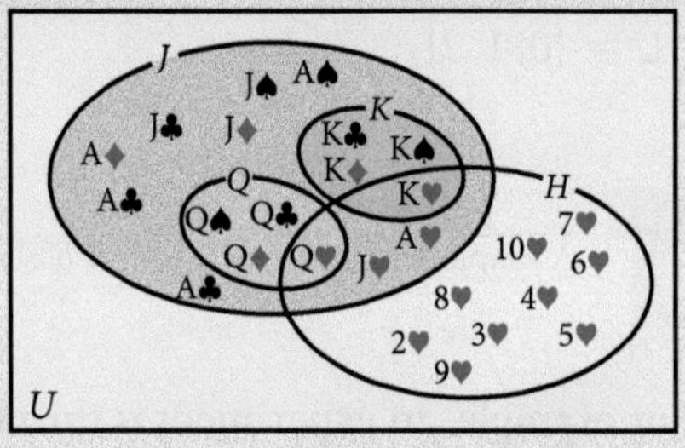

$U = \{$A♣, 2♣, ..., K♣, A♦, 2♦, ..., K♦, A♥, 2♥, ..., K♥, A♠, 2♠, ..., K♠$\}$

Some other events of interest are:

K = event of king = {K♣, K♦, K♥, K♠}

H = event of heart = {A♥, 2♥, ..., K♥}

J = event of jack or better

= {J♣, J♦, J♥, J♠, Q♣, Q♦, Q♥, Q♠, K♣, K♦, K♥, K♠, A♣, A♦, A♥, A♠}

Q = event of queen = {Q♣, Q♦, Q♥, Q♠}

Example 8.5

Throw a coin three times and record the results. Show the event 'observing two heads' as a Venn diagram.

Solution

The sample space is made up of eight possible outcomes: hhh, hht, tht, etc. Observing exactly two heads is an event with three elements: {hht, hth, thh}

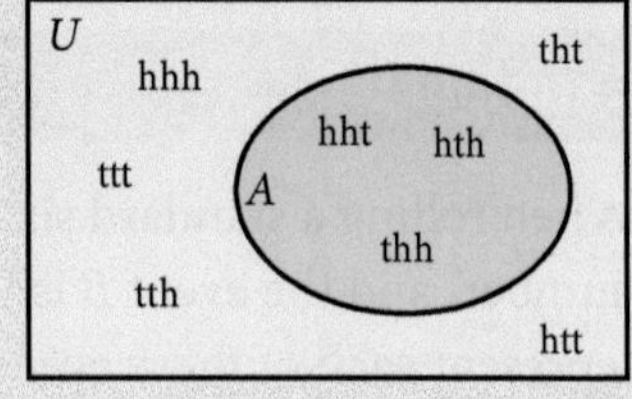

Set operations have a number of properties, which are basic consequences of the definitions. Some examples are:

$A \cup B = B \cup A$	$(A')' = A$	$A \cap U = A$
$A \cup U = U$	$A \cap A' = \varnothing$	$A \cup A' = U$

U is the sample space and $\varnothing$ is the empty set.
Two valuable properties are known as De Morgan's laws, which state that
$(A \cup B)' = A' \cap B'$
$(A \cap B)' = A' \cup B'$
Finally
$A \cap (B \cup C) = (A \cap B) \cup (A \cap C)$,
$A \cup (B \cap C) = (A \cup B) \cap (A \cup C)$

Tree diagrams, tables, and grids

In an experiment to check the blood types of patients, the experiment can be thought of as a two-stage experiment: first we identify the type of the blood and then we classify the Rh factor + or –.

The simple events in this experiment can be counted using another tool, the **tree diagram**, which is extremely powerful and helpful in solving probability problems.

Our sample space in this experiment is the set
$\{A+, A-, B+, B-, AB+, AB-, O+, O-\}$.
These can be read from the outcomes in the last column.

The same simple events can also be arranged in a **probability table**, as shown in Table 8.2

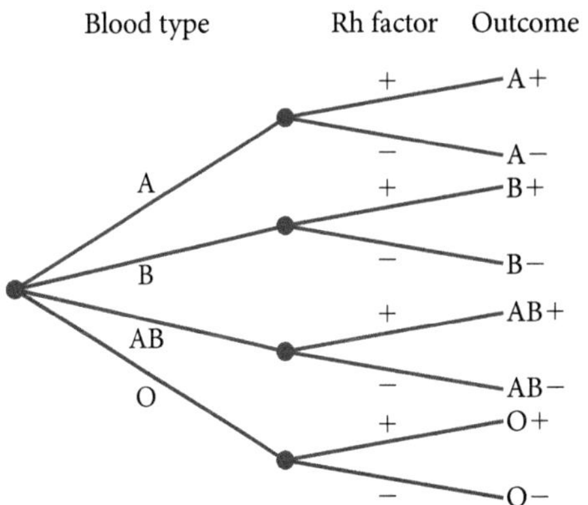

Figure 8.6 Blood types tree diagram

	Blood type			
Rh factor	A	B	AB	O
Positive	A+	B+	AB+	O+
Negative	A–	B–	AB–	O–

Table 8.2 Blood types probability table

They can also be shown using a 2-dimensional grid as in Figure 8.7.

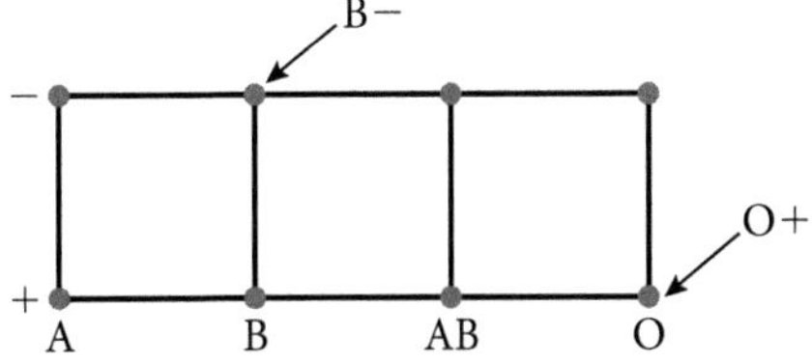

Figure 8.7 2D grid for the probability of simple events

Example 8.6

Two tetrahedral dice, with sides numbered 1 to 4, one blue and one yellow, are rolled. List the elements of the following events when both dice are thrown.

$T = \{$3 is face down on at least one dice$\}$

$B = \{$the 3 on the blue dice is face down$\}$

$S = \{$sum of the face-down faces on both dice is a six$\}$

Solution

$T = \{(1, 3), (2, 3), (3, 3), (4, 3), (3, 2), (3, 1), (3, 4)\}$

$B = \{(1, 3), (2, 3), (3, 3)\}$

$S = \{(2, 4), (3, 3), (4, 2)\}$

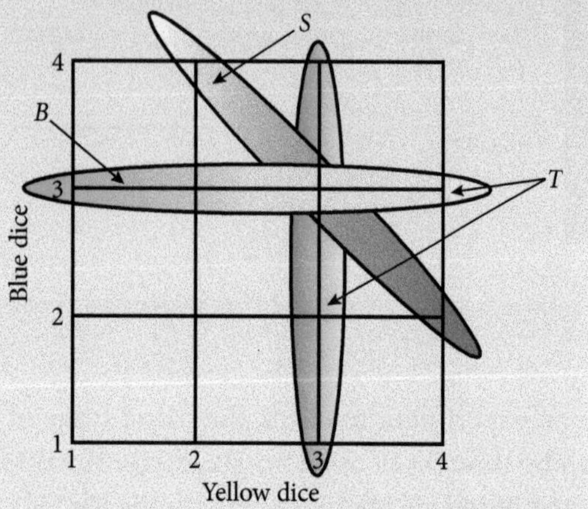

Exercise 8.1

1. In a large school, a student is selected at random. Give a reasonable sample space for answers to each of the following questions:
   (a) Are you left-handed or right-handed?
   (b) What is your height in centimetres?
   (c) For how many minutes did you study last night?
2. You throw a standard six-sided dice and a coin and record the number and the face that appear in that order. For example, (5, h) represents a 5 on the dice and a head on the coin. List the sample space.
3. You draw cards from a standard deck of 52 playing cards.
   (a) List the sample space if you draw one card at a time.
   (b) List the sample space if you draw two cards at a time.
   (c) How many outcomes do you have in each of the experiments above?
4. Tim carried out an experiment where he flipped 20 coins together and observed the number of heads showing. He repeated this experiment 10 times and got the following results:

   11, 9, 10, 8, 13, 9, 6, 7, 10, 11

   (a) Use Tim's data to calculate the probability of obtaining a head.
   (b) Tim flipped the 20 coins for the 11th time. How many heads should he expect to observe?
   (c) If he flipped the coins 1000 times, how many heads should he expect to observe?
5. In a game, a four-sided dice with sides marked 1, 2, 3, and 4 is used. The intelligence of the player is determined by rolling the dice twice and adding 1 to the sum of the two rolls.
   (a) What is the sample space for rolling the dice twice?
   (b) What is the sample space for the intelligence of a player?

6. A box contains a yellow ball, a green ball, and a blue ball. You run an experiment where you draw a ball at random, look at its colour and then replace it and draw a second ball at random.
   (a) What is the sample space of this experiment?
   (b) List the elements of the event of drawing yellow first.
   (c) List the elements of the event of drawing the same colour twice.

7. Repeat the same exercise as in question 6, except this time without replacing the first ball.

8. Nick flips a coin three times and each time he notes whether it is heads or tails.
   (a) What is the sample space of this experiment?
   (b) What is the event that heads occur more often than tails?

9. Franz lives in Vienna. He and his family decided that their next vacation will be to either Italy or Hungary. If they go to Italy, they can fly, drive, or take the train. If they go to Hungary, they will drive or take a boat. Letting the outcome of the experiment be the location of their vacation and their mode of travel, list all the points in the sample space. Also list the sample space of the event 'fly to destination.'

10. A hospital codes patients according to whether they have or do not have health insurance, and according to their condition. The condition of the patient is rated as good (g), fair (f), serious (s), or critical (c). The hospital clerk marks a 0 for a non-insured patient and a 1 for an insured patient, and one of the above letters for the patient's condition. For example, (1, c) means an insured patient with critical condition.
    (a) List the sample space of this experiment.
    (b) What is the event: not insured, in serious or critical condition?
    (c) What is the event: patient in good or fair condition?
    (d) What is the event: patient has insurance?

11. A social study investigates people for different characteristics. One part of the study classifies people according to gender (G_1 = female, G_2 = male), drinking habits (K_1 = abstain, K_2 = drinks occasionally, K_3 = drinks frequently), and marital status (M_1 = married, M_2 = single, M_3 = divorced, M_4 = widowed).
    (a) List the elements of an appropriate sample space for observing a person in this study.
    (b) Three events are defined as: A = the person is a male, B = the person drinks, and C = the person is single.
    List the elements of each event A, B, and C.
    (c) Interpret each event in the context of this situation:
    (i) $A \cup B$ **(ii)** $A \cap C$ **(iii)** C' **(iv)** $A \cap B \cap C$ **(v)** $A' \cap B$

12. Cars leaving the highway can make a right turn (R), left turn (L), or go straight on (S). You are collecting data on traffic patterns at this intersection and you group your observations by taking four cars at a time every 5 minutes.
    (a) List three outcomes in your sample space U. How many are there?
    (b) List the outcomes in the event that all cars go in the same direction.
    (c) List the outcomes that only two cars turn right.
    (d) List the outcomes that only two cars go in the same direction.

13. You are collecting data on traffic at an intersection for vehicles leaving a highway. Your task is to collect information about the type of vehicle: truck (T), bus (B), or car (C). You are also recording whether the driver is wearing a safety belt (SY) or is not wearing safety belt (SN), as well as whether the headlights are on (O) or off (F).
    (a) List the outcomes of your sample space, U.
    (b) List the outcomes of the event SY (the driver is wearing the safety belt).
    (c) List the outcomes of the event C (the vehicle is a car).
    (d) List the outcomes of the event in $C \cap SY$, C', and $C \cup SY$.

14. Many electric systems use a built-in backup system so that the equipment using the system will work even if some parts fail. Such a system in given in the diagram.

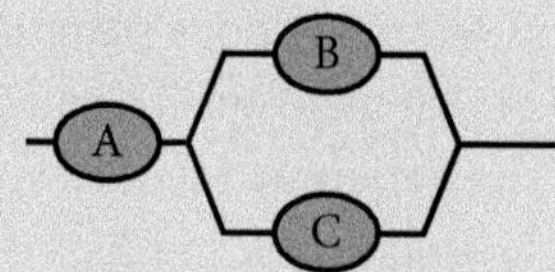

    Two parts of this system are installed in parallel, so that the system will work if at least one of them works. If we code a working part by 1 and a failing part by 0, then one of the outcomes would be (1, 0, 1), which means parts A and C work while B failed.
    (a) List the outcomes of your sample space, U.
    (b) List the outcomes of the event X, that exactly two of the parts work.

8.2 Probability assignments

Probability theories

There are a few theories of probability that assign meaning to statements such as 'the probability that A occurs is p%.' In this book we will primarily examine only the **relative frequency theory**. In essence, we follow the idea that probability is 'the long-run proportion of repetitions on which an event occurs.' This allows us to merge two concepts into one.

Equally likely outcomes

In the theory of equally likely outcomes, probability has to do with symmetries and the indistinguishability of outcomes. If a given experiment or trial has n possible outcomes among which there is no preference, they are equally likely. The probability of each outcome is then $\frac{100\%}{n}$ or $\frac{1}{n}$. For example, if a coin is balanced well, there is no reason for it to land heads in preference to tails when it is thrown, so the probability that the coin lands heads is equal to the probability that it lands tails, and both are $\frac{100\%}{2} = 50\%$. Similarly, if a dice is fair, the chance that when it is rolled it lands with the side showing 1 on top is the same as the chance that it shows 2, 3, 4, 5, or 6: $\frac{100\%}{6}$ or $\frac{1}{6}$. In the theory of equally likely outcomes, probabilities are between 0% and 100%. If an event consists of more than one possible outcome, the chance of the event is the number of ways it can occur, divided by the total number of things that could occur. For example, the chance that a dice lands showing an even number on top is the number of ways it could land showing an even number (2, 4, or 6), divided by the total number of things that could occur (6, namely showing 1, 2, 3, 4, 5, or 6).

In all theories, probability is on a scale of 0% to 100%. 'Probability' and 'chance' are synonymous.

Frequency theory

In frequency theory, probability is the limit of the relative frequency with which an event occurs in repeated trials. Relative frequencies are always between 0% and 100%. According to the frequency theory of probability, 'the probability that A occurs is p%' means that if you repeat the experiment over and over again, independently and under essentially identical conditions, the percentage of the time that A occurs will converge to p. For example, to say that the chance that a coin lands heads is 50% means that if you throw the coin over and over again, independently, the ratio of the number of times the coin lands heads to the total number of throws approaches a limiting value of 50% as the number of throws grows. Because the ratio of heads to throws is always between 0% and 100%, when the probability exists, it must be between 0% and 100%.

Using Venn diagrams and the equally likely concept, we can say that the probability of any event is the number of elements in an event A divided by the total number of elements in the sample space U. This is equivalent to saying: $\mathrm{P}(A) = \frac{n(A)}{n(U)}$, where $n(A)$ represents the number of outcomes in A and $n(U)$ represents the total number of outcomes. So in Example 8.5, the probability of observing exactly two heads is $\mathrm{P}(2 \text{ heads}) = \frac{3}{8}$.

Probability rules

Regardless of which theory we subscribe to, the probability rules apply.

Rule 1

No matter how little a chance you think an event has, there is no such thing as negative probability.

Any probability is a number between 0 and 1 – that is, the probability $P(A)$ of any event A satisfies $0 \leqslant P(A) \leqslant 1$. If the probability of any event is 0, the event *never* occurs. Likewise, if the probability is 1, it *always* occurs. In rolling a standard dice, it is impossible to get the number 9, so $P(9) = 0$. Also, the probability of observing any integer between 1 and 6 inclusive is 1.

Rule 2

No matter how large a chance you think an event has, there is no such thing as a probability larger than 1.

All possible outcomes together must have probability 1 – that is, the probability of the sample space U is 1, or $P(U) = 1$. Informally, this is sometimes called the 'something has to happen' rule.

Rule 3

If two events have no outcomes in common, the probability that one or the other occurs is the sum of their individual probabilities. Two events that have no outcomes in common, and hence can never occur together, are called **disjoint** events or **mutually exclusive** events.

The addition rule for mutually exclusive events is

$$P(A \text{ or } B) = P(A) + P(B)$$

For example, in throwing three coins, the events of getting exactly two heads or exactly two tails are disjoint, and hence the probability of getting exactly two heads or two tails is

$$\frac{3}{8} + \frac{3}{8} = \frac{6}{8} = \frac{3}{4}$$

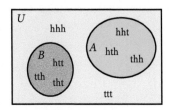

Figure 8.8 Events of two tails or two heads in throwing three dice

Additionally, we can always add the probabilities of **outcomes** because they are always disjoint. A trial cannot come out in two different ways at the same time. This will give you a way to check whether the probabilities you assigned are legitimate.

Rule 4

Suppose that the probability that you receive a 7 on your IB exam is 0.2. Then the probability of *not* receiving a 7 on the exam is 0.8. The event that contains the outcomes **not in** A is called the **complement** of A and is denoted by A'.

$$P(A') = 1 - P(A), \text{ or } P(A) = 1 - P(A')$$

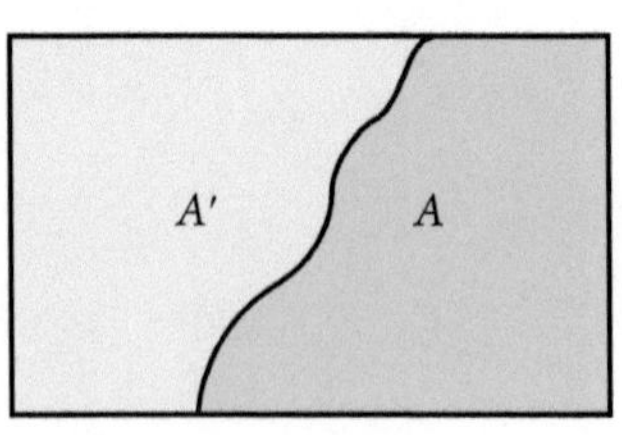

Figure 8.9 The complement of A is denoted by A'

The 'something has to happen' rule

You have to be careful with these rules. By the 'something has to happen' rule, the total of the probabilities of all possible outcomes must be 1. This is because they are disjoint, and their sum covers all the elements of the sample space. Suppose someone reports the following probabilities for students in your high school, which has four grades: 'The probability that a grade 1, 2, 3 or 4 student is chosen at random from the high school is 0.24, 0.24, 0.25, and 0.19 respectively.' You should know immediately that there is something wrong because these probabilities add up to 0.92.

Similarly, if someone claims that the probabilities are 0.24, 0.28, 0.25, 0.26 respectively, there is also something wrong. These probabilities add up to 1.03, which is more than 1.

Example 8.7

Data for traffic violations was collected in a certain country. A summary is given below.

Age group	18–20 years	21–30 years	31–40 years	Over 40 years
Probability	0.06	0.47	0.29	0.18

Estimate the probability that the offender is:

(a) in the youngest age group

(b) between 21 and 40

(c) 40 or younger.

Solution

Each probability is between 0 and 1, and the probabilities add up to 1. Therefore this is a legitimate assignment of probabilities.

(a) The probability that the offender is in the youngest group is 0.06.

(b) The probability that the driver is between 21 and 40 years is $0.47 + 0.29 = 0.76$.

(c) The probability that a driver is 40 or younger is $1 - 0.18 = 0.82$.

Example 8.8

When people create passcodes for their cell phones, the first digits follow distributions very similar to that shown in the table.

First digit	0	1	2	3	4	5	6	7	8	9
Probability	0.009	0.300	0.174	0.122	0.096	0.078	0.067	0.058	0.051	0.045

(a) Find the probabilities of the following three events:

$A = \{\text{first digit is 1}\}$

$B = \{\text{first digit is more than 5}\}$

$C = \{\text{first digit is an odd number}\}$

(b) Find the probability that the first digit is:

(i) 1 or greater than 5

(ii) not 1

(iii) an odd number or a number larger than 5.

Solution

(a) From the table:

$P(A) = 0.300$

$P(B) = P(6) + P(7) + P(8) + P(9)$

$= 0.067 + 0.058 + 0.051 + 0.045 = 0.221$

$P(C) = P(1) + P(3) + P(5) + P(7) + P(9)$

$= 0.300 + 0.122 + 0.078 + 0.058 + 0.045 = 0.603$

(b) (i) Since A and B are mutually exclusive, by the addition rule, the probability that the first digit is 1 or greater than 5 is

$\text{P}(A \text{ or } B) = 0.300 + 0.\ 221 = 0.521$

(ii) Using the complement rule, the probability that the first digit is not 1 is

$\text{P}(A') = 1 - \text{P}(A) = 1 - 0.300 = 0.700$

(iii) The probability that the first digit is an odd number or a number larger than 5 is

$$\begin{aligned}\text{P}(B \text{ or } C) &= \text{P}(1) + \text{P}(3) + \text{P}(5) + \text{P}(6) + \text{P}(7) + \text{P}(8) + \text{P}(9)\\ &= 0.300 + 0.122 + 0.078 + 0.067 + 0.058 + 0.051 + 0.045\\ &= 0.721\end{aligned}$$

Notice here that P(B or C) **is not** the sum of P(B) and P(C) because B and C are not disjoint.

Equally likely outcomes

In some cases, we are able to assume that individual outcomes are equally likely because of some balance in the experiment. Throwing a balanced coin renders heads or tails equally likely, each with a probability of 50%, and rolling a standard balanced dice gives the numbers from 1 to 6 as equally likely, each having a probability of $\frac{1}{6}$.

Suppose in Example 8.8, we consider all the digits to be equally likely to happen. Then the probabilities would be as shown in Table 8.3.

First digit	0	1	2	3	4	5	6	7	8	9
Probability	0.1	0.1	0.1	0.1	0.1	0.1	0.1	0.1	0.1	0.1

Table 8.3 All digits are equally likely to happen

$\text{P}(A) = 0.1$

$\text{P}(B) = \text{P}(6) + \text{P}(7) + \text{P}(8) + \text{P}(9) = 4 \times 0.1 = 0.4$

$\text{P}(C) = \text{P}(1) + \text{P}(3) + \text{P}(5) + \text{P}(7) + \text{P}(9) = 5 \times 0.1 = 0.5$

Also, by the complement rule, the probability that the first digit is not 1 is

$\text{P}(A') = 1 - \text{P}(A) = 1 - 0.1 = 0.9.$

Two-dimensional grids are also very helpful tools that are used to visualise two-stage or sequential probability models. For example, consider rolling a normal unbiased cubic dice twice. Figure 8.10 shows a two-dimensional grid and a number of possible events.

If we are interested in the probability that at least one roll shows a 6, we count the points on the column corresponding to 6 on the first roll and the points on the row corresponding to 6 on the second roll, observing that the point in the corner should not be counted twice.

If we are interested in the number showing on both rolls being the same, then we count the points on the diagonal as shown.

Finally, if we are interested in the probability that the first roll shows a number larger than the second roll, then we pick the points below the diagonal.

Hence

$$P(\text{first number} > \text{second number}) = \frac{15}{36}$$

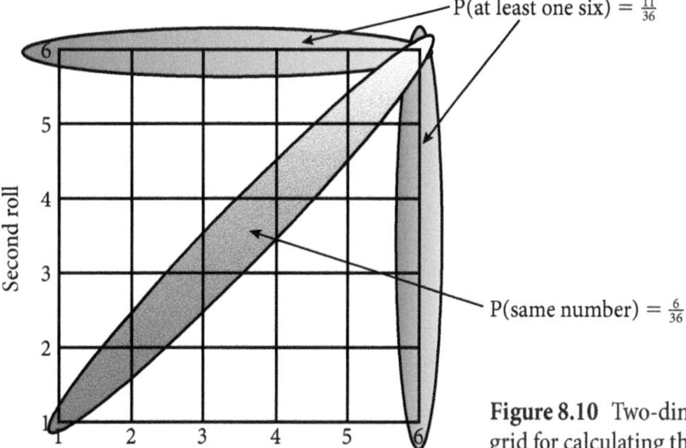

Figure 8.10 Two-dimensional grid for calculating the probability of events when rolling a dice

Geometric probability

Some cases give rise to interpreting events as areas in the plane. Take, for example, shooting at a circular target at random. What is the probability of hitting the central part?

The probability of hitting the central part is given by

$$P = \frac{\pi\left(\frac{R}{4}\right)^2}{\pi R^2} = \frac{1}{16}$$

This calculation comes from the area of the small circle over the area of the whole target.

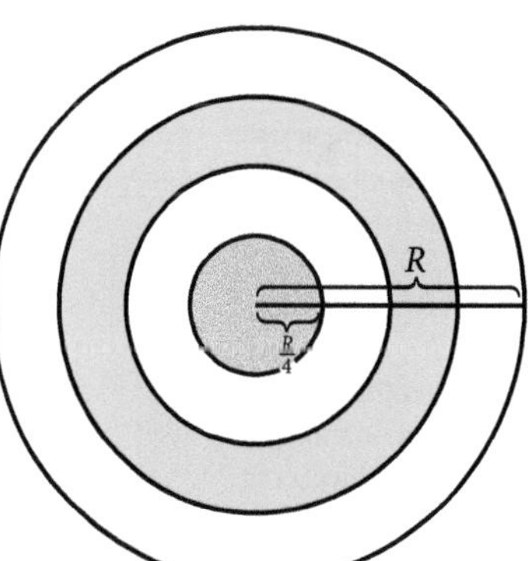

Figure 8.11 What is the probability of hitting the central part?

Example 8.9

Lydia and Rania agreed to meet at the museum between 12:00 and 13:00. The first person to arrive will wait 15 minutes. If the second person does not show up, the first person will leave. Assuming that their arrival times are random, what is the probability that they meet?

Solution

If Lydia arrives x minutes after 12:00 and Rania arrives y minutes after 12:00, then the conditions for them to meet are $|x - y| \leqslant 15$ and $x \leqslant 60, y \leqslant 60$.

Geometrically, the shaded region in the diagram on the right shows the arrival times that will allow them to meet.

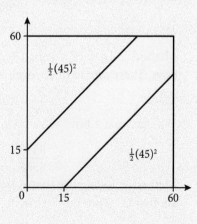

The area for each triangle is $\frac{1}{2}bh = \frac{1}{2}(45)^2$, so, the shaded area is $60^2 - 45^2$.

The probability they meet is therefore $\frac{60^2 - 45^2}{60^2} = \frac{7}{16}$.

Probability calculation for equally likely outcomes using counting principles (Optional – recommended!)

Counting principles are not part of the SL syllabus, so for the purposes of this section we mention only the tools you need. For a more in-depth discussion of counting principles, look at Chapter 3 of the HL version of this book.

In an experiment where all outcomes are equally likely, the theoretical probability of an event A is given by

$$\mathrm{P}(A) = \frac{n(A)}{n(U)}$$

where $n(A)$ is the number of outcomes that make up the event A, and $n(U)$ is the total number of outcomes in the sample space.

The new ideas we want to discuss here involve the calculation of $n(A)$ and $n(U)$. Such calculations will involve counting principles.

A **combination** of r objects out of n objects is a subset of the set of n objects. For example, consider the letters ABCDE. There are 10 combinations of 3 letters chosen from these 5: ABC, ABD, ABE, ACD, ACE, ADE, BCD, BCE, BDE, CDE. The general rule is that the number of combinations or subsets of r objects out of n objects is given by the binomial coefficient ${}^nC_r = \frac{n!}{(n-r)!r!}$. Thus the number of ways of choosing 3 letters from ABCDE is ${}^5C_3 = \frac{5!}{2!3!} = 10.$ Your GDC can do the work for you too!

Example 8.10

In a group of 18 students, 8 are females. If five students are chosen at random, what is the probability of choosing:

(a) all girls (b) 3 girls and 2 boys (c) at least 1 boy?

Solution

The total number of outcomes is the number of ways we can choose 5 out of the 18 students.

$$n(U) = {}^{18}C_5 = 8568$$

(a) This event requires that we pick our group from among the 8 girls.

$$n(A) = {}^8C_5 = 56 \Rightarrow \mathrm{P(all\ girls)} = \frac{56}{8568} = 0.0065$$

(b) This event requires that we pick 3 out of the 8 girls, and at the same time we pick 2 out of the 10 boys. So, using the multiplication principle,

$$n(\text{3 girls and 2 boys}) = {}^8C_3 \cdot {}^{10}C_2 = 56 \cdot 45 = 2520$$

$$\Rightarrow \mathrm{P}(\text{3 girls and 2 boys}) = \frac{2520}{8568} = 0.294$$

(c) This event can be approached in two ways.

Method 1: To have at least 1 boy means that we can have, 1, 2, 3, 4, or 5 boys. These are mutually exclusive, so the probability in question is the sum

$$\mathrm{P}(\text{at least 1 boy}) = \frac{{}^{10}C_1 \cdot {}^8C_4 + {}^{10}C_2 \cdot {}^8C_3 + \ldots + {}^{10}C_5 \cdot {}^8C_0}{{}^{18}C_5}$$

$$= \frac{8512}{8568} = 0.9935$$

Method 2: Recognise that at least 1 boy is the complement of no boys at all; that is, 0 boys or 5 girls.

$$\mathrm{P}(\text{at least 1 boy}) = 1 - \mathrm{P}(\text{all girls}) = 1 - 0.0065 = 0.9935$$

Did you observe that ${}^8C_5 = {}^8C_3$? Can you think of a reason for this?

Exercise 8.2

1. In a simple experiment, 20 chips labelled with integers 1 to 20 inclusive were placed in a box and one chip was picked at random.
   (a) What is the probability that the number drawn is a multiple of 3?
   (b) What is the probability that the number drawn is not a multiple of 4?

2. The probability that an event A happens is 0.37
   (a) What is the probability that it does not happen?
   (b) What is the probability that it may or may not happen?

3. You are playing with an ordinary deck of 52 cards by drawing cards at random and looking at them.
   (a) Find the probability that the card you draw is:
   (i) the ace of hearts **(ii)** the ace of hearts or any spade
   (iii) an ace or any heart **(iv)** not a face card.
   (b) Now you draw the ten of diamonds and put it on the table and draw a second card. What is the probability that the second card
   (i) is the ace of hearts? **(ii)** is not a face card?
   (c) Now you draw the ten of diamonds and return it to the deck and draw a second card. What is the probability that the second card
   (i) is the ace of hearts? **(ii)** is not a face card?

4. On Monday morning, my class wanted to know how many hours students spent studying on Sunday night. They stopped schoolmates at random as they arrived and asked each: 'How many hours did you study last night?' Here are the answers of the sample they chose on Monday 15 January 2018.

Hours spent studying	0	1	2	3	4	5
Number of students	4	12	8	3	2	1

   (a) Estimate the probability that a student spent less than three hours studying on Sunday night.
   (b) Estimate the probability that a student studied for exactly two or three hours.
   (c) Estimate the probability that a student studied for less than six hours.

5. You throw a coin and a standard six-sided dice and record the number and the face that appear.
   (a) Find the probability of getting a number larger than 3.
   (b) Find the probability of getting a head and a 6.

6. A dice is constructed in such a way that a 1 has a chance of occurring twice as often as any other number.
   (a) Find the probability that a 5 appears.
   (b) Find the probability an odd number appears.

7. You are given two fair dice to roll in an experiment.
   (a) Your first task is to report the numbers you observe.
      (i) What is the sample space of your experiment?
      (ii) What is the probability that the two numbers are the same?
      (iii) What is the probability that the two numbers differ by 2?
      (iv) What is the probability that the two numbers are not the same?
   (b) Your second task is to report the sum of the numbers that appear.
      (i) What is the probability that the sum is 1?
      (ii) What is the probability that the sum is 9?
      (iii) What is the probability that the sum is 8?
      (iv) What is the probability that the sum is 13?

8. The blood types of people can be one of the four types: O, A, B, or AB. The distribution of people with these blood types differs from one group of people to another. Here are the distributions of blood types for randomly chosen people in the US, China, and Russia.

Country \ Blood Type	O	A	B	AB
US	0.43	0.41	0.12	?
China	0.36	0.27	0.26	0.11
Russia	0.39	0.34	?	0.09

   (a) What is the probability of type AB in the US?
   (b) Dirk lives in the US and has type B blood. People with type B blood can receive blood only from people with type O or type B. What is the probability that a randomly chosen US citizen can donate blood to Dirk?
   (c) What is the probability of randomly choosing a US citizen and a Chinese citizen with type O blood?
   (d) What is the probability of randomly choosing a US, a Chinese and a Russian citizen with type O blood?
   (e) What is the probability of randomly choosing a US, a Chinese and a Russian citizen with the same blood type?

9. In each of the following situations, state whether or not the given assignment of probabilities to individual outcomes is legitimate. Give reasons for your answer.
   (a) A dice is loaded such that the probability of each face is according to the following assignment, where x is the number on the upper face and $P(x)$ is its probability.

x	1	2	3	4	5	6
$P(x)$	0	$\frac{1}{6}$	$\frac{1}{3}$	$\frac{1}{3}$	$\frac{1}{6}$	0

(b) Students at a school are categorised in terms of gender and whether or not they are diploma candidates.
P(female, diploma candidate) = 0.57
P(female, not a diploma candidate) = 0.23
P(male, diploma candidate) = 0.43
P(male, not a diploma candidate) = 0.18

(c) Draw a card from a deck of 52 cards (x is the suit of the card and P(x) is its probability).

x	Hearts	Spades	Diamonds	Clubs
P(x)	$\frac{12}{52}$	$\frac{15}{52}$	$\frac{12}{52}$	$\frac{13}{52}$

10. In Switzerland, there are three official languages: German, French, and Italian. You choose a Swiss citizen at random and ask, 'What is your mother tongue?' Here is the distribution of responses:

Language	German	French	Italian	Other
Probability	0.58	0.24	0.12	?

(a) What is the probability that a Swiss citizen's mother tongue is not one of the official languages?

(b) What is the probability that a Swiss citizen's mother tongue is not German?

(c) What is the probability that you choose two Swiss citizens independently of each other and they both have German as their mother tongue?

(d) What is the probability that you choose two Swiss citizens independently of each other and they both have the same mother tongue?

11. Lots of the email messages we receive are spam. Choose a spam email message at random. Here is the distribution of topics.

Topic	Adult	Financial	Health	Leisure	Products	Scams
Probability	0.165	0.142	0.075	0.081	0.209	0.145

(a) What is the probability of choosing a spam message and it does not concern these topics?

(b) Parents are usually concerned about spam messages with adult content and scams. What is the probability that a randomly chosen spam email falls into one of the other categories?

12. An experiment involves rolling a pair of dice, 1 white and 1 red, and recording the numbers that come up. Find the probability:

(a) that the sum is greater than 8

(b) that a number greater than 4 appears on the white dice

(c) that at most a total of 5 appears.

13. (a) A box contains 8 chips numbered 1 to 8. Two are chosen at random and their numbers are added together. What is the probability that their sum is 7?
    (b) A box contains 20 chips numbered 1 to 20. Two are chosen at random. What is the probability that the numbers on the two chips differ by 3?
    (c) A box contains 20 chips numbered 1 to 20. Two are chosen at random. What is the probability that the numbers on the two chips differ by more than 3?
14. Tim and Val want to meet for dinner at their favourite restaurant. They agree on meeting between 20:00 and 21:00. The first person to arrive will order a salad and spend 30 minutes before ordering the main meal. If the second person does not arrive within the 30 minutes, the first person will pay the bill and leave. What is the probability they manage to eat dinner together at the restaurant?
15. Bus 48A and Tram 49 serve different routes in the city. They share one stop next to the library. They stop at this station every 20 minutes. Every stay is 3 minutes long. Assuming their arrivals at the hour are random, find the probability that both are at the stop together in any 20 minute interval.
16. A wooden cube has its faces painted green. We cut the cube into 1000 small cubes of equal size. We mix the small cubes thoroughly. We draw one cube at random. What is the probability that the cube:
    (a) has exactly two faces coloured green
    (b) has exactly three faces coloured green
    (c) has no faces coloured green?

8.3 Operations with events

The **intersection** of two events, B and C, denoted by the symbol $B \cap C$ or simply BC, is the event containing all outcomes common to B and C.

In Example 8.8, we talked about the events

$B = \{\text{first digit is more than 5}\}$

$C = \{\text{first digit is an odd number}\}$

We also claimed that these two events are not disjoint. This brings us to another concept for looking at combined events.

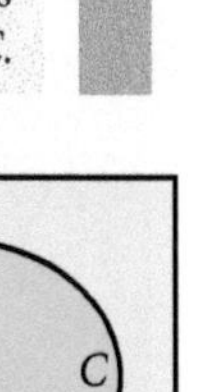

Figure 8.12 Event of B 'intersection' $C = \{7, 9\}$

Here $B \cap C = \{7, 9\}$ because these outcomes are in both B and C. Since the intersection has outcomes common to the two events, B and C, they are not mutually exclusive.

The probability of $B \cap C$ is $0.058 + 0.045 = 0.103$. Recall from Example 8.8 that we said the probability of B or C is not simply the sum of the two probabilities. That brings us to the next concept. How can we find the probability of B or C when they are not mutually exclusive? To answer this question, we need to define another operation.

Here $B \cup C = \{1, 3, 5, 6, 7, 8, 9\}$. In calculating the probability of $B \cup C$, we observe that the outcomes 7 and 9 are counted twice. To remedy the situation, if we decide to add the probabilities of B and C, we subtract one of the incidents of double counting. So, $P(B \cup C) = 0.221 + 0.603 - 0.103 = 0.721$, which is the result we received with direct calculation. In general, we can state the following probability rule.

The **union** of two events, B and C, denoted by the symbol $B \cup C$, is the event containing all the outcomes that belong to B or to C or to both.

Rule 5

For any two events A and B, $P(A \cup B) = P(A) + P(B) - P(A \cap B)$.

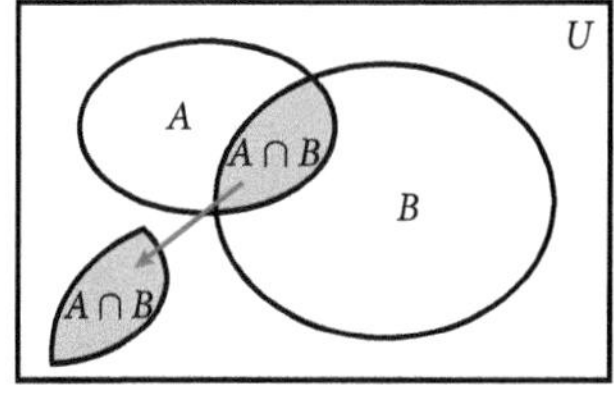

Figure 8.13 $P(A \cup B)$

As we see from Figure 8.13, $P(A \cap B)$ has been added twice, so the 'extra' one is subtracted to give the probability of $(A \cup B)$.

This general probability addition rule applies to the case of mutually exclusive events too. Consider any two mutually exclusive events A and B. The probability of A or B is given by

$$P(A \cup B) = P(A) + P(B) - P(A \cap B) = P(A) + P(B)$$

since $P(A \cap B) = 0$.

Some useful results

1. $P(A) = P(A \cap B) + P(A \cap B')$

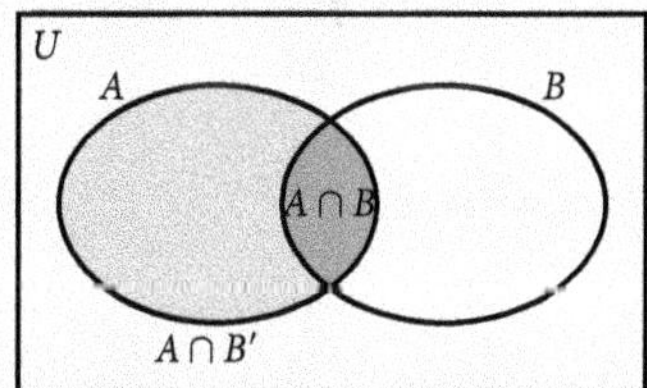

2. $P(B) = P(A \cap B) + P(A' \cap B)$

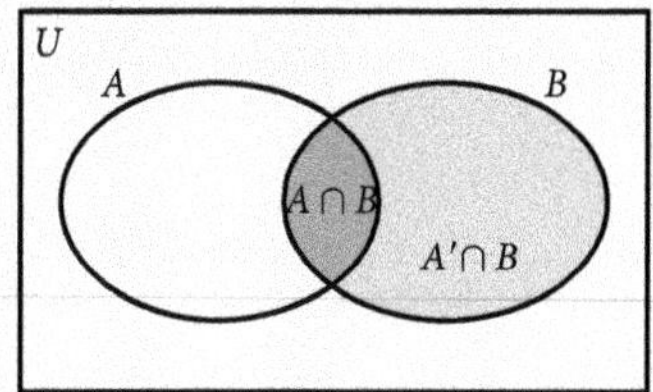

3. $P(A' \cap B') = 1 - P(A \cup B)$

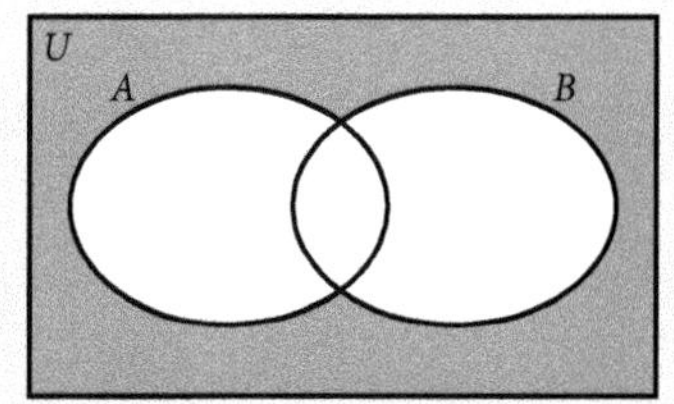

Rule 6 – The simple multiplication rule

Consider the following situation. In a large school, 55% of the students are male. It is also known that the percentage of cyclists among males and females in this school are the same, 22%. What is the probability of selecting a male cyclist when a student is selected at random from this population?

Applying common sense, we can think of the problem in the following manner. Since the proportion of cyclists is the same in both groups, cycling and gender are independent of each other in the sense that knowing that the student is a male does not influence the probability that he is a cyclist.

The chance of picking a male student is 55%. From that 55% of the population, we know that 22% are cyclists, so by simple arithmetic the chance that we select a male cyclist is $0.22 \times 0.55 = 12.1\%$.

This is an example of the multiplication rule for independent events.

Two events A and B are independent if knowing that one of them occurs does not change the probability that the other occurs.

If two events A and B are **independent**, then $P(A \cap B) = P(A) \times P(B)$.

Independent or disjoint? Do not confuse independent with disjoint. Disjoint means that if one of the events occurs, then the other does not occur. Independent means that knowing one of the events occurs does not influence whether the other occurs.

Example 8.11

At the start of this chapter, we found that the probability of encountering a green light on my drive to school is 30%. What is the probability that I encounter a green light on two consecutive days?

Solution

Assume that encountering a light green is a random event and that a green light on one day does not influence the event on the next day. In that case our calculation is very simple:

$$P(\text{green both days}) = P(\text{green first day}) \times P(\text{green second day})$$
$$= 0.30 \times 0.30 = 0.09$$

This rule can be extended to more than two independent events. For example, on the assumption of independence, what is the chance that I find the light green every day of the week?

$$P(\text{green every day}) = 0.3 \times 0.3 \times 0.3 \times 0.3 \times 0.3 = 0.002\,43$$

Example 8.12

Computers bought from a well-known producer require repairs quite frequently. It is estimated that in the first month after purchase, 17% of computers require one repair job, 7% will need two repairs, and 4% require three or more repairs.

(a) What is the probability that in the first month after purchase a computer chosen at random from this producer will need:
  (i) no repairs
  (ii) no more than one repair
  (iii) one or more repairs?

(b) If you buy two such computers, what is the probability that in the first month after purchase:
  (i) neither will require repair
  (ii) both will require repair?

Solution

(a) Since all of the events listed are disjoint, the addition rule can be used.
  (i) $P(\text{no repairs}) = 1 - P(\text{some repairs})$
  $= 1 - (0.17 + 0.07 + 0.04) = 1 - (0.28) = 0.72$
  (ii) $P(\text{no more than one repair}) = P(\text{no repairs or one repair})$
  $= 0.72 + 0.17 = 0.89$
  (iii) $P(\text{one or more repairs}) = P(\text{one or two or three or more repairs})$
  $= 0.17 + 0.07 + 0.04 = 0.28$

(b) Since repairs on the two computers are independent of one another, the multiplication rule can be used. Use the probabilities of events from part (a) in the calculations.
  (i) $P(\text{neither will need repair}) = 0.72 \times 0.72 = 0.5184$
  (ii) $P(\text{both will need repair}) = 0.28 \times 0.28 = 0.0784$

Conditional probability

In probability, conditioning means incorporating new restrictions on the outcome of an experiment, updating probabilities to take into account new information. This section describes conditioning, and how conditional probability can be used to solve complicated problems. Let us start with an example.

Example 8.13

A public health department wanted to study the exercise behaviour of high school students. They interviewed 768 students from grades 10 to 12 and asked them about their exercise habits. They categorised the students into the following three categories: regular exercise (three or more times per week), occasional exercise (one or two times per week), and no exercise. The results are summarised in the table.

	Regular exercise	**Occasional exercise**	**No exercise**	**Total**
Male	127	73	214	414
Female	99	66	189	354
Total	226	139	403	768

If we select a student at random from this study, what is the probability that we select:

(a) a female

(b) a male who exercises regularly

(c) a student who doesn't exercise?

Solution

(a) $\text{P(female)} = \dfrac{354}{768} = 0.461$

(b) Since we have 127 males categorised as taking regular exercise, the chance of a male who takes regular exercise will be

$$\text{P(male, regular)} = \frac{127}{768} = 0.165.$$

(c) $\text{P(no exercise)} = \dfrac{403}{768} = 0.525$

In Example 8.13, what if we know that the selected student is a female? Does that influence the probability that the selected student takes no exercise? Yes it does!

Knowing that the selected student is a female changes our choices. The 'revised' sample space is not made up of all students anymore, only of the female students. The chance of finding a student who doesn't exercise among the females is $\dfrac{189}{354} = 0.534$; that is, 53.4% of the females take no exercise as compared to the 52.5% of students who take no exercise in the whole population.

This probability is called a **conditional probability**. We write this as

$$\text{P(no exercise}|\text{female)} = \frac{189}{354}$$

We read this as 'probability of selecting a student who takes no exercise **given that** we have selected a female'.

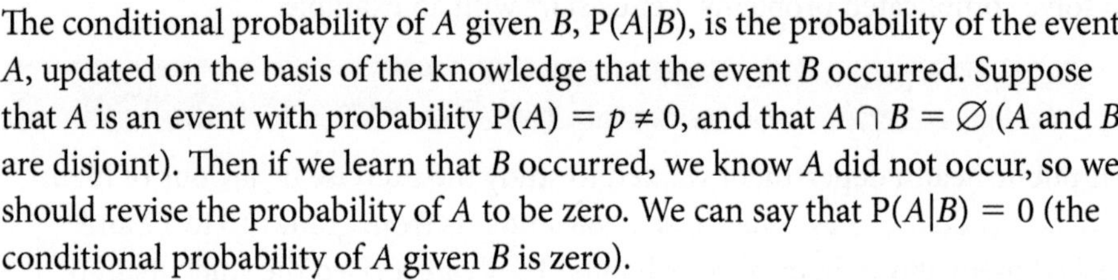

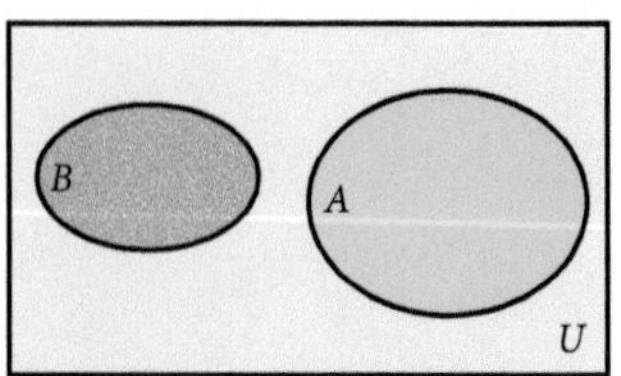

Figure 8.14 $A \cap B = \varnothing$

The conditional probability of A given B, $\text{P}(A|B)$, is the probability of the event A, updated on the basis of the knowledge that the event B occurred. Suppose that A is an event with probability $\text{P}(A) = p \neq 0$, and that $A \cap B = \varnothing$ (A and B are disjoint). Then if we learn that B occurred, we know A did not occur, so we should revise the probability of A to be zero. We can say that $\text{P}(A|B) = 0$ (the conditional probability of A given B is zero).

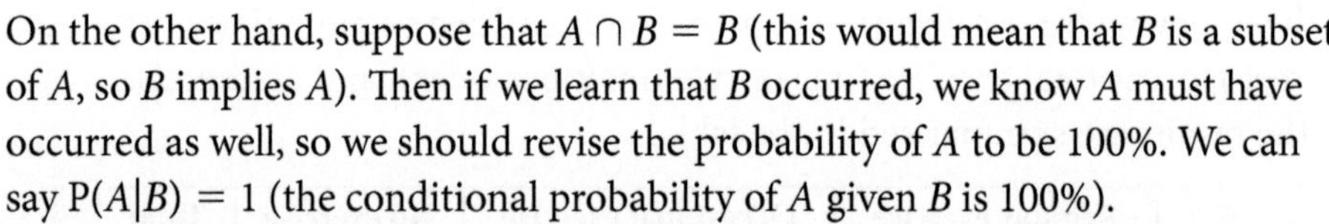

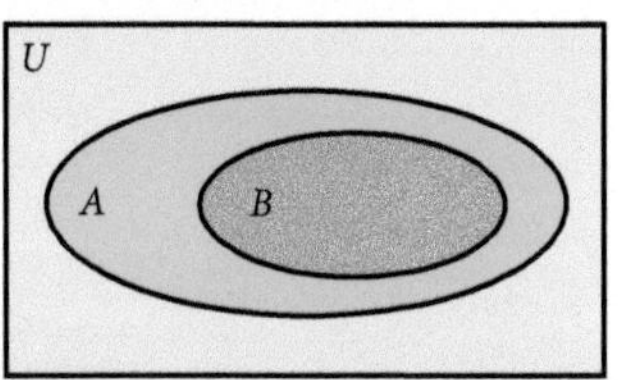

Figure 8.15 $A \cap B = B$

On the other hand, suppose that $A \cap B = B$ (this would mean that B is a subset of A, so B implies A). Then if we learn that B occurred, we know A must have occurred as well, so we should revise the probability of A to be 100%. We can say $\text{P}(A|B) = 1$ (the conditional probability of A given B is 100%).

Remember that the probability we assign to an event can change if we know that some other event has occurred. This idea is the key to understanding conditional probability.

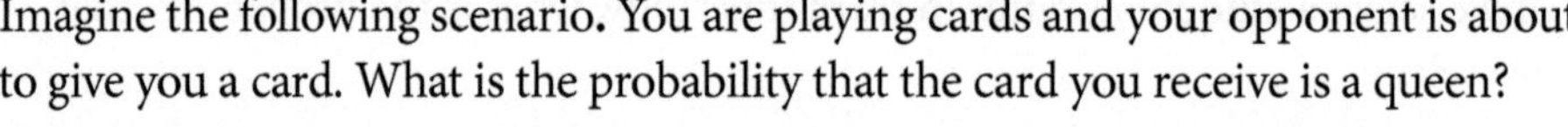

Imagine the following scenario. You are playing cards and your opponent is about to give you a card. What is the probability that the card you receive is a queen?

As you know, there are 52 cards in the deck, and 4 of these cards are queens. So, assuming that the deck was thoroughly shuffled, the probability of receiving a queen is

$$\text{P(queen)} = \frac{4}{52} = \frac{1}{13}$$

This calculation assumes that you know nothing about any cards already dealt from the deck.

Suppose now that you are looking at the five cards you have in your hand, and one of them is a queen. You know nothing about the other 47 cards except that exactly three queens are among them. The probability of being given a queen as the next card, given what you already know, is

$$\text{P(queen}|\text{1 queen in hand)} = \frac{3}{47} \neq \frac{1}{13}$$

So, knowing that there is one queen among your five cards changes the probability of the next card being a queen.

Consider Example 8.13 again. We want to express the table frequencies as relative frequencies or probabilities, as in Table 8.4.

	Regular exercise	Occasional exercise	No exercise
Male	0.165	0.095	0.279
Female	0.129	0.086	0.246

Table 8.4 Relative frequencies

To find the probability of selecting a student at random and finding that student is female and takes no exercise, we look at the intersection of the female row with the no exercise column and find that this probability is 0.246.

We can look at this calculation from a different perspective. We know that the percentage of females in our sample is 46.1, and among those females in Example 8.13, we found that 53.4% of them take no exercise. So, the percentage of females who take no exercise in the population is then 53.4% of those 46.1% females; that is, $0.534 \times 0.461 = 0.246$.

In terms of events, this can be read as

$$\text{P(no exercise}|\text{female)} \times \text{P(female)} = \text{P(female and no exercise) or P(female} \cap \text{no exercise)}$$

This is an example of the **multiplication rule** of any two events A and B.

Given any events A and B, the probability that both events happen is given by

$$P(A \cap B) = P(A|B) \times P(B)$$

Example 8.14

In a psychology lab, researchers are studying the colour preferences of young children. Six green toys and four red toys (identical apart from colour) are placed in a container. The child is asked to select two toys at random. What is the probability that the child chooses two red toys?

Solution

To solve this problem, we will use a tree diagram.

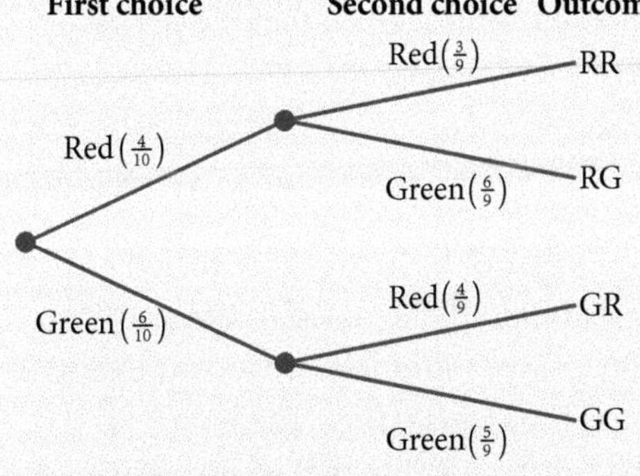

Every entry on each of the branches has a conditional probability. So, Red on the second choice is actually either Red|Red or Red|Green. We are interested in RR, so the probability is

$$P(RR) = P(R) \times P(R|R) = \frac{4}{10} \times \frac{3}{9} = 13.3\%$$

If $P(A \cap B) = P(A|B) \times P(B)$ as discussed above, and if $P(B) \neq 0$, we can rearrange the multiplication rule to produce a definition of the conditional probability $P(A|B)$ in terms of the 'unconditional' probabilities $P(A \cap B)$ and $P(B)$.

When $P(B) \neq 0$, the conditional probability of A given B is $P(A|B) = \frac{P(A \cap B)}{P(B)}$

Why does this formula make sense?

First of all, note that it agrees with the intuitive answers we found above: if $A \cap B = \varnothing$, then $P(A \cap B) = 0$, so $P(A|B) = \frac{0}{P(B)} = 0$. If $A \cap B = B$, then $P(A|B) = \frac{P(B)}{P(B)} = 100\%$.

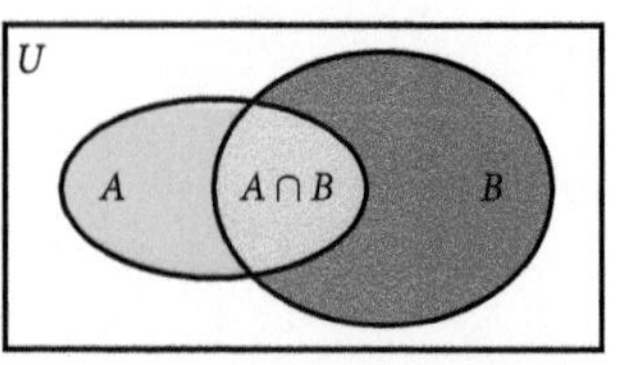

Figure 8.16 $P(A|B) = \frac{P(A \cap B)}{P(B)}$

Now, if we learn that B occurred, we can restrict attention to just those outcomes that are in B, and disregard the rest of U, so we have a new sample space that is just B (see Figure 8.16). For A to have occurred in addition to B requires that $A \cap B$ occurred, so the conditional probability of A given B is $\frac{P(A \cap B)}{P(B)}$, just as we defined it above.

Example 8.15

In an experiment to study the phenomenon of colour blindness, researchers collected information concerning 1000 people in a small town and categorised them according to colour blindness and gender. A summary of the findings is given in the table.

	Male	Female	Total
Colour blind	40	2	42
Not colour blind	470	488	958
Total	510	490	1000

What is the probability that a person is colour blind, given that the person is female?

Solution

To answer this question, we notice that we do not have to search the whole population for this event. We limit our search to the females. There are 490 females in the study. When we search for colour blindness, we look only for the females who are colour blind – that is, the intersection. There are two females who are colour blind. Therefore, the probability that a person is colour blind, given that the person is female, is

$$P(C|F) = \frac{P(C \cap F)}{P(F)} = \frac{n(C \cap F)}{n(F)} = \frac{2}{490} = 0.004$$

where C is the event of selecting a colour-blind person and F is the event of selecting a female.

Notice that in Example 8.15, we used the frequency rather than the probability. However, these are equivalent since dividing by $n(U)$ will transform the frequency into a probability.

$$\frac{n(C \cap F)}{n(F)} = \frac{\frac{n(C \cap F)}{n(U)}}{\frac{n(F)}{n(U)}} = \frac{P(C \cap F)}{P(F)} = P(C|F)$$

Example 8.16

A national airline is known for its punctuality. The probability that a regularly scheduled flight departs on time is $P(D) = 0.83$, the probability that it arrives on time is $P(A) = 0.92$, and the probability that it both arrives and departs on time is $P(A \cap D) = 0.78$. Find the probability that a flight:

(a) arrives on time given that it departed on time

(b) departed on time given that it arrived on time.

Solution

(a) The probability that a flight arrives on time given that it departed on time is

$$P(A|D) = \frac{P(A \cap D)}{P(D)} = \frac{0.78}{0.83} = 0.94$$

(b) The probability that a flight departed on time given that it arrived on time is

$$P(D|A) = \frac{P(D \cap A)}{P(A)} = \frac{0.78}{0.92} = 0.85$$

Independence

Two events are **independent** if learning that one occurred does not affect the chance that the other occurred. That is, if $P(A|B) = P(A)$, and vice versa.

If we apply this definition to the general multiplication rule, then

$$P(A \cap B) = P(A|B) \times P(B) = P(A) \times P(B)$$

which is the multiplication rule for independent events we studied earlier.

These results give us some helpful tools for checking the independence of events.

Two events are **independent** if and only if either $P(A \cap B) = P(A) \times P(B)$ or $P(A|B) = P(A)$. Otherwise, the events are **dependent**.

Example 8.17

Take another look at Example 8.16. Are the events of arriving on time (A) and departing on time (D) independent?

Solution

We can answer this question in two different ways.

Method 1: We are told that $P(A) = 0.92$, and we found that $P(A|D) = 0.94$. Since the two values are not the same, then we can say that the two events are not independent.

Method 2: $P(A \cap D) = 0.78$ and
$P(A) \times P(D) = 0.92 \times 0.83 = 0.76 \neq P(A \cap D)$.
Again we can say the events are not independent.

Example 8.18

If a doctor suspects that a patient is lactose intolerant, they might give the patient a breath test, or a blood test, or both in order to make a diagnosis. A study found the following results in a particular country: 81% of the patients suspected to be lactose intolerant were given a breath test, 40% a blood test, and 25% both tests.

(a) What is the probability that a patient suspected to be lactose intolerant is given:

(i) at least one test (ii) exactly one test (iii) no test?

(b) Are 'giving the patient a breath test' and 'giving the patient a blood test' independent?

Solution

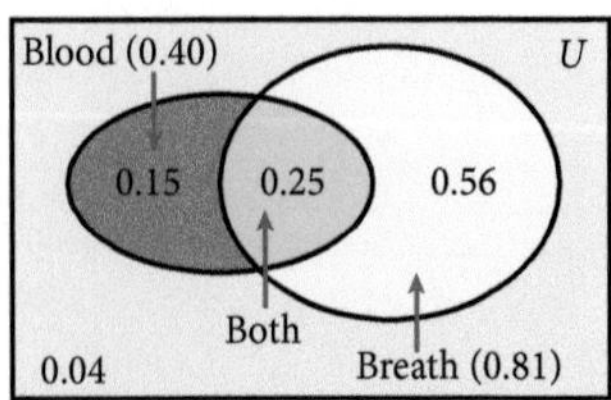

Figure 8.17 Venn diagram for the solution to Example 8.18

A Venn diagram can help explain the solution.

(a) (i) The probability that a patient receives a test means that they receive either a blood test, or a breath test, or both tests. This probability can be calculated directly from the diagram or by applying the addition rule.

The diagram shows that if 81% receive the breath test and 25% are also given the blood test, then the remaining 56% do not receive a blood test. Similarly 15% of the blood test receivers do not get a breath test. So, the probability of receiving a test is $0.56 + 0.25 + 0.15 = 0.96$.

Alternatively, if we apply the addition rule:

$$\text{P(at least one test)} = \text{P(breath)} + \text{P(blood)} - \text{P(both)}$$
$$= 0.81 + 0.40 - 0.25 = 0.96$$

(ii) To receive exactly one test is to receive a blood test or a breath test, but not both. From the Venn diagram it is clear that this probability is $0.15 + 0.56 = 0.71$.

Alternatively, since we know that the union of the two events still contains the intersection, we can subtract the probability of the intersection from that of the union. That is, $0.96 - 0.25 = 0.71$.

(iii) To receive no test is the complement of receiving at least one test. Hence, P(no test) = 1 − P(at least one test) = $1 - 0.96 = 0.04$.

(b) To check for independence, we can use either of the two methods we tried before.

Since all the necessary probabilities are given, we can use the product rule:

P(both tests) = P(breath) × P(blood) = $0.81 \times 0.40 = 0.324$, but P(both tests) = 0.25. Therefore, the events of receiving a breath and a blood test are not independent.

Example 8.19

Jane and Kate are long-time friends and frequently play tennis together. When Jane serves first, she wins 60% of the time. The same pattern is true for Kate. They alternate the serve. They usually play for a prize, which is a chocolate bar. The first one who loses on her serve has to buy the chocolate. Jane serves first.

(a) Find the probability that Jane pays on her second serve.

(b) Find the probability that Jane eventually pays for the chocolate.

(c) Find the probability that Kate pays for the chocolate.

Solution

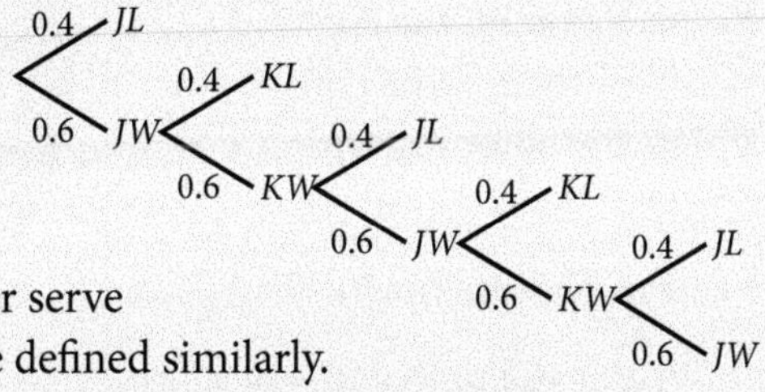

A tree diagram can help in understanding this problem. Let JW stand for Jane winning her serve and JL for Jane losing her serve and hence paying. KW and KL are defined similarly.

(a) For Jane to pay on her second serve, she should win her first serve. Kate must also win her first serve, and then Jane loses her second serve. See diagram above. The probability this happens is

$$P(JW) \cdot P(KW) \cdot P(JL) = 0.6 \times 0.6 \times 0.4 = 0.4 \cdot (0.6)^2 = 0.144$$

(b) For Jane to pay, she needs to be the first one to lose on her serve. This means, she loses on her first serve or the second or the third, and so on. So, the probability that she pays is

$$\begin{aligned} P(\text{Jane pays}) &= P(JL) + P(JW) \cdot P(KW) \cdot P(JL) + \\ &\quad P(JW) \cdot P(KW) \cdot P(JW) \cdot P(KW) \cdot P(JL) + \cdots \\ &= 0.4 + 0.4 \cdot (0.6)^2 + 0.4 \cdot (0.6)^4 + \cdots \end{aligned}$$

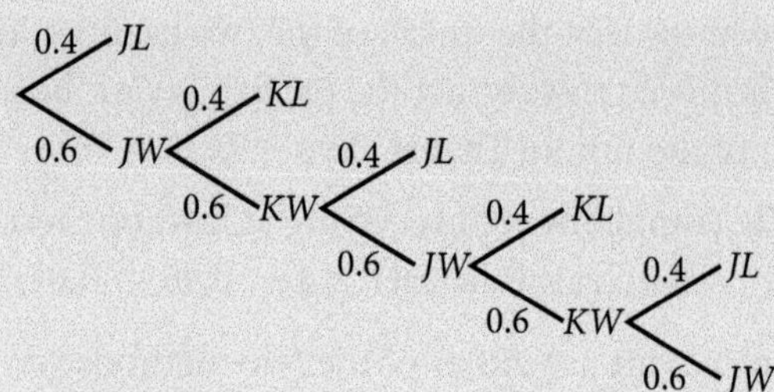

This appears to be the sum of an infinite geometric series with $(0.6)^2$ as common ratio, hence

$$\text{P(Jane pays)} = 0.4 + 0.4 \cdot (0.6)^2 + 0.4 \cdot (0.6)^4 + \cdots = \frac{0.4}{1 - (0.6)^2} = 0.625$$

(c) P(Kate pays) = 1 − P(Jane pays) = 1 − 0.625 = 0.375

Alternatively, for Kate to pay she needs to lose on her first serve, (0.6×0.4), or on her second, third, etc.,

$$\text{P(Kate pays)} = 0.6 \times 0.4 + 0.6 \times (0.6)^2 \times 0.4 + 0.6 \times (0.6)^4 \times 0.4 + \cdots$$

$$= \frac{0.6 \times 0.4}{1 - (0.6)^2} = 0.375$$

Example 8.20

A target for a dart game is shown here. The radius of the board is 40 cm and the board is divided into three regions as shown. You score 2 points if you hit the centre, 1 point for the middle region, and 0 points the outer region.

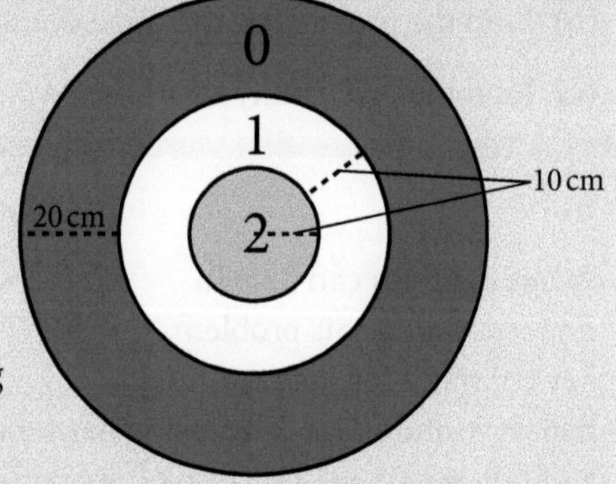

(a) What is the probability of scoring a 1 in one attempt?

(b) What is the probability of scoring a 2 in one attempt?

(c) How many attempts are necessary so that the probability of scoring at least one 2 is at least 50%?

Solution

(a) $\text{P}(1) = \dfrac{\pi(20^2 - 10^2)}{\pi(40^2)} = \dfrac{3}{16}$

(b) $\text{P}(2) = \dfrac{\pi(10^2)}{\pi(40^2)} = \dfrac{1}{16}$

(c) Let the number of attempts be n.

P(at least one 2) $= 1 - \text{P(no 2 in } n \text{ attempts)} = 1 - \left(\frac{15}{16}\right)^n$

For this probability to be at least 50%, then

$$1 - \left(\frac{15}{16}\right)^n \geqslant 0.5 \Leftrightarrow \left(\frac{15}{16}\right)^n \leqslant 0.5$$

$$\Rightarrow n \ln\left(\frac{15}{16}\right) \leqslant \ln(0.5)$$

$$\Rightarrow n \geqslant \frac{\ln(0.5)}{\ln\left(\frac{15}{16}\right)} \quad \left\{\text{since } \ln\left(\frac{15}{16}\right) < 0\right\}$$

$$\Rightarrow n \geqslant 10.74$$

So 11 attempts are required.

Exercise 8.3

1. Events A and B are given such that $P(A) = \frac{3}{4}$, $P(A \cup B) = \frac{4}{5}$ and $P(A \cap B) = \frac{3}{10}$. Find $P(B)$.

2. Events A and B are given such that $P(A) = \frac{7}{10}$, $P(A \cup B) = \frac{9}{10}$ and $P(A \cap B) = \frac{3}{10}$. Find:

(a) $P(B)$ **(b)** $P(B' \cap A)$ **(c)** $P(B \cap A')$

(d) $P(B' \cap A')$ **(e)** $P(B|A')$

3. Events A and B are given such that $P(A) = \frac{1}{3}$, $P(A \cup B) = \frac{4}{9}$ and $P(B) = \frac{2}{9}$. Show that A and B are neither independent nor mutually exclusive.

4. Events A and B are given such that $P(A) = \frac{3}{7}$, and $P(A \cap B) = \frac{3}{10}$. If A and B are independent, find $P(A \cup B)$.

5. Driving tests in a certain city are not easy to pass the first time. After going through training, the percentage of new drivers passing the test the first time is 60%. If a driver fails the first test, the chance of passing it on a second try, two weeks later, is 75%. Otherwise, the driver has to retrain and take the test after 6 months. Find the probability that a randomly chosen new driver will pass the test without having to retrain.

6. People with O^- blood type are universal donors. In other words, they can donate blood to individuals with any blood type. Only 8% of people have O^- blood.

(a) One person randomly arrives to give blood. What is the probability that the person does not have O^- blood?

(b) Two people arrive independently to give blood. Find the probability that:

(i) both have O^- blood

(ii) at least one of them has O^- blood

(iii) only one of them has O^- blood.

(c) Eight people arrive randomly to give blood. What is the probability that at least one of them has O^- blood?

7. PINs for cell phones usually consist of four digits that are not necessarily different.

(a) How many possible PINs are there?

(b) What is the probability that a PIN chosen at random does not start with a zero?

(c) What is the probability that a PIN contains at least one zero?

(d) Given a PIN with at least one zero, what is the probability that it starts with a zero?

8. An urn contains six red balls and two blue balls. We make two draws and each time we put the ball back after marking its colour.

(a) What is the probability that at least one of the balls is red?

(b) Given that at least one is red, what is the probability that the second one is red?

(c) Given that at least one is red, what is the probability that the second one is blue?

9. Two dice are rolled and the numbers on the top faces are observed.

(a) List the elements of the sample space.

(b) Let x represent the sum of the numbers observed. Copy and complete the following table.

x	2	3	4	5	6	7	8	9	10	11	12
P(x)		$\frac{1}{18}$									

(c) (i) What is the probability that at least one dice shows a 6?

(ii) What is the probability that the sum is at most 10?

(iii) What is the probability that a dice shows 4 or the sum is 10?

(iv) Given that the sum is 10, what is the probability that one of the dice is a 4?

10. A school has the following numbers categorised by class and gender:

	Grade 9	Grade 10	Grade 11	Grade 12	Total
Male	180	170	230	220	800
Female	200	130	190	180	700

(a) What is the probability that a student chosen at random will be a female?

(b) What is the probability that a student chosen at random is a male in grade 12?

(c) What is the probability that a female student chosen at random is a grade 12 student?

(d) What is the probability that a student chosen at random is a grade 12 or female student?

(e) What is the probability that a grade 12 student chosen at random is a male?

(f) Are gender and grade independent of each other? Explain.

11. Some young people do not like to wear glasses. A survey considered a large number of teenaged students as to whether they needed glasses to correct their vision and whether they used the glasses when they needed to. Here are the results.

		Used glasses when needed	
		Yes	**No**
Need glasses for correct vision	**Yes**	0.41	0.15
	No	0.04	0.40

(a) Find the probability that a randomly chosen young person from this group

(i) is judged to need glasses

(ii) needs to use glasses but does not use them.

(b) From those who are judged to need glasses, what is the probability that the student does not use them?

(c) Are the events of using and needing glasses independent?

12. Copy this table and fill in the missing entries.

$P(A)$	$P(B)$	Conditions for events A and B	$P(A \cap B)$	$P(A \cup B)$	$P(A \mid B)$
0.3	0.4	Mutually exclusive			
0.3	0.4	Independent			
0.1	0.5			0.6	
0.2	0.5		0.1		

13. In a large graduating class, there are 100 students taking the IB examination. 40 students are doing Economics SL, 30 students are doing Physics SL, and 12 are doing both.

(a) A student is chosen at random. Find the probability that this student is doing Physics, given that they are doing Economics SL.

(b) Are doing Physics and Economics SL independent events?

14. A market chain in Germany accepts only Mastercard and Visa. It estimates that 21% of its customers use Mastercard, 57% use Visa, and 13% use both cards.

(a) What is the probability that a customer will have an acceptable credit card?

(b) What proportion of their customers has neither card?

(c) What proportion of their customers has exactly one acceptable card?

15. 132 of 300 patients at a hospital are signed up for a special exercise programme that consists of a swimming class and an aerobics class. Each of these 132 patients takes at least one of the two classes. There are 78 patients in the swimming class and 84 in the aerobics class. Find the probability that a randomly chosen patient at this hospital is:

(a) not in the exercise programme

(b) enrolled in both classes.

16. An ordinary unbiased six-sided dice is rolled three times. Find the probability of rolling

(a) three twos

(b) at least one two

(c) exactly one two.

17. An athlete is shooting arrows at a target. She has a record of hitting the centre 30% of the time. Find the probability that she hits the centre:

(a) with her second shot but not with her first

(b) exactly once with her first three shots

(c) at least once with her first three shots.

18. Two unbiased dodecahedral (12 faces) dice, with faces numbered 1 to 12, are thrown. The scores are the numbers on the top side. Find the probability that:

(a) at least one 12 shows

(b) the total score is exactly 12

(c) there is a total score of at least 20

(d) a total score of at least 20 is achieved, given that a 12 shows on at least one dice.

19. The two dodecahedral dice in question 18 are thrown again. Two events are defined:

$A = \{\text{at least one of the numbers is a 10}\}$

$B = \{\text{the sum of the numbers is at most 15}\}$

Describe each event, list its elements, and find the probability.

(a) $A \cap B$ **(b)** $A \cup B$ **(c)** $A \cap B'$ **(d)** $A \cup B'$

(e) $A' \cup B'$ **(f)** $A' \cap B'$ **(g)** $(A \cap B') \cup (A' \cap B)$

20. Three fair six-sided dice are rolled.

  (a) Find the probability that a triple is rolled.

  (b) Given that the roll is a sum of 8 or less, find the probability that a triple is rolled.

  (c) Find the probability that at least one six appears.

  (d) Given that the dice all have different numbers, find the probability that at least one six appears.

21. You are given four coins: one has two heads, one has two tails, and the other two are normal. You choose a coin at random and flip it. The result is tails. What is the probability that the opposite face is heads?

22. George and Kassanthra play a game in which they roll two unbiased six-sided dice. The first one who rolls a sum of 6 wins. Kassanthra rolls the dice first.

  (a) What is the probability that Kassanthra wins on her second roll?

  (b) What is the probability that George wins on his second roll?

  (c) What is the probability that Kassanthra wins?

23. A small repair shop for washing machines has the following demand for their services:

  On 10% of the days, they have no requests; they have one request on 30% of the days, and two requests 50% of the time.

  (a) On Monday, what is the chance of more than two requests?

  (b) What is the chance of no requests for a whole week (of 5 days)?

24. A construction company is bidding on three projects: B_1, B_2, and B_3. From previous experience they have the following probabilities of winning the bids: $P(B_1) = 0.22$, $P(B_2) = 0.25$, and $P(B_3) = 0.28$. Winning the bids are not independent of each other. The joint probabilities are given in Table 8.5. Also, $P(B_1 \cap B_2 \cap B_3) = 0.01$. Find the following probabilities:

  (a) $P(B_1 \cup B_2)$ **(b)** $P(B_1' \cap B_2')$ **(c)** $P(B_1' \cap B_2') \cup B_3$

  (d) $P(B_1' \cap B_2' \cap B_3)$ **(e)** $P(B_2 \cap B_3 | B_1)$ **(f)** $P(B_2 \cup B_3 | B_1)$

	B_1	B_2	B_3
B_1		0.11	0.05
B_2	0.11		0.07
B_3	0.05	0.07	

Table 8.5 Data for question 24

25. Circuit boards used in electronic equipment go through more than one inspection. The process of finding faults in the solder joints on these boards is highly subjective and prone to disagreements among inspectors. In a batch of 20 000 joints, Nick found 1448 faulty joints while David found 1502 faulty ones. All in all, between both inspectors, 2390 joints were judged to be faulty. Find the probability that a randomly chosen joint is:

  (a) judged to be faulty by neither of the two inspectors

  (b) judged to be defective by David but not Nick.

Chapter 8 practice questions

1. Two independent events A and B are given such that:
   $P(A) = k$, $P(B) = k + 0.3$ and $P(A \cap B) = 0.18$.
   (a) Find k. **(b)** Find $P(A \cup B)$ **(c)** $P(A'|B')$

2. Many airport authorities test prospective employees for drug use. This procedure has plenty of opponents who claim that this procedure creates difficulties for some people and that it prevents others from getting these jobs even if they were not drug users. The claim depends on the fact that these tests are not 100% accurate. To test this claim, assume that a test is 98% accurate in that it correctly identifies a person as a user or non-user 98% of the time. Each job applicant takes this test twice. The tests are done at separate times and are designed to be independent of each other. What is the probability that:
   (a) a non-user fails both tests
   (b) a drug user is detected (i.e., they fail at least one test)
   (c) a drug user passes both tests?

3. Communications satellites are difficult to repair when something goes wrong. One satellite uses solar energy and it has two systems that provide electricity. The main system has a probability of failure of 0.002. It has a backup system that works independently of the main one and has a failure rate of 0.01. What is the probability that the systems do not fail at the same time?

4. In a group of 200 students taking the IB examination, 120 take Spanish, 60 take French, and 10 take both.
   (a) If a student is selected at random, what is the probability that the student:
   (i) takes either French or Spanish
   (ii) takes either French or Spanish, but not both
   (iii) does not take either French or Spanish?
   (b) Given that a student takes Spanish, what is the chance that the student takes French?

5. In a factory producing computer disk drives, there are three machines that work independently to produce one of the components. In any production process, machines are not 100% fault free. The production after one batch from each machine is listed in the table.

	Defective	**Non-defective**
Machine I	6	120
Machine II	4	80
Machine III	10	150

   (a) A component is chosen at random from the batches. Find the probability that the chosen component is:

(i) from machine I
(ii) a defective component from machine II
(iii) non-defective or from machine I
(iv) from machine I given that it is defective.

(b) Is the quality of the component dependent on the machine used?

6. At a school, the students are organising a lottery to raise money for their community. The lottery tickets that they have consist of small coloured envelopes containing a small note. The note either says: 'You won a prize!' or 'Sorry, try another ticket.' The envelopes have several colours. They have 70 red envelopes that contain two prizes, and the rest (130 envelopes) contain four other prizes.

(a) You want to help this class and you buy a ticket hoping that it does not have a prize. You pick your ticket at random by closing your eyes. What is the probability that your ticket does not have a prize?

(b) You picked a red envelope. What is the probability that you did not win a prize?

7. Two events A and B have the conditions:

$\text{P}(A|B) = 0.30$, $\text{P}(B|A) = 0.60$, $\text{P}(A \cap B) = 0.18$.

(a) Find $\text{P}(B)$

(b) Are A and B independent? Explain.

(c) Find $\text{P}(B \cap A')$

8. In several ski resorts in Austria and Switzerland, the local sports authorities use senior high school students as ski instructors to help deal with the surge in demand during vacations. However, to become an instructor, you have to pass a test and must be a senior at your school.

Here are the results of a survey of 120 students in a Swiss school who are training to become instructors. In this group, there are 70 boys and 50 girls. 74 students took the test, 32 boys and 16 girls passed the test, the rest, including 12 girls, failed the test. 10 of the students, including 6 girls, were too young to take the ski test.

(a) Copy and complete the table.

	Boys	**Girls**
Passed the ski test	32	16
Failed the ski test		12
Training, but did not take the test yet		
Too young to take the test		

(b) Find the probability that:
(i) a student chosen at random, has taken the test
(ii) a girl chosen at random has taken the test
(iii) a randomly chosen boy and randomly chosen girl have both passed the ski test.

9. Two events A and B are such that $P(A) = \frac{9}{16}$, $P(B) = \frac{3}{8}$, and $P(A|B) = \frac{1}{4}$. Find the probability that:

(a) both events will happen

(b) only one of the events will happen

(c) neither event will happen.

10. Martina plays tennis. When she serves, she has 60% chance of succeeding with her first serve and continuing the game. She has 95% chance on the second serve. Of course, if both serves are not successful, she loses the point.

(a) Find the probability that she misses both serves.

If Martina succeeds with the first serve, her probability of gaining the point against Steffy is 75%; if she is only successful with the second serve, the probability for that point goes down to 50%.

(b) Find the probability that Martina wins a point against Steffy.

11. For events X and Y, $P(X) = 0.6$, $P(Y) = 0.8$ and $P(X \cup Y) = 1$.

Find:

(a) $P(X \cap Y)$

(b) $P(X' \cup Y')$.

12. In a survey, 100 managers were asked 'Do you prefer to watch the news or play sport?' Of the 46 men in the survey, 33 said they would choose sport, while 29 women made this choice.

	Men	Women	Total
The news			
Sport	33	29	
Total	46		100

Find the probability that:

(a) a manager selected at random prefers to watch the news

(b) a manager prefers to watch the news, given that the manager is a man.

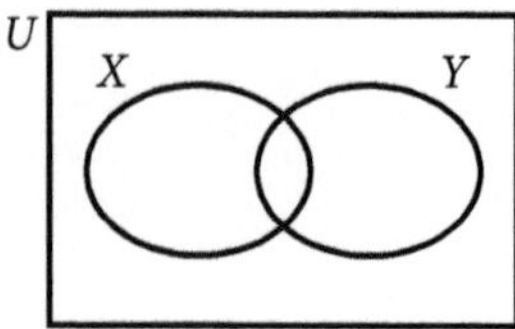

Figure 8.18 Venn diagram for question 13

13. The Venn diagram shows a sample space U and events X and Y.

$n(U) = 36$, $n(X) = 11$, $n(Y) = 6$ and $n(X \cup Y)' = 21$.

(a) Copy the diagram and shade the region $(X \cup Y)'$.

(b) Find:

(i) $n(X \cap Y)$

(ii) $P(X \cap Y)$.

(c) Are events X and Y mutually exclusive? Explain why or why not.

14. In a survey of 200 people, 90 of whom were female, it was found that 60 people were unemployed, including 20 males.

(a) Copy and complete the table, using this information.

	Males	Females	Totals
Unemployed			
Employed			
Totals			200

(b) If a person is selected at random from this group of 200, find the probability that this person is:

(i) an unemployed female

(ii) a male, given that the person is employed.

15. The Venn diagram shows the universal set U and the subsets M and N.

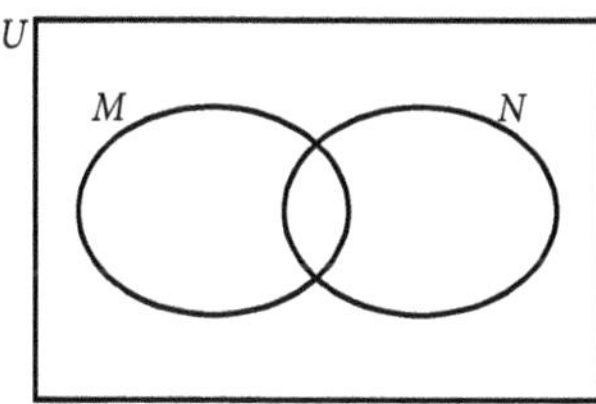

Figure 8.19 Venn diagram for question 15

(a) Copy the diagram and shade the area which represents the set $M \cap N'$.

$n(U) = 100$, $n(M) = 30$, $n(N) = 50$, and $n(M \cup N) = 65$.

(b) Find $n(N \cap M')$.

(c) An element is selected at random from U. What is the probability that this element is in $N \cap M'$?

16. Two fair dice are thrown and the number showing on each is noted. Find the probability that:

(a) the sum of the numbers is less than or equal to 7

(b) at least one dice shows a 3

(c) at least one dice shows a 3, given that the sum is less than 8.

17. For events A and B, the probabilities are $\text{P}(A) = \frac{3}{11}$, $\text{P}(B) = \frac{4}{11}$. Calculate the value of $\text{P}(A \cap B)$ if:

(a) $\text{P}(A \cup B) = \frac{6}{11}$

(b) events A and B are independent.

18. In a school of 88 boys, 32 study Economics (E), 28 study History (H), and 39 do not study either subject. This information is represented in the Venn diagram.

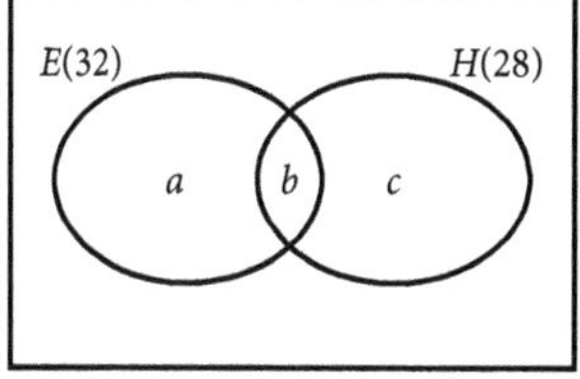

Figure 8.20 Venn diagram for question 18

(a) Calculate the values a, b, and c.

(b) A student is selected at random.

(i) Calculate the probability that he studies both Economics and history.

(ii) Given that he studies Economics, calculate the probability that he does not study History.

(c) A group of three students is selected at random from the school.

(i) Calculate the probability that none of these students studies Economics.

(ii) Calculate the probability that at least one of these students studies Economics.

19. Sophia is a student at an IB school.

The probability that she will be woken by her alarm clock is $\frac{7}{8}$.

If she is woken by her alarm clock, the probability she will be late for school is $\frac{1}{4}$.

If she is not woken by her alarm clock, the probability she will be late for school is $\frac{3}{5}$.

Let W be the event 'Sophia is woken by her alarm clock'.

Let L be the event 'Sophia is late for school'.

(a) Copy and complete the tree diagram.

(b) Calculate the probability that Sophia will be late for school.

(c) Given that Sophia is late for school, what is the probability that she was woken by her alarm clock?

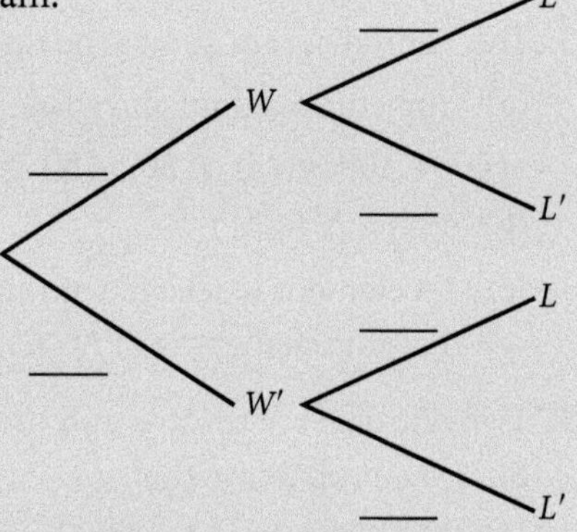

20. Two unbiased six-sided dice of different colours are rolled.

Find:

(a) P(the same number appears on both dice)

(b) P(the sum of the numbers is 10)

(c) P(the sum of the numbers is 10 or the same number appears on both dice).

21. The table below shows the subjects studied by 210 students at a college.

	Year 1	**Year 2**	**Totals**
History	50	35	85
Science	15	30	45
Art	45	35	80
Totals	110	100	210

(a) A student from the college is selected at random. Let A be the event the student studies art. Let B be the event the student is in Year 2.

(i) Find P(A).

(ii) Find the probability that the student is a Year 2 art student.

(iii) Are the events A and B independent? Justify your answer.

(b) Given that a history student is selected at random, calculate the probability that the student is in Year 1.

(c) Two students are selected at random from the college. Calculate the probability that one student is in Year 1 and the other in Year 2.

22. A new blood test has been shown to be effective in the early detection of a disease. The probability that the blood test correctly identifies someone with this disease is 0.99, and the probability that the blood test correctly identifies someone without the disease is 0.95. The incidence of this disease in the general population is 0.0001.

A doctor administered the blood test to a patient and the test result indicated that this patient had the disease. What is the probability that the patient has the disease?

23. Asha walks to school every day. If it is not raining, the probability that she is late is 0.2. If it is raining, the probability that she is late is $\frac{2}{3}$. The probability that it rains on any particular day is 0.25.

Last Friday, Asha was late. Find the probability that it was raining on that day.

24. The probability that Marco leaves his umbrella in any place he visits is $\frac{1}{3}$. After visiting two friends in succession, he finds he has left his umbrella at one of his friends' places. What is the probability that he left his umbrella at the second friend's place?

25. Two girls, Catherine and Lucy, play a game in which they take turns in throwing an unbiased six-sided dice. The first one to throw a 5 wins the game. Catherine is the first to throw.

(a) Find the probability that:

(i) Lucy wins on her first throw

(ii) Catherine wins on her second throw

(iii) Catherine wins on her nth throw.

(b) Let p be the probability that Catherine wins the game.

Show that $p = \frac{1}{6} + \frac{25}{36}p$.

(c) Find the probability that Lucy wins the game.

(d) Suppose that they play the game six times. Find the probability that Catherine wins more games than Lucy.

26. The chance of rain on any day during the summer in Schaditz, Austria, is 0.2. When it rains, the probability that the daily maximum temperature exceeds 25°C is 0.3, while it is 0.6 when it does not rain. Given that the maximum daily temperature exceeded 25°C on a certain summer's day, find the probability that it rained on that day.

27. The independent events A and B are such that $P(A) = 0.4$ and $P(A \cup B) = 0.88$ Find:

(a) $P(B)$

(b) the probability that either A occurs, or B occurs, but **not both**.

28. Roberto travels to school in a neighbouring town by bus every weekday from Monday to Friday. The probability that he catches the 08:00 bus on Friday is 0.66. The probability that he catches the 08:00 bus on any other weekday is 0.75. A weekday is chosen at random.

(a) Find the probability that he catches the 08:00 bus on that day.

(b) Given that he catches the 08:00 bus on that day, find the probability that the chosen day is Friday.

29. Antonio and Sarah play a game by throwing a dice in turn. If the dice shows a 3, 4, 5, or 6, the player who threw the dice wins the game. If the dice shows a 1 or 2, the other player has the next throw. Antonio plays first, and the game continues until there is a winner.

(a) Write down the probability that Antonio wins on his first throw.

(b) Calculate the probability that Sarah wins on her first throw.

(c) Calculate the probability that Antonio wins the game.

30. Six balls numbered 1, 2, 2, 3, 3, 3 are placed in a bag. Balls are taken one at a time from the bag at random and the number noted. Throughout the question a ball is always replaced before the next ball is taken.

(a) A single ball is taken from the bag. Let X denote the value shown on the ball. Find $E(X)$.

(b) Three balls are taken from the bag. Find the probability that:

(i) the total of the three numbers is 5

(ii) the median of the three numbers is 1.

(c) Ten balls are taken from the bag. Find the probability that fewer than four of the balls are numbered 2.

(d) Find the fewest number of balls that must be taken from the bag for the probability of taking out at least one ball numbered 2 to be greater than 0.95.

(e) Another bag also contains balls numbered 1, 2, or 3.

Eight balls are to be taken from this bag at random. It is calculated that the expected number of balls numbered 1 is 4.8, and the variance of the number of balls numbered 2 is 1.5.

Find the fewest possible number of balls numbered 3 in this bag.

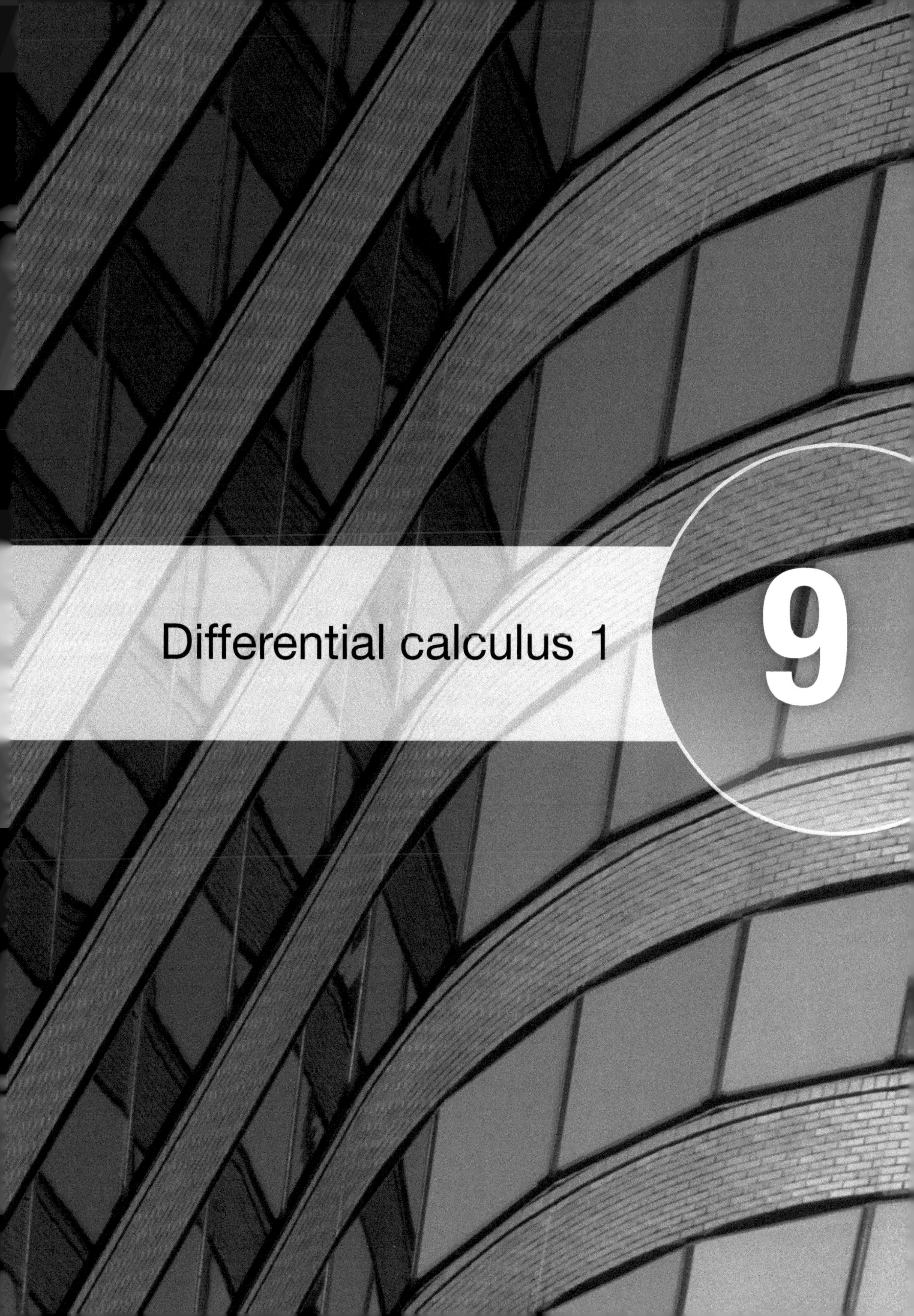

9 Differential calculus 1

9 Differential calculus 1

Learning objectives

By the end of this chapter, you should be familiar with...

- the concept of a limit
- the derivative as a gradient (slope) function and as a rate of change
- finding the derivative of polynomial functions
- the derivatives of $\sin x$ and $\cos x$
- the relationship between the graphs of a function, its first derivative and its second derivative
- identifying where a function is increasing or decreasing
- finding and testing for maximum, minimum, and inflection points
- kinematic problems involving displacement, velocity, and acceleration
- finding the equation of a tangent or a normal at a given point.

Figure 9.1 shows a distance–time graph for a 50-kilometre bicycle ride that included going up and then down a steep hill. There are four time intervals labelled A, B, C, and D. The cyclist's speed is the lowest in interval B. It is the highest in interval C. The cyclist's speed is about the same in intervals A and D. The shape of the distance–time graph gives information about the cyclist's speed during a certain interval and at a particular moment (instant) during the ride.

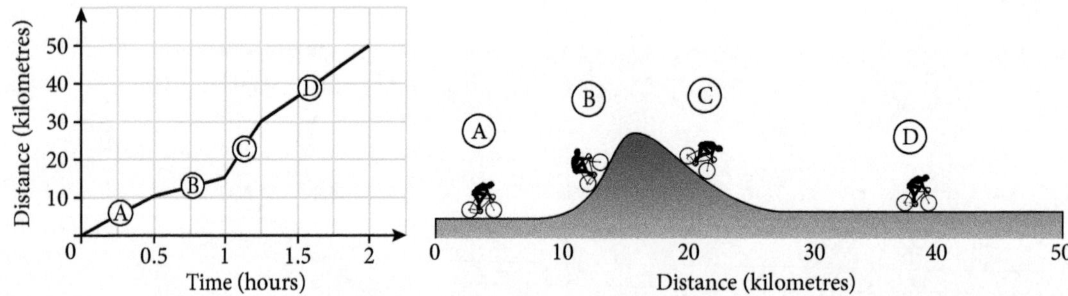

Figure 9.1 A distance–time graph for a cyclist

Calculus is the branch of mathematics that was developed to analyse and model changing quantities – such as velocity and acceleration. We can also apply it to study change in the context of slope, area, volume, and a wide range of other concepts that allow us to model real-life phenomena more precisely. Although mathematical techniques that we have previously studied dealt with many of these concepts, the ability to model change was restricted. For example, consider the curve in Figure 9.2 that illustrates the motion of an object by indicating the distance (y metres) travelled after a certain amount of time (t seconds). Without calculus, we can only compute the **average velocity** between two different times (Figure 9.3). With calculus, we can find the velocity of an object at a particular instant, known as its **instantaneous velocity** (Figure 9.4). The starting point for our study of calculus is the idea of a limit.

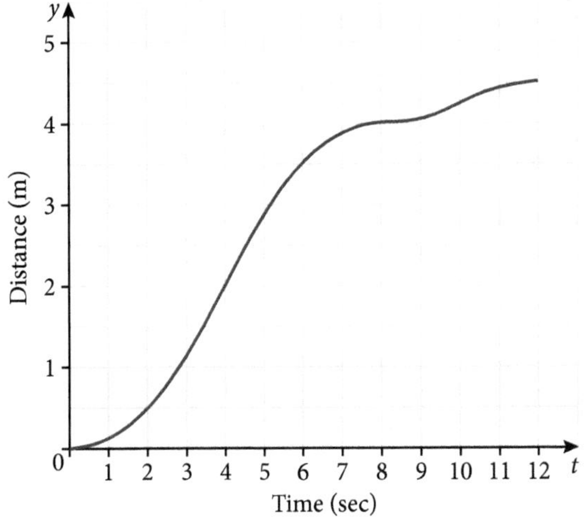

Figure 9.2 Distance–time graph for an object's motion

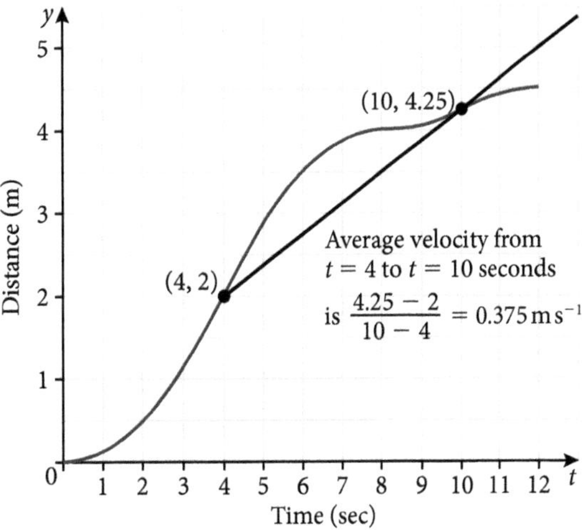

Figure 9.3 Average velocity from a distance–time graph

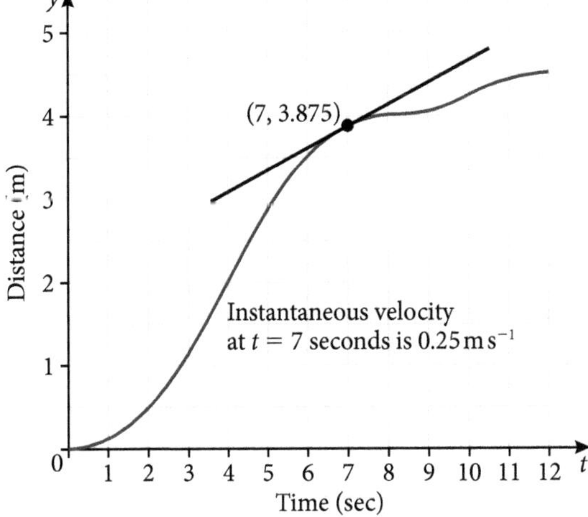

Figure 9.4 Instantaneous velocity from a distance–time graph

9.1 Limits of functions

A **limit** is one of the ideas that distinguishes calculus from algebra, geometry, and trigonometry. The notion of a limit is a fundamental concept of calculus. Limits are not new to us. We often use the idea of a limit in many non-mathematical situations. We have already used mathematical limits in this book – finding the sum of an infinite geometric series and computing the irrational number e.

In Chapter 3, we established that if the sequence of partial sums for an infinite series **converges** to a finite number L, we say that the infinite series has a sum of L. We used limits to confirm algebraically that the infinite series $2 + 1 + \frac{1}{2} + \frac{1}{4} + \frac{1}{8} + \cdots$ has a sum of 4. As part of the algebra for this, we reasoned that as the value of n increases in the positive direction without bound ($n \to +\infty$) the expression $\left(\frac{1}{2}\right)^n$ converges to zero – in other words, the **limit** of $\left(\frac{1}{2}\right)^n$ as n goes to positive infinity is zero. This result is expressed using limit notation as $\lim_{n\to\infty}\left(\frac{1}{2}\right)^n = 0$. It is beyond the requirements of this course to establish a precise formal definition of a limit, but a closer look at justifying a couple of limits can lead us to an informal understanding of the concept of a limit.

Example 9.1

Evaluate $\lim_{n\to\infty}\left(\frac{1}{2}\right)^n$ by using your GDC to analyse the behaviour of the function $f(x) = \left(\frac{1}{2}\right)^x$ for large positive x values.

Solution

The notation $\lim_{x\to+\infty} f(x)$ indicates the limit of the value of the function f as x takes on greater and greater positive values (also written simply as $\lim_{x\to\infty} f(x)$), and $\lim_{x\to-\infty} f(x)$ indicates the limit of the value of the function f as x takes on greater and greater (in magnitude) negative values.

(a)

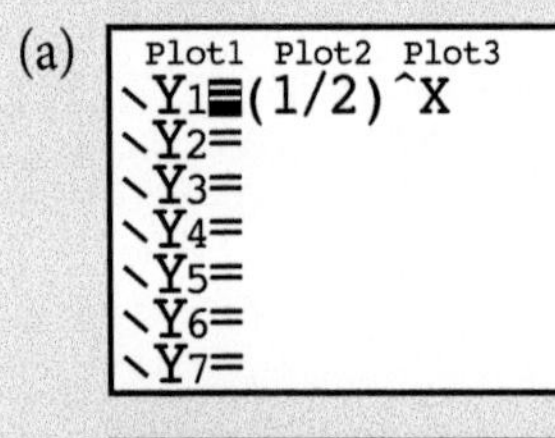

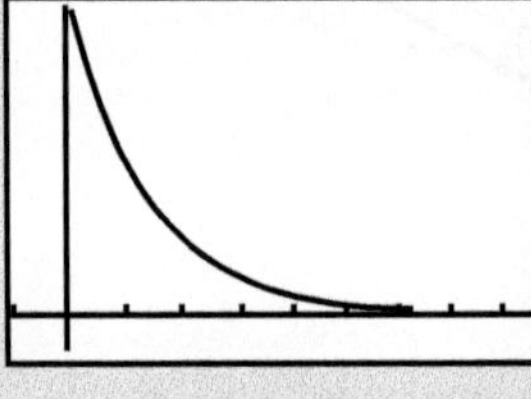

X	Y1
0	1
1	.5
2	.25
3	.125
4	.0625
5	.03125
6	.01563

X=0

X	Y1
7	.00781
8	.00391
9	.00195
10	9.8E-4
11	4.9E-4
12	2.4E-4
13	1.2E-4

Y1=1.22070313E-4

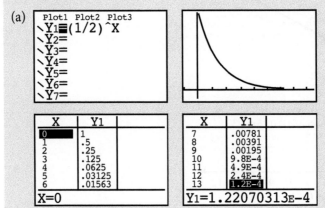

The GDC screen images show the graph and table of values for $y = \left(\frac{1}{2}\right)^x$

The larger the value of x, the closer y gets to zero. Although there is no value of x that will produce a value of y equal to zero, we can get as close to zero as we wish. For example, if we wish to produce a value of y within 0.001 of zero, then we could choose $x = 10$ and $y = \left(\frac{1}{2}\right)^{10} = \frac{1}{1024} \approx 0.00097656$

And if we want a result within 0.0 000 001 of zero, then we could choose $x = 24$ and $y = \left(\frac{1}{2}\right)^{24} = \frac{1}{16777216} \approx 0.000000059605$ and so on.

Therefore, we can conclude that $\lim_{n\to\infty}\left(\frac{1}{2}\right)^n = 0$

The line $y = c$ is a **horizontal asymptote** of the graph of a function $y = f(x)$ if either $\lim_{x\to\infty} f(x) = c$ or $\lim_{x\to-\infty} f(x) = c$.

For example, the line $y = 0$ (x-axis) is a horizontal asymptote of the graph of $y = \left(\frac{1}{2}\right)^x$ because $\lim_{n\to\infty}\left(\frac{1}{2}\right)^n = 0$.

In calculus, we are interested in limits of functions of real numbers. Although many of the limits of functions that we will encounter can only be approached and not actually reached (as in Example 9.1), this is not always the case. For example, if asked to evaluate the limit of the function $f(x) = \frac{x}{2} - 1$ as x approaches 6, then we evaluate the function for $x = 6$. Since $f(6) = 2$, then $\lim\limits_{x \to 6}\left(\frac{x}{2} - 1\right) = 2$. However, it is more common that we are unable to evaluate the limit of $f(x)$ as x approaches some number c because $f(c)$ does not exist.

Example 9.2

Find the value of the following two limits by using your GDC to analyse the graph of the relevant function.

(a) $\lim\limits_{x \to 0} \frac{\sin x}{x}$ (b) $\lim\limits_{x \to 0} \frac{\cos x - 1}{x}$

Solution

(a) We are not able to evaluate this limit by direct substitution because when $x = 0$, $\frac{\sin x}{x} = \frac{0}{0}$ and is therefore undefined. We use our GDC (in radian mode) to analyse the behaviour of the function $y = \frac{\sin x}{x}$ as x approaches zero from the right side and the left side.

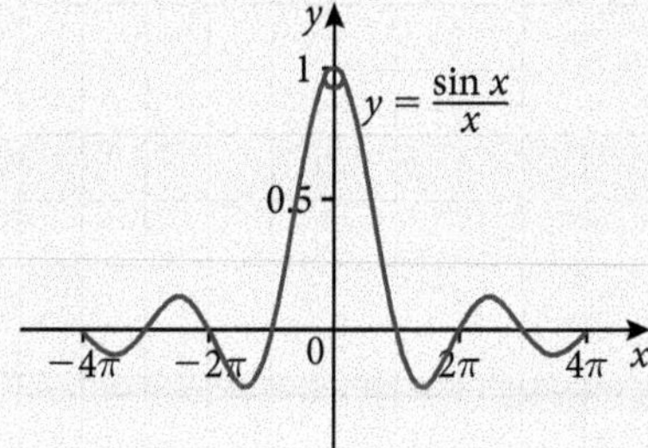


```
Plot1 Plot2 Plot3
\Y1 = sin(X)/X
```

```
Y1(-0.05)
        .9995833854
Y1(0.05)
        .9995833854
1-Ans
     4.166145864E-4
```

```
Y1(-0.002)
        .9999993333
Y1(0.002)
        .9999993333
1-Ans
        6.66666655E-7
```

Although there is no point on the graph of $y = \frac{\sin x}{x}$ corresponding to $x = 0$, it is clear from the graph that as x approaches zero (from either direction), the value of $\frac{\sin x}{x}$ converges to 1.

We can get the value of $\frac{\sin x}{x}$ arbitrarily close to 1 depending on our choice of x (see GDC images above). If we want $\frac{\sin x}{x}$ to be within 0.001 of 1, we choose $x = \pm 0.05$ giving $\frac{\sin 0.05}{0.05} \approx 0.999583$ and

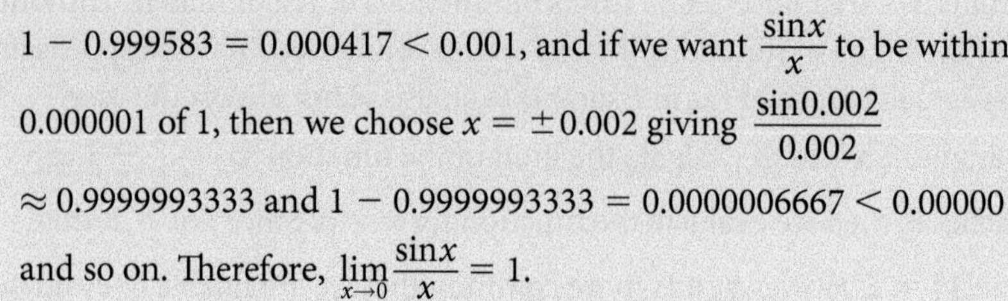

$1 - 0.999583 = 0.000417 < 0.001$, and if we want $\frac{\sin x}{x}$ to be within 0.000001 of 1, then we choose $x = \pm 0.002$ giving $\frac{\sin 0.002}{0.002}$ ≈ 0.9999993333 and $1 - 0.9999993333 = 0.0000006667 < 0.000001$ and so on. Therefore, $\lim\limits_{x \to 0} \frac{\sin x}{x} = 1$.

(b) As with $y = \frac{\sin x}{x}$, substituting $x = 0$ into the function $y = \frac{\cos x - 1}{x}$ produces $\frac{0}{0}$. The graph of $y = \frac{\cos x - 1}{x}$ shows that the function approaches 0 as x tends to 0.

A table produced on a GDC also shows that the function approaches zero from both directions.

```
Plot1 Plot2 Plot3
\Y1≣ (cos(X)-1)/X
```

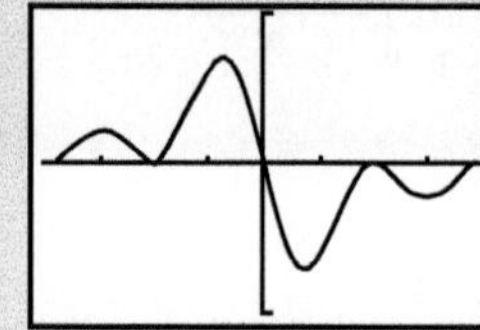

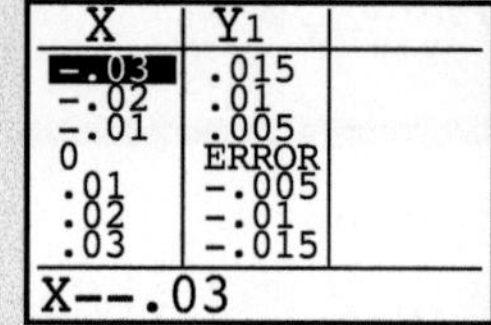

Therefore, $\lim\limits_{x \to 0} \frac{\cos x - 1}{x} = 0$

The analysis and result for Example 9.2 illustrates why it is preferred (and often necessary) that in calculus, the argument of a trigonometric function be in radian measure rather than degrees.

The limit of $\frac{\sin x}{x}$ as x tends to ∞ is **not** equal to 1 if x is in degrees.

Functions do not necessarily converge to a finite value at every point. It is possible that a limit does not exist.

Although we can describe the behaviour of the function $y = \frac{1}{x^2}$ by writing $\lim\limits_{x \to 0} \frac{1}{x^2} = \infty$, this does not mean that we consider ∞ to represent a number – it does not. This notation is simply a convenient way to indicate in what manner the limit does not exist.

Example 9.3

Find $\lim\limits_{x \to 0} \frac{1}{x^2}$, if it exists.

Solution

As x approaches zero, the value of $\frac{1}{x^2}$ becomes increasingly large in the positive direction. The graph of the function below seems to indicate that we can make the values of $y = \frac{1}{x^2}$ arbitrarily large by choosing x close enough to zero. Therefore, the values of $y = \frac{1}{x^2}$ do not approach a finite number, so $\lim\limits_{x \to 0} \frac{1}{x^2}$ does not exist.

The line $x = c$ is a **vertical asymptote** of the graph of a function $y = f(x)$ if either $\lim\limits_{x \to c} f(x) = \infty$ or $\lim\limits_{x \to c} f(x) = -\infty$.

For example, the line $x = 0$ (y-axis) is a vertical asymptote of the graph of $y = \frac{1}{x^2}$ because $\lim\limits_{x \to 0} \frac{1}{x^2} = \infty$.

If $f(x)$ becomes arbitrarily close to a unique finite number L as x approaches c from either side, then the limit of $f(x)$ as x approaches c is L. The notation for indicating this is $\lim\limits_{x \to c} f(x) = L$.

When a function $f(x)$ becomes arbitrarily close to a finite number L, we say that $f(x)$ **converges** to L.

Recall from Section 2.2 the end behaviour of rational functions. Since $\lim_{x\to\infty} \frac{3x^2+5x-1}{2x^2+1} = \frac{3}{2}$, the rational function $y = \frac{3x^2+5x-1}{2x^2+1}$ will have a horizontal asymptote of $y = \frac{3}{2}$. In other words, as $x \to +\infty$ or $x \to -\infty$, the limit of the function values approach $\frac{3}{2}$.

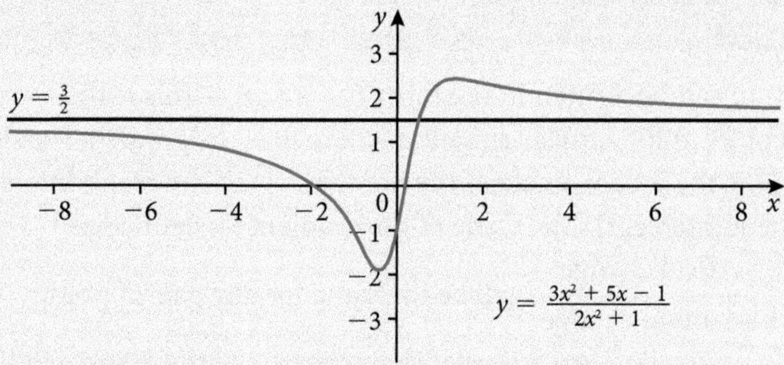

A rational function approaching a horizontal asymptote as $x \to +\infty$ or $x \to -\infty$

Exercise 9.1

1. Evaluate each limit. Confirm your result by means of a table or graph on your GDC.

(a) $\lim_{x\to\infty} \frac{4x+3}{x}$ **(b)** $\lim_{h\to 0}(3x^2 + 2hx + h^2)$

(c) $\lim_{x\to 0} \frac{2x^2-2x}{x}$ **(d)** $\lim_{x\to 3} \frac{x^2-9}{x-3}$

2. Investigate the limit (if it exists) of each expression as $x \to \infty$ by evaluating the expression for the following values of x: 10, 50, 100, 1000, 10 000, and 1 000 000. Hence, make a conjecture for the value of each limit.

(a) $\lim_{x\to\infty} \frac{3x+2}{x^2-3}$ **(b)** $\lim_{x\to\infty} \frac{5x-6}{2x+5}$ **(c)** $\lim_{x\to\infty} \frac{3x^2+2}{x-3}$

3. Use the graphing or table capabilities of your GDC to investigate the value of the expression $\left(1 + \frac{1}{c}\right)^c$ as c increases without bound (i.e. $c \to \infty$). Indicate the significance of the result.

4. If it is known that the line $y = 3$ is a horizontal asymptote for the function $f(x)$, then state the value of each of the following two limits: $\lim_{x\to\infty} f(x)$ and $\lim_{x\to -\infty} f(x)$.

5. If it is known that the line $x = a$ is a vertical asymptote for the function $g(x)$, and that $g(x) > 0$, then what conclusion can be made about $\lim_{x\to a} g(x)$?

6. State the equations of all horizontal and vertical asymptotes for the following functions.

(a) $f(x) = \frac{3x-1}{1+x}$ **(b)** $g(x) = \frac{1}{(x-2)^2}$ **(c)** $h(x) = \frac{1}{x-a} + b$

9.2 The derivative of a function: basic rules and facts

Tangent lines and the gradient (or slope) of a curve

Any linear function can be written in the form $y = mx + c$. This is the gradient–intercept form for a linear equation where m is the gradient (or slope) of the graph and c is the y-coordinate of the point at which the graph intersects the y-axis (i.e. the y-intercept). The value of the gradient m, defined as $m = \frac{y_2 - y_1}{x_2 - x_1} = \frac{\text{vertical change}}{\text{horizontal change}}$, will be the same for any pair of points, (x_1, y_1) and (x_2, y_2), on the line. An essential characteristic of the graph of a linear function is that it has a constant gradient. This is not true for the graphs of non-linear functions.

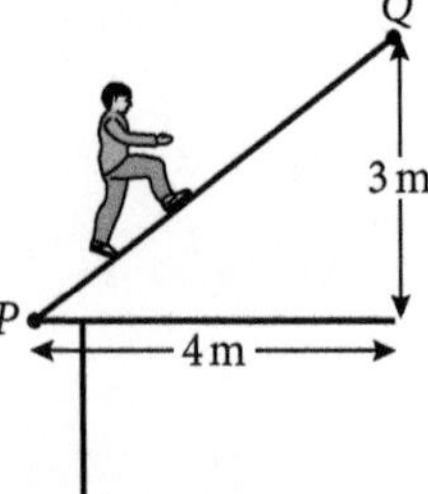

Figure 9.5 Gradient of a straight line

Consider a person walking up the side of a pitched roof as shown in Figure 9.5. At any point along the line segment PQ, the person is experiencing a gradient of $\frac{3}{4}$.

Now consider someone walking up the curve shown in Figure 9.6 that passes through the three points A, B, and C. As the person walks along the curve from A to C, they will experience a steadily increasing gradient. The gradient is continually changing from one point to the next along the curve. Therefore, it is incorrect to say that a non-linear function, whose graph is a curve, has a gradient – it has infinitely many gradients. We need a means to determine the gradient of a non-linear function at a specific point on its graph.

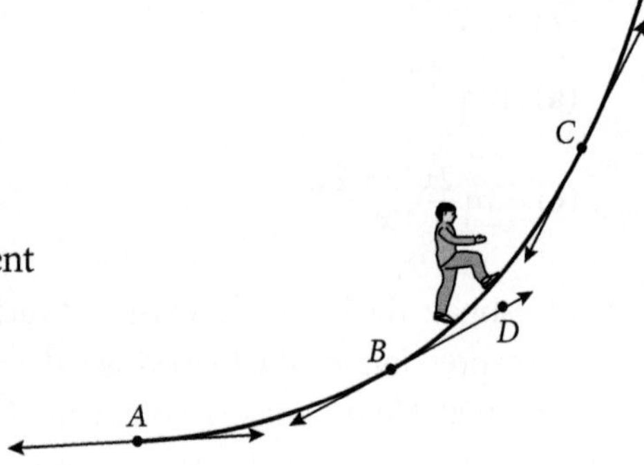

Figure 9.6 Gradient of a curve

The gradient (slope) of a curve at a point is the gradient of the line that is tangent to the curve at that point.

The word curve can often mean the same as function, even if the function is linear.

Imagine if the gradient of the curve in Figure 9.6 stopped increasing (remained constant) after point B. From that point on, a person walking up the curve would move along a line with a gradient equal to the gradient of the curve at point B. This line, containing point D in the diagram, only touches the curve once, at point B. Line (BD) is **tangent** to the curve at point B. Therefore, finding the gradient of the line that is tangent to a curve at a certain point will give us the gradient of the curve at that point.

Finding the gradient of a curve at a point – or better, finding a rule (function) that gives us the gradient at any point on the curve – is very useful information in many applications. The gradient of a line, or of a curve at a point, is a measure of how fast variable y is changing as variable x changes. The gradient represents the rate of change of y with respect to x. To find the gradient of a tangent, we first need to clarify what it means to say that a line is tangent to a curve at a point. Then we can establish a method to find the tangent at a point.

The three graphs in Figure 9.7 show different configurations of tangent lines. A tangent line may cross or intersect the graph at one or more points.

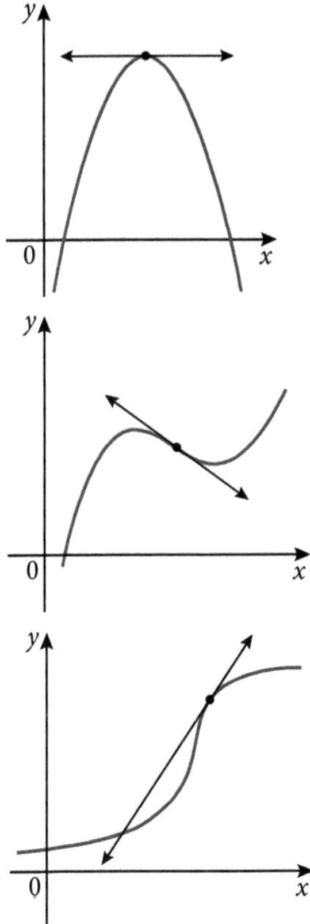

Figure 9.7 Different configurations of lines tangent to a curve

For many functions, the graph has a tangent at every point. Informally, a function is said to be smooth if it has this property. Any linear function is certainly smooth, since the tangent at each point coincides with the original graph. However, some graphs are not smooth at every point. Consider the point (0, 0) on the graph of the function $y = |x|$ (Figure 9.8). Zooming in on (0, 0) will always produce a V-shape rather than smoothing out to appear more and more linear. Therefore, there is no tangent to the graph at this point.

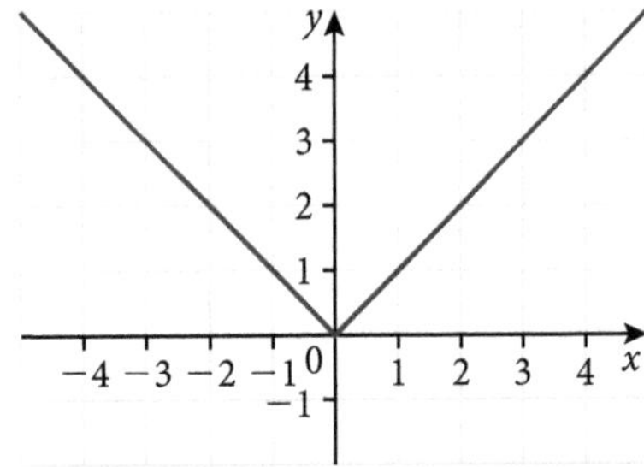

Figure 9.8 $y = |x|$

One way to find the tangent line of a graph at a particular point is to make a visual estimate. Figure 9.9 shows the distance–time graph for an object's motion. The gradient at any point (t, y) on the curve will give us the rate of change of the distance y with respect to time t, in other words the object's instantaneous velocity at time t. In the figure, an estimate of the tangent to the curve at (5, 3) has been drawn. Reading from the graph, the gradient appears to be $\frac{4}{6} = \frac{2}{3}$. Or, in other words, the object has a velocity of approximately 0.667 m s^{-1} at the instant when $t = 5$ seconds.

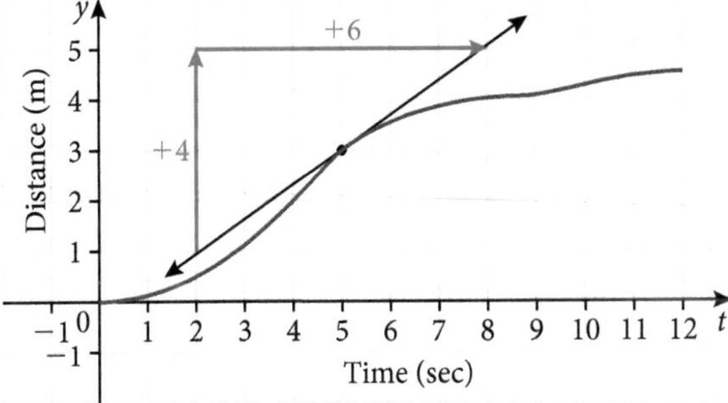

Figure 9.9 Estimating the gradient of a tangent

A more precise method of finding tangent lines makes use of a **secant line** and a **limit process**. Suppose that f is any smooth function, so the tangent to its graph exists at all points. A secant line (or chord) is drawn through the point for which we are trying to find a tangent to f and a second point on the graph of f, as shown in Figure 9.10. If P is the point of tangency with coordinates $(x, f(x))$, then choose a point Q to be a horizontal distance of h units away. Hence, the coordinates of point Q are $(x + h, f(x + h))$ and the gradient of the secant line (PQ) is $m_{\text{sec}} = \frac{f(x + h) - f(x)}{(x + h) - x} = \frac{f(x + h) - f(x)}{h}$

The right side of this equation is often referred to as a difference quotient. The numerator is the change in y, and the denominator h is the change in x. The limit process of achieving better and better approximations for the gradient of the tangent at P consists of finding the gradient of the secant (PQ) as Q

moves ever closer to P, as shown in the graphs in Figure 9.11 and Figure 9.12. In doing so, the value of h will approach zero.

By evaluating a limit of the gradient of the secant lines as h approaches zero, we can find the exact gradient of the tangent line at $P(x, f(x))$ (Figure 9.13).

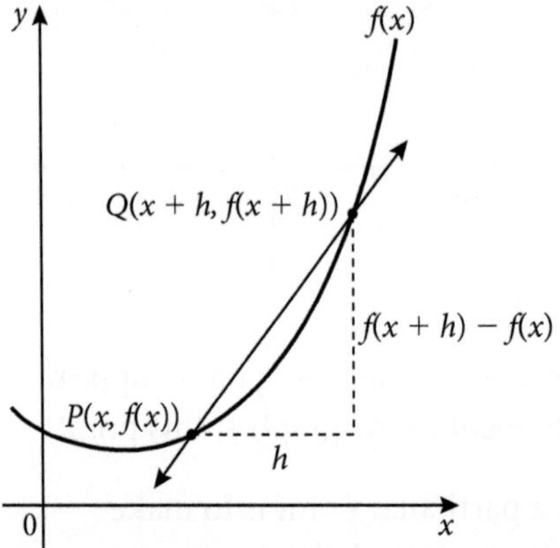

Figure 9.10 Secant line through points P and Q

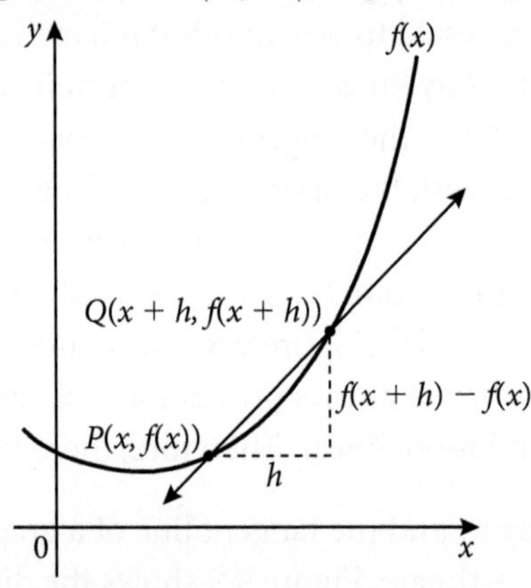

Figure 9.11 The horizontal distance h gets smaller as point Q moves closer to P

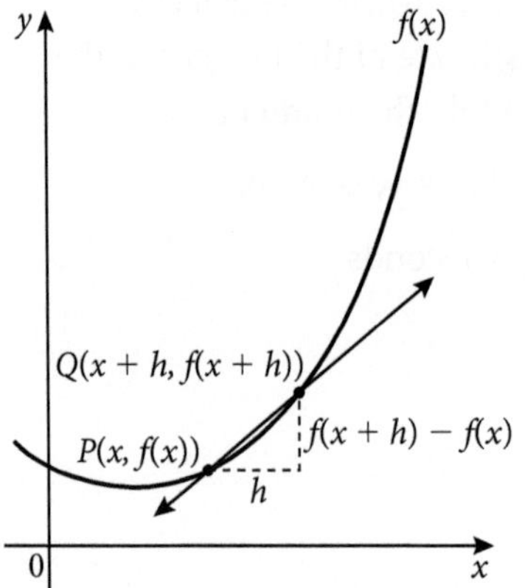

Figure 9.12 As h gets smaller, the secant line becomes a better approximation of the tangent to the graph of f at P

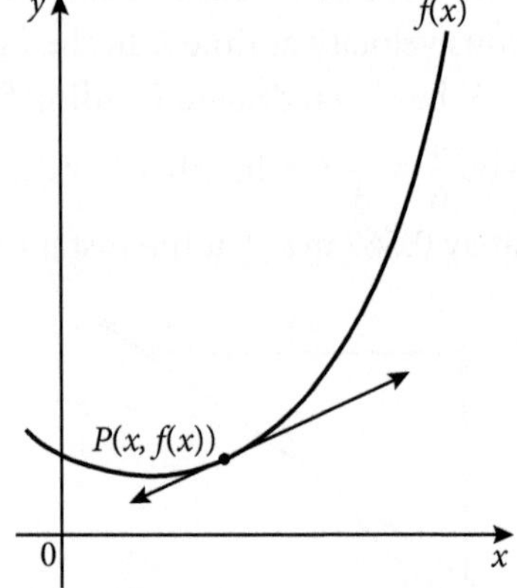

Figure 9.13 Tangent to f at point P

Let's apply this limit process to a specific curve. We will see that the gradient of a line through two points on a curve (a secant line) becomes a better and better approximation for the gradient of the tangent to the curve at a point. This occurs as one of the points that the secant line passes through moves closer and closer to the other (fixed) intersection point where we are trying to approximate the gradient of the tangent.

Consider the graph of the curve $f(x) = x^2 + 1$. We wish to calculate the gradient of the tangent to the curve at the point (1, 2). Let's compute the gradient of three secant lines that pass through (1, 2) and points on the curve where $x = -\frac{1}{2}$, then $x = 0$, and then $x = \frac{1}{2}$.

Given $f(x) = x^2 + 1$, we can show that $f\left(-\frac{1}{2}\right) = \frac{5}{4}$, $f(0) = 1$ and $f\left(\frac{1}{2}\right) = \frac{5}{4}$.

Figure 9.14 shows a secant line passing through the points $\left(-\frac{1}{2}, \frac{5}{4}\right)$ and (1, 2).

$f(x) = x^2 + 1$

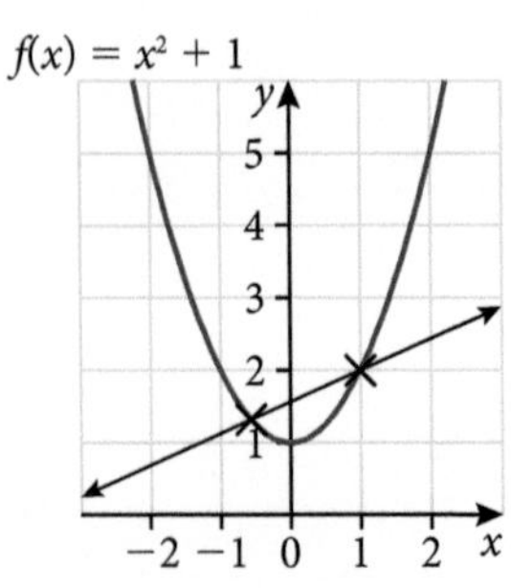

Figure 9.14 Secant line through $\left(-\frac{1}{2}, \frac{5}{4}\right)$ and (1, 2)

Its gradient is $\dfrac{2-\frac{5}{4}}{1-\left(-\frac{1}{2}\right)} = \dfrac{\frac{3}{4}}{\frac{3}{2}} = \dfrac{1}{2}$. Figure 9.15 shows a secant line passing through the points (0, 1) and (1, 2). Its gradient is 1. Figure 9.16 shows a secant line passing through the points $\left(\frac{1}{2}, \frac{5}{4}\right)$ and (1, 2). Its gradient is $\frac{3}{2}$. With the gradients of successive secant lines going from $\frac{1}{2}$, to 1, and then to $\frac{3}{2}$, it seems reasonable to conjecture that the gradient of the tangent to the curve at (1, 2) is 2, as shown in Figure 9.17.

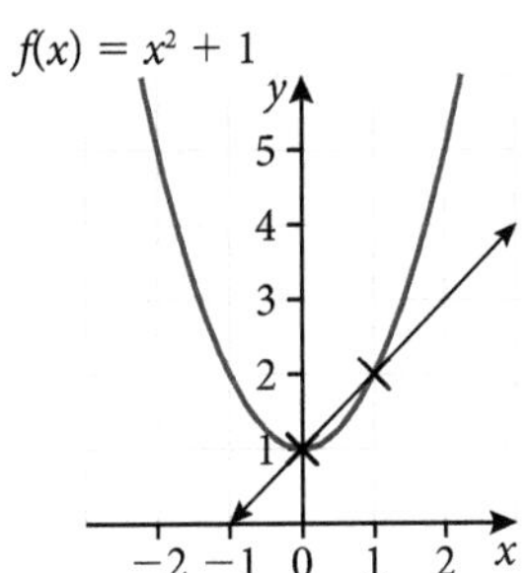

Figure 9.15 Secant line through (0, 1) and (1, 2)

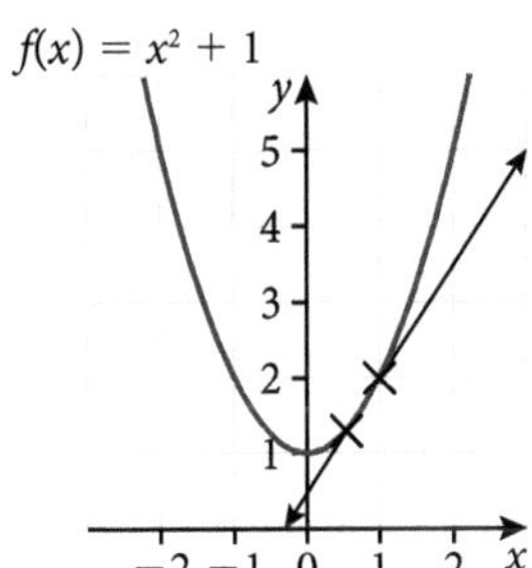

Figure 9.16 Secant line through $\left(\frac{1}{2}, \frac{5}{4}\right)$ and (1, 2)

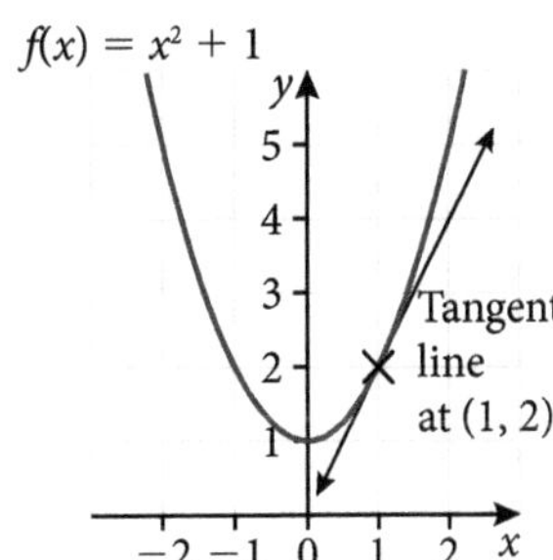

Figure 9.17 Tangent line at (1, 2)

Online

Demonstration of a secant line approximating line tangent to a curve at a point.

Ultimately, from the function $f(x) = x^2 + 1$, we wish to derive another function that will compute the gradient (slope) of the graph of f at a point by simply inputting the x-coordinate of the point. This derived function is called the **derivative** of $y = f(x)$ at x. There are two common ways to denote this function: either $f'(x)$ or $\dfrac{dy}{dx}$.

$\dfrac{dy}{dx}$ is not a fraction. If, for example, $y = x^2 + 1$, then the derivative can be denoted by writing $\dfrac{d}{dx}(x^2 + 1) = 2x$. This is read as, 'the derivative of $x^2 + 1$ with respect to x is $2x$.'

We could continue to apply the limit process (making secant lines closer to the tangent) to calculate the gradient of tangents at other points on the graph of $f(x) = x^2 + 1$ in order to make a conjecture for the derivative of the function, but that would be tedious. Let's use the power of a GDC to quickly compute the value of the derivative for $f(x) = x^2 + 1$ (the gradient of the tangent to the curve) at several selected points and use these results to make a reasonable conjecture for the function that is the derivative of $f(x) = x^2 + 1$.

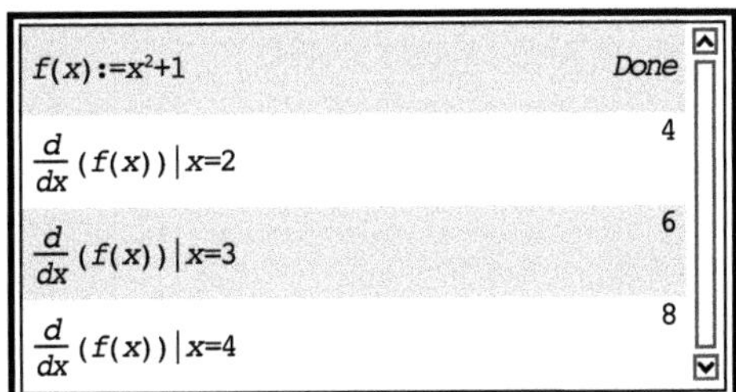

Figure 9.18 Using a GDC to make a conjecture for the function that is the derivative of $f(x) = x^2 + 1$

The command name and syntax for computing the value of a derivative at a point will vary from one GDC model to another.

The derivative of $f(x) = x^2 + 1$ seems to be $2x$. That is, $f'(x) = 2x$.

Now consider the function $g(x) = x^2$ and values of its derivative at $x = 1, x = 2, x = 3$, and $x = 4$.

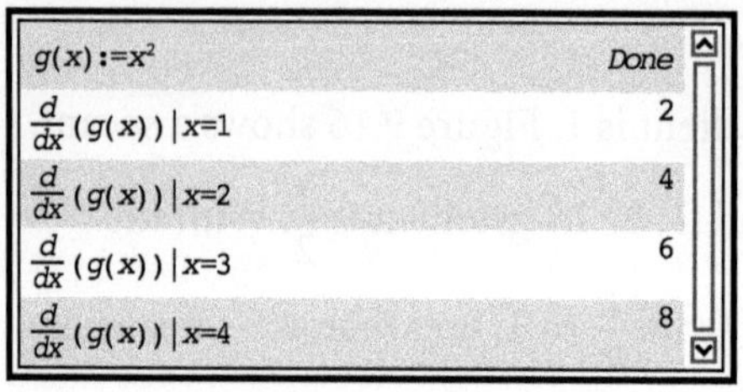

g(x):=x²	Done
$\frac{d}{dx}(g(x))\|x=1$	2
$\frac{d}{dx}(g(x))\|x=2$	4
$\frac{d}{dx}(g(x))\|x=3$	6
$\frac{d}{dx}(g(x))\|x=4$	8

Figure 9.19 Using a GDC to make a conjecture for the function that is the derivative of $f(x) = x^2$

It appears that the derivative of $g(x) = x^2$ is also $2x$. That is, $g'(x) = 2x$.

In order to develop some basic rules for finding the derivative of functions of the form $f(x) = ax^n + bx^{n-1} + \ldots$, where all powers are integers, let's continue to use a GDC to make convincing conjectures for the derivatives of three further functions: $f_1(x) = 3x^2 + 2x$, $f_2(x) = x^3$, and $f_3(x) = \frac{1}{x} = x^{-1}$.

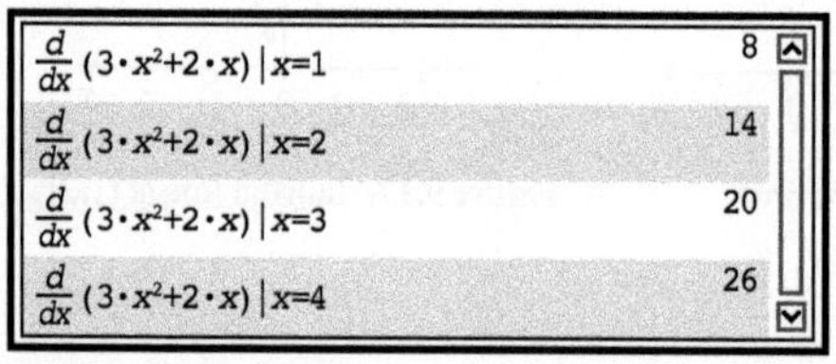

$\frac{d}{dx}(3\cdot x^2+2\cdot x)\|x=1$	8
$\frac{d}{dx}(3\cdot x^2+2\cdot x)\|x=2$	14
$\frac{d}{dx}(3\cdot x^2+2\cdot x)\|x=3$	20
$\frac{d}{dx}(3\cdot x^2+2\cdot x)\|x=4$	26

Figure 9.20 Using a GDC to make a conjecture for the function that is the derivative of $f(x) = 3x^2 + 2x$

A reasonable conjecture for the derivative of $f_1(x) = 3x^2 + 2x$ is $6x + 2$. That is, $f_1'(x) = 6x + 2$.

$\frac{d}{dx}(x^3)\|x=1$	3
$\frac{d}{dx}(x^3)\|x=2$	12
$\frac{d}{dx}(x^3)\|x=3$	27
$\frac{d}{dx}(x^3)\|x=4$	48

Figure 9.21 Using a GDC to make a conjecture for the function that is the derivative of $f(x) = x^3$

After some examination of the results here, it seems that the derivative of $f_2(x) = x^3$ is $3x^2$. That is, $f_2'(x) = 3x^2$.

$\frac{d}{dx}(x^{-1})\|x=1$	-1
$\frac{d}{dx}(x^{-1})\|x=2$	$\frac{-1}{4}$
$\frac{d}{dx}(x^{-1})\|x=3$	$\frac{-1}{9}$
$\frac{d}{dx}(x^{-1})\|x=4$	$\frac{-1}{16}$

Figure 9.22 Using a GDC to make a conjecture for the function that is the derivative of $f(x) = \frac{1}{x}$

These results make it clear that the derivative of $f_3(x) = \frac{1}{x} = x^{-1}$ is $-\frac{1}{x^2}$. That is, $f_3'(x) = -\frac{1}{x^2}$

Basic differentiation rules

We have now established the following results:

- When $f(x) = x^2$, then $f'(x) = 2x$
- When $f(x) = x^2 + 1$, then $f'(x) = 2x$
- When $f(x) = 3x^2 + 2x$, then $f'(x) = 6x + 2$
- When $f(x) = x^3$, then $f'(x) = 3x^2$
- When $f(x) = x^{-1}$, then $f'(x) = -x^{-2}$

In addition, we know that when $f(x) = x$, then $f'(x) = 1$, since the line $y = x$ has a constant gradient equal to 1, and that when $f(x) = 1$, then $f'(x) = 0$ because the line $y = 1$ is horizontal and thus has a constant gradient equal to 0. The graph of any function $f(x) = c$ where c is a constant is a horizontal line, confirming that if $f(x) = c$, $c \in \mathbb{R}$, then $f'(x) = 0$. Thus, the derivative of a constant is zero.

The constant rule

The derivative of a constant function is zero. That is, given c is a real number, and $f(x) = c$, then $f'(x) = 0$.

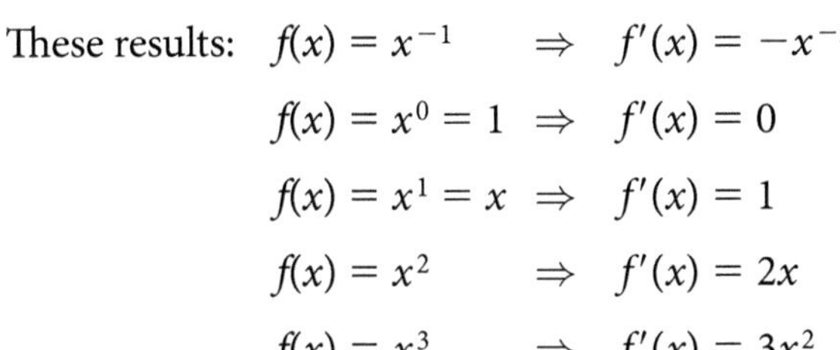

These results:

$$
\begin{aligned}
f(x) &= x^{-1} && \Rightarrow \quad f'(x) = -x^{-2} \\
f(x) &= x^0 = 1 && \Rightarrow \quad f'(x) = 0 \\
f(x) &= x^1 = x && \Rightarrow \quad f'(x) = 1 \\
f(x) &= x^2 && \Rightarrow \quad f'(x) = 2x \\
f(x) &= x^3 && \Rightarrow \quad f'(x) = 3x^2
\end{aligned}
$$

can be summarised in a single rule called the **power rule** that is true for any value of n that is a rational number ($n \in \mathbb{Q}$) (see key fact box on the right).

The power rule

Given n is a rational number, and if $f(x) = x^n$ then the derivative of x^n is $f'(x) = nx^{n-1}$.

Another basic rule of differentiation is suggested by our result that the derivative of $f(x) = x^2 + 1$ is $f'(x) = 2x$. The derivative of a sum of terms is obtained by differentiating each term separately (differentiating term by term) In this case, $\frac{d}{dx}(x^2 + 1) = \frac{d}{dx}(x^2) + \frac{d}{dx}(1) = 2x + 0 = 2x$.

The sum and difference rule

If $f(x) = g(x) \pm h(x)$ then $f'(x) = g'(x) \pm h'(x)$.

The sum rule for derivatives can help us give a very convincing justification of our first differentiation rule, the constant rule. The fact that the derivative of a constant must be zero can be verified by considering the transformation of the graph of a function (Section 2.4). The graph of the function $f(x) + c$, where $c \in \mathbb{R}$, is a vertical translation by c units of the graph of $f(x)$. As Figure 9.23 illustrates, when the graph of a function is translated vertically, its shape is preserved. Hence, the gradient of the tangent line to the graph of $f(x) + c$ will be the same as that for $f(x)$ at a particular value of x. This means that the derivatives for the two functions must be equal. That is,

$$\frac{d}{dx}[f(x) + c] = \frac{d}{dx}[f(x)]$$

$$\frac{d}{dx}[f(x)] + \frac{d}{dx}(c) = \frac{d}{dx}[f(x)]$$

This is only true if $\frac{d}{dx}(c) = 0$.

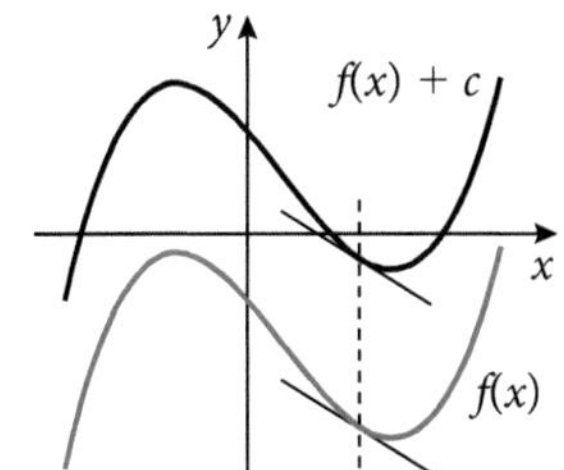

Figure 9.23 Translating the graph of a function vertically does not alter the gradient of the tangent line at a particular value of x. Hence, the derivatives of the two functions are equal.

Our final basic rule of differentiation is illustrated by the result that the derivative of $f(x) = 3x^2 + 2x$ is $f'(x) = 6x + 2$. Using the sum rule,

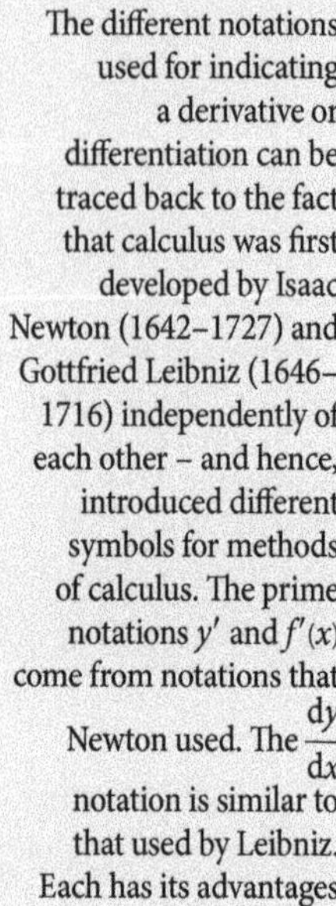

The constant multiple rule

If $f(x) = c \cdot g(x)$ then $f'(x) = c \cdot g'(x)$.

The different notations used for indicating a derivative or differentiation can be traced back to the fact that calculus was first developed by Isaac Newton (1642–1727) and Gottfried Leibniz (1646–1716) independently of each other – and hence, introduced different symbols for methods of calculus. The prime notations y' and $f'(x)$ come from notations that Newton used. The $\frac{dy}{dx}$ notation is similar to that used by Leibniz. Each has its advantages and disadvantages. For example, it is often easier to write our four basic rules of differentiation using Leibniz notation.

$f'(x) = \frac{d}{dx}(3x^2 + 2x) = \frac{d}{dx}(3x^2) + \frac{d}{dx}(2x) = 6x + 2$. The fact that $\frac{d}{dx}(3x^2) = 6x$ suggests that $3 \cdot \frac{d}{dx}(x^2) = 3 \cdot 2x = 6x$. In other words, the derivative of a function being multiplied by a constant is equal to the constant multiplying the derivative of the function.

Basic differentiation rules

Constant rule: $\frac{d}{dx}(c) = 0, c \in \mathbb{R}$

Power rule: $\frac{d}{dx}(x^n) = nx^{n-1}, n \in \mathbb{Q}$

Sum and difference rule: $\frac{d}{dx}[g(x) + h(x)] = \frac{d}{dx}[g(x)] + \frac{d}{dx}[h(x)]$

Constant multiple rule: $\frac{d}{dx}[c \cdot f(x)] = c \cdot \frac{d}{dx}[f(x)], c \in \mathbb{R}$

Example 9.4

For each function: (i) find the derivative using the basic differentiation rules; (ii) find the gradient of the graph of the function at the indicated points; and (iii) use your GDC to confirm your answer for (ii).

(a) $f(x) = x^3 + 2x^2 - 15x - 13$ at the points $(-3, 23)$, $(3, -13)$

(b) $f(x) = (2x - 7)^2$ at the points $(2, 9)$, $\left(\frac{7}{2}, 0\right)$

(c) $f(x) = 3\sqrt{x} - 6$ at the points $(4, 0)$, $(9, 3)$

(d) $f(x) = \frac{x^4}{4} - \frac{3x^3}{2} - 2x^2 + \frac{15x}{2} + \frac{3}{4}$ at the points $(5, -43)$, $(0, 0)$

Solution

(a) (i) $\frac{d}{dx}(x^3 + 2x^2 - 15x - 13) = \frac{d}{dx}(x^3) + 2 \cdot \frac{d}{dx}(x^2) - 15 \cdot \frac{d}{dx}(x) - \frac{d}{dx}(13)$

$$= 3x^2 + 2(2x) - 15(1) - 0$$
$$= 3x^2 + 4x - 15$$

Therefore the derivative of $f(x) = x^3 + 2x^2 - 15x - 13$ is $f'(x) = 3x^2 + 4x - 15$

(ii) Gradient of curve at $(-3, 23)$ is $f'(-3) = 3(-3)^2 + 4(-3) - 15$

$$= 27 - 12 - 15 = 0$$

We should observe a horizontal tangent (gradient = 0) to the curve at $(-3, 23)$

Gradient of curve at $(3, -13)$ is $f'(3) = 3(3)^2 + 4(3) - 15$

$$= 27 + 12 - 15 = 24$$

We should observe a very steep tangent (gradient = 24) to the curve at $(3, -13)$

(iii) Not only can we use the GDC to compute the value of the derivative at a particular value of x on the 'home' screen, but we can also do it on the graph screen.

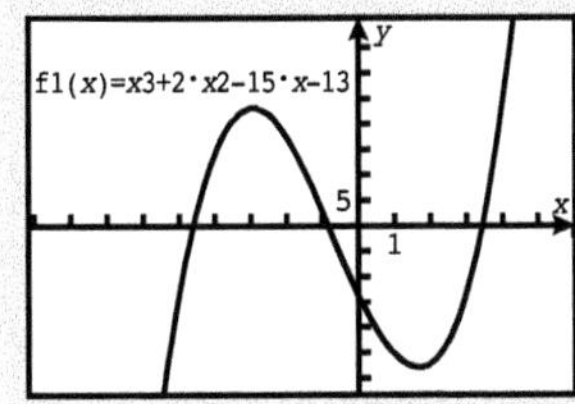

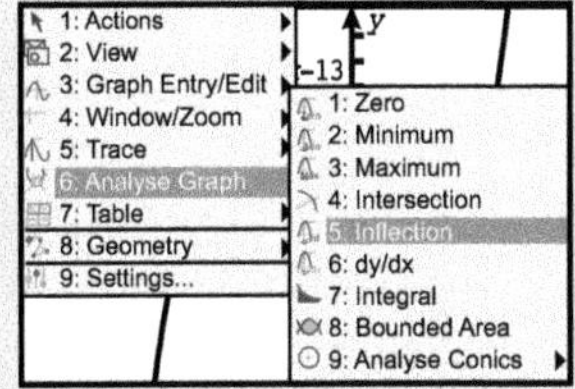

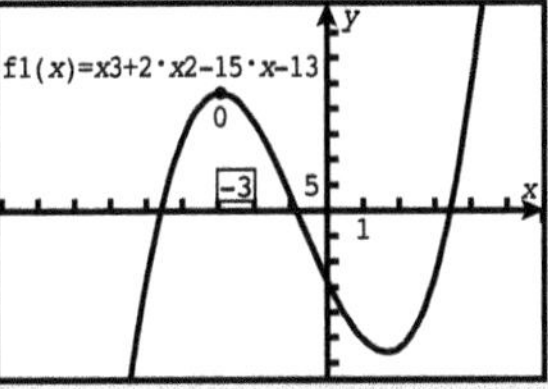

Observe that the graph of $y = x^3 + 2x^2 - 15x - 13$ appears to have a turning point at $(-3, 23)$, confirming that a tangent to the curve at that point would be horizontal.

'turning point'

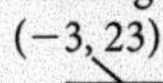

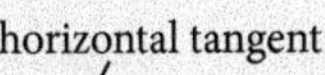

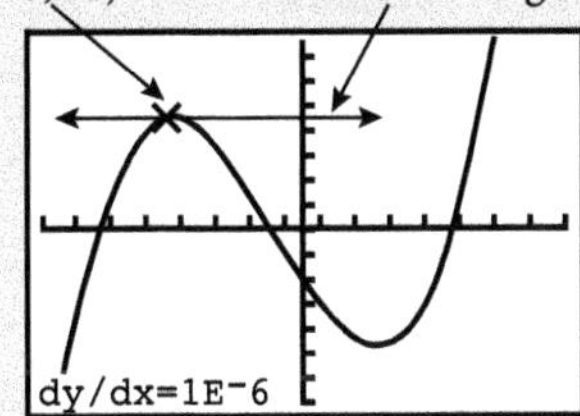

Let's check on our GDC that the gradient is 24 at $(3, -13)$.
Most GDCs are also capable of drawing a tangent at a point and displaying its equation, as shown in the screenshots.

```
CALCULATE
1:value
2:zero
3:minimum
4:maximum
5:intersect
6:dy/dx
7:∫f(x)dx
```

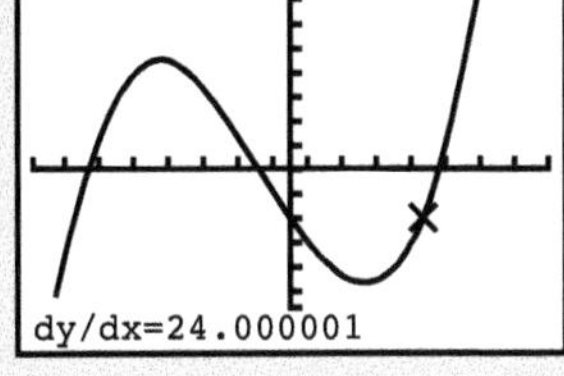

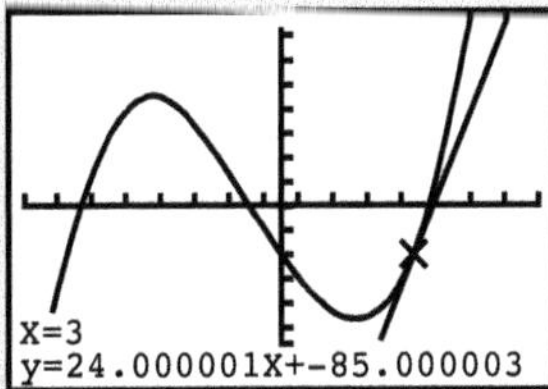

The equation of the tangent line at $(3, -13)$ is $y = 24x - 85$.
We will look at finding the equations of tangent lines analytically in Section 9.4.

(b) (i) Differentiate term by term after expanding:

$$\frac{d}{dx}[(2x - 7^2)] = \frac{d}{dx}[(2x - 7)(2x - 7)]$$

$$= \frac{d}{dx}(4x^2 - 28x + 49)$$

$$= 4\frac{d}{dx}(x^2) - 28\frac{d}{dx}(x) + \frac{d}{dx}(49)$$

$$= 8x - 28 + 0$$

Therefore the derivative of $f(x) = (2x - 7)^2$ is $f'(x) = 8x - 28$.

(ii) Gradient of curve at $(2, 9)$ is $f'(2) = 8(2) - 28 = -12$

Gradient of curve at $\left(\frac{7}{2}, 0\right)$ is $f'\left(\frac{7}{2}\right) = 8\left(\frac{7}{2}\right) - 28 = 0$

Thus, we should observe a horizontal tangent to the curve at $\left(\frac{7}{2}, 0\right)$.

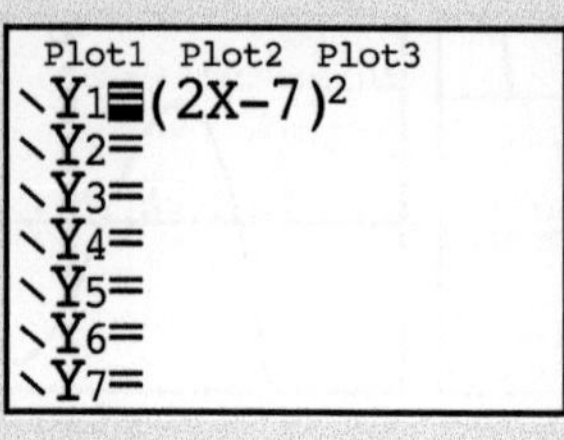

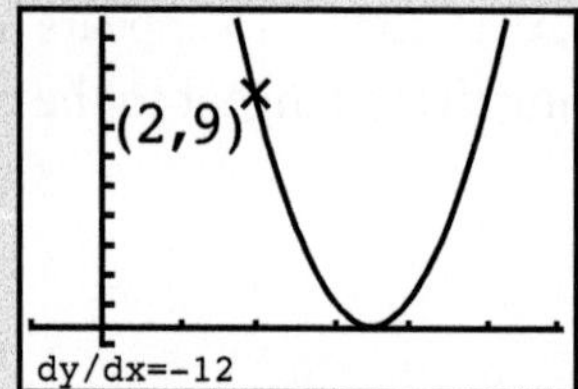

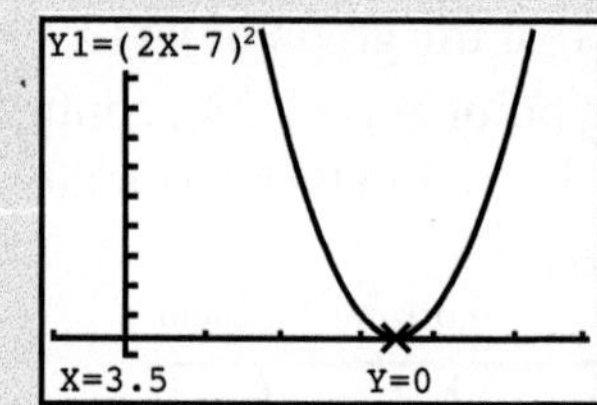

The GDC confirms that the gradient of the curve is −12 at (2, 9). The vertex of the parabola is at $\left(\frac{7}{2}, 0\right)$ confirming that it has a horizontal tangent at that point.

(c) (i) $\frac{\mathrm{d}}{\mathrm{d}x}(3\sqrt{x} - 6) = 3\frac{\mathrm{d}}{\mathrm{d}x}\left(x^{\frac{1}{2}}\right) - \frac{\mathrm{d}}{\mathrm{d}x}(6)$

$$= 3\left(\frac{1}{2}x^{-\frac{1}{2}}\right) - 0$$

$$= 3\left(\frac{1}{2}x^{-\frac{1}{2}}\right) - 0 = \frac{3}{2x^{\frac{1}{2}}}$$

Therefore, the derivative of $f(x) = 3\sqrt{x} - 6$ is $f'(x) = \frac{3}{2x^{\frac{1}{2}}}$ or $f'(x) = \frac{3}{2\sqrt{x}}$.

(ii) Gradient of curve at (4, 0) is $f'(4) = \frac{3}{2\sqrt{4}} = \frac{3}{4}$

Gradient of curve at (9, 3) is $f'(9) = \frac{3}{2\sqrt{9}} = \frac{1}{2}$

As the gradient at $x = 9$ is less than that at $x = 4$, we should observe the graph of the equation becoming less steep as we move along the curve from $x = 4$ to $x = 9$.

(iii) 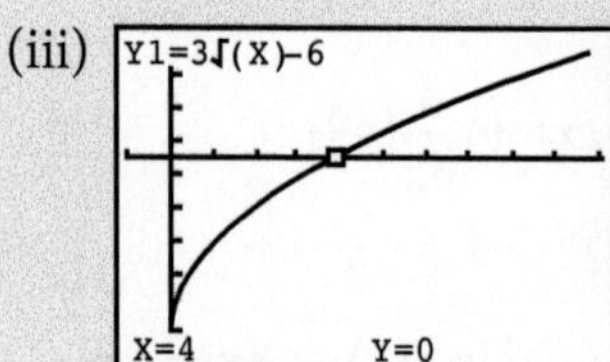

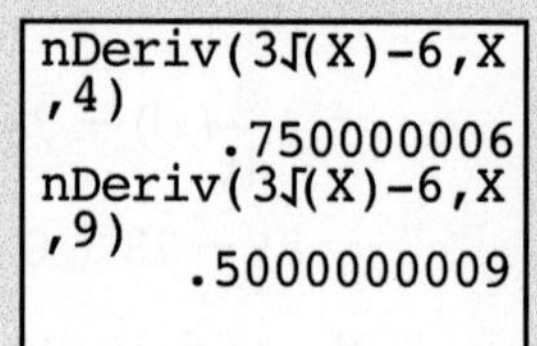

The gradient of the graph of $y = 3\sqrt{x} - 6$ appears to decrease steadily as x increases. Let's check the results for (ii) by evaluating the derivative at a point on the home screen.

The GDC confirms the gradients for the curve when $x = 4$ and $x = 9$, but the GDC computations have incorporated a small amount of error.

(d) (i) $\dfrac{\mathrm{d}}{\mathrm{d}x}\left(\dfrac{x^4}{4} - \dfrac{3x^3}{2} - 2x^2 + \dfrac{15x}{2} + \dfrac{3}{4}\right)$

$$= \frac{1}{4}(4x^3) - \frac{3}{2}(3x^2) - 2(2x) + \frac{15}{2}(1) + 0$$

$$= x^3 - \frac{9x^2}{2} - 4x + \frac{15}{2}$$

Therefore, the derivative of $f(x) = \dfrac{x^4}{4} - \dfrac{3x^3}{2} - 2x^2 + \dfrac{15x}{2} + \dfrac{3}{4}$ is

$f'(x) = x^3 - \dfrac{9x^2}{2} - 4x + \dfrac{15}{2}$

(ii) Gradient of curve at $(5, -43)$ is $f'(5) = 5^3 - \dfrac{9(5)^2}{2} - 4(5) + \dfrac{15}{2} = 0$.

Thus, there should a horizontal tangent to the curve at $(5, -43)$.

Gradient of curve at $(0, 0)$ is $f'(0) = \dfrac{15}{2}$.

(iii)

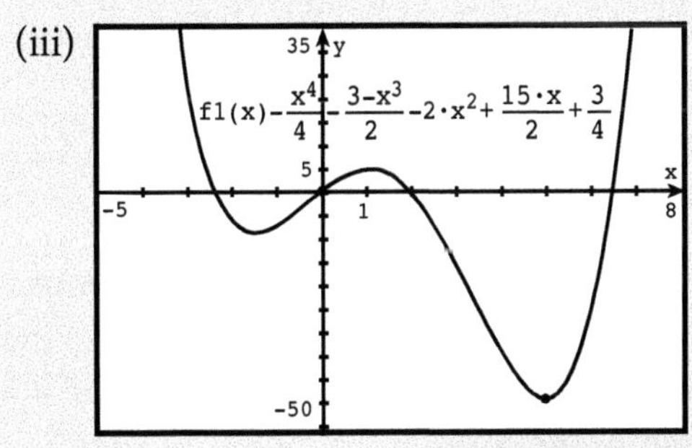

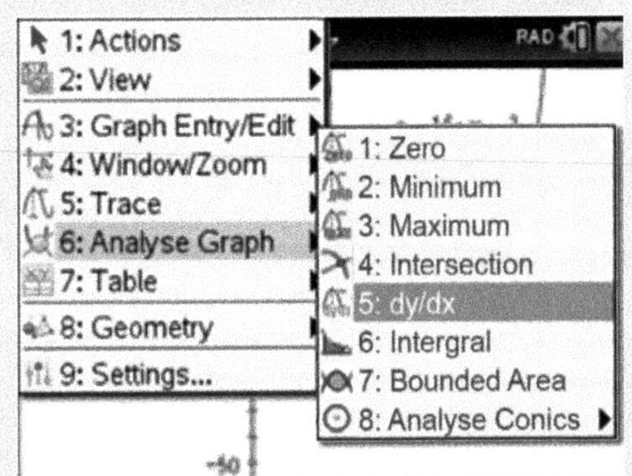

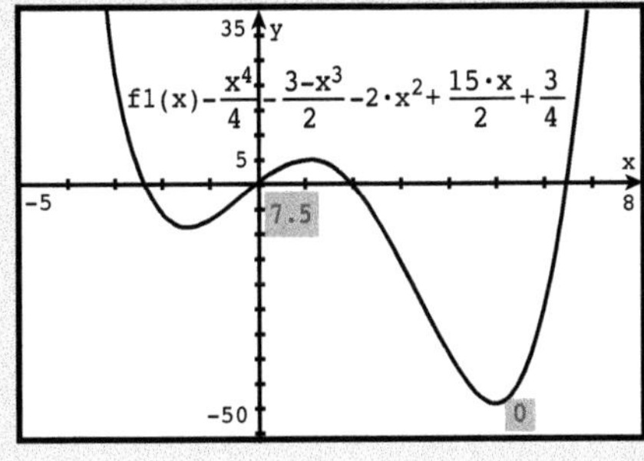

$\frac{d}{dx}(f1(x))\vert x=5$	0
$\frac{d}{dx}(f1(x))\vert x=0$	$\frac{15}{2}$

Example 9.5

The curve $y = ax^3 + 7x^2 - 8x - 5$ has a turning point at the point where $x = -2$. Determine the value of a.

Solution

A turning point on a graph is a point where the gradient changes from upwards to downwards, or vice versa.

There must be a horizontal tangent, and a gradient of zero, at the point where the graph has a turning point.

$$\frac{dy}{dx} = \frac{d}{dx}(ax^3 + 7x^2 - 8x - 5)$$
$$= a\frac{d}{dx}(x^3) + 7\frac{d}{dx}(x^2) - 8\frac{d}{dx}(x) + \frac{d}{dx}(-5)$$
$$= 3ax^2 + 14x - 8$$

$$\frac{dy}{dx} = 0 \text{ when } x = -2, \text{ so } 3a(-2)^2 + 14(-2) - 8 = 0$$
$$\Rightarrow 12a - 28 - 8 = 0$$
$$\Rightarrow 12a = 36$$
$$\Rightarrow a = 3$$

Online

Trace out the graph of the derivative of an inputted function.

Recall that the derivative of a function is a formula for the **instantaneous rate of change** of the dependent variable (commonly y) with respect to the independent variable (x). In other words, the gradient of the tangent at a point gives the gradient, or rate of change, of the curve at that point. The gradient of a **secant line** (that crosses the curve at two points) gives the **average rate of change** between the two points.

Example 9.6

Boiling water is poured into a cup. The temperature of the water in degrees Celsius, C, after t minutes is given by $C = 19 + \frac{182}{t^{\frac{3}{2}}}$, for $t \geqslant 1$ minute.

(a) Find the average rate of change of the temperature from $t = 2$ to $t = 6$.

(b) Find the rate of change of the temperature at the instant when $t = 4$.

Solution

(a) When $t = 2$, $C \approx 83.35°C$ and when $t = 6$, $C \approx 31.38°C$. The average rate of change from $t = 2$ to $t = 6$ is the gradient of the line through the points (2, 83.35) and (6, 31.38).

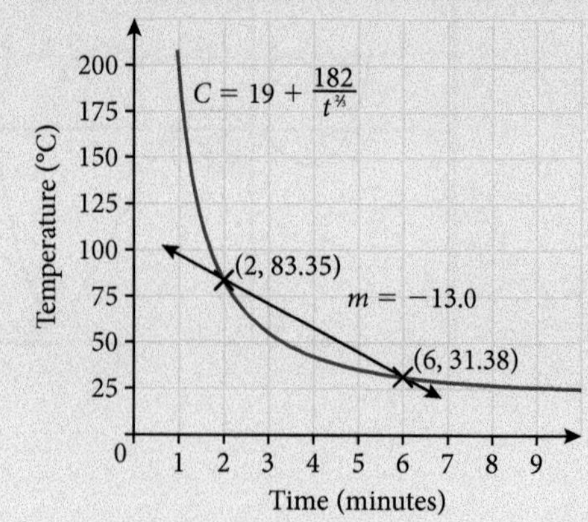

Average rate of change $= \frac{83.35 - 31.38}{2 - 6} = \frac{51.97}{-4} = -12.9925$

To an accuracy of three significant figures, the average rate of change from $t = 2$ to $t = 6$ is $-13.0°$C per minute. During that period of time, the water is, on average, getting 13.0° C cooler every minute.

(b) Work out the derivative $\frac{dC}{dt}$; that is, the rate of change of temperature with respect to time t, from which we can compute the rate at which the temperature is changing at the instant when $t = 4$.

$$\frac{dC}{dt} = \frac{d}{dt}\left(19 + \frac{182}{t^{\frac{3}{2}}}\right) = \frac{d}{dt}\left(19 + \frac{182}{t^{-\frac{3}{2}}}\right) = \frac{d}{dt}(19) + 182\frac{d}{dt}\left(t^{-\frac{3}{2}}\right)$$

$$= 0 + 182\left(-\frac{3}{2}t^{-\frac{5}{2}}\right) = -273t^{-\frac{5}{2}}$$

$$= -\frac{273}{t^{\frac{5}{2}}} = -\frac{273}{\sqrt{t^5}}$$

When $t = 4$:

$$\frac{dC}{dt} = -\frac{273}{\sqrt{4^5}} = -\frac{273}{32} \approx -8.53$$

Therefore, the instantaneous rate of change of temperature at $t = 4$ min is $-8.53°$C per min.

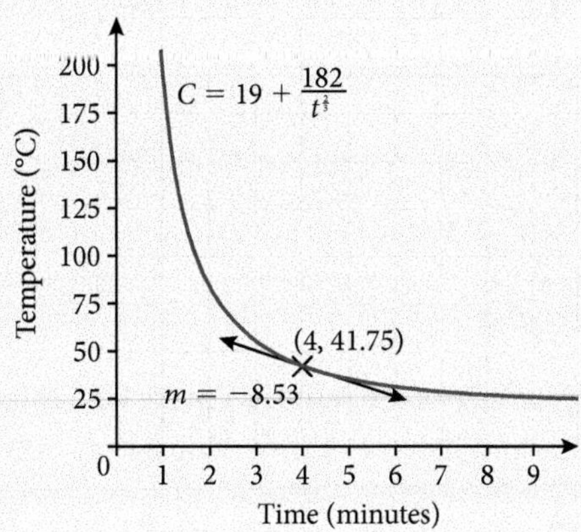

Derivatives of the sine and cosine functions

The graph of $y = \sin x$ (Figure 9.24) is periodic, with period 2π, so the same will be true of its derivative, which gives the gradient at each point on the graph. Therefore, we need to consider only the portion of the graph in the interval $0 \leqslant x \leqslant 2\pi$.

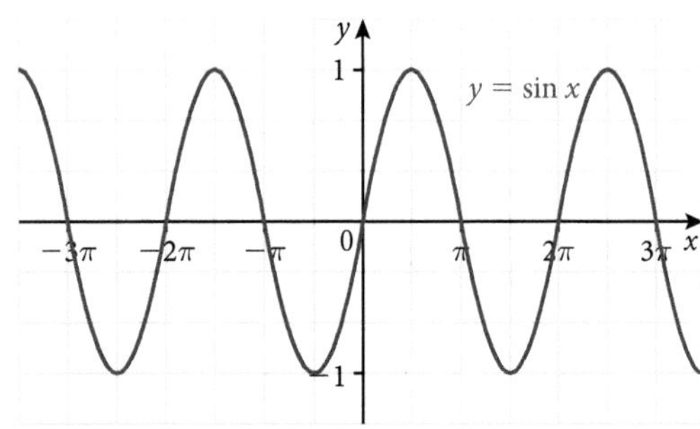

Figure 9.24 Graph of $y = \sin x$

Figure 9.25 shows two pairs of axes having equal scales on the x- and y-axes and corresponding x-coordinates aligned vertically. On the top pair of axes, $y = \sin x$ is graphed with tangent lines drawn at nine selected points. The points were chosen such that the gradients of the tangents at those points, in order, are equal to $1, \frac{1}{2}, 0, -\frac{1}{2}, -1, -\frac{1}{2}, 0, \frac{1}{2}, 1$. The values of these gradients were then plotted in the bottom graph, with the y-coordinate of each point indicating the gradient of the curve for that particular x value. Hence, the points in the bottom pair of axes should be on the graph of the derivative of $y = \sin x$.

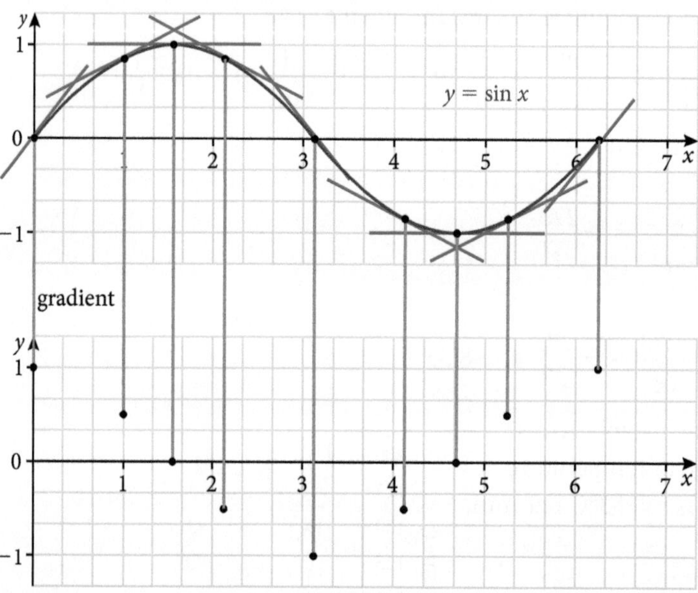

Figure 9.25 Values of gradient for selected tangents to $y = \sin x$ (shown in the top graph) plotted in the lower graph

Note that the graphs in Figures 9.24, 9.25, and 9.26 have x in radians. As mentioned previously, we must use radian measure when trigonometric functions are involved in calculus.

Figure 9.26 is the same as Figure 9.25 except that the graph of $y = \sin x$, the grid lines, and the lines connecting points between the two graphs have been removed.

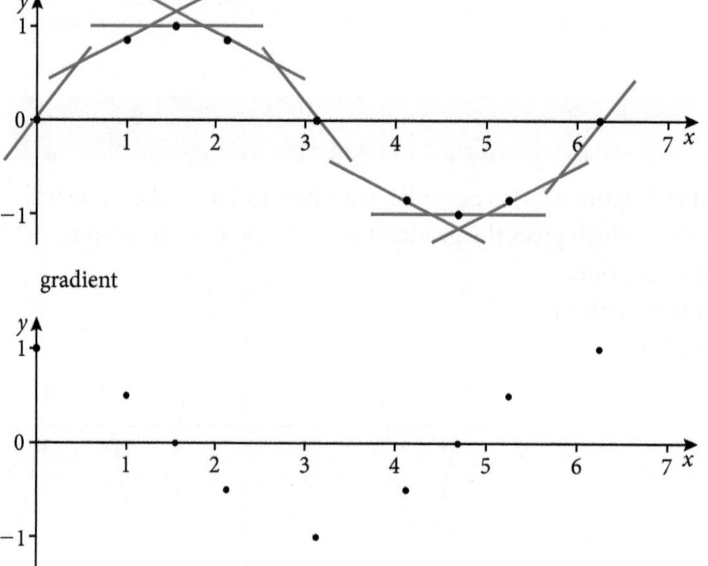

Figure 9.26 Tangent lines for $y = \sin x$ above and values of gradients of the tangent lines plotted below

This leads to an obvious choice for our conjecture for the derivative of the sine function. For $f(x) = \sin x$, it appears that $f'(x) = \cos x$. Let's use our GDC to provide confirmation of this conjecture.

The discrepancies beyond 6 significant figures are due to the small amount of error in the algorithm used by the GDC to compute the derivative of a function at a point.

The GDC screen images below show the derivative of $\sin x$ being evaluated for various values of x, along with $\cos x$ for the same x value. Note that the GDC must be in radian mode. You can see that the values are equivalent (to 6 significant figures).

```
nDeriv(sin(X),X,
π/4)
       .7071066633
cos(π/4)
       .7071067812
```

```
nDeriv(sin(X),X,
5π/6)
      -.8660252595
cos(5π/6)
      -.8660254038
```

```
nDeriv(sin(X),X,
5.25)
       .5120853919
cos(5.25)
       .5120854772
```

Figure 9.27 GDC screens showing the derivative of $\sin x$ being evaluated

Derivative of the sine function

If $f(x) = \sin x$, then $f'(x) = \cos x$. Or, in Leibniz notation, $\frac{d}{dx}(\sin x) = \cos x$.

This result is only true when x is in radian measure.

Now let's use a GDC to graph the derivative of $\cos x$, from which we should be able to make a conjecture.

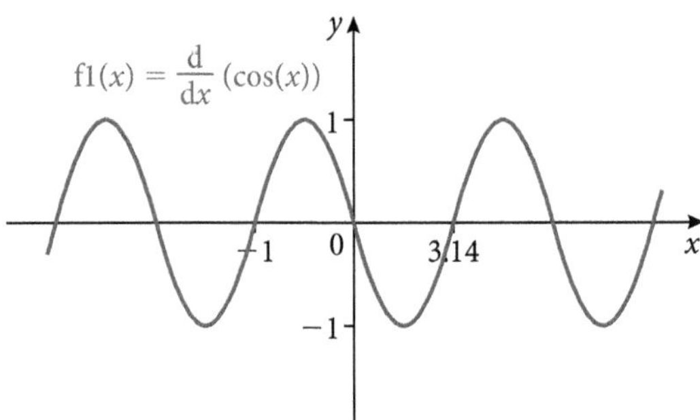

Figure 9.28 Graph of the derivative of $y = \cos x$; that is, graph of $y = \frac{d}{dx}(\cos x)$

At first glance, the graph of $y = \frac{d}{dx}(\cos x)$ looks like the graph of $y = \sin x$, but on closer inspection we see that it is the reflection about the x-axis of $y = \sin x$. Therefore, our conjecture is that the derivative of $\cos x$ is $-\sin x$.

That is $\frac{d}{dx}(\cos x) = -\sin x$

Again, let's use our GDC to help confirm our conjecture.

```
nDeriv(cos(X),X,
π/6)
      -.4999999167
-sin(π/6)
               -.5
```

```
nDeriv(cos(X),X,
3π/2)
       .9999998333
-sin(3π/2)
                 1
```

```
nDeriv(cos(X),X,
9)
      -.4121184166
-sin(9)
      -.4121184852
```

Figure 9.29 Using a GDC to confirm our conjecture

Derivative of the cosine function

If $f(x) = \cos x$, then $f'(x) = -\sin x$. Or, in Leibniz notation $\frac{d}{dx}(\cos x) = -\sin x$.

This result is only true when x is in radian measure.

Exercise 9.2

1. Find the gradient of the tangent line of the graph of each function at the point where $x = 1$. Sketch each function and draw a line tangent to the graph at $x = 1$.

 (a) $f(x) = 1 - x^2$ **(b)** $g(x) = x^3 + 2$

 (c) $h(x) = \sqrt{x}$ **(d)** $r(x) = \dfrac{1}{x^2}$

2. For each function:

 (i) find the derivative

 (ii) compute the gradient of the graph of the function at the indicated point.

 Use a GDC to confirm your results.

 (a) $y = 3x^2 - 4x$ point $(0, 0)$

 (b) $y = 1 - 6x - x^2$ point $(-3, 10)$

 (c) $y = \dfrac{2}{x^3}$ point $(-1, -2)$

 (d) $y = x^5 - x^3 - x$ point $(1, -1)$

 (e) $y = \sin x$ point $\left(\dfrac{\pi}{3}, \dfrac{\sqrt{3}}{2}\right)$

 (f) $y = (x + 2)(x - 6)$ point $(2, -16)$

 (g) $y = 2x + \dfrac{1}{x} - \dfrac{3}{x^3}$ point $(1, 0)$

 (h) $y = \dfrac{x^3 + 1}{x^2}$ point $(-1, 0)$

 (i) $y = \cos x$ point $\left(\dfrac{3\pi}{4}, \dfrac{\sqrt{2}}{2}\right)$

3. The gradient of the curve $y = x^2 + ax + b$ at the point $(2, -4)$ is -1. Find the value of a and the value of b.

4. Use the graph of f to answer each question.

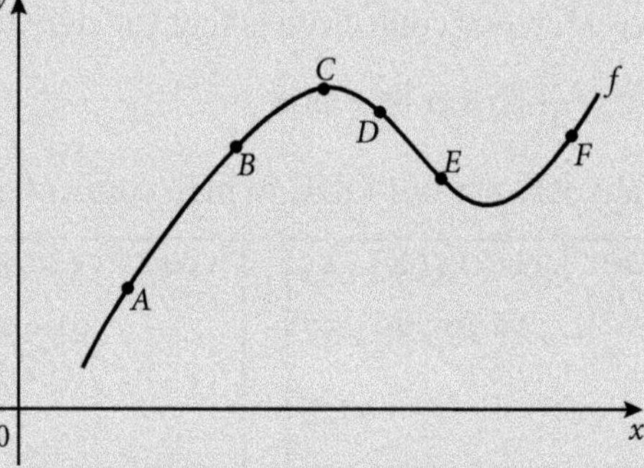

 (a) Between which two consecutive points is the average rate of change of the function greatest?

 (b) Indicate at which point(s) the instantaneous rate of change of f is:

 (i) positive **(ii)** negative **(iii)** zero

 (c) For what two pairs of consecutive points is the average rate of change approximately equal?

5. For each function, find the coordinates of any points on the graph of the function where the gradient is equal to the given value.

   (a) $y = x^2 + 3x$ gradient $= 3$

   (b) $y = x^3$ gradient $= 12$

   (c) $y = x^2 - 5x + 1$ gradient $= 0$

   (d) $y = x^2 - 3x$ gradient $= -1$

6. The gradient of the curve $y = x^2 - 4x + 6$ at the point $(3, 3)$ is equal to the gradient of the curve $y = 8x - 3x^2$ at (a, b). Find the value of a and the value of b.

7. The graph of the equation $y = ax^3 - 2x^2 - x + 7$ has a gradient of 3 at the point where $x = 2$. Find the value of a.

8. Find the coordinates of the point on the graph of $y = x^2 - x$ at which the tangent is parallel to the line $y = 5x$.

9. A car is parked with the windows and doors closed for five hours. The temperature inside the car in degrees Celsius, C, is given by $C = 2\sqrt{t^3} + 17$, with t representing the number of hours since the car was first parked.

   (a) Find the average rate of change of the temperature between $t = 1$ and $t = 4$.

   (b) Find the function which gives the instantaneous rate of change of the temperature for any time t, $0 < t < 5$.

   (c) Find the time t at which the instantaneous rate of change of the temperature is equal to the average rate of change from $t = 1$ to $t = 4$.

9.3 Maxima and minima: first and second derivatives

The relationship between a function and its derivative

The derivative is a function derived from a function f that gives the gradient (slope) of the graph of f at any x in the function's domain, given that the curve is differentiable at the value of x. The derivative is a gradient, or rate of change, function. Knowing the gradient of a function at different values in its domain tells us about properties of the function and the shape of its graph.

In the previous section, we observed that if a graph has a turning point at a particular point (for example, at the vertex of a parabola), then it has a horizontal tangent (gradient $= 0$) at the point. Hence, the derivative will equal zero at a turning point. In Section 2.1 (Quadratic functions), we found the vertex of the graph of a quadratic function by using the technique of completing the square to write its equation in vertex form. We can also find the

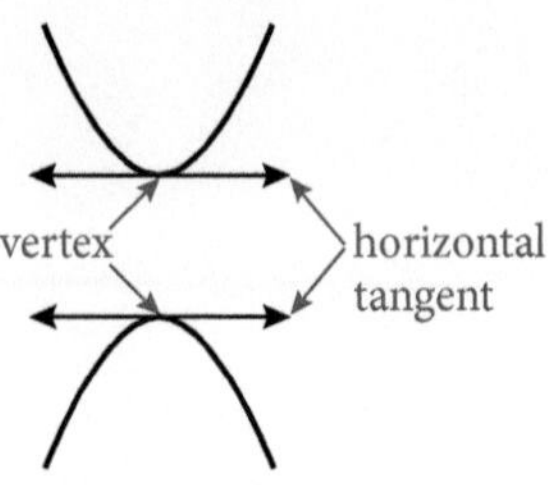

Figure 9.30 The tangent to a parabola at its vertex is horizontal

vertex by means of differentiation. As we look at the graph of a parabola moving from left to right (i.e. domain values increasing), it either turns from going down to going up (decreasing to increasing), or from going up to going down (increasing to decreasing), as in Figure 9.30.

Example 9.7

Using differentiation, find the coordinates of the vertex of the parabola with the equation $y = x^2 - 8x + 14$.

Solution

Find the value of x for which the derivative, $\frac{dy}{dx}$, is zero.

$$\frac{dy}{dx} = \frac{d}{dx}(x^2 - 8x + 14) = 2x - 8 = 0 \Rightarrow x = 4$$

The x-coordinate of the vertex is 4.

To find the y-coordinate of the vertex, substitute $x = 4$ into the equation: $y = 4^2 - 8(4) + 14 = -2$

Therefore, the vertex has coordinates $(4, -2)$

We know that the parabola in Example 9.7 will 'open upwards' because the coefficient of the quadratic term, x^2, is positive. The parabola has a negative gradient (decreasing) to the left of the vertex and a positive gradient (increasing) to the right of the vertex. As the values of x increase, the derivative of $y = x^2 - 8x + 14$ will change from negative, to zero, to positive, accordingly.

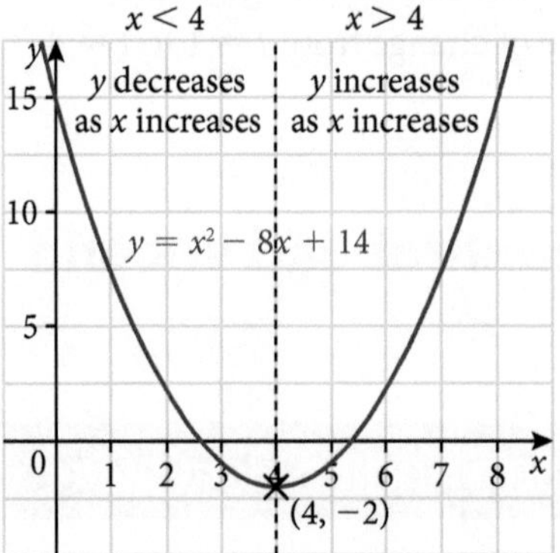

Figure 9.31 Gradient changes from negative to positive as x increases

$$\frac{dy}{dx} = 2x - 8 \Rightarrow \frac{dy}{dx} < 0 \text{ for } x < 4 \text{ and } \frac{dy}{dx} = 0 \text{ for } x = 4 \text{ and } \frac{dy}{dx} > 0 \text{ for } x > 4$$

In other words, the function $f(x) = x^2 - 8x + 14$ is decreasing for all $x < 4$; it is neither decreasing nor increasing at $x = 4$; and it is increasing for all $x < 4$. A point at which a function is neither increasing nor decreasing (where there is a horizontal tangent) is called a **stationary point**. A convenient way to demonstrate where a function is increasing or decreasing and the location of any stationary points is with a **sign chart** for the function and its derivative (Figure 9.32) for $f(x) = x^2 - 8x + 14$. The derivative $f'(x) = 2x - 8$ is zero only at

$x = 4$, thereby dividing the domain of f (i.e. $\mathbb{R}$) into two intervals: $x < 4$ and $x > 4$. $f'(x) = 2x - 8$ is a **continuous** function (i.e. no gaps in the domain) so it is only necessary to test one point in each interval in order to determine the sign of all the values of the derivative in that interval. $f'(x)$ can only change sign at $x = 4$. For example, the fact that $f'(3) = 2(3) - 8 = -2 < 0$ means that $f'(x) < 0$ for all x when $x < 4$. Therefore, f is decreasing for all x in the open interval $(-\infty, 4)$.

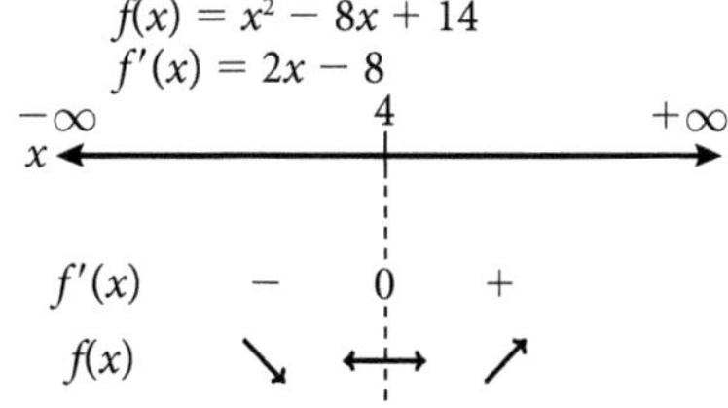

Figure 9.32 Sign chart for $f(x)$ and $f'(x)$

If $f'(x) > 0$ for $a < x < b$, then $f(x)$ is **increasing** on the interval $a < x < b$

If $f'(x) < 0$ for $a < x < b$, then $f(x)$ is **decreasing** on the interval $a < x < b$

If $f'(x) = 0$ for $a < x < b$, then $f(x)$ is **constant** on the interval $a < x < b$

If $f'(x) = 0$ for a single value $x = c$ on some interval $a < c < b$, then $f(x)$ has a **stationary point** at $x = c$. The corresponding point $(c, f(c))$ on the graph of f is called a stationary point.

It is at stationary points, or endpoints of the domain (if the domain is not all real numbers), where a function may have a maximum or minimum value. These points at which extreme values of a function may occur are often referred to as **critical points**. Whether a function is increasing or decreasing on either side of a stationary point will indicate whether the stationary point is a maximum, minimum or neither.

Example 9.8

Consider the function $f(x) = 2x^3 + 3x^2 - 12x - 4, x \in \mathbb{R}$

(a) Find the coordinates of any stationary points of f.

(b) Using the derivative of f, classify any stationary points as a maximum, a minimum, or neither.

Solution

(a) $f'(x) = 6x^2 + 6x - 12 = 0 \Rightarrow 6(x^2 + x - 2) = 0$

$$\Rightarrow 6(x + 2)(x - 1) = 0$$

$$\Rightarrow x = -2 \text{ or } x = 1$$

With a domain of all real numbers, there are no domain endpoints that may be an extreme value. Thus, f has two critical points: one at $x = -2$ and the other at $x = 1$.

When $x = -2$: $y = 2(-2)^3 + 3(-2)^2 - 12(-2) - 4 = 16$

So f has a stationary point at $(-2, 16)$

And when $x = 1$: $y = 2(1)^3 + 3(1)^2 - 12(1) - 4 = -11$

So f has a stationary point at $(1, -11)$

(b) Construct a sign chart for $f'(x)$ and $f(x)$ to show where f is increasing or decreasing. The derivative $f'(x)$ has two zeros, $x = -2$ and $x = -1$, thereby dividing the domain of f into three intervals that need to be tested.

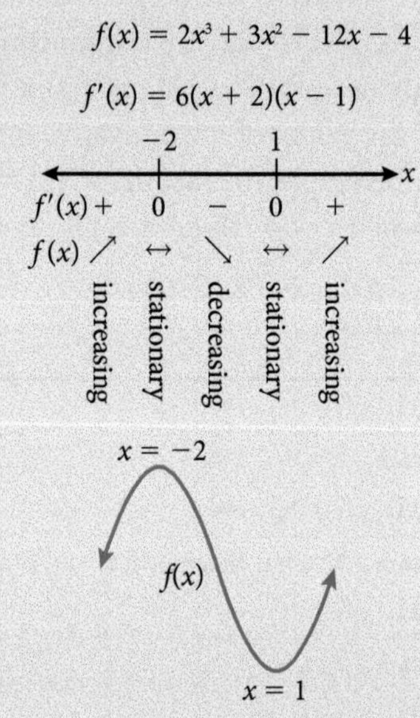

Since $f'(-3) = 6(-1)(-4) = 24 > 0$ then $f'(x) > 0$ for all $x < -2$. Likewise, since $f'(2) = 6(4)(1) = 24 > 0$ then $f'(x) > 0$ for all $x > 1$. Thus, f is increasing on the open intervals $(-\infty, -2)$ and $(1, \infty)$.

Since $f'(0) = -12 < 0$ then $f'(x) < 0$ for all x such that $-2 < x < 1$. Thus, f is decreasing on the open interval $(-2, 1)$, i.e. $-2 < x < 1$.

From this information, we can visualise for increasing values of x that the graph of f is going up for all $x < -2$, then turning down at $x = -2$, then going down for values of x from -2 to 1, then turning up at $x = 1$, and then going up for all $x > 1$. The basic shape of the graph of f will look something like the sketch in Figure 9.33. Clearly the stationary point $(-2, 16)$ is a maximum and the stationary point $(1, -11)$ is a minimum.

The graph of $f(x) = 2x^3 + 3x^2 - 12x - 4$ from Example 9.8 confirms the results acquired from analysing the derivative of f.

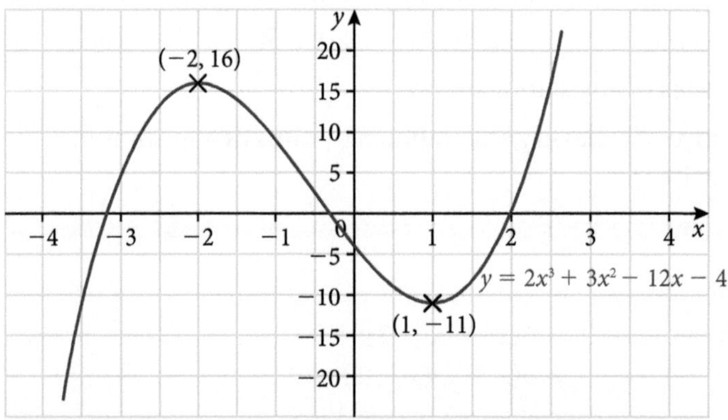

Figure 9.33 Gradient changes from negative to positive as x increases

The plural of maximum is maxima, and the plural of minimum is minima. Maxima and minima are collectively referred to as extrema – the plural of extremum. Extrema of a function that do not occur at domain endpoints will be turning points of the graph of the function.

For Example 9.8, we can express the result for part (b) most clearly by saying that $f(x)$ has a **relative maximum value** of 16 at $x = -2$, and a **relative minimum** value of -11 at $x = 1$. The reason that these extreme values are described as relative (or local) is because they are a maximum or minimum for the function in the immediate vicinity of the point but not for the entire domain of the function. A point that is a maximum/minimum for the entire domain is called an **absolute**, or **global**, **maximum/minimum**.

The first derivative test

From Example 9.8, we can see that a function f has a maximum at some $x = c$ if $f'(c) = 0$ and f is increasing immediately to the left of $x = c$ and decreasing immediately to the right of $x = c$. Similarly, f has a minimum at some $x = c$ if $f'(c) = 0$ and f is decreasing immediately to the left of $x = c$ and increasing immediately to the right of $x = c$. It is important to understand, however, that not all stationary points are either a maximum or a minimum.

Example 9.9

For the function $f(x) = x^4 - 2x^3$, find the coordinates of all stationary points and describe them completely.

Solution

$$f'(x) = \frac{d}{dx}(x^4 - 2x^3) = 4x^3 - 6x^2 = 0$$
$$\Rightarrow 2x^2(2x - 3) = 0$$
$$\Rightarrow x = 0 \text{ or } x = \frac{3}{2}$$

The implied domain is all real numbers, so $x = 0$ and $x = \frac{3}{2}$ are the critical points of f.

when $x = 0$, $y = f(0) = 0$; and when $x = \frac{3}{2}$, $y = f\left(\frac{3}{2}\right) = \left(\frac{3}{2}\right)^4 - 2\left(\frac{3}{2}\right)^3$

$$= \frac{81}{16} - \frac{54}{8} = -\frac{27}{16}$$

Therefore, f has stationary points at (0, 0) and $\left(\frac{3}{2}, -\frac{27}{16}\right)$.

Because f has two stationary points, there are three intervals for which to test the sign of the derivative. We could use a sign chart as shown previously, or we can use a more detailed table (see below) that summarises the testing of the three intervals and the two critical points.

Interval/point	$x < 0$	$x = 0$	$x < x < \frac{3}{2}$	$x = \frac{3}{2}$	$x > \frac{3}{2}$
Test value	$x = -1$		$x = 1$		$x = 2$
Sign of $f'(x)$	$f'(-1) = -10 < 0$	0	$f'(1) = -2 < 0$	0	$f'(2) = 8 > 0$
Conclusion	f decreasing ↘	none	f decreasing ↘	abs. min.	f increasing ↗

On either side of $x = 0$, f does not change from either decreasing to increasing or from increasing to decreasing. Although there is a horizontal tangent at (0, 0), it is **not** an extreme value (turning point). The function steadily decreases as x approaches zero, then at $x = 0$ the function has a rate of change (gradient) of zero for an instant and then continues on decreasing.

As x approaches $\frac{3}{2}$, f is decreasing and then switches to increasing at $x = \frac{3}{2}$.

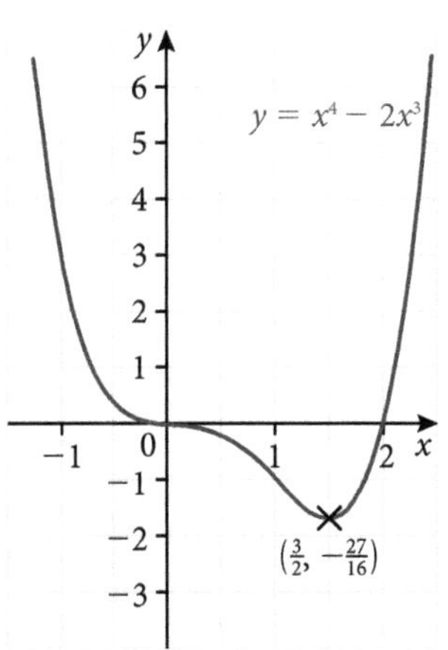

Figure 9.34 Graph for solution to Example 9.9

Therefore, the stationary point $(0, 0)$ is neither a maximum nor a minimum; and the stationary point $\left(\frac{3}{2}, -\frac{27}{16}\right)$ is an absolute minimum. In other words, f has an absolute (global) minimum value of $-\frac{27}{16}$ at $x = \frac{3}{2}$. The reason that an absolute, rather than a relative, minimum value occurs at $x = \frac{3}{2}$ is because for all $x < \frac{3}{2}$ the function f is either increasing or constant (at $x = 0$) and for all $x > \frac{3}{2}$ the function f is increasing.

First derivative test for maxima and minima of a function

Suppose that $x = c$ is a critical point of a continuous and smooth function f. That is, $f(c) = 0$ and $x = c$ is a stationary point or $x = c$ is an endpoint of the domain.

At a stationary point $x = c$:

If $f'(x)$ changes sign from positive to negative as x increases through $x = c$, then f has a relative maximum at $x = c$.

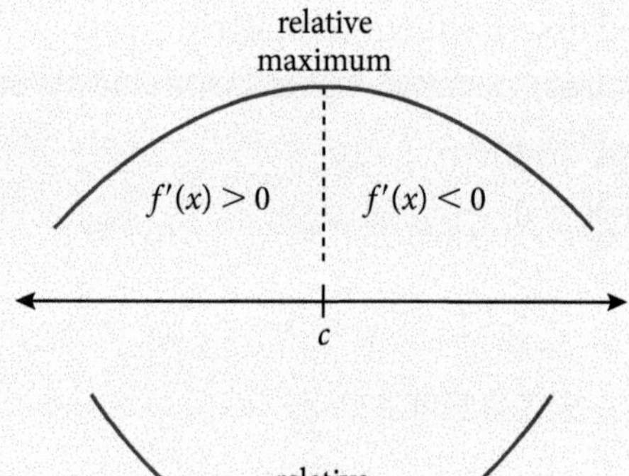

If $f'(x)$ changes sign from negative to positive as x increases through $x = c$, then f has a relative minimum at $x = c$.

relative
minimum
$f'(x) < 0$
$f'(x) > 0$
c

If $f'(x)$ does not change sign as x increases through $x = c$, then f has neither a relative maximum nor a relative minimum at $x = c$.

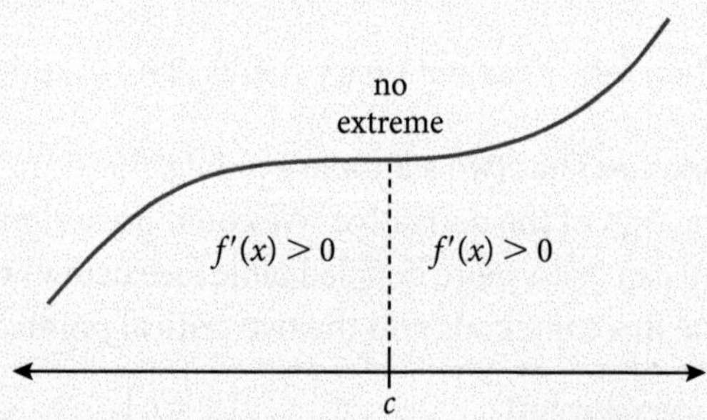

When $x = c$ is an endpoint of the domain, then $x = c$ will be a relative maximum or minimum of f if the sign of $f'(x)$ is always positive or always negative for $x > c$ (at a left endpoint), or for $x < c$ (at a right endpoint).

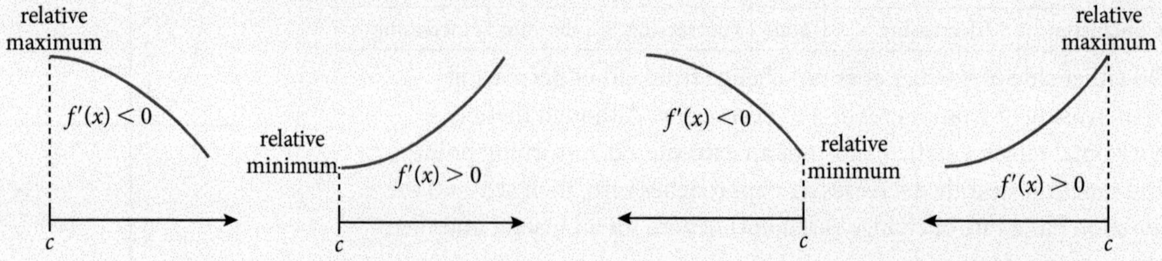

If it is possible to show that a relative maximum or minimum at $x = c$ is the greatest or least value for the entire domain of f, then it is classified as an absolute maximum or minimum.

Example 9.10

Apply the first derivative test to find any extreme values for $f(x) = 4x^3 - 9x^2 - 120x + 25$

Solution

$$f'(x) = \frac{d}{dx}(4x^3 - 9x^2 - 120x + 25) = 12x^2 - 18x - 120$$

$$f'(x) = 12x^2 - 18x - 120 = 0$$

$$\Rightarrow 6(2x^2 - 3x - 20) = 0$$

$$\Rightarrow 6(2x + 5)(x - 4) = 0$$

Thus, f has stationary points at $x = -\frac{5}{2}$ and $x = 4$

To classify the stationary point at $x = -\frac{5}{2}$, we need to choose test points on either side of $-\frac{5}{2}$, for example $x = -3$ (left) and $x = 0$ (right).

$$f'(-3) = 6(-1)(-7) = 42 > 0$$

$$f'(0) = 6(5)(-4) = -120 < 0$$

so f has a relative maximum at $x = -\frac{5}{2}$

$$f\left(-\frac{5}{2}\right) = 4\left(-\frac{5}{2}\right)^3 - 9\left(-\frac{5}{2}\right)^2 - 120\left(-\frac{5}{2}\right) + 25 = 206.25$$

Therefore, f has a relative maximum value of 206.25 at $x = -\frac{5}{2}$

To classify the stationary point at $x = 4$, we need to choose test points on either side of 4, for example $x = 0$ (left) and $x = 5$ (right).

$$f'(0) = 6(5)(-4) = -120 < 0$$

$$f'(5) = 6(15)(1) = 90 > 0$$

so f has a relative minimum at $x = 4$

$$f(4) = 4(4)^3 - 9(4)^2 - 120(4) + 25 = -343$$

Therefore, f has a relative minimum value of -343 at $x = 4$

Change in displacement and velocity

Consider the motion of an object such that its position s relative to a reference point or line as a function of time t is given by $s(t)$. The **displacement** of the object over the time interval from t_1 to t_2 is:

$$\text{change in } s = \text{displacement} = s(t_2) - s(t_1)$$

The **average velocity** of the object over the time interval is:

$$v_{avg} = \frac{\text{displacement}}{\text{change in time}} = \frac{s(t_2) - s(t_1)}{t_2 - t_1}$$

The object's **instantaneous velocity** at a particular time, t, is the value of the derivative of the position function, s, with respect to time at t.

$$\text{velocity} = \frac{ds}{dt} = s'(t)$$

Example 9.11

A toy rocket is launched upwards into the air. Its vertical position, s metres, above the ground at t seconds is given by $s(t) = -5t^2 + 18t + 1$

(a) Find the average velocity over the time interval from $t = 1$ second to $t = 2$ seconds.

(b) Find the instantaneous velocity at $t = 1$ second.

(c) Find the maximum height reached by the rocket and the time at which this occurs.

Solution

(a) $v_{\text{avg}} = \dfrac{s(2) - s(1)}{2 - 1} = \dfrac{[-5(2)^2 + 18(2) + 1] - [-5 + 18 + 1]}{1} = 3\,\text{m s}^{-1}$

(b) $s'(t) = -10t + 18 \Rightarrow s'(1) = -10 + 18 = 8\,\text{m s}^{-1}$

(c) $s'(t) = -10t + 18 = 0 \Rightarrow t = 1.8$

Thus, s has a stationary point at $t = 1.8$. t must be positive and ranges from time of launch ($t = 0$) until when the rocket hits the ground ($h = 0$).

$$s(t) = -5t^2 + 18t + 1 = 0 \Rightarrow t = \frac{-18 \pm \sqrt{18^2 - 4(-5)(1)}}{2(-5)}$$

$$\Rightarrow t \approx -0.5472 \text{ or } t \approx 3.655$$

So, the rocket hits the ground about 3.66 seconds after the time of launch. Hence, the domain for the position (s) and velocity (v) functions is $0 \leqslant t \leqslant 3.66$. Therefore, the function s has three critical points: $t = 0$, $t = 1.8$, and $t \approx 3.66$.

Applying the first derivative test, we determine the sign of the derivative, $s'(t) = 0$, for values on either side of $t = 1.8$; for example $t = 0$ and $t = 2$. $s'(0) = 18 > 0$ and $s'(2) = -2 < 0$. Neither of the domain endpoints, $t = 0$ and $t \approx 3.66$, are at a maximum or minimum because the function is not constantly increasing or constantly decreasing before or after the endpoint. Since $s'(t)$ changes from increasing to decreasing at $t = 1.8$ and $s(1.8) = -5(1.8)^2 + 18(1.8) + 1 = 17.2$, then the toy rocket reaches a maximum height of 17.2 metres 1.8 seconds after it was launched.

A function and its second derivative

There is another useful test for the purpose of analysing the stationary point of a function that makes use of the derivative of the derivative, the second derivative, of the function.

When we differentiate a function $y = f(x)$, we obtain the first derivative $f'(x)$ $\left(\text{also written as } \dfrac{dy}{dx}\right)$. We can often also differentiate the derivative, which is denoted in Newton notation as $f''(x)$ or in Leibniz notation as $\dfrac{d^2y}{dx^2}$ and called the second derivative of f with respect to x. For example, if $f(x) = x^3$, then $f'(x) = 3x^2$ and $f''(x) = 6x$.

Second derivatives, like first derivatives, occur often in methods of applying calculus. In Example 9.11, the function $s(t)$ gave the position, in metres above the ground, of a projectile (toy rocket) where t, in seconds, is the time since the projectile was launched. The function $s'(t)$, the first derivative of the position function, gives the rate of change of the object's position – its velocity, in metres per second (m s^{-1}). Differentiation of this function gives the rate of change of the object's velocity – its acceleration, measured in metres per second per second (m s^{-2}).

The graphs of the position, velocity, and acceleration functions for Example 9.11 aligned vertically (Figure 9.35) nicely illustrate the relationships between a function, its first derivative, and its second derivative. The gradient of the graph of $s(t)$ is initially a large positive value (graph is steep) but steadily decreases until it is zero (horizontal tangent) at $t = 1.8$ and then continues to decrease becoming a large negative value (again, steep, but in the other direction). This corresponds to the real-life situation in which the rocket is launched with a high initial velocity ($v(0) = 18\ \text{m s}^{-1}$) and then its velocity decreases steadily due to gravity. The rocket's velocity is zero for just an instant when it reaches its maximum height at $t = 1.8$ and then its velocity becomes more and more negative because it has changed direction and is moving back (negative direction) to the ground. The rate of change of the velocity, $v'(t)$, is constant and it is negative because the velocity is decreasing from positive values to zero to negative values. This is clear from the fact that the graph of the velocity function, $v(t)$, is a straight line with a negative gradient. It follows then that the acceleration function – the rate of change of velocity – is a negative constant, $a = -10$ in this case, and its graph is a horizontal line.

Position function:
$s(t) = -5t^2 + 18t + 1$

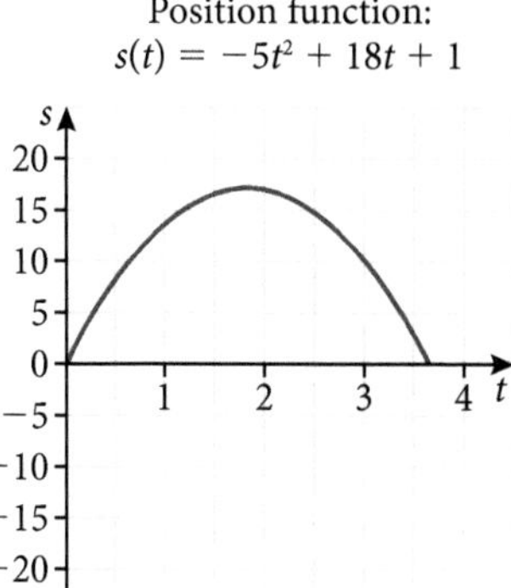

Velocity function:
$v(t) = s'(t) = -10t + 18$

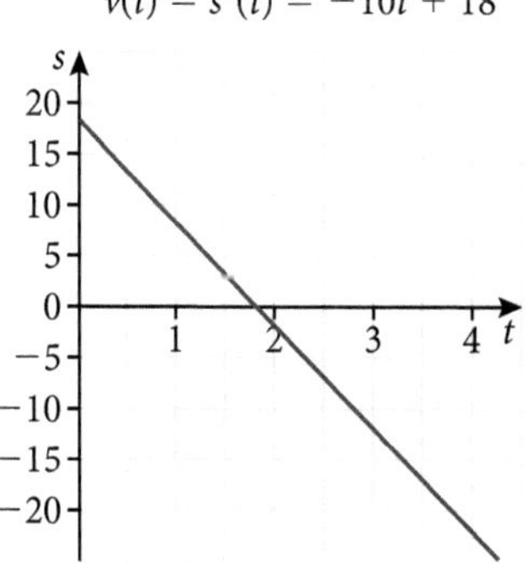

Acceleration function:
$a(t) = v'(t) = s''(t) = -10$

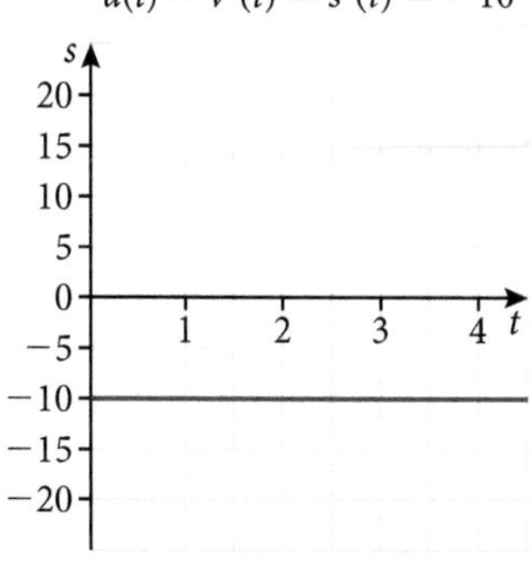

Figure 9.35 Graphs of the position, velocity and acceleration functions for Example 9.11

It would be incorrect to graph a function and its first and/or second derivative on the same axes because they have different units. For example, in Figure 9.35 the units on each vertical axis are: metres for $s(t)$, metres per second for $v(t)$ and metres per second per second for $a(t)$.

In Example 9.11, it is not possible to have a negative function value for $s(t)$ because the rocket's position is always above, or at, ground level. In many motion problems in calculus, we consider a simplified version by limiting an object's motion to a line with its position given as its **displacement** from a fixed point (usually the origin). At a position left of the fixed point the object's displacement is negative and at a position right of the fixed point the displacement is positive. Velocity can also be positive or negative depending on the direction of travel (i.e. the sign of the rate of change of the object's displacement). Likewise, acceleration is positive if velocity is increasing (i.e. rate of change of velocity is positive) and negative if velocity is decreasing.

If an object moves in a straight line such that at time t its displacement (position) from a fixed point is $s(t)$, then the first derivative $s'(t)$, also written as $\frac{ds}{dt}$, gives the velocity $v(t)$ at time t.

The second derivative $s''(t)$, also written as $\frac{d^2s}{dt}$, is the first derivative of $v(t)$. Hence the second derivative of the displacement, or position, function is a measure of the rate at which the velocity is changing; that is, it represents the acceleration of the object, which we express as

$$a(t) = v'(t) = s''(t) \text{ or } a(t) = \frac{dv}{dt} = \frac{d^2s}{dt^2}$$

Displacement can be negative, positive, or zero. **Distance** is the absolute value of displacement. **Velocity** can be negative, positive, or zero. **Speed** is the absolute value of velocity.

A common misconception is that acceleration is positive for motion in the positive direction (usually right or up) and negative for motion in the negative direction (usually left or down). Acceleration indicates how velocity is changing. Even though an object may be moving in a positive direction, e.g. to the right, if it is slowing down then its acceleration is acting in the opposite direction and is negative. In Example 9.11, the rocket was always accelerating in the negative direction, -10 m s^{-2}, due to the force of gravity.

Online

Simultaneous displacement and velocity graphs produced for linear motion.

Example 9.12

An object moves along a straight line so that after t seconds its displacement from the origin is s metres. Given that $s(t) = -2t^3 + 6t^2$, find:

(a) expressions for the (i) velocity and (ii) acceleration at time t seconds.

(b) the (i) initial velocity and (ii) initial acceleration of the object (when $t = 0$)

(c) the (i) maximum displacement and (ii) maximum velocity for the interval $0 \leqslant t \leqslant 3$

Solution

(a) (i) $v(t) = \dfrac{ds}{dt} = \dfrac{d}{dt}(-2t^3 + 6t^2) = -6t^2 + 12t$

(ii) $a(t) = \dfrac{d^2s}{dt^2} = \dfrac{dv}{dt} = \dfrac{d}{dt}(-6t^2 + 12t) = -12t + 12$

(b) (i) $v(0) = -6(0)^2 + 12(0) = 0$

The object's initial velocity is 0 m s^{-1}

(ii) $a(0) = -12(0) + 12 = 12$

The object's initial acceleration is 12 m s^{-2}

(c) (i) To find the maximum displacement, we can apply the first derivative test to $s(t)$. Since the first derivative of displacement, $s(t)$, is velocity, $v(t)$, then the critical points of $s(t)$ are where the velocity is zero (stationary points) and domain endpoints.

$s'(t) = v(t) = -6t^2 + 12t = 0 \Rightarrow 6t(-t + 2) = 0$

$v(t) = 0$ when $t = 0$ or $t = 2$

For the interval $0 \leqslant t \leqslant 3$ the critical points to be tested for finding the maximum displacement are at $t = 0$, $t = 2$ and $t = 3$. Check whether the velocity is increasing or decreasing on either side of the stationary point at $t = 2$ by finding the sign of $v(t)$ for $t = 1$ and $t = 2.5$

$v(1) = -6(1)^2 + 12(1) = 6$

$v(2.5) = -6(2.5)^2 + 12(2.5) = -7.5$

Hence, the displacement s is increasing for $0 < t < 2$ and decreasing for $2 < t < 3$. This indicates that the stationary point at $t = 2$ must be an absolute maximum for s in the interval $0 \leqslant t \leqslant 3$

$s(2) = -2(2)^3 + 6(2)^2 = 8$

Therefore, the object has a maximum displacement of 8 metres at $t = 2$ seconds.

(ii) To find the maximum velocity, we can apply the first derivative test to $v(t)$. The first derivative of $v(t)$ is acceleration $a(t)$, which is the second derivative of $s(t)$. Hence, where $s''(t) = 0$ (acceleration is zero) indicates critical points for $v(t)$; that is, where velocity may change from increasing to decreasing or vice versa.

$$s''(t) = a(t) = \frac{d}{dt}(-6t^2 + 12t) = -12t + 12 \Rightarrow 12(-t + 1) = 0$$

$$a(t) = 0 \text{ when } t = 1$$

For the interval $0 \leqslant t \leqslant 3$, the critical points to be tested for finding the maximum velocity are at $t = 0$, $t = 1$ and $t = 3$. Check whether the velocity is increasing or decreasing on either side of $t = 1$ by finding the sign of $a(t)$ for $t = 0.5$ and $t = 2$

$$a(0.5) = -12(0.5) + 12 = 6$$

$$a(2) = -12(2) + 12 = -12$$

Hence, the velocity v is increasing for $0 < t < 1$ and decreasing for $1 < t < 3$. This indicates that the point at $t = 1$ must be an absolute maximum for v in the interval $0 \leqslant t \leqslant 3$

$$v(1) = -6(1)^2 + 12(1) = 6$$

Therefore, the object has a maximum velocity of $6\ \text{m s}^{-1}$ at $t = 1$ s

The second derivative of a function tells us how the first derivative of the function changes. From this we can use the second derivative, as we did the first derivative, to reveal information about the shape of the graph of a function. Note in Example 9.12 that the object's velocity changed from increasing to decreasing when the object's acceleration was zero at $t = 1$. Let's examine graphically the significance of the point where acceleration is zero (i.e. velocity changing from increasing to decreasing) in connection to the displacement graph for Example 9.12. In other words, what can the second derivative of a function tell us about the shape of the function's graph?

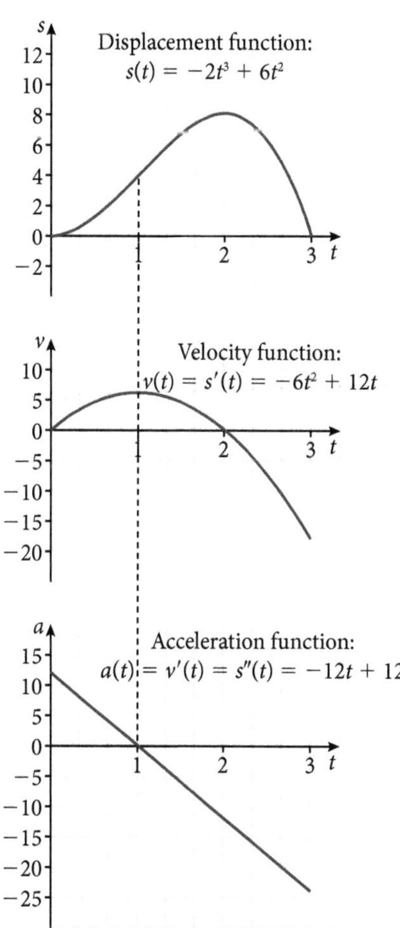

Figure 9.36 Graphs of the displacement, velocity, and acceleration functions for the motion of the object in Example 9.12

Figure 9.36 shows the graphs of the displacement, velocity and acceleration functions for the motion of the object in Example 9.12. A dashed vertical line shows the displacement, velocity and acceleration at $t = 1$. At this point velocity has a maximum value and acceleration is zero. It is also where velocity changes from increasing to decreasing, which has a corresponding effect on the shape of the displacement function $s(t)$. At $t = 1$, the graph of $s(t)$ changes from curving upwards (concave up) to curving downwards (concave down) because its gradient (corresponding to velocity) changes from increasing to decreasing. This can only occur when velocity (first derivative) has a maximum and, hence, where acceleration (second derivative) is zero. We can see from this illustration that for a general function $f(x)$, finding intervals where the first derivative $f'(x)$ is increasing (positive acceleration) or decreasing (negative acceleration) can be used to determine where the graph of $f(x)$ is curving upwards or curving downwards. A point at which a function's curvature (concavity) changes – as at $t = 1$ for the graph of $s(t)$ in Figure 9.36 – is called a **point of inflection.**

Online

The relationship between the values of the first and second derivative and the graph of a function.

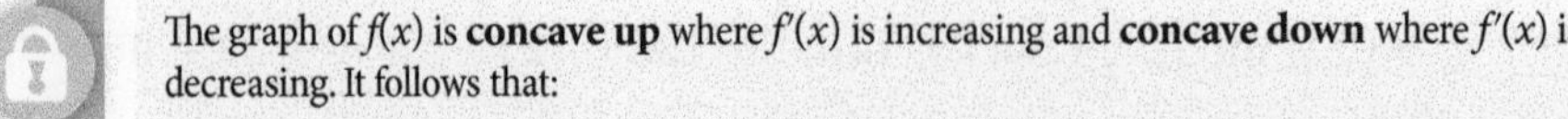

The graph of $f(x)$ is **concave up** where $f'(x)$ is increasing and **concave down** where $f'(x)$ is decreasing. It follows that:

if $f''(x) > 0$ for all x in some interval of the domain of f, then the graph of f is concave up in the interval; if $f''(x) < 0$ for all x in some interval of the domain of f, then the graph of f is concave down in the interval.

if $f(x)$ is a continuous function, its graph can only change concavity where $f''(x) = 0$. Hence, for a continuous function, an **inflection point** may only occur where $f''(x) = 0$.

Example 9.13

Determine the intervals on which the graph of $y = x^4 - 4x^3$ is concave up or concave down and identify any inflection points.

Solution

We first note that the function is continuous for its domain of all real numbers. To locate points of inflection, we then find for what value(s) the second derivative is zero.

$$\frac{dy}{dx} = \frac{d}{dx}(x^4 - 4x^3) = 4x^3 - 12x^2$$

$$\frac{d^2y}{dx^2} = \frac{d}{dx}(4x^3 - 12x^2) = 12x^2 - 24x = 12x(x - 2)$$

Setting $\frac{d^2y}{dx^2} = 0$, it follows that inflection points may occur at $t = 0$ and $t = 2$.

These two values divide the domain of the function into three intervals that we need to test. Let's choose $t = -1$, $t = 1$ and $t = 3$ as our test values. At $t = -1$, $\frac{d^2y}{dx^2} = 36 > 0$; at $t = 1$, $\frac{d^2y}{dx^2} = -2 < 0$; and at $t = 3$, $\frac{d^2y}{dx^2} = 36 > 0$. These results can be organised in a sign chart illustrating that the graph of $y = x^4 - 4x^3$ is concave up for the open intervals $(-\infty, 0)$ and $(2, \infty)$, and is concave down on the open interval $(0, 2)$.

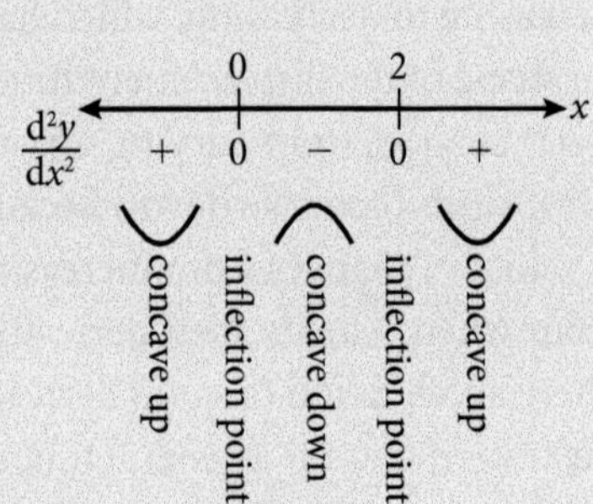

At $t = 0$, $y = 0$ and at $t = 2$, $y = 2^4 - 4(2)^3 = -16$.
Therefore, $(0, 0)$ and $(2, -16)$ are inflection points because it is at these points that the concavity of the graph changes.

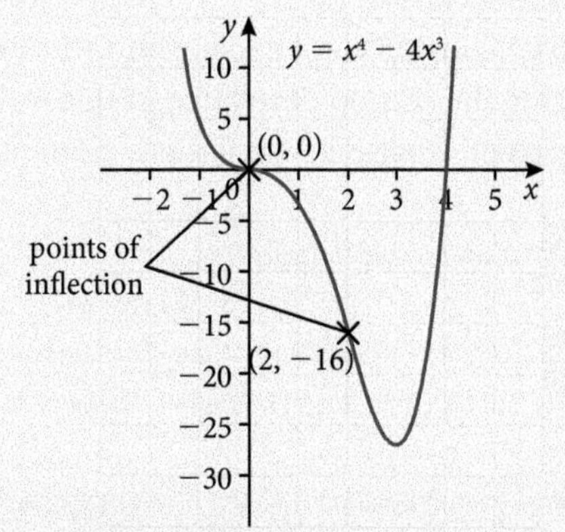

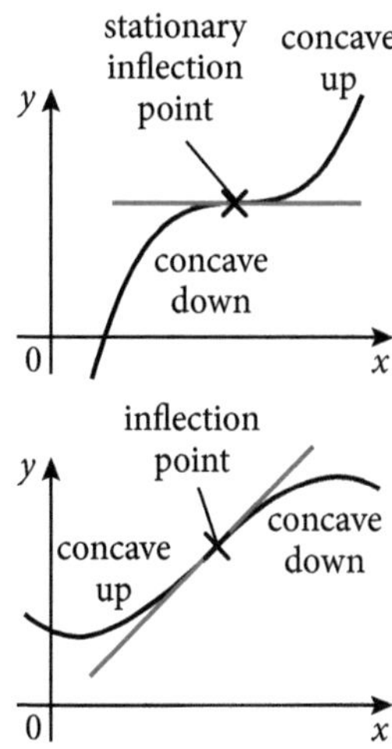

Figure 9.37 For either type of inflection point the graph crosses its tangent line at the point of inflection

The graph of the function from Example 9.13 reveals two different types of inflection point. The gradient of the curve at $(0, 0)$ is zero – that is, it is a stationary point. The gradient of the curve at the other inflection point $(2, -16)$ is negative.

For either type of inflection point the graph crosses its tangent line at the point of inflection as shown in Figure 9.37.

The fact that the second derivative of a function is zero at a certain point does not guarantee that an inflection point exists at the point. The functions $y = x^3$ and $y = x^4$ (Figures 9.38 and 9.39) show that $\frac{d^2y}{dx^2} = 0$ is a necessary but not sufficient condition for the existence of an inflection point.

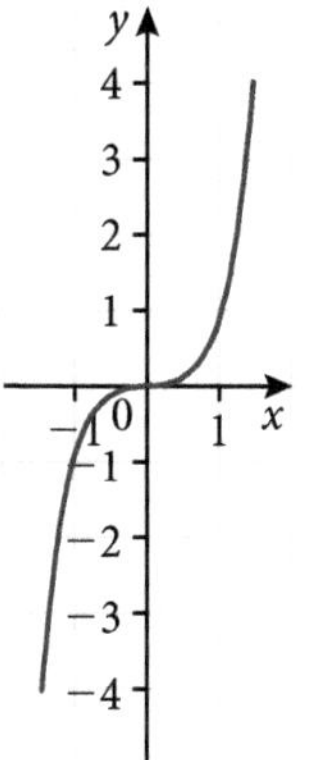

Figure 9.38 Graph of $y = x^3$

For $y = x^3$: $\frac{dy}{dx} = \frac{d}{dx}(x^3) = 3x^2 \Rightarrow \frac{d^2y}{dx^2} = \frac{d}{dx}(3x^2) = 6x \Rightarrow \frac{d^2y}{dx^2} = 0$ at $x = 0$.
We can conclude from this that there *may* be an inflection point at $x = 0$.
We need to investigate further by checking to see if $\frac{d^2y}{dx^2}$ changes sign at $x = 0$.
At $x = -1$, $\frac{d^2y}{dx^2} = -6$ and at $x = 1$, $\frac{d^2y}{dx^2} = 6$. Thus, there is an inflection point at $x = 0$ (Figure 9.38) because the second derivative changes sign at $x = 0$.

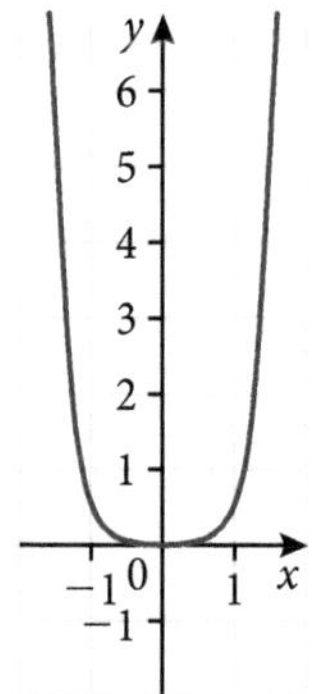

Figure 9.39 Graph of $y = x^4$

For $y = x^4$: $\frac{dy}{dx} = \frac{d}{dx}(x^4) = 4x^3 \Rightarrow \frac{d^2y}{dx^2} = \frac{d}{dx}(4x^3) = 12x^2 \Rightarrow \frac{d^2y}{dx^2} = 0$ at $x = 0$.

Again, we need to see if $\frac{d^2y}{dx^2}$ changes sign at $x = 0$. At $x = -1$, $\frac{d^2y}{dx^2} = 12$ and at $x = 1$, $\frac{d^2y}{dx^2} = 12$. Thus, there is no inflection point at $x = 0$ (Figure 9.39) because the second derivative does not change sign at $x = 0$.

The second derivative test

Instead of using the first derivative to check whether a function changes from increasing to decreasing (maximum) or decreasing to increasing (minimum) at a stationary point, we can simply evaluate the second derivative at the stationary point. If the graph is concave up at the stationary point then it will be a minimum, and if it is concave down then it will be a maximum.

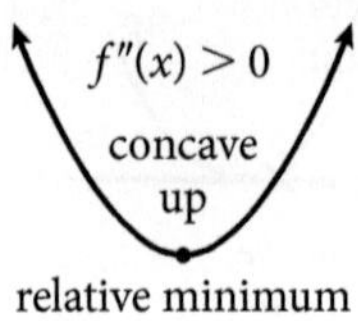

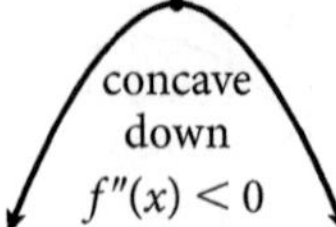

Figure 9.40 Relative minimum and relative maximum

If the second derivative is zero at a stationary point (as for $y = x^3$ and $y = x^4$), no conclusion can be made and we need to go back to the first derivative test. Using the second derivative in this way is a very efficient method for telling us whether a stationary point is a relative maximum or minimum.

The second derivative test

If $f'(c) = 0$ and $f''(c) < 0$, then f has a relative maximum at $x = c$.

If $f'(c) = 0$ and $f''(c) > 0$, then f has a relative minimum at $x = c$.

If $f'(c) = 0$, the test fails and the first derivative test should be applied.

Example 9.14

Find any relative extrema for $f(x) = 3x^5 - 25x^3 + 60x + 20$

Solution

The implied domain of f is all real numbers. Solve $f'(x) = 0$ to obtain possible extrema.

$$f'(x) = 15x^4 - 75x^2 + 60 = 0$$
$$15(x^4 - 5x^2 + 4) = 0$$
$$15(x^2 - 4)(x^2 - 1) = 0$$
$$15(x + 2)(x - 2)(x + 1)(x - 1) = 0$$

Therefore, f has four stationary points: $x = -2$, $x = 2$, $x = -1$ and $x = 1$.

Applying the second derivative test:

$$f''(x) = 60x^3 - 150x = 30x(2x - 5)$$
$f''(-2) = -180 < 0 \Rightarrow f$ has a relative maximum at $x = -2$

$f''(2) = 180 > 0 \Rightarrow f$ has a relative minimum at $x = 2$

$f''(-1) = 90 > 0 \Rightarrow f$ has a relative minimum at $x = -1$

$f''(1) = -90 < 0 \Rightarrow f$ has a relative maximum at $x = 1$

Exercise 9.3

1. Find the vertex of each parabola using differentiation.

(a) $y = x^2 - 2x - 6$ **(b)** $y = 4x^2 + 12x + 17$

(c) $y = -x^2 + 6x - 7$

2. For each function:

(i) find the derivative, $f'(x)$

(ii) indicate the interval(s) for which $f(x)$ is increasing

(iii) indicate the interval(s) for which $f(x)$ is decreasing.

(a) $y = x^2 - 5x + 6$ **(b)** $y = 7 - 4x - 3x^2$

(c) $y = \frac{1}{3}x^3 - x$ **(d)** $y = x^4 - 4x^3$

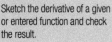

Online

Sketch the derivative of a given or entered function and check the result.

3. For each function:

(i) find the coordinates of any stationary points for the graph of the equation

(ii) state, with reasoning, whether each stationary point is a minimum, maximum, or neither

(iii) sketch a graph of the equation and indicate the coordinates of each stationary point on the graph.

(a) $y = 2x^3 + 3x^2 - 72x + 5$ (b) $y = \frac{1}{6}x^3 - 5$

(c) $y = x(x - 3)^2$ (d) $y = x^4 - 2x^3 - 5x^2 + 6$

(e) $y = x^3 - 2x^2 - 7x + 10$ (f) $y = x - \sqrt{x}$

4. An object moves along a line such that its displacement s metres at time t seconds from the origin O is given by $s(t) = t^3 - 4t^2 + t$.

(a) Find expressions for the object's velocity and acceleration in terms of t.

(b) For the interval $-1 \leqslant t \leqslant 3$, sketch the graphs of the displacement–time, velocity–time, and acceleration–time graphs of separate sets of axes vertically.

(c) For the interval $-1 \leqslant t \leqslant 3$, find the time at which the displacement is a maximum and find its value.

(d) For the interval $-1 \leqslant t \leqslant 3$, find the time at which the velocity is a minimum and find its value.

(e) Describe the motion of the object during the interval $-1 \leqslant t \leqslant 3$ accurately.

5. For each function $f(x)$:

(i) find any relative extrema and points of inflection

(ii) state the coordinates of any such points.

Use your GDC to assist you in sketching the function.

(a) $f(x) = x^3 - 12x$ (b) $f(x) = \frac{1}{4}x^4 - 2x^2$

(c) $f(x) = x + \frac{4}{x}$ (d) $f(x) = -3x^5 + 5x^3$

(e) $f(x) = 3x^4 - 4x^3 - 12x^2 + 5$

6. Consider the function $g(x) = x + 2\cos x$. For the interval $0 \leqslant x \leqslant 2\pi$:

(a) find the exact x-coordinates of any stationary points.

(b) determine whether each stationary point is a maximum, minimum, or neither and give a brief explanation.

7. An object moves along a line such that its displacement s metres from a fixed point P at time t seconds is given by $s(t) = t(t - 3)(8t - 9)$.

(a) Find the initial velocity and initial acceleration of the object.

(b) Find the velocity and acceleration of the object at $t = 3$ seconds.

(c) Find the values of t for which the object changes direction. What significance do these times have in connection to the displacement of the object?

(d) Find the value of t for which the object's velocity is a minimum. What significance does this time have in connection to the acceleration of the object?

8. The delivery cost per tonne of bananas, D (in thousands of dollars), when x tonnes of bananas are shipped is given by $D = 3x + \frac{100}{x}$, $x > 0$.

Find the value of x for which the delivery cost per tonne of bananas is a minimum, and find the value of the minimum delivery cost. Explain why this cost is a minimum rather than a maximum.

9. The curve $y = x^4 + ax^2 + bx + c$ passes through the point $(-1, -8)$ and at that point $\frac{dy}{dx} = \frac{d^2y}{dx^2} = 6$. Find the values of a, b, and c and sketch the curve.

10. Find any maxima, minima or stationary points of inflection of the function $f(x) = \frac{x^3 + 3x - 1}{x^2}$, stating, with explanation, the nature of each point. Sketch the curve, indicating clearly what happens as $x \to \pm\infty$.

Remember that the derivative of an even function is odd and the derivative of an odd function is even.

11. For each graphed function, sketch its derivative on a separate pair of axes. Do not use your GDC.

(a) semicircle

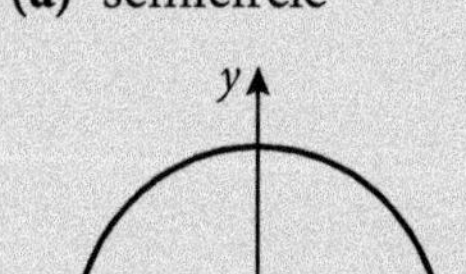

(b) parabola

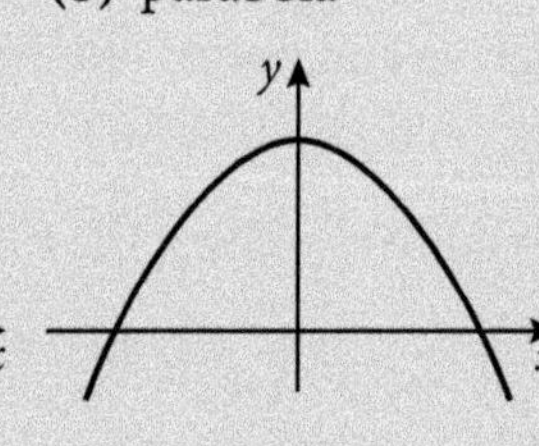

(c) odd function

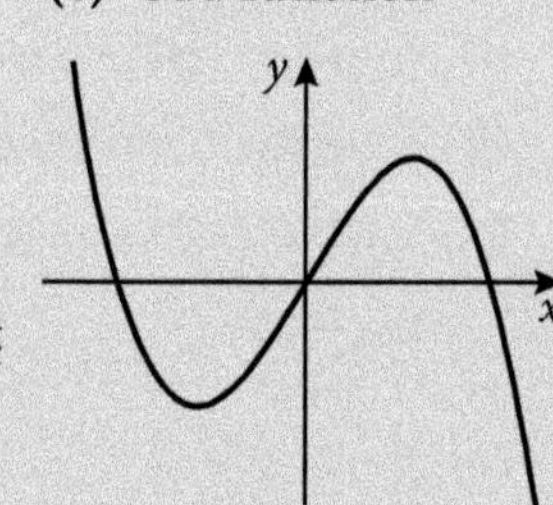

(d) even function

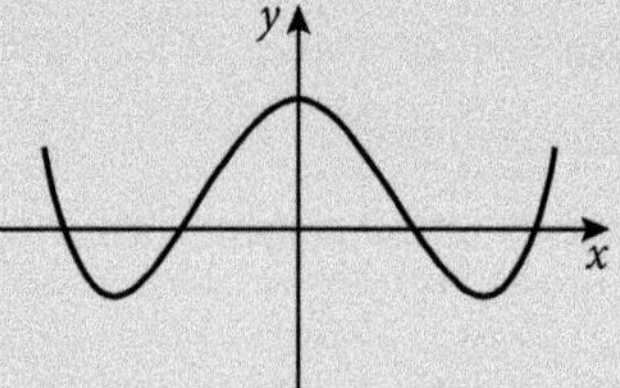

(e) periodic function

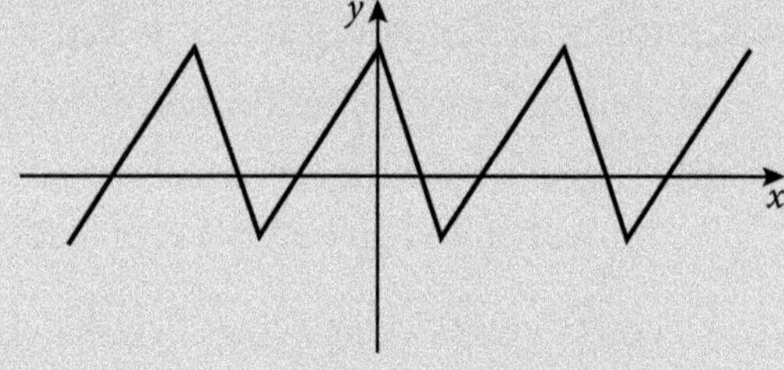

12. The graph of the derivative of a function f is shown.

(i) On what intervals is f increasing or decreasing?

(ii) For what value(s) of x does f have a local maximum or minimum?

(a)

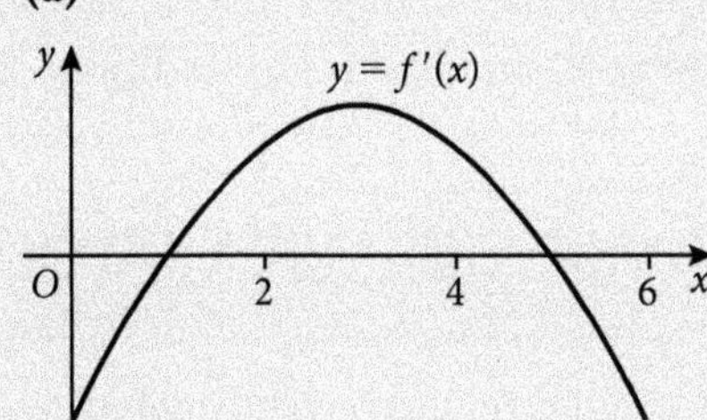

(b)

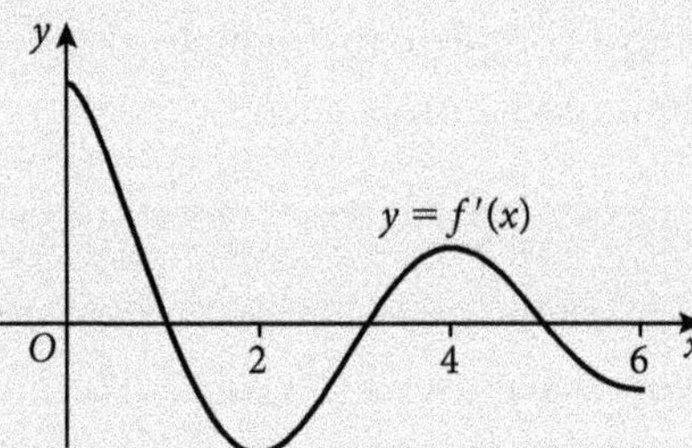

13. The graph of the second derivative f'' of a function f is shown. State the x-coordinates of the inflection points of f. Give reasons for your answers.

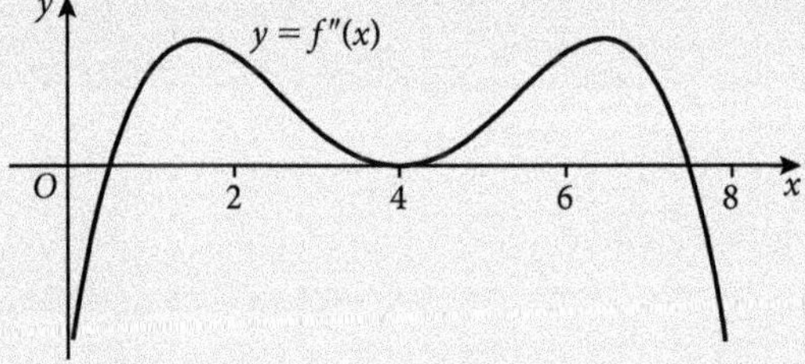

14. Sketch a continuous curve $y = f(x)$ with the following properties. Label coordinates where possible.

(a) $f(-2) = 8$

(b) $f(0) = 4$

(c) $f(2) = 0$

(d) $f'(2) = f'(-2) = 0$

(e) $f'(x) > 0$ for $|x| > 2$

(f) $f'(x) < 0$ for $|x| < 2$

(g) $f''(x) < 0$ for $x < 0$

(h) $f''(x) > 0$ for $x > 0$

15. An object moves along a horizontal line such that its displacement, s metres, from its starting position at any time $t \geqslant 0$ is given by the function $s(t) = -2t^3 + 15t^2 - 24t$. The positive direction is to the right.

(a) Find the intervals of time when the object is moving to the right, and the intervals when it is moving to the left.

(b) Find the (i) initial velocity, and (ii) initial acceleration of the object.

16. **(a)** Use your GDC to approximate to three significant figures the maximum and minimum values of the function $f(x) = x - \sqrt{2}\sin x$ in the interval $0 \leqslant x \leqslant 2\pi$

(b) Find $f'(x)$ and find the exact minimum and maximum values for $f(x)$ in the interval $0 \leqslant x \leqslant 2\pi$

9.4 Tangents and normals

In many areas of mathematics and physics, it is useful to have an accurate description of a line that is tangent or normal (perpendicular) to a curve. The most complete mathematical description we can obtain is to find the algebraic equation of such lines.

Finding equations of tangents

We now make use of the basic differentiation rules that we established earlier to determine the equation of lines that are tangent to a curve at a point. Example 9.15 shows how we can approximate the square root of a number quite accurately without a calculator by making use of a tangent line.

Example 9.15

Find the equation of the line tangent to $y = \sqrt{x}$ at $x = 9$.

Use this tangent line to approximate $\sqrt{10}$.

Solution

We can find the equation of any line if we know its gradient and a point it passes through.

Since $y = 3$ when $x = 9$, then the point the tangent passes through is (9, 3). We differentiate to find the gradient of the curve at $x = 9$, thus giving us the gradient of the tangent line.

$$\frac{\mathrm{d}y}{\mathrm{d}x} = \frac{\mathrm{d}}{\mathrm{d}x}(\sqrt{x}) = \frac{\mathrm{d}}{\mathrm{d}x}x^{\frac{1}{2}} = \frac{1}{2}x^{-\frac{1}{2}} = \frac{1}{2\sqrt{x}}$$

at $x = 9$: $\frac{\mathrm{d}y}{\mathrm{d}x} = \frac{1}{2\sqrt{9}} = \frac{1}{6} \Rightarrow$ The gradient of the curve and tangent at $x = 9$ is $\frac{1}{6}$.

Now that we have a point and a gradient for the line, we can substitute in the point-gradient form for the equation of a line.

$$y - 3 = \frac{1}{6}(x - 9) \Rightarrow y = \frac{1}{6}x + \frac{3}{2}$$

The equation of the line tangent to $y = \sqrt{x}$ at $x = 9$ is $y = \frac{x}{6} + \frac{3}{2}$.

For values of x near 9, $y = \sqrt{x} \approx \frac{x}{6} + \frac{3}{2}$.

$$\sqrt{10} \approx \frac{10}{6} + \frac{3}{2} = \frac{19}{6} = 3.1\overline{6}$$

The actual value of $\sqrt{10}$ to four significant figures is 3.162. Our approximation expressed to four significant figures is 3.167.
The percentage error is less than 0.2%.

Figure 9.41 shows the graphs of $y = \sqrt{x}$ and its tangent at $x = 9$, $y = \frac{x}{6} + \frac{3}{2}$, and illustrates that the tangent is a very good approximation to the curve in the interval $5 < x < 13$ centred on the point the tangent passes through (9, 3).

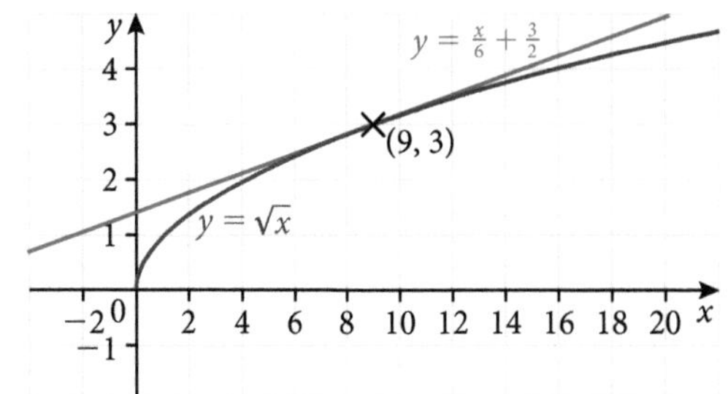

Figure 9.41 Graphs of $y = \sqrt{x}$ and its tangent at $x = 9$

Example 9.16

Find the equation of the tangent to $f(x) = x + \frac{1}{x}$ at the point $\left(\frac{1}{2}, \frac{5}{2}\right)$.

Solution

$$f(x) = x + \frac{1}{x} = x + x^{-1}$$

$$f'(x) = 1 - x^{-2} = 1 - \frac{1}{x^2}$$

When $x = \frac{1}{2}$, $f'\left(\frac{1}{2}\right) = 1 - \frac{1}{\left(\frac{1}{2}\right)^2} = -3$

Hence, the gradient of the tangent at this point is -3

$$y - \frac{5}{2} = -3\left(x - \frac{1}{2}\right) \Rightarrow y = -3x + \frac{3}{2} + \frac{5}{2} \Rightarrow y = -3x + 4$$

The equation of the line tangent to $f(x) = x + \frac{1}{x}$ at $x = \frac{1}{2}$ is $y = -3x + 4$

Example 9.17

Consider the function $g(x) = x^2(x - 1)$

(a) Find the two points on the graph of g at which the gradient of the curve is 8.

(b) Find the equations of the tangents to the curve at both of these points.

Solution

(a) In order to differentiate by applying the power rule term by term, we need to write $g(x)$ in expanded form: $g(x) = x^2(x - 1) = x^3 - x^2$

$$g'(x) = \frac{d}{dx}(x^3 - x^2) = 3x^2 - 2x$$

$$g'(x) = 3x^2 - 2x = 8 \Rightarrow 3x^2 - 2x - 8 = 0$$

$$(3x + 4)(x - 2) = 0 \Rightarrow x = -\frac{4}{3} \text{ or } x = 2$$

$$g\left(-\frac{4}{3}\right) = \left(-\frac{4}{3}\right)^3 - \left(-\frac{4}{3}\right)^2 = -\frac{112}{27}$$

$$g(2) = 2^3 - 2^2 = 4$$

Thus, the gradient of the curve is equal to 8 at $\left(-\frac{4}{3}, -\frac{112}{27}\right)$ and $(2, 4)$

(b) tangent at $\left(-\frac{4}{3}, -\frac{112}{27}\right)$:

$$y - \left(-\frac{112}{27}\right) = 8\left[x - \left(-\frac{4}{3}\right)\right] \Rightarrow y = 8x + \frac{32}{3} - \frac{112}{27}$$

$$\Rightarrow y = 8x + \frac{176}{27}$$

Therefore, the equation of the tangent at $\left(-\frac{4}{3}, -\frac{112}{27}\right)$ is $y = 8x + \frac{176}{27}$

tangent at (2, 4):

$$y - 4 = 8(x - 2) \Rightarrow y = 8x - 16 + 4 \Rightarrow y = 8x - 12$$

Therefore, the equation of the tangent at (2, 4) is $y = 8x - 12$

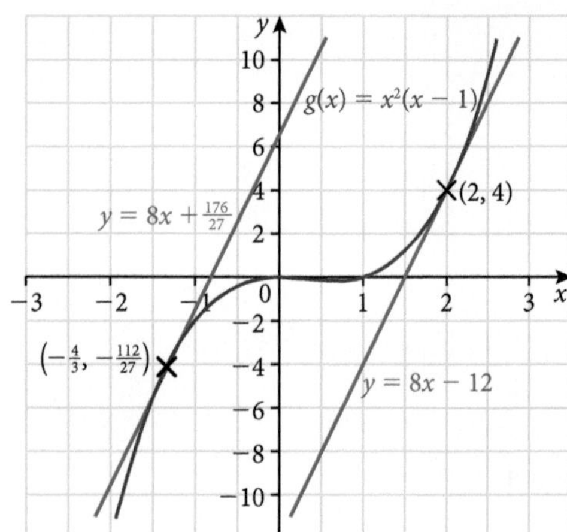

Figure 9.42 Results for Example 9.17

Figure 9.42 shows the results for Example 9.17 – the graph of the function g and the two tangent lines to the graph of the function that have a gradient of 8. Note that the scales on the x- and y-axes are not equal, which causes the gradient of the tangent lines to appear less than 8 for this particular graph.

A **normal** to a graph of a function at a point is the line through the point that is at a right angle to the tangent at the point. In other words, the tangent and normal to a curve at a certain point are perpendicular.

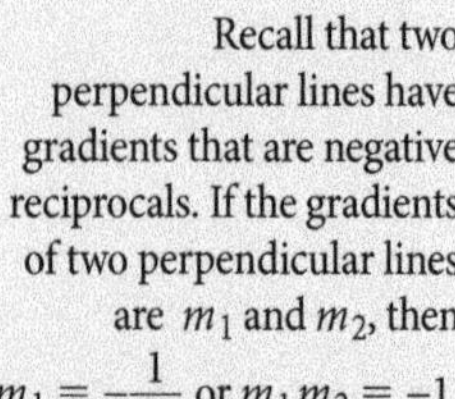

Recall that two perpendicular lines have gradients that are negative reciprocals. If the gradients of two perpendicular lines are m_1 and m_2, then $m_1 = -\frac{1}{m_2}$ or $m_1 m_2 = -1$. The exception is when one of the lines is horizontal (gradient is zero), and the other is vertical (gradient is undefined).

The normal to a curve at a point

We often need to find the line that is perpendicular to a curve at a certain point, which we define to be the line that is perpendicular to the tangent at that point. In this particular context, we apply the adjective **normal** rather than perpendicular to denote that two lines are at right angles to one another.

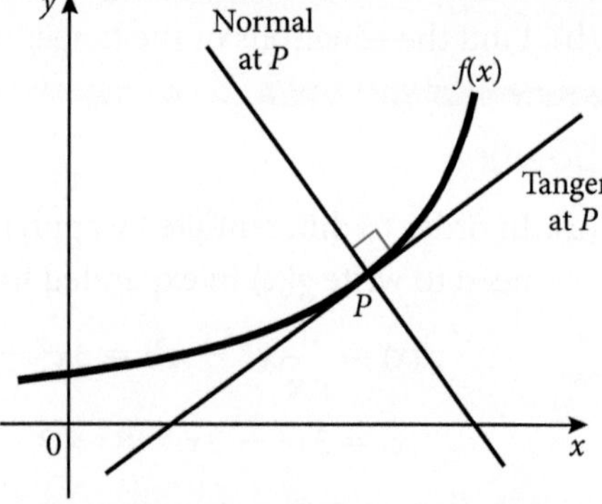

Figure 9.43 The normal to a curve at a point

Example 9.18

Find the equation of the normal to the graph of $y = 2x^2 - 6x + 3$ at the point $(1, -1)$.

Solution

$$\frac{dy}{dx} = \frac{d}{dx}(2x^2 - 6x + 3)$$
$$= 4x - 6$$

Gradient of tangent at $(1, -1)$ is $4(1) - 6 = -2$. Hence, gradient of normal is $+\frac{1}{2}$.

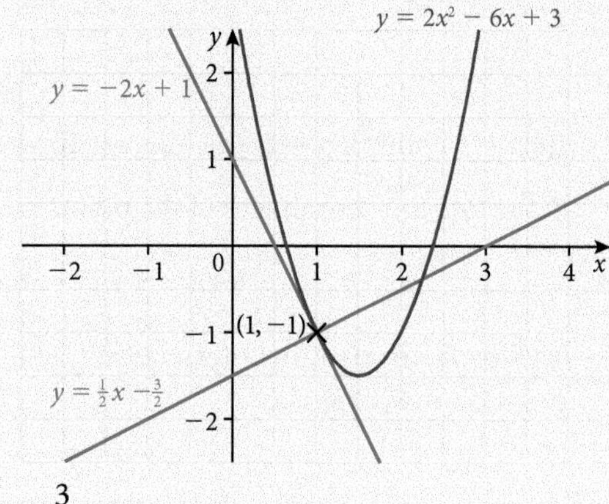

Equation of normal:

$$y - (-1) = \frac{1}{2}(x - 1) \Rightarrow y = \frac{1}{2}x - \frac{3}{2}$$

The graph in Example 9.18 shows the curve with both its tangent and normal at the point $(1, -1)$. Remember that if you graph a function with its tangent and normal at a certain point, the normal will only appear perpendicular if the scales on the x- and y-axes are equal. However, the tangent will always appear tangent to the curve.

Example 9.19

Consider the parabola with the equation $y = \frac{1}{4}x^2$

(a) Find the equation of the normals at the points $(-2, 1)$ and $(-4, 4)$.

(b) Show that the point of intersection of these two normals lies on the parabola.

Solution

(a) $\frac{dy}{dx} = \frac{1}{2}x$

Gradient of tangent at $(-2, 1)$ is $\frac{1}{2}(-2) = -1$, so the gradient of the normal at that point is $+1$

The equation of the normal at $(-2, 1)$ is:

$$y - 1 = x - (-2) \Rightarrow y = x + 3$$

Gradient of tangent at $(-4, 4)$ is $\frac{1}{2}(-4) = -2$, so the gradient of the normal at that point is $\frac{1}{2}$

The equation of the normal at $(-4, 4)$ is:

$$y - 4 = \frac{1}{2}[x - (-4)] \Rightarrow y = \frac{1}{2}x + 6$$

(b) Set the equations of the two normals equal to each other to find their intersection.

$$x + 3 = \frac{1}{2}x + 6 \Rightarrow \frac{1}{2}x = 3 \Rightarrow x = 6$$

then $y = 9$ implies that the intersection point is (6, 9)

Substitute the coordinates of the points into the equation for the parabola.

$$y = \frac{1}{4}x^2 \Rightarrow 9 = \frac{1}{4}(6)^2 \Rightarrow 9 = \frac{1}{4} \cdot 36 \Rightarrow 9 = 9$$

This confirms that the intersection point, (6, 9), of the normals is also a point on the parabola.

Exercise 9.4

1. Find an equation of the tangent line to the graph of the equation at the indicated value of x.

(a) $y = x^2 + 2x + 1$ $\quad x = -3$

(b) $y = x^3 + x^2$ $\quad x = -\frac{2}{3}$

(c) $y = 3x^2 - x + 1$ $\quad x = 0$

(d) $y = 2x + \frac{1}{x}$ $\quad x = \frac{1}{2}$

2. Find the equations of the normal to the functions in question 1 at the indicated value of x.

3. Find the equations of the lines tangent to the curve $y = x^3 - 3x^2 + 2x$ at any point where the curve intersects the x-axis.

4. Find the equation of the tangent to the curve $y = x^2 - 2x$ that is perpendicular to the line $x - 2y = 1$.

5. Using your GDC for assistance make accurate sketches of the curves $y = x^2 - 6x + 20$ and $y = x^3 - 3x^2 - x$ on the same set of axes. The two curves have the same gradient at an integer value for x somewhere in the interval $0 \leqslant x \leqslant 7$.

(a) Find this value of x.

(b) Find the equation for the tangent to each curve at this value of x.

6. Find the equation of the normal to the curve $y = x^2 + 4x - 2$ at the point where $x = -3$. Find the coordinates of the other point where this normal intersects the curve again.

7. Consider the function $g(x) = \frac{1 - x^3}{x^4}$. Find the equation of both the tangent and the normal to the graph of g at the point (1, 0).

8. The normal to the curve $y = ax^{\frac{1}{2}} + bx$ at the point where $x = 1$ has a gradient of 1 and intersects the y-axis at $(0, -4)$. Find the value of a and the value of b.

9. **(a)** Find the equation of the tangent to the function $f(x) = x^3 + \frac{1}{2}x^2 + 1$ at the point $\left(-1, \frac{1}{2}\right)$.

   (b) Find the coordinates of another point on the graph of f where the tangent is parallel to the tangent found in **(a)**.

10. Find the equation of both the tangent and the normal to the curve $y = \sqrt{x}(1 - \sqrt{x})$ at the point where $x = 4$.

11. Consider the function $f(x) = (1 + x)^2(5 - x)$.

    (a) Show that the tangent to the graph of f where $x = 1$ does not intersect the graph of the function again.

    (b) Also show that the tangent line at $(0, 5)$ intersects the graph of f at a turning point.

    (c) Sketch the graph of f and the two tangents from **(a)** and **(b)**.

12. Find equations of both lines through the point $(2, -3)$ that are tangent to the parabola $y = x^2 + x$

13. Find all tangent lines through the origin to the graph of $y = 1 + (x - 1)^2$

14. **(a)** Find the equation of the tangent line to $y = \sqrt[3]{x}$ at $x = 8$.

    (b) Use the equation of this tangent line to approximate $\sqrt[3]{9}$ to three significant figures.

15. Find the equation of the tangent line for $f(x) = \frac{1}{\sqrt{x}}$ at $x = a$.

16. The tangent to the graph of $y = x^3$ at a point P intersects the curve again at another point Q. Find the coordinates of Q in terms of the coordinates of P.

Chapter 9 practice questions

1. The function f is defined as $f(x) = x^2$

   (a) Find the gradient (slope) of f at the point P, where $x = 1.5$

   (b) Find an equation for the tangent to f at the point P.

   (c) Draw a diagram to show clearly the graph of f and the tangent at P.

   The tangent from part **(b)** intersects the x-axis at the point Q, and the y-axis at the point R.

   (d) Find the coordinates of Q and R.

   (e) Verify that Q is the midpoint of $[PR]$.

   (f) Find an equation, in terms of a, for the tangent to f at the point $S(a, a^2)$, $a \neq 0$.

The tangent from part **(f)** intersects the x-axis at the point T, and the y-axis at the point U.

(g) Find the coordinates of T and U.

(h) Prove that whatever the value of a, T is the midpoint of SU.

2. The curve with equation $y = Ax + B + \frac{C}{x}, x \in \mathbb{R}, x \neq 0$, has a minimum at $P(1, 4)$ and a maximum at $Q(-1, 0)$. Find the value of each of the constants A, B and C.

3 Differentiate:

(a) $x^2(2 - 3x^3)$ **(b)** $\frac{1}{x}$

4. Consider the function $f(x) = \frac{8}{x} + 2x, x > 0$

(a) Solve the equation $f'(x) = 0$. Show that the graph of f has a turning point at $(2, 8)$.

(b) Find the equations of the asymptotes to the graph of f, and hence sketch the graph.

5. Find the coordinates of the stationary point on the curve with equation $y = 4x^2 + \frac{1}{x}$

6. The curve $y = ax^3 - 2x^2 - x + 7$ has a gradient (slope) of 3 at the point where $x = 2$. Determine the value of a.

7. If $f(2) = 3$ and $f'(2) = 5$, find an equation of **(a)** the tangent to the graph of f at $x = 2$, and **(b)** the normal to the graph of f at $x = 2$.

8. The function $g(x)$ is defined for $-3 \leqslant x \leqslant 3$. The behaviour of $g'(x)$ and $g''(x)$ is given in the tables.

x	$-3 < x < -2$	-2	$-2 < x < 1$	1	$1 < x < 3$
$g'(x)$	negative	0	positive	0	negative

x	$-3 < x < -\frac{1}{2}$	$-\frac{1}{2}$	$-\frac{1}{2} < x < 3$
$g''(x)$	positive	0	negative

Use the information above to answer the following. In each case, justify your answer.

(a) Write down the value of x for which g has a maximum.

(b) On which intervals is the value of g decreasing?

(c) Write down the value of x for which the graph of g has a point of inflection.

(d) Given that $g(-3) = 1$, sketch the graph of g. On the sketch, clearly indicate the position of the maximum point, the minimum point, and the point of inflection.

9. Given the function $f(x) = x^2 - 3bx + (c + 2)$, determine the values of b and c such that $f(1) = 0$ and $f'(3) = 0$.

10. The graphs show the functions f_1, f_2, f_3, f_4, and their derivatives. Match each function to its derivative.

f_1

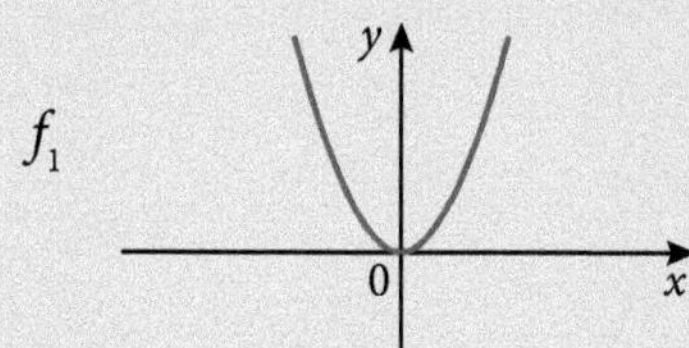

(a)

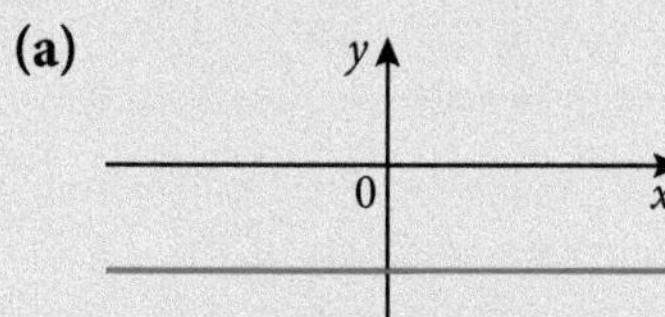

f_2

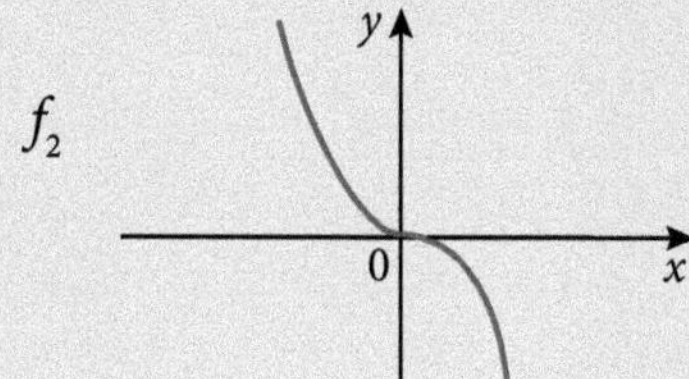

(b)

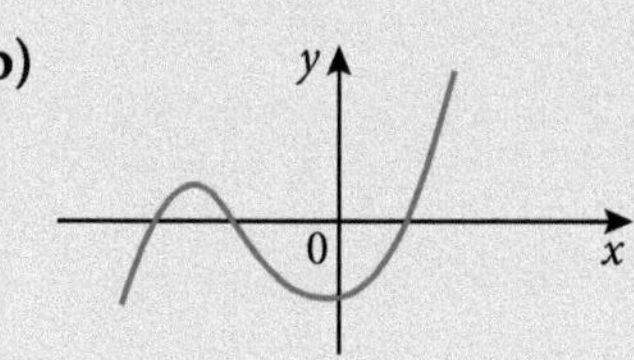

f_3

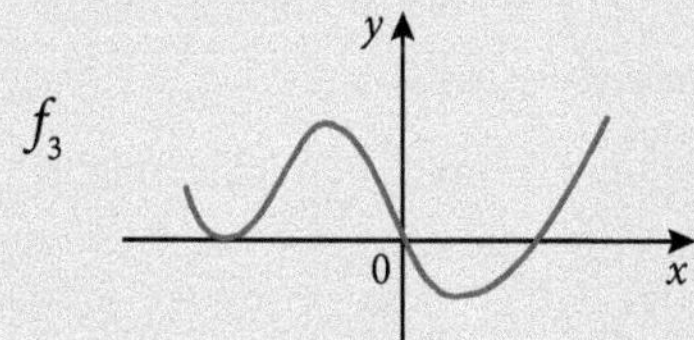

(c)

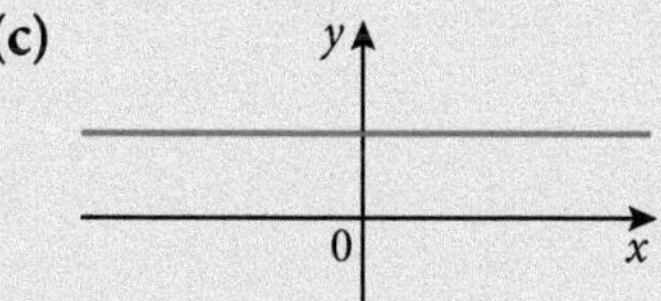

f_4

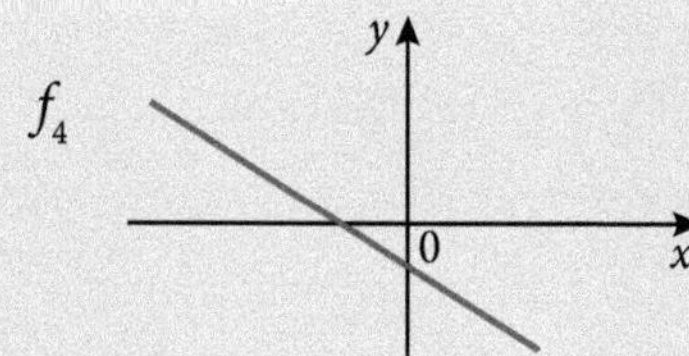

(d)

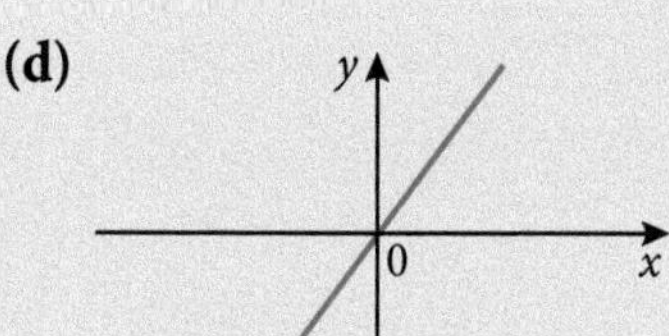

(e)

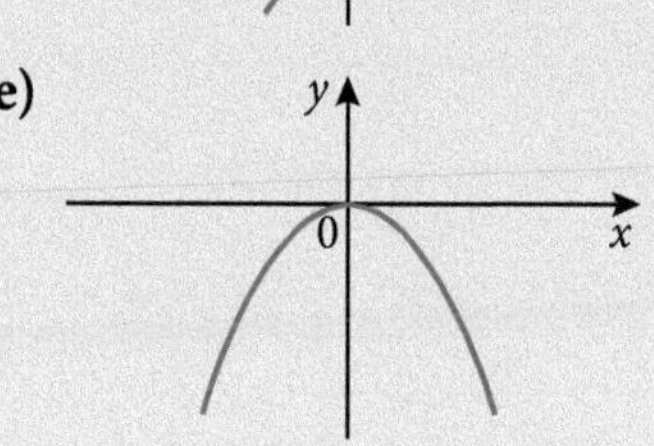

11. Consider the function $f(x) = 1 + \sin x$

(a) Find the average rate of change of f from $x = 0$ to $x = \frac{\pi}{2}$.

(b) Find the instantaneous rate of change of f at $x = \frac{\pi}{4}$

(c) At what value of x in the interval $0 < x < \frac{\pi}{2}$ is the instantaneous rate of change of f equal to the average rate of change of f from $x = 0$ to $x = \frac{\pi}{2}$ (answer to part (a))?

12. Consider the function $y = \frac{3x - 2}{x}$. The graph of this function has a vertical and a horizontal asymptote.

(a) Write down the equation of:

(i) the vertical asymptote **(ii)** the horizontal asymptote.

(b) Find $\frac{dy}{dx}$

(c) Indicate the intervals for which the curve is increasing or decreasing.

(d) How many stationary points does the curve have? Explain using your result to (c).

13. Show that there are two points at which the function $h(x) = 2x^2 - x^4$ has a maximum value, and one point at which h has a minimum value. Find the coordinates of these three points, indicating whether each is a maximum or minimum.

14. The normal to the curve $y = x^{\frac{1}{2}} + x^{\frac{1}{3}}$ at the point (1, 2) meets the axes at $(a, 0)$ and $(0, b)$. Find a and b.

15. The displacement s metres of a car, t seconds after leaving a fixed point A, is given by $s(t) = 10t - \frac{1}{2}t^2$

(a) Calculate the velocity when $t = 0$.

(b) Calculate the value of t when the velocity is zero.

(c) Calculate the displacement of the car from A when the velocity is zero.

16. A ball is thrown vertically upwards from ground level such that its height h at t seconds is given by $h = 14t - 4.9t^2$

(a) Write expressions for the ball's velocity and acceleration.

(b) Find the maximum height the ball reaches and the time it takes to reach the maximum.

(c) At the moment the ball reaches its maximum height, what is the ball's velocity and acceleration?

17. Find the exact coordinates of the inflection point on the curve $y = x^3 + 12x^2 - x - 12$.

18. Given the function $f(x) = 2\cos x - 3$. At the point on the curve where $x = \frac{\pi}{3}$, find:

(a) the equation of the tangent to f

(b) the equation of the normal to f.

Express both equations exactly.

19. The curve $y = ax^2 + bx + c$ has a maximum point at (2, 18) and passes through the point (0, 10). Find a, b and c.

20. For the function $f(x) = \frac{1}{2}x^2 - 5x + 3$, find:

(a) the equation of the tangent line at $x = -2$

(b) the equation of the normal line at $x = -2$.

21. Consider the function $f(x) = x^4 - x^3$.

(a) Find the coordinates of any maximum or minimum points. Identify each as relative or absolute.

(b) State the domain and range of f.

(c) Find the coordinates of any inflection point(s).

(d) Sketch the function clearly indicating any maximum, minimum or inflection points.

22. Evaluate each limit.

(a) $\lim_{x \to \infty} \frac{2 - 3x + 5x^2}{8 - 3x^2}$ (b) $\lim_{x \to 1} \frac{x^3 - 1}{x - 1}$

23. Find the derivative $f'(x)$ for each function.

(a) $f(x) = \frac{x^2 - 4x}{\sqrt{x}}$ (b) $f(x) = x^3 - 3\sin x$

(c) $f(x) = \frac{1}{x} + \frac{x}{2}$ (d) $f(x) = \frac{7}{3x^{13}}$

24. A point (p, q) is on the graph of $y = x^3 + x^2 - 9x - 9$, and the line tangent to the graph at (p, q) passes through the point $(4, -1)$. Find p and q.

25. For what values of c, such that $c \geqslant 0$, is the line $y = -\frac{1}{12}x + c$ normal to the graph of $y = x^3 + \frac{1}{3}$?

26. Find the points on the curve $y = \frac{1}{3}x^3 - x$ where the tangent line is parallel to the line $y = 3x$.

27. At what point does the line that is normal to the graph of $y = x - x^2$ at the point $(1, 0)$ intersect the graph of the curve a second time?

28. An object moves along a line according to the position function $s(t) = t^3 - 9t^2 + 24t$. Find the positions of the object when

(a) its velocity is zero

(b) its acceleration is zero.

29. A particle moves along a straight line in the time interval $0 \leqslant t \leqslant 2\pi$ such that its displacement from the origin O is s metres given by the function $s = t + \sin t$.

(a) Find the value(s) of t in the interval $0 \leqslant t \leqslant 2\pi$ when the particle changes direction.

(b) Show that the particle always remains on the same side of the origin O.

(c) Find the value(s) of t in the interval $0 \leqslant t \leqslant 2\pi$ when the acceleration of the particle is zero.

(d) Sketch a graph of the displacement of the particle from O for $0 \leqslant t \leqslant 2\pi$, and state the maximum value of s in this interval.

30. The curve whose equation is $y = ax^3 + bx^2 + cx + d$ has a point of inflection at $(-1, 4)$, a turning point when $x = 2$, and passes through the point $(3, -7)$. Find the values of a, b, c, and d, and the y-coordinate of the turning point.

31. Find the stationary values of the function $f(x) = 1 - \frac{9}{x^2} + \frac{18}{x^4}$ and determine their nature.

32. **(a)** Find the equation of the tangent to the curve $y = \frac{1}{x}$ at the point $(1, 1)$

(b) Find the equation of the tangent to the curve $y = \cos x$ at the point $\left(\frac{\pi}{2}, 0\right)$

(c) Deduce that $\frac{1}{x} > \cos x$ for $0 \leqslant x \leqslant \frac{\pi}{2}$

33. Show that there is just one tangent to the curve $y = x^3 - x + 2$ that passes through the origin. Find:

(a) the equation of the tangent

(b) the coordinates of the point of tangency.

34. The displacement s metres of a moving body B from a fixed point O at time t seconds is given by $s = 50t - 10t^2 + 1000$

(a) Find the velocity of B in $\mathrm{m\,s^{-1}}$

(b) Find its maximum displacement from O.

35. The diagram shows a sketch of the graph of $y = f'(x)$ for $a \leqslant x \leqslant b$

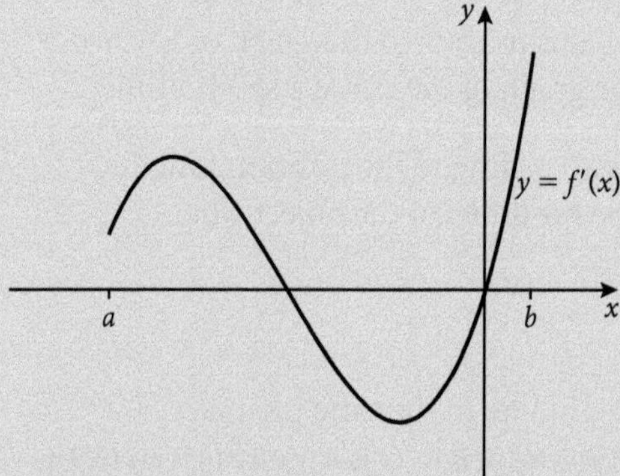

Sketch the graph of $y = f(x)$ for $a \leqslant x \leqslant b$, given that $f(0) = 0$, $f(a) = 0$ and $f(x) \geqslant 0$ for all x. On your graph, you should clearly indicate any minimum or maximum points, or points of inflection.

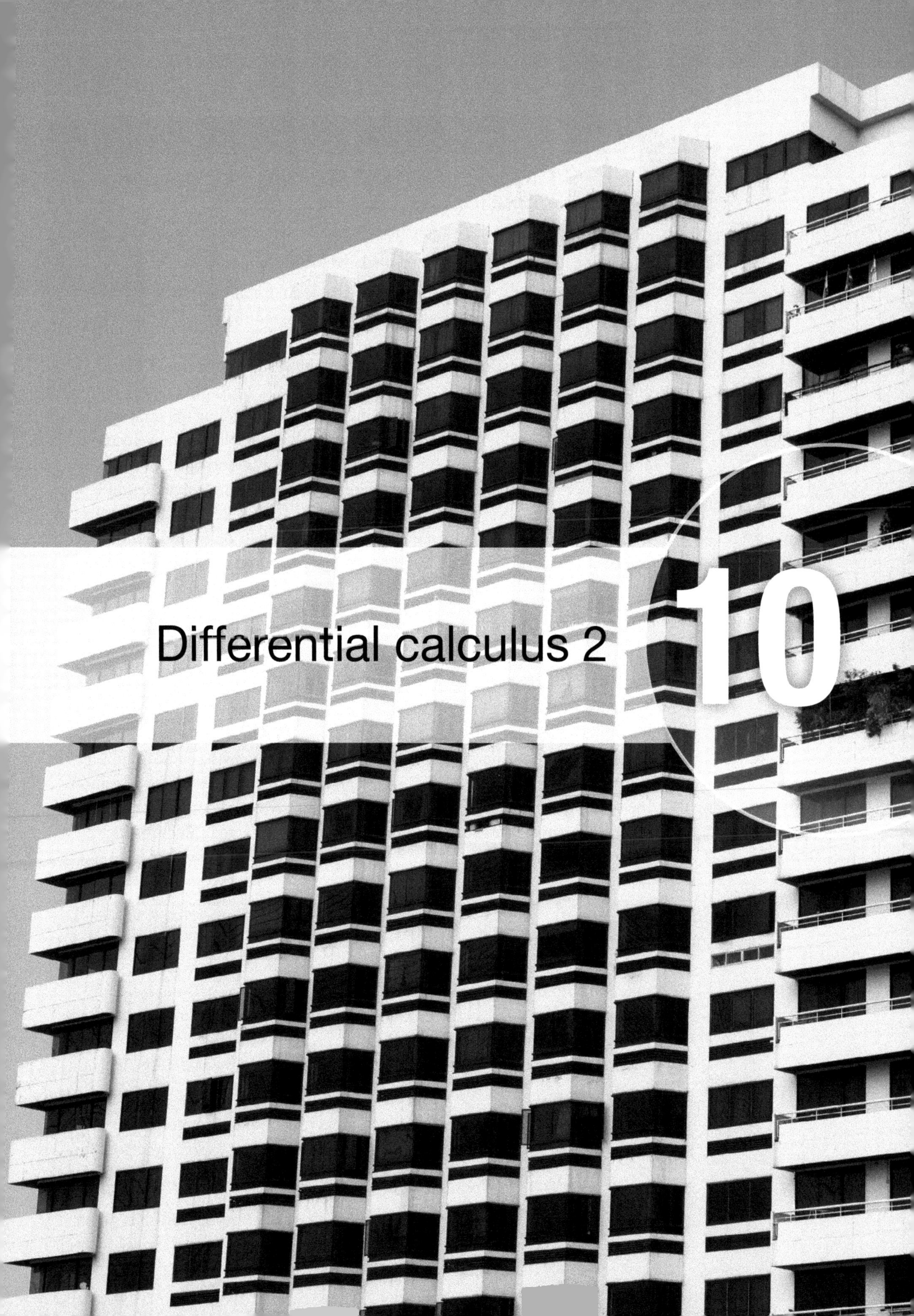

10 Differential calculus 2

Learning objectives

By the end of this chapter, you should be familiar with...

- finding the derivative of a composite function
- finding the derivative of a function that is in the form of a product or quotient
- finding the derivative of exponential and logarithmic functions
- solving problems requiring a solution that is an optimum; that is, a maximum or minimum (optimisation).

The primary purpose of Chapter 9 was to establish some fundamental concepts and techniques of differential calculus. Chapter 9 also introduced some applications involving the differentiation of functions: finding maxima and minima of a function; kinematic problems involving displacement, velocity, and acceleration; and finding equations of tangents and normals. The focus of this chapter is to expand our set of differentiation rules and techniques and to deepen and extend the applications introduced in Chapter 9 – particularly using methods of finding extrema in the context of finding an 'optimum' solution to a problem. We start by investigating the derivatives of two important functions.

10.1 Derivatives of exponential and logarithmic functions

To make a conjecture for the derivatives of the functions $y = e^x$ and $y = \ln x$, we will use the same informal approach that worked in the previous chapter for determining the derivative of $y = \cos x$. We start by using some graphing technology (for example, a GDC) to graph the derivative of the function and then examine the graph's shape to make a persuasive conjecture for the rule for the derivative. We then check the rule for a few selected values to help to confirm our conjecture.

Derivative of the exponential function $y = e^x$

Let's review some important facts about exponential functions. An exponential function with base b is defined as $f(x) = b^x$, $b > 0$ and $b \neq 1$. The graph of f passes through (0, 1), has the x-axis as a horizontal asymptote, and, depending on the value of the base of the exponential function b, will either be a continually increasing exponential growth curve, as shown in Figure 10.1, or a continually decreasing exponential decay curve, as shown in Figure 10.2.

In Chapter 4 we learned that the exponential function e^x, sometimes written as 'exp x', is a particularly important function for modelling exponential growth and decay. The number e was defined in Section 4.2 as the limit of $\left(1 + \frac{1}{x}\right)^x$ as $x \to \infty$. Let's make a conjecture for the derivative of e^x by looking at its graph on our GDC.

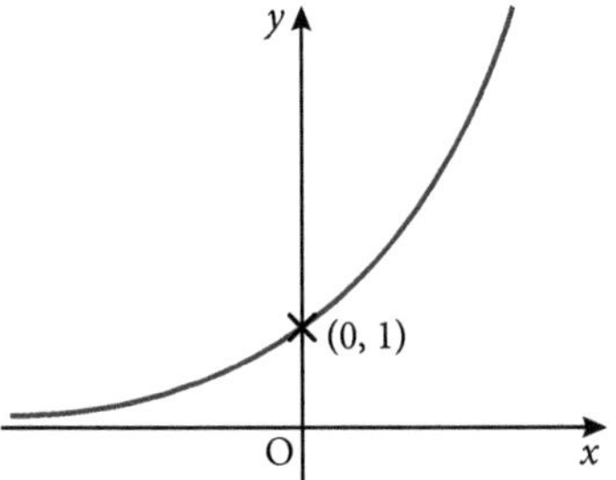

Figure 10.1 Exponential growth curve: $f(x) = b^x$ for $b > 1$. As $x \to \infty, f(x) \to \infty$; f is an increasing function

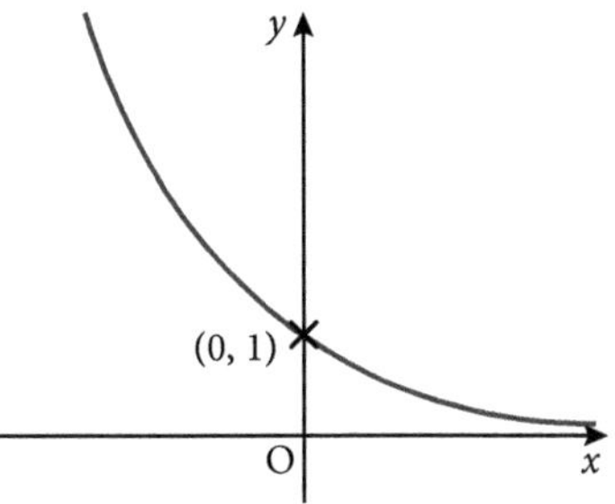

Figure 10.2 Exponential decay curve: $f(x) = b^x$ for $0 < b < 1$. As $x \to \infty, f(x) \to 0$; f is a decreasing function

Look at Figure 10.3. The graph of the derivative of e^x appears to be identical to e^x itself. That is, $\frac{d}{dx}(e^x) = e^x$. Let's make further use of a GDC to compare results for the derivative's value computed by the GDC with the value of the rule from our conjecture for selected values of x.

$\frac{d}{dx}(e^x)\|_{x=1}$	2.71828
e^1	2.71828
$\frac{d}{dx}(e^x)\|_{x=3}$	20.0855
e^3	20.0855

$\frac{d}{dx}(e^x)\|_{x=8.37}$	4315.64
$e^{8.37}$	4315.64
$\frac{d}{dx}(e^x)\|_{x=-4.5}$	0.011109
$e^{-4.5}$	0.011109

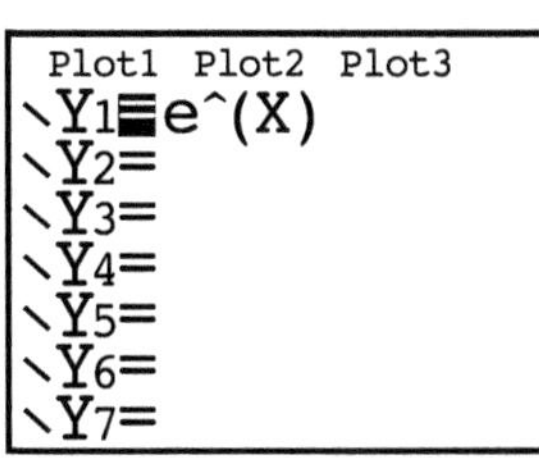

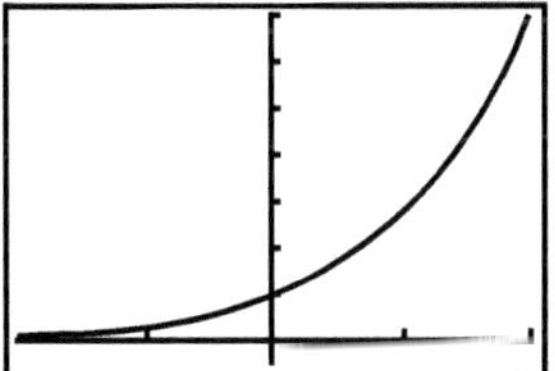

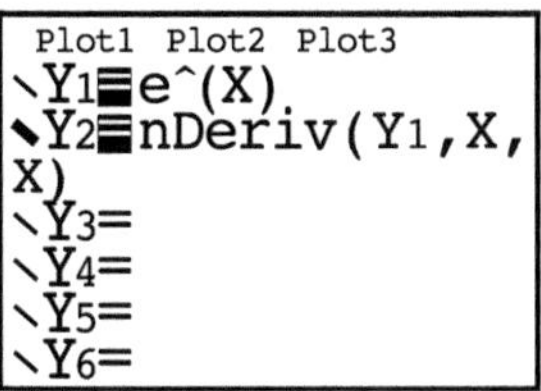

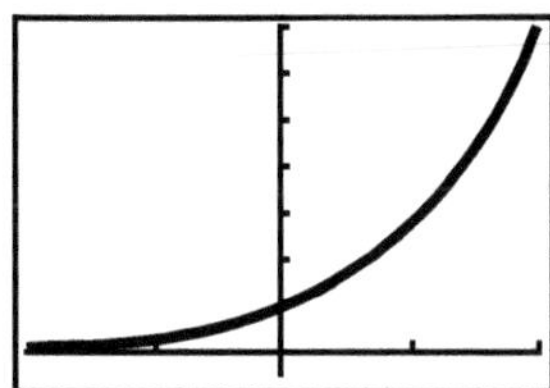

Figure 10.3 Graphing the derivative of e^x

Derivative of the exponential function

If $f(x) = e^x$, then $f'(x) = e^x$. In Leibniz notation, $\frac{d}{dx}(e^x) = e^x$.

The derivative of the exponential function **is** the exponential function. More precisely, the slope of the graph of $f(x) = e^x$ at any point (x, e^x) is equal to the y-coordinate of the point.

Derivative of the natural logarithm function $y = \ln x$

Now that we have found the derivative of $y = e^x$, let's find the derivative of its inverse, $y = \ln x$, $x > 0$. We start by using our GDC to view a graph of the derivative of $f(x) = \ln x$ and also construct a table of ordered pairs $(x, f'(x))$.

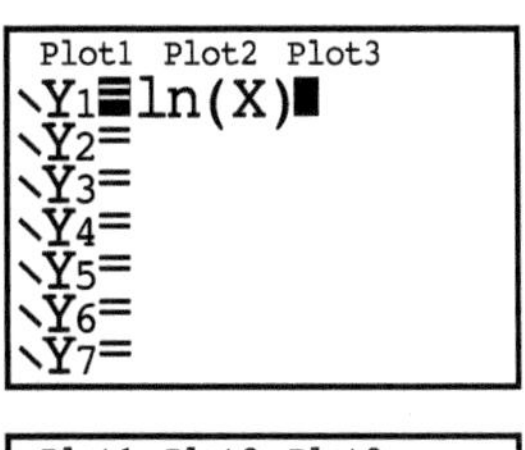

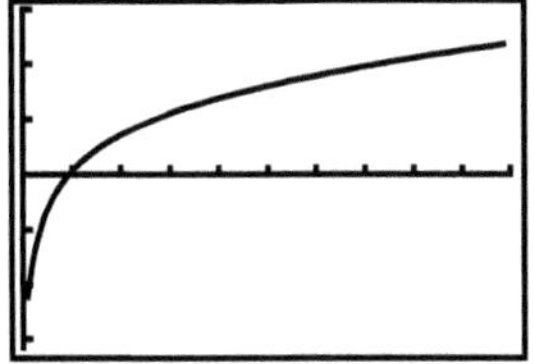

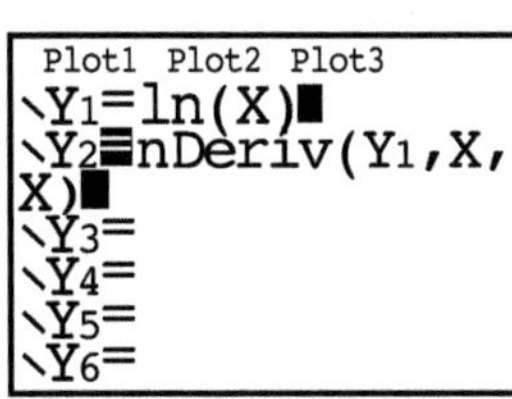

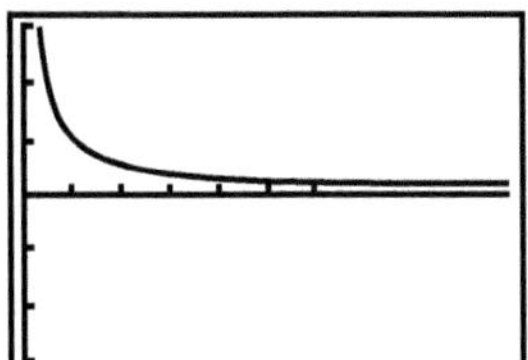

X	Y2
0	ERROR
1	1
2	.5
3	.33333
4	.25
5	.2
6	.16667

X=0

Figure 10.4 Graphing the derivative of $y = \ln x$

Note that in the GDC images above, the graph of $y = e^x$ is displayed in thin style and the graph of $y = \frac{d}{dx}(e^x)$ is displayed in bold style.

Note that in the GDC images on the left, the second graph screen has the graph of $y = \ln x$ turned off, so that only the graph of the derivative of $y = \ln x$ is displayed.

In the table, each value in the Y_2 column is the gradient of the tangent to $y = \ln x$ at the given value for x. From the graph of the derivative, and especially from the table, we conjecture that the derivative of $\ln x$ is $\frac{1}{x}$. Let's use a GDC to evaluate the derivative of $\ln x$ and graph the function of $\frac{1}{x}$ and see if they match.

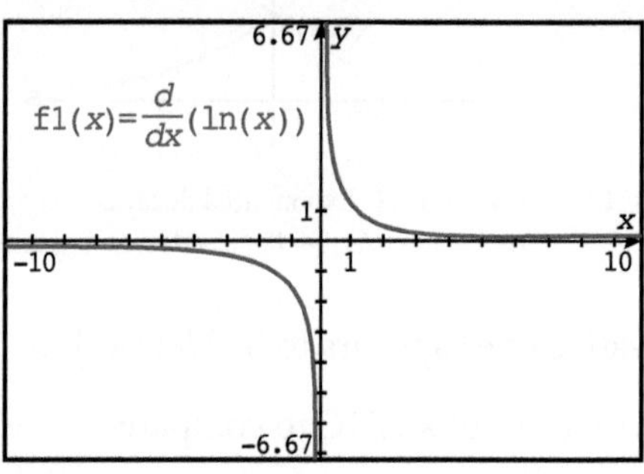

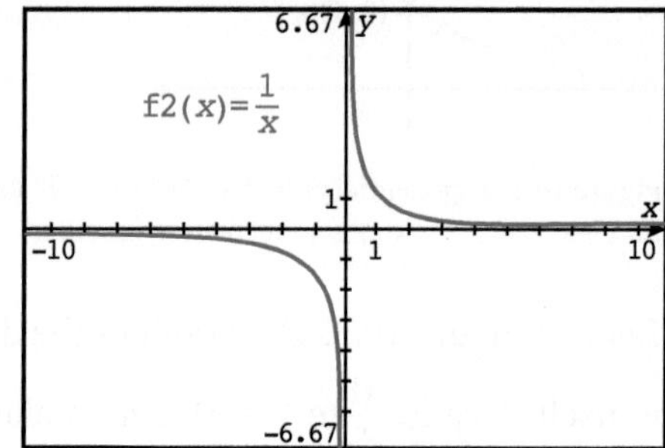

Figure 10.5 GDC screens showing the derivative of $\ln x$ and $\frac{1}{x}$

Therefore, $\frac{\mathrm{d}}{\mathrm{d}x}(\ln x) = \frac{1}{x}$

Derivative of the natural logarithm function
If $f(x) = \ln x$, then $f'(x) = \frac{1}{x}$. Or, in Leibniz notation $\frac{\mathrm{d}}{\mathrm{d}x}(\ln x) = \frac{1}{x}$

Summary of differentiation rules
Derivative of x^n $f(x) = x^n \Rightarrow f'(x) = nx^{n-1}$
Derivative of $\sin x$ $f(x) = \sin x \Rightarrow f'(x) = \cos x$
Derivative of $\cos x$ $f(x) = \cos x \Rightarrow f'(x) = -\sin x$
Derivative of e^x $f(x) = \mathrm{e}^x \Rightarrow f'(x) = \mathrm{e}^x$
Derivative of $\ln x$ $f(x) = \ln x \Rightarrow f'(x) = \frac{1}{x}$

Example 10.1

Differentiate each of the following functions.

(a) $f(x) = 5 - 2\mathrm{e}^x$ (b) $g(x) = \frac{\ln x}{2}$ (c) $y = 6 - \ln\left(\frac{\mathrm{e}^3}{x}\right)$

Solution

(a) $f'(x) = \frac{\mathrm{d}}{\mathrm{d}x}(5) - 2\frac{\mathrm{d}}{\mathrm{d}x}(\mathrm{e}^x)$

$= 0 - 2\mathrm{e}^x$

$= -2\mathrm{e}^x$

(b) $g'(x) = \frac{1}{2}\frac{\mathrm{d}}{\mathrm{d}x}(\ln x)$

$= \frac{1}{2} \cdot \frac{1}{x}$

$= \frac{1}{2x}$

(c) $\frac{\mathrm{d}y}{\mathrm{d}x} = \frac{\mathrm{d}}{\mathrm{d}x}(6) - \frac{\mathrm{d}}{\mathrm{d}x}\left(\ln\left(\frac{\mathrm{e}^3}{x}\right)\right)$

$= 0 - \frac{\mathrm{d}}{\mathrm{d}x}(\ln \mathrm{e}^3 - \ln x)$

$= -\frac{\mathrm{d}}{\mathrm{d}x}(\ln \mathrm{e}^3) + \frac{\mathrm{d}}{\mathrm{d}x}(\ln x)$

$= -\frac{\mathrm{d}}{\mathrm{d}x}(3) + \frac{1}{x}$

$= 0 + \frac{1}{x}$

$= \frac{1}{x}$

Example 10.2

Find the equation of the line tangent to each function at the specified value of x. Express the equation exactly.

(a) $y = e^x + 1 \quad x = 1$ (b) $y = \ln x \quad x = 4$

Solution

(a) When $x = 1$, $y = e^1 + 1 = e + 1$. The point of tangency is $(1, e + 1)$.

$\frac{dy}{dx} = e^x \Rightarrow$ slope of tangent line $= e^1 = e$

Substitute into point-slope form of a line: $y - y_1 = m(x - x_1)$

$y - (e + 1) = e(x - 1) \Rightarrow$ equation of tangent line is $y = ex + 1$

A graph of the curve and the result for the tangent line on a GDC provides evidence that the result is correct.

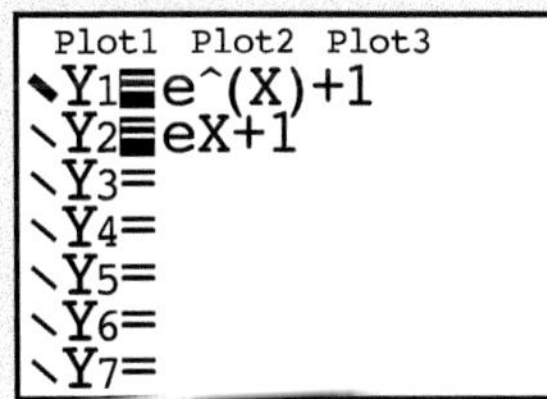

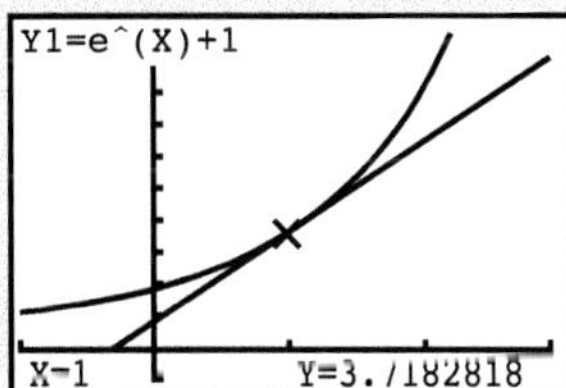

(b) When $x = 4$, $y = \ln 4$. The point of tangency is $(4, \ln 4)$.

$\frac{dy}{dx} = \frac{1}{x} \Rightarrow$ slope of tangent line $= \frac{1}{4}$

Substitute into point-slope form of a line: $y - y_1 = m(x - x_1)$

$y - \ln 4 = \frac{1}{4}(x - 4) \Rightarrow$ equation of tangent line is $y = \frac{1}{4}x - 1 + \ln 4$

Again, a graph of the curve and the result for the tangent line provides evidence that the result is correct.

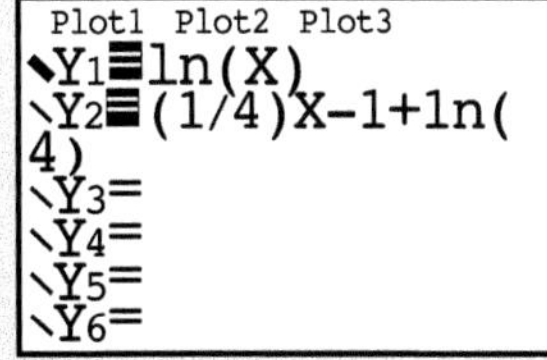

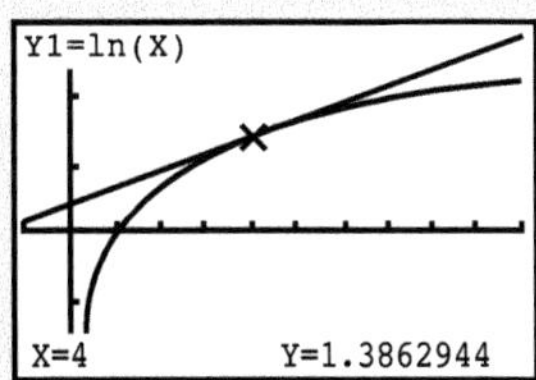

Exercise 10.1

1. Write down the derivative of each function.

(a) $y = 5 - e^x$ **(b)** $y = x + \ln x$ **(c)** $y = \frac{2e^x}{5}$

(d) $y = 2e\ln x$ **(e)** $y = \frac{1}{2}(e^x + 2\cos x)$ **(f)** $y = 2e + \ln x$

2. Find the equation of the tangent to the given curve at the specified value of x. Express the equation exactly, in the form $y = mx + c$.

   (a) $y = \dfrac{e^x - 1}{4}$ $\quad x = 0$

   (b) $y = x + e^x$ $\quad x = 0$

   (c) $y = \frac{2}{3}\ln x$ $\quad x = e$

3. Find the coordinates of any stationary points on the curve $y = x - e^x$. Classify any such points as a maximum, minimum, or neither, giving reasons for your answer.

4. Show that the curve $y = x - \ln x$ has no points of inflection.

5. Find the equation of the normal line to the curve $y = 3 + \sin x$ at the point where $x = \dfrac{\pi}{2}$

6. Consider the function $y = e^x - x^3$.

   (a) Find $f'(x)$ and $f''(x)$.

   (b) Find the x-coordinates (accurate to three significant figures) for any points where $f'(x) = 0$.

   (c) Indicate the intervals for which $f(x)$ is increasing and the intervals for which $f(x)$ is decreasing.

   (d) For the values of x found in part **(b)**, state whether that point on the graph of f is a maximum, minimum, or neither.

   (e) Find the x-coordinate of any inflection point(s) for the graph of f.

   (f) Indicate the intervals for which $f(x)$ is concave up, and indicate the intervals for which $f(x)$ is concave down.

7. A line with gradient m passes through the origin and is tangent to the graph of $y = \ln x$. Find the value of m.

10.2 The chain rule

We know how to differentiate functions such as $f(x) = \sqrt{x}$ and $g(x) = x^3 + 2x - 3$, but how do we differentiate the composite function $f(g(x)) = \sqrt{x^3 + 2x - 3}$? The rule for computing the derivative of the composite of two functions (a 'function of a function'), is called the **chain rule**. Because most functions that we encounter in applications are composites of other functions, it can be argued that the chain rule is the most important, and most widely used, differentiation rule.

Table 10.1 shows some examples of functions that we can differentiate with the rules that we have learned thus far, and further examples of functions that are best differentiated with the chain rule.

Differentiate without the chain rule	Differentiate with the chain rule
$y = \cos x$	$y = \cos 2x$
$y = 3x^2 + 5x$	$y = \sqrt{3x^2 + 5x}$
$y = \ln x$	$y = \ln(1 - 3x)$
$y = \frac{1}{3x^2}$	$y = \frac{1}{3x^2 + x}$

Table 10.1 Functions that can be differentiated with and without the chain rule

Differentiating a composite function

The chain rule says, in a very basic sense, that given two functions, the derivative of their composite is the product of their derivatives – remembering that a derivative is a rate of change of one quantity (variable) with respect to another quantity (variable). For example, the function $y = 8x + 6 = 2(4x + 3)$ is the composite of the functions $y = 2u$ and $u = 4x + 3$. Note that the function y is in terms of u, and the function u is in terms of x. How are the derivatives of these three functions related? Clearly, $\frac{dy}{dx} = 8$, $\frac{dy}{du} = 2$ and $\frac{du}{dx} = 4$. Since $8 = 2 \cdot 4$, the derivatives relate such that $\frac{dy}{dx} = \frac{dy}{du} \cdot \frac{du}{dx}$. In other words, rates of change multiply.

If we think of derivatives as rates of change, the relationship $\frac{dy}{dx} = \frac{dy}{du} \cdot \frac{du}{dx}$ can be illustrated by a practical example. Consider the pair of levers in Figure 10.6, with lever endpoints U and U′ connected by a segment that can shrink and stretch but always remains horizontal. Hence, points U and U′ are always the same distance, u, from the ground.

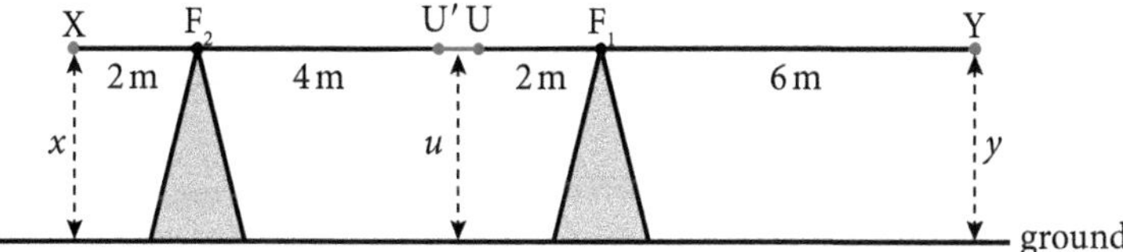

Figure 10.6 Two levers with horizontal connection between U and U′

As point Y moves down, points U and U′ move up, and point X moves down but at different rates. Let dy, du, and dx represent the change in distance from the ground for the points Y, U, and X, respectively. Because $YF_1 = 6$ and $UF_1 = 2$, then if point Y moves such that $dy = 3$, then $du = 1$. Since $U'F_2 = 4$ and $XF_2 = 2$, then if point U′ moves so that $du = 2$, then $dx = 1$.

Hence, $\frac{dy}{du} = 3$ and $\frac{du}{dx} = 2$.

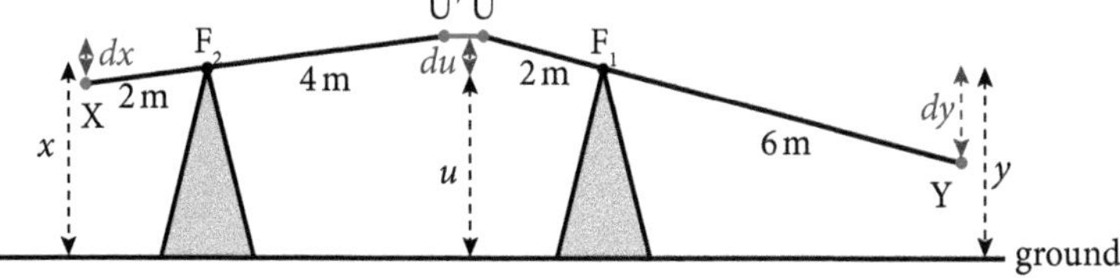

Figure 10.7 dx, du, and dy represent the change in distance from the ground for X, U, and Y

Combining these two results, we can see that for every 6 units that Y's distance changes, X's distance will change 1 unit. That is, $\frac{dy}{dx} = 6$. Therefore, we can write $\frac{dy}{dx} = \frac{dy}{du} \cdot \frac{du}{dx} = 3 \cdot 2 = 6$. In other words, the rate of change of y with respect to x is the product of the rate of change of y with respect to u and the rate of change of u with respect to x.

For the polynomial function in Example 10.3, we could have differentiated the function in expanded form by differentiating term by term rather than differentiating the factored form by the chain rule.
$\frac{dy}{dx} = \frac{d}{dx}(16x^4 - 8x^2 + 1)$
$= 64x^3 - 16x$
This confirms the result. It is not always easier to differentiate powers of polynomials by expanding and then differentiating term by term. For example, it is far better to find the derivative of $y = (3x + 5)^8$ by the chain rule.

Example 10.3

The polynomial function $y = 16x^4 - 8x^2 + 1 = (4x^2 - 1)^2$ is the composite of $y = u^2$ and $u = 4x^2 - 1$. Use the chain rule to find $\frac{dy}{dx}$, the derivative of y with respect to x.

Solution

$y = u^2 \Rightarrow \frac{dy}{du} = 2u$

$u = 4x^2 - 1 \Rightarrow \frac{du}{dx} = 8x$

Applying the chain rule: $\frac{dy}{dx} = \frac{dy}{du} \cdot \frac{du}{dx} = 2u \cdot 8x$

$= 2(4x^2 - 1) \cdot 8x$

$= 64x^3 - 16x$

We often write composite functions using nested function notation. For example, the notation $f(g(x))$ denotes a function composed of functions f and g such that g is the 'inside' function and f is the 'outside' function. For the composite function $y = (4x^2 - 1)^2$ in Example 10.3, the inside function is $g(x) = 4x^2 - 1$ and the outside function is $f(u) = u^2$. Looking again at the solution for Example 10.3, we see that we can choose to express and work out the chain rule in function notation rather than in Leibniz notation.

Leibniz notation	Function notation
$\frac{dy}{dx} = \frac{dy}{du} \cdot \frac{du}{dx} = 2u \cdot 8x$ $= 2(4x^2 - 1) \cdot 8x$ $= 64x^3 - 16x$	$\frac{d}{dx}[f(g(x))] = f'(u) \cdot g'(x) = 2u \cdot 8x$ $= f'(g(x)) \cdot g'(x) = 2(4x^2 - 1) \cdot 8x$ $= 64x^3 - 16x$

Table 10.2 Leibniz and function notation for using the chain rule to differentiate $y = (4x^2 - 1)^2$

This leads us to formally state the chain rule in the two different notations.

Chain rule

If $y = f(u)$ is a function in terms of u and $u = g(x)$ is a function in terms of x, then the function $y = f(g(x))$ is differentiated as follows:

$\frac{dy}{dx} = \frac{dy}{du} \cdot \frac{du}{dx}$ (Leibniz form)

or, equivalently,

$\frac{dy}{dx} = \frac{d}{dx}[f(g(x))] = f'(g(x)) \cdot g'(x)$ (function notation form)

The chain rule needs to be applied carefully. Consider the function notation form for the chain rule $\frac{d}{dx}[f(g(x))] = f'(g(x)) \cdot g'(x)$. Although it is the product of two derivatives, it is important to point out that the first derivative involves the function f differentiated at $g(x)$ and the second is function g differentiated at x. The chain rule written in Leibniz form, $\frac{dy}{dx} = \frac{dy}{du} \cdot \frac{du}{dx}$, is easily remembered because it appears to be an obvious statement about fractions – but they are **not** fractions. The expressions $\frac{dy}{dx}$, $\frac{dy}{du}$, and $\frac{du}{dx}$ are derivatives or, more precisely, limits, and although du and dx essentially represent very small changes in the variables u and x, we cannot guarantee that they are non-zero.

The function notation form of the chain rule offers a very useful way of saying the rule 'in words' and thus, a very useful structure for applying it.

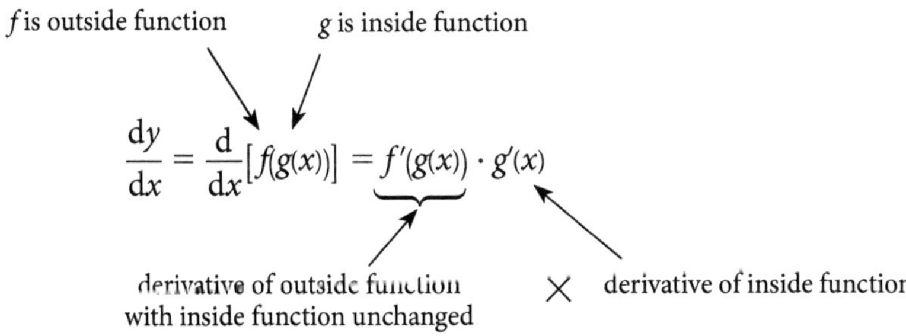

The chain rule in words:

$$\text{derivative of composite} = \text{derivative of outside function with inside function unchanged} \times \text{derivative of inside function}$$

Although this is taking some liberties with mathematical language, the mathematical interpretation of the phrase 'with inside function unchanged' is that the derivative of the outside function, f, is evaluated at $g(x)$, the inside function.

The chain rule is our most important rule of differentiation. It is an indispensable tool in differential calculus. Forgetting to apply the chain rule when it needs to be applied, or applying it improperly, is a common source of errors in calculus computations. It is important to understand it, practise it, and master it.

Example 10.4

Differentiate each function by applying the chain rule. Start by decomposing the composite function into the outside function and the inside function.

(a) $y = \cos 2x$ (b) $y = \sqrt{3x^2 + 5x}$

(c) $y = \ln(1 - 3x)$ (d) $y = \frac{1}{3x^2 + x}$

Solution

(a) $y = f(g(x)) = \cos 2x$ outside function is $f(u) = \cos u$

inside function is $g(x) = 2x$

In Leibniz form: $\frac{dy}{dx} = \frac{dy}{du} \cdot \frac{du}{dx} = (-\sin u) \cdot 2 = -2\sin(2x)$

Alternatively, in function notation form:

$$\frac{dy}{dx} = f'(g(x)) \cdot g'(x) = \underbrace{[-\sin(2x)]} \cdot 2 = -2\sin(2x)$$

derivative of outside function with inside function unchanged × derivative of inside function

(b) $y = f(g(x)) = \sqrt{3x^2 + 5x}$ outside function is $f(u) = \sqrt{u} = u^{\frac{1}{2}}$ $f'(u) = \frac{1}{2}u^{-\frac{1}{2}}$
inside function is $g(x) = 3x^2 + 5x$

$$\frac{dy}{dx} = f'(g(x)) \cdot g'(x) = \frac{1}{2}(3x^2 + 5x)^{-\frac{1}{2}} \cdot (6x + 5)$$

$$\frac{dy}{dx} = \frac{6x + 5}{2(3x^2 + 5x)^{\frac{1}{2}}} \text{ or } \frac{6x + 5}{2\sqrt{3x^2 + 5x}}$$

(c) $y = f(g(x)) = \ln(1 - 3x)$ outside function is $f(u) = \ln u$ $f'(u) = \frac{1}{u}$
inside function is $g(x) = 1 - 3x$

$$\frac{dy}{dx} = f'(g(x)) \cdot g'(x) = \frac{1}{1 - 3x} \cdot (-3)$$

$$\frac{dy}{dx} = -\frac{3}{1 - 3x} \text{ or } \frac{3}{3x - 1}$$

(d) $y = f(g(x)) = \frac{1}{3x^2 + x}$ outside function is $f(u) = \frac{1}{u} = u^{-1}$ $f'(u) = -u^{-2}$
inside function is $g(x) = 3x^2 + x$

$$\frac{dy}{dx} = f'(g(x)) \cdot g'(x) = -(3x^2 + x)^{-2} \cdot (6x + 1)$$

$$\frac{dy}{dx} = -\frac{6x + 1}{(3x^2 + x)^2}$$

Example 10.5

Find the derivative of the function $y = (2x + 3)^3$ by:
(a) expanding the binomial and differentiating term by term
(b) using the chain rule.

Solution

(a) $y = (2x + 3)^3 = (2x + 3)(2x + 3)^2$

$= (2x + 3)(4x^2 + 12x + 9)$

$= 8x^3 + 24x^2 + 18x + 12x^2 + 36x + 27$

$= 8x^3 + 36x^2 + 54x + 27$

$$\frac{dy}{dx} = 24x^2 + 72x + 54$$

(b) $y = f(g(x)) = (2x + 3)^3$

$y = f(u) = u^3 \Rightarrow f'(u) = 3u^2$

$u = g(x) = 2x + 3 \Rightarrow g'(x) = 2$

$$\frac{dy}{dx} = \frac{dy}{du} \cdot \frac{du}{dx} = 3u^2 \cdot 2 = 6u^2$$
$$= 6(2x + 3)^2$$
$$= 6(4x^2 + 12x + 9)$$
$$= 24x^2 + 72x + 54$$

Example 10.6

For each function $f(x)$, find $f'(x)$.

(a) $f(x) = \sin^2 x$ (b) $f(x) = \sin x^2$ (c) $f(x) = e^{\sin x}$ (d) $f(x) = \sqrt[3]{(7 - 5x)^2}$

Solution

(a) The expression $\sin^2 x$ is an abbreviated way of writing $(\sin x)^2$.

Hence, if $f(x) = g(h(x)) = (\sin x)^2$, then the outside function is $g(u) = u^2$, and the inside function is $h(x) = \sin x$.

By the chain rule, $f'(x) = g'(h(x)) \cdot h'(x)$
$$= 2(\sin x)^1 \cdot \cos x$$

Therefore, $f'(x) = 2\sin x \cos x$

(b) The expression $\sin x^2$ is equivalent to $\sin(x^2)$, and is *not* $(\sin x)^2$.

Hence, if $f(x) = g(h(x)) = \sin(x^2)$, then the outside function is $g(u) = \sin u$, and the inside function is $h(x) = x^2$.

By the chain rule, $f'(x) = g'(h(x)) \cdot h'(x)$
$$= \cos(x^2) \cdot 2x$$

Therefore, $f'(x) = 2x\cos(x^2)$

(c) $f(x) = g(h(x)) = e^{\sin x}$ outside function is $g(u) = e^u$

inside function is $h(x) = \sin x$

By the chain rule, $f'(x) = g'(h(x)) \cdot h'(x)$

Therefore, $f'(x) = e^{\sin x} \cdot \cos x$

(d) First change from radical form to rational power form.

$f(x) = \sqrt[3]{(7 - 5x)^2} = (7 - 5x)^{\frac{2}{3}}$

$f(x) = g(h(x)) = (7 - 5x)^{\frac{2}{3}}$ outside function is $g(u) = u^{\frac{2}{3}}$

inside function is $h(x) = 7 - 5x$

By the chain rule, $f'(x) = g'(h(x)) \cdot h'(x)$
$$= \frac{2}{3}(7 - 5x)^{-\frac{1}{3}} \cdot (-5)$$

Therefore, $f'(x) = -\dfrac{10}{3(7 - 5x)^{\frac{1}{3}}}$ or $-\dfrac{10}{3(\sqrt[3]{7 - 5x})}$

Endeavour to write functions in a way that eliminates any confusion regarding the argument of the function. For example, write $\sin(x^2)$ rather than $\sin x^2$; $1 + \ln x$ rather than $\ln x + 1$; $5 + \sqrt{x}$ rather than $\sqrt{x} + 5$; $\ln(4 - x^2)$ rather than $\ln 4 - x^2$.

Exercise 10.2

1. Find the derivative of each function.

(a) $y = (3x - 8)^4$ (b) $y = \sqrt{1 - x}$ (c) $y = \ln x^2$

(d) $y = 2\sin\left(\frac{x}{2}\right)$ (e) $y = (x^2 + 4)^{-2}$ (f) $y = e^{-3x}$

(g) $y = \frac{1}{\sqrt{x + 2}}$ (h) $y = \cos^2 x$ (i) $y = e^{x^2} - 2x$

(j) $y = \frac{1}{3x^2 - 5x + 7}$ (k) $y = \sqrt[3]{2x + 5}$ (l) $y = \ln(x^2 - 9)$

2. Find the equation of the tangent to the given curve at the specified value of x. Express the equation exactly in the form $y = mx + c$.

(a) $y = (2x^2 - 1)^3$ $x = -1$

(b) $y = \sqrt{3x^2 - 2}$ $x = 3$

(c) $y = \sin 2x$ $x = \pi$

3. An object moves along a line so that its position, s, relative to a starting point at any time $t \geqslant 0$ is given by $s(t) = \cos(t^2 - 1)$.

(a) Find the velocity of the object as a function of t.

(b) Find the object's velocity at $t = 0$.

(c) In the interval $0 < t < 2.5$, find any times (values of t) for which the object is stationary.

(d) Describe the object's motion during the interval $0 < t < 2.5$.

4. Find $\frac{dy}{dx}$ for each function. Use your GDC to check your answer.

(a) $y = \sqrt{x^2 + 2x + 1}$ (b) $y = \frac{1}{\sin x}$ (c) $y = \left(x + \sqrt{x}\right)^3$

(d) $y = e^{\cos x}$ (e) $y = (\ln x)^2$ (f) $y = \frac{3}{\sqrt{2x + 1}}$

5. Find the equation of **(i)** the tangent, and **(ii)** the normal to each curve at the given point.

(a) $y = \frac{2}{x^2 - 8}$ at (3, 2)

(b) $y = \sqrt{1 + 4x}$ at (2, 3)

(c) $y = \ln(4x - 3)$ at (1, 0)

6. Consider the trigonometric curve $y = \sin\left(2x - \frac{\pi}{2}\right)$

(a) Find $\frac{dy}{dx}$ and $\frac{d^2y}{dx^2}$.

(b) Find the exact coordinates of any inflection points for the curve in the interval $0 < x < \pi$.

10.3 The product and quotient rules

The product rule

With the differentiation rules that we have learned thus far, we can differentiate some functions that are products. For example, we can differentiate the function $f(x) = (x^2 + 3x)(2x - 1)$ by expanding and then differentiating the polynomial term by term. In doing so, we are applying the sum and difference, constant multiple, and power rules from the previous chapter.

$$f(x) = (x^2 + 3x)(2x - 1) = 2x^3 + 5x^2 - 3x$$

$$f'(x) = 2\frac{d}{dx}(x^3) + 5\frac{d}{dx}(x^2) - 3\frac{d}{dx}(x)$$

$$f'(x) = 6x^2 + 10x - 3$$

The sum and difference rule states that the derivative of a sum or difference of two functions is the sum or difference of their derivatives. Perhaps the derivative of the product of two functions is the product of their derivatives. If we try this with the above example we get

$$f'(x) = \frac{d}{dx}(x^2 + 3x) \cdot \frac{d}{dx}(2x - 1) = (2x + 3) \cdot 2 = 4x + 6.$$

However, $4x + 6 \neq 6x^2 + 10x - 3$, so this is clearly incorrect.

The derivative of a product of two functions is not the product of their derivatives. However, there are many products, such as $y = (4x - 3)^3(x - 1)^4$ and $f(x) = x^2\sin x$, for which it is either difficult or impossible to write the function as a polynomial. In order to differentiate functions like this we need a **product rule**.

Product rule

If y is a function in terms of x that can be expressed as the product of two functions, u and v, that are also in terms of x, then the product $y = uv$ can be differentiated as follows:

$$\frac{dy}{dx} = \frac{d}{dx}(uv) = u\frac{dv}{dx} + v\frac{du}{dx}$$

Equivalently, if $y = f(x) \cdot g(x)$, then

$$\frac{dy}{dx} = \frac{d}{dx}\left[f(x) \cdot g(x)\right] = f(x) \cdot g'(x) + g(x) \cdot f'(x)$$

A formal proof of the product rule is beyond the scope of this book. The following examples show how to apply the product rule in a variety of situations.

Example 10.7

Use the product rule to compute the derivative of the function $y = (x^2 + 3x)(2x - 1)$.

Solution

Let $u(x) = x^2 + 3x$ and $v(x) = 2x - 1$. Then $y = u(x) \cdot v(x)$ or $y = uv$

$\frac{du}{dx} = 2x + 3$ and $\frac{dv}{dx} = 2$

By the product rule (in Leibniz form)

$$\frac{dy}{dx} = \frac{d}{dx}(uv) = u\frac{dv}{dx} + v\frac{du}{dx} = (x^2 + 3x) \cdot 2 + (2x - 1) \cdot (2x + 3)$$
$$= (2x^2 + 6x) + (4x^2 + 4x - 3)$$
$$= 6x^2 + 10x - 3$$

This result agrees with the derivative we obtained earlier from differentiating the expanded polynomial.

Example 10.8

Given $y = x^2\sin x$, find $\frac{dy}{dx}$

Solution

Let $y = f(x) \cdot g(x) = x^2\sin x \Rightarrow f(x) = x^2$ and $g(x) = \sin x$

$f'(x) = 2x$ and $g'(x) = \cos x$

By the product rule (function notation form),

$$\frac{dy}{dx} = \frac{d}{dx}[f(x) \cdot g(x)] = f(x) \cdot g'(x) + g(x) \cdot f'(x)$$
$$= x^2 \cdot \cos x + (\sin x) \cdot 2x$$
$$\frac{dy}{dx} = x^2\cos x + 2x\sin x$$

As with the chain rule, it is helpful to remember the structure of the product rule in words.

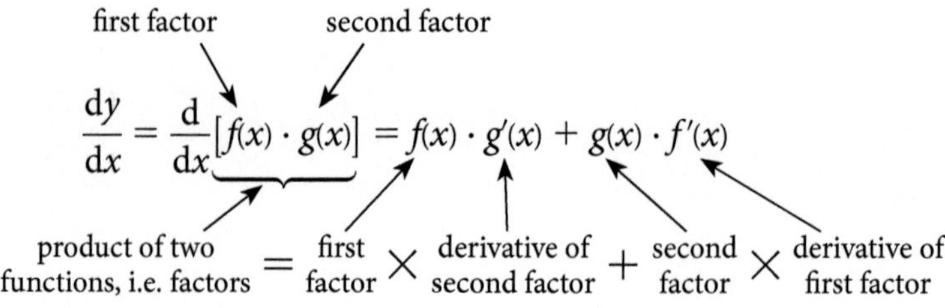

Example 10.9

Find the exact coordinates of any stationary points, and any inflection points, for the curve $y = xe^x$. Classify any stationary points as a maximum, minimum, or neither.

Solution

Recall from Chapter 9 that stationary points occur where the first derivative is zero and that inflection points (where concavity changes) may occur where the second derivative is zero.

$$\frac{dy}{dx} = \frac{d}{dx}\left[f(x) \cdot g(x)\right] = f(x) \cdot g'(x) + g(x) \cdot f'(x)$$

$$= \frac{d}{dx}(xe^x) = x\frac{d}{dx}(e^x) + e^x\frac{d}{dx}(x)$$

$$\frac{dy}{dx} = xe^x + e^x = 0 \Rightarrow e^x(x + 1) = 0 \Rightarrow \frac{dy}{dx} = 0 \text{ when } x = -1$$

When $x = -1, y = -e^{-1} = -\frac{1}{e}$

Therefore the curve has a stationary point at $\left(-1, -\frac{1}{e}\right)$.

$$\frac{d^2y}{dx^2} = \frac{d}{dx}(xe^x + e^x) = \frac{d}{dx}(xe^x) + \frac{d}{dx}e^x$$

$$= (xe^x + e^x) + e^x$$

$$\frac{d^2y}{dx^2} = xe^x + 2e^x = 0 \Rightarrow e^x(x + 2) = 0 \Rightarrow \frac{d^2y}{dx^2} = 0 \text{ when } x = -2$$

An inflection point will occur at $x = -2$ if the sign of the second derivative changes (i.e. concavity changes) at that point. Find the sign of $\frac{d^2y}{dx^2}$ at test points $x = -3$ and $x = 0$.

At $x = -3, \frac{d^2y}{dx^2} = e^{-3}(-3 + 2) = -\frac{1}{e^3} < 0$

At $x = 0, \frac{d^2y}{dx^2} = e^0(0 + 2) = 2 > 0$

The second derivative undergoes a sign change at $x = -2$, hence there is an inflection point on the curve at that point.

When $x = -2, y = -2e^{-2} = -\frac{2}{e^2}$

Therefore, the curve has an inflection point at $\left(-2, -\frac{2}{e^2}\right)$.

We can use the second derivative test to classify the stationary point $\left(-1, -\frac{1}{e}\right)$.

At $x = -1, \frac{d^2y}{dx^2} = e^{-1}(-1 + 2) = \frac{1}{e} > 0 \Rightarrow$ curve is concave up at $x = -1$.

Therefore, the stationary point $\left(-1, -\frac{1}{e}\right)$ is a minimum point for the curve.

It's good practice to perform a graphical check of our results on a GDC.

The graph on the GDC not only visually confirms our results but also informs us that $\left(-1, -\frac{1}{e}\right)$ is an absolute minimum.

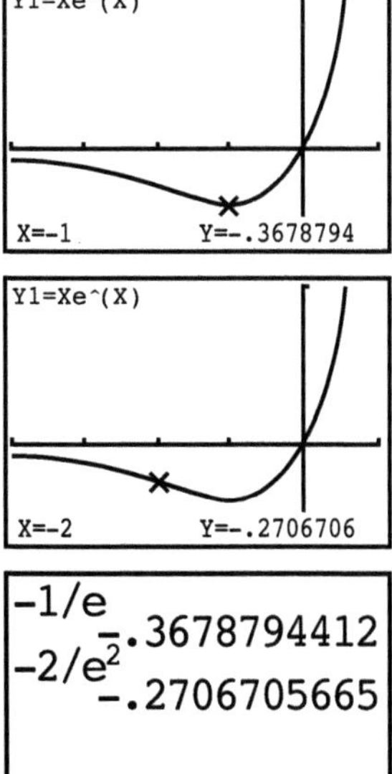

Figure 10.8 GDC screens for solution to Example 10.9

The quotient rule

In part (d) of Example 10.4 in Section 10.2, we used the chain rule to find the derivative of the rational function $y = \frac{1}{3x^2 + x}$ by expressing the function as $y = (3x^2 + x)^{-1}$. Can we apply the same approach to differentiating $y = \frac{x+3}{2x-1}$?

Although we can write the function as $y = (x + 3)(2x - 1)^{-1}$ and then apply both the product rule and the chain rule, it seems worthwhile to have a **quotient rule** to differentiate such functions more directly.

Quotient rule

If y is a function in terms of x that can be expressed as the quotient of two functions, u and v, that are also in terms of x, then the quotient $y = \frac{u}{v}$ can be differentiated as follows:

$$\frac{dy}{dx} = \frac{d}{dx}\left(\frac{u}{v}\right) = \frac{v\frac{du}{dx} - u\frac{dv}{dx}}{v^2}$$

Equivalently, if $y = \frac{f(x)}{g(x)}$, then

$$\frac{dy}{dx} = \frac{d}{dx}\left[\frac{f(x)}{g(x)}\right] = \frac{g(x) \cdot f'(x) - f(x) \cdot g'(x)}{[g(x)]^2}$$

A full proof is beyond the scope of this course. As with the chain rule and the product rule, it is helpful to recognise the structure of the quotient rule by remembering it in words.

$$\begin{pmatrix}\text{derivative of}\\ \text{quotient}\end{pmatrix} = \frac{(\text{denominator}) \times \begin{pmatrix}\text{derivative of}\\ \text{numerator}\end{pmatrix} - (\text{numerator})\begin{pmatrix}\text{derivative of}\\ \text{denominator}\end{pmatrix}}{(\text{denominator})^2}$$

Example 10.10

Given $y = \frac{x+3}{2x-1}$, find $\frac{dy}{dx}$

Solution

$y = \frac{u}{v} = \frac{x+3}{2x-1} \Rightarrow u = x + 3$ and $v = 2x - 1$

$\frac{du}{dx} = 1$ and $\frac{dv}{dx} = 2$

By the quotient rule (Leibniz form),

$$\frac{dy}{dx} = \frac{d}{dx}\left(\frac{u}{v}\right) = \frac{v\frac{du}{dx} - u\frac{dv}{dx}}{v^2} = \frac{(2x-1) \cdot 1 - (x+3) \cdot 2}{(2x-1)^2}$$

$$= \frac{2x - 1 - 2x - 6}{(2x-1)^2}$$

$$\frac{dy}{dx} = \frac{-7}{(2x-1)^2}$$

Example 10.11

Given $y = \frac{1}{2x-3}$, find $\frac{dy}{dx}$ by using:

(a) the quotient rule (b) the chain rule.

Solution

(a) $y = \frac{f(x)}{g(x)} = \frac{1}{2x-3} \Rightarrow f(x) = 1$ and $g(x) = 2x - 3$

$f'(x) = 0$ and $g'(x) = 2$

By the quotient rule (function notation form),

$$\frac{dy}{dx} = \frac{d}{dx}\left[\frac{f(x)}{g(x)}\right] = \frac{g(x) \cdot f'(x) - f(x) \cdot g'(x)}{[g(x)]^2}$$

$$= \frac{(2x-3) \cdot 0 - 1 \cdot (2)}{(2x-3)^2}$$

$$\frac{dy}{dx} = -\frac{2}{(2x-3)^2}$$

(b) $y = f(g(x)) = \frac{1}{2x-3} = (2x-3)^{-1}$

outside function is $f(u) = u^{-1}$
$\Rightarrow f'(u) = -u^{-2}$
inside function is $g(x) = 2x - 3$

By the chain rule (function notation form),

$$\frac{dy}{dx} = f'(g(x)) \cdot g'(x) = -(2x-3)^{-2} \cdot 2$$

$$\frac{dy}{dx} = -\frac{2}{(2x-3)^2}$$

As Example 10.11 illustrates, when required to differentiate a quotient you can choose to rewrite the quotient $y = \frac{u}{v}$ as $y = uv^{-1}$, and then the chain rule and/or the product rule instead of the quotient rule.

Example 10.12

For each function, find its derivative by using
(i) the quotient rule (ii) another method.

(a) $f(x) = \frac{3x-2}{2x-5}$ (b) $g(x) = \frac{5x-1}{3x^2}$

Solution

(a) (i) $f(x) = y = \frac{u}{v} = \frac{3x - 2}{2x - 5}$

$$f'(x) = \frac{dy}{dx} = \frac{v\frac{du}{dx} - u\frac{dv}{dx}}{v^2} = \frac{(2x-5)\cdot 3 - (3x-2)\cdot 2}{(2x-5)^2}$$

$$= \frac{6x - 15 - 6x + 4}{(2x-5)^2}$$

$$f'(x) = \frac{-11}{(2x-5)^2}$$

$v = (2x-5)^{-1}$ is a composite function, so we'll need the chain rule to find $\frac{dv}{dx}$

(ii) Rewrite $f(x)$ as a product and apply the product rule with the chain rule.

$f(x) = y = \frac{3x-2}{2x-5} = (3x-2)(2x-5)^{-1}$

$\Rightarrow u = 3x - 2$ and $v = (2x-5)^{-1}$

$$f'(x) = \frac{d}{dx}(uv) = u\frac{dv}{dx} + v\frac{du}{dx}$$

$$= (3x-2)\cdot\frac{d}{dx}[(2x-5)^{-1}] + (2x-5)^{-1}\cdot 3$$

$$= (3x-2)[-(2x-5)^{-2}\cdot 2] + 3(2x-5)^{-1}$$ Apply chain rule to $\frac{d}{dx}[(2x-5)^{-1}]$

$$= (-6x+4)(2x-5)^{-2} + 3(2x-5)^{-1}$$

$$= (2x-5)^{-2}[(-6x+4) + 3(2x-5)]$$ Factorise out HCF of $(2x-5)^{-2}$

$$= (2x-5)^{-2}[-6x + 4 + 6x - 15]$$

$$f'(x) = \frac{-11}{(2x-5)^2}$$

Note the use of brackets. When using the quotient rule, it is a good idea to enclose all factors and derivatives in brackets. Also, be extra careful with the subtraction in the numerator, which causes many errors when using the quotient rule.

(b) (i) $g(x) = y = \frac{u}{v} = \frac{5x-1}{3x^2}$

$$g'(x) = \frac{dy}{dx} = \frac{v\frac{du}{dx} - u\frac{dv}{dx}}{v^2} = \frac{3x^2\cdot 5 - (5x-1)\cdot 6x}{(3x^2)^2}$$

$$= \frac{15x^2 - 30x^2 + 6x}{9x^4}$$

$$= \frac{3x(-5x+2)}{9x^4}$$

$$g'(x) = \frac{-5x+2}{3x^3}$$

(ii) Use algebra to split the numerator:

$$g(x) = \frac{5x-1}{3x^2} = \frac{5x}{3x^2} - \frac{1}{3x^2} = \frac{5}{3x} - \frac{1}{3x^2} = \frac{5}{3}x^{-1} - \frac{1}{3}x^{-2}$$

Now, differentiate term-by-term using the power rule.

$$g'(x) = \frac{5}{3}\frac{d}{dx}(x^{-1}) - \frac{1}{3}\frac{d}{dx}(x^{-2})$$

$$= \frac{5}{3}(-x^{-2}) - \frac{1}{3}(-2x^{-3})$$

$$g'(x) = -\frac{5}{3x^2} + \frac{2}{3x^3} = -\frac{5}{3x^2}\cdot\frac{x}{x} + \frac{2}{3x^3} = -\frac{5x}{3x^3} + \frac{2}{3x^3} = \frac{-5x+2}{3x^3}$$

As Example 10.12 demonstrates, before differentiating a quotient, consider whether it is worthwhile to perform some algebra that may allow you to differentiate more efficiently.

The function $h(x) = \frac{3x^2}{5x-1}$ initially looks similar to the function g in Example 10.12 part (b) (they are reciprocals). However, it is not possible to split the denominator and express as two fractions. Recognise that $\frac{3x^2}{5x-1}$ is not equivalent to $\frac{3x^2}{5x} - \frac{3x^2}{1}$. Hence, in order to differentiate $h(x) = \frac{3x^2}{5x-1}$ we would apply either the quotient rule, or the product rule with the function rewritten as $h(x) = 3x^2(5x-1)^{-1}$, using the chain rule to differentiation the factor $(5x-1)^{-1}$.

Exercise 10.3

1. Find the derivative of each function.

(a) $y = x^2e^x$

(b) $y = x\sqrt{1-x}$

(c) $y = x\ln x$

(d) $y = \sin x \cos x$

(e) $y = \frac{e^x}{x}$

(f) $y = \frac{x+1}{x-1}$

(g) $y = (2x-1)^3(x^4+1)$

(h) $y = \frac{\sin x}{x}$

(i) $y = \frac{x}{e^x - 1}$

(j) $y = \frac{6x-7}{3x+2}$

(k) $y = (x^2-1)\ln(3x)$

(l) $y = \frac{1}{\sin^2 x + \cos^2 x}$

2. Find the equation of the tangent to the given curve at the specified value of x. Express the equation exactly in the form $y = mx + c$.

(a) $y = \frac{8}{4+x^2}$ $\quad x = 2$

(b) $y = \frac{x^3+1}{2x}$ $\quad x = 1$

(c) $y = x\sqrt{x^2-3}$ $\quad x = 2$

3. Consider the function $h(x) = \frac{x^2 - 3}{e^x}$

(a) Find the exact coordinates of any stationary points.

(b) Determine whether each stationary point is a maximum, minimum, or neither.

(c) What do the function values approach as **(i)** $x \to \infty$ **(ii)** $x \to -\infty$?

(d) Write down the equation of any asymptotes for the graph of $h(x)$.

(e) Make an accurate sketch of the curve, indicating any extrema and points where the graph intersects the x- and y-axes.

4. Find the equation for the tangent and the normal to the graph of $y = \frac{1}{\sqrt{3 - 2x}}$ at the point where $x = -3$.

5. A curve has equation $y = x(x - 4)^2$.

(a) For this curve, find:

(i) the x-intercepts

(ii) the coordinates of the maximum point

(iii) the coordinates of the point of inflection.

(b) Use your answers to part **(a)** to sketch a graph of the curve for $0 \leqslant x \leqslant 4$, clearly indicating the features you have found in part **(a)**.

6. Find the equation of **(i)** the tangent **(ii)** the normal to each curve at the given point.

(a) $y = \frac{2}{x^2 - 8}$ at (3, 2) **(b)** $y = x\sqrt{1 + x}$ at (0, 0)

(c) $y = \frac{x}{x + 1}$ at $\left(1, \frac{1}{2}\right)$

7. The tangent to the graph of $y = 3x\sqrt{1 + 2x}$ at the point (4, 36) intersects the x-axis at point A and intersects the y-axis at point B. Find the exact coordinates of A and B.

8. The function h is defined as $h(x) = \frac{2x - 4}{x^2 - 4x + 5}$

(a) Find the derivative of h, $h'(x)$.

(b) Without using your GDC, find the exact coordinates of any points on the graph of h where there is a horizontal tangent.

(c) Use your GDC to confirm your results for part **(b)**.

To find the derivative of $\tan x$, express it as $\frac{\sin x}{\cos x}$ and apply the quotient rule.

9. Find the equation of both the tangent and the normal to the curve $y = x \tan x$ at the point where $x = \frac{\pi}{4}$.

10. Consider the function $f(x) = \frac{x^2 - 3x + 4}{(x + 1)^2}$

(a) Show that $f'(x) = \frac{5x - 11}{(x + 1)^3}$ **(b)** Show that $f''(x) = \frac{-10x + 38}{(x + 1)^4}$

(c) Does the graph of f have an inflection point at $x = 3.8$? Give reasons for your answer.

10.4 Optimisation

Many problems in science and mathematics involve finding the maximum or minimum value (**optimum** value) of a function over a specified or implied domain. The development of calculus in the 17th century was motivated to a large extent by maxima and minima (**optimisation**) problems. One such problem led Pierre de Fermat (1601–1665) to develop his principle of least time: a ray of light will follow the path that takes the least (or minimum) time. The solution to Fermat's principle lead to Snell's law, or the law of refraction. The solution is found by applying techniques of differential calculus, which can also be used to solve other optimisation problems involving ideas such as least cost, maximum profit, minimum surface area, and greatest volume.

We have learned the theory of how to use the derivative of a function to locate points where the function has a maximum or minimum (i.e. extreme) value. It is important to remember that if the derivative of a function is zero at a certain point, it does not necessarily follow that the function has an extreme value (relative or absolute) at that point – it only means that the function has a horizontal tangent (stationary point) at that point. An extreme value may occur where the derivative is zero or at the endpoints of the function's domain.

Figure 10.9 shows the graph of $f(x) = x^4 - 8x^3 + 18x^2 - 16x - 2$. The derivative of $f(x)$ is $f'(x) = 4x^3 - 24x^2 + 36x - 16 = 4(x - 4)(x - 1)^2$. The function has horizontal tangents at both $x = 1$ and $x = 4$ since the derivative is zero at these points. However, an extreme value (absolute minimum) occurs only at $x = 4$. It is important to confirm – graphically or algebraically – the precise nature of a point on a function where the derivative is zero. Some different algebraic methods for confirming that a value is a maximum or minimum will be illustrated in the examples that follow.

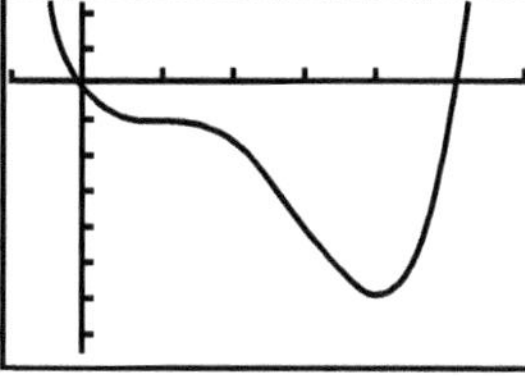

Figure 10.9 GDC screen showing the graph of $f(x) = x^4 - 8x^3 + 18x^2 - 16x - 2$

It is also useful to not ignore that one can often find extreme values (extrema) without calculus (e.g. using a minimum command on a graphics calculator). Calculator or computer technology can be very helpful in modelling, solving or confirming solutions to optimisation problems. However, it is important to learn how to apply algebraic methods of differentiation to optimisation problems because it may be the only efficient way to obtain an accurate solution.

Let's start with a relatively straightforward example. We can use the steps in the solution to develop a general strategy that can be applied to more sophisticated problems.

Example 10.13

Find the maximum area of a rectangle inscribed in an isosceles right-angled triangle whose hypotenuse is 20 cm long.

General strategy for solving optimisation problems

Step 1: Draw a diagram that accurately illustrates the problem. Label all known parts of the diagram. Using variables, label the important unknown quantity (or quantities), for example, x for base and y for height in Example 10.13.

Step 2: For the quantity that is to be optimised (area in Example 10.13), express this quantity as a function in terms of a single variable. From the diagram and/or information provided, determine the domain of this function.

Step 3: Find the derivative of the function from Step 2, and determine where the derivative is zero. This value (or values) of the derivative, along with any domain endpoints, are the critical values to be tested – in Example 10.13: $x = 0$, $x = 10$, and $x = 20$).

Step 4: Using algebraic (e.g. second derivative test) or graphical (e.g. GDC) methods, analyse the nature (maximum, minimum or neither) of the points at the critical values for the optimised function. Make sure you answer the precise question that was asked in the problem.

Solution

Step 1: Draw an accurate diagram. Let the base of the rectangle be x cm and the height y cm.

The area of the rectangle is $A = xy\,\text{cm}^2$.

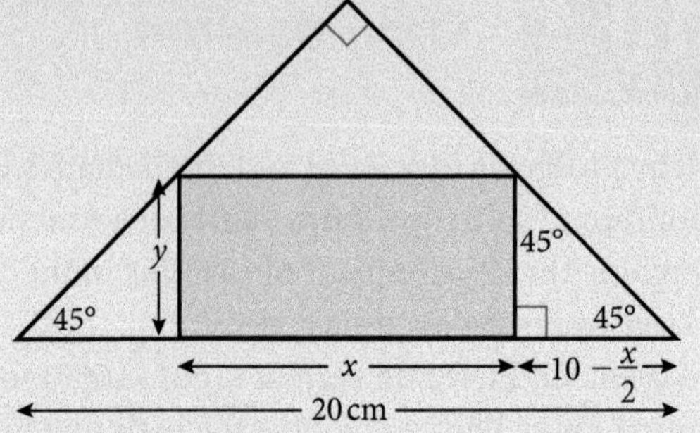

Step 2: Express the area as a function in terms of only one variable.

It can be deduced from the diagram that $y = 10 - \frac{1}{2}x$.

Therefore, $A(x) = x\left(10 - \frac{1}{2}x\right) = 10x - \frac{1}{2}x^2$

x must be positive; from the diagram it is clear that x must be less than 20.

Step 3: Find the derivative of the area function and find for what value(s) of x it is zero.

$A'(x) = 10 - x$

$A'(x) = 0$ when $x = 10$

Step 4: Analyse $A(x)$ at $x = 10$ and also at the endpoints of the domain, $x = 0$ and $x = 20$.

The second derivative test (Chapter 12) provides information about the concavity of a function. The second derivative is $A''(x) = -1$ and since $A''(x)$ is always negative, $A(x)$ is always concave down, indicating that $A(x)$ has a maximum at $x = 10$.

$A(0) = 0$ and $A(20) = 0$. Therefore $A(x)$ has an absolute maximum at $x = 10$.

Example 10.14

Two vertical posts, with heights of 7 m and 13 m, are secured by a rope going from the top of one post to a point on the ground that is between the posts and then to the top of the other post (see diagram). The distance between the two posts is 25 m. Where should the point at which the rope touches the ground be located so that the least amount of rope is used?

Online

An interactive diagram for the optimisation problem in Example 10.14.

Solution

Step 1: Draw an accurate diagram.
Draw the posts as line segments PQ and TS and the point where the rope touches the ground is labeled R. The optimum location of point R can be given as a distance from the base of the shorter post, QR, or from the taller post, SR.

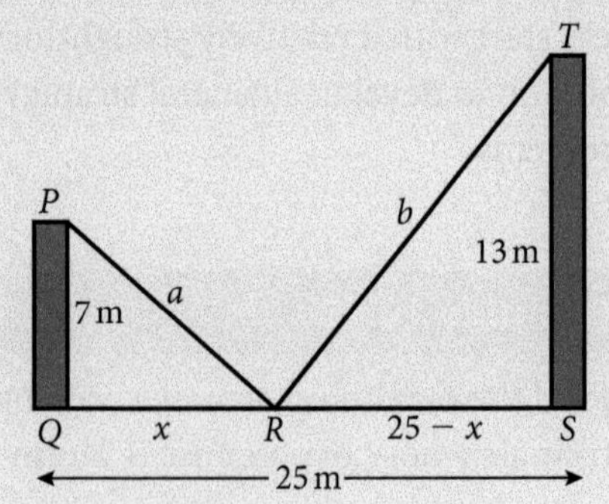

It is decided to give the answer as the distance from the shorter post, and this is labeled x. There are two other important unknown quantities – the lengths of the two portions of the rope, PR and TR. These are labeled a and b, respectively.

Step 2: The quantity to be minimised is the length L of the rope which is the sum of a and b. From Pythagoras' theorem, $a = \sqrt{x^2 + 49}$ and $b = \sqrt{(25 - x)^2 + 169}$. Therefore, the function for length (L) can be expressed in terms of the single variable x as

$$\begin{aligned} L(x) &= \sqrt{x^2 + 49} + \sqrt{(25 - x)^2 + 169} \\ &= \sqrt{x^2 + 49} + \sqrt{x^2 - 50x + 625 + 169} \\ &= \sqrt{x^2 + 49} + \sqrt{x^2 - 50x + 794} \end{aligned}$$

From the given information and diagram, the domain of $L(x)$ is $0 \leqslant x \leqslant 25$

Step 3: To facilitate differentiation, express $L(x)$ using fractional exponents.

$$L(x) = (x^2 + 49)^{\frac{1}{2}} + (x^2 - 50x + 794)^{\frac{1}{2}}$$

Apply the chain rule for differentiation:

$$\frac{dL}{dx} = \frac{1}{2}(x^2 + 49)^{-\frac{1}{2}}(2x) + \frac{1}{2}(x^2 - 50x + 794)^{-\frac{1}{2}}(2x - 50)$$

$$\frac{dL}{dx} = \frac{x}{\sqrt{x^2 + 49}} + \frac{x - 25}{\sqrt{x^2 - 50x + 794}}.$$

By setting $\frac{dL}{dx} = 0$, we obtain

$$\begin{aligned} x\sqrt{x^2 - 50x + 794} &= -(x - 25)\sqrt{x^2 + 49} \\ x^2(x^2 - 50x + 794) &= (25 - x)^2(x^2 + 49) \\ x^4 - 50x^3 + 794x^2 &= x^4 - 50x^3 + 674x^2 - 2450x + 30\,625 \\ 120x^2 + 2450x - 30\,625 &= 0 \\ 5(4x - 35)(6x + 175) &= 0 \end{aligned}$$

$$x = \frac{35}{4} \quad \text{or} \quad x = -\frac{175}{6}$$

Step 4: Since $x = -\frac{175}{6}$ is not in the domain for $L(x)$, then the critical values are $x = 0$, $x = \frac{35}{4}$ and $x = 25$. Simply evaluate $L(x)$ for these critical values.

$$L(0) = 49 + \sqrt{794} \approx 35.18$$

$$L(25) = \sqrt{674} + 13 \approx 38.96$$

$$L\left(\frac{35}{4}\right) = 5\sqrt{41} \approx 32.02$$

Therefore, the rope should touch the ground at a distance of $\frac{35}{4} = 8.75$ m from the base of the shorter post to give a minimum rope length of approximately 32.02 m.

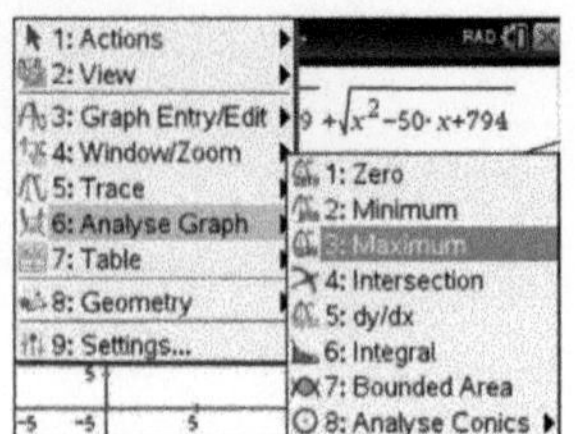

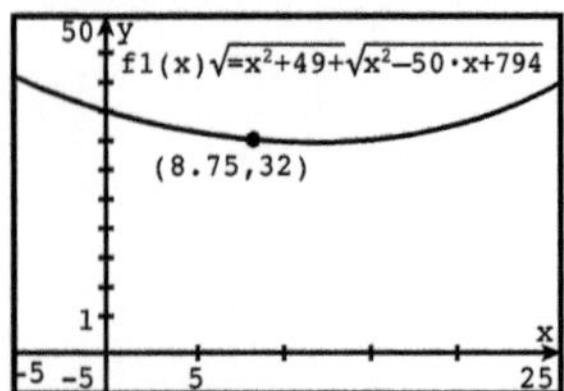

Figure 10.10 GDC screens for steps 3 and 4 of the solution to Example 10.14

The minimum value could also be confirmed from the graph of $L(x)$, but it would be difficult to confirm using the second derivative test because of the tedious algebra required. From this example, we can see that applied optimisation problems can involve a high level of algebra. If you have access to suitable graphing technology, you could perform steps 3 and 4 graphically rather than algebraically (as shown in Figure 10.10).

In both Examples 10.13 and 10.14, the extreme value occurred at a point where the derivative was zero. Although this often happens, an extreme value may occur at the endpoint of the domain.

Example 10.15

Four metres of wire is to be used to form a square and a circle. How much of the wire should be used to make the square and how much should be used to make the circle in order to enclose the greatest amount of total area?

Solution

Step 1: Let x = length of each edge of the square, and r = radius of the circle.

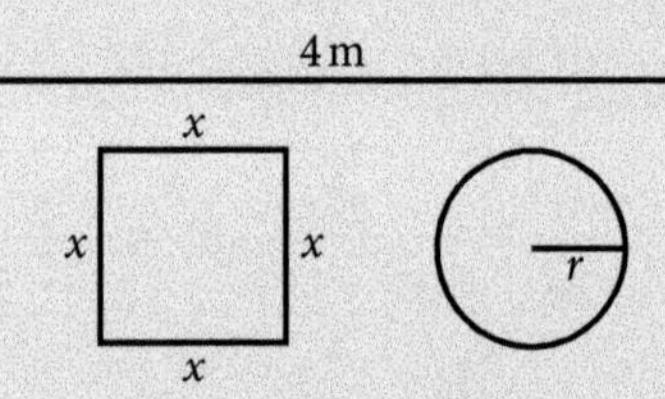

Step 2: The total area is given by $A = x^2 + \pi r^2$. The task is to write the area A as a function of a single variable.
Therefore it is necessary to express r in terms of x or vice-versa and perform a substitution.

The perimeter of the square is $4x$ and the circumference of the circle is $2\pi r$. The total amount of wire is 4 m which gives:

$$4 = 4x + 2\pi r \Rightarrow 2\pi r = 4 - 4x \Rightarrow r = \frac{2(1-x)}{\pi}$$

Substitute for r:

$$A(x) = x^2 + \pi\left[\frac{2(1-x)}{\pi}\right]^2 = x^2 + \frac{4(1-x)^2}{\pi} = \frac{1}{\pi}[(\pi+4)x^2 - 8x + 4]$$

Because the square's perimeter is $4x$, then the domain for $A(x)$ is $0 \leqslant x \leqslant 1$

Step 3: Differentiate the function $A(x)$, set equal to zero, and solve.

$$\frac{d}{dx}\left(\frac{1}{\pi}[(\pi+4)x^2 - 8x + 4]\right)$$

$$= \frac{1}{\pi}[2(\pi+4)x - 8] = 0$$

$$2(\pi+4)x - 8 = 0 \Rightarrow \Rightarrow x = \frac{4}{\pi+4} \approx 0.5601$$

The critical values are $x = 0$, $x \approx 0.5601$, and $x = 1$

Step 4: Evaluate $A(x)$: $A(0) \approx 1.273$, $A(0.5601) \approx 0.5601$, and $A(1) = 1$

Therefore, the maximum area occurs when $x = 0$ which means all the wire is used for the circle.

Example 10.16

A pipeline needs to be constructed to link an offshore drilling rig to an onshore refinery depot. The oil rig is located at a distance (perpendicular to the coast) of 140 km from the coast. The depot is located inland at a perpendicular distance of 60 km from the coast. For modelling purposes, the coastline is assumed to follow a straight line. The rate at which crude oil is pumped through the pipeline varies according to several variables, including pipe dimensions, materials, temperature, and so on. On average, oil flows through the offshore section of the pipeline at a rate of 9 km h^{-1} and through the onshore section at a rate of 5 km h^{-1}. Assume that both sections of pipeline can travel straight from one point to another. At what point should the pipeline intersect with the coastline in order for the oil to take a minimum amount of time to flow from the rig to the depot?

Solution

Step 1: The optimum location of the point, C, where the pipeline comes ashore will be designated by the distance it is from the point on the coast that is a minimum distance (perpendicular) from the rig, R (140 km).

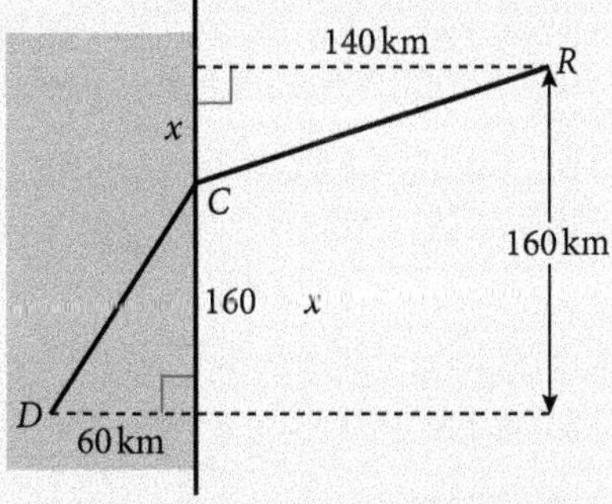

The distance from R to C is $\sqrt{x^2 + 140^2}$ and the distance from D (depot) to C is $\sqrt{(160 - x)^2 + 60^2}$.

Step 2: The quantity to be minimised is time, so it is necessary to express the total time it takes the oil to flow from R to D in terms of a single variable.

$$\text{time} = \frac{\text{distance}}{\text{rate}}$$

$$\text{time (offshore)} = \frac{\sqrt{x^2 + 19\,600}}{9}$$

$$\text{time (onshore)} = \frac{\sqrt{x^2 - 320x + 29\,200}}{5}$$

The function for time T in terms of x is:

$$T(x) = \frac{\sqrt{x^2 + 19600}}{9} + \frac{\sqrt{x^2 - 320x + 29200}}{5}$$

and the domain for $T(x)$ is $0 \leqslant x \leqslant 160$

Steps 3 and 4: The algebra for finding the derivative of $T(x)$ is similar to that of step 3 in Example 10.14.

Use a GDC to find the value of x that produces a minimum for $T(x)$.

Therefore, the optimum point for the pipeline to intersect with the coast is approximately 134.9 km from the point on the coast nearest to the drilling rig.

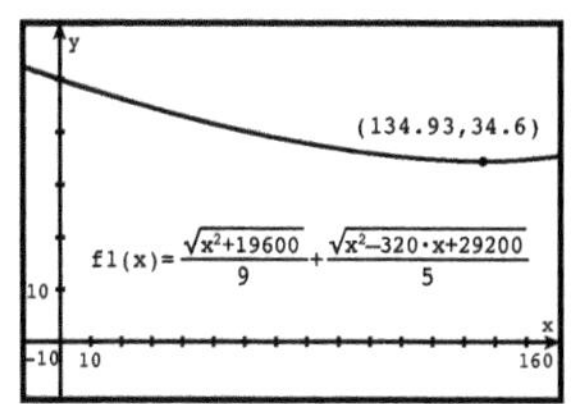

Figure 10.11 GDC screens for steps 3 and 4 of the solution to Example 10.16

Exercise 10.4

1. Find the points on the graph of the equation $y = 4 - x^2$ that are nearest to the point (0, 2).

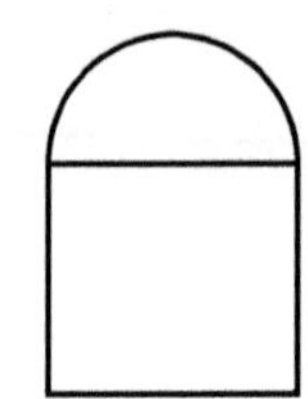

Figure 10.12 Diagram for question 2

2. A window is in the shape of a rectangle with a semicircle on top. Find the dimensions of the rectangular section of the window when the perimeter of the entire window is 4 metres and the area of the entire window is a maximum.

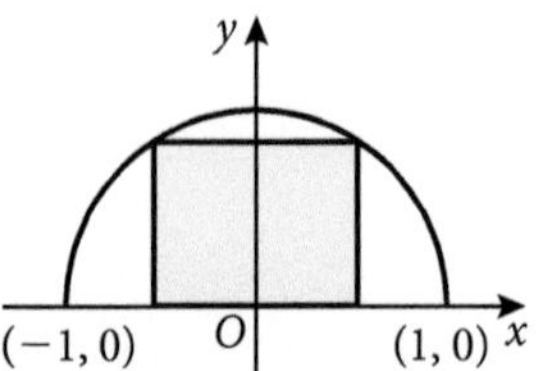

Figure 10.13 Diagram for question 3

3. Find the dimensions of the rectangle with maximum area that is inscribed in a semicircle with radius 1 cm. Two vertices of the rectangle are on the semicircle and the other two vertices are on the x-axis, as shown in Figure 10.13.

4. A rectangular piece of aluminium is to be rolled to make a cylinder with open ends (a tube). Regardless of the dimensions of the rectangle, the perimeter of the rectangle must be 40 cm. Find the dimensions (length and width) of the rectangle that gives a maximum volume for the cylinder.

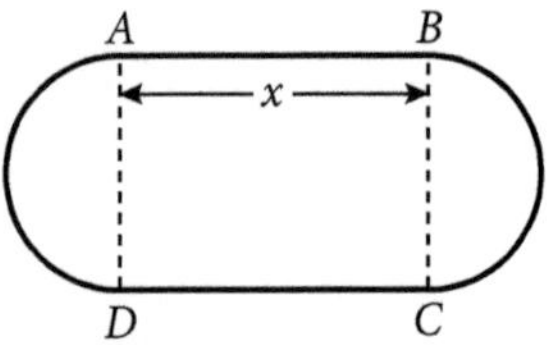

Figure 10.14 Diagram for question 7

5. Find the minimum distance from the graph of the function $y = \sqrt{x}$ to the point $\left(\frac{3}{2}, 0\right)$.

6. A rectangular box has height h cm, width x cm, and length $2x$ cm. It is designed to have a volume equal to 1 litre (1000 cm^3).

   (a) Show that $h = \frac{500}{x^2}$ cm.

   (b) Find an expression for the total surface area, s cm^2, of the box in terms of x.

   (c) Find the dimensions of the box that produces a minimum surface area, giving your answers to 3 significant figures.

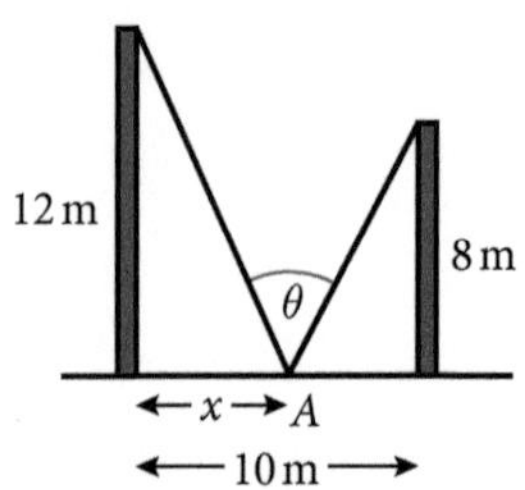

Figure 10.15 Diagram for question 8

7. The shape in Figure 10.14 consists of a rectangle $ABCD$ and two semicircles on either end. The rectangle has an area of 100 cm^2. If x represents the length of the rectangle AB, then find the value of x that makes the perimeter of the entire figure a minimum.

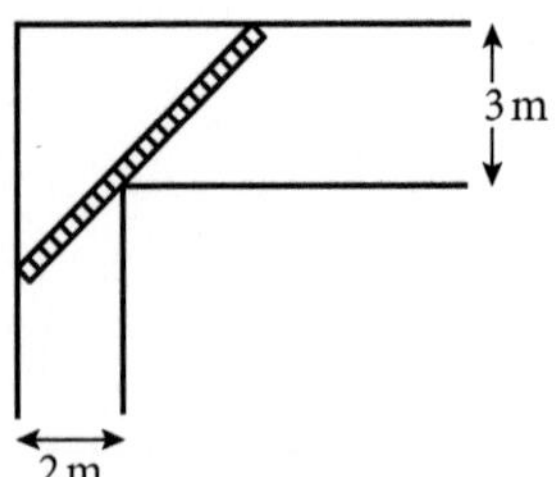

Figure 10.16 Diagram for question 9

8. Two vertical posts, with heights 12 metres and 8 metres, are 10 metres apart on horizontal ground (Figure 10.15). A rope that stretches is attached to the top of both posts and is stretched down so that it touches the ground at point A between the two posts. The distance from the base of the taller post to point A is represented by x and the angle between the two sections of rope is θ. What value of x makes θ a maximum?

9. A ladder is to be carried horizontally down an L-shaped hallway. The first section of the hallway is 2 metres wide and then there is a right-angled turn into a 3-metre wide section of the hallway. What is the longest ladder that can be carried around the corner?

Online 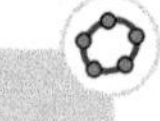

An interactive diagram for the optimisation problem in question 9.

10. Erica is walking from the wildlife observation tower (T) to the Big Desert Park office (O). The tower is 7 km due west and 10 km due south of the office. There is a road that goes to the office that Erica can get to if she walks 10 km due north from the tower. Erica can walk at a rate of 2 km h^{-1} through the sandy terrain of the park, but she can walk at a faster rate of 5 km h^{-1} on the road. To what point A on the road should Erica walk to in order to take the least time to walk from the tower to the office? Find the value of d such that point A is d km from the office

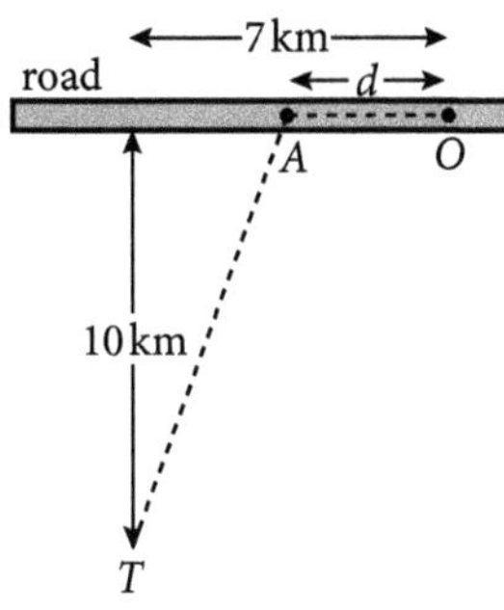

Figure 10.17 Diagram for question 10

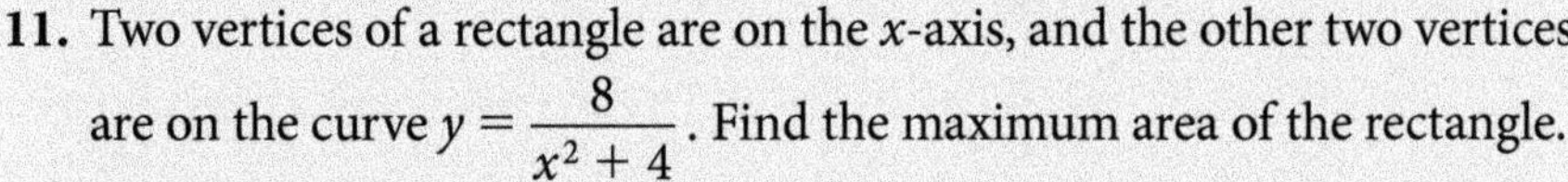

11. Two vertices of a rectangle are on the x-axis, and the other two vertices are on the curve $y = \dfrac{8}{x^2 + 4}$. Find the maximum area of the rectangle.

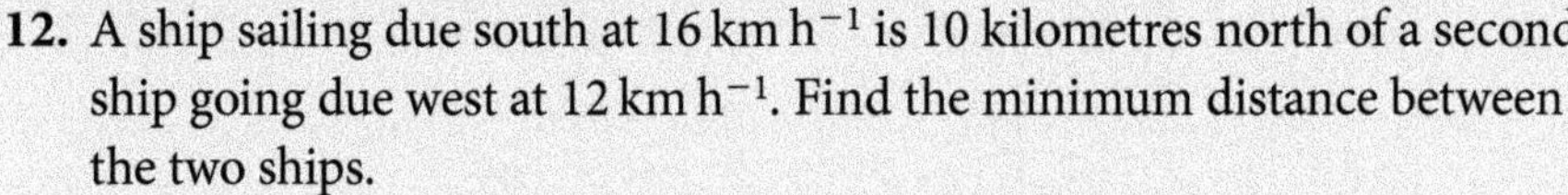

12. A ship sailing due south at 16 km h^{-1} is 10 kilometres north of a second ship going due west at 12 km h^{-1}. Find the minimum distance between the two ships.

13. Find the height, h, and the base radius, r, of the largest right circular cylinder that can be made by cutting it away from a sphere with a radius of R.

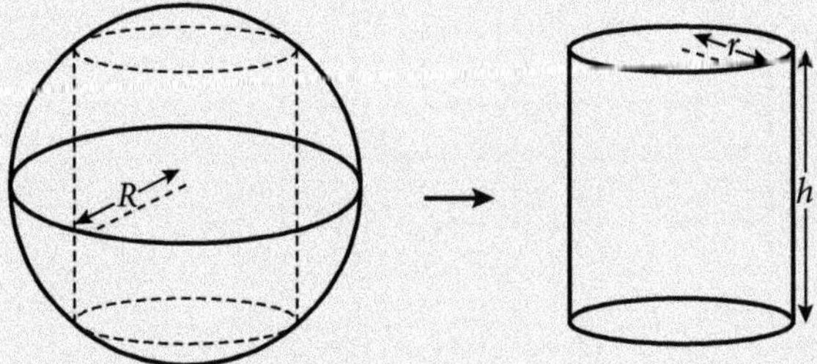

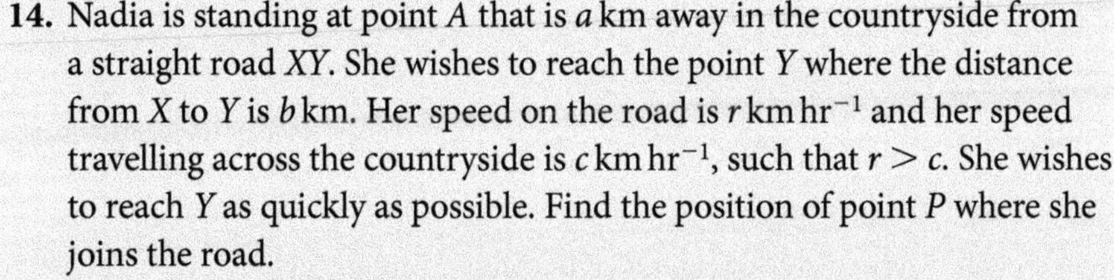

14. Nadia is standing at point A that is a km away in the countryside from a straight road XY. She wishes to reach the point Y where the distance from X to Y is b km. Her speed on the road is r km hr^{-1} and her speed travelling across the countryside is c km hr^{-1}, such that $r > c$. She wishes to reach Y as quickly as possible. Find the position of point P where she joins the road.

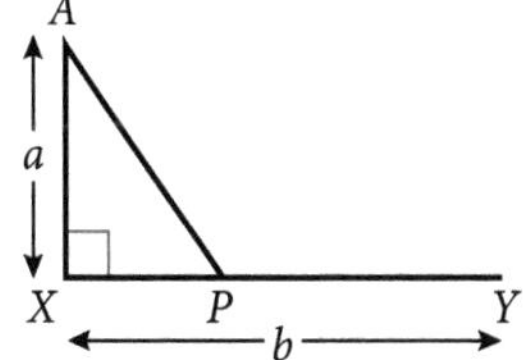

Figure 10.18 Diagram for question 14

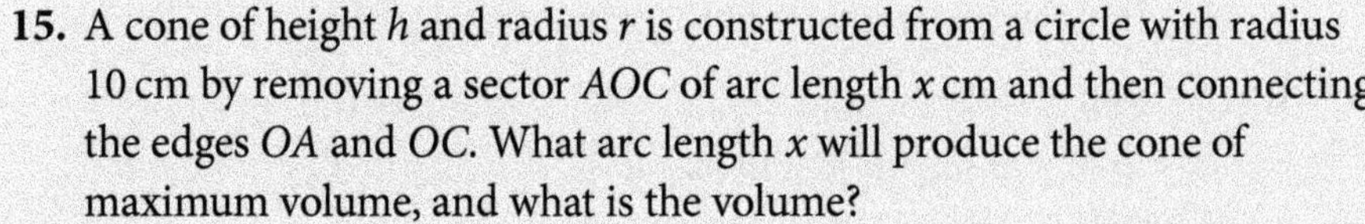

15. A cone of height h and radius r is constructed from a circle with radius 10 cm by removing a sector AOC of arc length x cm and then connecting the edges OA and OC. What arc length x will produce the cone of maximum volume, and what is the volume?

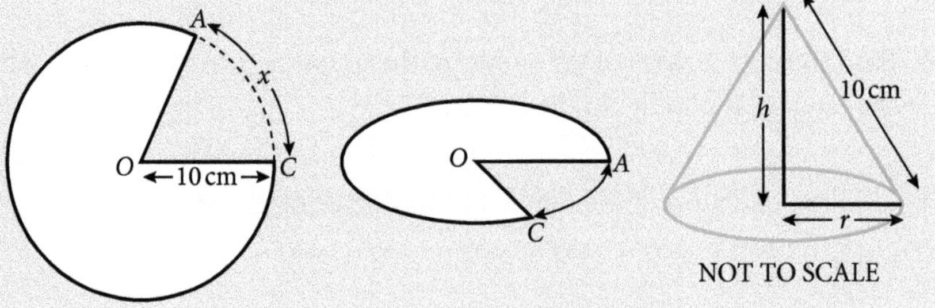

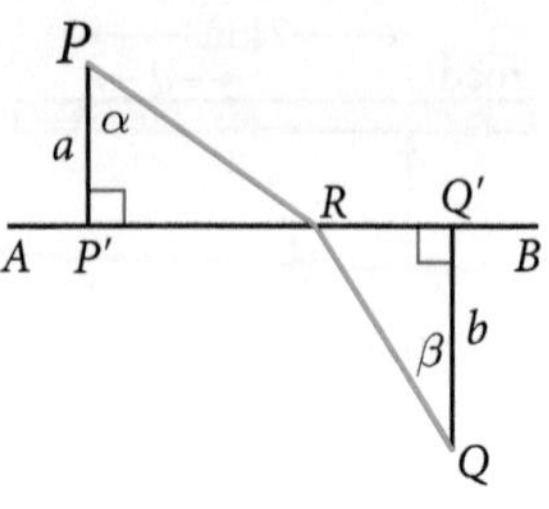

Figure 10.19 Diagram for question 16

16. Point P is a units above the line AB, and point Q is b units below line AB. The velocity of light is u units/second above AB and v units/second below AB, and $u > v$. The angles α and β are the angles that a ray of light makes with a perpendicular to line AB above and below AB, respectively. Show that the following relationship must hold true.

$$\frac{\sin\alpha}{\sin\beta} = \frac{u}{v}$$

Chapter 10 practice questions

1. The diagram shows the graph of $y = f(x)$.

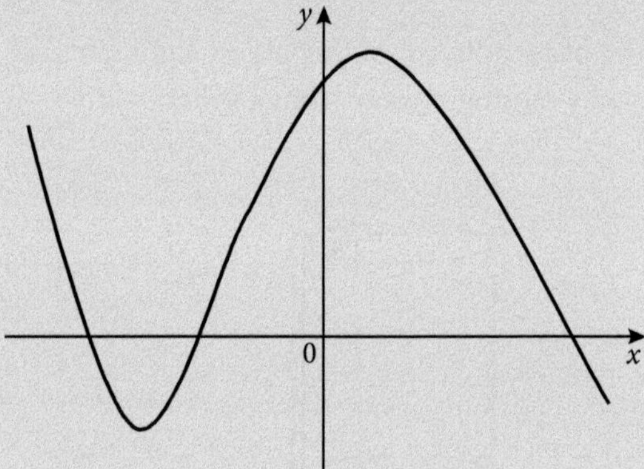

Sketch the graph of $y = f'(x)$.

2. A curve has equation $y = -x(x + 5)^2$.

(a) For this curve find:

(i) the x-intercepts

(ii) the exact coordinates of the maximum point

(iii) the exact coordinates of the point of inflection.

(b) Use your answers to part (a) to sketch a graph of the curve for $-5 \leqslant x \leqslant 0$, clearly indicating the features you have found in part (a).

3. Find the coordinates of the point on the graph of $y = 3x^2 + 2x$ at which the tangent is parallel to the line $y = 4x$.

4. Find the equation of the tangent to the curve of $y = \sin(3x + 1)$ at the point $\left(-\frac{1}{3}, 0\right)$.

5. The diagram shows part of the graph of the function $f{:}x \mapsto -x^3 - 2x^2 + 8x$.

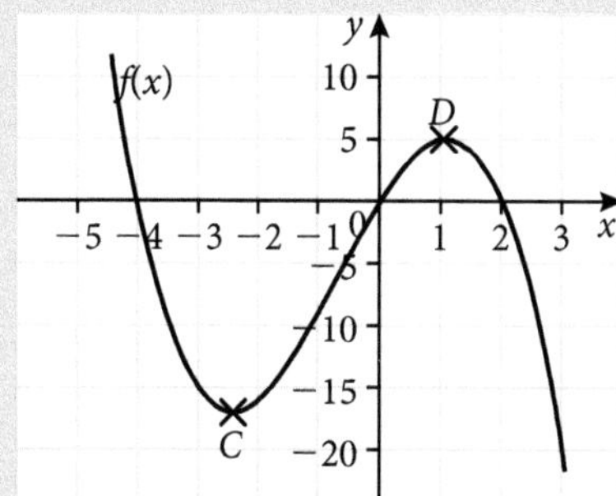

The graph intersects the x-axis at $(-4, 0)$, $(0, 0)$, and $(2, 0)$. There is a minimum point at C and a maximum point at D.

(a) The function may also be written in the form $f{:}x \mapsto -x(x - a)(x - b)$, where $a < b$. Write down the value of:

(i) a

(ii) b.

(b) Find:

(i) $f'(x)$

(ii) the exact values of x at which $f'(x) = 0$

(iii) the value of the function at D.

(c) **(i)** Find the equation of the tangent to the graph of $f(x)$ at $(0, 0)$.

(ii) This tangent cuts the graph of $f(x)$ at another point. Give the x-coordinate of this point.

6. In a controlled experiment, a tennis ball is dropped from the uppermost observation deck (447 metres high) of the CN Tower in Toronto. The ball's velocity is given by

$$v(t) = 66 - 66e^{-0.15t}$$

where v is in metres per second and t is in seconds.

(a) Find the value of v when:

(i) $t = 0$

(ii) $t = 10$.

(b) **(i)** Find an expression for the acceleration, a, as a function of t.

(ii) What is the value of a when $t = 0$?

(c) **(i)** As t becomes large, what value does v approach?

(ii) As t becomes large, what value does a approach?

(iii) Explain the relationship between the answers to parts (i) and (ii).

7. Given the function $f(x) = x^3 + 7x^2 + 8x - 3$:

(a) identify any points as a relative maximum or minimum, and find their exact coordinates

(b) find the exact coordinates of any inflection point(s).

8. Consider the function $g(x) = 2 + \frac{1}{e^{3x}}$.

(a) (i) Find $g'(x)$.

(ii) Explain briefly how this shows that $g(x)$ is a decreasing function for all values of x (i.e. that $g(x)$ always decreases in value as x increases).

Let P be the point on the graph of g where $x = -\frac{1}{3}$.

(b) Find an expression in terms of e for:

(i) the y-coordinate of P

(ii) the gradient of the tangent to the curve at P.

(c) Find the equation of the tangent to the curve at P, giving your answer in the form $y = mx + c$.

9. Consider the function f given by $f(x) = \frac{2x^2 - 13x + 20}{(x-1)^2}$, $x \neq 1$.

(a) Show that $f'(x) = \frac{9x - 27}{(x-1)^3}$, $x \neq 1$.

The second derivative is given by $f''(x) = \frac{72 - 18x}{(x-1)^4}$, $x \neq 1$

(b) Using values of $f'(x)$ and $f''(x)$, explain why a minimum must occur at $x = 3$.

(c) There is a point of inflection on the graph of $f(x)$. Write down the coordinates of this point.

10. Differentiate with respect to x:

(a) $\frac{1}{(2x+3)^2}$ (b) $e^{\sin 5x}$

11. The curve with equation $y = Ax + B + \frac{C}{x}$, $x \in \mathbb{R}$, $x \neq 0$, has a minimum at $P(1, 4)$ and a maximum at $Q(-1, 0)$. Find the value of each of the constants A, B, and C.

12. (a) Differentiate:

(i) $\ln x$ (ii) $\frac{1}{x}$

(b) The curve C has equation $y = \frac{\ln x}{x}$, $0 < x < \infty$.

(i) Show that $\frac{dy}{dx} = \frac{1}{x^2}(1 - \ln x)$.

(ii) Show that y has a maximum value of $\frac{1}{e}$ and justify that this is a maximum value.

(c) Assuming $y \to 0$ as $x \to \infty$, draw a sketch of the graph of the curve C.

(d) Find the two values of x for which $\frac{\ln x}{x} = \frac{1}{2}\ln 2$.

13. Differentiate with respect to x:

(a) $\dfrac{x^3}{x^2 + 1}$ **(b)** $e^x \sin 2x$

14. The curve $y = ax^3 - 2x^2 - x + 7$ has a gradient of 3 at the point where $x = 2$. Determine the value of a.

15. Let $y = h(x)$ be a function of x for $0 \leqslant x \leqslant 6$. The graph of h has an inflection point at P, and a maximum point at M.

Partial sketches of the curves of $h'(x)$ and $h''(x)$ are shown below.

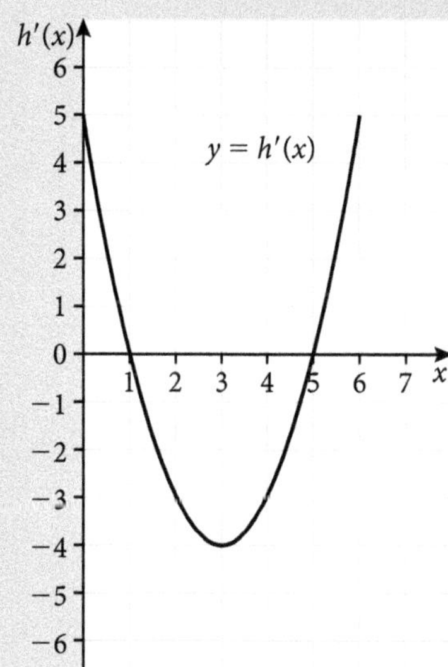

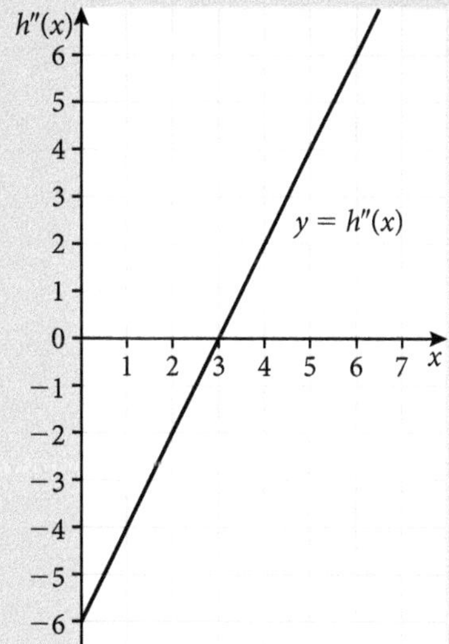

Use the above information to answer the following.

(a) Write down the x-coordinate of P, and justify your answer.

(b) Write down the x-coordinate of M, and justify your answer.

(c) Given that $h(3) = 0$, sketch the graph of h. On the sketch, mark the points P and M.

16. Find the equation of the tangent to the curve $y = xe^x$ at the point on the curve where $x = 1$.

17. A cylinder is to be made with an exact volume of $128\pi\,\text{cm}^3$. Find the height, h, of the cylinder and the radius, r, of the cylinder's base so that the cylinder's surface area is a minimum?

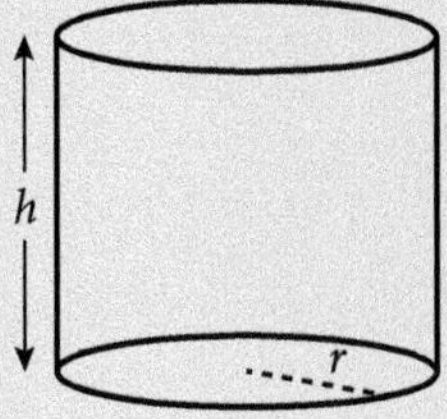

18. A rectangle has its base on the x-axis and its upper two vertices on the parabola $y = 12 - x^2$. Calculate the largest area that the rectangle can have, and the dimensions (length and width) that give this area.

19. The figure shows the graph of a function $y = f(x)$.

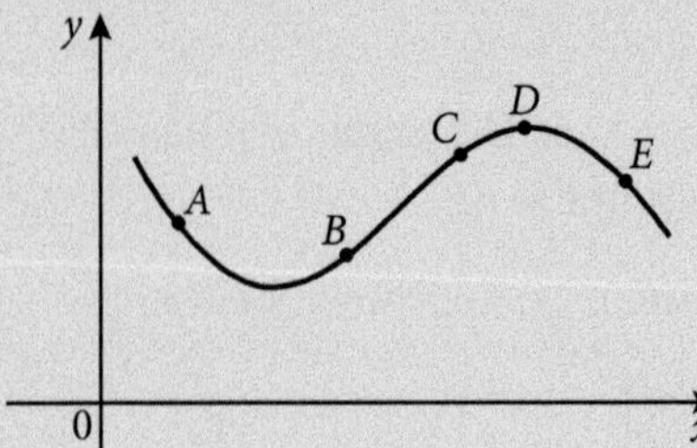

State at which one of the five points marked on the graph:

(a) $f'(x)$ and $f''(x)$ are both negative

(b) $f'(x)$ is negative and $f''(x)$ is positive

(c) $f'(x)$ is positive and $f''(x)$ is negative.

20. Find the equation of the normal to the curve with equation $y = \dfrac{2x - 1}{x + 2}$ at the point $(-3, 7)$.

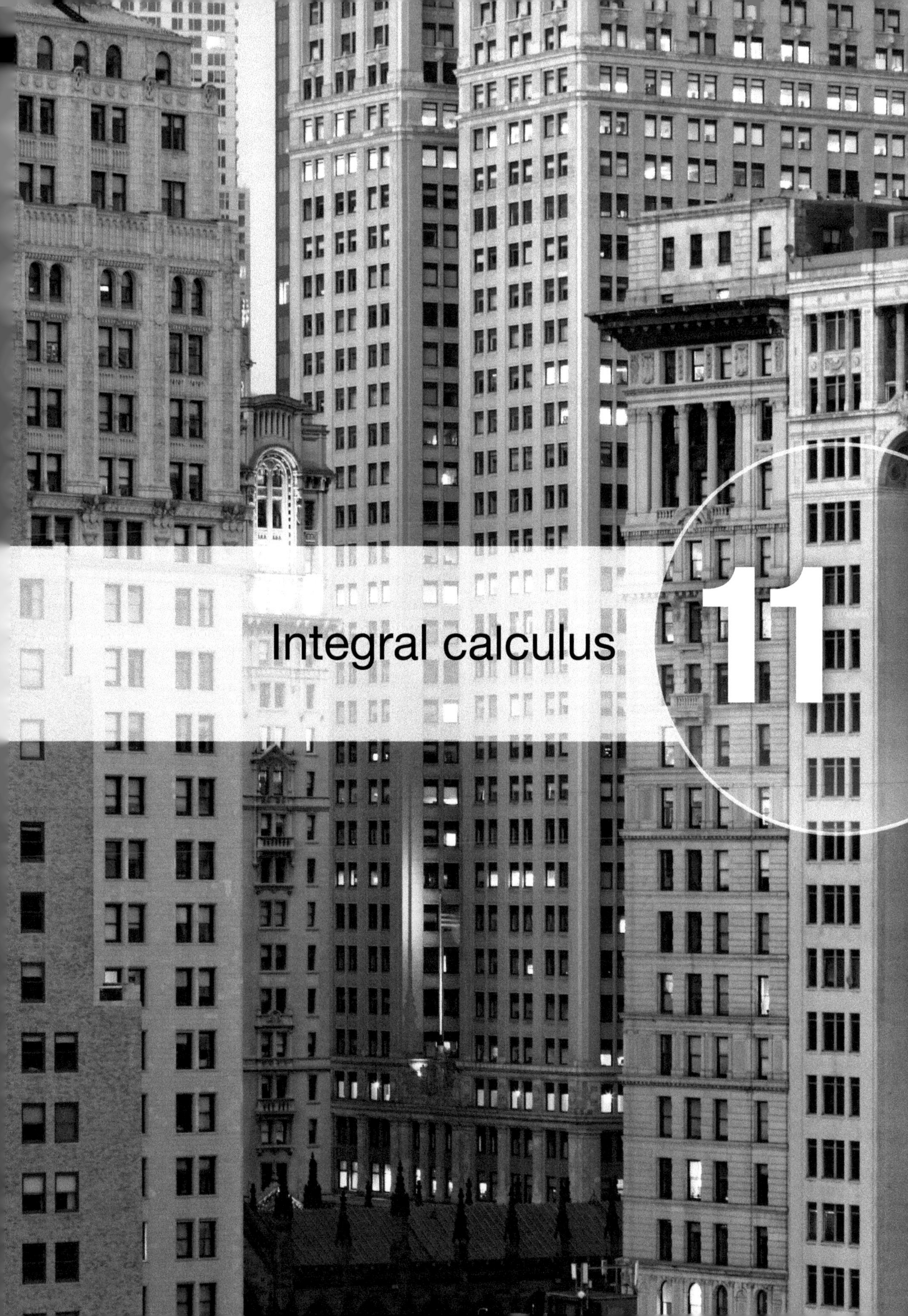

11 Integral calculus

11 Integral calculus

Learning objectives

By the end of this chapter, you should be familiar with...

- integration as antidifferentiation of functions
- calculating and applying definite integrals
- finding areas under curves (between the curve and the x-axis), and areas between curves
- antidifferentiation with a boundary condition
- solving kinematic problems involving displacement s, velocity v, acceleration a, and total distance travelled
- working with integration of polynomial functions, trigonometric functions and their inverses, and exponential functions
- integration by inspection (reverse chain rule) and integration by substitution.

In Chapters 9 and 10 we learned about the process of differentiation. That is, finding the derivative of a given function. In this chapter, we will reverse the process. That is, given a function $f(x)$ how can we find a function $F(x)$ whose derivative is $f(x)$? This process is the opposite of differentiation and is therefore called **antidifferentiation** or **integration**.

11.1 Antiderivative

An **antiderivative** of the function $f(x)$ is a function $F(x)$ such that $\frac{d}{dx}F(x) = F'(x) = f(x)$ wherever $f(x)$ is defined.

For instance, let $f(x) = x^2$. It is not difficult to discover an antiderivative of $f(x)$. Keep in mind that this is a power function. Since the power rule reduces the power of the function by 1, we examine the derivative of x^3: $\frac{d}{dx}(x^3) = 3x^2$

This derivative, however, is 3 times $f(x)$. To 'compensate' for the 'extra' 3 we have to multiply by $\frac{1}{3}$ so that the antiderivative is $\frac{1}{3}x^3$. Now $\frac{d}{dx}\left(\frac{1}{3}x^3\right) = x^2$, and therefore $\frac{1}{3}x^3$ is an antiderivative of x^2.

Table 11.1 shows some examples of functions, each paired with one of its antiderivatives. The diagrams show the relationship between the derivative and the integral as opposite operations.

Function $f(x)$	Antiderivative $F(x)$
1	x
x	$\frac{x^2}{2}$
$3x^2$	x^3
x^4	$\frac{x^5}{5}$
$\cos x$	$\sin x$
$\cos 2x$	$\frac{1}{2}\sin 2x$
e^x	e^x
$\sin x$	$-\cos x$

Table 11.1 Examples of functions paired to antiderivatives

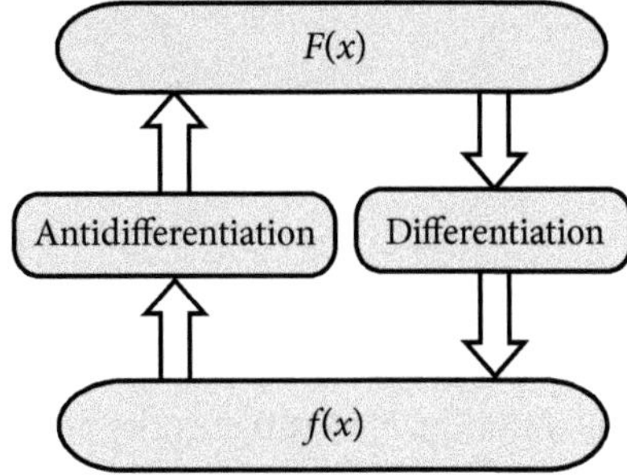

Figure 11.1 The relationship between a derivative and its integral

Example 11.1

Given the function $f(x) = 3x^2$. Find an antiderivative of $f(x)$.

Solution

$F_1(x) = x^3$ is one such antiderivative because $\frac{d}{dx}(F_1(x)) = 3x^2$

The following functions are also antiderivatives because the derivative of each one of them is also $3x^2$.

$$F_2(x) = x^3 + 27 \quad F_3(x) = x^3 - \pi \quad F_4(x) = x^3 + \sqrt{5}$$

Indeed, $F(x) = x^3 + c$ is an antiderivative of $f(x) = 3x^2$ for any constant c.

This is simply because

$$(F(x) + c)' = F'(x) + c' = F'(x) + 0 = f(x)$$

Thus we can say that any single function $f(x)$ has many antiderivatives, whereas a function has only one derivative.

If $F(x)$ is an antiderivative of $f(x)$, then so is $F(x) + c$ for any choice of the constant c. This statement is an indirect conclusion of one of the results of the mean value theorem. Two functions with the same derivative on an interval differ only by a constant on that interval.

The **mean value theorem** states that a function $H(x)$, continuous over an interval $[a, b]$ and differentiable over $]a, b[$ satisfies:

$H(b) - H(a) = (b - a)H'(c)$ for some $c \in]a, b[$

Let $F(x)$ and $G(x)$ be any antiderivatives of $f(x)$; that is, $F'(x) = G'(x)$.

Take $H(x) = F(x) - G(x)$ and any two numbers x_1 and x_2 in the interval $[a, b]$ such that $x_1 < x_2$, then

$$\begin{aligned} H(x_2) - H(x_1) &= (x_2 - x_1)H'(c) = (x_2 - x_1) \cdot (F'(c) - G'(c)) \\ &= (x_2 - x_1) \cdot 0 = 0 \Rightarrow H(x_1) = H(x_2) \end{aligned}$$

which means $H(x)$ is a constant function. Hence $H(x) = F(x) - G(x) = $ constant. That is, any two antiderivatives of a function differ by a constant.

11 Integral calculus

Note that if we differentiate an antiderivative of $f(x)$, we obtain $f(x)$. Thus

$$\frac{d}{dx}\left(\int f(x)\,dx\right) = f(x)$$

The expression $\int f(x)dx$ is called an **indefinite integral** of $f(x)$. The function $f(x)$ is called the **integrand**, and the constant c is called the **constant of integration**.

The integral symbol $\int$ is a medieval S, used by Leibniz as an abbreviation for the Latin word *summa* ('sum').

We think of the combination $\int [\ ]dx$ as a single symbol; we fill in the blank with the formula of the function whose antiderivative we seek. We may regard the differential dx as specifying the independent variable x both in the function $f(x)$ and in its antiderivatives. This is true for any independent variable, say t, with the notation adjusted appropriately. Thus

$$\frac{d}{dt}\left(\int f(t)\,dt\right) = f(t)$$

and

$$\int f(t)dt = F(t) + c$$

are equivalent statements.

Notation

The notation

$$\int f(x)dx = F(x) + c \qquad \textbf{(1)}$$

where c is an arbitrary constant, means that $F(x) + c$ is an antiderivative of $f(x)$.

Equivalently, $F(x)$ satisfies the condition that

$$\frac{d}{dx}F(x) = F'(x) = f(x) \qquad \textbf{(2)}$$

for all x in the domain of $f(x)$.

It is important to note that **(1)** and **(2)** are just different notations to express the same fact. For example

$$\int x^2\,dx = \frac{1}{3}x^3 + c \text{ is equivalent to } \frac{d}{dx}\left(\frac{1}{3}x^3\right) = x^2$$

Derivative formula	Equivalent integration formula
$\frac{d}{dx}(x^3) = 3x^2$	$\int 3x^2\,dx = x^3 + c$
$\frac{d}{dx}(\sqrt{x}) = \frac{1}{2\sqrt{x}}$	$\int \frac{1}{2\sqrt{x}}\,dx = \sqrt{x} + c$
$\frac{d}{dt}(\tan t) = \sec^2 t$	$\int \sec^2 t\,dt = \tan t + c$
$\frac{d}{dv}\left(v^{\frac{3}{2}}\right) = \frac{3}{2}v^{\frac{1}{2}}$	$\int \frac{3}{2}v^{\frac{1}{2}}\,dv = v^{\frac{3}{2}} + c$

Table 11.2 Derivative formulae and their equivalent integration formulae

Basic integration formulae

Many basic integration formulae can be obtained directly from their companion differentiation formulae. Some of the most important are given in Table 11.3.

	Derivative formula	Integration formula
1	$\frac{d}{dx}(x) = 1$	$\int dx = x + c$
2	$\frac{d}{dx}(x^{n+1}) = (n+1)x^n,\ n \neq -1$	$\int x^n\,dx = \frac{x^{n+1}}{n+1} + c,\ n \neq -1$
3	$\frac{d}{dx}(\sin x) = \cos x$	$\int \cos x\,dx = \sin x + c$
4	$\frac{d}{dv}(\cos v) = -\sin v$	$\int \sin v\,dv = -\cos v + c$
5	$\frac{d}{dt}(\tan t) = 1/\cos^2 t = \sec^2 t$	$\int \sec^2 t\,dt = \tan t + c$
6	$\frac{d}{dv}(e^v) = e^v$	$\int e^v\,dv = e^v + c$
7	$\frac{d}{dx}(\ln\lvert x\rvert) = \frac{1}{x}$	$\int \frac{1}{x}\,dx = \ln\lvert x\rvert + c$

Table 11.3 Many basic integration formulae can be obtained directly from their companion differentiation formulae

The integral sign and differential serve as delimiters, adjoining the integrand on the left and right, respectively. In particular we do not write $\int dx\, f(x)$ when we mean $\int f(x)dx$

Formula 7 is a special case of the 'power' rule shown in formula 2, but needs some modification.

If we are asked to integrate $\frac{1}{x}$, we may attempt to do it using the power rule:

$$\int \frac{1}{x}\,dx = \int x^{-1}\,dx = \frac{1}{(-1)+1}x^{(-1)+1} + c = \frac{1}{0}x^{0} + c, \text{ which is undefined.}$$

However, the solution is found by observing what we learned in Chapter 10:

$\frac{d}{dx}(\ln x) = \frac{1}{x}, x > 0$, implies $\int \frac{1}{x}\,dx = \ln x + c, x > 0$.

The function $\frac{1}{x}$ is differentiable for $x < 0$ too. So, we must be able to find its integral.

The solution lies in the chain rule.

If $x < 0$, then we can write $x = -u$ where $u > 0$. Then $dx = -du$, and

$$\int \frac{1}{x}\,dx = \int \frac{1}{-u}(-du) = \int \frac{1}{u}\,du = \ln u + c, u > 0$$

But $u = -x$, therefore when $x < 0$:

$\int \frac{1}{x}\,dx = \ln u + c = \ln(-x) + c$, and combining the two results, we have

$$\int \frac{1}{x}\,dx = \ln|x| + c, x \neq 0$$

Suppose that $f(x)$ and $g(x)$ are differentiable functions and k is a constant. Then:

A constant factor can be moved through an integral sign; that is,

$$\int kf(x)\,dx = k\int f(x)\,dx$$

An antiderivative of a sum (difference) is the sum (difference) of the antiderivatives; i.e.,

$$\int (f(x) + g(x))dx = \int f(x)\,dx \pm \int g(x)\,dx$$

Example 11.2

Evaluate

(a) $\int 3\cos x\,dx$ (b) $\int (x^3 + x^2)dx$

Solution

(a) $\int 3\cos x\,dx = 3\int \cos x\,dx = 3\sin x + c$

(b) $\int (x^3 + x^2)dx = \int x^3\,dx + \int x^2\,dx = \frac{x^4}{4} + \frac{x^3}{3} + c$

Sometimes it is useful to rewrite the integrand in a different form before performing the integration.

Example 11.3

Evaluate

(a) $\int \frac{t^3 - 3t^5}{t^5}\,dt$ (b) $\int \frac{x + 5x^4}{x^2}\,dx$

Solution

(a) $\int \frac{t^3 - 3t^5}{t^5}\,dt = \int \frac{t^3}{t^5}\,dt - \int \frac{3t^5}{t^5}\,dt = \int t^{-2}\,dt - \int 3\,dt = \frac{t^{-1}}{-1} - 3t + c$

$= -\frac{1}{t} - 3t + c$

(b) $\int \frac{x + 5x^4}{x^2}\,dx = \int \frac{x}{x^2}\,dx + \int \frac{5x^4}{x^2}\,dx = \int \frac{1}{x}\,dx + \int 5x^2\,dx$

$= \ln|x| + \frac{5x^3}{3} + c$

Integration by simple substitution – change of variables

In this section, we will study substitution, a technique that can often be used to transform complex integration problems into simpler ones.

The method of substitution depends on our understanding of the chain rule as well as the use of variables in integration. Two facts to recall:

When we find an antiderivative, we can use any other variable.

That is, $\int f(u)\,du = F(u) + c$, where u is a dummy variable in the sense that it can be replaced by any other variable.

Using the chain rule $\frac{d}{dx}(F(u(x))) = F'(u(x)) \cdot u'(x)$

Which can be written in integral form as $\int F'(u(x)) \cdot u'(x)\,dx = F(u(x)) + c$

Or equivalently, since $F(x)$ is an antiderivative of $f(x)$,

$\int f(u(x)) \cdot u'(x)\,dx = F(u(x)) + c$

For our purposes, it will be useful and simpler to let $u(x) = u$ and to write $\frac{du}{dx} = u'(x)$ in its differential form as $du = u'(x)\,dx$ or simply $du = u'\,dx$.

We can now write the integral as

$$\int f(u(x)) \cdot u'(x)\,dx = \int f(u)du = F(u(x)) + c$$

Example 11.4 demonstrates how the method works.

Example 11.4

Evaluate

(a) $\int (x^3 + 2)^{10} \cdot 3x^2\,dx$ (b) $\int \tan x\,dx$ (c) $\int \cos 5x\,dx$

(d) $\int \cos x^2 \cdot x\,dx$ (e) $\int e^{3x+1}\,dx$

Solution

(a) To integrate this function, it is simplest to make the substitution $u = x^3 + 2$, and so $du = 3x^2 dx$. Now we can write the integral as

$$\int (x^3+2)^{10} \cdot 3x^2\,dx = \int u^{10}\,du = \frac{u^{11}}{11} + c = \frac{(x^3+2)^{11}}{11} + c$$

(b) This integrand has to be rewritten first and then we make the substitution:

$$\int \tan x\,dx = \int \frac{\sin x}{\cos x}\,dx = \int \frac{1}{\cos x} \cdot \sin x\,dx$$

We now let $u = \cos x \Rightarrow du = -\sin x\,dx$, and

$$\int \tan x\,dx = \int \frac{1}{\cos x} \cdot \sin x\,dx = \int \frac{1}{u} \cdot (-du) = -\int \frac{1}{u}\,du = -\ln|u| + c$$

This last result can be then expressed in two ways:

$$\int \tan x\,dx = -\ln|\cos x| + c, \text{ or}$$

$$\int \tan x\,dx = -\ln|\cos x| + c = \ln|(\cos x)^{-1}| + c = \ln\left|\frac{1}{(\cos x)}\right| + c$$
$$= \ln|\sec x| + c$$

(c) We let $u = 5x$, then $du = 5dx \Rightarrow dx = \frac{1}{5}du$, and so

$$\int \cos 5x\,dx = \int \cos u \cdot \frac{1}{5}du = \frac{1}{5}\int \cos u\ du = \frac{1}{5}\sin u + c$$
$$= \frac{1}{5}\sin 5x + c$$

Another method can be applied here:

The substitution $u = 5x$ requires $du = 5dx$. As there is no factor of 5 in the integrand, and since 5 is a constant, we can multiply and divide by 5 so that we group the 5 and dx to form the du required by the substitution:

$$\int \cos 5x\,dx = \frac{1}{5}\int \cos x \cdot 5dx = \frac{1}{5}\int \cos u\ du = \frac{1}{5}\sin u + c$$
$$= \frac{1}{5}\sin 5x + c$$

(d) By letting $u = x^2$, $du = 2x\,dx$ and so

$$\int \cos x^2 \cdot x\,dx = \frac{1}{2}\int \cos x^2 \cdot 2x\,dx = \frac{1}{2}\int \cos u\,du$$
$$= \frac{1}{2}\sin u + c = \frac{1}{2}\sin x^2 + c$$

(e) $\int e^{3x+1}dx = \frac{1}{3}\int e^{3x+1}3dx = \frac{1}{3}\int e^u\,du = \frac{1}{3}e^u + c = \frac{1}{3}e^{3x+1} + c$

The main challenge in using the substitution rule is to think of an appropriate substitution. You should try to select u to be a part of the integrand whose differential is also included (except for the constant). In Example 11.4 (a), we selected u to be $(x^3 + 2)$ knowing that $du = 3x^2 dx$. Then we compensated for the absence of 3. Finding the right substitution is a subtle art, which you will acquire with practice. It is often the case that your first guess may not work.

In integration, multiplying by a constant inside the integral and compensating for that with the reciprocal outside the integral depends on formula 2 from Table 11.3.

However, we cannot do this with a variable.

For example,

$\int e^x dx = \frac{1}{2x}\int e^x 2x\,dx$

is **not** valid because $2x$ is not a constant.

Example 11.5

Evaluate each integral.

(a) $\int e^{-3x}\,dx$ (b) $\int \sin^2 x \cos x\,dx$ (c) $\int 2\sin(3x-5)\,dx$

(d) $\int e^{mx+n}\,dx$ (e) $\int x\sqrt{x}\,dx$ and $F(1) = 2$

Solution

The main challenge in using the substitution rule is to think of an appropriate substitution. You should try to select u to be a part of the integrand whose differential is also included (except for the constant). In Example 11.4 (a), we selected u to be (x^3+2) knowing that $du = 3x^2\,dx$. Then we compensated for the absence of 3. Finding the right substitution is a subtle art, which you will acquire with practice. It is often the case that your first guess may not work.

(a) Let $u = -3x$, then $du = -3dx$

$$\int e^{-3x}\,dx = -\frac{1}{3}\int e^{-3x}(-3\,dx) = -\frac{1}{3}\int e^{u}\,du = -\frac{1}{3}e^{u} + c$$

$$= -\frac{1}{3}e^{-3x} + c$$

(b) Let $u = \sin x \Rightarrow du = \cos x\,dx$, and hence

$$\int \sin^2 x \cos x\,dx = \int u^2\,du = \frac{1}{3}u^3 + c = \frac{1}{3}\sin^3 x + c$$

(c) Let $u = 3x - 5$, then $du = 3dx$

$$\int 2\sin(3x-5)\,dx = 2\cdot\frac{1}{3}\int \sin(3x-5)\,3\,dx = \frac{2}{3}\int \sin u\,du$$

$$= -\frac{2}{3}\cos u + c = -\frac{2}{3}\cos(3x-5) + c$$

(d) Let $u = mx + n$, then $du = m\,dx$

$$\int e^{mx+n}\,dx = \frac{1}{m}\int e^{mx+n}\,m\,dx = \frac{1}{m}\int e^{u}\,du = \frac{1}{m}e^{u} + c = \frac{1}{m}e^{mx+n} + c$$

(e) $F(x) = \int x\sqrt{x}\,dx = \int x^{\frac{3}{2}}\,dx = \dfrac{x^{\frac{5}{2}}}{\left(\frac{5}{2}\right)} + c = \dfrac{2}{5}x^{\frac{5}{2}} + c$, but $F(1) = 2$

$$F(1) = \frac{2}{5}1^{\frac{5}{2}} + c = \frac{2}{5} + c = 2 \Rightarrow c = \frac{8}{5}$$

Therefore $F(x) = \frac{2}{5}x^{\frac{5}{2}} + \frac{8}{5}$

Examples 11.4 and 11.5 make it clear that Table 11.3 is limited in scope because we cannot use the integrals directly to evaluate composite functions. We therefore need to revise some of the derivative formulae.

	Derivative formula	Integration formula
1	$\frac{d}{dx}(u(x)) = u'(x) \Rightarrow du = u'(x)dx$	$\int du = u + c$
2	$\frac{d}{dx}\left(\frac{u^{n+1}}{n+1}\right) = u^n u'(x), n \neq -1 \Rightarrow d\left(\frac{u^{n+1}}{n+1}\right) = u^n u'(x)\,dx$	$\int u^n\,du = \frac{u^{n+1}}{n+1} + c, n \neq -1$
3	$\frac{d}{dx}(\sin(u)) = \cos(u)u'(x) \Rightarrow d(\sin(u)) = \cos(u)u'(x)\,dx$	$\int \cos u\,du = \sin u + c$
4	$\frac{d}{dx}(\cos(u)) = -\sin(u)u'(x) \Rightarrow d(\cos(u)) = -\sin(u)u'(x)dx$	$\int \sin u\,du = -\cos u + c$
5	$\frac{d}{dt}(\tan u) = \sec^2 u\,u'(t) \Rightarrow d(\tan u) = \sec^2 u\,u'(t)\,dt$	$\int \sec^2 u\,du = \tan u + c$
6	$\frac{d}{dx}(e^u) = e^u u'(x)\,dx \Rightarrow d(e^u) = e^u u'(x)\,dx$	$\int e^u\,du = e^u + c$
7	$\frac{d}{dx}(\ln\lvert u\rvert) = \frac{1}{u}u'(x) \Rightarrow d(\ln\lvert u\rvert) = \frac{1}{u}u'(x)\,dx$	$\int \frac{1}{u}\,du = \ln\lvert u\rvert + c$

Table 11.4 More advanced derivative and integration formulae.

Example 11.6

Evaluate each integral.

(a) $\int \sqrt{6x + 11}\,dx$

(b) $\int (5x^3 + 2)^8 x^2\,dx$

(c) $\int \frac{x^3 - 2}{\sqrt[5]{x^4 - 8x + 13}}\,dx$

(d) $\int \sin^4(3x^2)\cos(3x^2)\,x\,dx$

Solution

(a) We let $u = 6x + 11$ and calculate du:

$$u = 6x + 11 \Rightarrow du = 6\,dx$$

Since du contains the factor 6, the integral is not in the form $\int f(u)\,du$. However, here we can use one of two approaches.

Introduce the factor 6, as we have done before; that is,

$$\int \sqrt{6x + 11}\,dx = \frac{1}{6}\int \sqrt{6x + 11}\,6\,dx$$

$$= \frac{1}{6}\int \sqrt{u}\,du = \frac{1}{6}\int u^{\frac{1}{2}}\,du$$

$$= \frac{1}{6}\frac{u^{\frac{3}{2}}}{\frac{3}{2}} + c = \frac{2}{18}u^{\frac{3}{2}} + c$$

$$= \frac{1}{9}(6x + 11)^{\frac{3}{2}} + c$$

Or since $u = 6x + 11 \Rightarrow du = 6\,dx \Rightarrow dx = \frac{du}{6}$, then

$\int\sqrt{6x + 11}\,dx = \int\sqrt{u}\frac{du}{6} = \frac{1}{6}\int u^{\frac{1}{2}}du$, then we follow the same steps as before.

(b) We let $u = 5x^3 + 2$, so $du = 15x^2\,dx$. This means that we need to introduce the factor 15 into the integrand

$$\int(5x^3 + 2)^8 x^2\,dx = \frac{1}{15}\int(5x^3 + 2)^8\,15x^2\,dx$$

$$= \frac{1}{15}\int u^8\,du = \frac{1}{15}\frac{u^9}{9} + c$$

$$= \frac{1}{135}(5x^3 + 2)^9 + c$$

(c) We let $u = x^4 - 8x + 13 \Rightarrow du = (4x^3 - 8)dx = 4(x^3 - 2)dx$

$$\int\frac{x^3 - 2}{\sqrt[5]{x^4 - 8x + 13}}dx = \frac{1}{4}\int\frac{4(x^3 - 2)dx}{\sqrt[5]{x^4 - 8x + 13}} = \frac{1}{4}\int\frac{du}{u^{\frac{1}{5}}}$$

$$= \frac{1}{4}u^{\frac{1}{5}}du = \frac{1}{4}\frac{u^{\frac{4}{5}}}{\frac{4}{5}} + c$$

$$= \frac{5}{16}(x^4 - 8x + 13)^{\frac{4}{5}} + c$$

(d) We let $u = \sin(3x^2) \Rightarrow du = \cos(3x^2)6x\,dx$ using the chain rule.

$$\int\sin^4(3x^2)\cos(3x^2)\,x\,dx = \frac{1}{6}\int\sin^4(3x^2)\cos(3x^2)\,6x\,dx$$

$$= \frac{1}{6}\int u^4\,du = \frac{1}{6}\frac{u^5}{5} + c$$

$$= \frac{1}{30}\sin^5(3x^2) + c$$

Exercise 11.1

1. Find the most general antiderivative of each function.

(a) $f(x) = x + 2$

(b) $f(t) = 3t^2 - 2t + 1$

(c) $g(x) = \frac{1}{3} - \frac{2}{7}x^3$

(d) $f(t) = (t - 1)(2t + 3)$

(e) $g(u) = u^{\frac{2}{5}} - 4u^3$

(f) $f(x) = 2\sqrt{x} - \frac{3}{2\sqrt{x}}$

(g) $h(\theta) = 3\sin\theta + 4\cos\theta$

(h) $f(t) = 3t^2 - 2\sin t$

(i) $f(x) = \sqrt{x}(2x - 5)$

(j) $g(\theta) = 3\cos\theta - 2\sec^2\theta$

(k) $h(t) = e^{3t-1}$

(l) $f(t) = \frac{2}{t}$

(m) $h(u) = \frac{t}{3t^2 + 5}$

(n) $h(\theta) = e^{\sin\theta} \cos\theta$

(o) $f(x) = (3 + 2x)^2$

2. Find f.

(a) $f''(x) = 4x - 15x^2$

(b) $f''(x) = 1 + 3x^2 - 4x^3, f'(0) = 2, f(1) = 2$

(c) $f''(t) = 8t - \sin t$

(d) $f'(x) = 12x^3 - 8x + 7, f(0) = 3$

(e) $f'(\theta) = 2\cos\theta - \sin(2\theta)$

3. Evaluate each integral.

(a) $\int x(3x^2 + 7)^5 \, dx$

(b) $\int \frac{x}{(3x^2 + 5)^4} dx$

(c) $\int 2x^2 \sqrt[4]{5x^3 + 2} \, dx$

(d) $\int \frac{(3 + 2\sqrt{x})^5}{\sqrt{x}} dx$

(e) $\int t^2 \sqrt{2t^3 - 7} \, dt$

(f) $\int \left(2 + \frac{3}{x}\right)^5 \left(\frac{1}{x^2}\right) dx$

(g) $\int \sin(7x - 3) dx$

(h) $\int \frac{\sin(2\theta - 1)}{\cos(2\theta - 1) + 3} d\theta$

(i) $\int \sec^2(5\theta - 2) d\theta$

(j) $\int \cos(\pi x + 3) dx$

(k) $\int \sec 2t \tan 2t \, dt$

(l) $\int x e^{x^2+1} dx$

(m) $\int \sqrt{t} \, e^{2t\sqrt{t}} dt$

(n) $\int \frac{2}{\theta} (\ln \theta)^2 d\theta$

4. Evaluate each integral.

(a) $\int t\sqrt{3 - 5t^2} \, dt$

(b) $\int \theta^2 \sec^2 \theta^3 \, d\theta$

(c) $\int \frac{\sin\sqrt{t}}{2\sqrt{t}} dt$

(d) $\int \tan^5 2t \sec^2 2t \, dt$

(e) $\int \frac{dx}{\sqrt{x}(\sqrt{x} + 2)}$

(f) $\int (x^2 + 1) e^{x^3+3x+1} dt$

(g) $\int \frac{x + 3}{x^2 + 6x + 7} dx$

(h) $\int \frac{k^3 x^3}{\sqrt{a^2 - a^4 x^4}} dx$

(i) $\int 3x\sqrt{x - 1} \, dx$

(j) $\int \sqrt{1 + \cos\theta} \sin\theta \, d\theta$

(k) $\int t^2 \sqrt{1 - t} \, dt$

(l) $\int \frac{r^2 - 1}{\sqrt{2r - 1}} dr$

(m) $\int \frac{e^{x^2} - e^{-x^2}}{e^{x^2} + e^{-x^2}} x dx$

11.2 Area and the definite integral

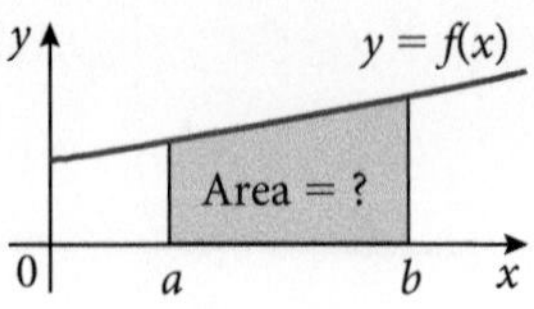

Figure 11.2 How do we find the area?

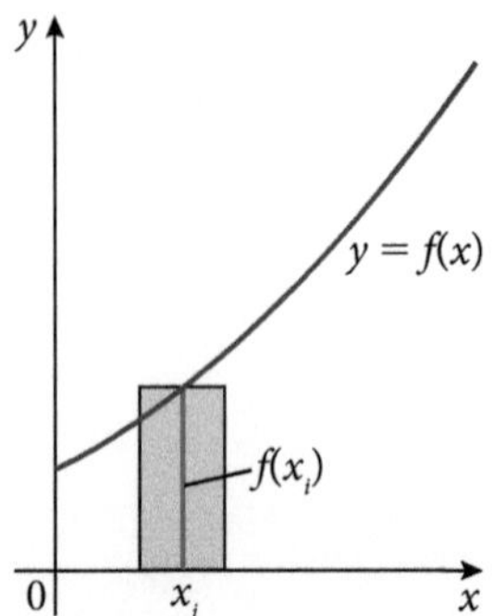

Figure 11.3 Dividing the base interval into subintervals

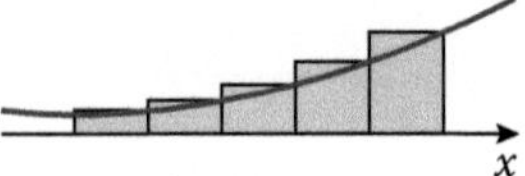

Figure 11.4 The total area of the rectangles can be viewed as an approximation

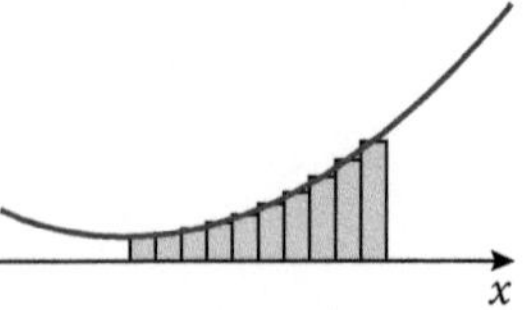

Figure 11.5 As n increases, the approximations get better

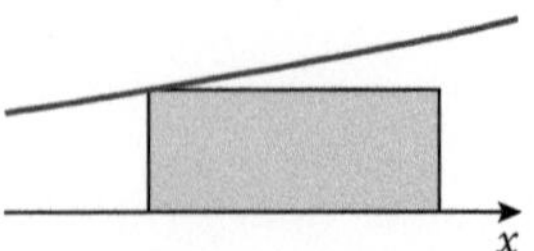

Figure 11.6 Underestimation of area

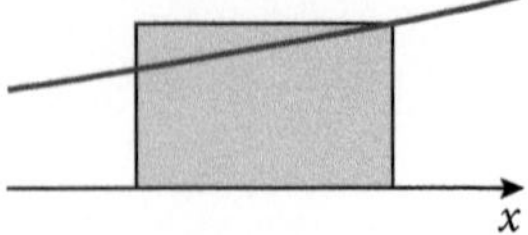

Figure 11.7 Overestimation of area

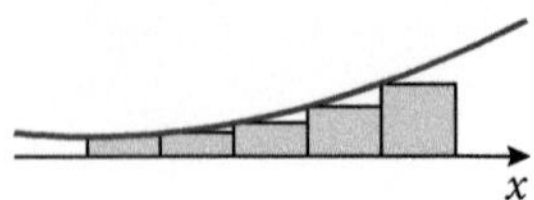

Figure 11.8 n inscribed rectangles

The function $f(x)$ is continuous and non-negative on an interval $[a, b]$. How do we find the area between the graph of $f(x)$ and the interval $[a, b]$ on the x-axis? (Figure 11.2)

We divide the base interval $[a, b]$ into n equal subintervals, and over each subinterval, we construct a rectangle that extends from the x-axis to any point on the curve $y = f(x)$ that is above the subinterval; the particular point does not matter – it can be above the centre, above one endpoint, or above any other point in the subinterval. In Figure 11.3 it is at the centre.

For each n, the total area of the rectangles can be viewed as an approximation to the exact area in question. As n increases, these approximations will get better and better and will eventually approach the exact area as a limit. See Figures 11.3–11.5

A traditional approach would be to study how the choice of where to put the rectangular strip does not affect the approximation as the number of intervals increases. We can construct inscribed rectangles that, at the start, give us an underestimate of the area (Figure 11.6). On the other hand we can construct circumscribed rectangles that, at the start, overestimate the area (Figure 11.7).

As the number of intervals increases, the difference between the overestimates and the underestimates will approach 0.

Figures 11.8 and 11.9 show n inscribed and circumscribed rectangles and Figure 11.10 shows the difference between the overestimates and underestimates.

Figure 11.10 shows that as the number n increases, the difference between the estimates will approach 0. Because we set up our rectangles by choosing a point inside the interval, the areas of the rectangles will lie between the overestimates and underestimates, and hence, as the difference between the extremes approaches zero, the rectangles we construct will give the area of the region required.

If we consider the width of each interval to be $\triangle x$, then the area of any rectangle is given as

$$A_i = f(x_i^*)\triangle x$$

The total area of the rectangles so constructed is

$$A_n = \sum_{i=0}^{n} f(x_i^*)\triangle x$$

where x_i^* is an arbitrary point within any subinterval $[x_{i-1}, x_i]$, $x_0 = a$, and $x_n = b$.

In the case of a function $f(x)$ that has both positive and negative values on $[a, b]$, it is necessary to consider the signs of the areas in the following sense.

On each subinterval, we have a rectangle with width Δx and height $f(x^*)$. If $f(x^*) > 0$, then this rectangle is above the x-axis; if $f(x^*) < 0$, then this rectangle is below the x-axis. We will consider the sum defined above as the sum of the signed areas of these rectangles. That means the total area on the interval is the sum of the areas above the x-axis minus the sum of the areas of the rectangles below the x-axis.

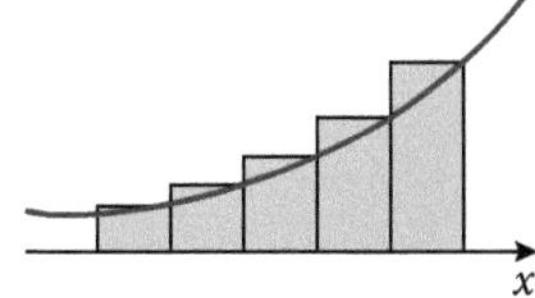

Figure 11.9 n circumscribed rectangles

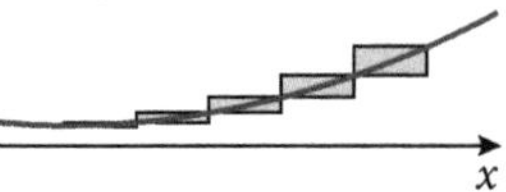

Figure 11.10 difference between over- and under-estimates

We are now ready to look at a loose definition of the definite integral:

If $f(x)$ is a continuous function defined for $a \leqslant x \leqslant b$, we divide the interval $[a, b]$ into n subintervals of equal width $\Delta x = \dfrac{(b-a)}{n}$. We let $x_0 = a$, and $x_n = b$ and we choose $x_1^*, x_2^*, \dots, x_n^*$ in these subintervals, so that x_i^* lies in the ith subinterval $[x_{i-1}, x_i]$. Then the definite integral of $f(x)$ from a to b is

$$\int_a^b f(x)\,dx = \lim_{n\to\infty} \sum_{i=1}^{n} f(x_i^*)\,\Delta x$$

In the notation $\int_a^b f(x)\,dx$, a and b are called the limits of integration: a is the lower limit and b is the upper limit.

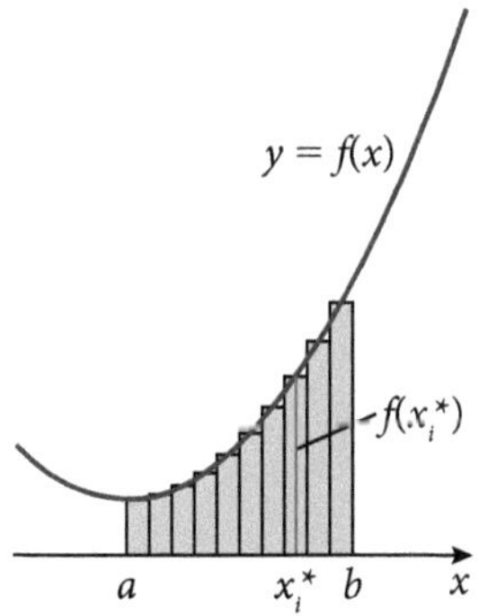

Figure 11.11 Area of each circumscribed rectangle

Because we have assumed that $f(x)$ is continuous, it can be proved that the limit definition above always exists and gives the same value no matter how we choose the points x_i^*. If we take these points at the centre, at two-thirds the distance from the lower endpoint, or at the upper endpoint, the value is the same. This why we will state the definition of the integral from now on as

$$\int_a^b f(x)\,dx = \lim_{n\to\infty} \sum_{i=1}^{n} f(x_i)\,\Delta x$$

Figure 11.12 Areas above and below the x-axis

For a more rigorous treatment of the definition of definite integrals using Riemann sums, refer to university calculus books. Such a treatment is beyond the scope of the IB syllabus and this book.

Online 

Using an interactive graph, guess the rule for the definite integral of an inputted function.

Calling the area under the function an integral is no coincidence. To make the point, let us take the following example:

Example 11.7

Find the area $A(x)$ between the graph of the function $f(x) = 3$ and the interval $[-1, x]$, and find the derivative $A'(x)$ of this area function.

Solution

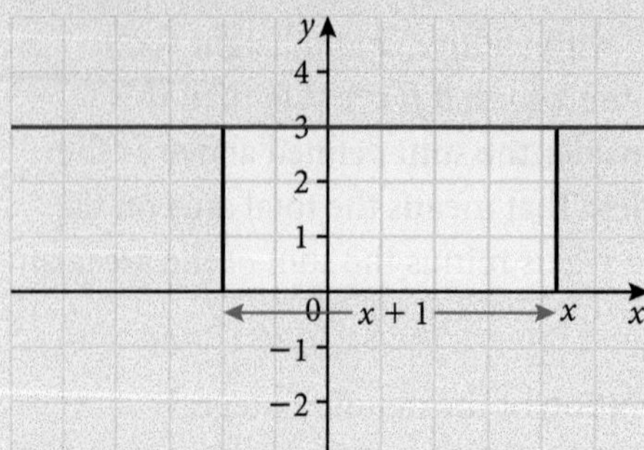

The area in question is

$$A(x) = 3(x - (-1)) = 3x + 3$$

$$A'(x) = 3 = f(x)$$

Example 11.8

Find the area $A(x)$ between the graph of the function $f(x) = 3x + 2$ and the interval $\left[-\frac{2}{3}, x\right]$, and find the derivative $A'(x)$ of this area function.

Solution

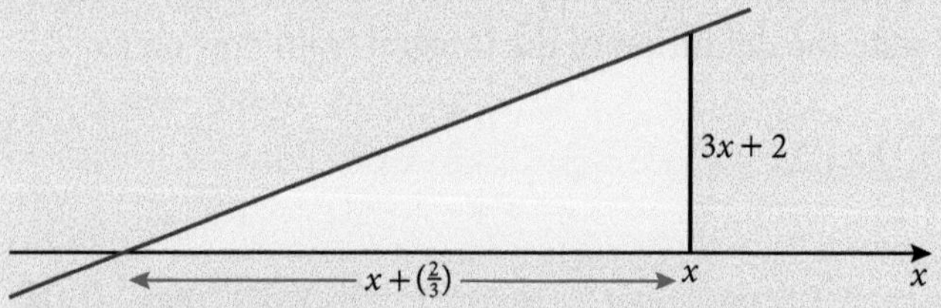

The area in question is

$$A(x) = \frac{1}{2}\left(x + \frac{2}{3}\right)(3x + 2) = \frac{1}{6}(3x + 2)^2$$

since this is the area of a triangle. Hence

$$A'(x) = \frac{1}{6} \times 2(3x + 2) \times 3 = 3x + 2 = f(x)$$

Example 11.9

Find the area $A(x)$ between the graph of the function $f(x) = x + 2$ and the interval $[-1, x]$, and find the derivative $A'(x)$ of this area function.

Solution

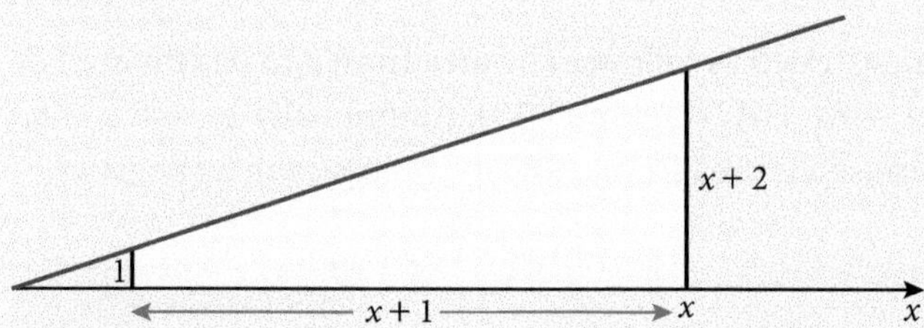

This is a trapezium, so the area is

$$A(x) = \frac{1}{2}(1 + (x + 2))(x + 1) = \frac{1}{2}(x^2 + 4x + 3), \text{ and}$$

$$A'(x) = \frac{1}{2} \times (2x + 4) = x + 2 = f(x)$$

Note that in every case, $A'(x) = f(x)$

That is, the derivative of the area function $A(x)$ is the function whose graph forms the upper boundary of the region. It can be shown that this relation is true not only for linear functions but for all continuous functions. Thus, to find the area function $A(x)$, we can look instead for a particular function whose derivative is $f(x)$. This is, of course, nothing but the antiderivative of $f(x)$.

So, intuitively, as we have seen above, we define the area function as

$$A(x) = \int_a^x f(t)\,dt, \text{ that is } A'(x) = f(x)$$

This is the trigger to the **fundamental theorem of calculus.**

We will now look at some of the properties of the definite integral.

Basic properties of the definite integral

$$\int_a^b f(x)\,dx = -\int_b^a f(x)\,dx$$

When we defined the definite integral $\int_a^b f(x)\,dx$, we implicitly assumed that $a < b$. When we reverse a and b, then Δx changes from $\frac{(b-a)}{n}$ to $\frac{(a-b)}{n}$. Therefore the result above follows.

$$\int_a^b f(x)\,dx = 0$$

When $a = b$, then $\Delta x = 0$, and so, the result above follows.

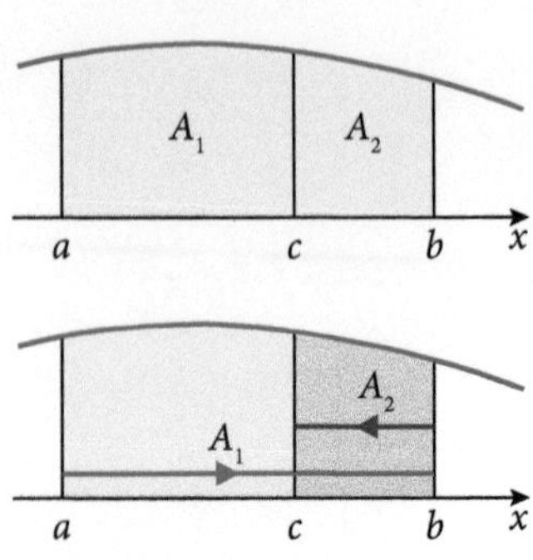

Figure 11.13 $A(x) = A_1 + A_2$

$$\int_a^b c\,dx = c(b - a)$$

$$\int_a^b [f(x) \pm g(x)]\,dx = \int_a^b f(x)\,dx \pm \int_a^b g(x)\,dx$$

$$\int_a^b cf(x)\,dx = c\int_a^b f(x)\,dx, \text{ where } c \text{ is any constant.}$$

$$\int_a^b f(x)\,dx = \int_a^c f(x)\,dx + \int_c^b f(x)\,dx$$

This property can be demonstrated as follows. The area from a to b is the sum of the two areas, that is $A(x) = A_1 + A_2$ (Figure 11.13). Additionally, even if $c > b$, the relationship holds because the area from c to b in this case will be negative.

The first fundamental theorem of integral calculus

Our understanding of the definite integral as the area under the curve for $f(x)$ helps us establish the basis for the fundamental theorem of integral calculus.

In the definition of definite integral, we'll make the upper limit a variable, say x. Then we will call the area between a and x, $A(x)$; that is,

$$A(x) = \int_a^x f(t)\,dt$$

This equation is stating that

$$\frac{d}{dx}(A(x)) = A'(x) = \frac{d}{dx}\left(\int_a^x f(t)\,dt\right) = f(x)$$

It is important to remember that $\int_a^x f(t)\,dt$ is a function of x.

This very powerful statement is called the **first fundamental theorem of integral calculus**. In essence, it says that the processes of integration and differentiation are inverses of one another.

Example 11.10

Find each derivative.

(a) $\frac{d}{dx}\int_5^x 7t^3\,dt$ (b) $\frac{d}{dx}\int_0^x \frac{dt}{1 + t^4}$ (c) $\frac{d}{dx}\int_x^\pi \frac{1}{1 + t^4}\,dt$

Solution

(a) This is a direct application of the fundamental theorem:

$$\frac{d}{dx}\int_5^x 7t^3\,dt = 7x^3$$

(b) This is also straightforward:

$$\frac{d}{dx}\int_0^x \frac{dt}{1 + t^4} = \frac{1}{1 + x^4}$$

(c) We need to rewrite the expression before we perform the calculation.

$$\frac{d}{dx}\int_x^{\pi}\frac{1}{1+t^4}dt = \frac{d}{dx}\int_{\pi}^{x} -\frac{1}{1+t^4}dt = -\frac{d}{dx}\int_x^{\pi}\frac{1}{1+t^4}dt = \frac{-1}{1+x^4}$$

The second fundamental theorem of integral calculus

Recall that $A(x) = \int_a^x f(t)\,dt$. If $F(x)$ is any antiderivative of $f(x)$, then applying what we learned earlier

$F(x) = A(x) + c$ where c is an arbitrary constant.

Now

$$F(b) = A(b) + c = \int_a^b f(t)\,dt + c, \text{ and}$$

$$F(a) = A(a) + c = \int_a^a f(t)\,dt + c = 0 + c, \text{ and hence}$$

$$F(b) - F(a) = \int_a^b f(t)\,dt + c - c$$

$$= \int_a^b f(t)\,dt$$

The second fundamental theorem of calculus states:

$$\int_a^b f(t)\,dt = F(b) - F(a)$$

The theorem is also known as the **evaluation theorem**. Also, since we know that $F'(x)$ is the rate of change in $F(x)$ with respect to x, and that $F(b) - F(a)$ is the change in y when x changes from a to b, we can reformulate the theorem in words to read:

The integral of a rate of change is the **total change**:

$$\int_a^b F'(x)\,dx = F(b) - F(a)$$

Here are a few instances where this applies:

- If $V'(t)$ is the rate at which a liquid flows into or out of a container at time t, then $\int_{t_1}^{t_2} V'(t)\,dt = V(t_2) - V(t_1)$ is the change in the amount of liquid in the container between time t_1 and t_2.
- If the rate of growth of a population is $n'(t)$, then $\int_{t_1}^{t_2} n'(t)\,dt = n(t_2) - n(t_1)$ is the increase (or decrease) in population during the period from t_1 to t_2.

This theorem has many other applications in calculus and several other fields. It is a very powerful tool that allows us to deal with problems of area, volume, and work. In this book, we will apply it to finding areas between functions, volumes of revolution, and in displacement problems.

Notation

We will use the following notation in evaluating definite integrals. If we know that $F(x)$ is an antiderivative of $f(x)$, then we will write

$$\int_a^b f(t)\,dt = F(x)\Big|_a^b = F(b) - F(a)$$

Example 11.11

Evaluate each integral

(a) $\int_{-1}^{3} x^5 \, dx$ (b) $\int_{0}^{4} \sqrt{x} \, dx$

(c) $\int_{\pi}^{2\pi} \cos\theta \, d\theta$ (d) $\int_{1}^{2} \frac{4 + u^2}{u^3} du$

Solution

(a) $\int_{-1}^{3} x^5 \, dx = \frac{x^6}{6}\Big|_{-1}^{3} = \frac{3^6}{6} - \frac{1}{6} = \frac{364}{3}$

(b) $\int_{0}^{4} \sqrt{x} \, dx = \frac{2}{3}x^{\frac{3}{2}}\Big|_{0}^{4} = \frac{2}{3}4^{\frac{3}{2}} - 0 = \frac{16}{3}$

(c) $\int_{\pi}^{2\pi} \cos\theta \, d\theta = \sin\theta\Big|_{\pi}^{2\pi} = 0 - 0 = 0$

(d) $\int_{1}^{2} \frac{4 + u^2}{u^3} du = \int_{1}^{2} \left(\frac{4}{u^3} + \frac{1}{u}\right) du = 4 \cdot \frac{u^{-2}}{-2} + \ln|u|\Big|_{1}^{2}$

$$= -2u^{-2} + \ln u\Big|_{1}^{2}$$

$$= (-2 \cdot 2^{-2} + \ln 2) - (-2 \cdot 1 + \ln 1)$$

$$= -\frac{1}{2} + \ln 2 + 2 = \frac{3}{2} + \ln 2$$

Using substitution with the definite integral

In Section 11.1, we discussed the use of substitution to evaluate integrals in cases that are not easily recognised. We established that

$$\int f(u(x)) \cdot u'(x) \, dx = \int f(u) du = F(u(x)) + c$$

When evaluating definite integrals by substitution, two methods are available.

- Evaluate the indefinite integral first, revert to the original variable, then use the fundamental theorem. For example, to evaluate

$$\int_{0}^{\frac{\pi}{3}} \tan^5 x \sec^2 x \, dx$$

we find the indefinite integral

$$\int \tan^5 x \sec^2 x \, dx = \int u^5 \, du = \frac{1}{6}u^6 = \frac{1}{6}\tan^6 x$$

then we use the fundamental theorem

$$\int_{0}^{\frac{\pi}{3}} \tan^5 x \sec^2 x \, dx = \frac{1}{6}\tan^6 x\Big|_{0}^{\frac{\pi}{3}} = \frac{1}{6}(\sqrt{3})^6 = \frac{27}{6} = \frac{9}{2}$$

- Or we can use the following substitution rule for definite integrals

$$\int_a^b f(u(x))u'(x)\,dx = \int_{u(a)}^{u(b)} f(u)\,du$$

The change of variable is possible because:

if $F(x)$ is an antiderivative of $f(x)$, then by the fundamental theorem

$$\int_a^b f(u(x))u'(x)\,dx = F(u(x))\Big|_a^b = F(u(b)) - F(u(a))$$

also

$$\int_{u(a)}^{u(b)} f(u)\,du = F(u)\Big|_{u(a)}^{u(b)} = F(u(b)) - F(u(a))$$

therefore, to evaluate

$$\int_0^2 x^2(2x^3+5)^4\,dx$$

Let $u = 2x^3 + 5 \Rightarrow du = 6x^2\,dx \Rightarrow dx = \dfrac{du}{6x^2}, u(2) = 21, u(0) = 5$, and so

$$\int_0^2 (2x^3+5)^4 x^2\,dx = \int_5^{21} u^4 x^2 \frac{du}{6x^2} = \frac{1}{6}\int_5^{21} u^4\,du = \frac{u^5}{30}\Big]_5^{21} = \frac{2040488}{15}$$

Example 11.12

Evaluate $\int_2^6 \sqrt{4x+1}\,dx$

Solution

Let $u = 4x + 1$, then $du = 4dx$. The limits of integration are $u(2) = 9$, and $u(6) = 25$. Therefore

$$\int_2^6 \sqrt{4x+1}\,dx = \frac{1}{4}\int_9^{25} \sqrt{u}\,du = \frac{1}{4}\left(\frac{2}{3}u^{3/2}\right)\Big|_9^{25} = \frac{1}{6}(125-27) = \frac{49}{3}$$

Note that, using this method, we do not return to the original variable of integration. We simply evaluate the new integral between the appropriate values of u.

Notice that the substitution $u = 4x + 1$ stretched the interval $[2, 6]$ by a factor of 4, and shifted it by 1 unit to the right. But the areas are the same.

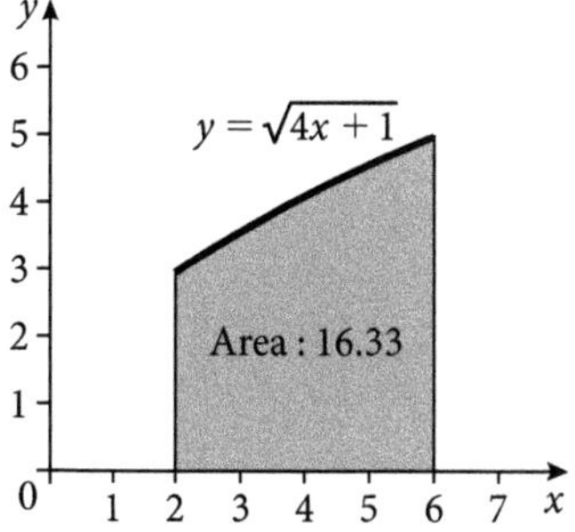

Figure 11.14 The area under the curve $y = \sqrt{4x+1}$ between $x = 2$ and $x = 6$

Exercise 11.2

1. Evaluate each integral.

(a) $\int_{-2}^{1} (3x^2 - 4x^3)\,dx$

(b) $\int_2^7 8\,dx$

(c) $\int_1^5 \frac{2}{t^3}\,dt$

(d) $\int_2^2 (\cos t - \tan t)\,dt$

(e) $\int_1^7 \frac{2x^2 - 3x + 5}{\sqrt{x}}\,dx$

(f) $\int_0^{\pi} \cos\theta\,d\theta$

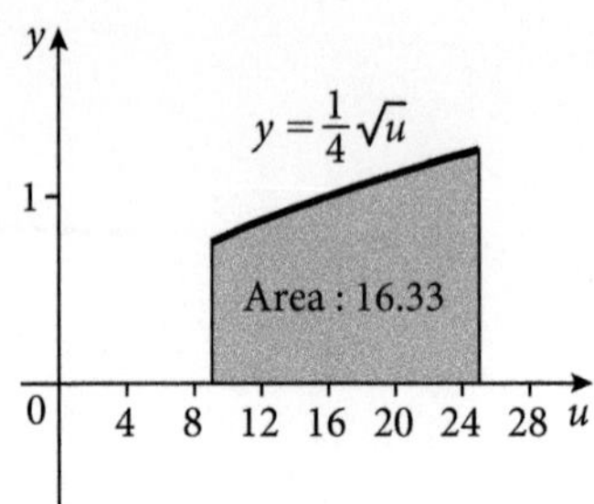

Figure 11.15 The area under the curve $y = \frac{1}{4}\sqrt{u}$ between $u = 9$ and $u = 25$

(g) $\int_0^{\pi} \sin\theta \, d\theta$ **(h)** $\int_3^1 (5x^4 + 3x^2)\, dx$

(i) $\int_1^3 \frac{u^5 + 2}{u^2}\, du$ **(j)** $\int_1^e \frac{2\, dx}{x}$

(k) $\int_1^3 \frac{2x}{x^2 + 2}\, dx$ **(l)** $\int_1^3 (2 - \sqrt{x})^2\, dx$

(m) $\int_0^1 (8x^7 + \sqrt{\pi})\, dx$

(n) **(i)** $\int_0^2 |3x|\, dx$ **(ii)** $\int_{-2}^0 |3x|\, dx$ **(iii)** $\int_{-2}^2 |3x|\, dx$

(o) $\int_0^{\frac{\pi}{2}} \sin 2x\, dx$ **(p)** $\int_1^9 \frac{1}{\sqrt{x}}\, dx$

(q) $\int_{-2}^2 (e^x - e^{-x})dx$

2. Evaluate each integral.

(a) $\int_0^4 \frac{x^3\, dx}{\sqrt{x^2 + 1}}$ **(b)** $\int_1^{\sqrt{e}} \frac{\sin(\pi \ln x)}{x}\, dx$

(c) $\int_e^{e^2} \frac{dt}{t \ln t}$ **(d)** $\int_{-1}^2 3x\sqrt{9 - x^2}\, dx$

(e) $\int_{-\frac{\pi}{3}}^{\frac{2\pi}{3}} \frac{\sin x}{\sqrt{3 + \cos x}}\, dx$ **(f)** $\int_e^{e^2} \frac{\ln x}{x}\, dx$

(g) $\int_{-\ln 2}^{\ln 2} \frac{e^{2x}}{e^{2x} + 9}\, dx$ **(h)** $\int_0^{\sqrt{\pi}} 7x \cos x^2\, dx$

(i) $\int_{\pi^2}^{4\pi^2} \frac{\sin\sqrt{x}}{\sqrt{x}}\, dx$ **(j)** $\int_0^{\frac{\pi}{6}} (1 - \sin 3t)\cos 3t\, dt$

(k) $\int_0^{\frac{\pi}{4}} e^{\sin 2\theta} \cos 2\theta\, d\theta$ **(l)** $\int_0^{\sqrt{\ln \pi}} 4t\, e^{t^2} \sin(e^{t^2})dt$

3. Find the indicated derivative.

(a) $\frac{d}{dx}\int_2^x \frac{\sin t}{t}\, dt$ **(b)** $\frac{d}{dt}\int_t^3 \frac{\sin x}{x}\, dx$

(c) $\frac{d}{dt}\int_{-\pi}^t \frac{\cos y}{1 + y^2}\, dy$

4. **(a)** Find $\int_0^k \frac{dx}{3x + 2}$, giving your answer in terms of k.

(b) Given that $\int_0^k \frac{dx}{3x + 2} = 1$, calculate the value of k.

5. Given that $p, q \in \mathbb{N}$, show that

$$\int_0^1 x^p(1-x)^q\,dx = \int_0^1 x^q(1-x)^p\,dx$$

Do not attempt to evaluate the integrals.

6. Given that $k \in \mathbb{N}$, evaluate each integral:

(a) $\int x(1-x)^k\,dx$ **(b)** $\int_0^1 x(1-x)^k\,dx$

7. Let $F(x) = \int_3^x \sqrt{5t^2+2}\,dt$, find:

(a) $F(3)$ **(b)** $F'(3)$ **(c)** $F''(3)$

11.3 Areas

We have seen how the area between a curve defined by $y = f(x)$ and the x-axis can be computed by the integral $\int_a^b f(x)\,dx$ on an interval $[a, b]$ where $f(x) > 0$. In this section, we shall use integration to find the area of more general regions between curves.

Areas between curves of functions of the form $y = f(x)$ and the x-axis

If the function $y = f(x)$ is always above the x-axis, finding the area is a straightforward computation of the integral $\int_a^b f(x)\,dx$.

Example 11.13

Find the area between the curve $f(x) = x^3 - x + 1$ and the x-axis over the interval $[-1, 2]$

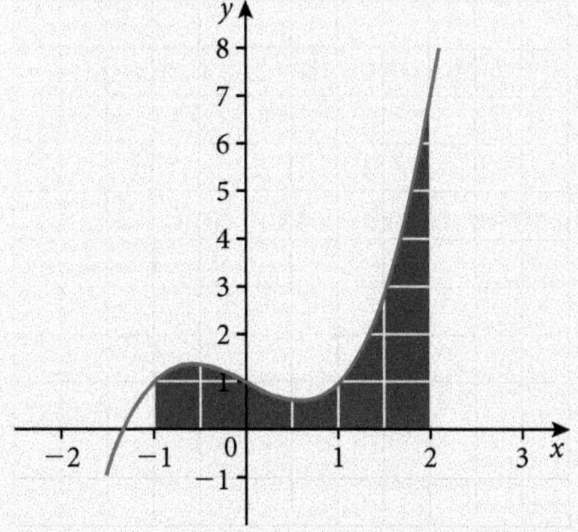

$\int_{-1}^{2}$ (x³-x+1)dx
21/4

Figure 11.16 Using a GDC to find the area

Solution

This area is:

$$\int_{-1}^{2}(x^3 - x + 1)\,dx = \left[\frac{x^4}{4} - \frac{x^2}{2} + x\right]_{-1}^{2} = (4 - 2 + 2) - \left(\frac{1}{4} - \frac{1}{2} - 1\right) = 5\frac{1}{4}$$

You can use your GDC to work out the area. The calculation is usually straightforward.

In some cases, we will have to adjust how to work. This is the case when the graph intersects the x-axis. Since we are interested in the area bounded by the curve and the interval $[a, b]$ on the x-axis, we do not want the two areas to cancel each other. This is why we have to split the process into subintervals where we take the absolute values of the areas found and add them.

Online 

Explore the relationship between the definite integral and the area between the function and the x-axis.

Example 11.14

Find the area under the curve $f(x) = x^3 - x - 1$ and the x-axis over the interval $[-1, 2]$

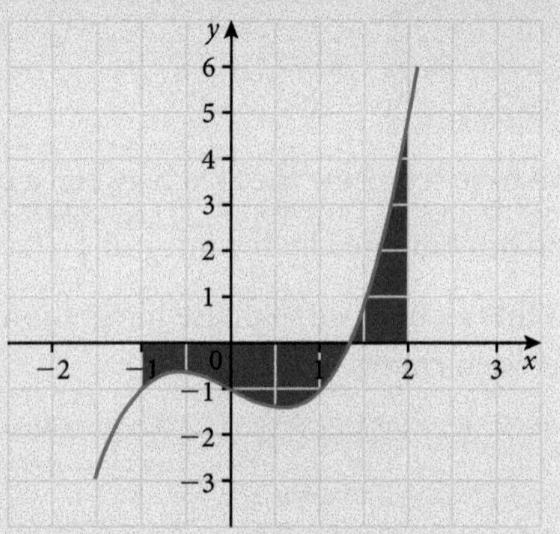

Solution

As we see from the diagram, a part of the graph is below the x-axis, and its area will be negative. If we try to integrate this function without paying attention to the intersection with the x-axis, here is what we get:

$$\int_{-1}^{2}(x^3 - x - 1)\,dx = \frac{x^4}{4} - \frac{x^2}{2} - x\Big|_{-1}^{2} = (4 - 2 - 2) - \left(\frac{1}{4} - \frac{1}{2} + 1\right)$$
$$= -\frac{3}{4}$$

This integration has to be split before we start. However, this is a function where we cannot find the intersection point. So, we either use a GDC to find the intersection or we just take the absolute values of the different parts of the region. This is done by integrating the absolute value of the function:

$$Area = \int_{a}^{b}|f(x)|\,dx$$

$\int_{-1}^{2}$ |x³-x-1|dx
3.614515769

Figure 11.17 It is best to use a GDC

As we said earlier, this is not easy to find given the difficulty with the x-intercept. It is best if we use a GDC.

Hence, $Area = \int_{-1}^{2}|x^3 - x - 1|\,dx \approx 3.6145$

Example 11.15

Find the area enclosed by the graph of the function $f(x) = x^3 - 4x^2 + x + 6$ and the x-axis.

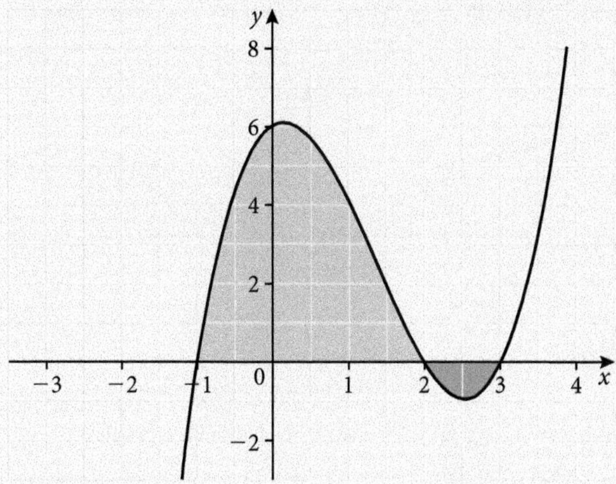

Solution

This function intersects the x-axis at three points, where $x = -1, 2$, and 3. To find the area, we split it into two and then add the absolute values:

$$Area = \int_{-1}^{3} |f(x)|\,dx = \int_{-1}^{2} f(x)\,dx + \int_{2}^{3} (-f(x))\,dx$$

$$= \int_{-1}^{2} (x^3 - 4x^2 + x + 6)\,dx + \int_{2}^{3} (-x^3 + 4x^2 - x - 6)\,dx$$

$$= \frac{x^4}{4} - \frac{4x^3}{3} + \frac{x^2}{2} + 6x\Big|_{-1}^{2} + -\frac{x^4}{4} + \frac{4x^3}{3} - \frac{x^2}{2} - 6x\Big|_{2}^{3}$$

$$= \frac{45}{4} + \frac{7}{12} = \frac{71}{6}$$

Area between curves

In some practical problems, we may have to compute the area between two curves. Let $f(x)$ and $g(x)$ be functions such that $f(x) \geqslant g(x)$ on the interval $[a, b]$ (Figure 11.18). We do not insist that both functions are non-negative.

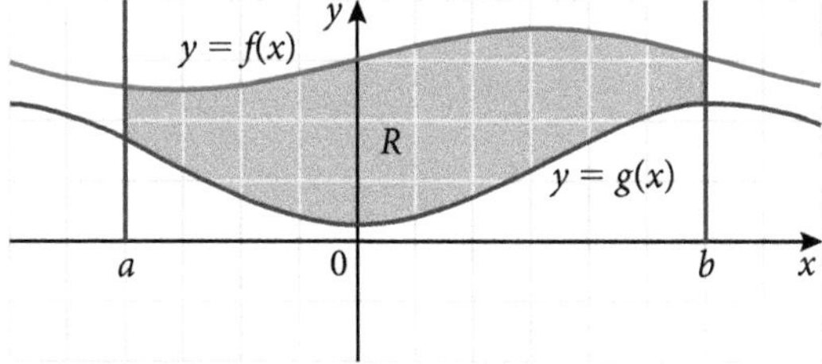

Figure 11.18 Area between two curves

To find the area of the region R between the curves from $x = a$ to $x = b$, we subtract the area between the lower curve $g(x)$ and the x-axis from the area between the upper curve $f(x)$ and the x-axis; that is

$$\text{Area of } R = \int_a^b f(x)\,dx - \int_a^b g(x)\,dx = \int_a^b [f(x) - g(x)]\,dx$$

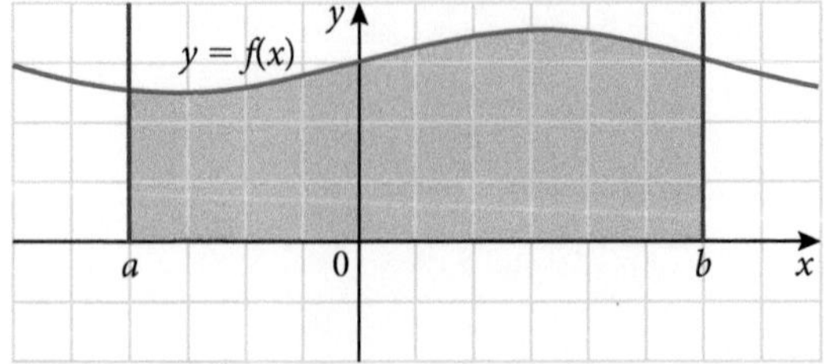

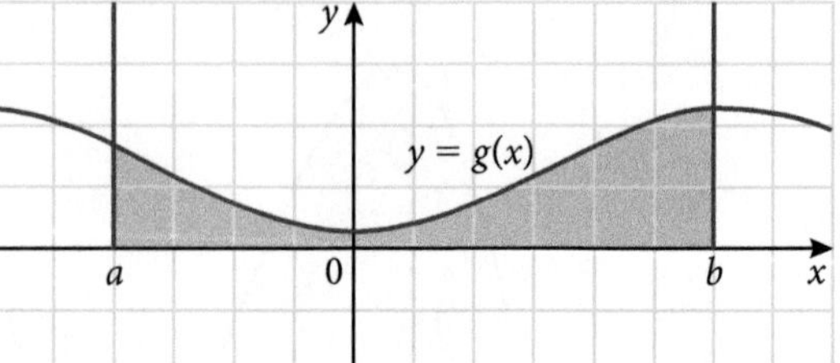

Figure 11.19 Areas under functions f and g

If $f(x)$ and $g(x)$ are functions such that $f(x) \geqslant g(x)$ on the interval $[a, b]$, then the area between the two curves is given by $A = \sum_a^b [f(x) - g(x)]\,dx$

This fact applies to all functions, not only to positive functions. These facts are used to define the area between curves.

Example 11.16

Find the area of the region between the curves $y = x^3$ and $y = x^2 - x$ on the interval $[0, 1]$.

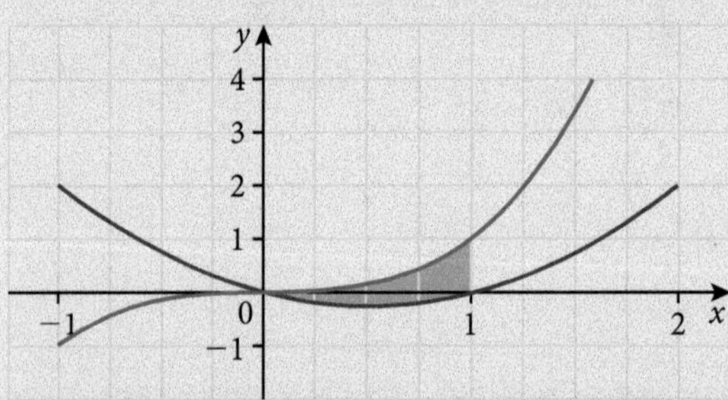

Solution

$y = x^3$ appears to be higher than $y = x^2 - x$ with one intersection at $x = 0$. Thus, the required area is

$$A = \int_0^1 [x^3 - (x^2 - x)]\,dx = \frac{x^4}{4} - \frac{x^3}{3} + \frac{x^2}{2}\Big|_0^1 = \frac{5}{12}$$

In some cases, we must be very careful how we calculate the area. This is the case where the two functions intersect at more than one point.

Example 11.17

Find the area of the region bounded by the curves $y = x^3 + 2x^2$ and $y = x^2 + 2x$.

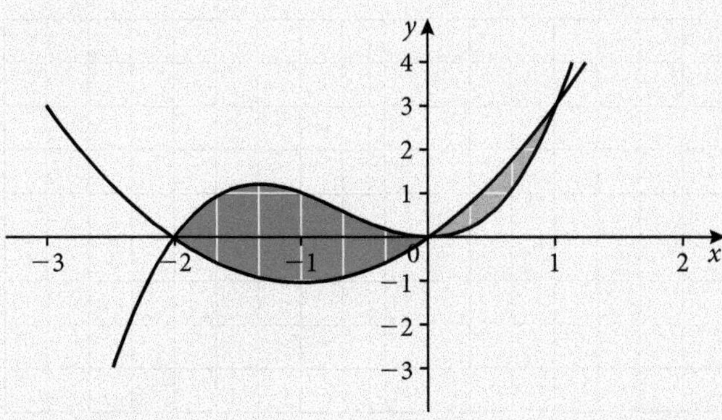

Solution

The two curves intersect when:

$$x^3 + 2x^2 = x^2 + 2x \Rightarrow x^3 + x^2 - 2x = 0 \Rightarrow x(x + 2)(x - 1) = 0$$

That is, when $x = -2, 0$, or 1

The area is equal to:

$$A = \int_{-2}^{0} [x^3 + 2x^2 - (x^2 + 2x)]\, dx + \int_{0}^{1} [x^2 + 2x - (x^3 + 2x^2)]\, dx$$

$$= \int_{-2}^{0} [x^3 + x^2 - 2x]\, dx + \int_{0}^{1} [-x^2 + 2x - x^3]\, dx$$

$$= \left[\frac{x^4}{4} + \frac{x^3}{3} - x^2\right]_{-2}^{0} + \left[-\frac{x^4}{4} - \frac{x^3}{3} + x^2\right]_{0}^{1}$$

$$= 0 - \left[\frac{16}{4} - \frac{8}{3} - 4\right] + \left[-\frac{1}{4} - \frac{1}{3} + 1\right] - 0 = \frac{37}{12}$$

This discussion leads us to stating the general expression we should use in evaluating areas between curves.

If $f(x)$ and $g(x)$ are functions that are continuous on the interval $[a, b]$, then the area between the two curves is given by

$$A = \int_{a}^{b} |f(x) - g(x)|\, dx$$

We can do this on our GDC.

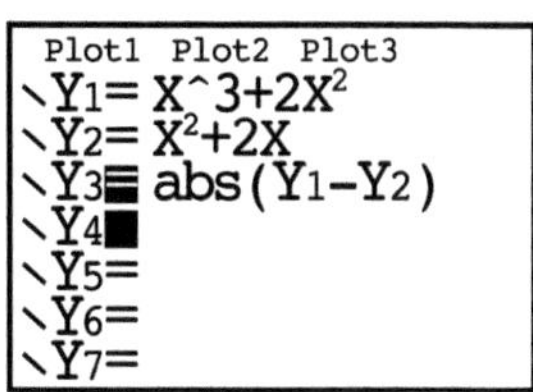

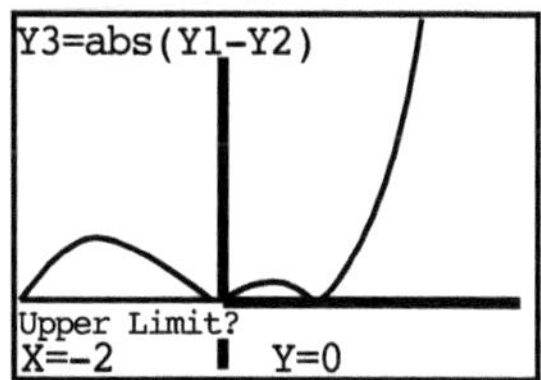

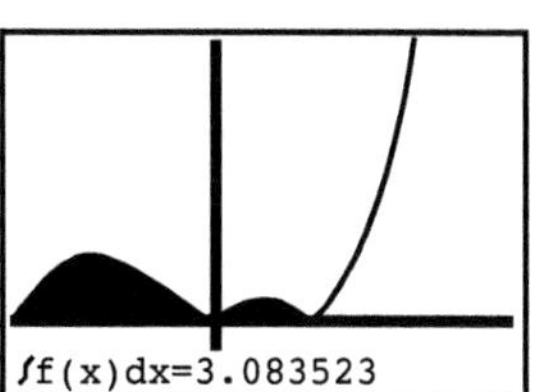

Figure 11.20 Using a GDC to find the area between two curves

Areas along the y-axis (optional)

To find the area enclosed by $y = 1 - x$ and $y^2 = x + 1$, it is best to treat the region between them by regarding x as a function of y (Figure 11.21).

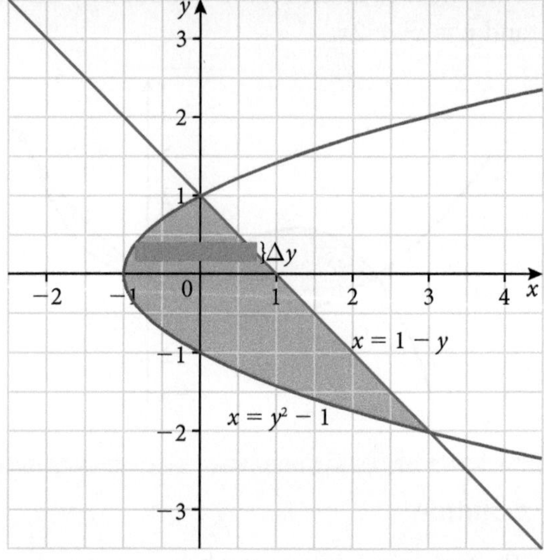

Figure 11.21 Area between two curves expressed by regarding x as a function of y

The area of the shaded region can be calculated using the integral:

$$A(y) = \int_{-2}^{1} \left|(1 - y) - (y^2 - 1)\right| \,dy$$

$$= \int_{-2}^{1} |2 - y - y^2| \,dy = \left|2y - \frac{y^2}{2} - \frac{y^3}{3}\right|_{-2}^{1} = \frac{9}{2}$$

If we used y as a function of x, then the calculation would involve calculating the area by dividing the interval into two: $[-1, 0]$ and $[0, 3]$.

In the first part, the area is enclosed between $y = \sqrt{x + 1}$ and $y = -\sqrt{x + 1}$, and the area in the second part is enclosed by $y = 1 - x$ and $y = -\sqrt{x + 1}$:

$$A(x) = 2\int_{-1}^{0} \sqrt{x + 1} \,dx + \int_{0}^{3} \left((1 - x) - (-\sqrt{x + 1})\right) dx = \frac{4}{3} + \frac{19}{6} = \frac{27}{6}$$

Exercise 11.3

1. Find the area of the region bounded by the given curves. Sketch the region and then compute the required area.

   (a) $y = x + 1, y = 7 - x^2$

   (b) $y = \cos x, y = x - \frac{\pi}{2}, x = -\pi$

   (c) $y = 2x, y = x^2 - 2$

   (d) $y = x^3, y = x^2 - 2, x = 1$

   (e) $y = x^6, y = x^2$

   (f) $y = 5x - x^2, y = x^2$

   (g) $y = 2x - x^3, y = x - x^2$

   (h) $y = \sin x, y = 2 - \sin x$ (one period)

   (i) $y = \frac{x}{2}, y = \sqrt{x}, x = 9$

   (j) $y = \frac{x^4}{10}, y = 3x - x^3$

   (k) $y = \frac{1}{x}, y = \frac{1}{x^3}, x = 8$

   (l) $y = 2\sin x, y = \sqrt{3}\tan x, -\frac{\pi}{4} \leqslant x \leqslant \frac{\pi}{4}$

(m) $y = x^3 + 2x^2$, $y = x^3 - 2x$, $x = -3$, and $x = 2$

(n) $y = x^3 + 1$, $y = (x + 1)^2$

(o) $y = x^3 + x$, $y = 3x^2 - x$

(p) $y = 3 - \sqrt{x}$, $y = \dfrac{2\sqrt{x} + 1}{2\sqrt{x}}$

2. Find the area of the shaded region.

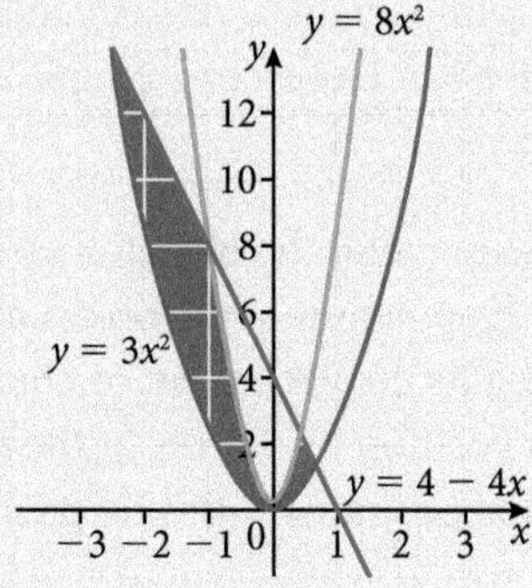

3. Find the area of the region enclosed by $y = e^x$, $x = 0$, and the tangent to $y = e^x$ at $x = 1$

4. Find the area of the region enclosed by $y = (x - 2)^2$ and $y = x(x - 4)^2$

5. Find a value for $m > 0$ such that the area under the graph of $y = e^{2x}$ over the interval $[0, m]$ is 3 square units.

6. Find the area of the region bounded by $y = x^3 - 4x^2 + 3x$ and the x-axis.

11.4 Modelling linear motion

So far, our mathematical models considered the motion of an object only along a straight line. For example, projectile motion (e.g. a ball being thrown) is often modelled by a position function that simply gives the height (displacement) of the object. In this way, we are modelling the motion as if it was restricted to a vertical line.

In this section, we will again analyse the motion of an object as if its motion takes place along a straight line in space. This makes sense only if the mass (and thus, size) of the object is not taken into account. Hence, the object is modelled by a particle whose mass is considered to be zero. This study of motion, without reference either to the forces that cause it or to the mass of the object, is known as **kinematics**.

Displacement and total distance travelled

Recall from Chapter 10 that given time t, displacement s, velocity v, and acceleration a, we have:

$$v = \frac{\mathrm{d}s}{\mathrm{d}t} \text{ and } a = \frac{\mathrm{d}v}{\mathrm{d}t}, \text{ also } a = \frac{\mathrm{d}}{\mathrm{d}t}\left(\frac{\mathrm{d}s}{\mathrm{d}t}\right) = \frac{\mathrm{d}^2 s}{\mathrm{d}t^2}$$

It is important to understand the difference between displacement and distance travelled. Consider a couple of simple examples of an object moving along the x-axis.

Assume that the object does not change direction during the interval $0 \leqslant t \leqslant 5$. If the position of the object at $t = 0$ is $x = 2$, and at $t = 5$ its position is $x = -3$, then its displacement, or change in position, is -5 because the object changed its position by 5 units in the negative x-direction. This can be calculated by: (final position) − (initial position) $= -3 - 2 = -5$.

However, the distance travelled would be the absolute value of displacement, calculated by |final position − initial position| $= |-3 - 2| = 5$.

Assume that another object's initial and final positions are the same as in the first example; that is, at $t = 0$ its position is $x = 2$, and at $t = 5$ its position is $x = -3$. However, the object changed direction in that it first travelled to the left (negative velocity) from $x = 2$ to $x = -5$ during the interval $0 \leqslant t \leqslant 3$, and then travelled to the right (positive velocity) from $x = -5$ to $x = -3$. The object's displacement is -5, the same as in the first example because its net change in position is just the difference between its final and initial positions. However, it's clear that the object has travelled further than in the first example. But, we cannot calculate it the same way as we did in the first example. We will have to make a separate calculation for each interval where the direction changed. Hence, total distance travelled $= |-5 - 2| + |-3 - (-5)| = 7 + 2 = 9$.

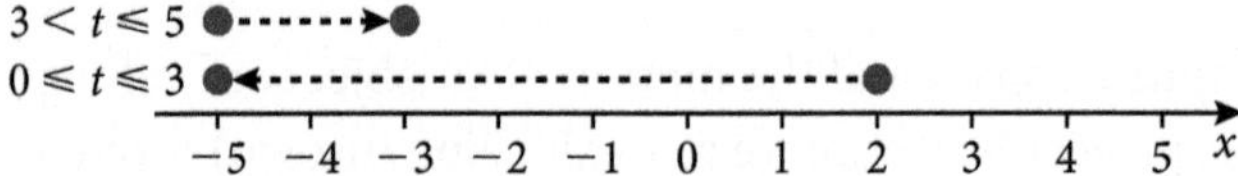

Figure 11.22 Travelled distances

- The **velocity** $v = \frac{\mathrm{d}s}{\mathrm{d}t}$ of a particle is a measure of how fast it is moving and of its direction of motion relative to a fixed point.
- The **speed** $|v|$ of a particle is a measure of how fast it is moving that does not indicate direction. Thus, speed is the magnitude of velocity and is always positive.
- The **acceleration** $a = \frac{\mathrm{d}v}{\mathrm{d}t}$ of a particle is a measure of how fast its velocity is changing.

Example 11.18

The displacement s of a particle on the x-axis, relative to the origin, is given by the position function $s(t) = -t^2 + 6t$ where s in centimetres and t is in seconds.

(a) Find a function for the particle's velocity $v(t)$ in terms of t. Graph the functions $s(t)$ and $v(t)$ on separate axes.

(b) Find the particle's position at the following times: $t = 0, 1, 3,$ and 6 seconds

(c) Find the particle's displacement for the following intervals: $0 \leqslant t \leqslant 1$, $1 \leqslant t \leqslant 3$, $3 \leqslant t \leqslant 6$, and $0 \leqslant t \leqslant 6$

(b) Find the particle's total distance travelled for the following intervals: $0 \leqslant t \leqslant 1$, $1 \leqslant t \leqslant 3$, $3 \leqslant t \leqslant 6$, and $0 \leqslant t \leqslant 6$

Solution

(a) $v(t) = \frac{d}{dt}(-t^2 + 6t) = -2t + 6$

Position function: $s(t) = -t^2 + 6t$

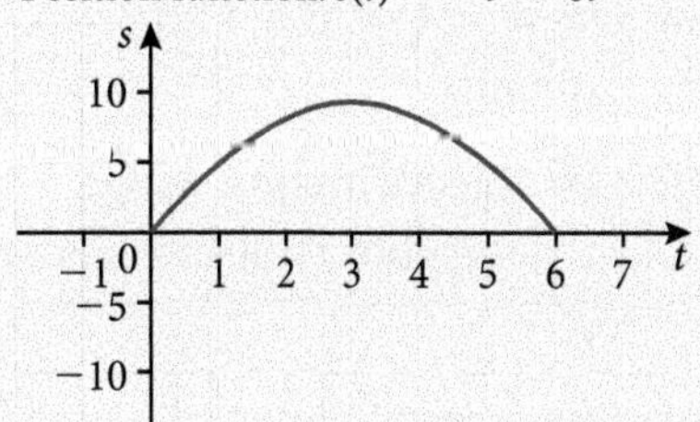

Velocity function: $v(t) = s'(t) = -2t + 6$

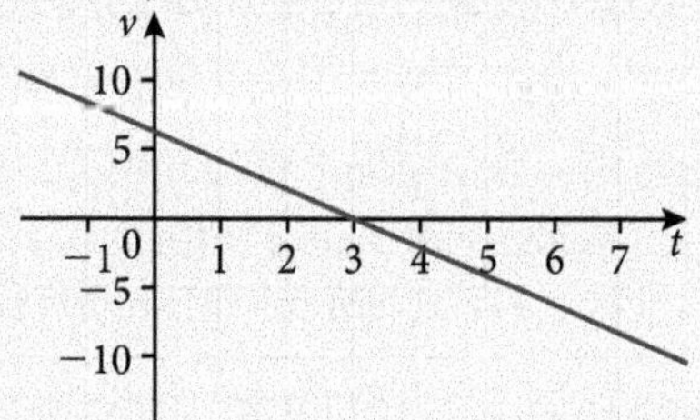

(b) The particle's position at:

$t = 0$ is $s(0) = -(0)^2 + 6(0) = 0$ cm

$t = 1$ is $s(1) = -(1)^2 + 6(1) = 5$ cm

$t = 3$ is $s(3) = -(3)^2 + 6(3) = 9$ cm

$t = 6$ is $s(6) = -(6)^2 + 6(6) = 0$ cm

(c) The particle's displacement for the interval:

$0 \leqslant t \leqslant 1$: Δ position $= s(1) - s(0) = 5 - 0 = 5$ cm

$1 \leqslant t \leqslant 3$: Δ position $= s(3) - s(1) = 9 - 5 = 4$ cm

$3 \leqslant t \leqslant 6$: Δ position $= s(6) - s(3) = 0 - 9 = -9$ cm

$0 \leqslant t \leqslant 6$: Δ position $= s(6) - s(0) = 0 - 0 = 0$ cm

This last result makes sense considering the particle moved to the right 9 cm then, at $t = 3$, it turned around and moved to the left 9 cm, ending where it started – thus, no change in net position.

(d) The particle's total distance travelled for the interval:

$0 \leqslant t \leqslant 1$ is $|s(1) - s(0)| = |5 - 0| = 5$

$1 \leqslant t \leqslant 3$ is $|s(3) - s(1)| = |9 - 5| = 4$

$3 \leqslant t \leqslant 6$ is $|s(6) - s(3)| = |0 - 9| = |-9| = 9$

The object's motion changed direction (velocity$= 0$) at $t = 3$

$0 \leqslant t \leqslant 6$ is $|s(3) - s(0)| + |s(6) - s(3)| = |9 - 0| + |0 - 9|$
$= 9 + 9 = 18$

Since differentiation of the position function gives the velocity function $\left(\text{i.e. } v = \frac{\mathrm{d}s}{\mathrm{d}t}\right)$, we expect that the inverse of differentiation (integration) will lead us in the reverse direction – that is, from velocity to position. When velocity is constant, we can find the displacement with the formula:

displacement = velocity × change in time

If we drove a car at a constant velocity of $50\,\text{km h}^{-1}$ for 3 hours, then our displacement (same as distance travelled in this case) is 150 km. If a particle travelled to the left on the x-axis at a constant rate of -4 units s^{-1} for 5 seconds, then the particle's displacement is -20 units.

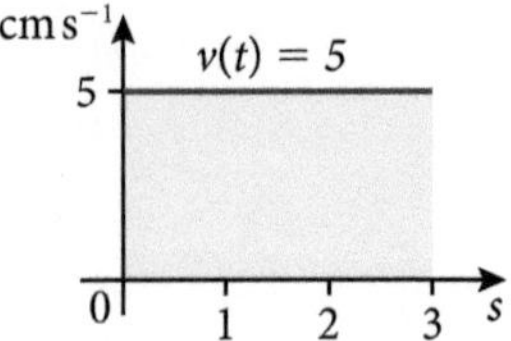

Figure 11.23 Velocity–time graph

The velocity–time graph (Figure 11.23) depicts an object's motion with a constant velocity of $5\,\text{cm s}^{-1}$ for $0 \leqslant t \leqslant 3$. Clearly, the object's displacement is $5\,\text{cm s}^{-1} \times 3\,\text{s} = 15\,\text{cm}$ for this interval.

The area ($3 \times 5 = 15$) under the velocity curve for a certain interval is equal to the object's displacement. We can argue that just as the total area can be found by summing the areas of narrow rectangular strips, the displacement can be found by summing small displacements ($v \cdot \Delta t$). Consider:

displacement = velocity × change in time $\Rightarrow s = v \cdot \Delta t \Rightarrow s = v \cdot \mathrm{d}t$

We already know that when $f(x) \geqslant 0$, the definite integral $\int_a^b f(x)\,\mathrm{d}x$ gives the area between $y = f(x)$ and the x-axis from $x = a$ to $x = b$. And if $f(x) \leqslant 0$, then $\int_a^b f(x)\,\mathrm{d}x$ gives a number that is the opposite of the area between $y = f(x)$ and the x-axis from a to b.

Given that $v(t)$ is the velocity function for a particle moving along a line, then:

$\int_a^b v(t)\,\mathrm{d}t$ gives the displacement from $t = a$ to $t = b$

$\int_a^b |v(t)|\,\mathrm{d}t$ gives the total distance travelled from $t = a$ to $t = b$

Let's apply integration to find the displacement and distance travelled for the two intervals $3 \leqslant t \leqslant 6$ and $0 \leqslant t \leqslant 6$ in Example 11.18

For $3 \leqslant t \leqslant 6$:

$$\text{Displacement} = \int_3^6 (-2t + 6)\mathrm{d}t = -t^2 + 6t\Big|_3^6 = 0 - 9 = -9$$

$$\text{Distance travelled} = \int_3^6 |(-2t + 6)|\mathrm{d}t = \left|-t^2 + 6t\right|\Big|_3^6 = |0 - 9| = 9$$

For $0 \leqslant t \leqslant 6$:

$$\text{Displacement} = \int_0^6 (-2t + 6)\mathrm{d}t = -t^2 + 6t\Big|_0^6 = 0$$

$$\text{Distance travelled} = \int_0^3 |(-2t + 6)|\mathrm{d}t + \int_3^6 |(-2t + 6)|\mathrm{d}t$$

(particle changed direction at $t = 3$)

$$= \left|-t^2 + 6t\right|\Big|_0^3 + \left|-t^2 + 6t\right|\Big|_3^6 = 9 + 9 = 18$$

Note when using a GDC or your computer, you do not need to separate the integrals as we did here.

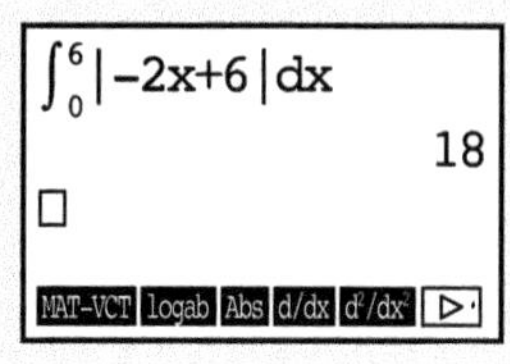

Example 11.19

The function $v(t) = \sin(\pi t)$ gives the velocity in $\mathrm{m\,s^{-1}}$ of a particle moving along the x-axis.

(a) Determine when the particle is moving to the right, to the left, and stopped. If it stops, determine if it changes direction at that time.

(b) Find the particle's displacement for the time interval $0 \leqslant t \leqslant 3$.

(c) Find the particle's total distance travelled for the time interval $0 \leqslant t \leqslant 3$.

Solution

(a) $v(t) = \sin(\pi t) = 0 \Rightarrow \sin(k \cdot \pi) = 0$ for $k \in \mathbb{Z} \Rightarrow \pi t = k\pi \Rightarrow t = k, k \in \mathbb{Z}$ for $0 \leqslant t \leqslant 3$, $t = 0, 1, 2, 3$. Therefore, the particle is stopped at $t = 0, 1, 2, 3$.

Since $t = 0$ and $t = 3$ are endpoints of the interval, the particle can change direction only at $t = 1$ or $t = 2$.

$v\left(\frac{1}{2}\right) = \sin\left(\pi \cdot \frac{1}{2}\right) = 1$; $v\left(\frac{3}{2}\right) = \sin\left(\pi \cdot \frac{3}{2}\right) = -1$

$\Rightarrow$ direction changes at $t = 1$

$v\left(\frac{3}{2}\right) = \sin\left(\pi \cdot \frac{3}{2}\right) = -1$; $v\left(\frac{5}{2}\right) = \sin\left(\pi \cdot \frac{5}{2}\right) = 1$

$\Rightarrow$ direction changes again at $t = 2$

(b) displacement $= \int_0^3 \sin(\pi t)\,\mathrm{d}t = -\frac{1}{\pi}\cos(\pi t)\Big|_0^3$

$$= -\frac{1}{\pi}\cos(3\pi) - \left(-\frac{1}{\pi}\cos(0)\right) = \frac{2}{\pi} \approx 0.637 \text{ metres}$$

(c) total distance travelled $= \int_0^1 |\sin(\pi t)|\,\mathrm{d}t + \int_1^2 |\sin(\pi t)|\,\mathrm{d}t + \int_2^3 |\sin(\pi t)|\,\mathrm{d}t$

$$= \left|\frac{2}{\pi}\right| + \left|-\frac{2}{\pi}\right| + \left|\frac{2}{\pi}\right| = \frac{6}{\pi} \approx 1.91 \text{ metres}$$

Note that in Example 11.20, the position function is not known precisely. The position function can be obtained by finding the antiderivative of the velocity function.

$$s(t) = \int v(t)\,\mathrm{d}t = \int \sin(\pi t)\,\mathrm{d}t = -\frac{1}{\pi}\cos(\pi t) + C$$

We can determine the constant of integration c only if we know the particle's initial position (or position at any other specific time). However, the particle's initial position will not affect displacement or distance travelled for any interval.

Position and velocity from acceleration

If we can obtain position from velocity by applying integration, then we can also obtain velocity from acceleration by integrating. Consider the next example.

Example 11.20

The motion of a falling parachutist is modelled as linear motion by considering that the parachutist is a particle moving along a line whose positive direction is vertically downwards. The parachute is opened at $t = 0$, at which time the parachutist's position is $s = 0$. According to the model, the acceleration function for the parachutist's motion for $t > 0$ is given by:

$$a(t) = -54\mathrm{e}^{-1.5t}$$

(a) At the moment the parachute opens, the parachutist has a velocity of 42 m s^{-1}. Find the velocity function of the parachutist for $t > 0$. What does the model say about the parachutist's velocity as $t \to \infty$?

(b) Find the position function of the parachutist for $t > 0$.

Solution

(a) $v(t) = \int a(t)\,\mathrm{d}t = \int (-54\mathrm{e}^{-1.5t})\,\mathrm{d}t$

$$= -54\left(\frac{1}{-1.5}\right)\mathrm{e}^{-1.5t} + C$$

$$= 36\mathrm{e}^{-1.5t} + C$$

Since $v = 42$ when $t = 0$, then $42 = 36e^0 + C \Rightarrow 42 = 36 + C \Rightarrow C = 6$

Therefore, after the parachute opens ($t > 0$), the velocity function is $v(t) = 36e^{-1.5t} + 6$

Since $\lim_{t\to\infty} e^{-1.5t} = \lim_{t\to\infty} \frac{1}{e^{1.5t}} = 0$, then as $t \to 0$, $\lim_{t\to\infty} v(t) = 6\,\text{m}\,\text{s}^{-1}$

(b) $s(t) = \int v(t)\,dt = \int (36e^{-1.5t} + 6)\,dt$

$$= 36\left(\frac{1}{-1.5}\right)e^{-1.5t} + 6t + C$$

$$= -24e^{-1.5t} + 6t + C$$

Since $s = 0$ when $t = 0$, then $0 = -24e^0 + 6(0) + C \Rightarrow 0 = -24 + C$ $\Rightarrow C = 24$

Therefore, after the parachute opens ($t > 0$), the position function is $s(t) = -24e^{-1.5t} + 6t + 24$

The limit of the velocity as $t \to \infty$, for a falling object, is called the **terminal velocity** of the object. While the limit $t \to \infty$ is never attained as the parachutist eventually lands on the ground, the velocity gets close to the terminal velocity very quickly. For example, after just 8 seconds, the velocity is $v(8) = 36e^{-1.5(8)} + 6$ $\approx 6.0002\,\text{m}\,\text{s}^{-1}$

Uniformly accelerated motion

Motion under the effect of gravity in the vicinity of Earth (or other planets) is an important case of rectilinear motion. This is called **uniformly accelerated motion.**

If a particle moves with constant acceleration along the s-axis, and if we know the initial speed and position of the particle, then it is possible to have specific formulas for the position and speed at any time t.

Assume acceleration is constant; that is, $a(t) = a$, $v(0) = v_0$ and $s(0) = s_0$.

$v(t) = \int a(t)dt = at + c$; however, we know that $v(0) = v_0$, so

$v(0) = v_0 = a \times 0 + c \Rightarrow c = v_0$, hence $v(t) = at + v_0$

$s(t) = \int v(t)dt = \int (at + v_0)dt = \frac{1}{2}at^2 + v_0t + c$, and, as above, substituting $s(0) = s_0$ into the equation, we have

$$s(t) = \frac{1}{2}at^2 + v_0t + s_0$$

When this is applied to the free-fall model (s-axis vertical), then

$v(t) = -gt + v_0$ and

$s(t) = -\frac{1}{2}gt^2 + v_0t + s_0$, where $g = 9.8\,\text{m}\,\text{s}^{-2}$

Example 11.21

A ball is hit directly upwards from a point 2 m above the ground with initial velocity of $45\,\text{m}\,\text{s}^{-1}$. How high will the ball travel?

Solution

$$v(t) = -9.8t + 45$$

$$s(t) = -\frac{1}{2}(9.8)\,t^2 + 45t + 2 = -4.9t^2 + 45t + 2$$

The ball will rise until $v(t) = 0, \Rightarrow 0 = -9.8t + 45, \Rightarrow t \approx 4.6\text{ s}$

At this time

$$s(4.6) = -4.9(4.6)^2 + 45(4.6) + 2 \approx 105.32\text{ m}$$

Exercise 11.4

1. The velocity of a particle along a rectilinear path is given by each equation for $v(t)$ in m s^{-1}. Find both the net distance and the total distance it travels between the times $t = a$ and $t = b$.

(a) $v(t) = t^2 - 11t + 24, a = 0, b = 10$

(b) $v(t) = t - \frac{1}{t^2}, a = 0.1, b = 1$

(c) $v(t) = \sin 2t, a = 0, b = \frac{\pi}{2}$

(d) $v(t) = \sin t + \cos t, a = 0, b = \pi$

(e) $v(t) = t^3 - 8t^2 + 15t, a = 0, b = 6$

(f) $v(t) = \sin\left(\frac{\pi t}{2}\right) + \cos\left(\frac{\pi t}{2}\right), a = 0, b = 1$

2. The acceleration of a particle along a rectilinear path is given by each equation for $a(t)$ in m s^{-2} and the initial velocity v_0 in m s^{-1} is also given. Find the velocity of the particle as a function of t, and both the net distance and the total distance travelled between times $t = a$ and $t = b$.

(a) $a(t) = 3, v_0 = 0, a = 0, b = 2$

(b) $a(t) = 2t - 4, v_0 = 3, a = 0, b = 3$

(c) $a(t) = \sin t, v_0 = 0, a = 0, b = \frac{3\pi}{2}$

(d) $a(t) = \frac{-1}{\sqrt{t+1}}, v_0 = 2, a = 0, b = 4$

(e) $a(t) = 6t - \frac{1}{(t+1)^3}, v_0 = 2, a = 0, b = 2$

3. The velocity and initial position of an object moving along a coordinate line are given. Find the position of the object at time t.

(a) $v = 9.8t + 5, s(0) = 10$

(b) $v = 32t - 2, s(0.5) = 4$

(c) $v = \sin \pi t, s(0) = 0$

(d) $v = \frac{1}{t+2}, t > -2, s(-1) = \frac{1}{2}$

4. The acceleration, initial velocity, and initial position of an object moving on a coordinate line are given. Find the position of the object at time t.

(a) $a = e^t$, $v(0) = 20$, $s(0) = 5$

(b) $a = 9.8$, $v(0) = -3$, $s(0) = 0$

(c) $a = -4\sin 2t$, $v(0) = 2$, $s(0) = -3$

(d) $a = \frac{9}{\pi^2}\cos\frac{3t}{\pi}$, $v(0) = 0$, $s(0) = -1$

5. An object moves with a speed of $v(t)$ m s^{-1} along the s-axis. Find the displacement and the distance travelled by the object during the given time interval.

(a) $v(t) = 2t - 4; 0 \leqslant t \leqslant 6$

(b) $v(t) = |t - 3|; 0 \leqslant t \leqslant 5$

(c) $v(t) = t^3 - 3t^2 + 2t; 0 \leqslant t \leqslant 3$

(d) $v(t) = \sqrt{t} - 2, 0 \leqslant t \leqslant 3$

6. An object moves with an acceleration $a(t)$ m s^{-2} along the s-axis. Find the displacement and the distance travelled by the object during the given time interval.

(a) $a(t) = t - 2$, $v_0 = 0$, $1 \leqslant t \leqslant 5$

(b) $a(t) = \frac{1}{\sqrt{5t+1}}$, $v_0 = 2$, $0 \leqslant t \leqslant 3$

(c) $a(t) = -2$, $v_0 = 3$, $1 \leqslant t \leqslant 4$

7. The velocity of an object moving along the s-axis is $v = 9.8t - 3$.

(a) Find the object's displacement between $t = 1$ and $t = 3$ given that $s(0) = 5$

(b) Find the object's displacement between $t = 1$ and $t = 3$ given that $s(0) = -2$

(c) Find the object's displacement between $t = 1$ and $t = 3$ given that $s(0) = s_0$

8. The displacement s metres of a moving object from a fixed point O at time t seconds is given by $s(t) = 50t - 10t^2 + 1000$.

(a) Find the velocity of the object in m s^{-1}.

(b) Find its maximum displacement from O.

9. A particle moves along a line so that its speed v at time t is given by

$$v(t) = \begin{cases} 5t & 0 \leqslant t < 1 \\ 6\sqrt{t} - \frac{1}{t} & t \geqslant 1 \end{cases}$$

where t is in seconds and v is in cm s^{-1}. Estimate the time(s) at which the particle is 4 cm from its starting position.

10. A projectile is fired vertically upwards with an initial velocity of 49 m s^{-1} from a platform 150 m high.

(a) How long will it take the projectile to reach its maximum height?

(b) What is the maximum height of the projectile?

(c) How long will it take the projectile to pass its starting point on the way down?

(d) What is the velocity of the projectile when it passes the starting point on the way down?

(e) How long will it take the projectile to hit the ground?

(f) What will its speed be at impact?

Chapter 11 practice questions

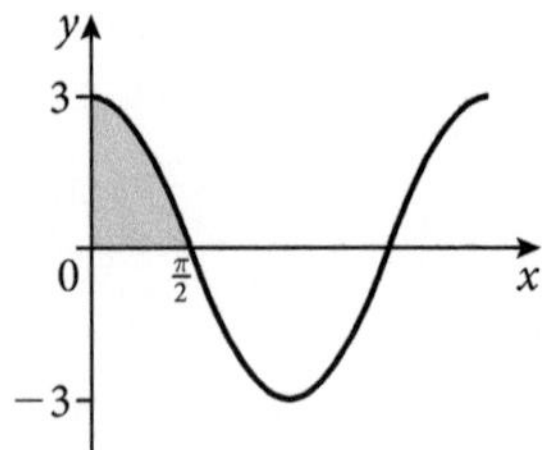

Figure 11.24 Graph for question 1

1. The graph in Figure 11.24 represents the function

$f: x \mapsto p\cos x,\ p \in \mathbb{N}$.

Find:

(a) the value of p

(b) the area of the shaded region.

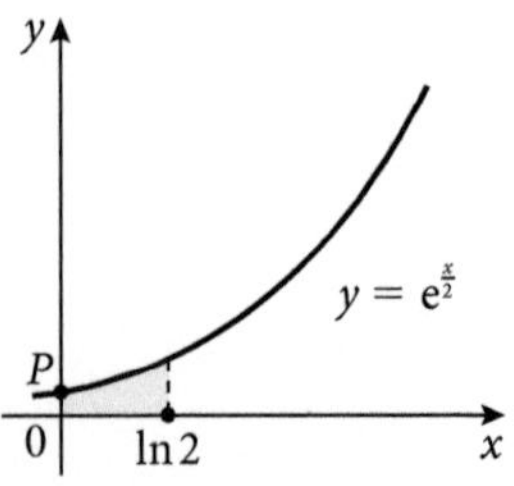

Figure 11.25 Diagram for question 2

2. The diagram in Figure 11.25 shows part of the graph of $y = e^{\frac{x}{2}}$.

(a) Find the coordinates of the point P, where the graph meets the y-axis.

(b) Find the exact value of the area of the shaded region between the graph and the x-axis, bounded by $x = 0$ and $x = \ln 2$.

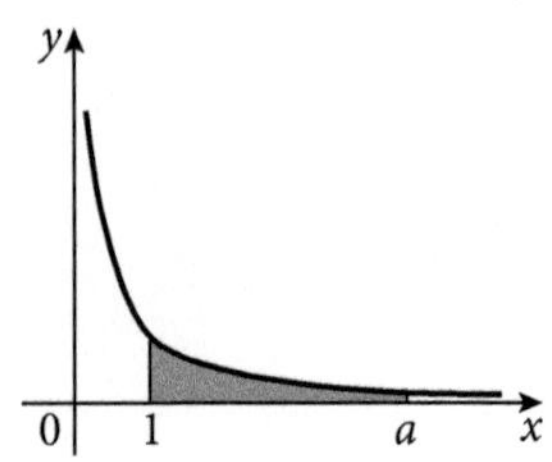

Figure 11.26 Diagram for question 3

3. The diagram in Figure 11.26 shows part of the graph of $y = \frac{1}{x}$.
The area of the shaded region is 2 units.
Find the exact value of a.

4. (a) Find the equation of the tangent to the curve $y = \ln x$ at the point $(\mathrm{e}, 1)$, and verify that the origin is on this line.

(b) Show that $(x \ln x - x)' = \ln x$

(c) The graph in Figure11.27 shows the region enclosed by the curve $y = \ln x$, the tangent in part (a), and the line $y = 0$.

Use the result of part (b) to show that the area of this region is $\frac{1}{2}\mathrm{e} - 1$

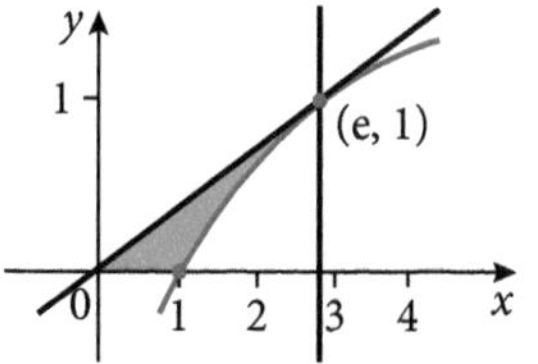

Figure 11.27 Graph for question 4

5. The main runway at Concordville airport is 2 km long. An aeroplane landing at Concordville touches down at point T and immediately starts to slow down. The point A is at the southern end of the runway. A marker is located at point P on the runway.

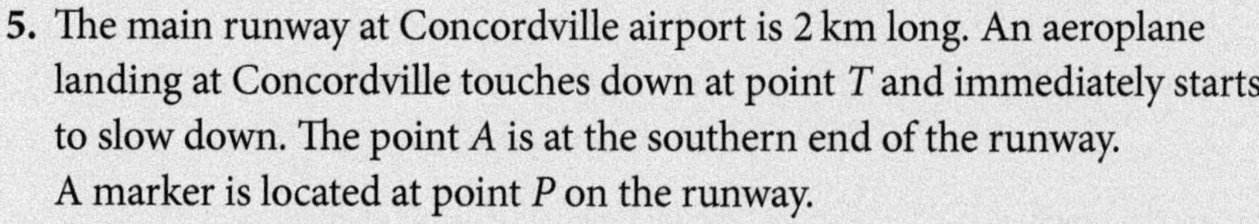

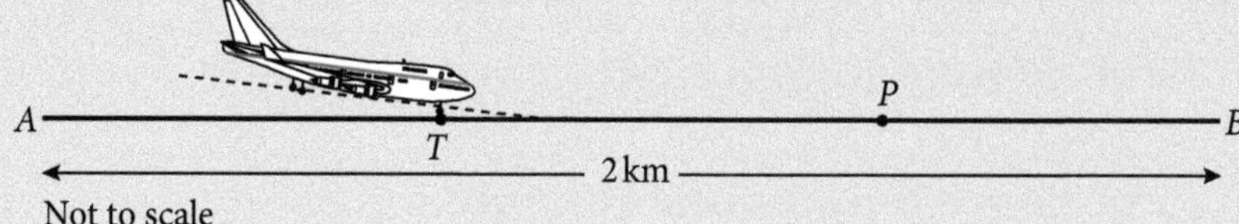

As the aeroplane slows down, its distance, s, from A, is given by

$$s = c + 100t - 4t^2$$

where t is the time in seconds after touchdown, and c metres is the distance of T from A.

(a) The aeroplane touches down 800 m from A, (i.e. $c = 800$).

(i) Find the distance travelled by the aeroplane in the first 5 seconds after touchdown.

(ii) Write down an expression for the velocity of the aeroplane at time t seconds after touchdown, and hence find the velocity after 5 seconds.

The aeroplane passes the marker at P with a velocity of $36\ \mathrm{m\,s^{-1}}$. Find:

(iii) how many seconds after touchdown it passes the marker

(iv) the distance from P to A.

(b) Show that if the aeroplane touches down before reaching point P, it can stop before reaching the northern end, B, of the runway.

6. (a) Sketch the graph of $y = \pi \sin x - x$, $-3 \leqslant x \leqslant 3$, on millimetre square paper, using a scale of 2 cm per unit on each axis.

Label and number both axes and indicate clearly the approximate positions of the x-intercepts and the local maximum and minimum points.

(b) Find the solution of the equation $\pi \sin x - x = 0$, $x > 0$.

(c) Find the indefinite integral $\int (\pi \sin x - x)\mathrm{d}x$ and hence, or otherwise, calculate the area of the region enclosed by the graph, the x-axis, and the line $x = 1$.

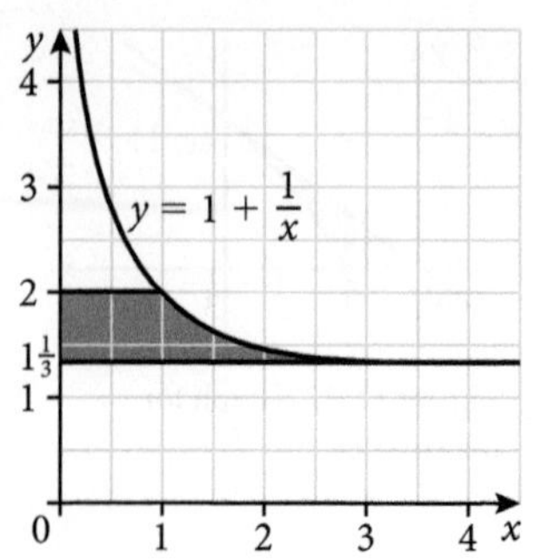

Figure 11.28 Diagram for question 7

7. Figure 11.28 shows the graph of the function $y = 1 + \frac{1}{x}, 0 < x \leqslant 4$.

   Find the exact value of the area of the shaded region.

8. Note that radians are used throughout this question.

   (a) (i) Sketch the graph of $y = x^2 \cos x$, for $0 \leqslant x \leqslant 2$, making clear the approximate positions of the positive intercept, the maximum point, and the endpoints.

   (ii) Write down the approximate coordinates of the positive x-intercept, the maximum point and the endpoints.

   (b) Find the exact value of the positive x-intercept for $0 \leqslant x \leqslant 2$.

   Let R be the region in the first quadrant enclosed by the graph and the x-axis.

   (c) (i) Shade R on your sketch.

   (ii) Write down an integral which represents the area of R.

   (d) Evaluate the integral in part (c) (ii), either by using a graphic display calculator or by using:

$$\frac{d}{dx}(x^2 \sin x + 2x \cos x - 2 \sin x) = x^2 \cos x.$$

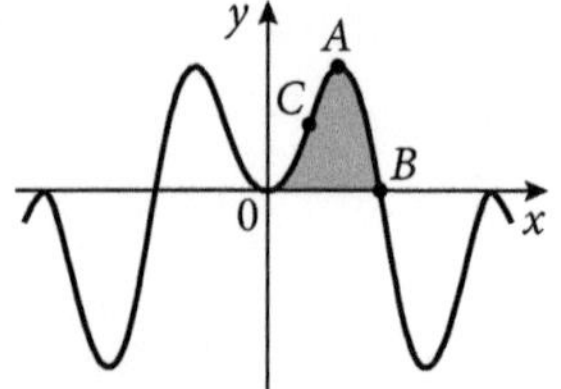

Figure 11.29 Diagram for question 9

9. Note that radians are used throughout this question.
   The function f is given by $f(x) = (\sin x)^2 \cos x$

   Figure 11.29 shows part of the graph of $y = f(x)$.

   The point A is a maximum point, the point B lies on the x-axis, and the point C is a point of inflection.

   (a) Give the period of f.

   (b) From consideration of the graph of $y = f(x)$, find the range of f, accurate to 1 significant figure.

   (c) (i) Find $f'(x)$

   (ii) Hence, show that at the point A, $\cos x = \sqrt{\frac{1}{3}}$

   (iii) Find the exact maximum value.

   (d) Find the exact value of the x-coordinate at the point B.

   (e) (i) Find $\int f(x)dx$

   (ii) Find the area of the shaded region in the diagram.

   (f) Given that $f''(x) = 9(\cos x)^3 - 7 \cos x$, find the x-coordinate at the point C.

10. Note that radians are used throughout this question.

(a) Draw the graph of $y = \pi + x\cos x$, $0 \leqslant x \leqslant 5$, on millimetre square graph paper, using a scale of 2 cm per unit. Make clear:

(i) the integer values of x and y on each axis

(ii) the approximate positions of the x-intercepts and the turning points.

(b) Without the use of a calculator, show that π is a solution of the equation $\pi + x\cos x = 0$

(c) Find another solution of the equation $\pi + x\cos x = 0$ for $0 \leqslant x \leqslant 5$, giving your answer to 6 significant figures.

(d) Let R be the region enclosed by the graph and the axes for $0 \leqslant x \leqslant \pi$. Shade R on your diagram, and write down an integral which represents the area of R.

(e) Evaluate the integral in part (d) to an accuracy of 6 significant figures. If considered necessary, you can make use of the result

$$\frac{d}{dx}(x\sin x + \cos x) = x\cos x$$

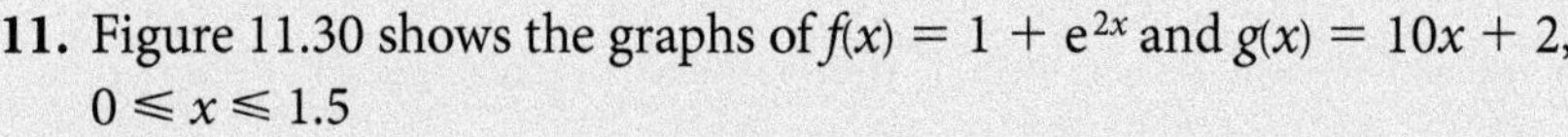

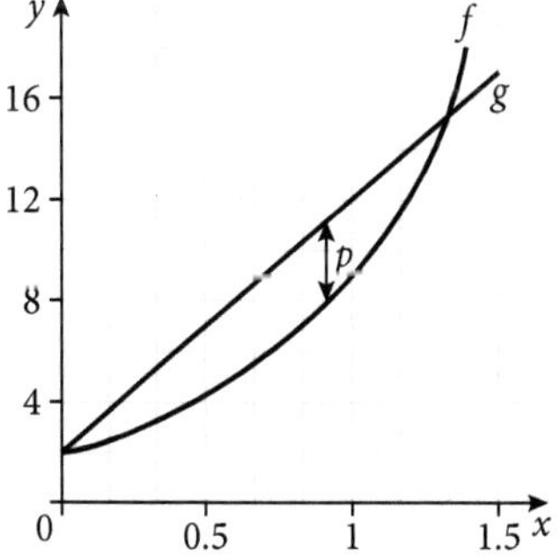

Figure 11.30 Diagram for question 11

11. Figure 11.30 shows the graphs of $f(x) = 1 + e^{2x}$ and $g(x) = 10x + 2$, $0 \leqslant x \leqslant 1.5$

(a) (i) Write down an expression for the vertical distance p between the graphs of f and g.

(ii) Given that p has a maximum value for $0 \leqslant x \leqslant 1.5$, find the value of x at which this occurs.

The graph of $y = f(x)$ only is shown Figure 11.31.
When $x = a$, $y = 5$.

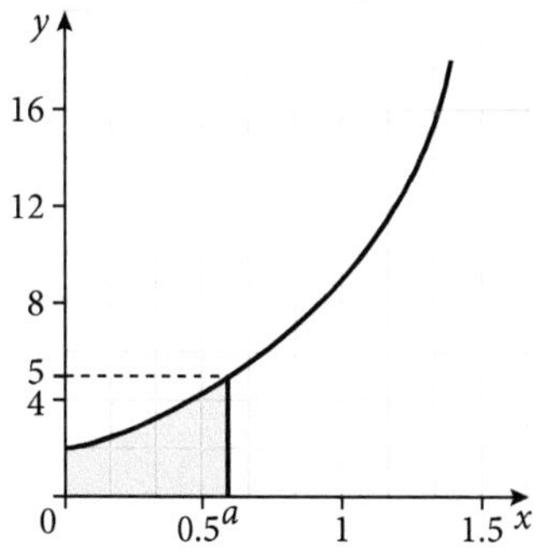

Figure 11.31 Second diagram for question 11

(b) (i) Find $f^{-1}(x)$

(ii) Hence, show that $a = \ln 2$

(c) The region shaded in Figure 11.31 is rotated through 360° about the x-axis. Write down an expression for the volume obtained.

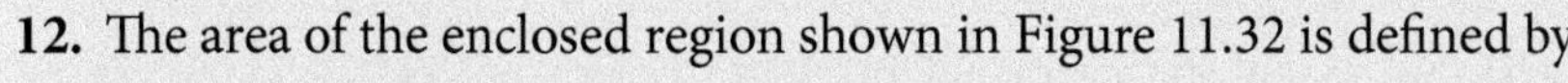

12. The area of the enclosed region shown in Figure 11.32 is defined by

$$y \geqslant x^2 + 2,\ y \leqslant ax + 2, \text{ where } a > 0$$

Find, in terms of a, the area of the region.

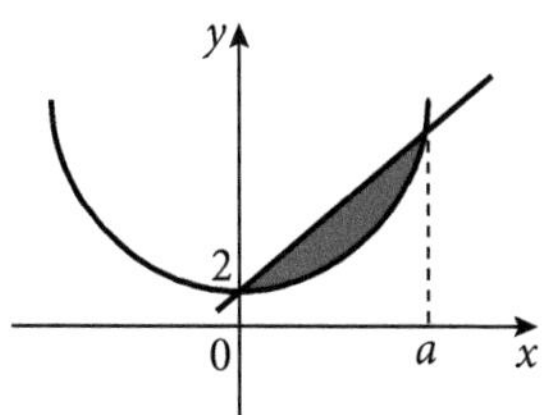

Figure 11.32 Diagram for question 12

13. Using the substitution $u = \frac{1}{2}x + 1$, or otherwise, find the integral

$$\int x\sqrt{\frac{1}{2}x + 1}\,dx$$

14. A particle moves along a straight line. When it is a distance s from a fixed point, where $s > 1$, the velocity v is given by $v = \frac{3s + 2}{2s - 1}$

Find the acceleration when $s = 2$.

15. The area between the graph of $y = xe^x$ and the x-axis from $x = 0$ to $x = k$ $(k > 0)$ is equal to 1. Find the exact value of k.

16. Find the real number $k > 1$ for which $\int_1^k \left(1 + \frac{1}{x^2}\right)dx = \frac{3}{2}$

17. The acceleration, $a(t)$ m s^{-2}, of a fast train during the first 80 seconds of motion is given by

$$a(t) = -\frac{1}{20}t + 2$$

where t is the time in seconds. If the train starts from rest at $t = 0$, find the distance travelled by the train in the first minute.

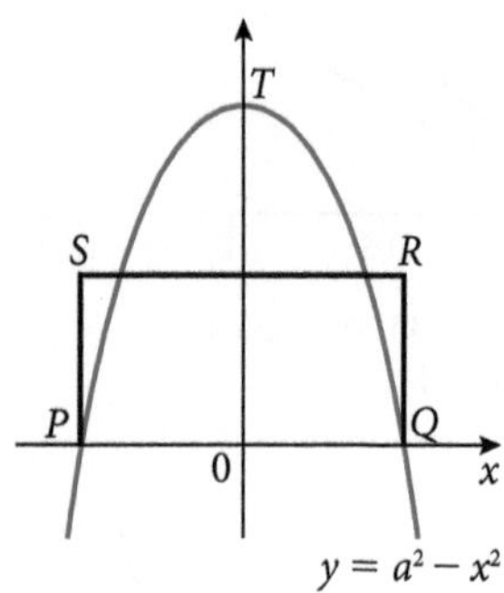

Figure 11.33 Diagram for question 18

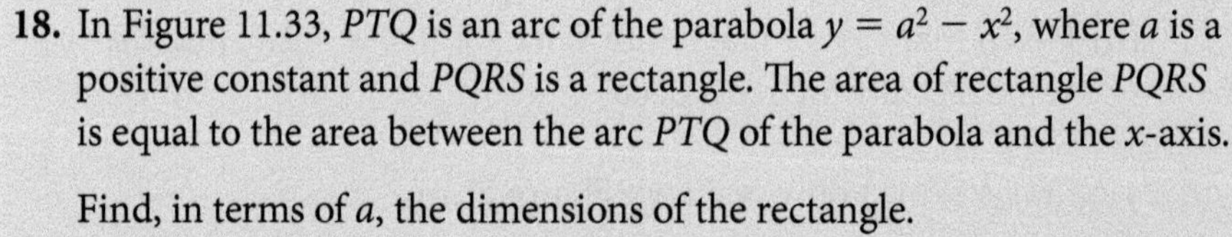

18. In Figure 11.33, PTQ is an arc of the parabola $y = a^2 - x^2$, where a is a positive constant and $PQRS$ is a rectangle. The area of rectangle $PQRS$ is equal to the area between the arc PTQ of the parabola and the x-axis.

Find, in terms of a, the dimensions of the rectangle.

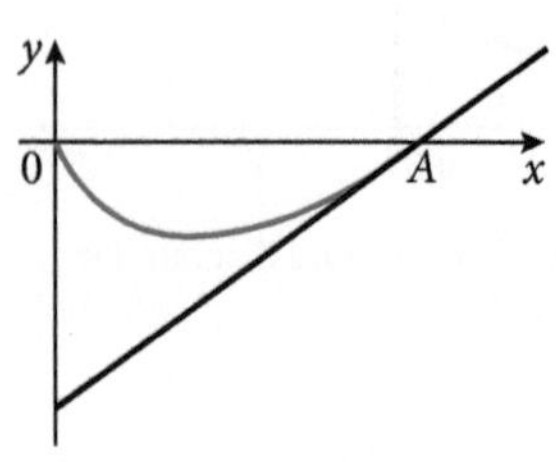

Figure 11.34 Diagram for question 19

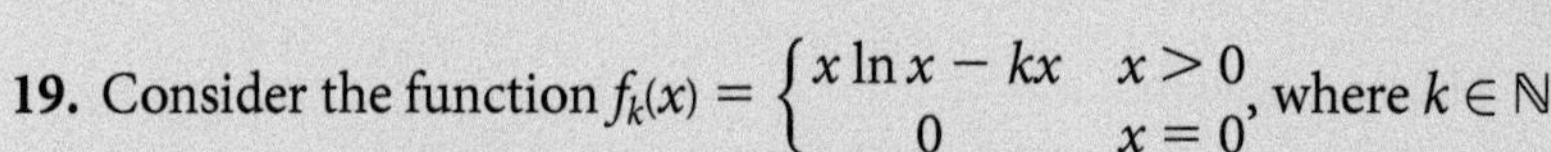

19. Consider the function $f_k(x) = \begin{cases} x\ln x - kx & x > 0 \\ 0 & x = 0 \end{cases}$, where $k \in \mathbb{N}$

(a) Find the derivative of $f_k(x)$, $x > 0$.

(b) Find the interval over which $f(x)$ is increasing.

The graph of the function $f_k(x)$ is shown in Figure 11.34.

(c) **(i)** Show that the stationary point of $f_k(x)$ is at $x = e^{k-1}$.

(ii) One x-intercept is at $(0, 0)$. Find the coordinates of the other x-intercept.

(d) Find the area enclosed by the curve and the x-axis.

(e) Find the equation of the tangent to the curve at A.

(f) Show that the area of the triangular region created by the tangent and the coordinate axes is twice the area enclosed by the curve and the x-axis.

(g) Show that the x-intercepts of $f_k(x)$ for consecutive values of k form a geometric sequence.

20. Consider the graphs of the functions $f(x) = a - |x - a|$ and $g(x) = |x - a|$, where $a > 0$. Find the value of a if the two graphs enclose an area of 12.5 square units.

21. The equation of motion of a particle with mass m subjected to a force kx can be written as $kx = mv\frac{dv}{dx}$, where x is the displacement and v is the velocity. When $x = 0$, $v = v_0$. Find v, in terms of v_0, k, and m, when $x = 2$.

22. (a) Sketch and label the graphs of $f(x) = e^{-x^2}$ and $g(x) = e^{-x^2} - 1$ for $0 \leqslant x \leqslant 1$, and shade the region A that is bounded by the graphs and the y-axis.

(b) Let the x-coordinate of the point of intersection of the curves $y = f(x)$ and $y = g(x)$ be p. Without finding the value of p, show that $\frac{p}{2} <$ area of region $A < p$.

(c) Find the value of p correct to 4 decimal places.

(d) Express the area of region A as a definite integral and calculate its value.

23. Let $f(x) = x\cos 3x$

(a) Use integration by parts to show that

$\int f(x)\,dx = \frac{1}{3}x\sin 3x + \frac{1}{9}\cos 3x + c$

(b) Use your answer to part (a) to calculate the exact area enclosed by $f(x)$ and the x-axis in each of the following cases. Give your answers in terms of π.

(i) $\frac{\pi}{6} \leqslant x \leqslant \frac{3\pi}{6}$ **(ii)** $\frac{3\pi}{6} \leqslant x \leqslant \frac{5\pi}{6}$ **(iii)** $\frac{5\pi}{6} \leqslant x \leqslant \frac{7\pi}{6}$

(c) Given that the above areas are the first three terms of an arithmetic sequence, find an expression for the total area enclosed by $f(x)$ and the x-axis for $\frac{\pi}{6} \leqslant x \leqslant \frac{(2n+1)\pi}{6}$, where $n \in \mathbb{Z}$. Give your answers in terms of n and π.

24. A particle is moving along a straight line so that t seconds after passing through a fixed point O on the line, its velocity $v(t)$ m s^{-1} is given by $v(t) = t\sin\left(\frac{\pi}{3}t\right)$.

(a) Find the values of t for which $v(t) = 0$, given that $0 \leqslant t \leqslant 6$.

(b) (i) Write down a mathematical expression for the total distance travelled by the particle in the first six seconds after passing through O.

(ii) Find this distance.

25. A particle is projected along a straight-line path. After t seconds, its velocity v in metres per second is given by $v = \dfrac{1}{2 + t^2}$

(a) Find the distance travelled in the first second.

(b) Find an expression for the acceleration at time t.

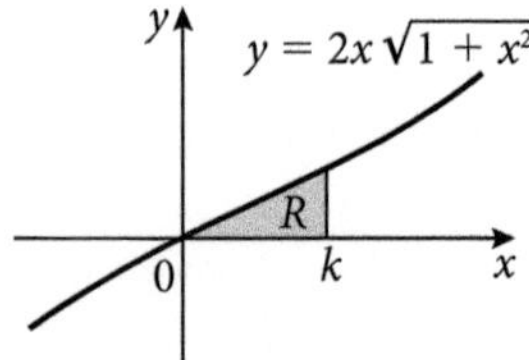

Figure 11.35 Diagram for question 26

26. Figure 11.35 shows the shaded region R enclosed by the graph of $y = 2x\sqrt{1 + x^2}$, the x-axis, and the vertical line $x = k$.

(a) Find $\dfrac{dy}{dx}$

(b) Using the substitution $u = 1 + x^2$ or otherwise, show that
$\int 2x\sqrt{1 + x^2}\,dx = \frac{2}{3}(1 + x^2)^{\frac{3}{2}} + c$

(c) Given that the area of R equals 1, find the value of k.

27. A particle moves in a straight line with velocity v m s^{-1}, at time t seconds, given by $v(t) = 6t^2 - 6t$, $t \geqslant 0$.

Calculate the total distance travelled by the particle in the first two seconds of motion.

28. A particle moves in a straight line. Its velocity v m s^{-1} after t seconds is given by $v = e^{-\sqrt{t}}\sin t$.
Find the total distance travelled in the time interval $[0, 2\pi]$.

29. The temperature T °C of an object in a room after t minutes satisfies the differential equation $\dfrac{dT}{dt} = k(T - 22)$, where k is a constant.

(a) Show that $T = Ae^{kt} + 22$, where A is a constant.

(b) When $t = 0$, $T = 100$, and when $t = 15$, $T = 70$.

(i) Use this information to find the values of A and k.

(ii) Hence, find the value of t when $T = 40$.

30. Consider the function $f(x) = \dfrac{1}{x^2 + 5x + 4}$

(a) Sketch the graph of the function, indicating the equations of the asymptotes, intercepts, and extreme values.

(b) Find $\int_0^1 f(x)\,dx$ and express it in the form $\ln k$.

(c) Sketch the graph of $f(|x|)$ and hence determine the area of the region between this graph, the x-axis, and the lines $x = -1$, and $x = 1$.

31. Use the substitution $u = x + 2$ to find $\int \dfrac{x^3\,dx}{(x + 2)^2}$

32. (a) On the same axes, sketch the graphs of the functions, $f(x)$ and $g(x)$, where

$$f(x) = 4 - (1 - x)^2, \text{ for } -2 \leqslant x \leqslant 4$$

$$g(x) = \ln(x + 3) - 2, \text{ for } -3 \leqslant x \leqslant 5$$

(b) (i) Write down the equation of any vertical asymptotes.

(ii) State the x-intercept and y-intercept of $g(x)$.

(c) Find the values of x for which $f(x) = g(x)$.

(d) Let A be the region where $f(x) \geqslant g(x)$ and $x \geqslant 0$.

(i) On your graph, shade the region A.

(ii) Write down an integral that represents the area of A.

(iii) Evaluate this integral.

(e) In the region A, find the maximum vertical distance between $f(x)$ and $g(x)$.

33. Consider the functions $f(x) = \dfrac{3}{x + 2}$ and $g(x) = x^2$ where $x \geqslant 0$.

(a) Sketch the graphs of f and g on the same set of axes.

(b) Find the point of intersection of the two graphs.

(c) Find the equation of the tangent to $g(x)$ at the point of intersection as well as the normal to $f(x)$ at the same point.

(d) Find the area of the region bounded by the two lines in (c) and the x-axis.

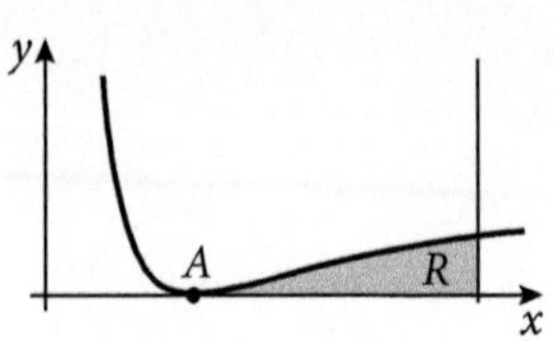

Figure 11.36 Diagram for question 34

34. Figure 11.36 shows part of the graph of $y = \frac{(\ln x)^2}{x}$, $x > 0$

(a) Find the extreme points of the curve.

(b) The region R is enclosed by the curve, the x-axis, and the line $x = \mathrm{e}$. Find the area of the region R.

12 Probability distributions

12 Probability distributions

Learning objectives

By the end of this chapter, you should be familiar with...

- discrete random variables and their probability distributions
- the effect of linear transformations of X on the values of its parameters
- the normal distribution: properties, normal probability calculations, and inverse normal calculations
- standardising normal variables (z-values)
- inverse normal calculations where mean and standard deviation are unknown
- the binomial distribution including its mean and variance.

Investing in securities, calculating premiums for insurance policies, or overbooking policies used in the airline industry are only a few of the many applications of probability and statistics. Actuaries, for example, calculate the expected claims that an insurance company will incur and decide on how high the premiums should be. These applications depend mainly on what we call probability distributions. A probability distribution describes the behaviour of a population in that it lists the distribution of possible outcomes to an event, along with the probability of each potential outcome. This can be done by a table of values with their corresponding probabilities, or by using a mathematical model.

In this chapter, we will get an understanding of the basic ideas of distributions and will study two specific ones: the binomial and normal distributions.

12.1 Random variables

A **random variable** is a variable that takes on numerical values determined by the outcome of a random experiment.

Random variables are customarily denoted by upper case letters, such as X and Y. Lower case letters are used to represent particular values of the random variable. That is, if X represents the numbers resulting from the throw of a dice, then $x = 2$, represents the case when the outcome is 2.

In Chapter 7, variables were defined as characteristics that change or vary over time and/or for different objects under consideration. A numerically valued variable x will vary or change depending on the outcome of the experiment we are performing. For example, suppose we are counting the number of smartphones owned by families in a certain city. The variable of interest, X, can take any of the values 0, 1, 2, 3, and so on, depending on the random outcome of the experiment. For this reason, we call the variable X a random variable.

When a probability experiment is performed, we are often not interested in all the details of the outcomes, but only in the value of some numerical quantity determined by the result. For instance, in tossing two dice (used in plenty of games), we care often only about their sum and not the values on the individual dice. A sample space for which the points are equally likely is given in Table 12.1. It consists of 36 ordered pairs (a, b) where a is the number on the first dice, and b is the number on the second dice. For each sample point, we can let the random variable X stand for the sum of the numbers. The resulting values of x are also presented in the table.

$(1, 1); x = 2$	$(2, 1); x = 3$	$(3, 1); x = 4$	$(4, 1); x = 5$	$(5, 1); x = 6$	$(6, 1); x = 7$
$(1, 2); x = 3$	$(2, 2); x = 4$	$(3, 2); x = 5$	$(4, 2); x = 6$	$(5, 2); x = 7$	$(6, 2); x = 8$
$(1, 3); x = 4$	$(2, 3); x = 5$	$(3, 3); x = 6$	$(4, 3); x = 7$	$(5, 3); x = 8$	$(6, 3); x = 9$
$(1, 4); x = 5$	$(2, 4); x = 6$	$(3, 4); x = 7$	$(4, 4); x = 8$	$(5, 4); x = 9$	$(6, 4); x = 10$
$(1, 5); x = 6$	$(2, 5); x = 7$	$(3, 5); x = 8$	$(4, 5); x = 9$	$(5, 5); x = 10$	$(6, 5); x = 11$
$(1, 6); x = 7$	$(2, 6); x = 8$	$(3, 6); x = 9$	$(4, 6); x = 10$	$(5, 6); x = 11$	$(6, 6); x = 12$

Table 12.1 Sample space and the values of the random variable X in the two-dice experiment.

Notice that events can be more accurately and concisely defined in terms of the random variable X; for example, the event of getting a sum greater than or equal to 5 but less than 9 can be replaced by $5 \leqslant x < 9$.

We can think of many examples of random variables:

- $X =$ the number of calls received by a household on a Friday night
- $X =$ the number of beds available at hotels in a large city
- $X =$ the number of customers a sales person contacts on a working day
- $X =$ the length of a metal bar produced by a certain machine
- $X =$ the weight of newborn babies in a large hospital.

As you have seen in Chapter 7, these variables are classified into discrete or continuous, according to the values that X can assume. In the examples above, the first three are discrete and the last two are continuous. A random variable is discrete if its set of possible values is isolated points on the number line; that is, there is a countable number of possible values for the variable. The variable is continuous if its set of possible values is an entire interval on the number line; that is, it can take any value in an interval. Consider the number of times you flip a coin until the head side appears. The possible values are $x = 1, 2, 3, \ldots$. This is a discrete variable, even though the number of times may be infinite. On the other hand, consider the time it takes a student at your school to eat lunch. This can be anywhere between zero and the length of the lunch period at your school.

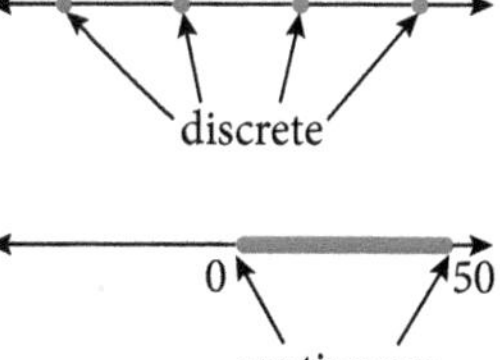

Figure 12.1 Discrete and continuous varables

Example 12.1

State whether each of the following is a discrete or a continuous random variable.

(a) The number of hairs on a Scottish terrier

(b) The height of a building

(c) The amount of fat in a steak

(d) A high school student's grade on a history test

(e) The number of fish in the Atlantic Ocean

(f) The temperature of an electric kettle

Solution

(a) Even though the number of hairs is incredibly large, it is countable. So, it is a discrete random variable.

(b) This can be any real number. Even when we say this building is 15 m high, the number could be 15.1 m, 15.02 m, and so on. So, it is continuous.

(c) This is continuous as the amount of fat could be zero up to the maximum amount of fat that can be held in one piece.

(d) Grades are discrete. No matter how detailed a score the teacher gives, the grades are isolated points on a scale.

(e) This is unfathomably large, but still theoretically countable, hence discrete.

(f) This is continuous, as the temperature can take any value from room temperature up to 100 degrees Celsius.

Discrete probability distribution

In Chapter 7, we learned how to work with the frequency distribution and relative or percentage frequency distribution for a set of numerical measurements on a variable X. The distribution gave the following information about X:

- what value of X occurred
- how often each value occurred

We also learned how to use the mean and standard deviation to measure the centre and variability of the data set.

Here is an example of the frequency distribution of 25 families in Lower Austria that were polled in a marketing survey to list the number of litres of milk consumed during a particular week. Table 12.2 lists the number of litres consumed, to the nearest litre, along with the relative frequency that number is observed. One of the interpretations of probability is that it is understood to be the long-term relative frequency of the event.

The **probability distribution** for a discrete random variable is a table, graph, or formula that gives the possible values of x, and the probability $P(x) = P(X = x)$ associated with each value of x. This is also called the **probability mass function (PMF)** and in many sources it is called the **probability distribution function (PDF)**.

Number of litres to the nearest litre	Relative frequency
0	0.08
1	0.20
2	0.36
3	0.20
4	0.12
5	0.04

Table 12.2 Number of litres of milk consumed by families during a particular week

A table like this, where we replace the relative frequency with probability, is called a probability distribution of the random variable.

In other words, for every possible value x of the random variable X, the probability mass function specifies the probability of observing that value when the experiment is performed.

Note: we write $P(X = x)$ as $P(x)$ for convenience.

If X is the number of litres of milk (to the nearest litre) consumed by a family, the **probability distribution** of X is as follows:

x	0	1	2	3	4	5
$P(x)$	0.08	0.20	0.36	0.20	0.12	0.04

Table 12.3 Probability distribution of milk consumption

The other way of representing the probability distribution is with a histogram, as shown in Figure 12.2. Every bar corresponds to the probability of the associated value of x. The values of x naturally represent mutually exclusive events. Summing $P(x)$ over all values of X is equivalent to adding the probabilities of all simple events in the sample space, and hence the total is 1.

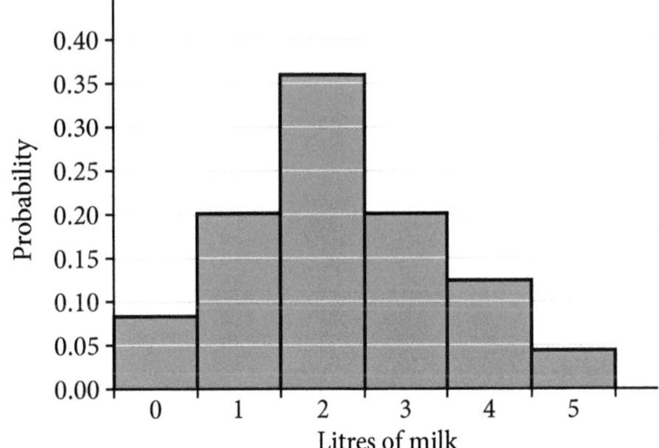

Figure 12.2 Milk consumption probability distribution

This result can be generalised for all probability distributions.

Probability distribution functions of discrete random variables

Let X be a discrete random variable with probability distribution function $P(x)$. Then

- $0 \leq P(x) \leq 1$ for any value x
- The individual probabilities sum to 1; that is $\sum_x P(x) = 1$, where the notation indicates summation over all possible values of x.

For some value x of the random variable X, we often wish to compute the probability that the observed value of X is at most x. This gives rise to the **Cumulative Distribution Function (CDF)**.

The **cumulative distribution function (CDF)** of a random variable X expresses the probability that X does not exceed the value x as a function of x. That is

$F(x) = P(X \leq x) = \sum_{y: y \leq x} p(y)$

It is also known as the cumulative probability function $F(x)$

The notation in CDF indicates that summation is over all possible values of y that are less than or equal to x. The choice of the variable name to be y is arbitrary – we can use any letter.

For example, in the milk consumption case, the CDF will look like Table 12.4.

x	$F(x)$
0	0.08
1	0.28
2	0.64
3	0.84
4	0.96
5	1.00

Table 12.4 CDF for milk consumption

So, $F(3) = 0.84$ is the probability of a family consuming up to 3 litres of milk. This result can be achieved by adding the probabilities corresponding to $x = 0$, 1, 2, and 3.

In many cases, we use the cumulative distribution to find individual probabilities,

$$P(X = x) = P(X \leq x) - P(X < x)$$

For example, to find the probability that $x = 3$, we can use Table 12.4

$$P(x = 3) = P(x \leq 3) - P(x < 3) = 0.84 - 0.64 = 0.20$$

This property is of great value when studying the binomial and Poisson distributions.

Example 12.2

Radon is a major cause of lung cancer. It is a radioactive gas produced by the natural decay of radium in rocks that contain small amounts of uranium. Studies in areas with high levels of radon revealed that one third of houses in these areas have dangerous levels of radon. Suppose that two houses are randomly selected and we define the random variable X to be the number of houses with dangerous levels. Find the probability distribution of X by a table, a graph, and a formula.

Solution

Since two houses are selected, the possible values of X are 0, 1, or 2. The assumption here is that we are choosing the houses randomly and independently of each other.

$$p(x = 2) = p(2) = p(\text{1st house with dangerous levels and 2nd house with dangerous levels})$$
$$= p(\text{1st house with dangerous levels}) \times p(\text{2nd house with dangerous levels})$$
$$= \frac{1}{3} \times \frac{1}{3} = \frac{1}{9}$$

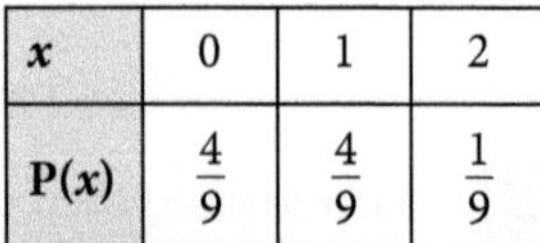

x	0	1	2
P(x)	$\frac{4}{9}$	$\frac{4}{9}$	$\frac{1}{9}$

Table 12.5 Probability distribution table for Example 12.2

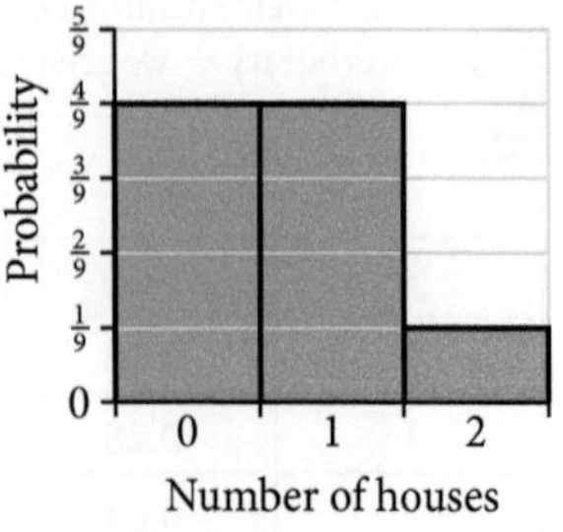

Figure 12.3 Probability distribution graph for Example 12.2

$$p(x = 0) = p(0) = p(\text{1st house without dangerous levels and 2nd house without dangerous levels})$$
$$= p(\text{1st house without dangerous levels}) \times p(\text{2nd house without dangerous levels})$$
$$= \frac{2}{3} \times \frac{2}{3} = \frac{4}{9}$$

$$p(x = 1) = 1 - [p(0) + p(2)] = 1 - \left[\frac{4}{9} + \frac{1}{9}\right] = \frac{4}{9}$$

Any type of graph can be used to give the probability distribution as long as it shows the possible values of X and the corresponding probabilities. The probability here is graphically displayed as the height of a rectangle. Moreover, the rectangle corresponding to each value of X has an area equal to the probability P(x). The histogram is the preferred tool because of its connection to the continuous distributions discussed later in the chapter.

The probability distribution of x can also be given by

$$P(x) = {}^2C_x \cdot \left(\frac{1}{3}\right)^x \cdot \left(\frac{2}{3}\right)^{2-x}$$

where $\binom{2}{x} = {}^2C_x$ represents the binomial coefficient (see Chapter 3).

Note that when x is replaced by 0, 1, or 2, we obtain the results we are looking for

$$p(0) = {}^2C_0 \cdot \left(\frac{1}{3}\right)^0 \cdot \left(\frac{2}{3}\right)^{2-0} = 1 \cdot 1 \cdot \frac{4}{9} = \frac{4}{9}$$

$$p(1) = {}^2C_1 \cdot \left(\frac{1}{3}\right)^1 \cdot \left(\frac{2}{3}\right)^{2-1} = 2 \cdot \frac{1}{3} \cdot \frac{2}{3} = \frac{4}{9}$$

$$p(2) = {}^2C_2 \cdot \left(\frac{1}{3}\right)^2 \cdot \left(\frac{2}{3}\right)^{2-2} = 1 \cdot \frac{1}{9} \cdot 1 = \frac{1}{9}$$

Don't be concerned now with how we came up with this formula as we will discuss it later in the chapter. The only reason we are looking at it now is to illustrate the fact that a formula/rule can sometimes be used to give the probability distribution.

Example 12.3

Many universities have a policy of posting the grade distributions for their courses. Several of the universities have a grade-point average that codes the grades in the following manner: A = 4, B = 3, C = 2, D = 1, and F = 0. During the spring term at a certain large university, 13% of the students in an introductory Statistics course received grade A, 37% B, 45% C, 4% D, and 1% F. A student is chosen at random and the grade noted. The student's grade on the 4-point scale is a random variable X.

Here is the probability distribution of X:

x	0	1	2	3	4
P(x)	0.01	0.04	0.45	0.37	0.13

(a) Is this a probability distribution?

(b) What is the probability that a randomly chosen student receives a grade B or better?

Solution

(a) Yes, it is. Each probability is between 0 and 1, and the sum of all probabilities is 1.

(b) $p(x \geqslant 3) = p(x = 3) + p(x = 4) = 0.37 + 0.13 = 0.40$

Example 12.4

When people choose codes for their smartphones, the first digits follow a probability distribution similar to the one below.

First digit	0	1	2	3	4	5	6	7	8	9
Probability	0.009	0.300	0.174	0.122	0.096	0.078	0.067	0.058	0.051	0.045

Here, X represents the first digit chosen.

What is the probability that you pick a first digit that is more than 5?

Show a probability histogram for the distribution.

Solution

$P(x > 5) = p(x = 6) + p(x = 7) + p(x = 8) + p(x = 9) = 0.221$

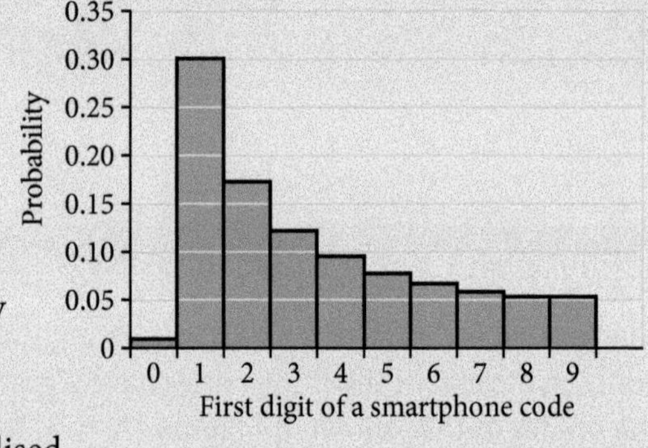

Note that the height of each bar shows the probability of the outcome at its base. The heights add up to 1, of course. The bars have the same width, namely 1. So the areas also display the probability assignments of the outcomes. Think of such histograms (probability histograms) as idealised pictures of the results of very many repeated trials.

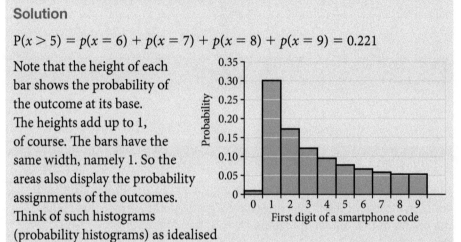

Expected values

The probability distribution for a random variable looks very similar to the relative frequency distribution discussed in Chapter 7. The difference is that the relative frequency distribution describes a sample of measurements, whereas the probability distribution is constructed as a model for the entire population. Just as the mean and standard deviation give us measures for the centre and spread of the sample data, we can calculate similar measures to describe the centre and spread of the population.

The population mean, which measures the average value of X in the population, is also called the **expected value** of the random variable X. It is the value that we would expect to observe on average if we repeated the experiment an infinite number of times. The formula we use to determine the expected value can be simply understood with an example.

Let's revisit the milk consumption example. Here is the table of probabilities:

x	0	1	2	3	4	5
$P(x)$	0.08	0.20	0.36	0.20	0.12	0.04

Table 12.6 Milk consumption probabilities

Suppose we choose a large number of families, say 100 000. Intuitively, using the relative frequency concept of probability, we would expect to observe 8000 families consuming no milk, 20 000 consuming 1 litre, and the rest consuming: 36 000, 20 000, 12 000, and 4000.

The average (mean) value of X as defined in Chapter 7 would then be equal to

$$\frac{\text{Sum of all measurements}}{n} = \frac{0 \cdot 8000 + 1 \cdot 20\,000 + 2 \cdot 36\,000 + 3 \cdot 20\,000 + 4 \cdot 12\,000 + 5 \cdot 4000}{100\,000}$$

$$= 0 \cdot 0.08 + 1 \cdot 0.20 + 2 \cdot 0.36 + 3 \cdot 0.20 + 4 \cdot 0.12 + 5 \cdot 0.04$$

$$= 0 \cdot p(0) + 1 \cdot p(1) + 2 \cdot p(2) + 3 \cdot p(3) + 4 \cdot p(4) + 5 \cdot p(5) = 2.2$$

That is, we expect to see families, on average, consuming 2.2 litres of milk. This does not mean that we know what a family will consume, but we can say what we expect to happen.

Let X be a discrete random variable with probability distribution $P(x)$. The mean or **expected value** of X is given by $\mu = E(X) = \sum x \cdot P(x)$

Insurance companies make extensive use of expected value calculations. Here is a simplified example.

An insurance company offers a policy that pays you €10,000 when your car is damaged beyond repair or €5000 for major damages (50%). They charge you €50 per year for this service. Can they make a profit?

Suppose in any year that 1 out of every 1000 cars is damaged beyond repair, and that another 2 out of 1000 will have serious damage. Then we can display the probability model for this policy in a table.

Type of accident	Amount paid x	Probability $P(X = x)$
Total damage	10 000	$\frac{1}{1000}$
Major damage	5000	$\frac{2}{1000}$
Minor or no damage	0	$\frac{997}{1000}$

Table 12.7 Probability table for car insurance policy

The expected amount the insurance company pays is given by

$$\mu = E(X) = \sum xP(x) = €10{,}000\left(\frac{1}{1000}\right) + €5000\left(\frac{2}{1000}\right) + €0\left(\frac{997}{1000}\right)$$

$$= €20$$

This means that the insurance company expects to pay, on average, an amount of €20 per insured car. Since it is charging people €50 for the policy, the company expects to make a profit of €30 per car. Thinking about the problem from a different perspective, suppose they insure 1000 cars. Then the company would expect to pay €10,000 for 1 car, and €5000 to each of two cars with major damage. This is a total of €20,000 for all cars, or an average of €20 per car.

Of course, this expected value is not what actually happens to any particular policy. No individual policy actually costs the insurance company €20. We are dealing with random events, so a few car owners may require a payment of €10,000 or €5000, while most of the others receive nothing. Because of the need to anticipate such variability, the insurance company needs to know a measure of this variability; this is the **standard deviation**.

Variance and standard deviation

In Chapter 7, we calculated the variance by computing the deviation from the mean, $x - \mu$, and then squaring it. We do that with random variables as well.

Let X be a discrete random variable with probability distribution $P(x)$ and mean μ. The **variance of X**, $Var(X)$, given by

$\sigma^2 = E((X - \mu)^2)$
$= \sum(x - \mu)^2 \cdot P(x)$

This is sometimes called $Var(X)$, or $V(X)$.

The standard deviation σ of a random variable x is equal to the positive square root of its variance.

We can use similar arguments to justify the formulas for the population variance σ^2 and consequently the population standard deviation σ. These measures describe the spread of the values of the random variable around the centre. We similarly use the idea of the average or expected value of the squared deviations of the x-values from the mean μ or $E(X)$.

It can also be shown that there is another formula for the variance:

$$\sigma^2 = \sum(x - \mu)^2 \cdot P(x) = \sum x^2 \cdot P(x) - \mu^2 = \sum x^2 \cdot P(x) - [E(X)]^2$$
$$= \sum E(X^2) - [E(X)]^2 = \sum x^2 \cdot P(x) - [\sum xP(x)]^2$$

Let's go back to the milk consumption example. We calculated the expected mean value to be 2.2 litres. To calculate the variance, we can tabulate our work to make the manual calculation simple.

x	$P(x)$	Deviation $(x - \mu)$	Squared deviation $(x - \mu)^2$	$(x - \mu)^2 \cdot P(x)$
0	0.08	−2.2	4.84	0.3872
1	0.20	−1.2	1.44	0.2880
⋮	⋮	⋮	⋮	⋮
5	0.04	2.8	7.84	0.3136
		Total	$\sum(x - \mu)^2 \cdot P(x)$	**1.52**

Table 12.8 Calculating variance for milk consumption

So, the variance of the milk consumption is 1.52 litres^2, or the standard deviation is 1.233 litres.

GDC notes

You can do these calculations using your GDC. The method will depend on which GDC you are using; some may require that you store your data in lists and perform the calculations as described by the formulas above, and some may give you the results after you enter your data in lists, making sure that the probability is given as a frequency. For discrete random variable calculations, take the σx values and not the sx values. Figure 12.4 shows a sample of a GDC output.

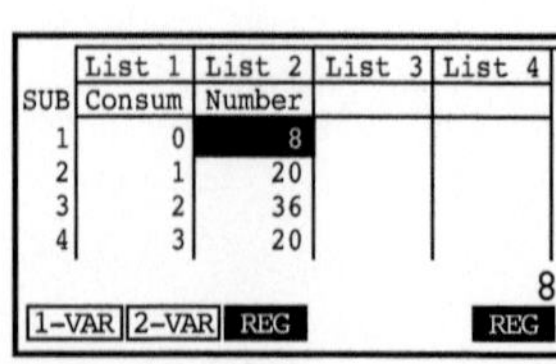

List 1 List 2 List 3 List 4
SUB Consum Number
1 0 8
2 1 20
3 2 36
4 3 20
8
1-VAR 2-VAR REG
REG


```
1-Variable
x̄    =2.2
Σx   =220
Σx2  =636
σx   =1.2328828
sx   =1.23909383
n    =100
```

Figure 12.4 GDC output

You can also do the calculation using a spreadsheet.

x	$P(x)$	$xP(x)$	$x - \mu$	$(x - \mu)^2$	$(x - \mu)^2P(x)$
0	0.08	0	−2.2	4.84	0.3872
1	0.2	0.2	−1.2	1.44	0.288
⋮	⋮	⋮	⋮	⋮	⋮
Totals	1	2.2			1.52

Figure 12.5 Spreadsheet calculations

Example 12.5

A computer store sells a particular type of laptop. The number of laptops sold each day is given in the table; x is the number of laptops sold each day. The store has only four laptops left in stock and would like to know how well they are prepared for all eventualities. Find the expected value of the demand as well as the standard deviation.

x	$P(X = x)$
0	0.08
1	0.40
2	0.24
3	0.15
4	0.08
5	0.05

Table 12.9 Table for Example 12.5

Solution

$$E(X) = \sum xP(x) = 0 \times 0.08 + 1 \times 0.40 + 2 \times 0.24 + 3 \times 0.15 + 4 \times 0.08 + 5 \times 0.05 = 1.90$$

$$\begin{aligned} Var(X) = \sigma^2 &= \sum (x - \mu)^2 P(x) \\ &= (0 - 1.9)^2 \cdot 0.08 + (1 - 1.9)^2 \cdot 0.40 + (2 - 1.9)^2 \cdot 0.24 \\ &\quad + (3 - 1.9)^2 \cdot 0.15 + (4 - 1.9)^2 \cdot 0.08 + (5 - 1.9)^2 \cdot 0.05 \\ &= 1.63 \end{aligned}$$

$\sigma = 1.28$

The graph of the probability distribution is given. As an approximation, we can use the empirical rule to see where most of the demand is expected to be. Recall that the empirical rule tells us that about 95% of the values would lie within 2 standard deviations of the mean.

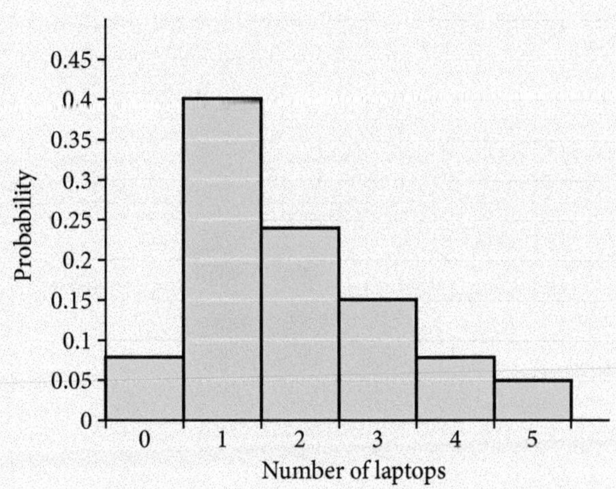

In this case $\mu \pm 2\sigma = 1.9 \pm 2 \times 1.28 \Rightarrow (-0.66, 4.46)$. This interval does not contain the 5 units sold in a day. We can say that it is unlikely that 5 or more customers of this shop will want to buy a laptop today.

Using a GDC, enter the demand in L1 and the probabilities in L2. We then find the sum of their product.

```
sum(L1*L2)
                1.9
```

For the variance, we follow the same procedure as described in Example 12.4

Notice here that we combined several steps in one.

```
L1*L2→L3
(0 .4 .48 .45 ....
(L1-1.9)²*L2→L5
(.2888 .324 .00...
sum(L5)
               1.63
```

Exercise 12.1

1. Classify each of the following as discrete or continuous random variables.
   - **(a)** The number of words spelled correctly by a student on a spelling test
   - **(b)** The volume of water flowing through the Niagara Falls per year
   - **(c)** The length of time a student is late to class
   - **(d)** The number of bacteria per ml of drinking water in Geneva
   - **(e)** The amount of carbon dioxide produced per litre of fuel
   - **(f)** The amount of a flu vaccine in a syringe
   - **(g)** The heart rate of a lab mouse
   - **(h)** The barometric pressure at the top of Mount Everest
   - **(i)** The distance travelled by a taxi driver per day
   - **(j)** The total score of football teams in national leagues
   - **(k)** The height of ocean tides on the shores of Portugal
   - **(l)** The tensile breaking strength (in newtons per square metre) of a 5-cm diameter steel cable
   - **(m)** The number of overdue books at a public library

2. A random variable Y has this probability distribution:

y	0	1	2	3	4	5
P(y)	0.1	0.3		0.1	0.05	0.05

   - **(a)** Find P(2).
   - **(b)** Construct a probability histogram for this distribution.
   - **(c)** Find μ and σ.
   - **(d)** Locate the interval $\mu \pm \sigma$ as well as $\mu \pm 2\sigma$ on the histogram.
   - **(e)** We create another random variable $Z = Y + 1$. Find μ and σ of Z.
   - **(f)** Compare your results for **(c)** and **(e)** and generalise for $Z = Y + b$, where b is a constant.

3. A discrete random variable X can assume five possible values: 12, 13, 15, 18, and 20. Its probability distribution is shown below.

x	12	13	15	18	20
P(x)	0.14	0.11		0.26	0.23

   - **(a)** What is P(15)?
   - **(b)** What is the probability that x equals 12 or 20?
   - **(c)** What is $p(X \leqslant 18)$?
   - **(d)** Find E(X).
   - **(e)** Find V(X).
   - **(f)** Let $Y = 0.5X - 4$. Find E(Y) and V(Y).
   - **(g)** Compare your results in (d), (e) and (f) and generalise for $Y = aX + b$, where a and b are constants.

4. Medical research has shown that a certain type of chemotherapy is successful 70% of the time when used to treat skin cancer. In a study to check the validity of such a claim, researchers chose different treatment centres and selected five of their patients at random. Here is the probability distribution of the number of successful treatments for groups of five:

x	0	1	2	3	4	5
P(x)	0.002	0.029	0.132	0.309	0.360	0.168

(a) Find the probability that at least two patients would benefit from the treatment.

(b) Find the probability that the majority of the group does not benefit from the treatment.

(c) Find E(X) and interpret the result.

(d) Show that $\sigma(X) = 1.02$

(e) Graph P(x). Locate μ, $\mu \pm \sigma$, and $\mu \pm 2\sigma$ on the graph. Use the empirical rule to approximate the probability that X falls in this interval. Compare this with the actual probability.

5. The probability function of a discrete random variable X is given by

$\mathrm{P}(X = x) = \dfrac{kx}{2}$ for $x = 12, 14, 16, 18$

Set up a table showing the probability distribution and find the value of k.

6. X has probability distribution as shown in the table.

x	5	10	15	20	25
P(x)	$\frac{3}{20}$	$\frac{7}{30}$	k	$\frac{3}{10}$	$\frac{13}{60}$

(a) Find the value of k

(b) Find $p(X > 10)$

(c) Find $p(5 < X \leqslant 20)$

(d) Find the expected value and the standard deviation.

(e) Let $Y = \frac{1}{5}X - 1$. Find E(Y) and V(Y).

7. The discrete random variable Y has a probability density function

$p(Y = y) = k(16 - y^2)$ for $y = 0, 1, 2, 3, 4$

(a) Find the value of the constant k.

(b) Draw a histogram to illustrate the distribution.

(c) Find $p(1 \leqslant Y \leqslant 3)$

(d) Find the mean and variance.

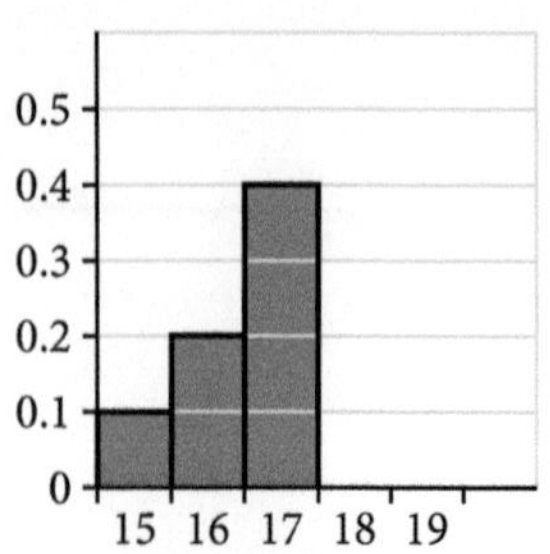

Figure 12.6 Probability distribution for question 8

8. The probability distribution of students categorised by age who visit a certain cinema on weekends is given in Figure 12.6. The probabilities for 18-year-olds and 19-year-olds are missing. We know that:

$P(X = 18) = 2P(X = 19)$

(a) Complete the graph and describe the distribution.

(b) Find the expected value and the variance.

9. In a small town, a computer store sells laptops to the local residents. However, because of low demand, they like to keep their stock at a manageable level. The data they have indicate that the weekly demand for the laptops they sell follows a distribution given in the table below.

***x*: laptops sold**	0	1	2	3	4	5
P(*x*)	0.10	0.40	0.20	0.15	0.10	0.05

(a) Find the mean and standard deviation of this distribution.

(b) Use the empirical rule to find the approximate number of computers sold about 95% of the time.

10. The discrete random variable X has probability function given by

$$P(x) = \begin{cases} \left(\frac{1}{4}\right)^{x-1} & x = 2, 3, 4, 5, 6 \\ k & x = 7 \\ 0 & \text{otherwise} \end{cases}$$

where k is a constant.

Determine the value of k and the expected value of X.

11. The following is a probability distribution for a random variable Y.

y	0	1	2	3
P($Y = y$)	0.1	0.40	k	$(k-1)^2$

(a) Find the value of k.

(b) Find the expected value.

12. A closed box contains 8 red balls and four white ones. A ball is taken out at random, its colour noted, and it is then returned. This is done three times. Let X represent the number of red balls drawn.

(a) Set up a table to show the probability distribution of X.

(b) What is the expected number of red balls drawn in this experiment?

13. A discrete random variable Y has the following probability distribution function:

$P(Y = y) = k(4 - y)$ for $y = 0, 1, 2, 3,$ and 4

(a) Find the value of k

(b) Find $P(1 \leqslant y < 3)$

14. Airlines sometimes overbook flights. Suppose, for a 50-seat plane, that 55 tickets were sold. Let X represent the number of ticketed passengers that show up for the flight. From records, the airline has the following PMF for this flight.

x	45	46	47	48	49	50	51	52	53	54	55
P(x)	0.05	0.08	0.12	0.15	0.25	0.20	0.05	0.04	0.03	0.02	0.01

(a) Construct a CDF table for this distribution.

(b) What is the probability that the flight will accommodate all ticket holders that show up?

(c) What is the probability that not all ticket holders will have a seat on the flight?

(d) Calculate the expected number of passengers who will show up.

(e) Calculate the standard deviation of the number of passengers who will show up.

(f) Calculate the probability that the number of passengers showing up will be within one standard deviation of the expected number.

15. A small internet provider has 6 telephone service lines operating 24 hours daily. Defining X as the number of lines in use at any specific 10-minute period of the day, the PMF of X is given in the table.

x	0	1	2	3	4	5	6
P(x)	0.08	0.15	0.22	0.27	0.20	0.05	0.03

(a) Construct a CDF table.

(b) Calculate the probability that at most three lines are in use.

(c) Calculate the probability that a customer calling for service will have a free line.

(d) Calculate the expected number of lines in use.

(e) Calculate the standard deviation of the number of lines in use.

16. Some torches use one AA-type battery to work. The voltage in any new battery is considered acceptable if it is at least 1.3 volts. 90% of the AA batteries from a specific supplier have an acceptable voltage. Batteries are usually tested until an acceptable one is found, then it is installed in the torch. Let X represent the number of batteries that must be tested.

(a) Find P(1), i.e., P($X = 1$)

(b) Find P(2)

(c) Find P(3)

(d) To have $X = 5$, what must be true of the

(i) fourth battery tested

(ii) fifth battery tested?

(e) Use your observations above to obtain a general model for P(x).

17. Repeat question 16 for a torch that needs two batteries.

18. A biased dice with four faces is used in a game. A player pays 10 counters to roll the dice. The table below shows the possible scores on the dice, the probability of each score, and the number of counters the player receives in return for each score.

Score	1	2	3	4
Probability	$\frac{1}{2}$	$\frac{1}{5}$	$\frac{1}{5}$	$\frac{1}{10}$
Number of counters player receives	4	4	15	n

Find the value of n in order for the player to get an expected return of 9 counters per roll.

19. Two children, Alan and Belle, each throw two fair six-sided dice simultaneously. The score for each child is the sum of the two numbers shown on their respective dice.

(a) (i) Calculate the probability that Alan obtains a score of 9.

(ii) Calculate the probability that Alan and Belle both obtain a score of 9.

(b) (i) Calculate the probability that Alan and Belle obtain the same score.

(ii) Deduce the probability that Alan's score exceeds Belle's score.

(c) Let X represent the largest number shown on the four dice.

(i) Show that $P(X \leqslant x) = \left(\frac{x}{6}\right)^4$, for $x = 1, 2, \ldots, 6$

(ii) Copy and complete the following probability distribution table.

x	1	2	3	4	5	6
$P(X = x)$	$\frac{1}{1296}$	$\frac{15}{1296}$				$\frac{671}{1296}$

(ii) Calculate E(X)

20. Consider the 10 data items $x_1, x_2, \ldots, x_{10}$. Given that $\sum_{i=1}^{10} x_i^2 = 1341$ and the standard deviation is 6.9, find the value of $\bar{x}$.

12.2 The binomial distribution

Examples of discrete random variables are abundant in everyday situations. However, there are a few discrete probability distributions that are widely applied and serve as models for a great number of the applications. One of them is the **binomial** distribution.

The binomial distribution

We start with an example.

Example 12.6

A cereal company puts miniature figures in boxes of cornflakes to make them attractive for children and thus boost sales. The manufacturer claims that 20% of the boxes contain a figure. You buy three boxes of this cereal.

(a) Find the probability that you will get

  (i) exactly three figures

  (ii) exactly 2 figures.

(b) If you bought five boxes, what is the probability that you will get exactly 2 figures?

Solution

(a) (i) To get three figures means that the first box contains a figure (0.20 chance), as does the second (also 0.20), and the third (0.20). We want three figures, so this is the intersection of three events and the probability is simply $0.20^3 = 0.008$.

(ii) To get exactly 2 figures, the situation becomes more complicated. A tree diagram can help us to visualise it better.

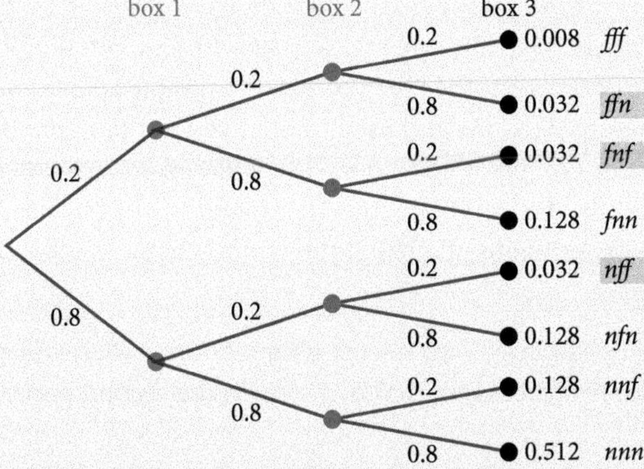

Let f stand for figure, and n for no figure. There are three events of interest to us. Since we are interested in two figures, we want to see *ffn*, which has a probability of

$$0.2 \times 0.2 \times 0.8 = 0.2^2 \times 0.8 = 0.032,$$

the other events of interest are *fnf* and *nff*, both with probabilities

$$0.2 \times 0.8 \times 0.2 = 0.032 \text{ and } 0.8 \times 0.2 \times 0.2 = 0.032$$

A **binomial experiment** has the characteristics:

- The experiment consists of n identical trials.
- Each trial has one of two outcomes. We call one of them success, S, and the other failure, F.
- The probability of success on a single trial, p, is constant throughout the whole experiment. The probability of failure is $1 - p$, which is sometimes denoted by q. That is $p + q = 1$.
- The trials are independent.
- We are interested in the number of successes x that are possible during the n trials. That is: $x = 0, 1, 2, \ldots, n$.

Since the order of multiplication is not important, the three probabilities are the same. These three events are disjoint, as can be seen from the tree diagram, so the probability of exactly two figures is the sum of the three numbers, $0.032 + 0.032 + 0.032$. Of course you may realise that it would be much simpler if we wrote $3(0.032)$, since there are three events with the same probability.

(b) The situation is similar, of course. However, a tree diagram would not be useful in this case as there is too much information to construct to see the solution. No matter how we succeed in finding a figure, whether it is in the first box, the second, or the third, it has the same probability, 0.2. So, to have two successes (finding figures) in the five boxes, we need the other three to be failures (no figures) with a probability of 0.8 for each failure. Therefore the chance of having a case like *ffnnn* is $0.2^2 \times 0.8^3$. However, this can happen in several disjoint ways. There are 10 possibilities: *ffnnn*, *fnfnn*, *fnnfn*, *fnnnf*, *nffnn*, *nnffn*, *nnnff*, *nfnfn*, *nnfnf*, *nfnnf*. Which means the probability of having exactly two figures in five boxes is

$$10 \times 0.2^2 \times 0.8^3 = 0.2048$$

Suppose that a random experiment can result in two possible mutually exclusive and collectively exhaustive outcomes: success and failure, and that p is the probability of a success from a single trial. When n independent trials are carried out, the distribution of the number of successes x is called the **binomial distribution**. The probability distribution function for the binomial random variable x is:

$P(x \text{ successes in } n \text{ independent trials}) = P(x) = {}^nC_x p^x (1-p)^{n-x} = {}^nC_x p^x q^{n-x}$, for $x = 0, 1, 2, \ldots n$.

The notation used to indicate that a variable has a binomial probability distribution with n trials and success probability of p is: $X \sim B(n, p)$

The number 10 is the binomial coefficient (Pascal's entry) that you saw in Chapter 3. This is also the combination of three events out of five.

The previous result can be written as $\binom{5}{2} 0.2^2 \cdot 0.8^3$, where $\binom{5}{2}$ is the binomial coefficient.

You can find experiments like this one in many situations. Coin flipping is one simple example of this. Another very common example is opinion polling that is conducted before elections and used to predict voter preferences. Each sampled person can be compared to a coin – but a biased coin. A voter you sample who is in favour of a candidate can correspond to either a head or a tail on a coin. Such experiments all exhibit the typical characteristics of the binomial experiment.

In the cereal company's example, we started with $n = 3$, $p = 0.2$, and asked for the probability of two successes; that is, $x = 2$. In the second part we have $n = 5$.

Imagine repeating a binomial experiment n times. If the probability of success is p, then the probability of having x successes is $pppp\ldots$, x times, (p^x), because the order is not important. However, in order to have exactly x successes, the other $(n - x)$ trials must be failures – that is, with probability of $qqqq\ldots$, $(n - x)$ times, (q^{n-x}). This is only one order (combination) where the successes happen the first x times and the rest are failures. We have to count the number of orders (combinations) possible. This is given by the binomial coefficient $\binom{n}{x} = {}^nC_x$.

Example 12.7

A computer shop orders its notebooks from a supplier that has a rate of defective items of 10%. The shop usually takes a sample of 10 computers and checks them for defects. If they find two computers defective, they return the shipment. What is the probability that their random sample will contain two defective computers?

Solution

We will consider this to be a random sample and the shipment large enough to render the trials independent of each other. The probability of finding two defective computers in a sample of 10 is given by

$$P(x = 2) = {}^{10}C_2(0.1)^2(0.9)^{10-2} = 45 \times 0.01 \times 0.43047 = 0.194$$

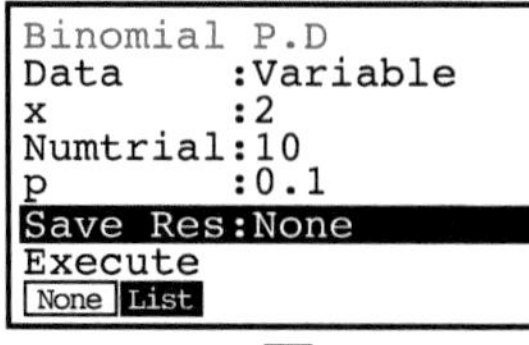

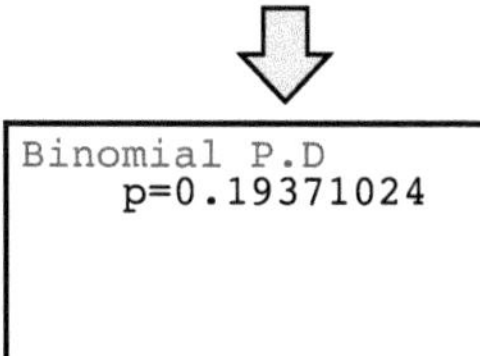

Figure 12.7 A GDC can do this calculation for you

Of course it is a daunting task to do all the calculations by hand. A GDC can do this calculation for you. You need to learn how your GDC performs such calculations. Figure 12.7 shows a sample from one GDC.

Using a spreadsheet, we can also produce this result or even a set of probabilities covering all the possible values. The formula used for this example for Excel is =BINOMDIST(B1:G1,10,0.1,FALSE).

Similarly, a GDC can also give us a list of the probabilities (see Figure 12.8).

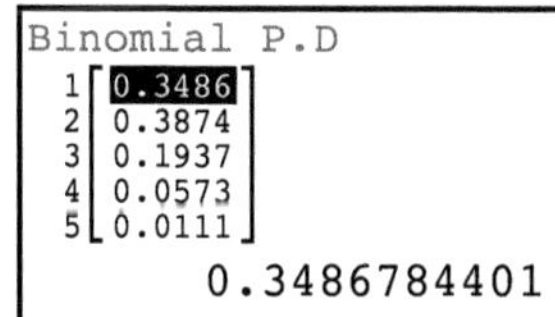

Figure 12.8 GDC list of probabilities

Like other distributions, when we look at the binomial distribution, we want to look at its expected value and standard deviation.

Using the formula we developed for the expected value, $\sum xP(x)$, we can, of course, add $xP(x)$ for all the values involved in the experiment. The process would be long and tedious for something we intuitively know. For example, in the defective items sample, if we know that the defect rate of the computer manufacturer is 10%, then it is natural to expect to have $10 \times 0.1 = 1$ defective computer. If we have 100 computers with a defective rate of 10%, how many would you expect to be defective? Can you think of a reason why it would not be 10?

The binomial probability model

n = number of trials

p = probability of success

$q = 1 - p$ probability of failure

x = number of successes in n trials

$P(x) = {}^nC_x p^x(1-p)^{n-x} = {}^nC_x p^x q^{n-x}$, for $x = 0, 1, 2, \dots n$.

Expected value $= \mu = np$

Variance $= \sigma^2 = npq$

$\sigma = \sqrt{npq}$

The expected value of the successes in the binomial is the number of trials n multiplied by the probability of success, np.

So, in the defective notebooks case, the expected number of defective items in the sample of 10 is $np = 10 \times 0.1 = 1$

And the standard deviation is $\sigma = \sqrt{npq} = \sqrt{10 \times 0.1 \times 0.9} = 0.949$

How do we know that the binomial distribution is a probability distribution?

We can easily verify that the binomial distribution as developed satisfies the probability distribution conditions.

1. $0 \leqslant \text{P}(x) \leqslant 1$ **2.** $\sum_x \text{P}(x) = 1$

1. Since $p > 0$ by definition, then $p^x > 0$, for $x = 0, 1, 2, \ldots$. Similarly, $q^{n-x} > 0$.

We also know that ${}^nC_x > 0$. Therefore

$$\text{P}(x) = {}^nC_x p^x q^{n-x} > 0$$

$\text{P}(x) \leqslant 1$ will be a natural result of proving the second condition. If the sum of n positive parts is equal to 1, none of the parts can be greater than 1.

2. $\sum_{x=0}^{n} \text{P}(x) = \sum_{x=0}^{n} {}^nC_x p^x q^{n-x}$

The binomial theorem states

$$(p + q)^n = \sum_{x=0}^{n} {}^nC_x p^x q^{n-x} = \sum_{x=0}^{n} \text{P}(x)$$

Since $p + q = 1$, then $(p + q)^n = 1$, and therefore

$$\sum_{x=0}^{n} \text{P}(x) = \sum_{x=0}^{n} {}^nC_x p^x q^{n-x} = (p + q)^n = 1$$

Online 

Explore binomial distribution with a dynamic plot.

Example 12.8

A study to examine the effectiveness of advertising on the internet reported that 4 out of 10 users remember advertisement banners after seeing them.

(a) 20 users are chosen at random and shown an advertisement. What is the expected number of users who will remember the advertisement?

(b) What is the chance that 5 of those 20 will remember the advertisement?

(c) What is the probability that at most 1 user will remember the advertisement?

(d) What is the chance that at least two users will remember the advertisement?

Solution

(a) $X \sim \text{B}(20, 0.4)$. The expected number is simply $20 \times 0.4 = 8$.
We expect 8 of the users to remember the advertisement.
Notice that on the histogram, the area in red corresponds to the expected value 8.

```
Binomial P.D
     p=0.0746470195
```

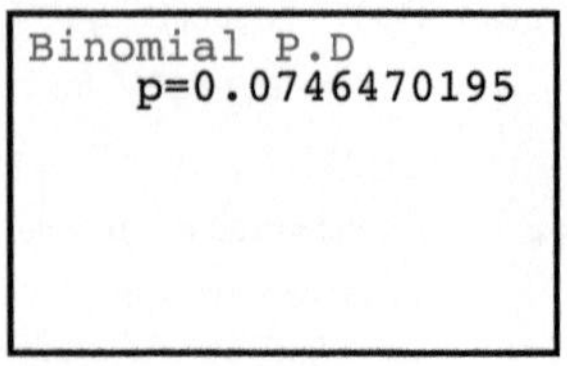

Figure 12.9 GDC solution to Example 12.8 (b)

(b) $\text{P}(5) = \binom{20}{5}(0.4)^5(0.6)^{15} = 0.0746$, or see the output from a GDC.

This area is shown as the green area above 5 in the histogram.

(c) $\text{P}(x \leqslant 1) = \text{P}(x = 0) + \text{P}(x = 1) = 0.000524$

(d) $P(x \geq 2) = 1 - P(x \leq 1)$

$= 1 - 0.000524 = 0.999475$

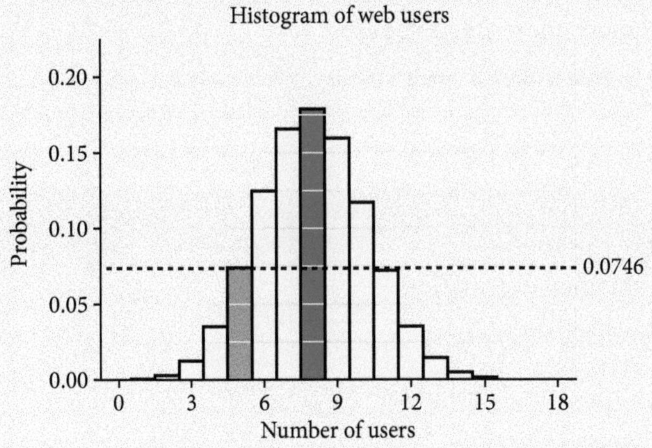

Histogram of web users

Probability 0.20 0.15 0.10 0.05 0.00

0.0746

0 3 6 9 12 15 18

Number of users

The cumulative binomial distribution function

The cumulative distribution function F(x) of a random variable X expresses the probability that X does not exceed the value x. That is

$$F(x) = P(X \leq x) = \sum_{y:\, y \leq x} p(y)$$

So, for the binomial distribution, the cumulative distribution function (CDF) is given by

$$F(x) = P(X \leq x) = \sum_{y:\, y \leq x} p(y)$$

$$= \sum_{y:\, y \leq x} {}^{n}C_{y} p^{y} q^{n-y}$$

The cumulative distribution is very helpful when we need to find the probability that a binomial variable assumes values over a certain interval.

You will not be required to perform calculations manually. A GDC can produce the values requested.

Example 12.9

A large shipment of light bulbs contains 4% defective bulbs. In a sample of 20 randomly selected bulbs from the shipment, find the probability that

(a) there are at most 3 defective bulbs

(b) there are at least 6 defective bulbs.

Solution

(a) This can be considered as a binomial distribution with $n = 20$ and $p = 0.04$

We need $P(x \leq 3)$, which we can calculate either by finding the probabilities for $x = 0, 1, 2$, and 3, and adding them, or by using the cumulative function. In both cases, we can use a GDC.

Using the CDF is a much more straightforward procedure.

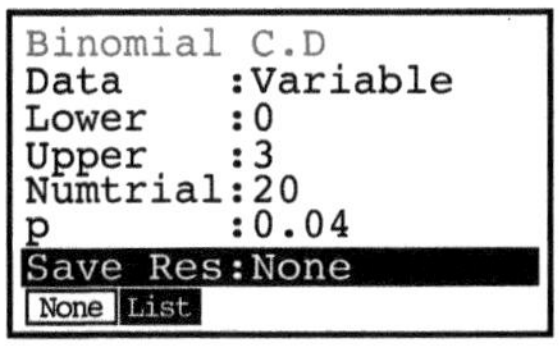

```
Binomial C.D
Data     :Variable
Lower    :0
Upper    :3
Numtrial:20
p        :0.04
Save Res:None
None List
```

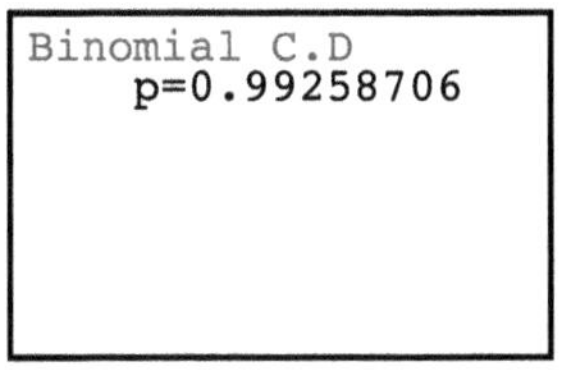

```
Binomial C.D
    p=0.99258706
```

Figure 12.10 (a) GDC solution to Example 12.9 (a)

```
Binomial C.D
Data     :Variable
Lower    :6
Upper    :20
Numtrial:20
p        :0.04
Save Res:None
```

```
Binomial C.D
    p=9.7654E-05
```

Figure 12.10 (b) GDC solution to Example 12.9 (b)

(b) Here we need $P(X \geqslant 6)$. The first approach is not feasible at all as we need to calculate 15 individual probabilities and add them. However, setting the problem as a complement and then using the cumulative distribution is much more efficient. i.e.,

$P(X \geqslant 6) = 1 - P(X < 6) = 1 - P(X \leqslant 5)$

Exercise 12.2

1. Consider the binomial distribution

$P(x) = {}^5C_x(0.6)^x(0.4)^{5-x}, x = 0, 1, \ldots, 5$

(a) Make a table for this distribution.

(b) Graph this distribution.

(c) Find the mean and standard deviation:

(i) using a formula

(ii) by using the table of values you created in part (a).

(d) Locate the mean μ and the two intervals $\mu \pm \sigma$ and $\mu \pm 2\sigma$ on the graph.

(e) Find the actual probabilities for x to lie within each of the intervals $\mu \pm \sigma$ and $\mu \pm 2\sigma$ and compare them to the empirical rule.

2. A poll of 20 adults is taken in a large city. The purpose is to determine whether they support banning smoking in restaurants. It is known that approximately 60% of the population supports the decision.
Let X represent the number of respondents in favour of the decision.

(a) What is the probability that 5 respondents support the decision?

(b) What is the probability that none of the 20 support the decision?

(c) What is the probability that at least 1 supports the decision?

(d) What is the probability that at least 2 respondents support the decision?

(e) Find the mean and standard deviation of the distribution.

3. Consider the binomial random variable with $n = 6$ and $p = 0.3$

(a) Copy the table and fill in the probabilities.

k	0	1	2	3	4	5	6
$P(x \leqslant k)$							

(b) Copy and complete the table. Some cells have been filled to guide you.

Number of successes x	List the values of x	Write the probability statement	Explain it, if needed	Find the required probability
At most 3				
At least 3				
More than 3	4, 5, 6	$p(x > 3)$	$1 - p(x \leq 3)$	0.07047
Fewer than 3				
Between 3 and 5 (inclusive)				
Exactly 3				

4. Repeat question 3 with $n = 7$ and $p = 0.4$

5. A box contains eight balls: five are green, one is white, one red, and one yellow. Three balls are chosen at random without replacement, and the number of green balls y is recorded.

(a) Explain why y is not a binomial random variable.

(b) Explain why, when we repeat the experiment with replacement, then y is a binomial random variable.

(c) Give the values of n and p and display the probability distribution in tabular form.

(d) What is the probability that at most two green balls are drawn?

(e) What is the expected number of green balls drawn?

(f) What is the variance of the number of balls drawn?

(g) What is the probability that some green balls will be drawn?

6. On a multiple-choice test, there are 10 questions, each with 5 possible answers, one of which is correct. Nick is unaware of the content of the material and so he guesses on all questions. Find the probability that:

(a) Nick does not answer any question correctly

(b) Nick answers at most half of the questions correctly

(c) Nick answers at least one question correctly.

(d) How many questions should Nick expect to answer correctly?

7. Houses in a large city are equipped with alarm systems to protect them from burglary. A company claims their system to be 98% reliable. That is, it will trigger an alarm in 98% of the cases. In a certain neighbourhood, 10 houses equipped with this system experience an attempted burglary.

(a) Find the probability that all the alarms work properly.

(b) Find the probability that at least half of the alarms are triggered.

(c) Find the probability that at most 8 alarms will work properly.

8. Graphic novels are purchased by readers of all ages. 40% of graphic novels were purchased by readers who were 30 years of age or older. Fifteen readers are chosen at random. Find the probability that

   (a) at least 10 of them are 30 years or older

   (b) exactly 10 of them are 30 or older

   (c) at most 10 of them are younger than 30.

9. A factory makes computer hard disks. Over a long period, 1.5% of them are found to be defective. A random sample of 50 hard disks is tested.

   (a) Write down the expected number of defective hard disks in the sample.

   (b) Find the probability that three hard disks are defective.

   (c) Find the probability that more than one hard disk is defective.

10. Car colour preferences change over time and according to the area the customer lives in and the car model they are interested in. In a certain city, a car dealer noticed that 10% of the cars he sells are metallic grey. Twenty of his customers are selected at random and their car orders are checked for colour.

    Find the probability that:

    (a) at least 5 cars are metallic grey

    (b) at most 6 cars are metallic grey

    (c) more than 5 cars are metallic grey

    (d) between 4 and 6 cars are metallic grey

    (e) more than 15 cars are not metallic grey.

    In a sample of 100 customer records, find:

    (f) the expected number of metallic grey car orders

    (g) the standard deviation of metallic grey car orders.

    According to the empirical rule, 95% of the orders of metallic grey orders are between a and b.

    (h) Find a and b.

11. Dog owners in many countries buy health insurance for their dogs. 3% of all dogs have health insurance. In a random sample of 100 dogs in a large city, find:

    (a) the expected number of dogs with health insurance

    (b) the probability that 5 of the dogs have health insurance

    (c) the probability that more than 10 dogs have health insurance.

12. A balanced coin is flipped five times. Let x be the number of heads observed.

    (a) Using a table, construct the probability distribution of x.

    (b) What is the probability that no heads are observed?

(c) What is the probability that all observations are heads?

(d) What is the probability that at least one head is observed?

(e) What is the probability that at least one tail is observed?

(f) Another coin is unbalanced so that it shows 2 heads in every 10 flips. Repeat parts (a) to (e) for this coin.

13. On a television channel, the news is shown at the same time each day. The probability that Alice watches the news on a given day is 0.4. Calculate the probability that on five consecutive days, she watches the news on at most three days.

14. A satellite relies on solar cells for its power and will operate provided that at least one of the cells is working. Cells fail independently of each other, and the probability that an individual cell fails within one year is 0.8.

(a) For a satellite with ten solar cells, find the probability that all ten cells fail within one year.

(b) For a satellite with ten solar cells, find the probability that the satellite is still operating at the end of one year.

(c) For a satellite with n solar cells, write down the probability that the satellite is still operating at the end of one year. Hence, find the smallest number of solar cells required so that the probability of the satellite still operating at the end of one year is at least 0.95.

12.3 Continuous distributions – The normal distribution

When a random variable X is discrete, we assign a positive probability to each value that X can take and get the probability distribution for X. The sum of all the probabilities associated with the different values of X is 1.

We have seen, in the discrete variable case, that we can graphically represent the probabilities corresponding to the different values of the random variable X with a probability histogram (relative frequency histogram), where the area of each bar corresponds to the probability of the specific value it represents.

Consider now a continuous random variable X, such as height, weight, or length of life of a particular product – a TV set, for example. Because it is continuous, the possible values of X are over an interval. Moreover, there are an infinite number of possible values of X. Hence, we cannot find a probability distribution function for X by listing all the possible values of x along with their probabilities. If we try to assign probabilities to each of these uncountable values, the probabilities will no longer sum to 1. Thus we must use a different approach to generate the probability distribution for such random variables.

Suppose that we have a set of measurements on a continuous random variable, and we create a relative frequency histogram to describe their distribution. For a small number of measurements, we can use a small number of classes, but as more and more measurements are collected, we can use more classes and reduce the class width.

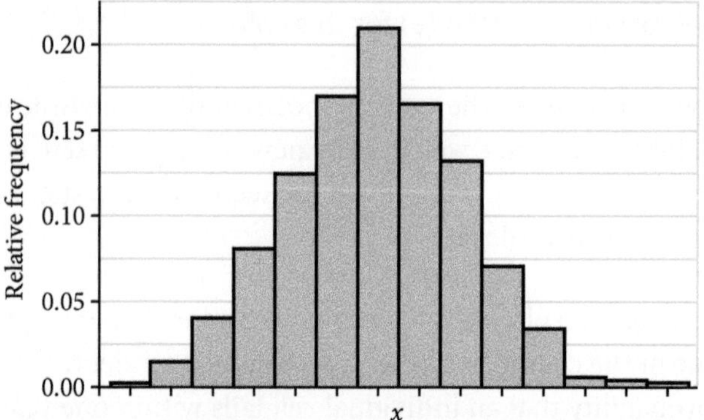

Figure 12.11 Relative frequency histogram

The histogram will slightly change as the class width becomes smaller and smaller as shown in Figure 12.12. As the number of measurements becomes very large and the class width becomes very narrow, the relative frequency histogram appears more and more like the smooth curve we see in Figure 12.13. This is what happens in the continuous case, and the smooth curve describing the probability distribution of the continuous random variable becomes the **probability density function** of X, represented by a curve $y = f(x)$. This curve is such that the entire area under the curve is 1 and the area between any two points is the probability that x falls between those 2 points.

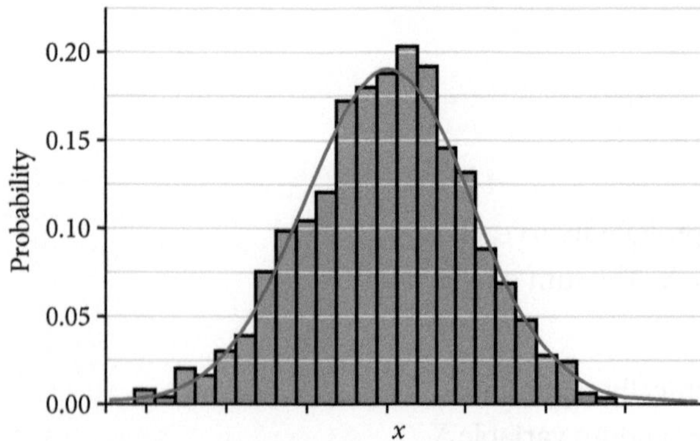

Figure 12.12 As the number of measurements increases, class width becomes narrower

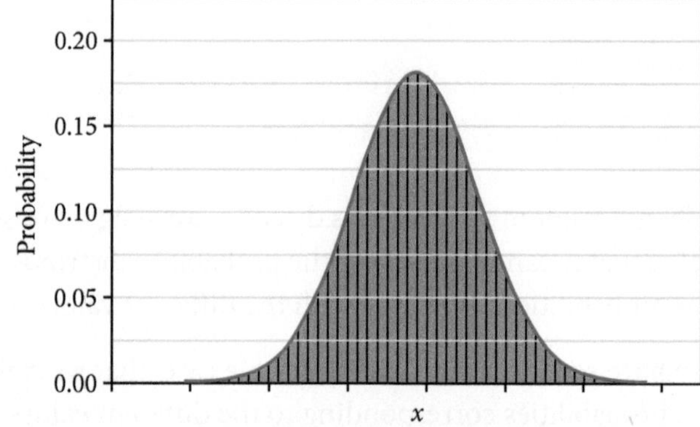

Figure 12.13 Smooth curve

Probability density function

Let X be a continuous random variable. The probability density function, $f(x)$, of the random variable is a function with the properties:

- $f(x) > 0$ for all values of X.
- The area under the probability density function $f(x)$ over all values of the random variable x is equal to 1.0; that is, $\int_{-\infty}^{\infty} f(x)\mathrm{d}x = 1$.

- A graph is drawn of the density function (Figure 12.14). Let a and b be two possible values of the random variable X, with $a < b$. The probability that x lies between a and b [$p(a<x<b)$] is the area under the density function between these points, i.e., $p(a < x < b) = \int_a^b f(x)\mathrm{d}x$.

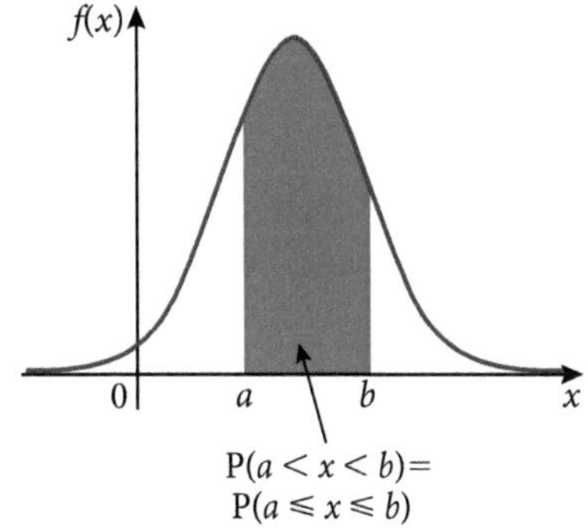

Figure 12.14 Density function graph

For example, Figure 12.15 shows the graph of a model for the PDF f of a random variable X defined to be the height, in cm, of an adult female in Spain. The probability that the height of a female chosen at random from this population is between 160 and 175 is equal to the area under the curve between 160 and 175.

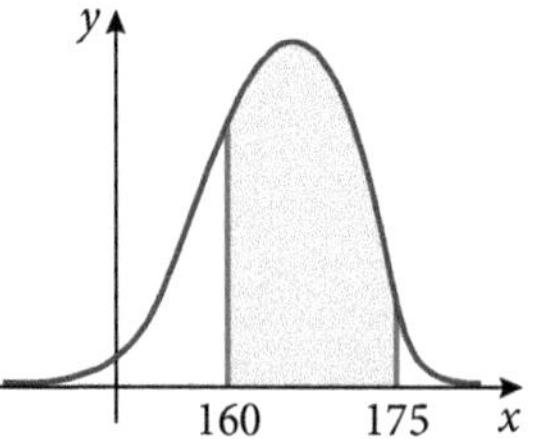

Figure 12.15 Green area = probability that a female has a height between 160 and 175 cm

Based on this definition of a continuous PDF, the probability that x equals any point a, is 0. This is so because the area above a value, say a, is a rectangle of width 0, or equivalently

$$p(X = a) = \int_a^a f(x)\mathrm{d}x = 0$$

So, for the continuous case, regardless of whether the endpoints a and b are themselves included, the area included between a and b is the same.

$$p(a < x < b) = p(a \leqslant x \leqslant b) = p(a \leqslant x < b) = p(a < x \leqslant b)$$

The function represented here is

$$f(x) = \int_{160}^{175} \frac{e^{-\frac{(x-165)^2}{50}}}{5\sqrt{2\pi}}\mathrm{d}x$$

```
∫175 ( e((-(x-165)²)÷50))÷(5/
 180
                    0.8185946141
```

Figure 12.16 Approximating using a GDC

This is not an integral you can calculate exactly. So, we use a GDC to approximate it.

So, the chance to choose a female at random with a height between 160 cm and 175 cm is approximately 81.9%.

Continuous probability distributions can assume a variety of shapes. However, for reasons of staying within (with some extensions) the boundaries of the IB syllabus, we will focus on one distribution. In fact, a large number of random variables observed in our surroundings possess a frequency distribution that is approximately bell-shaped. We call that distribution the **normal probability distribution.**

The normal distribution

The most important type of continuous random variable is the **normal** random variable. The probability density function of a normal random variable x is determined by two parameters: the mean or expected value μ, and the standard deviation σ of the variable.

The normal probability density function is a bell-shaped density curve that is symmetric about the mean μ. Its variability is measured by σ. The larger the value of σ, the more variability there is in the curve – that is, the higher the probability of finding values of the random variable further away from the mean. Figure 12.17 represents three different normal density functions with the same mean but different standard deviations. Note how the curves flatten as σ increases. This is because the area under the curve has to stay equal to 1.

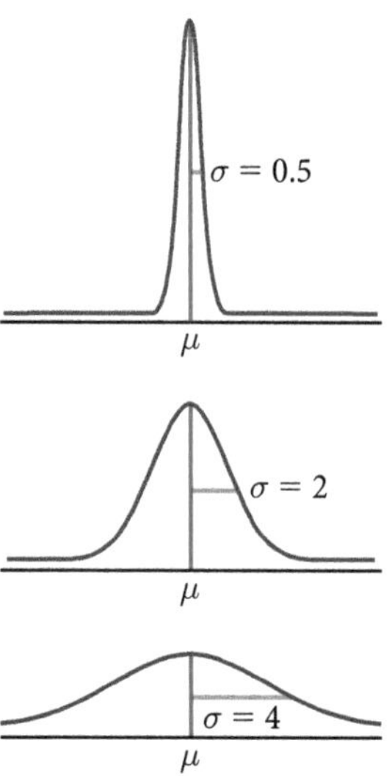

Figure 12.17 Three different normal density functions with the same mean but different standard deviations

The probability density function for a normally distributed random variable x is

$$f(x) = \frac{1}{\sigma\sqrt{2\pi}}e^{-\frac{(x-\pi)^2}{2\sigma^2}} = \frac{1}{\sigma\sqrt{2\pi}}e^{-\frac{1}{2}\left(\frac{x-\pi}{\sigma}\right)^2} \text{ for } -\infty < x < \infty$$

where μ and σ^2 are any number such that $-\infty < \mu < \infty$ and $0 < \sigma^2 < \infty$ and where e and π are the constants, e = 2.71828….. and π = 3.14159….

When a variable is normally distributed, we write:

$$X \sim N(\mu, \sigma^2)$$

Although we will not make direct use of the formula, it is interesting to note its properties because they help us understand how the normal distribution works.

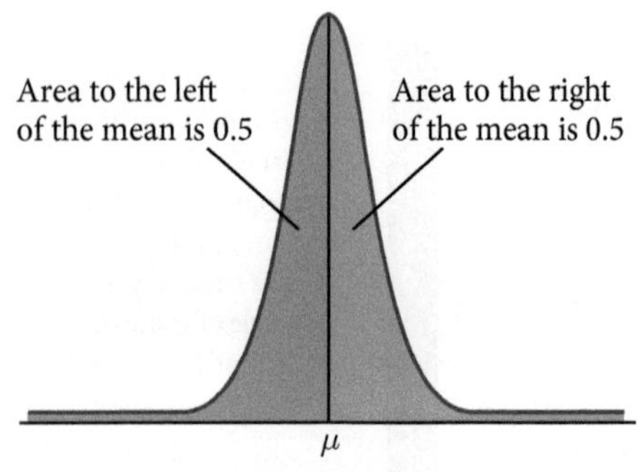

Figure 12.18 Normal probability distribution

The graph of a normal probability distribution is shown in Figure 12.18. The mean or expected value locates the centre of the distribution and the distribution is symmetric about this mean. Since the total area under the curve is 1, the symmetry of the curve implies that the area to the right of the mean and the area to the left are both equal to 0.5. Large values of σ tend to reduce the height of the curve and increase the spread, and small values of σ increase the height to compensate for the narrowness of the distribution.

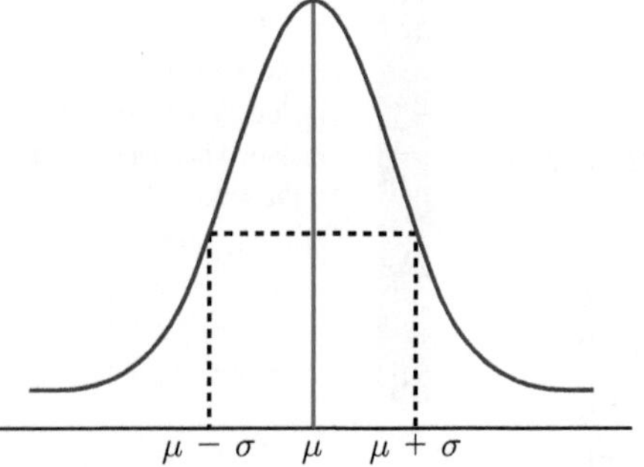

Figure 12.19 One σ to the right or left of the mean μ marks the point where the curvature of the curve changes.

So, the normal distribution is fully determined by its mean μ and its standard deviation, σ. Changing μ without changing σ moves the normal curve along the horizontal axis without changing its spread. The standard deviation σ controls the spread of the curve. You can also locate the standard deviation by eye on the curve. One σ to the right or left of the mean μ marks the point where the curvature of the curve changes. That is, as you move right from the mean, at the point where $x = \mu + \sigma$, the curve changes its curvature from downwards to upwards, and similarly as you move one σ to the left from the mean.

The empirical rule for the normal distribution

- Approximately 68% of the observations fall within σ of the mean
- Approximately 95% of the observations fall within 2σ of the mean
- Approximately 99.7% of the observations fall within 3σ of the mean.

Although there are many normal curves, they all have common properties.

Figure 12.20 illustrates this rule. Later in this section, you will learn how to find these areas from a table or from your GDC.

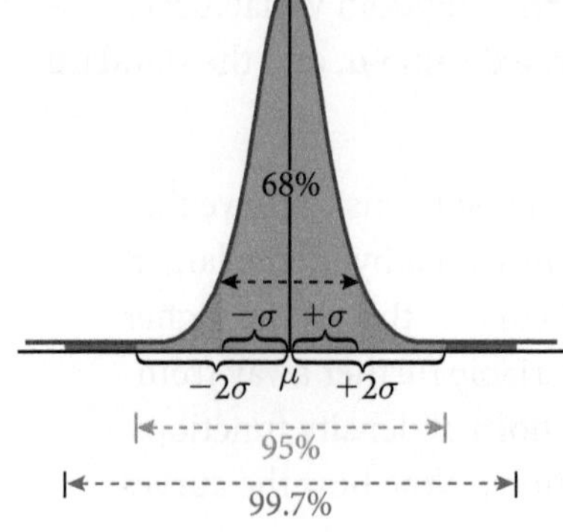

Figure 12.20 Empirical rule

Example 12.10

Heights of young German men between 18 and 19 years of age follow a distribution that is approximately normal, with mean 181 cm and a standard deviation of approximately 8 cm. Describe this population of young men.

Solution

According to the empirical rule, we find that approximately 68% of those young men have a height between 173 cm and 189 cm, 95% of them between 165 cm and 197 cm, and 99.7% between 157 cm and 205 cm. You can say that only 0.15% are taller than 205 cm or shorter than 157 cm.

As the empirical rule suggests, all normal distributions are the same if we measure in units of size σ about the mean μ as centre. Changing to these units is called **standardising**. To standardise a value, measure how far it is from the mean and express that distance in terms of σ.

Online

Explore normal distribution with a dynamic graph.

If x is a normal random variable, with mean μ and standard deviation σ, the standardised value of x is $z = \dfrac{x - \mu}{\sigma}$

A standardised value is also called the **z-score**.

The quantity $x - \mu$ tells us how far a value is from the mean; dividing by σ then tells us how many standard deviations that distance is equal to.

The standardising process is a transformation of the normal curve. For discussion purposes, assume the mean μ to be positive. The transformation $x - \mu$ shifts the graph back μ units. So, the new centre is shifted from μ back μ units. That is, the new centre is 0.

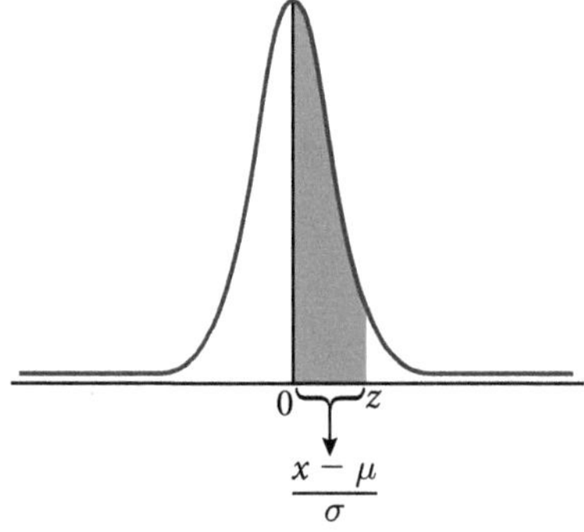

Figure 12.21 Transformation of the normal curve

Dividing by σ is going to scale the distances from the mean and express everything in terms of σ. So, a point that is one standard deviation from the mean is going to be 1 unit above the new mean; that is, it will be represented by $+1$. Now, if you look at the empirical rule discussed earlier, points that are within one standard deviation from the mean will be within a distance of 1 in the new distribution. Instead of being at $\mu + \sigma$ and $\mu - \sigma$, they will be at $0 + 1$ and $0 - 1$ respectively; that is, -1 and $+1$.

The probability density function for the standard normal distribution is $f(z) = \dfrac{1}{\sqrt{2\pi}} e^{-\frac{1}{2}z^2}$ for $-\infty < z < \infty$

The new distribution created by this transformation is called the **standard normal distribution**. It has a mean of 0 and a standard deviation of 1. It is a very helpful distribution because it will enable us to read the areas under any normal distribution through the standardisation process.

Since linear transformations can transform all normal functions to standard, this becomes a very convenient and efficient way of finding the area under any normal distribution.

The proof that the mean and the variance of the standard normal variable are 0 and 1 respectively is straightforward.

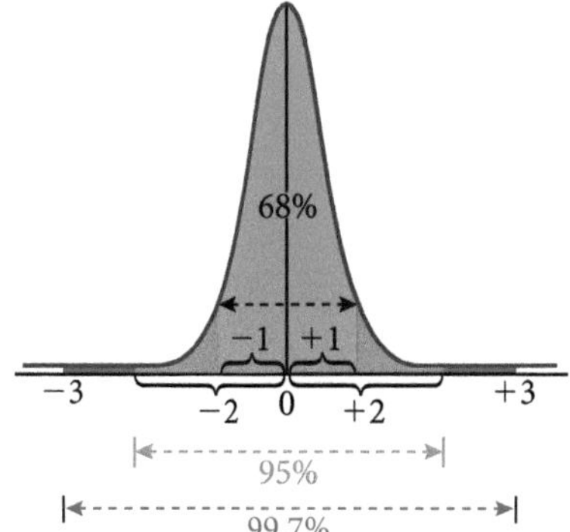

Figure 12.22 The standard normal distribution

Let $z = \frac{x - \mu}{\sigma}$ be the standard variable corresponding to a normal variable x.

$$\mathrm{E}(z) = \mathrm{E}\left(\frac{x - \mu}{\sigma}\right) = \mathrm{E}\left(\frac{1}{\sigma}(x - \mu)\right) = \frac{1}{\sigma}\mathrm{E}(x - \mu) = \frac{1}{\sigma}(\mu - \mu) = 0$$

$$\mathrm{V}(z) = \mathrm{V}\left(\frac{x - \mu}{\sigma}\right) = \mathrm{V}\left(\frac{1}{\sigma}(x - \mu)\right) = \frac{1}{\sigma^2}\mathrm{V}(x - \mu) = \frac{1}{\sigma^2}\mathrm{V}(x) = \frac{1}{\sigma^2}\sigma^2 = 1$$

Example 12.11

Using the data in Example 12.10, work out the z-score of a young German man with a height of

(a) 192 cm (b) 175 cm.

Solution

(a) z-score:

$$z = \frac{x - \mu}{\sigma} = \frac{192 - 181}{8} = 1.375$$

or 1.375 standard deviations above the mean.

(b) $z = \frac{x - \mu}{\sigma} = \frac{175 - 181}{8} = -0.75$

or 0.75 standard deviations below the mean.

To find the probability that a normal variable x lies in the interval a to b, we need to find the area under the normal curve $\mathrm{N}(\mu, \sigma^2)$ between the points a and b. However, there is an infinitely large number of normal curves – one for each mean and standard deviation. A separate table of areas for each of these curves is obviously not practical. Instead, we use one table for the standard normal distribution that gives us the required areas.

When we standardise a and b, we get two standard numbers z_1 and z_2 such that the area between z_1 and z_2 is the same as the area we need.

In this example, we are interested in the proportion of young German men whose height is between 175 cm and 192 cm.

To find the required area, we can use software or a GDC as shown in Figure 12.23.

With a GDC, we do not need to standardise our variables. However, there are cases where we need to understand standardisation in order to use it in solving some problems where the mean or the standard deviation, or both, are not given.

If we want to use the standard normal, our commands will be the same, but we do not need to include the mean and standard deviation. They are the default.

If we need the probability that a young man is taller than 175 cm, we can also read it either by looking at the distribution with the original data or by standardising.

```
normalcdf(175,19
2,181,8)
          .6888069418
```

```
normalcdf(-.75,1
.375)
          .6888069418
```

Figure 12.23 GDC output

Example 12.12

The age of graduate students in engineering programmes throughout the USA is normally distributed with mean $\mu = 24.5$ and standard deviation $\sigma = 2.5$

A student is chosen at random.

(a) What is the probability that the student is younger than 26 years old?

(b) What proportion of students are older than 23.7 years?

(c) What percentage of students are between 22 and 28 years old?

(d) What percentage of the ages falls within

(i) 1 standard deviation of the mean

(ii) 2 standard deviations of the mean

(iii) 3 standard deviations of the mean?

Solution

Let X = age of students, then $X \sim N(\mu = 24.5, \sigma^2 = 6.25)$

(a) We can either standardise and then read the table for the area left of 0.6 or use a GDC:

$$P\left(z < \frac{26 - 24.5}{2.5}\right) = P(z < 0.6) = 0.7257$$

Notice here that we put 0 as a lower limit. We can put a number as a lower limit far enough from the mean to make sure we are receiving the correct cumulative distribution.

```
Normal C.D
p    =0.72574688
z:Low=-9.8
z:Up =0.6
```

Figure 12.24 GDC screen for Example 12.12 (a)

(b) This can be done in a similar way:

$$P(x > 23.7) = P\left(z > \frac{23.7 - 24.5}{2.5}\right) = -0.32, \text{ so by symmetry we know}$$

$$P(z > -0.32) = P(z < 0.32) = 0.6255$$

```
Normal C.D
p    =0.62551583
z:Low=-0.32
z:Up =30.2
```

Figure 12.25 GDC screen for Example 12.12 (b)

(c) $P(22 < X < 28) = P\left(\frac{22 - 24.5}{2.5} < z < \frac{28 - 24.5}{2.5}\right) = P(-1 < z < 1.4)$

Find the area to the left of 1.4 and to the left of -1 and subtract them.

$$P(-1 < z < 1.4) = 0.9192 - 0.1587 = 0.7606 = 76.06\%$$

```
Normal C.D
p    =0.76058808
z:Low=-1
z:Up =1.4
```

Figure 12.26 GDC screen for Example 12.12 (c)

(d) This is the empirical rule. Find what percentage of the approximately normal data will lie within 1, 2, and 3 standard deviations.

(i) $P(-1 \leqslant z \leqslant 1) = 0.6826$

(ii) $P(-2 \leqslant z \leqslant 2) = 0.9544$

(iii) $P(-3 \leqslant z \leqslant 3) = 0.9973$

These are the exact values corresponding to the empirical rule's 68%, 95% and 99.7%.

The inverse normal distribution

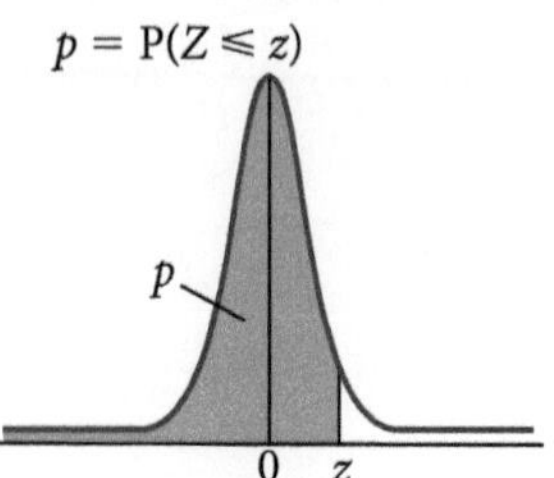

Figure 12.27 The z-score corresponding to Q_3 is 0.6745

Another type of problem arises when we are given a cumulative probability and would like to find the value in our data that has this cumulative probability. For example, what age marks the 95th percentile? That is, what age is higher than or equal to 95% of the population? To answer this question, we need to reverse our steps. So far, we are given a value and then we look for the area corresponding to it. Now, we are given the area and we have to look for the number. That is why this is called the **inverse normal distribution**. Again, the approach is to find the standard inverse normal number and then to de-standardise it. That is, to find the value from the original data that corresponds to the z-value at hand.

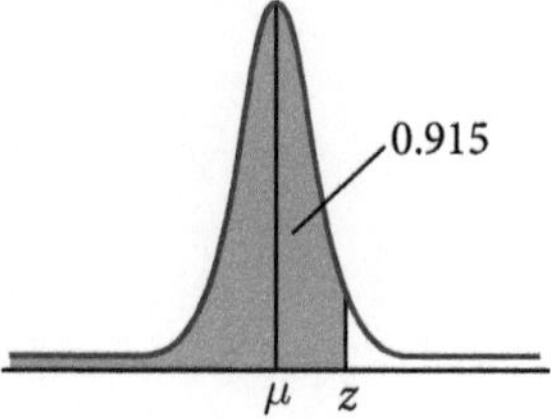

Figure 12.28
$P(Z < 1.3722) = 0.915$

For example, if we need to know what z-score the third quartile Q_3 is, we need to look up 0.75. The z-score corresponding to Q_3 is 0.6745 (Figure 12.27).

Suppose we want to find the z-score that leaves an area of 0.915 below it.

The z-score corresponding to 0.915 is 1.3722. That is

$$P(Z < 1.3722) = 0.915 \text{ (Figure 12.28)}$$

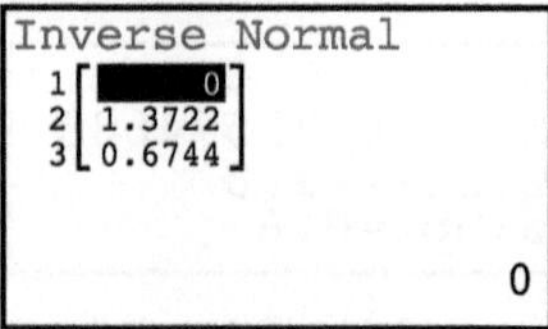

Inverse Normal
xInv=194.158829

Figure 12.29 Using invNorm

We could also use a GDC. The process is similar to the normal calculation, but choosing invNorm instead.

So, 95% of the young German men are shorter than 194.16 cm.

Example 12.13

The average time it takes fast trains to travel between London and Paris is 2 hours 15 minutes with a standard deviation of 4 minutes. Assume a normal distribution.

(a) What is the probability that a randomly chosen trip will take longer than 2 hours 20 minutes?

(b) What is the probability that a randomly chosen trip will take less than 2 hours 10 minutes?

(c) What is the IQR of the length of a trip?

Solution

We will do each problem using a GDC.

(a) $\mu = 2.25$ and $\sigma = 0.0667$

2 hours 20 minutes = 2.333, using our GDC. (We use z here for demonstration only. We don't need to standardise when using a GDC.)

$$P(x > 2.333) = P\left(z > \frac{2.333 - 2.25}{0.0667}\right) = P(z > 1.244) = 0.1067$$

(b) 2 hours 10 minutes = 2.167

$$P(x < 2.167) = P\left(z < \frac{2.167 - 2.25}{0.067}\right) = P(z < -1.244) = 0.106718$$

(c) To find the IQR, we need to find Q_1 and Q_3.

Q_1 is the number that leaves 25% of the data before it. Q_3 is the number that leaves 75% of the data before it. So, we need to find the inverse normal variable that has an area of 0.25 or 0.75 before it.

Using a GDC and the inverse normal, we find $Q_1 = 2.205$ and $Q_3 = 2.295$.

IQR $= 2.295 - 2.205 = 0.090$ of an hour (5.4 minutes)

Example 12.14

The age at which babies develop the ability to walk can be described by a normal distribution model. It is known that 5% of the babies learn how to walk by the age of 10 months, while 25% need more than 13 months. Find the mean and standard deviation of the distribution.

Solution

There are several approaches to this problem. Here is one.

Look at the distance between 10 and 13 months in two different ways. First, 10 and 13 months are 3 months apart. When standardised, the respective z-scores are -1.645 and 0.674. The z-scores are 2.319 standard deviations apart. So, 3 months must be the same as 2.319 standard deviations.

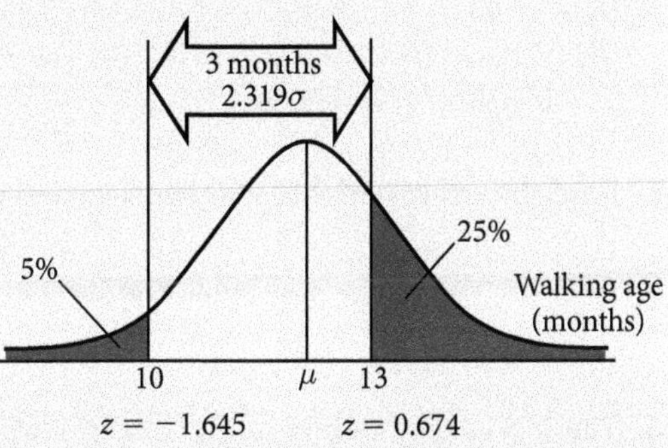

To find the z-score for the lowest 5% and the highest 25%, we can use a GDC.

Here is the calculation:

$2.319\,\sigma = 3$, $\sigma = 1.294$, and

$$z = \frac{x - \mu}{\sigma} \Rightarrow 0.674 = \frac{13 - \mu}{1.294} \Rightarrow \mu = 12.128, \text{ or alternatively}$$

$$-1.645 = \frac{10 - \mu}{\sigma} \Rightarrow \mu - 1.645\sigma = 10$$

$$0.674 = \frac{13 - \mu}{\sigma} \Rightarrow \mu + 0.674\sigma = 13$$

Solving the system of two equations in μ and σ will give the same result.

Exercise 12.3

1. The time it takes to change the batteries of your GDC is approximately normal with mean 50 hours and standard deviation of 7.5 hours.

   Find the probability that your newly equipped GDC will last

   (a) at least 50 hours
   (b) between 50 and 75 hours
   (c) less than 42.5 hours
   (d) between 42.5 and 57.5 hours
   (e) more than 65 hours
   (f) 47.5 hours.

2. Find each of the following probabilities.

   (a) $p(|z| < 1.2)$
   (b) $p(|z| > 1.4)$
   (c) $p(X < 3.7)$, where $X \sim \mathrm{N}(3, 3)$
   (d) $p(X > -3.7)$, where $X \sim \mathrm{N}(3, 3)$

3. A car manufacturer introduces a new model that has a fuel consumption of 11.4 litres per 100 km in urban areas. Tests show that this model has a standard deviation of 1.26. The distribution is assumed to be normal.

   A car is chosen at random from this model.

   (a) What is the probability that it will have consumption less than 8.4 litres per 100 km?
   (b) What is the probability that the consumption is between 8.4 and 14.4 litres per 100 km?

4. Find the value of z that will be exceeded only 10% of the time.

5. Find the value of $z = z_0$ such that 95% of the values of z lie between $-z_0$ and $+z_0$.

6. The scores on a public school's examination are normally distributed with a mean of 550 and a standard deviation of 100.

   (a) What is the probability that a randomly chosen student from this population scores below 400?
   (b) What is the probability that a student will score between 450 and 650?
   (c) What score should you have in order to be in the 90th percentile?
   (d) Find the IQR of this distribution.

7. A company producing and packaging sugar for home consumption put labels on their sugar bags noting the weight to be 500 g. Their machines are known to fill the bags with weights that are normally distributed with a standard deviation of 5.7 g. A bag that contains less than 500 g is considered to be underweight.

(a) The company decides to set their machines to fill the bags with a mean of 512 g. What fraction will be underweight?

(b) The company wants the percentage of underweight bags to be a maximum of 4%. What should the mean be?

(c) The company decides that they do not want to set the mean as high as 512 g, but instead at 510 g. What standard deviation gives them a maximum of 4% underweight bags?

8. In a large school, heights of students who are 13 years old are normally distributed with a mean of 151 cm and a standard deviation of 8 cm.

Find the probability that a randomly chosen child is:

(a) shorter than 166 cm

(b) within 6 cm of the average.

9. The time it takes Kevin to get to school every day is normally distributed with a mean of 12 minutes and a standard deviation of 2 minutes. Estimate the number of days when Kevin takes:

(a) longer than 17 minutes

(b) less than 10 minutes

(c) between 9 and 13 minutes.

There are 180 school days in Kevin's school year.

10. X has a normal distribution with mean 16. Given that the probability that X is less than 16.56 is 64%, find the standard deviation σ of this distribution.

11. X has a normal distribution with mean 91. Given that the probability that X is larger than 104 is 24.6%, find the standard deviation σ of this distribution.

12. X has a normal distribution with variance of 9. Given that the probability that X is more than 36.5 is 2.9%, find the mean μ of this distribution.

13. X has a normal distribution with standard deviation of 32. Given that the probability that X is more than 63 is 87.8%, find the mean μ of this distribution.

14. X has a normal distribution with variance of 25. Given that the probability that X is less than 27.5 is 0.312, find the mean μ of this distribution.

15. X has a normal distribution such that the probability that X is larger than 14.6 is 93.5% and $P(x > 29.6) = 2.2\%$. Find the mean μ and the standard deviation σ of this distribution.

16. $X \sim N(\mu, \sigma^2)$. $P(X > 19.6) = 0.16$ and $P(X < 17.6) = 0.012$.
Find μ and σ.

17. $X \sim N(\mu, \sigma^2)$. $P(X > 162) = 0.122$ and $P(X < 56) = 0.0276$.
Find μ and σ.

18. Wooden poles produced for electricity networks in rural areas have lengths that are normally distributed.

2% of the poles are rejected because they are considered too short, and 5% are rejected because they are too long.

(a) Find the mean and standard deviation of these poles if the acceptable range is between 6.3 m and 7.5 m.

(b) In a randomly selected sample of 20 poles, find the probability of finding 2 rejected poles.

19. Bottles of mineral water sold by a company are advertised to contain 1 litre of water. The company adjusts its filling process to fill the bottles with an average of 1012 ml to ensure that there is a minimum of 1 litre. The process follows a normal distribution with standard deviation of 5 ml.

(a) Find the probability that a randomly chosen bottle contains more than 1010 ml.

(b) Find the probability that a bottle contains less than the advertised volume.

(c) In a shipment of 10 000 bottles, what is the expected number of under-filled bottles?

20. Cholesterol plays a major role in a person's heart health. High blood cholesterol is a major risk factor for coronary heart disease and stroke. The level of cholesterol in the blood is measured in milligrams per decilitre (mg/dL). According to the WHO, in general, less than 200 mg/dL is a desirable level, 200 to 239 is borderline high, and above 240 is a high risk level and the person with this level has more than twice the risk of heart disease as a person with less than 200 mg/dL.

In a certain country, it is known that the average cholesterol level of the adult population is 184 mg/dL with a standard deviation of 22 mg/dL. It can be modelled by a normal distribution.

(a) What percentage do you expect to be borderline high?

(b) What percentage do you consider are high risk?

(c) Estimate the interquartile range of the cholesterol levels in this country.

(d) Above what value are the highest 2% of adults' cholesterol levels in this country?

21. A manufacturer of car tyres claims that its winter tyres can be described by a normal model with an average life of 52 000 km and a standard deviation of 4000 km.

(a) If you buy a set of tyres from this manufacturer, is it reasonable for you to hope they last more than 64 000 km?

(b) What percentage of these tyres do you expect to last less than 48 000 km?

(c) What percentage of these tyres do you expect to last between 48 000 km and 56 000 km?

(d) What is the IQR of the life of this type of tyre?

(e) The company wants to guarantee a minimum life for these tyres. They will refund customers whose tyres last less than a specific distance. What should their minimum life guarantee be so that they do not end up refunding more than 2% of their customers?

22. Chicken eggs are graded by size for the purpose of sales. In Europe, modern egg sizes are defined as follows: very large eggs have a mass of 73 g or more, large eggs are between 63 and 73 g, medium eggs are between 53 g and 63 g, and small eggs are less than 53 g.

(a) Mature hens (older than 1 year) produce eggs with an average mass of 67 g. 98% of the eggs produced by mature hens are 53 g or above. What is the standard deviation if the egg production can be modelled by a normal distribution?

(b) Young hens produce eggs with a mean of 51 g. Only 28% of their eggs exceed 53 g. What is the standard deviation?

(c) A farmer finds that 7% of his farm's eggs are small, and 12% are very large. Estimate the mean and standard deviation of these eggs.

23. A machine produces bearings with diameters that are normally distributed with mean 3.0005 cm and standard deviation 0.0010 cm. Specifications require the bearing diameters to lie in the interval 3.000 ± 0.0020 cm. Those outside the interval are considered scrap and must be disposed of. What percentage of the production will be scrap?

24. A soft-drink machine can be regulated so that it discharges an average μ ml per bottle. The amount of fill is normally distributed with a standard deviation 9 ml.

(a) Give the setting for μ so that 237 ml bottles will overflow only 1% of the time.

(b) The standard deviation σ of the machine can be adjusted to the required levels when needed. What is the largest value of σ that will allow the actual amount dispensed to fall within 30 ml of the mean with probability at least 95%?

25. The speeds of cars on a main highway are approximately normal. Data collected at a certain point show that 95% of the cars travel at a speed less than 140 km h^{-1}, and 10% travel at a speed less than 90 km h^{-1}.

(a) Find the average speed and the standard deviation for the cars travelling on that specific stretch of the highway.

(b) Find the proportion of cars that travel at speeds exceeding 110 km h^{-1}.

26. The random variable X is normally distributed and

$$P(X \leqslant 10) = 0.670; P(X \leqslant 12) = 0.937$$

Find $E(X)$.

27. A machine is set to produce bags of salt, whose weights are distributed normally, with a mean of 110 g and standard deviation of 1.142 g. If the weight of a bag of salt is less than 108 g, the bag is rejected. With these settings, 4% of the bags are rejected.

The settings of the machine are altered, and it is found that 7% of the bags are rejected.

(a) **(i)** If the mean has not changed, find the new standard deviation, correct to 3 decimal places.

The machine is adjusted to operate with this new value of the standard deviation.

(ii) Find the value, correct to 2 decimal places, at which the mean should be set so that only 4% of the bags are rejected.

(b) With the new settings from part (a), it is found that 80% of the bags of salt have a weight that lies between A g and B g, where A and B are symmetric about the mean. Find the values of A and B, giving your answers correct to 2 decimal places.

Chapter 12 practice questions

1. Residents of a small town have savings that are normally distributed with a mean of \$3000 and a standard deviation of \$500.

(a) What percentage of townspeople have savings greater than \$3200?

(b) Two townspeople are chosen at random. What is the probability that both of them have savings between \$2300 and \$3300?

(c) The percentage of townspeople with savings less than d dollars is 74.22%. Find the value of d.

2. A box contains 35 red discs and 5 black discs. A disc is selected at random and its colour noted. The disc is then replaced in the box.

(a) In eight such selections, find the probability that a black disc is selected:

(i) exactly once **(ii)** at least once.

(b) The process of selecting and replacing is carried out 400 times. What is the expected number of black discs that would be drawn?

3. The graph shows a normal curve for the random variable X, with mean μ and standard deviation σ.

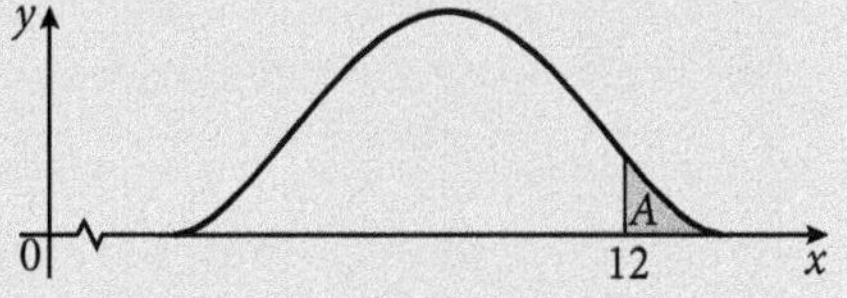

It is known that $P(X \geqslant 12) = 0.1$

(a) The shaded region A is the region under the curve where $X \geqslant 12$. Write down the area of the shaded region A.

It is also known that $P(X \leqslant 8) = 0.1$

(b) Find the value of μ, explaining your method in full.

(c) Show that $\sigma = 1.56$, correct to 3 significant figures.

(d) Find $P(X \leqslant 11)$.

4. A fair coin is tossed eight times. Calculate

(a) the probability of obtaining exactly 4 heads

(b) the probability of obtaining exactly 3 heads

(c) the probability of obtaining 3, 4, or 5 heads.

5. The lifespan of a particular species of insect is normally distributed with a mean of 57 hours and a standard deviation of 4.4 hours.

The probability that the lifespan of an insect of this species lies between 55 and 60 hours is represented by the shaded area in the diagram. This diagram represents the standard normal curve.

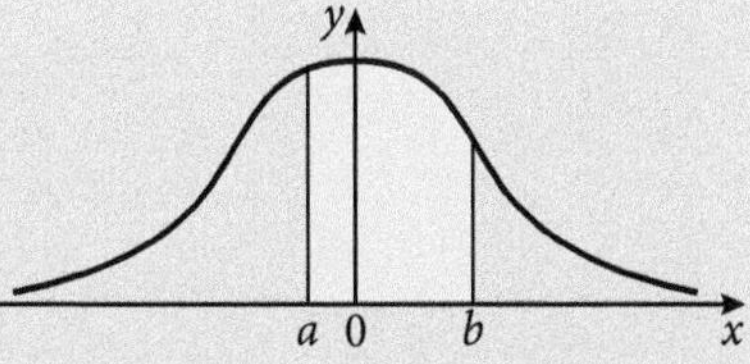

(a) Write down the values of a and b.

(b) Find the probability that the lifespan of an insect of this species is:

(i) more than 55 hours **(ii)** between 55 and 60 hours.

90% of the insects die after t hours.

(c) Represent this information on a standard normal curve diagram, similar to the one given in part (a), indicating clearly the area representing 90%.

(d) Find the value of t.

6. Intelligence quotient (IQ) in a certain population is normally distributed with a mean of 100 and a standard deviation of 15.

   (a) What percentage of the population has an IQ between 90 and 125?

   (b) If two people are chosen at random from the population, what is the probability that both have an IQ greater than 125?

7. Bags of cement are labelled as weighing 25 kg. The bags are filled by a machine and the actual weights are normally distributed with mean 25.7 kg and standard deviation 0.50 kg.

   (a) What is the probability a bag selected at random will weigh less than 25.0 kg?

   In order to reduce the number of underweight bags (bags weighing less than 25 kg) to 2.5% of the total, the mean is increased without changing the standard deviation.

   (b) Show that the increased mean is 26.0 kg.

   It is decided to purchase a more accurate machine for filling the bags. The requirements for this machine are that only 2.5% of bags be under 25 kg and that only 2.5% of bags be over 26 kg.

   (c) Calculate the mean and standard deviation that satisfy these requirements.

   The cost of the new machine is $5000. Cement sells for $0.80 per kg.

   (d) Compared to the cost of operating with a 26 kg mean, how many bags must be filled in order to recover the cost of the new equipment?

8. The mass of packets of a breakfast cereal is normally distributed with a mean of 750 g and standard deviation of 25 g.

   (a) Find the probability that a packet chosen at random has mass:

      (i) less than 740 g

      (ii) at least 780 g

      (iii) between 740 g and 780 g.

   (b) Two packets are chosen at random. What is the probability that both packets have a mass that is less than 740 g?

   (c) The mass of 70% of the packets is more than x g. Find the value of x.

9. In Tallopia, the height of adults is normally distributed with a mean of 187.5 cm and a standard deviation of 9.5 cm.

   (a) What percentage of adults in this village are taller than than 197 cm?

   (b) A standard doorway in the village is designed so that 99% of adults have a space of at least 17 cm over their heads when going through the doorway. Find the height of a standard doorway in the village. Give your answer to the nearest cm.

10. It is claimed that the masses of a population of lions are normally distributed with a mean mass of 310 kg and a standard deviation of 30 kg.

(a) Calculate the probability that a lion selected at random will have a mass of 350 kg or more.

(b) The probability that the mass of a lion lies between a and b is 0.95, where a and b are symmetric about the mean. Find the values of a and b.

11. Reaction times of human beings are normally distributed with a mean of 0.76 seconds and a standard deviation of 0.06 seconds.

The graph is that of the standard normal curve. The shaded area represents the probability that the reaction time of a person chosen at random is between 0.70 and 0.79 seconds.

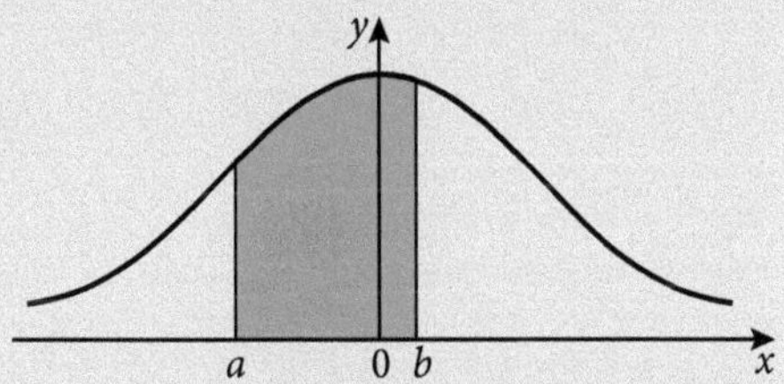

(a) Write down the values of a and b.

(b) Calculate the probability that the reaction time of a person chosen at random is:

(i) greater than 0.70 seconds

(ii) between 0.70 and 0.79 seconds.

Three per cent of the population have a reaction time less than c seconds.

(c) **(i)** Represent this information on a diagram similar to the one above. Indicate clearly the area representing 3%.

(ii) Find c.

12. A factory makes calculators. Over a long period, 2% of them are found to be faulty. A random sample of 100 calculators is tested.

(a) Write down the expected number of faulty calculators in the sample.

(b) Find the probability that three calculators are faulty.

(c) Find the probability that more than one calculator is faulty.

13. The speeds of cars at a certain point on a straight road are normally distributed with mean μ and standard deviation σ. 15% of the cars travelled at speeds greater than 90 km h^{-1}, and 12% of them at speeds less than 40 km h^{-1}. Find μ and σ.

14. Bag A contains 2 red balls and 3 green balls. Two balls are chosen at random from the bag without replacement. Let X denote the number of red balls chosen. The table shows the probability distribution for X.

x	0	1	2
$\text{P}(X = x)$	$\frac{3}{10}$	$\frac{6}{10}$	$\frac{1}{10}$

(a) Calculate $\text{E}(X)$, the mean number of red balls chosen.

Bag B contains 4 red balls and 2 green balls. Two balls are chosen at random from bag B.

(b) (i) Draw a tree diagram to represent this information, including the probability of each event.

(ii) Hence find the probability distribution for Y, where Y is the number of red balls chosen.

A standard dice with six faces is rolled. If a 1 or 6 is obtained, two balls are chosen from bag A, otherwise two balls are chosen from bag B.

(c) Calculate the probability that two red balls are chosen.

(d) Given that two red balls are obtained, find the conditional probability that a 1 or 6 was rolled on the dice.

15. Ball bearings are used in engines in large quantities. A car manufacturer buys these bearings from a factory. They agree on the following terms: The car company chooses a sample of 50 ball bearings from the shipment. If they find more than 2 defective bearings, the shipment is rejected. It is a fact that the factory produces 4% defective bearings.

(a) What is the probability that the sample is free of defects?

(b) What is the probability that the shipment is accepted?

(c) What is the expected number of defective bearings in the sample of 50?

16. Each DVD produced by a certain company is guaranteed to function properly with a probability of 98%. The company sells these DVDs in packs of 10 and offers a money-back guarantee that all the DVDs in a package will function.

(a) What is the probability that a package is returned?

(b) You buy three packages. What is the probability that exactly 1 of them must be returned?

17. It is estimated that 2.3% of the cherry tomato fruits produced at a certain farm are considered to be small and cannot be sold for commercial purposes. The farmers have to separate such fruits and use them for domestic consumption instead.

(a) 12 tomatoes are randomly selected from the produce.

(i) Calculate the probability that 3 are not fit for selling.

(ii) Calculate the probability that at least 4 are not fit for selling.

(b) It is known that the sizes of such tomatoes are normally distributed with a mean of 3 cm and a standard deviation of 0.5 cm. Tomatoes that are categorised as large will have to be larger than 2.5 cm. What proportion of the produce is large?

18. A factory has a machine designed to produce 1 kg bags of sugar. It is found that the average weight of sugar in the bags is 1.02 kg. Assuming that the weights of the bags are normally distributed, find the standard deviation if 1.7% of the bags weigh below 1 kg.

Give your answer correct to the nearest 0.1 gram.

19. Ian and Karl have been chosen to represent their countries in the Olympic discus throw. Assume that the distance thrown by each athlete is normally distributed. The mean distance thrown by Ian in the past year was 60.33 m with a standard deviation of 1.95 m.

(a) In the past year, 80% of Ian's throws have been longer than x metres. Find x, correct to two decimal places.

(b) In the past year, 80% of Karl's throws have been longer than 56.52 m. If the mean distance of his throws was 59.39 m, find the standard deviation of his throws, correct to two decimal places.

(c) This year, Karl's throws have a mean of 59.50 m and a standard deviation of 3.00 m. Ian's throws still have a mean of 60.33 m and standard deviation 1.95 m. In a competition, an athlete must have at least one throw of 65 m or more in the first round to qualify for the final round. Each athlete is allowed three throws in the first round.

(i) Determine which of these two athletes is more likely to qualify for the final on their first throw.

(ii) Find the probability that both athletes qualify for the final.

20. A company buys 44% of its stock of bolts from manufacturer A and the rest from manufacturer B. The diameters of the bolts produced by each manufacturer follow a normal distribution with a standard deviation of 0.16 mm.

The mean diameter of the bolts produced by manufacturer A is 1.56 mm. 24.2% of the bolts produced by manufacturer B have a diameter less than 1.52 mm.

(a) Find the mean diameter of the bolts produced by manufacturer B.

A bolt is chosen at random from the company's stock.

(b) Show that the probability that the diameter is less than 1.52 mm is 0.312, to 3 significant figures.

(c) The diameter of the bolt is found to be less than 1.52 mm. Find the probability that the bolt was produced by manufacturer B.

(d) Manufacturer B makes 8000 bolts in one day. It makes a profit of \$1.50 on each bolt sold, on condition that its diameter measures between 1.52 mm and 1.83 mm. Bolts whose diameters measure less than 1.52 mm must be discarded at a loss of \$0.85 per bolt.

Bolts whose diameters measure over 1.83 mm are sold at a reduced profit of \$0.50 per bolt.

Find the expected profit for manufacturer B.

Internal assessment

Internal assessment (IA) is an important component of the Analysis and Approaches SL course and contributes 20% to your final grade. It is a significant part of the overall assessment for the course and should be taken seriously. It should also be pointed out that your work in completing the IA component differs in important ways from the written exams (external assessment) for the course.

- Unlike written examinations, you do *not* perform IA work under strict time constraints.
- You have some freedom to decide which mathematical topic you wish to explore.
- Your IA work involves writing about mathematics, not just using mathematical procedures.
- Regular discussion with, and feedback from, your teacher will be essential.
- You should endeavour to explore a topic in which you have a genuine personal interest.
- You will be rewarded for evidence of creativity, curiosity, and independent thinking.

Mathematical exploration

To satisfy the IA component, you are required to complete a piece of written work on a mathematical topic that you choose in consultation with your teacher. This piece of written work is formally referred to as the mathematical exploration. It will be referred to simply as the 'exploration' throughout this chapter. Your primary objective is to *explore* a mathematical topic in which you are *genuinely interested* and that is at an *appropriate level* for the course. A fundamental aspect of your exploration must be the *use of mathematics* in a manner that clearly demonstrates your knowledge and understanding of the relevant mathematics. Your teacher may provide you with a list of ideas (or 'stimuli') to help you in the process of finding a suitable topic.

It is your responsibility to determine whether or not you are sufficiently interested in a particular topic – and it is your teacher's responsibility to determine if an exploration of the topic can be conducted at a mathematical level that is suitable for the course. Your teacher will help you determine if an exploration of a certain topic can potentially address the five assessment criteria satisfactorily. Your exploration should be approximately 12 to 20 pages long with double line spacing.

See the list of 200 ideas included in the eBook. You may find a suitable topic in the list, or the list may help you find or develop your own ideas for a mathematical topic to explore.

Internal assessment criteria

Your exploration will be assessed by your teacher according to the following five criteria.

A Presentation

This criterion assesses the organisation and coherence of the exploration. A well-organised exploration has an introduction, a rationale (which includes a brief explanation of why the topic was chosen), a description of the aim of the exploration, and a conclusion.

B Mathematical communication

This criterion assesses to what extent you are able to:

- use appropriate mathematical language (notation, symbols, terminology)
- clearly define key terms, variables, and parameters
- use multiple forms of mathematical representation, such as formulae, diagrams, tables, charts, graphs, and models
- apply a deductive approach in general, and present any proofs in a logical manner.

C Personal engagement

This criterion assesses the extent to which you engage with the exploration and present it in such a way that clearly shows *your own personal approach*. Personal engagement may be recognised in several different ways. These may include – but are not limited to – thinking independently and/or creatively, addressing personal interest, presenting mathematical ideas in your own words and diagrams, developing your own ideas and testing them, and creating your own examples to illustrate important results.

D Reflection

This criterion assesses how well you *review*, *analyse*, and *evaluate* the exploration. Although reflection may be seen in the conclusion to the exploration, you should also give evidence of reflective thought throughout the exploration. Reflection can be demonstrated by consideration of limitations and/or extensions, commenting on what you've learned, or comparing different mathematical methods and approaches.

E Use of mathematics

This criterion assesses to what extent and how well you use mathematics in your exploration. The mathematical working in your exploration needs to be *sufficiently sophisticated* and *rigorous*. The chosen topic should involve mathematics in the Analysis and Approaches SL syllabus or at a similar level. Sophistication and rigour can include understanding and use of challenging mathematical concepts, looking at a problem from different perspectives, mathematical arguments expressed clearly in a logical manner, or seeing underlying structures to link different areas of mathematics.

Your exploration will earn a score out of a total of 20 possible marks. The five criteria do not contribute equally to the overall score for your exploration. For example, criterion E (Use of mathematics) is 30% of the overall score, whereas crtieria C (Personal engagement) and D (Reflection) contribute 15% each.

It is very important that you familiarise yourself with the assessment criteria for the Analysis and Approaches SL exploration and refer to them while you are writing your exploration. The achievement levels for each criteria and associated descriptors are as follows:

A Presentation	
0	The exploration does not reach the standard described by the descriptors below.
1	The exploration has some coherence or some organisation.
2	The exploration has some coherence and shows some organisation.
3	The exploration is coherent and well organised.
4	The exploration is coherent, well organised, and concise.

B Mathematical communication	
0	The exploration does not reach the standard described by the descriptors below.
1	The exploration contains some relevant mathematical communication that is partially appropriate.
2	The exploration contains some relevant appropriate mathematical communication.
3	The mathematical communication is relevant, appropriate, and is mostly consistent.
4	The mathematical communication is relevant, appropriate, and consistent throughout.

C Personal engagement	
0	The exploration does not reach the standard described by the descriptors below.
1	There is evidence of some personal engagement.
2	There is evidence of significant personal engagement.
3	There is evidence of outstanding personal engagement.

D Reflection	
0	The exploration does not reach the standard described by the descriptors below.
1	There is evidence of limited reflection.
2	There is evidence of meaningful reflection.
3	There is substantial evidence of critical reflection.

E Use of mathematics	
0	The exploration does not reach the standard described by the descriptors below.
1	Some relevant mathematics is used.
2	Some relevant mathematics is used. Limited understanding is demonstrated.
3	Relevant mathematics commensurate with the level of the course is used. The mathematics explored is correct. Limited understanding is demonstrated.

4	Relevant mathematics commensurate with the level of the course is used. The mathematics is partially correct. Some knowledge and understanding are demonstrated.
5	Relevant mathematics commensurate with the level of the course is used. The mathematics explored is mostly correct. Good knowledge and understanding are demonstrated.
6	Relevant mathematics commensurate with the level of the course is used. The mathematics explored is correct. Thorough knowledge and understanding are demonstrated.

Guidance

Your teacher will provide oral and/or written advice on a draft of your exploration pertaining to how it can be improved. Your teacher will also write thorough and descriptive comments on the final version of your exploration to assist IB moderators in confirming the criteria scores they've awarded.

Conducting an in-depth individual exploration into the mathematics of a particular topic can be an interesting and very rewarding experience. It is important to take all stages of your work on the exploration seriously – not only because it is worth 20% of your final grade for the course but also because of the opportunity to pursue your own personal interests without the pressure of examination conditions. The exploration should *not* be approached as simply an extended homework assignment. The task of writing the exploration will require you to analyse, think, write, edit, and use mathematics in a readable and focused manner. Hopefully, it will also be enjoyable, thought-provoking, and satisfying, and it should give you the opportunity to gain a deeper appreciation for the beauty, power, and usefulness of mathematics.

Warning: Failure to properly cite any text, diagrams, images, mathematical working, or ideas that are not your own may result in your exploration being reviewed for malpractice, which could have serious consequences.

Although it is required that your exploration is completely your own work, you should consult with your teacher on a regular basis. You are allowed to work collaboratively with fellow students, but this should be limited to the following: selecting a topic, finding resources, understanding relevant mathematical knowledge and skills, and receiving peer feedback on your writing. While you are encouraged to *talk* through your ideas with others, it is not appropriate for you to *work* with others on your exploration. Your teacher should provide support and advice during the planning and writing stages of your exploration. Both you and your teacher will need to verify the authenticity of your exploration.

If you are uncertain about the formatting and style of citations and a bibliography (not the same thing), then you should consult with teacher(s) at your school who have expertise in this area – such as an English teacher or librarian. A bibliography is required but it does not replace the need for appropriate citations (inline or footnotes) at the pertinent location in the exploration.

Any text, diagrams, images, mathematical working, or ideas that are not your own must be cited where they appear in your exploration. Otherwise, all of the work connected with your exploration must be your own. Your exploration must provide the reader with the exact sources of quotations, ideas, and points of view with a complete and accurate bibliography. There are a number of acceptable bibliographic styles. Whichever style you choose, it must include all relevant source information and be applied consistently. Group work is not allowed. Also, if you are writing an extended essay for mathematics, you are not allowed to submit the same or similar piece of work for the exploration – and you should not write about the same mathematical topic for both.

In organising a successful exploration, consider the following suggestions:

1. Select a topic in which you are *genuinely interested*. Include a brief explanation in the early part of your exploration about why you chose your topic – including why you find it interesting.

2. Consult with your teacher to confirm that the topic is at the *appropriate level of mathematics* – namely, that it is at the same or similar level of the mathematics in the SL syllabus.

3. Find as much *information* about the topic as possible. Although information found on websites can be very helpful, try to also find information in books, journals, textbooks, and other printed material.

4. Although there is no requirement that you present your exploration to your classmates, it should be written so that they can follow it without trouble. Your exploration needs to be *logically organised* and use appropriate mathematical terminology and notation.

5. The most important aspects of your exploration should be about *mathematical communication and using mathematics.* Although other aspects of your topic – for example, historical, personal, cultural – can be discussed, be careful to keep focus on the mathematical features.

6. Two of the assessment criteria – Personal engagement and Reflection – are about *what you think about the topic* you are exploring. Don't hesitate to pose your own relevant and insightful questions as part of your exploration – and then to address these questions using mathematics at a suitably sophisticated level along with sufficient written commentary.

7. Although your teacher will expect and require you to work independently, you are allowed to *consult with your teacher* – and your teacher is allowed to give you advice and feedback to a certain extent while you are working on your exploration. It is especially important to check with your teacher that any *mathematics in your exploration is correct.* Your teacher will not give mathematical answers or corrections but can indicate where any errors have been made or where improvement is needed.

Warning: Although you will need to conduct some research, your exploration is *not* a research paper where you simply report what you've learned about a mathematical topic. You must discuss your thoughts about the mathematics and you must apply the mathematics in a way that clearly demonstrates your knowledge and understanding of the mathematics.

Keep in mind that you should write your exploration so that a student in your Analysis and Approaches SL class can understand it. That is, *your audience is your fellow students.*

Mathematical exploration - SL student checklist

- Is your exploration written entirely by yourself? Have you avoided simply replicating work and ideas from sources you found? ☐ Yes ☐ No
- Have you strived to apply your personal interest, develop your own ideas, and use critical thinking skills during your exploration? ☐ Yes ☐ No
- Did you refer to the five assessment criteria while writing your exploration? ☐ Yes ☐ No
- Does your exploration focus on good mathematical communication – and does it read like an article from a mathematical journal? ☐ Yes ☐ No
- Does your exploration have a clearly identified introduction and conclusion? ☐ Yes ☐ No
- Have you provided appropriate citation for any ideas, mathematical working, images, graphs, etc. that are not your own at the point they appear in your exploration? ☐ Yes ☐ No
- Not including the bibliography, is your exploration 12 to 20 pages? ☐ Yes ☐ No
- Are graphs, tables, and diagrams sufficiently described and labelled? ☐ Yes ☐ No
- To the best of your knowledge, have you used mathematics that is at the same level, or similar, to that studied in Analysis and Approaches SL? ☐ Yes ☐ No
- Have you attempted to discuss mathematical ideas, and use mathematics, with a sufficient level of sophistication and rigour? ☐ Yes ☐ No
- Are formulae, graphs, tables, and diagrams in the main body of text? (Preferably no full-page graphs, and no separate appendices.) ☐ Yes ☐ No
- Have you used technology – such as a GDC, spreadsheet, mathematics software, drawing and word-processing software – to enhance mathematical communication? ☐ Yes ☐ No
- Have you used appropriate mathematical language (notation, symbols, terminology) and defined key terms? ☐ Yes ☐ No
- Is the mathematics in your exploration performed precisely and accurately? ☐ Yes ☐ No
- Has calculator/computer notation and terminology been used? ($y = x^2$, not $y = x$^2; π, not <pi>; $\approx$, not approximately equal to; $|x|$, not abs(x); etc) ☐ Yes ☐ No
- Have you included reflective and explanatory comments about the topic being explored throughout your exploration? ☐ Yes ☐ No

Finding, developing, and choosing a topic for your exploration

It is fair to say that the *most important stage* of completing your exploration is determining the mathematical topic you are going to investigate, write about, and apply. Your exploration is much more likely to be successful – and gratifying – if it focuses on a mathematical topic in which you have a genuine interest, is at a suitable level for the Analysis and Approaches SL course, and for which you are confident that you can discuss and use the relevant mathematics in a manner that demonstrates thorough knowledge and understanding. There is no single approach for determining an exploration topic that is guaranteed to be successful for all students. Your teacher will provide helpful advice and support. Your teacher may supply you with a short list of some broad stimuli to start the process of finding a much narrower topic. Many teachers have found that starting with a sufficiently narrow topic is often more successful than starting with a very broad topic that requires a significant effort to reduce to the extent that it can be explored in less than 20 pages (double spaced).

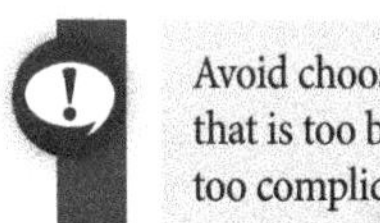

Avoid choosing a topic that is too broad and/or too complicated.

In the eBook for this textbook you will find a list of 200 mathematical topics. Some of the topics in the list are broad but many are already quite narrow in scope. It is possible that some of these 200 topics could be the focus of an exploration, while others will require you to investigate further to develop a narrower focus to explore. Do not restrict yourself to the topics in the list. This list is only the tip of the iceberg with regard to potential topics for your exploration. Reading through this list may stimulate you to think of some other topic(s) that you may find interesting to explore. Many of the items in the list may be unfamiliar to you. A quick search on the internet should give you a better idea what each is about and help you determine if you're interested enough to investigate further – and to see if it might be a suitable topic for your exploration.

Theory of knowledge

At the start of his wonderful book *Nature's Numbers*, the mathematician Ian Stewart writes:

> *'We live in a universe of patterns. Every night the stars move in circles across the sky. The seasons cycle at yearly intervals. No two snowflakes are ever exactly the same, but they all have sixfold symmetry. Tigers and zebras are covered in patterns of stripes, leopards and hyenas are covered in patterns of spots. Intricate trains of waves march across the oceans; very similar trains of sand dunes march across the desert. Coloured arcs of light adorn the sky in the form of rainbows, and a bright circular halo sometimes surrounds the moon on winter nights. Spherical drops of water fall from clouds.'*

We could add to Stewart's list. Wallpaper is patterned (there are surprisingly only 17 different distinct groups of possible patterns); buildings often exhibit mirror symmetry and their structure is carefully proportioned; the digital traces on memory sticks or hard drives are patterned in a way that makes them suitable for storing data; mechanical devices such as clocks and engines depend on symmetry and patterning for their smooth movement; the day is divided into equal parts that are represented using angles or digits; music possesses horizontal and vertical symmetries – and human behaviour is patterned.

It is no accident that the world is full of patterns. Symmetry in a building is not only easy on the eye but it ensures that the design is simple. Pattern is a labour-saving strategy. The same plan can be used for each window, or the plan for one side of a building can be used in reverse for the other side. These informational shortcuts can be found both in the man-made world and in nature. The same blueprint for generating twig patterns can be used for bigger branches, or one plan can be used for all the petals in a flower. It is a sort of design efficiency. The wealth of patterns in the world is a series of cost-effective solutions to problems – and that is why these patterns are worth studying.

Mathematics is one way in which human beings formally study patterns. While the natural sciences study patterns by going out into the world, collecting examples and analysing them, mathematics studies patterns in the abstract. Mathematics in its purest form is not fieldwork or experiment. Its raw materials are abstract structures specified by symbols, and mathematicians arrive at conclusions through their manipulation. In this sense, mathematics is a little 'other-wordly' – a characteristic that makes it interesting from a ToK perspective. It means that in some sense, mathematics is more like an art than a science. There is in this suggestion more than a hint of a deep reliance on creativity and imagination. A comparison with the arts and the sciences is instructive and reveals the truly special place that mathematics occupies in human knowledge.

In this chapter, we will investigate mathematics using the basic structure of the knowledge framework: Perspectives, methods and tools, and the link to the individual.

Under 'Perspectives', we will look at the orientation of mathematics within the academy. There are a number of key questions to be answered here:

- What is mathematics about?
- How should we think of mathematics: as a human construction or something in the world?
- Why is mathematics useful?

Under 'Methods and Tools' we will discuss exactly what mathematicians do – how they arrive at mathematical knowledge and what counts as facts and truth in mathematics. This is where we unpack the key conceptual building blocks of mathematical thought.

The final section deals with mathematics and the individual. What is the link between mathematics and supposedly subjective phenomena such as beauty? How reliable are our mathematical intuitions? Is mathematics a personal journey or is it something that we collaborate on?

On the way, we will have fun with infinite numbers, self-similar patterns and security codes. While it might be removed from the physical world, the world of mathematics is just as fascinating, if not more so. Enjoy!

What role does mathematics play in your life?

Perspectives

Mathematics and number

As a first definition, let's say that mathematics is the formal study of patterns. In this section we will see how far this basic idea takes us.

Imagine a simple pattern in the world – a set of similar objects, for example, a field of animals. Let's say that the animals are of the same kind – they are cows. To recognise that a group of different things all belong to the same kind is already remarkable. It means ignoring all the things that mark out individual animals and focusing only on what they have in common. Grouping a set of things together by common characteristics is a powerful technique in the sciences. If such a classification is effective, it might yield understanding, generalisations and predictions. We call groups that have these properties **natural kinds** – it is something that might be expected to happen in biology. But mathematics goes one step further. Suppose that we make a mark 'I' on a clay tablet for every cow in the field. We end up with a mark 'IIIIIII'. What we have done now is to abstract away everything about the animals in the field: the fact that they are animals, that they are cows, that they are eating grass. What is left is their number.

So, the simplest pattern that we can deal with abstractly is number. In a somewhat magical way, the inscriptions of the tablet **represent** the cows in the field. They are a convenient stand-in for the real world. If we want to find out what happens when we remove 'III' cows from the field. We can either move them physically or we simply separate the 'cow' symbols: 'IIIII III'. Manipulating the symbols is clearly easier to perform. Mathematics manipulates representations rather than the real world because it is easier.

We do not know if something like this story is accurate at the beginning of the long history of mathematics. But we do know that imprints on a Sumerian clay tablet led eventually to the astounding sophistication of the proof of Fermat's last theorem and to modern algebra, analysis, and geometry. Mathematics has been shaped by the job it is expected to perform and through countless quirks of culture. Improvised methods designed to deliver a temporary solution to an unforeseen problem become permanent. If they work well, they get passed on and take on a life of their own. Less good solutions eventually fall into disuse in a sort of Darwinian selection of competing ideas. We could call histories like this **cultural evolution**.

But has the counting of cows in a field really got anything to do with modern mathematics? Let's examine the example more closely. We add an 'I' on the tablet for each cow in the field, subject to two strict rules: no cow should be 'counted' more than once and all the cows in the field are counted. Although these rules are quite natural to us, they are mathematically sophisticated. Mathematically, we are establishing a mapping between the marks on the tablet and the cows in the field that is a **one-to-one correspondence**. This means a mapping links a mark to a unique cow (injective) and that all cows in the field are linked (surjective). While these early users of mathematics might not have understood it quite in these terms, they nonetheless needed to use these properties when counting. But there is something else at work here. The compound symbol 'IIIIIIII' stands for the whole field of cows. It is a property of the whole set. It expresses the size of the set or its **cardinality**. The counting of cows in a field has a lot to do with the deep nature of mathematics itself.

Indeed, there are three more ideas illustrated by this simple example. The first is the power of numbers to create ordering: I II III IIII is such an ordering. This is called the **ordinal** property of number. Second, it illustrates the special place of sets and mappings in mathematics. We focused on the set of cows and the set of marks on the tablet. Third, we counted the first set by establishing a one-to-one correspondence with the second. This is a technique that works with any sets, including those that have infinitely many members. Mathematics is truly about sets and the mappings between them.

By representing the real world by marks bearing a special relation to their targets, human beings initiated perhaps the most extraordinary technical advance in their history: the invention of symbolic representation. Manipulating symbols is easier than manipulating objects in the world. Moreover, symbols allow this information to be communicated over distance and time. But the most powerful feature of symbols is that they can be used to represent states of affairs that are not physically present. Symbols can represent past worlds, possible worlds, and desired future worlds. Symbols allow us to tell stories, write histories, and make plans. Symbols that do not actually correspond with the world are called **counterfactuals.** They describe 'what if' situations. What if the Allies had lost World War II? What if we add sulfuric acid to copper? What if we wake up one morning to discover that we have been transformed into a giant insect? What if parallel lines could actually meet? What if there was a solution to the equation $x^2 = -1$? The power of symbolic representation is that it allows us to build abstract worlds – virtual realities where the 'what if' conditions are true.

There is a sense in which the world of mathematics is one such virtual universe, containing all manner of exciting and weird things. Mathematicians discuss 11-dimensional hypercubes, infinite sets of numbers, infinite numbers, surfaces that turn you from being right-handed to left-handed as you traverse them, spaces where the angles of a triangle add up to more than 180 degrees, spaces where parallel lines diverge, systems where the order of the operation matters (where $A * B$ is not the same as $B * A$), vectors in infinite-dimensional space, series that go on forever, and geometric figures that are self-similar called fractals (where you can take a small piece of the original figure then enlarge it and it looks identical – truly identical – to the original). And all this started with the making of a simple mark on a clay tablet.

Mathematics uses symbols to describe these amazing structures in the basic language of sets and the mappings between them. Because symbols are abstract and not limited to representing things in the world, mathematicians can use their imaginations to create a virtual reality following its own rule system unhindered by what the world is really like, a **counterfactual world**. In this world, mathematicians can explore the patterns they encounter.

Yet mathematics is remarkably useful in this world. From building bridges to controlling strategy in football, mathematics lies at the heart of the modern world. If mathematics really is so other-worldy, how come it has so much to say about this one?

This is an important question that motivates much of what follows.

If symbolic representation is the most significant technical advance in history, what would you put in second place?

Purpose: mathematics for its own sake

ToK uses the map metaphor; knowledge is taken to be like a map that is used for a particular purpose, such as solving a particular problem or answering a question. The map is a simplified picture of the world and its simplicity is its strength. It ensures that we get the job done with the least cognitive cost. If this is right, then it is natural to ask about the purpose of this particular map. What problems does it solve or what questions does it answer? There seem to be two categories: those questions that occur strictly within the virtual reality of mathematics itself (mathematics for its own sake) and those that occur in the world outside (mathematics as a tool). These categories broadly correspond to two subdivisions of mathematics that are often two different departments within a university: pure mathematics and applied mathematics.

Let's start with pure mathematics. A typical example of a problem in this category is how to solve a particular type of equation.

An example of a problem in pure mathematics might be how to solve the equation

(1) $x^3 - 2x^2 - x + 2 = 0$

The task is to find a value for x that satisfies the equation. In books like this, there are many such equations and, in this context, they often have simple integer solutions. An initial strategy might be to try a value for x to see if it fits. If we try $x = 0$, then equation (1) gives us:

$0^3 - 2 \cdot 0^2 - 0 + 2 = 0$, i.e. $2 = 0$

which is clearly not true. So, we can say that $x = 0$ is not a solution to the equation.

But if we try $x = 1$, then equation (1) gives us:

$1^3 - 2 \cdot 1^2 - 1 + 2 = 0$

In other words, $1 - 2 - 1 + 2 = 0$ is true. So, $x = 1$ is a solution to the equation.

The trick now, as you know, is to factor out $(x - 1)$ from equation (1) to give:

(2) $(x - 1)(x^2 - x - 2) = 0$

We can now try to find values of x that make the second bracket in (2) equal to 0. This can be done either by trying out hopeful values of x (2 seems to be a good bet, for example) or using the quadratic formula. We end up with $x = 2$ or $x = -1$

The equation therefore has three solutions: $x = 1$ or $x = -1$ or $x = 2$

The history of these problems illustrates the great attraction of pure mathematics. Certainly, these problems were of interest from the 7th century in what is now the Middle East – the home of algebra. The great 11th century Persian mathematician and poet Omar Khayyam wrote a treatise about similar so-called cubic equations and realised they could have more than one solution. By the 16th century, cubic equations were of public interest. In Italy, contests were held to showcase the ability of mathematicians to solve cubic equations, often with a great deal of money at stake. One such contest took place in 1635 between Antonio Fior and Niccolò Tartaglia. Fior was a student of Scipione del Ferro, who had found a method for solving equations of the type $x^3 + ax = b$, which is known as the 'unknowns and cubes problem' (where a and b are given numbers).

Del Ferro kept his method secret until just before his death when he passed the method on to his student. Fior began to boast that he knew how to solve cubics. Tartaglia also announced that he had been able to solve a number of cubic equations successfully. Fior immediately challenged Tartaglia to a contest. Each was to give the other a set of 30 problems and put up a sum of money. The person who had solved the most after 30 days would take all the money.

Tartaglia had produced a method to solve a different type of cubic $x^3 + ax^2 = b$. Fior was confident that his ability to solve cubic equations would defeat Tartaglia and submitted 30 problems of the 'unknowns and cubes' type, but Tartaglia submitted a variety of different problems. Although Tartaglia could not initially solve the 'unknowns and cubes' type of equation, he worked hard and discovered a method to solve this type of problem. He then managed to solve all of Fior's problems in less than two hours. In the meantime, Fior had made little progress with Tartaglia's problems and it was obvious who was the winner. Tartaglia did not take Fior's money though; the honour of winning was enough.

Tartaglia represents the essence of the pure mathematician: someone who is intrigued by puzzles and has a deep desire to solve them. It is the problem itself that is the motivation, not possible real-world applications.

What other knowledge is worth pursuing for its own sake?

A modern example is the solution of Fermat's conjecture by Andrew Wiles. The French mathematician Pierre de Fermat wrote the conjecture in 1627 as a short observation in his copy of *The Arithmetics of Diophantus*.

The conjecture is that the equation

$$A^n + B^n = C^n$$

where A, B, C are positive integers and $n > 2$ has no solution. Despite a large number of attempts to prove it, the conjecture remained unproved for 358 years until Wiles published his successful proof in 1995. The proof is way beyond the scope of this book, but there have been a number of interesting books and TV programmes made about it, including Simon Singh *Fermat's Last Theorem* (1997) and the BBC Horizon programme *Fermat's Last Theorem* (1996). As mathematician Roger Penrose remarked, '*QED: how to solve the greatest mathematical puzzle of your age. Lock self in room. Emerge 7 years later*'.

Purpose: mathematical models

Unlike pure mathematics, which is about the solution of exclusively mathematical puzzles, applied mathematics is about solving real-world problems. The mathematics it produces can be just as interesting from an insider's viewpoint as the problems of pure mathematics (and often the two are inseparable), but a piece of applied mathematics is judged by whether it can be usefully applied in the world.

Here is an example of applied mathematics at work. This is a problem that could have been posed in this book or, indeed (and this is the point), in a physics course.

A stone is dropped down a 30 m well. How long will it take the stone to reach the bottom of the well, neglecting the effect of air resistance?

The typical way to solve this type of problem is to use what we call a **mathematical model**. The essence of mathematical modelling is to produce a description of the problem where the main physical features become variables in an equation which is then solved and translated back into the real world.

To model the situation above:

We know that the acceleration due to gravity is 9.8 m s^{-2}, and we also know that the distance travelled s is given by the equation:

$$s = \frac{1}{2}at^2, \text{ where } a = \text{acceleration and } t = \text{time}$$

So we substitute the known values into the equation and get:

$$30 = \frac{1}{2}(9.8)t^2$$

Rearranging the equation gives us:

$$\frac{60}{9.8} = t^2, \text{ so } t = \sqrt{\frac{60}{9.8}} = 2.47 \text{ seconds (3 s.f.)}$$

There are a number of points to make about the process here that are typical of mathematical models.

(1) The model neglects factors that are known to operate in the real-world situation. There are two big assumptions made: that the stone will not experience air resistance, which will act as a significant drag force, and that the acceleration due to gravity is constant.

(2) The model appeals to a law of nature. In this case, the law of acceleration due to gravity.

(3) The model uses values for constants that are established empirically. In this case, the acceleration due to gravity at the Earth's surface.

We know that neither of the assumptions in (1) is true. The effect of air resistance can be highly significant. We know that if you have the misfortune to fall from an airplane above 100 m or so, the height does not matter – the speed of impact with the ground will be the same, around 150 km h^{-1}, because of the effect of air resistance (of course, it matters how you fall). The changing strength of gravitational force is a less important factor for normal wells.

But if we are dealing with a well that is 4000 km deep, then this factor would be significant. The point is that the model is actually fictional (it even breaks a major law of physics). It could never be true in the sense of exactly corresponding to reality. However, it is a sort of idealisation that we accept because the model provides an approximation to the behaviour of the stone (although not such a good one for deeper wells) and more importantly it gives us understanding of the system. If we were to make the modelling assumptions more realistic, the mathematics in the model would become too complicated to solve easily. Points (2) and (3) show us that the actual content of the model depends on something outside mathematics – namely some well-established results in physics. The mathematics is only a tool, albeit an important one. A model is a mathematical map – a simplified picture of reality that is useful.

Another beautiful example is the Lotka–Volterra model of prey–predator population dynamics in biology. This model was proposed by Alfred Lotka in 1925 and independently by Vito Volterra in 1926.

The model assumes a closed environment where there are only two species, prey and predator, and no other factors. The rate of growth of prey is assumed to be a constant proportion A of the population. The rate at which predators eat prey is B, which is assumed to be a constant proportion of the product of predators and prey. The death rate of predators, C, is assumed to be a constant proportion of the population, and there is a rate of generation of new predators, D, dependent on the product of prey and predators.

These modelling assumptions give rise to a pair of coupled differential equations:

(1) $\frac{dx}{dt} = Ax - Bxy$

(2) $\frac{dy}{dt} = -Cy + Dxy$

A modern computer package gives the following evolution of prey and predators over time:

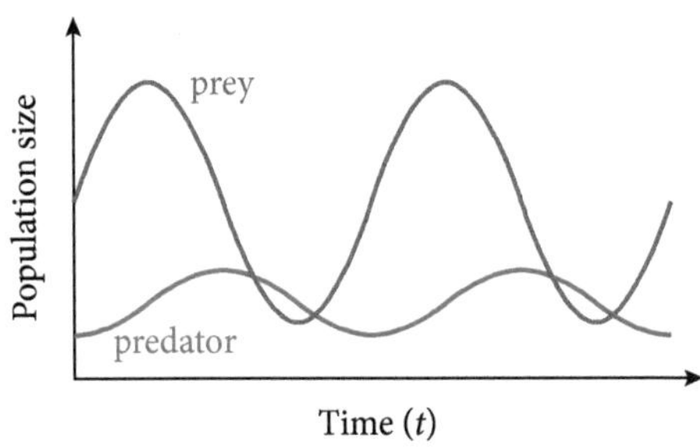

Figure 1 Evolution of prey and predator populations over time

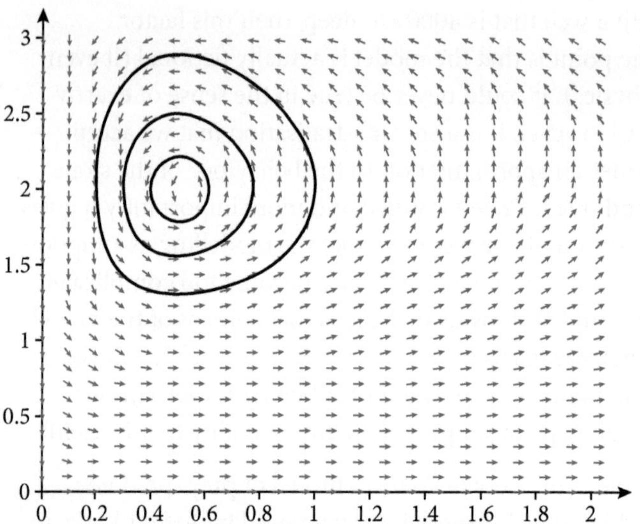

Figure 2 A phase space diagram. Number of prey (in units of 1000) on the x-axis, number of predators on the y-axis

It is interesting to look at a phase space diagram that represents each point (x, y) as a combination of numbers of prey and predators. Here the evolution of the system over time appears as a closed loop around the stationary point $\left(\frac{C}{D}, \frac{A}{B}\right)$, which is an 'attractor' of the dynamical system. (You could try to prove that this is a stationary point – it is not hard.) The position of an orbit around the attractor depends on the initial numbers of prey and predator. Notice that starting the model with too great a population of prey could end up with an extinction of predators (Figure 2) because the very high prey numbers leads to overpopulation of predators for whom there is not enough prey left to eat. The system itself is a nice example of circular causality.

As with the previous example, the modelling assumptions ensure that the mathematics of the model remains tractable, but the cost is that the model is not realistic. It is assumed that the prey do not die from natural causes or that the predators do not come into existence except through the provision of food. There is no competition between either prey or predators. Nonetheless, the model provides some important and powerful insights about the nature of population dynamics. As the model becomes more sophisticated and more factors are taken into consideration, not only does the mathematics become rapidly more difficult, but we lose sight of clear trends in the model (such as orbits around stationary points in phase space). We gain accuracy but lose understanding. This is a characteristic of both models and maps. A map that is as detailed as the territory it depicts is no use to anyone. It is precisely the simplification (literally what makes it false) that makes it useful. Virginia Woolf said about art, '*Art is not a copy of the world; one of the damn things is enough*', and the same could be said about models.

The distinction between pure and applied mathematics becomes blurred in the hands of someone like the great Carl Friedrich Gauss (1777–1855). He was perhaps happiest in the realm of number theory which he called the 'queen of mathematics', and the idea that queens stay in their rarified towers and do not dirty their hands in the ways of the world was perhaps not so far from his thinking. He found great satisfaction in working with patterns and sequences of numbers. It is the same Gauss who, as a young man, enabled astronomers to rediscover the minor planet Ceres after they had lost it in the glare of the sun, by calculating its orbit from the scant data that had been collected on its initial discovery in 1801 and then predicting where in the sky it would be found more than a year later. This feat immediately brought Gauss to the attention of the scientific community. His skills as a number theorist presented him with the opportunity of solving a very real scientific problem.

Who would have guessed that recent work in prime number theory would give rise to a system of encoding data that is used by banks all over the world? The system is called 'dual key cryptography'. The key to the code is a very large number that is the product of two primes. The bank holds one of the primes and the client's computer the other. The key can be made public because in order for it to work it has to be split up into its component prime factors. This task is virtually impossible for large numbers. For example, present computer programs would take longer than the 13.8 billion years since the big bang to find the two prime factors of the number:

25 195 908 475 657 893 494 027 183 240 048 398 571 429 282 126 204 032
027 777 137 836 043 662 020 707 595 556 264 018 525 880 784 406 918 290
641 249 515 082 189 298 559 149 176 184 502 808 489 120 072 844 992 687
392 807 287 776 735 971 418 347 270 261 896 375 014 971 824 691 165 077
613 379 859 095 700 097 330 459 748 808 428 401 797 429 100 642 458 691
817 195 118 746 121 515 172 654 632 282 216 869 987 549 182 422 433 637
259 085 141 865 462 043 576 798 423 387 184 774 447 920 739 934 236 584
823 824 281 198 163 815 010 674 810 451 660 377 306 056 201 619 676 256
133 844 143 603 833 904 414 952 634 432 190 114 657 544 454 178 424 020
924 616 515 723 350 778 707 749 817 125 772 467 962 926 386 356 373 289
912 154 831 438 167 899 885 040 445 364 023 527 381 951 378 636 564 391
212 010 397 122 822 120 720 357

But this number is indeed of the form of the product of two large primes. If you know one of them, it takes an ordinary computer a fraction of a second to do the division and find the other.

Just as pure research in the natural sciences produced results that could also be used for technological or engineering applications, so in mathematics, problems motivated purely from within the most abstract recesses of the subject (pure mathematics) give rise to very useful techniques for solving problems with strong applications in the world outside of mathematics. Mathematicians often practise their art as art for its own sake. They are motivated by the internal beauty and elegance of their subject. Nevertheless, it often happens that pure mathematics created for no other purpose than solving internal mathematical problems turns out to have some extraordinary and very practical applications.

Can you think of an example of a model that does not represent the world well but is nonetheless useful?

What other examples are there of pure research that end up having immense practical benefit?

Constructivist view of mathematics

Having thought a little about what the purpose of mathematics could be, let's move on to the question of whether it is best thought of as an invention or as something out there in the world.

Broadly speaking, the **constructivist** views mathematics as a human invention. The vision we had of mathematics as a vast virtual reality limited only by the imagination and the rules that are installed there is a constructivist view. However, we are then bound to ask why mathematics has so many useful applications in the real world. Why is mathematics important when it comes to building bridges, doing science and medicine, economics and even playing basketball? Chess is also a game invented by humans, but it does not have very much use in the outside world. Constructivism cannot account for the success of mathematics in the outside world.

On this view, mathematics is what might be called a **social fact**. A social fact is true by virtue of the role that it plays in our social lives. Social facts do have real causal power in the world. That a particular piece of paper is money is a social fact that does make things happen. That piece of paper acquires its status ultimately from a whole set of social agreements. In the end, social facts are produced by **language acts** – performances that change the social world. A language act would be a registry officer saying 'I pronounce you married'. The use of language in a **performative** manner creates social facts. Social facts are no less real or definite than those about the natural world. The statement 'John is married' is definitely either true or false. One is reminded of the story about the little boy who, when asked by his grandmother what day it will be tomorrow, replies, 'Let's wait and see'. Social facts do not require us to wait and see. They rely on social agreements, not on empirical evidence.

The mathematician Reuben Hersh argues for a type of constructivism that he calls **Humanism**. For Hersh, numbers and other mathematical objects are social facts. Hersh defends this view on the Edge website:

'[Mathematics] ... is neither physical nor mental, it's social. It's part of culture, it's part of history, it's like law, like religion, like money, like all those very real things, which are real only as part of collective human consciousness. Being part of society and culture, it's both internal and external: internal to society and culture as a whole, external to the individual, who has to learn it from books and in school. That's what math is.'

Hersh called his theory of mathematics humanism because it's saying that mathematics is something human. *'There's no math without people. Many people think that ellipses and numbers and so on are there whether or not any people know about them; I think that's a confusion.'*

Hersh points out that we do use numbers to describe physical reality and that this seems to contradict the idea that numbers are a social construction.

It is important to note here that we use numbers in two distinct ways: as nouns and as adjectives. When we say nine apples, nine is an adjective.

If it's an objective fact that there are nine apples on the table, that's just as objective as the fact that the apples are red, or that they're ripe or anything else about them; that's a fact. The problem occurs when we make a subconscious switch to 'nine' as an abstract noun in the sort of problems we deal with in Mathematics class. Hersh thinks that this is not really the same nine. They are connected, but the number nine is an abstract object as part of a number system. It is a result of our mathematics game – our deduction from axioms. It is a human creation.

Hersh sees a political and pedagogical dimension to his thinking about mathematics. He thinks that a humanistic vision of mathematics chimes in with more progressive politics. How can politics enter mathematics? As soon as we think of mathematics as a social construction then the exact arrangements by which this comes about – the institutions that build and maintain it – become important. These arrangements are political. Particularly interesting for us here is how a different view of mathematics can bring about changes in teaching and learning.

'Humanism sees mathematics as part of human culture and human history. It's hard to come to rigorous conclusions about this kind of thing, but I feel it's almost obvious that Platonism and Formalism are anti educational, and interfere with understanding, and Humanism at least doesn't hurt and could be beneficial. Formalism is connected with rote, the traditional method, which is still common in many parts of the world. Here's an algorithm; practise it for a while; now here's another one. That's certainly what makes a lot of people hate mathematics (...) There are various kinds of Platonists. Some are good teachers, some are bad. But the Platonist idea, that, as my friend Phil Davis puts it, Pi is in the sky, helps to make mathematics intimidating and remote. It can be an excuse for a pupil's failure to learn, or for a teacher's saying "some people just don't get it". The humanistic philosophy brings mathematics down to earth, makes it accessible psychologically, and increases the likelihood that someone can learn it, because it's just one of the things that people do.'

There is a possibility that the arguments explored in this section might cast light on an aspect of mathematics learning that has seemed puzzling – why it is that mathematical ability is seen to be closely correlated with a certain type of intelligence. There is a widespread view that mathematics polarises society into two distinct groups: those who can do it and those who cannot. Those who cannot do it often feel the stigma of failure and that there is an exclusive club whose membership they have been denied. Those who can do it often find themselves labelled as 'nerds' or as people who are, in some sense, socially deficient. Is Hersh correct in attributing this to a formalistic or Platonic view? Is he right to suggest that if mathematics is just a meaningless set of formal exercises, then it will not be valued by society? If we deny that mathematics is out there to be discovered, it takes the stigma away from the particular individual who does not make the discovery. It is interesting to speculate on how consequences in the classroom flow from a humanist view of mathematics.

What are the strengths and weaknesses of mathematical humanism?

Platonic view of mathematics

One way to explain why mathematics applies so well to things like bridges and planets is simply to take mathematics as being out there in the world, independent of human beings. As with other things in the natural world, it is our task to discover it (literally to 'lift the cover'). This is called the **Platonic** view because the philosopher Plato (427–347 BC) took the view that mathematical objects belonged to the real world, underlying the world of appearances in which we lived. Mathematical objects such as perfect circles and numbers existed in this real world; circles on Earth were mere inferior shadows. Many mathematicians have at least some sympathy with this view. They talk about mathematical objects as though they had an existence independent of us and that we are accountable to mathematical truths in the same way as we are accountable to physical facts about the universe. They feel that there really is a mathematical world out there and that they are trying to discover truths about it, much like natural science discovers truth about the physical world.

This view is itself not entirely without problems. In ToK we might want to ask: 'If mathematics is out there in the world, where is it?' We do not see circles, triangles, $\sqrt{2}\,\pi$, i, e, and other mathematical objects obviously floating around in the world. We have to do a great deal of work to find them through inference and abstraction.

While this might be true, there is some evidence that mathematics is hidden not too far below the surface of our reality. Take prime numbers as an example. The Platonist might want to try to find them somewhere in nature. One place where she might start is in Tennessee. In the summer of 2016, the forests were alive with a cicada that exploits a property of prime numbers for its own survival. These cicadas have a curious life cycle. They stay in the ground for 13 years. Then they emerge and enjoy a relatively brief period courting and mating before laying eggs in the ground and dying. There is another species of cicada that has the same cycle and no fewer than 12 types that have a cycle of 17 years. There are, to add to the puzzle, none that have cycles of 12, 14, 15, 16 or 18 years. The clue is that 13 and 17 are prime numbers. There is a predator wasp that has evolved to have a similar life cycle. But if a predator had a life cycle of 6 years, the prey and the predator would only meet every $6 \times 17 = 112$ years. Whereas, if the cicada had a life cycle of 12 years, the prey and predator would meet every cicada cycle. Nature has discovered prime numbers through the cicada life cycles by evolutionary trial and error.

The relationship of nature to geometry was explored by the Scottish biologist D'Arcy Wentworth Thompson in his magnificent book of 1917, *On Growth and Form.* He explored the formation of shells and the wings of dragonflies, and examined the skeletons of dinosaurs through the eyes of a civil engineer constructing bridges and wondered about the formation of bee cells and the arrangement of sunflower seeds.

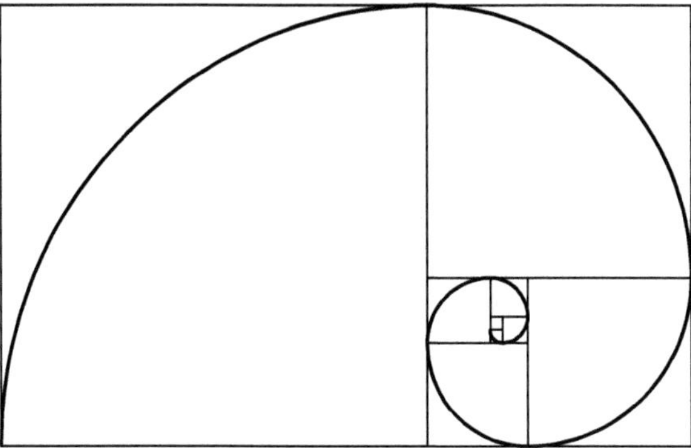

Figure 3 Spirals in nature

Many spirals in nature are formed, like the one in Figure 3, from the sequence:

1, 1, 2, 3, 5, 8, 13, 21, 34, …

This is called the Fibonacci sequence after the Italian mathematician Leonardo Pisano Bigolio (1170–1250), known as Fibonacci. The Fibonacci sequence is related to the golden number φ. The interested reader is referred to the many excellent sources on the internet.

Do you think that the mathematics teaching you have experienced reflects a Platonist or constructivist view of mathematics?

The methods and tools of mathematics

The language and concepts of mathematics

Knowledge in mathematics is like a map representing some aspect of the world. Like other areas of knowledge, it possesses a specialised vocabulary naming important concepts to build this map. Unlike some areas, this vocabulary is very precisely defined. This makes sense. If the world of mathematics is populated by some rather esoteric objects that are literally like nothing on Earth, then it is very important that these objects are precisely specified.

The other chapters in this book are all about establishing and using this very special vocabulary and becoming fluent in the methods that connect mathematical concepts into meaningful mathematical sentences. We will not spend too much time on these matters here, but there are a few aspects to highlight.

Notation

Since mathematical objects are abstract and we cannot point to them, we have to represent them with symbols. But the symbol and the idea are different things – there is a danger that we confuse them. Take representations of fractions. The symbols $\frac{1}{3}$, $\frac{2}{6}$, $\frac{3}{9}$, 0.3333… all represent the same number despite appearing to be quite different. (Perhaps the infinite number of ways of representing fractions is one of the reasons why some students have so much difficulty with them.) Some symbols such as $\frac{1}{0}$, $\sin^{-1}(1.2)$ or $\log(-2)$ have no meaning at all. More worrying is that an expression such as 'the smallest real number larger than 1' doesn't actually mean anything either. This is because there is no smallest real number larger than 1. (Think carefully about this.)

In a similar vein, the fact that there are different conventions for writing mathematics does not mean that the mathematics is different. Some conventions represent the number $\frac{3}{10}$ by the decimal 0.3, others by 0,3. Either way, the mathematics is the same and these do not really count as different mathematical cultures. Carl Friedrich Gauss, one of the greatest mathematicians of all time, said '*non notations, sed notions*' – not notations but notions.

Algebra

A staple method used in mathematics is the substitution of letters for numbers. In fact, mathematicians use letters for many sorts of mathematical objects, not just numbers. The reason is that they want to make generalised statements. By using a letter, they do not have to commit to making a statement about a specific number, but instead can make one about all numbers of a particular kind at once. This is a very powerful tool.

This is illustrated by a worked example. Imagine we want to prove that if we add an odd number to another odd number we get an even number. We hope to show that this is true for any choice of odd numbers. We could proceed by trying out different pairs of odd numbers and checking that the result is even:

$1 + 3 = 4$ even

$5 + 7 = 12$ even

$13 + 9 = 22$ even

$131 + 257 = 388$ even

You can see that this method will not serve as a proof because we would have to check every possible pair of odd numbers and, since this set is infinite, we would never finish. What we need is to define a general odd number without committing to a particular one. For example, we can define 'odd' by being 'one more than an even number'.

If k is an even number, then we can write $k = 2j$ for some whole number j.

If m is an odd number, then we can write $m = 2j + 1$ for some whole number j.

All we have to do now is to add two of these general odd numbers together.

So, we want to take two odd numbers, let's say $m = 2j + 1$ and $n = 2i + 1$ where j and i are whole numbers. There is a subtlety here because we use different letters j and i for the whole numbers in the expressions above because we want to allow m and n the possibility of being different odd numbers.

If we used the same letter, say j, in the expressions for m and n then we would be making our odd numbers equal and we would only have proved that if we add together two equal odd numbers, then the result is even.

Now we have to use some symbolic rules.

$$m + n = (2j + 1) + (2i + 1)$$

We can remove the brackets and rearrange to give:

$$m + n = 2j + 2i + 2$$

Finally, we can use the fact that 2 is a common factor of all terms in the expression to place it outside a bracket.

$$m + n = 2(j + i + 1)$$

But j, i and 1 are all whole numbers so $j + i + 1$ is also a whole number. Technically, this comes from the fact that the whole numbers are **closed under the operation of addition** because they form an important structure called a **group**. Let's call this whole number p.

So, we have that $m + n = 2p$. But this is precisely the definition of an even number that we started with. An even number is 2 times a whole number. This proves that any two odd numbers added together gives an even number.

The big chain of reasoning above is called a proof. It is immensely powerful because it covers an infinite number of situations. There is an infinite number of possible pairs of odd numbers to which the result applies. This is the power and beauty of using letters for numbers — a practice that was developed in Baghdad and Damascus about 1000 years ago. In one sense, mathematicians have a god-like ability when it comes to dealing with infinite sets.

Proof

Proof is the central concept in mathematics because it guarantees mathematical truth. When something is proved, we can say that it is true.

This type of truth is independent of place and time. In contrast to the science of the day, the mathematical truths of Pythagoras are just as true today as they were then – indeed his famous theorem is still taught today as can be seen in this book. But the science of the time has long been rejected. There were four chemical elements in the 4th century BC, and Aristotle thought that the heart was the organ for thinking. Actually, we do not have to go far back in time to find textbooks in the natural sciences that contain statements that we would dispute today. The truths of the natural sciences are always subject to revision, but mathematical truths are eternal.

But there is something even more striking about mathematical truths — that is, mathematical statements that have been proved. A statement such as 'odd + odd = even' has such power that we can say that it is certain. This is not just a matter of confidence – we are not talking about psychological certainty here.

It is certain because it cannot be otherwise. The negation of a mathematical truth (like 'odd + odd = odd') is to utter a self-contradiction or absurdity. Let's reflect on the power of this statement. This means that there is no possible world in which 'odd + odd = odd' (given the standard meanings of these terms). A story that makes this statement is describing a world that is self-contradictory – that is, an absurd and unintelligible world. Such a story is just not credible. But this means that mathematics is really radically different from other areas of knowledge, including the natural sciences. It is not a contradiction to say that the moon is composed of green cheese. There could be universes where this is true, but it just happens not to be in ours. Mathematics deals in what we call **necessary truths**, while the sciences deal mainly in **contingent truth**.

This is something that students of ToK should think about carefully. What is it about mathematical truth that makes it immune to revision and provides the basis for certainty and makes the negation of a mathematical truth a contradiction?

Recall that the constructivist sees mathematics as a big abstract game played by human beings according to invented rules. The hero of *The Glass Bead Game,* a novel by the German writer Hermann Hesse, must learn music, mathematics, and cultural history to play the game. On this view, mathematics is just like the glass bead game. There are parallels we can draw between a game like chess and mathematical proof. First, chess is played on a special board with pieces that can move in a particular way. The pieces must be set up on the board in a particular fashion before the game can begin. The same is true of mathematical proof. It starts with a collection of statements in mathematical language called **axioms**. They themselves cannot be proved. They are simply taken as self-evidently true and form the starting point for mathematical reasoning.

Once the game is set up, we can start playing. A move in chess means transforming the position of the pieces on the board by applying one of the game's rules that govern movement. Typically in chess, a move involves the movement of only one piece. (Can you think of an exception?) If the state of the pieces before the move was legitimate and the move was made according to the rules of the game, then the state of the pieces after the move is also legitimate. The same is true of a mathematical proof. One applies the rules (these are rules of algebra typically) to a line in the proof to get the next line. The whole proof is a chain of such moves.

Finally, the chess game ends. Either one of the players has achieved checkmate, or a stalemate (a draw) has been agreed. Similarly, a mathematical proof has an end. This is a point where the proof arrives at the required result at the end of the chain of reasoning. This result is called a **theorem**.

Once a proof of a mathematical statement is produced, we have a logical duty to believe the result, however unlikely. This is illustrated with a famous example.

Many people do not believe that $1 = 0.99999999\ldots$
(The three dots indicate that the 9's continue indefinitely).

The proof is straightforward.

Let	$x = 0.9999999\ldots$
Then	$10x = 9.9999999\ldots$
Subtract both equations	$10x - x = 9.99999999\ldots - 0.9999999\ldots$
This implies	$9x = 9$

Giving $x = 1$ as required.

0.999999... really does look very different to 1 but if the proof works then we are forced to believe that they are the same.

Are you happy with every stage of this proof?

Sets

A set is a collection of elements that can themselves be sets. They can be combined in various ways to produce new sets. The concepts of a set and membership of a set are **primitive.** This means that they cannot be explained in terms of more simple ideas. These seem to be rather modest beginnings on which to build the complexities of modern mathematics. Nevertheless, in the 20th century there were a number of projects that were designed to do just that: reduce the whole of mathematics to set theory. The most important work here was by Quine, von Neumann and Zermelo, and Bertrand Russell and Alfred North Whitehead in the three volumes of their *Principia Mathematica* of 1910–1913. Starting out with the notion of the empty set and the idea that no set can be a member of itself, we can construct the whole number system.

Mappings between sets

Once we have established sets in our mathematical universe, we want to do something useful with them. One of the most important ideas in the whole of mathematics is that of a mapping. A mapping is a rule that associates every member of a set with a member of a second set. This is what we were doing when we started this chapter by counting cows. We set up a one-to-one correspondence between a set of numbers and a set of cows.

Infinite sets

Consider the function $f(x) = 2x$ defined over the natural numbers.

Clearly it sets up a one-to-one correspondence between the set of natural numbers and the set of even numbers (check this yourself). So, this means that there are as many even numbers as there are natural numbers.

This is rather strange because we would think intuitively that there were more natural numbers than even numbers – they are after all the result of taking away an infinite number of odd numbers from the original set. But we are saying that the set that is left over has as many members as the original set. This strangeness is characteristic of infinite sets (indeed it can be used to define what we mean by infinite). Infinite sets can be put in a one-to-one correspondence with a proper subset of themselves.

But the story doesn't stop here. Using sets and mappings we can show that there are many different types of infinity. The set of natural numbers contains the smallest type of infinity, usually denoted by $\aleph_0$, which we call 'aleph nought'. In the 19th century, the German mathematician Georg Cantor showed by an ingenious argument that the number of numbers between 0 and 1 is a bigger type of infinity than aleph nought.

What is it about the difference between the methods of the natural sciences and mathematics that accounts for the radical difference in types of knowledge produced?

It turns out that there is an infinity of different types of infinity — a whole hierarchy of infinities, in fact — and this probably does not surprise you anymore, there are more infinities than finite cardinal numbers.

The methods and concepts of mathematics, therefore, are quite unlike anything to be found in the sciences, although they do seem to bear a strong resemblance to the arts in terms of the setting of the rules of the game and the use of the imagination. This is something we will explore in the next section.

Mathematics and the knower

English poet John Keats said, *"Beauty is truth, truth beauty – that is all / Ye know on earth, and all ye need to know."*

In this section we will see how mathematics impinges on our personal thinking about the world. One of the more surprising aspects of mathematics is the two-way link to the arts and beauty.

Beauty by the numbers

There is a long-held view that we find certain things beautiful because of their special proportions or some other intrinsic mathematical feature. This is the thinking that has inspired architects since the times of ancient Egypt and generations of painters, sculptors, musicians, and writers. Mathematics seems to endow beauty with a certain eternal objectivity. Things are beautiful because of the mathematical relationships between their parts. Moreover, this is a very public beauty because it can be dissected and discussed.

Let's take the example of the builders of the Parthenon. They were deeply interested in symmetry and proportion. In particular, they were interested in how to divide a line so that the proportion of the shorter part to the longer part is the same as that of the longer part to the whole. You can check that you get the quadratic equation $x^2 + x - 1 = 0$. One solution to this equation is the golden ratio $x = \frac{-1 + \sqrt{5}}{2} = 0.61803398875... = \varphi$.
This proportion features significantly in the design of the Parthenon and many other buildings of the period. Since it is also related to the Fibonacci sequence, you will find φ turning up anywhere where there are spirals. It is used quite self-consciously in painting (Piet Mondrian, for example) and in music (particularly the music of Debussy). There are those who go as far as saying that it is present in the proportions of the perfect human figure and that we have a predisposition towards this ratio.

See if you can spot the connection between the golden ratio and the Fibonacci sequence. Hint: write down a difference equation for generating the sequence.

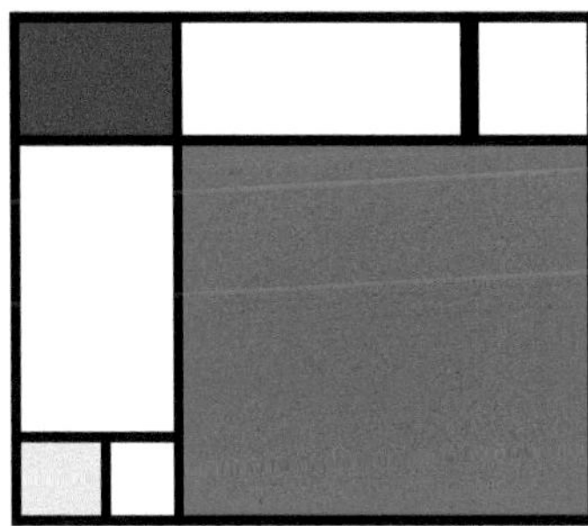

Figure 4 *Composition with Red, Blue and Yellow* (1926) Piet Mondrian. The proportions of some of the rectangles in this painting is φ

Beauty in numbers

Keats also put it the other way around: the beautiful is the true. Could we allow ourselves to be guided to truth in mathematics because of the beauty of the equations? This is a position taken by surprisingly many mathematicians. They look for beauty and elegance as an indicator of truth. Many mathematical physicists were guided in the 20th century by considerations of beauty and elegance.

Einstein suggested that the most incomprehensible thing about the universe was that it was comprehensible. From a ToK point of view, the most incomprehensible thing about the universe is that it is comprehensible in the language of mathematics. Galileo wrote, '*Philosophy is written in this grand book, the universe... It is written in the language of mathematics, and its characters are triangles, circles and other geometric figures...*'

Perhaps what is more puzzling is not just that we can describe the universe in mathematical terms, but that the mathematics we need to do this is mostly simple, elegant, and even beautiful.

To illustrate this, let's look at some of the famous equations of physics. Most people will be familiar with Einstein's field equations and Maxwell's equations.

EINSTEIN'S FIELD EQUATION
for general relativity

$$R_{\mu v} - Rg_{\mu v} = \frac{8\pi G}{C^4} T_{\mu v}$$

Figure 5 Einstein's field equation

1. $\nabla \cdot \mathbf{D} = \rho_V$
2. $\nabla \cdot \mathbf{B} = 0$
3. $\nabla \times \mathbf{E} = -\dfrac{\partial \mathbf{B}}{\partial t}$
4. $\nabla \times \mathbf{H} = \dfrac{\partial \mathbf{D}}{\partial t} + \mathbf{J}$

Figure 6 Maxwell's equations

It is perplexing that the whole crazy complex universe can be described by such simple, elegant, and even beautiful equations. It seems that our mathematics fits the universe rather well. It is difficult to believe that mathematics is just a mind game that we humans have invented.

But the argument from simplicity and beauty goes further. Symmetry in the underlying algebra led mathematical physicists to propose the existence of new fundamental particles, which were subsequently discovered. In some cases, beauty and elegance of the mathematical description have even been used as evidence of truth. The physicist Paul Dirac said, '*It seems that if one is working from the point of view of getting beauty in one's equations, and if one has really a sound insight, one is on a sure line of progress*'.

Dirac's own equation for the electron must rate as one of the most profoundly beautiful of all. Its beauty lies in the extraordinary neatness of the underlying mathematics – it all seems to fit so perfectly together:

$$(i\not{\partial} - m)\psi = 0$$

Figure 7 Dirac's equation of the electron

The physicist and mathematician Palle Jorgensen wrote:

> '*[Dirac] … liked to use his equation for the electron as an example stressing that he was led to it by paying attention to the beauty of the math, more than to the physics experiments.*'

It was because of the structure of the mathematics in particular that there were two symmetrical parts to the equation — one representing a negatively charged particle (the electron) and the other a similar particle but with a positive charge — that scientists were led to the discovery of the positron. It seems fair to say that the mathematics did really come first here.

We will leave the last word on this subject to Dirac himself, writing in Scientific American in 1963:

> *'I think there is a moral to this story, namely that it is more important to have beauty in one's equations than to have them fit experiment.'*

By any standards this is an extraordinary statement for a mathematical physicist to make.

Mathematics and personal intuitions

Sometimes our intuition can let us down badly when it comes to making judgments of probability. Here is an example to illustrate how we might have to correct our intuition by careful mathematical reasoning.

Consider the following case. There is a rare genetic disease among the population. Very few people have the disease. As a precaution, a test has been developed to detect whether particular individuals have the disease. Although the test is quite good, it is not perfect — it is only 99% accurate. Person X takes the test and it shows positive. The question for your mathematical intuition is: 'What is the probability that X actually has the disease?' (You should recognise this as being a problem of conditional probability.)

Think about this for a moment before we continue.

Many of the students (and teachers) we have worked with in the past give the same answer: the probability that X actually has the disease given a positive test result is about 99%. Did you say the same? If you did, then your mathematical intuition let you down – very badly.

Let's put some numbers into the problem to illustrate this. For the sake of simplicity, assume that the country in which the test takes place has a population of 10 million. We are told that the disease is very rare. Assume that only 100 people in the whole country have the disease. We are told that the test is 99% accurate so, of the 100 cases of the disease the test would show positive in 99 cases and negative in one. So far so good.

Now consider the 9 999 900 people who don't have the disease. In 99% of these cases the test does its job and records a negative result. In 1% of the cases however it gets it wrong and produces a positive result. 1% of 9 999 900 is 99 999. So, of the whole population tested there would be a total of $99\,999 + 99 = 100\,098$ positive results. But of these only 99 have the disease. Therefore, the probability of having the disease given a positive result is $\frac{99}{100\,098} = 0.0989\%$ or about 1 in 1000. This is quite a big difference from the 990 in 1000 that we expected intuitively. That is out by a whopping 99 000%.

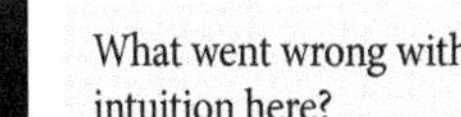

What went wrong with intuition here?

Mathematics and personal qualities

There are undoubtedly special qualities well-suited to doing mathematics. There are a host of great mathematicians from Archimedes, Euclid, Hypatia, through to Andrew Wiles, Grigori Perelman, and Maryam Mirzakhani, who contributed significantly to the area. Maryam was the first woman to receive the Fields medal (the equivalent of the Nobel prize in mathematics). Although mathematics is collaborative in the sense that mathematicians build on the work of others and take on the challenges that the area itself has recognised as being important, it is nevertheless largely a solitary pursuit. It requires great depth of thought, imaginative leaps, careful and sometimes laborious computations, innovative ways of solving very hard problems, and, most of all, great persistence. Mathematicians need to develop their intuition and their nose for a profitable strategy. They are guided by emotion and by hunches – they are a far cry from the stereotype of the coldly logical thinker who is closer to computer than human.

Conclusion

We have seen that mathematics is really one of the crowning achievements of human civilisation. Its ancient art has been responsible for some of the most extraordinary intellectual journeys taken by humankind, and its methods have allowed the building of great cities, and the production of great art, and it has been the language of great science.

From a ToK perspective, mathematics, with its absolute and unchanging notion of necessary truth, makes a good contrast to the natural sciences with their reliance on observation of the external world, experimental method, and provisional nature of its results.

Two countering arguments should be set against this view of mathematics. The idea that the axioms of mathematics (the rules of the game) are arbitrary both deprives mathematics of its status as something independent of human beings, and makes it vulnerable to the charge that its results cannot ever be entirely relevant to the world outside mathematics.

Platonists would certainly argue that mathematics is out there in the universe, with or without human beings. They would argue that it is built into the structure of the cosmos – a fact that explains why the laws of the natural sciences lend themselves so readily to mathematical expression.

Both views produce challenging questions in ToK. The constructivist is a victim of the success of mathematics in fields such as the natural sciences. She has to account for why mathematics is so supremely good at describing the outside world to which, according to this view, it should ultimately be blind. The Platonist, on the other hand, finds it hard to identify mathematical structures embedded in the world or has a hard time explaining why they are there.

We have seen how mathematics is closely integrated into artistic thinking; perhaps because both are abstract areas of knowledge indirectly linked to the world and not held to account through experiment and observation, but instead, open to thought experiment and leaps of imagination. Mathematics can challenge our intuitions and can push our cognitive resources as individual knowers. Infinity is not something that the human mind can fathom in its entirety. Instead, mathematics gives us the tools to deal with it in precisely this unfathomed state. We can be challenged by results that seem counter to our intuition, but ultimately, the nature of mathematical proof is that it forces us to accept them nonetheless. In turn, individuals can, through their insight and personal perspectives, make ground-breaking contributions that change the direction of mathematics forever. The history of mathematics is a history of great thinkers building on the work of previous generations to do ever more powerful things using ever more sophisticated tools.

The Greek thinkers of the 4th century BC thought that mathematics lay at the core of human knowledge. They thought that mathematics was one of the few areas in which humans could apprehend the eternal forms only accessible to pure unembodied intellect. They thought that in mathematics they could glimpse the very framework on which the world and its myriad processes rested. Maybe they were right.

Answers

Chapter 1

Exercise 1.1

1. (a) $x = h - \frac{n}{m}$ (b) $a = \frac{v^2 + t}{b}$
   (c) $b_1 = \frac{2A}{h} - b_2$ (d) $r = \sqrt{\frac{2A}{\theta}}$
   (e) $k = \frac{gh}{f}$ (f) $t = \frac{x}{a + b}$
   (g) $r = \sqrt[3]{\frac{3V}{\pi h}}$ (h) $k = \frac{g}{F(m_1 + m_2)}$
2. (a) $y = -\frac{2}{3}x - 5$ (b) $y = -4$
   (c) $y = \frac{5}{4}x + 6$ (d) $x = \frac{7}{3}$
   (e) $y = -4x + 11$ (f) $y = -\frac{5}{2}x - 7$
3. (a) (i) 17 (ii) $\left(0, \frac{5}{2}\right)$
   (b) (i) $\sqrt{40}$ (ii) $(2, 3)$
   (c) (i) $\frac{\sqrt{82}}{3}$ (ii) $\left(-1, \frac{7}{6}\right)$
   (d) (i) $\sqrt{533}$ (ii) $\left(1, \frac{11}{2}\right)$
4. (a) $k = 1$ or 9 (b) $k = -11$ or -3
5. (a) $(\sqrt{5})^2 + (\sqrt{45})^2 = (\sqrt{50})^2$
   (b) sides are: $\sqrt{29}, \sqrt{29}, \sqrt{58}$
   (c) sides are: $\sqrt{45}, \sqrt{10}, \sqrt{45}, \sqrt{10}$

Exercise 1.2

1. (a) G (b) L (c) H (d) K (e) J
   (f) C (g) A (h) I (i) F
2. $A = \frac{C^2}{4\pi}$ 3. $A = \frac{l^2\sqrt{3}}{4}$
4. $A = 4x^2 + 60x$ 5. $h = x\sqrt{2}$
6. (a) 9.4 (b) $V = \frac{3525}{P}$
7. (a) $F = kx$ (b) 6.25 (c) 37.5 N
8. (a) $\{-6.2, -1.5, 0.7, 3.2, 3.8\}$ (b) $r > 0$
   (c) $\mathbb{R}$ (d) $\mathbb{R}$ (e) $t \leqslant 3$
   (f) $\mathbb{R}$ (g) $x \neq \pm 3$
   (h) $-1 \leqslant x \leqslant 1$ and $x \neq 0$
9. no, $x = c$ is a vertical line
10. (a) (i) $\sqrt{17}$ (ii) 7 (iii) 0
    (b) $x < 4$
    (c) domain: $x \geqslant 4$, range: $h(x) \geqslant 0$
11. (a) (i) domain $\{x : x \in \mathbb{R}, x \neq 5\}$, range $\{y : y \in \mathbb{R}, y \neq 0\}$
    (ii) y-intercept $\left(0, -\frac{1}{5}\right)$, vertical asymptote $x = 5$, horizontal asymptote $y = 0$

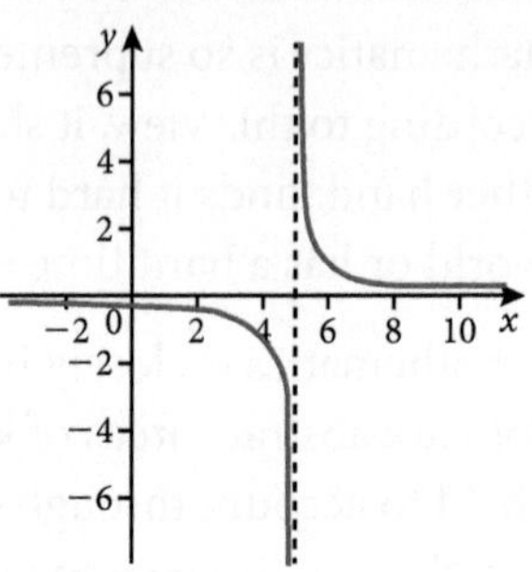

(b) (i) domain $\{x : x < -3, x > 3\}$, range $\{y : y > 0\}$
(ii) vertical asymptotes $x = -3$ and $x = 3$
horizontal asymptote $y = 0$

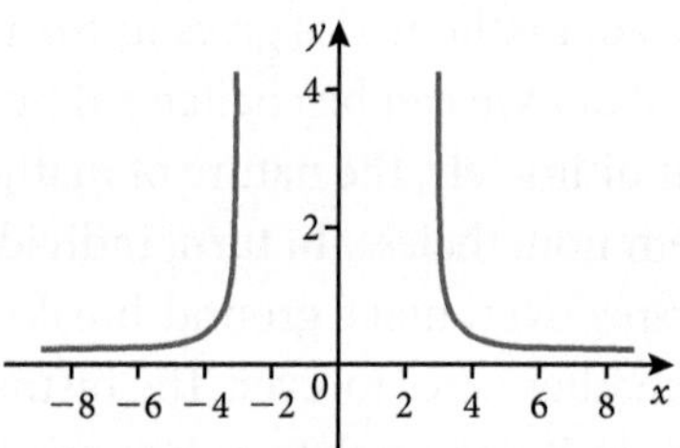

(c) (i) domain $\{x : x \in \mathbb{R}, x \neq -2\}$, range $\{y : y \in \mathbb{R}, y \neq 2\}$
(ii) y-intercept $\left(0, -\frac{1}{2}\right)$, vertical asymptote $x = -2$, horizontal asymptote $y = 2$

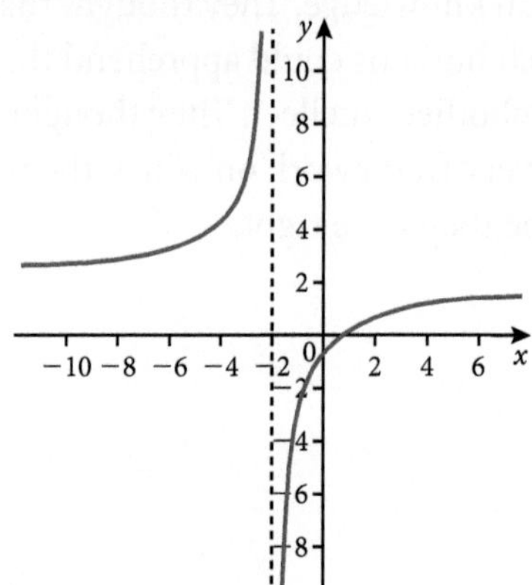

(d) (i) domain $\left\{x : -\frac{\sqrt{10}}{2} \leqslant x \leqslant \frac{\sqrt{10}}{2}\right\}$,
range $\{y : 0 < y \leqslant \sqrt{5}\}$
(ii) y-intercept $(0, \sqrt{5})$, x-intercepts $\left(-\frac{\sqrt{10}}{2}, 0\right)$ and $\left(\frac{\sqrt{10}}{2}, 0\right)$

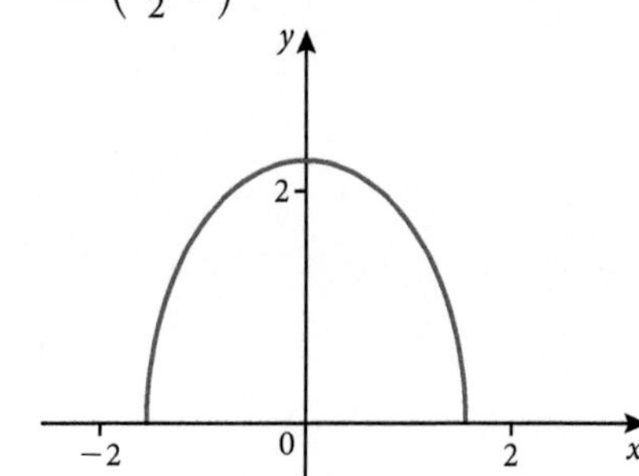

(e) (i) domain $\{x : x \in \mathbb{R}, x \neq 0\}$, range $\{y : y \in \mathbb{R}, y \neq -4\}$
(ii) vertical asymptote $x = 0$, horizontal asymptote $y = -4$

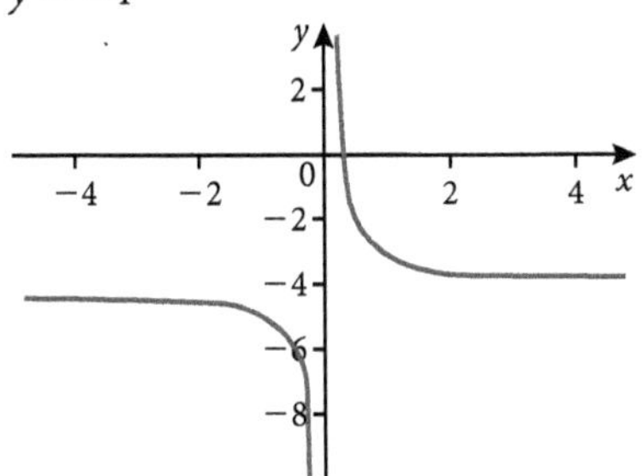

Exercise 1.3

1. (a) $(f \circ g)(5) = 1$ **(b)** $(g \circ f)(5) = \frac{1}{7}$
(c) $(f \circ g)(x) = \frac{2}{x-3}$ **(d)** $(g \circ f)(x) = \frac{1}{2x-3}$

2. (a) 1 **(b)** -7 **(c)** 7
(d) -47 **(e)** -1 **(f)** -79
(g) $1 - 2x^2$ **(h)** $-4x^2 + 12x - 7$ **(i)** $4x - 9$
(j) $-x^4 + 4x^2 - 2$

3. (a) $(f \circ g)(x) = 12x + 7$, domain: $x \in \mathbb{R}$; $(g \circ f)(x) = 12x - 1$, domain: $x \in \mathbb{R}$
(b) $(f \circ g)(x) = 4x^2 + 1$, domain: $x \in \mathbb{R}$; $(g \circ f)(x) = -2x^2 - 2$, domain: $x \in \mathbb{R}$
(c) $(f \circ g)(x) = \sqrt{x^2 + 2}$, domain: $x \in \mathbb{R}$; $(g \circ f)(x) = x + 2$, domain: $x \geqslant -1$
(d) $(f \circ g)(x) = \frac{2}{x+3}$, domain: $x \in \mathbb{R}, x \neq -3$; $(g \circ f)(x) = -\frac{x+2}{x+4}$, domain: $x \in \mathbb{R}, x \neq -4$
(e) $(f \circ g)(x) = x$, domain: $x \in \mathbb{R}$; $(g \circ f)(x) = x$, domain: $x \in \mathbb{R}$
(f) $(f \circ g)(x) = 1 + x^2$, domain: $x \in \mathbb{R}$; $(g \circ f)(x) = \sqrt[3]{-x^6 + 4x^3 - 3}$, domain: $x \in \mathbb{R}$
(g) $(f \circ g)(x) = \frac{2}{4x^2 - 1}$, domain: $x \neq 0, x \neq \pm\frac{1}{2}$; $(g \circ f)(x) = \frac{(4-x)^2}{4x^2}$, domain: $x \neq 0, x \neq 4$
(h) $(f \circ g)(x) = x$, domain: $x \neq -3$; $(g \circ f)(x) = x$, domain: $x \neq -3$
(i) $(f \circ g)(x) = \frac{x^2 - 1}{x^2 - 2}$, domain: $x \neq \pm\sqrt{2}$; $(g \circ f)(x) = \frac{2x-1}{(x-1)^2}$, domain: $x \neq 1$

4. (a) $(g \circ h)(x) = \sqrt{9 - x^2}$, domain: $-3 \leqslant x \leqslant 3$, range: $0 \leqslant y \leqslant 3$
(b) $(h \circ g)(x) = -x + 11$, domain: $x \geqslant 1$, range: $y \leqslant 10$

5. (a) $(f \circ g)(x) = \frac{1}{10 - x^2}$, domain: $x \neq \pm\sqrt{10}$, range: $y \neq 0$
(b) $(g \circ f)(x) = 10 - \frac{1}{x^2}$, domain: $x \neq 0$, range: $y < 10$

6. (a) $h(x) = x + 3, g(x) = x^2$
(b) $h(x) = x - 5, g(x) = \sqrt{x}$
(c) $h(x) = \sqrt{x}, g(x) = 7 - x$
(d) $h(x) = x + 3, g(x) = \frac{1}{x}$
(e) $h(x) = x + 1, g(x) = 10^x$
(f) $h(x) = x - 9, g(x) = \sqrt[3]{x}$
(g) $h(x) = x^2 - 9, g(x) = |x|$
(h) $h(x) = \sqrt{x - 5}, g(x) = \frac{1}{x}$

7. (a) (i) domain of f: $x \geqslant 0$
(ii) domain of g: $x \in \mathbb{R}$
(iii) $(f \circ g)(x) = \sqrt{x^2 + 1}$, domain: $x \in \mathbb{R}$
(b) (i) domain of f: $x \neq 0$
(ii) domain of g: $x \in \mathbb{R}$
(iii) $(f \circ g)(x) = \frac{1}{x+3}$, domain: $x \neq -3$
(c) (i) domain of f: $x \neq \pm 1$
(ii) domain of g: $x \in \mathbb{R}$
(iii) $(f \circ g)(x) = \frac{3}{x^2 + 2x}$, domain: $x \neq 0, -2$
(d) (i) domain of f: $x \in \mathbb{R}$
(ii) domain of g: $x \in \mathbb{R}$
(iii) $(f \circ g)(x) = x + 3$, domain $x \in \mathbb{R}$

Exercise 1.4

1. (a) 2 **(b)** 6
2. (a) -1 **(b)** b
3. 4
4. 6
5. (a) (i)
(ii)

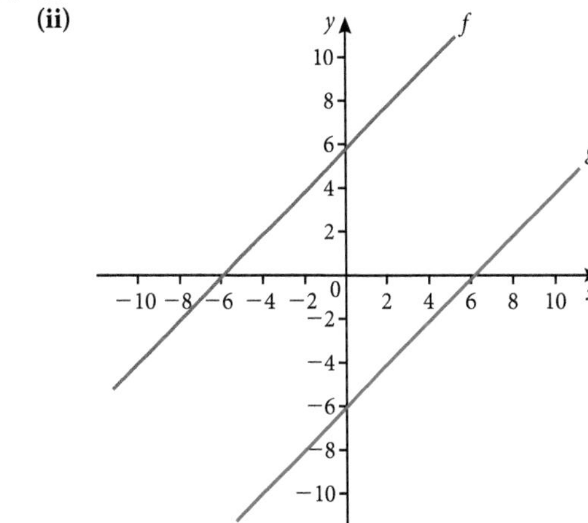

(b) (i)
(ii)

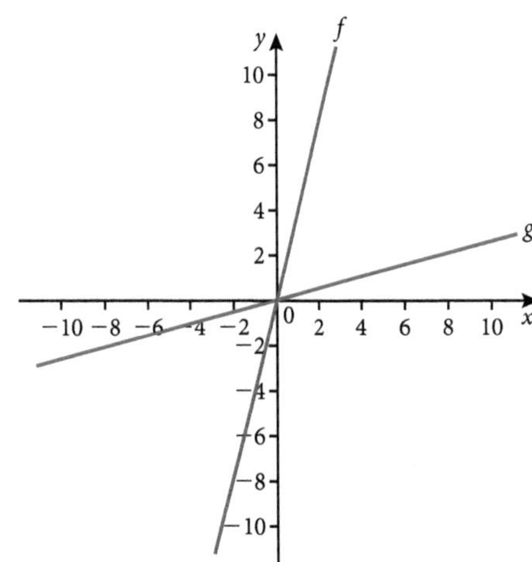

(c) (i)
(ii)

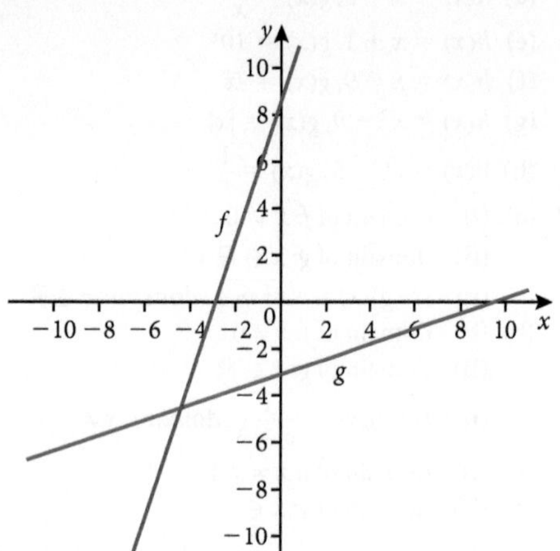

(d) (i)
(ii)

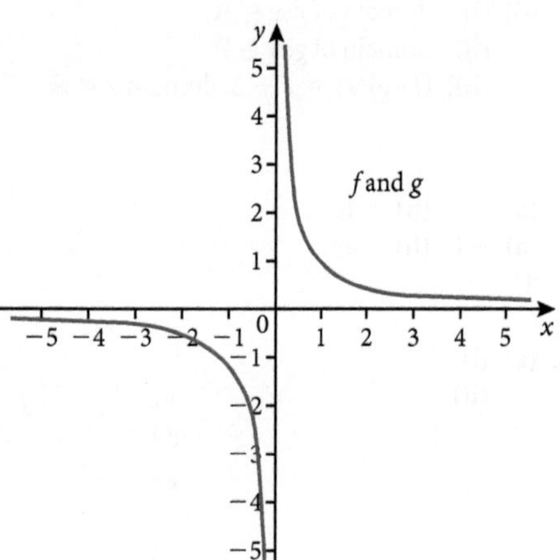

(e) (i)
(ii)

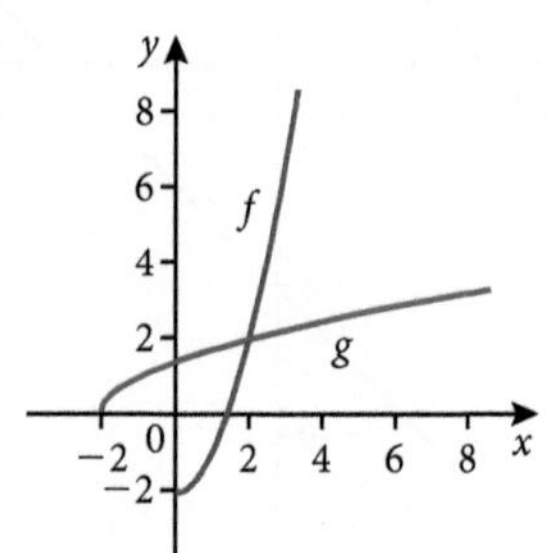

(f) (i)
(ii)

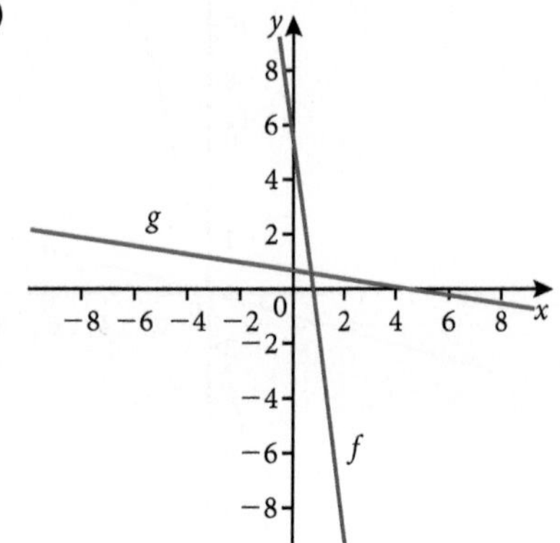

(g) (i)
(ii)

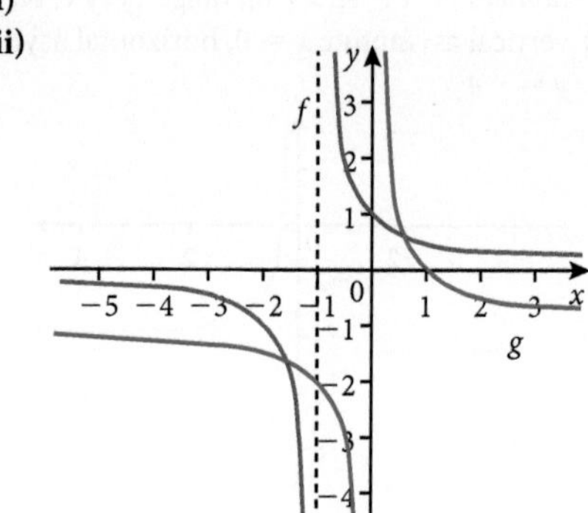

(h) (i)
(ii)

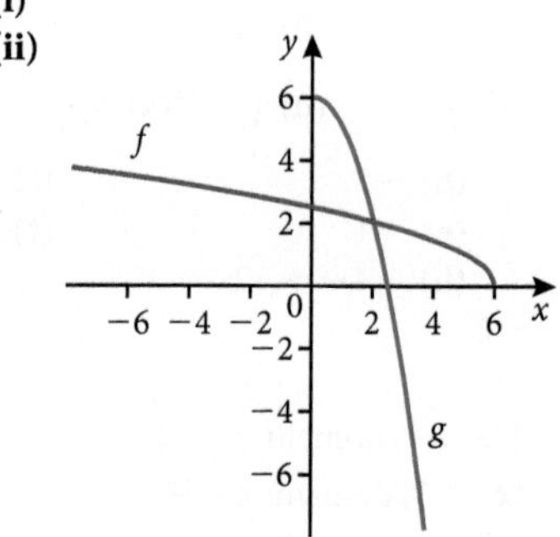

(i) (i)
(ii)

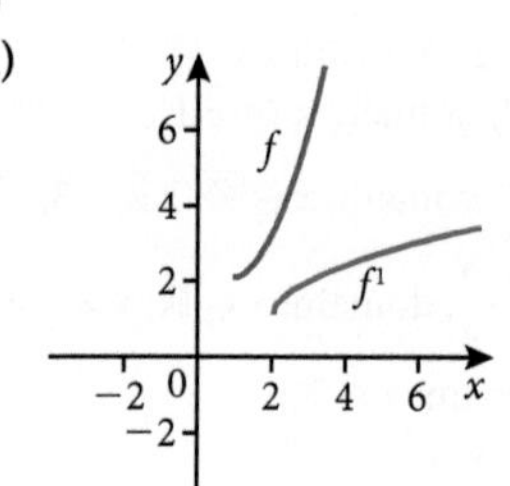

(j) (i)
(ii)

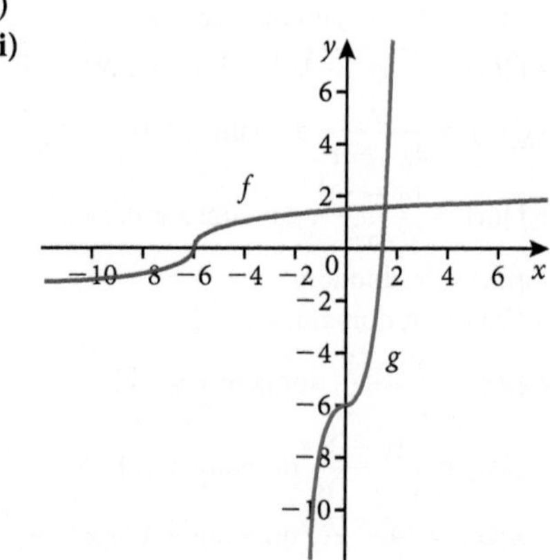

6. (a) $f^{-1}(x) = \frac{1}{2}x + \frac{3}{2}, x \in \mathbb{R}$
(b) $f^{-1}(x) = 4x - 7, x \in \mathbb{R}$
(c) $f^{-1}(x) = x^2, x \geqslant 0$
(d) $f^{-1}(x) = \frac{1}{x} - 2, x \in \mathbb{R}, x \neq 0$
(e) $f^{-1}(x) = \sqrt{4 - x}, x \leqslant 4$
(f) $f^{-1}(x) = x^2 + 5, x \geqslant 0$
(g) $f^{-1}(x) = \frac{1}{a}x - \frac{b}{a}, x \in \mathbb{R}$

(h) $f^{-1}(x) = -1 + \sqrt{x+1}, x \geqslant -1$

(i) $f^{-1}(x) = \sqrt{\dfrac{1+x}{1-x}}, -1 \leqslant x < 1$

(j) $f^{-1}(x) = \sqrt[3]{x-1}, x \in \mathbb{R}$

7. $x < -1, -1 \leqslant x \leqslant 1, x > 1$

8. (a) $\dfrac{3}{2}$ (b) 5

(c) -4 (d) $\dfrac{7}{2}$

(e) $g^{-1} \circ h^{-1} = \dfrac{1}{2}x - 1$ (f) $h^{-1} \circ g^{-1} = \dfrac{1}{2}x + \dfrac{1}{2}$

(g) $(g \circ h)^{-1} = \dfrac{1}{2}x + \dfrac{1}{2}$ (h) $(h \circ g)^{-1} = 2x + 2$

Exercise 1.5

1. (a)

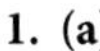

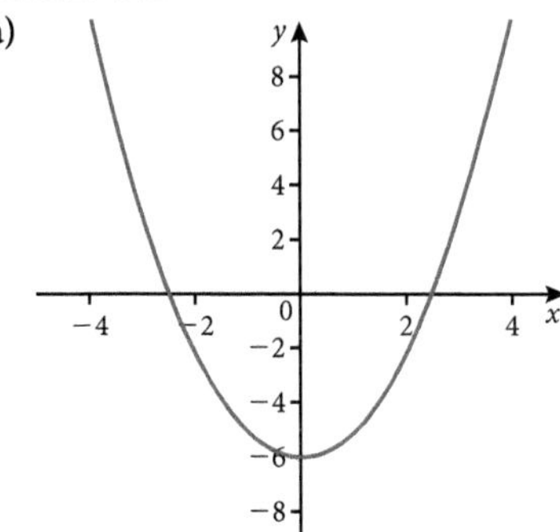

(b)

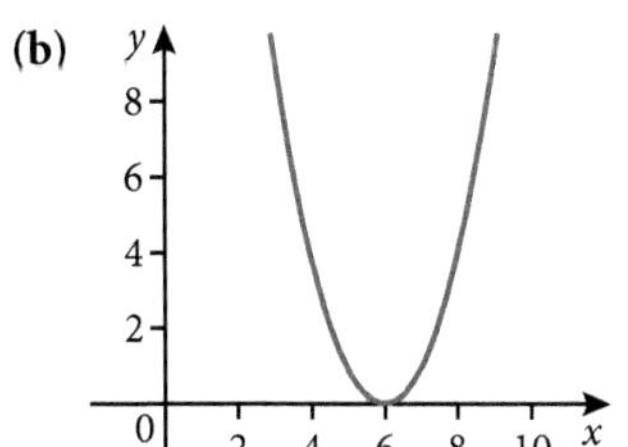

(c)

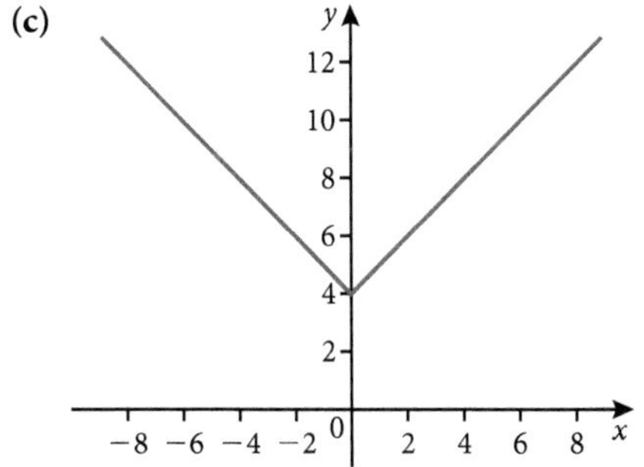

(d)

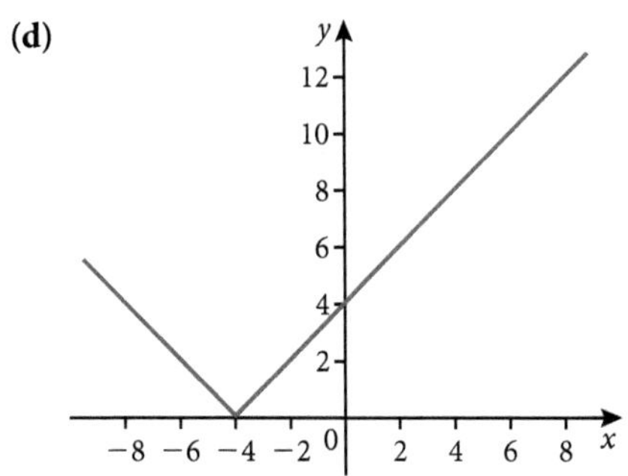

(e)

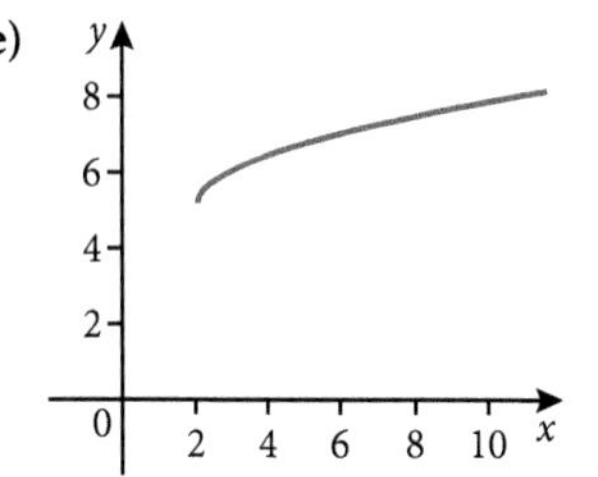

(f)

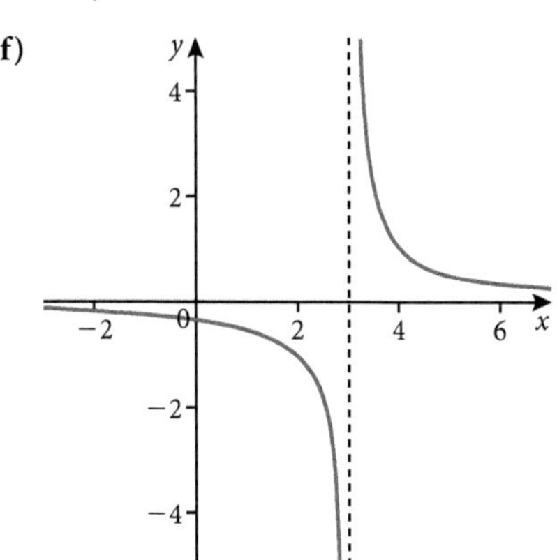

(g)

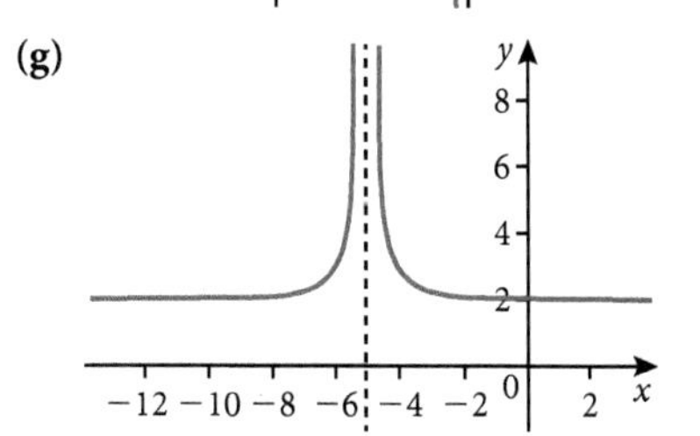

(h)

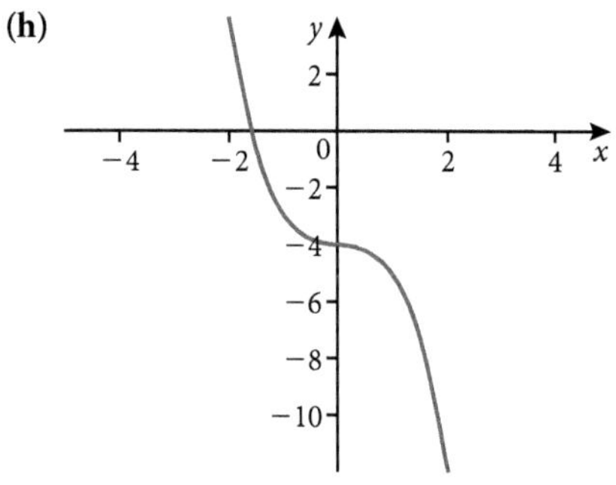

(i)

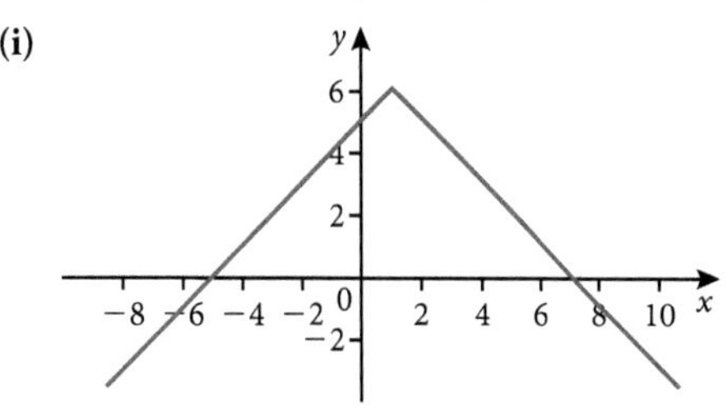

(j)

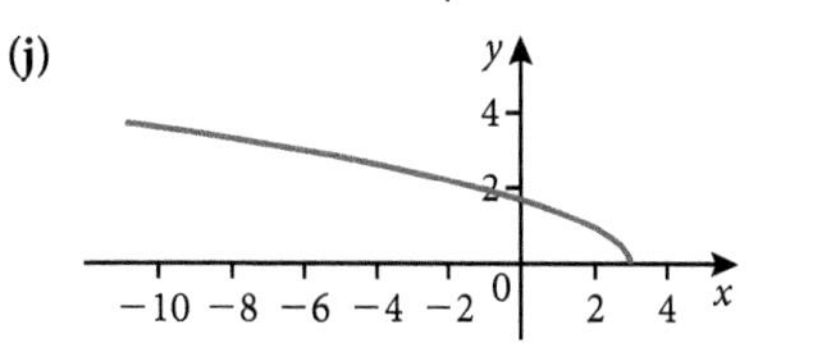

(k)

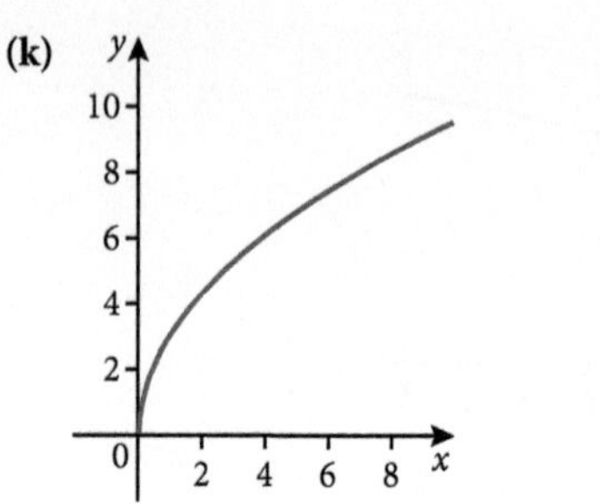

(l)

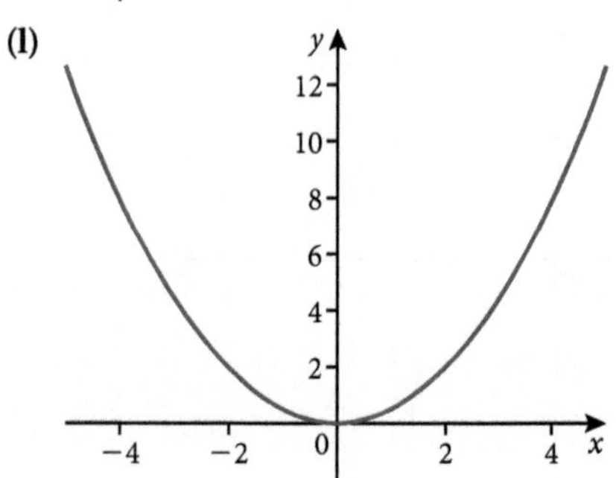

(m)

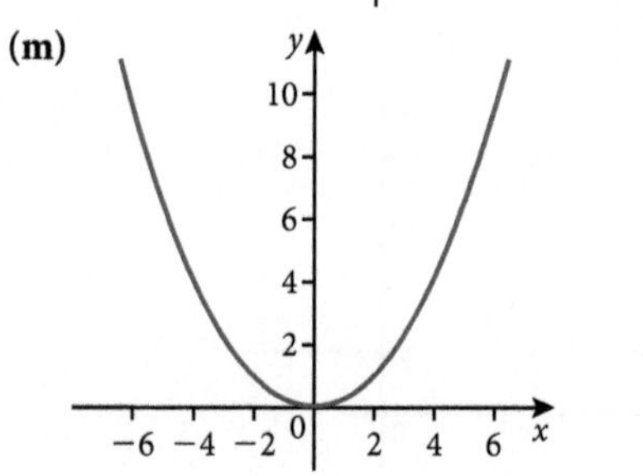

(n)

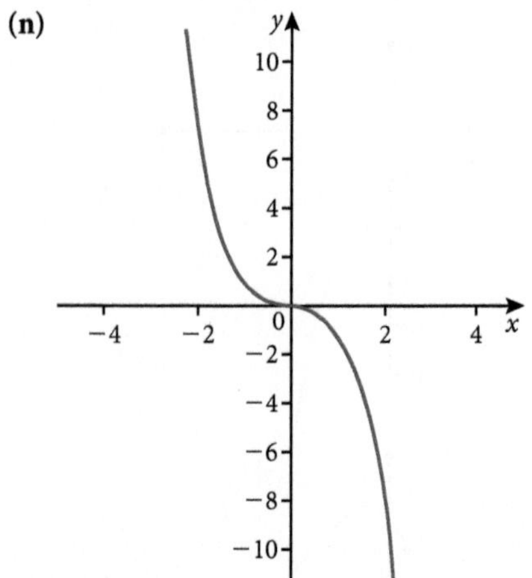

2. **(a)** $y = -x^2 + 5$ **(b)** $y = \sqrt{-x}$

(c) $y = -|x + 1|$ **(d)** $y = \dfrac{1}{x-2} - 3$

3. **(a)**

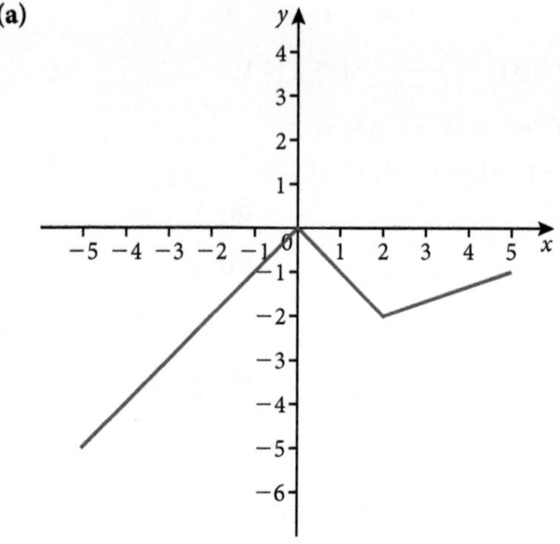

(b)

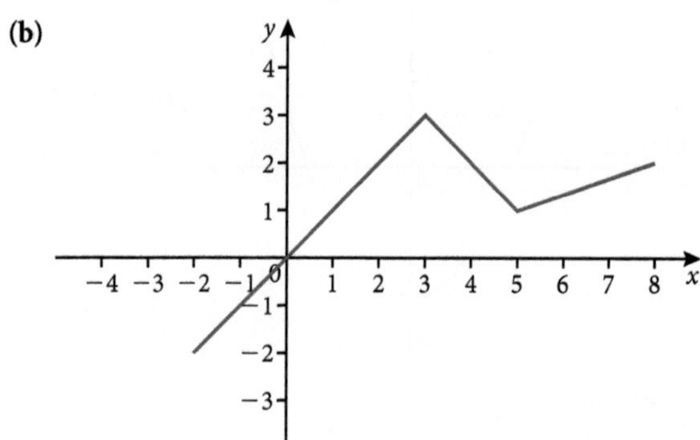

(c)

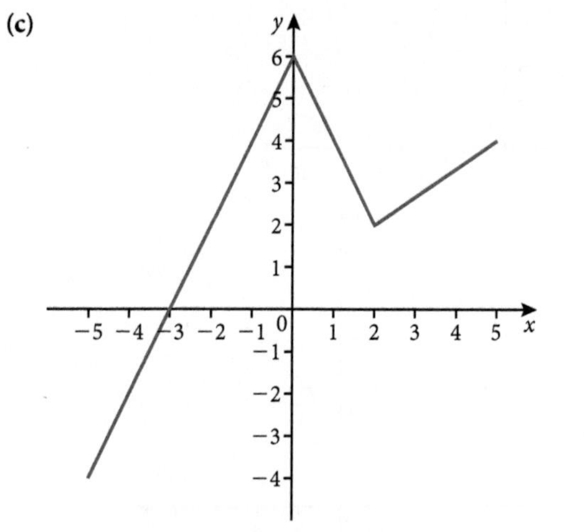

(d)

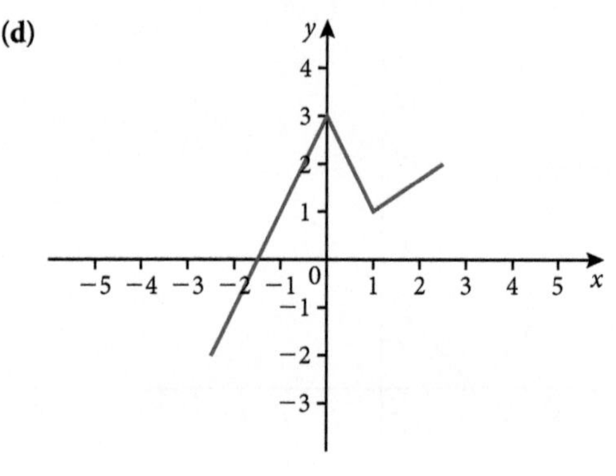

(e)

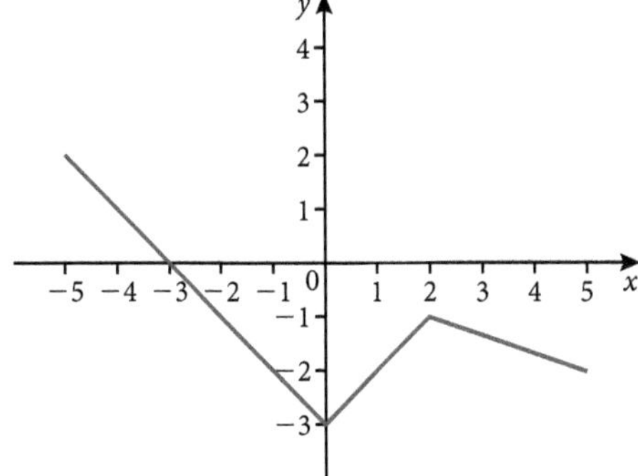

(f)

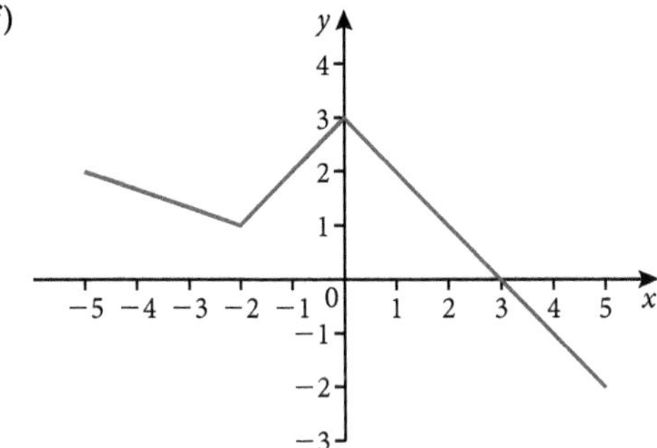

(g)

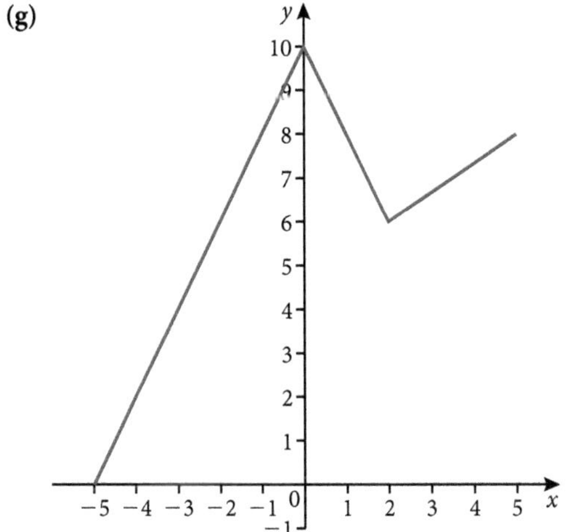

4. (a) horizontal translation 3 units right; vertical translation 5 units up (or reverse order)

(b) reflect over the x-axis; vertical translation 2 units up (or reverse order)

(c) horizontal translation 4 units left; vertical shrink by factor $\frac{1}{2}$ (or reverse order)

(d) horizontal shrink by factor $\frac{1}{3}$; horizontal translation 1 unit right; vertical translation 6 units down

Chapter 1 practice questions

1. (a) $a = -3, b = 1$ **(b)** range: $y \geqslant 0$

2. (a) 5 **(b)** -9

3. (a) $g^{-1}(x) = -3x + 4$ **(b)** $x = \frac{2}{3}$

4. (a) $(g \circ h)(x) = 2x - 3$ **(b)** See Worked Solutions

5. (a)

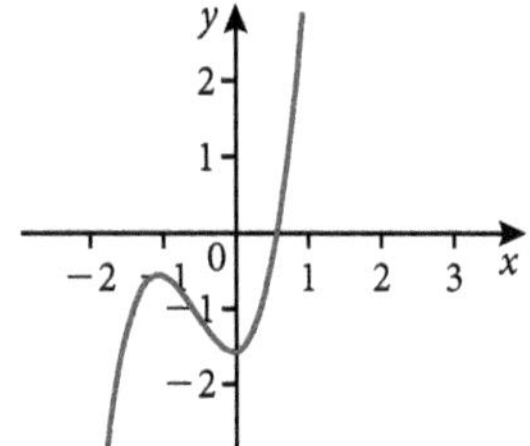

(b) maximum at $\left(-1, -\frac{1}{2}\right)$; minimum at $\left(0, -\frac{3}{2}\right)$

6. (a) $k = \frac{1}{2}$ **(b)** $p = -5$ **(c)** $q = 3$

7. (a)

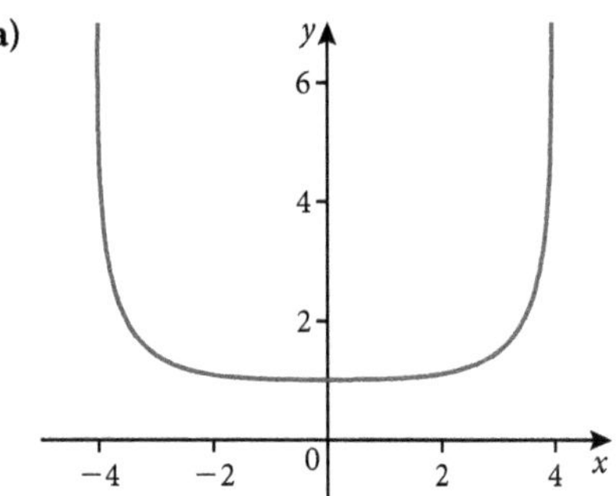

(b) $x = 4, x = -4$

(c) range: $y \geqslant 1$

8. (a)

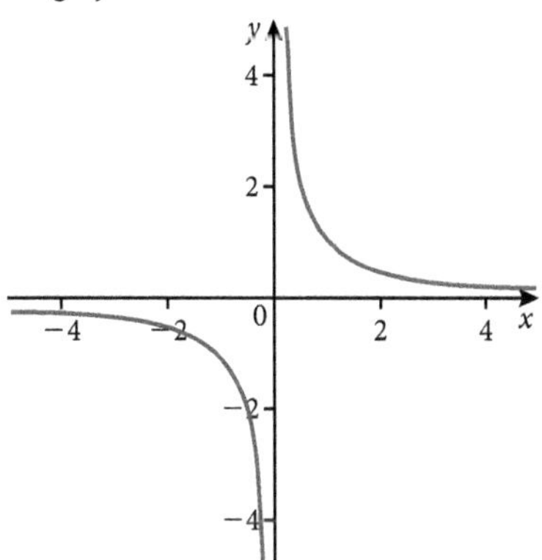

(b) $h(x) = \frac{1}{x+4} - 2$

(c) (i) x-intercept: $\left(-\frac{7}{2}, 0\right)$; y-intercept: $\left(0, -\frac{7}{4}\right)$

(ii) vertical asymptote: $x = -4$; horizontal asymptote: $y = -2$

(iii)

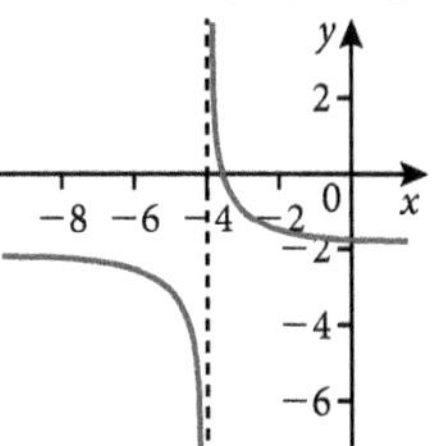

9. (a) (i) $\sqrt{11}$ **(ii)** 7 **(iii)** 0

(b) $x < -3$

(c) $(g \circ f)(x) = x - 2$

10. (a) 4 **(b)** $(g^{-1} \circ h)(x) = 2x^2 + 6$

(c) $x = \pm 2\sqrt{2}$

11. (a) $f^{-1}(x) = \frac{1}{3}x + \frac{1}{3}$ (b) $(f \circ g)(x) = \frac{12}{x} - 1$
(c) $(f \circ g)^{-1}(x) = \frac{12}{x+1}$ (d) $(g \circ g)(x) = x$

12. (a) (i) $a = 8$ (ii) $b = -3$
(b) reflection over x-axis

13. (a)

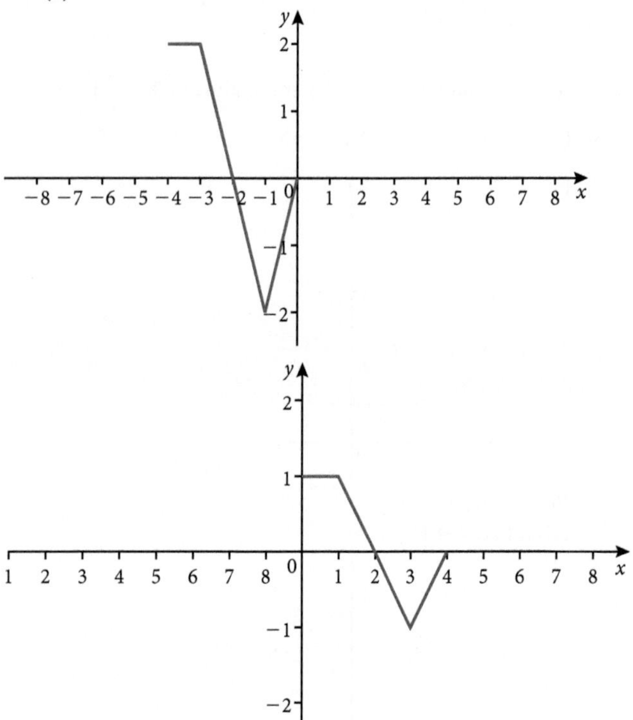

(b) $A'(-3, -2)$

14.

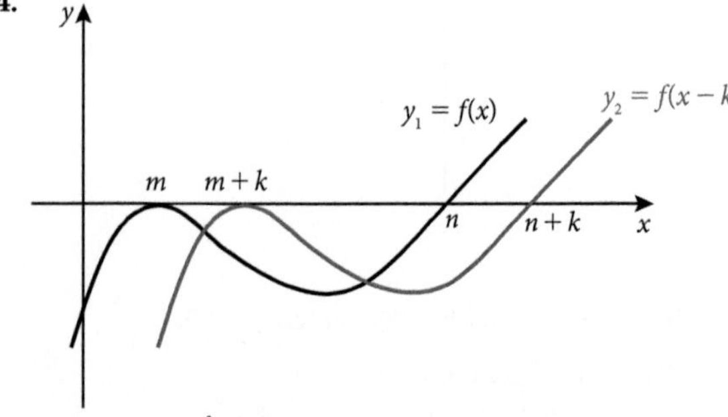

15. $(f \circ g)^{-1}(x) = \sqrt[3]{x-1}$

16. (a) $g(x) = \frac{x}{2x+1}$ (b) $\frac{2}{9}$

17. (a) $-\frac{\sqrt{2}}{2} \leqslant x \leqslant \frac{\sqrt{2}}{2}, x \neq 0$ (b) $f(x) \geqslant 0$

18. $f^{-1}(x) = \frac{x+1}{x-2}, x \neq 2$

19. (a) $-\frac{1}{2} < A < 2$ (b) $f^{-1}(x) = \frac{-2x-1}{x-2}$

20. (a) $g(x) = \sqrt[3]{x+1}$ (b) $g(x) = \sqrt[3]{x} + 1$

21. (a) $S = \{x: -\sqrt{3} < x < \sqrt{3}\}$ (b) $f(x) \geqslant \frac{\sqrt{3}}{3}$

22. $\frac{x}{4}$

23. (a) $A(1, 25), B(4, 0), C(7, -35), D(10, 0)$
(b) $A(1, 25), B(0, 0), C(-1, -35), D(-2, 0)$

Chapter 2

Exercise 2.1

1. (a) (i) $x = 5, (5, 7)$
(ii) horizontal translation 5 units right; vertical translation 7 units up
(iii) minimum value of f: 7
(b) (i) $x = -3, (-3, -1)$
(ii) horizontal translation 3 units left; vertical translation 1 unit down
(iii) minimum value of f: -1
(c) (i) $x = -1, (-1, 12)$
(ii) horizontal translation 1 unit left; reflection in x-axis; vertical stretch by factor 2; vertical translation 12 units up
(iii) maximum value of f:12
(d) (i) $x = \frac{1}{2}, \left(\frac{1}{2}, 8\right)$
(ii) horizontal translation $\frac{1}{2}$ unit right; vertical stretch by factor 4; vertical translation 8 units up
(iii) minimum value of f: 8
(e) (i) $x = -7, \left(-7, \frac{3}{2}\right)$
(ii) horizontal translation 7 units left; vertical stretch by factor $\frac{1}{2}$; vertical translation $\frac{3}{2}$ unit up
(iii) minimum value of f: $\frac{3}{2}$

2. (a) $x = 2, x = -4$ (b) $x = 5, x = -2$
(c) $x = \frac{3}{2}, x = 0$ (d) $x = 6, x = -1$
(e) $x = 3$ (f) $x = \frac{1}{3}, x = -4$
(g) $x = 3, x = 2$ (h) $x = 2, x = \frac{1}{4}$

3. (a) $x = -2 \pm \sqrt{7}$ (b) $x = 5, x = -1$
(c) no real solution (d) $x = -4 \pm \sqrt{13}$
(e) $x = 2, x = -4$ (f) $x = \frac{2 \pm \sqrt{22}}{2}$

4. (a) $x = 2 \pm \sqrt{5}$
(b) axis of symmetry: $x = 2$ (c) minimum value of f is -5

5. (a) two real solutions (b) no real solutions
(c) two real solutions (d) no real solutions

6. $p = \pm 2\sqrt{2}$

7. $k < 4$

8. $k < -1, k > 1$

9. $m < -3, m > 3$

10. $k > 12$

11. $x - 2 - x^2 \Rightarrow -(x^2 - x + 2) \Rightarrow -\left(x^2 - x + \frac{1}{4}\right) - \frac{7}{4}$
$\Rightarrow -\left(x - \frac{1}{2}\right)^2 - \frac{7}{4} \leqslant -\frac{7}{4}$ for all x

12. (a) $y = -2x^2 + 6x + 8$
(b) $y = \frac{2}{3}x^2 - \frac{7}{3}x + 1$

13. $-1 < k < 15$

14. $m < -2\sqrt{10}$ or $m > 2\sqrt{10}$

15. (a) (i) $y = (x + 2)^2 - 3$ (ii) $(-2, -3)$
(iii) minimum
(b) (i) $y = -2(x - 1)^2 + 5$ (ii) $(1, 5)$
(iii) maximum

(c) **(i)** $y = 3(x + 2)^2$ **(ii)** $(-2, 0)$
(iii) minimum
(d) **(i)** $y = -(x - 3)^2 + 12$
(ii) $(3, 12)$ **(iii)** maximum

16. **(a)** $x = 4$ **(b)** $h(x) = \frac{1}{2}x^2 - 4x + 6$

17. Answer shows that there is no real t for which $(2 - t)^2 - 4 \times 2 \times (t^2 + 3) > 0$

18. Answer shows that

$$\frac{1}{\frac{-b + \sqrt{(b^2 - 4a^2)}}{2a}} = \frac{-b - \sqrt{(b^2 - 4a^2)}}{2a}$$

Exercise 2.2

1. **(a)**

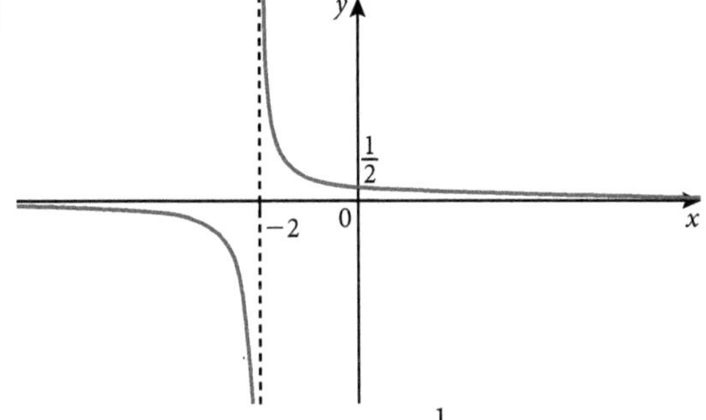

x-intercept: none, y-intercept: $(0, \frac{1}{2})$
vertical asymptote: $x = -2$
horizontal asymptote: $y = 0$

(b)

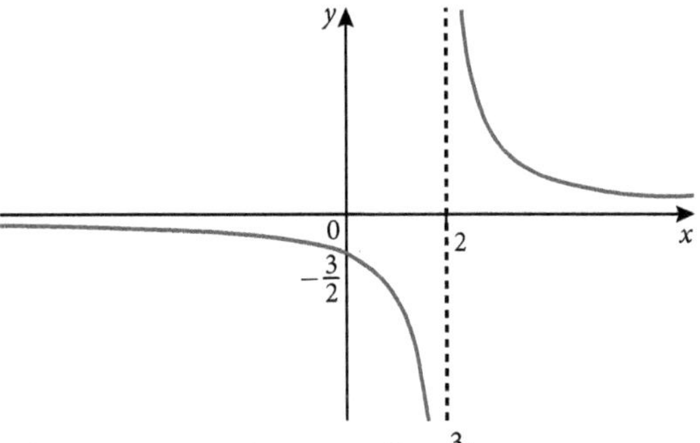

x-intercept: none, y-intercept: $(0, -\frac{3}{2})$
vertical asymptote: $x = 2$
horizontal asymptote: $y = 0$

(c)

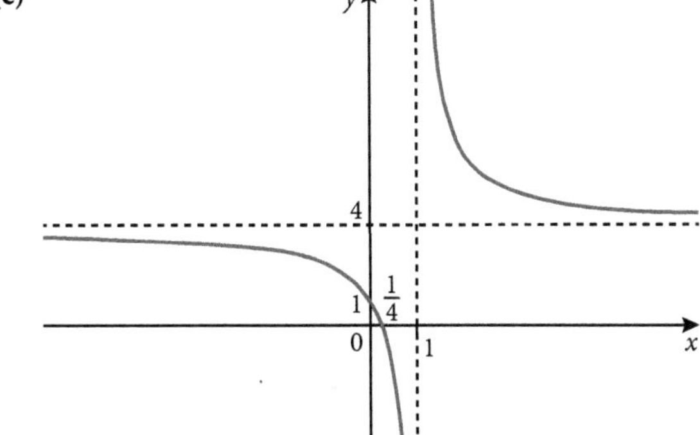

x-intercept: $\left(\frac{1}{4}, 0\right)$, y-intercept: $(0, 1)$
vertical asymptote: $x = 1$
horizontal asymptote: $y = 4$

(d)

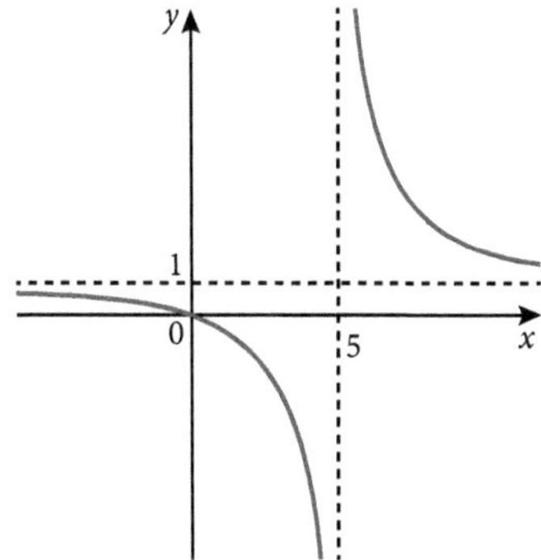

x-intercept: $(0, 0)$, y-intercept: $(0, 0)$
vertical asymptote: $x = 5$
horizontal asymptote: $y = 1$

(e)

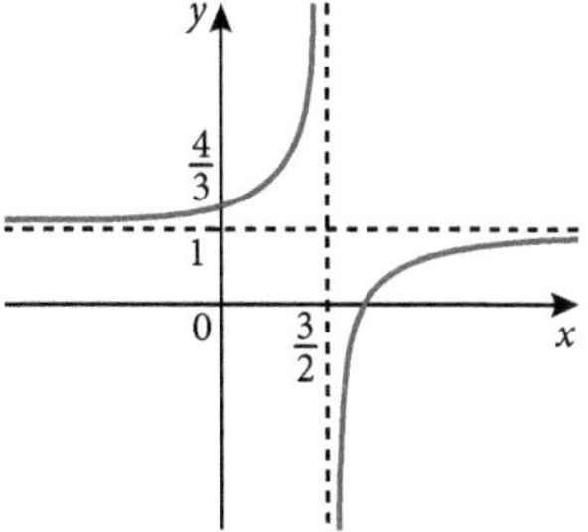

x-intercept: $(2, 0)$, y-intercept: $\left(0, \frac{4}{3}\right)$
vertical asymptote: $x = \frac{3}{2}$
horizontal asymptote: $y = 1$

(f)

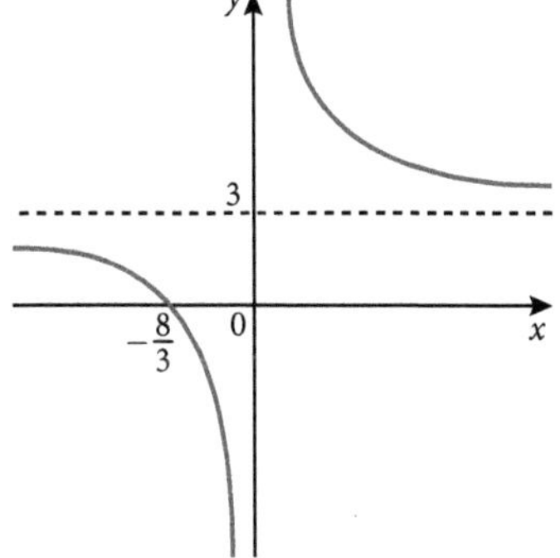

x-intercept: $\left(-\frac{8}{3}, 0\right)$, y-intercept: none
vertical asymptote: $x = 0$
horizontal asymptote: $y = 3$

2. **(a)**

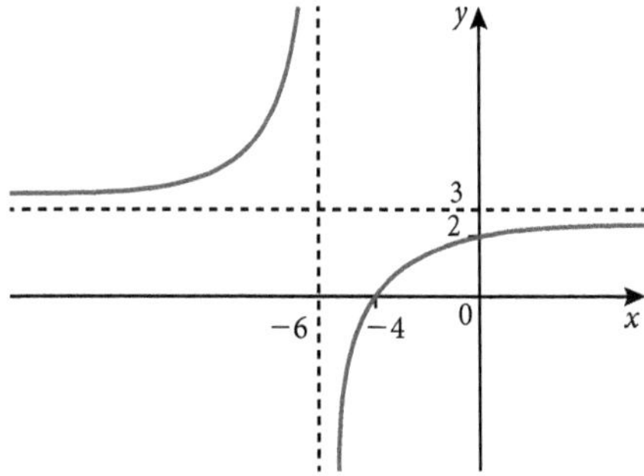

domain: $x \in \mathbb{R}, x \neq -6$
range: $y \in \mathbb{R}, y \neq 3$

(b)

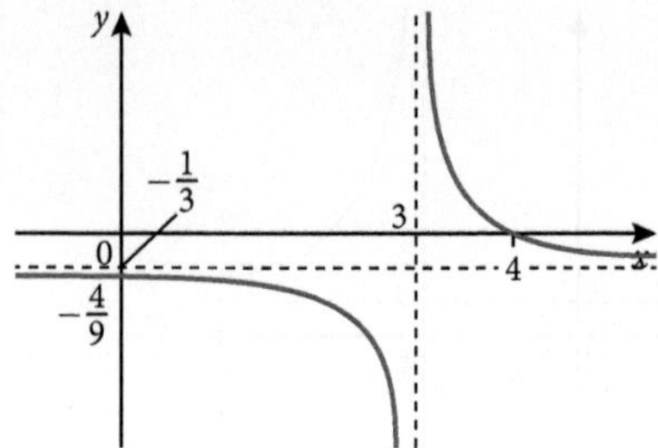

domain: $x \in \mathbb{R}, x \neq 3$

range: $y \in \mathbb{R}, y \neq -\frac{1}{3}$

(c)

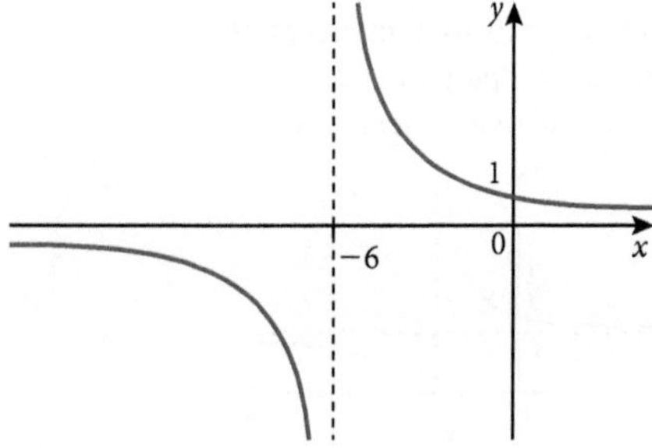

domain: $x \in \mathbb{R}, x \neq -6$

range: $y \in \mathbb{R}, y \neq 0$

(d)

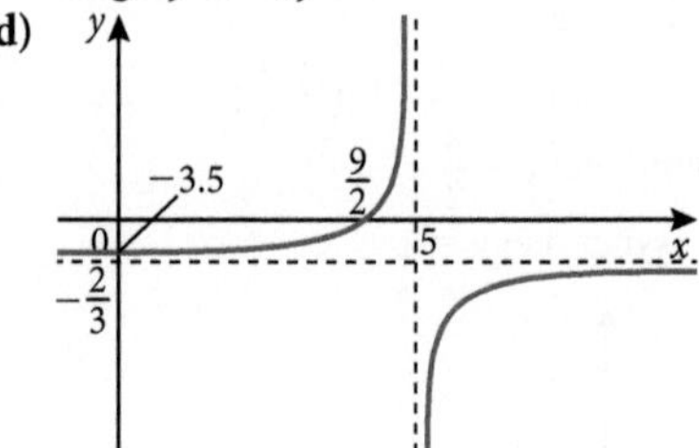

domain: $x \in \mathbb{R}, x \neq 5$

range: $y \in \mathbb{R}, y \neq -\frac{2}{3}$

3. **(a)**

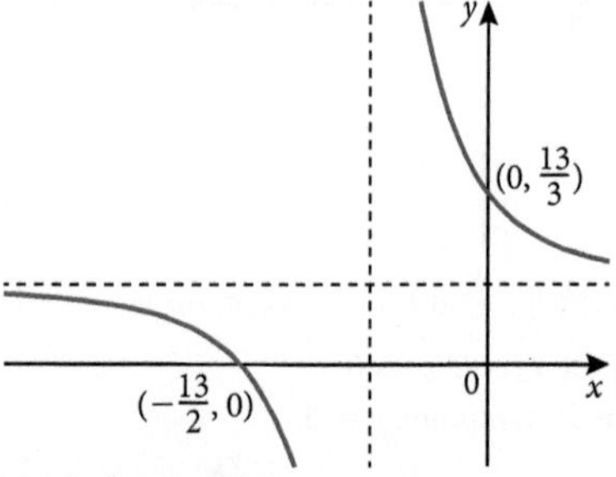

(b) vertical asymptote: $x = -3$;
horizontal asymptote: $y = 2$

(c) $g(x) = \dfrac{2x + 13}{x + 3}$

4. **(a)**

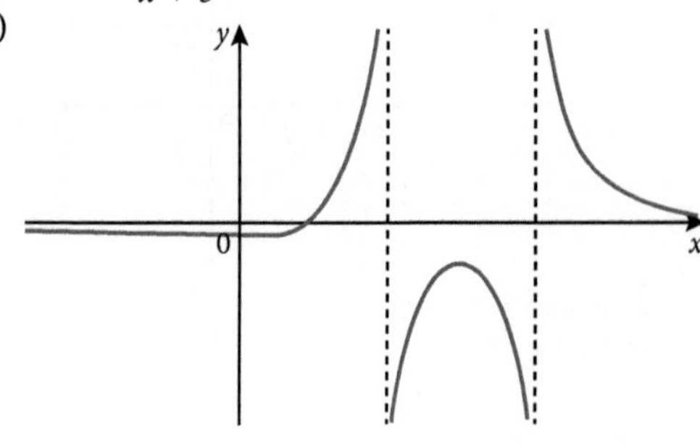

(b)

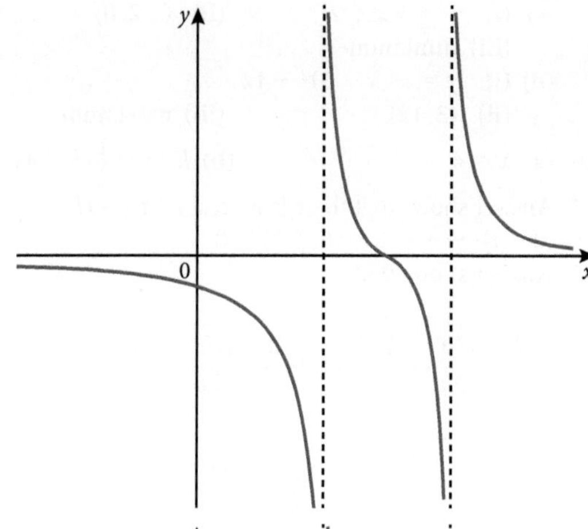

(c)

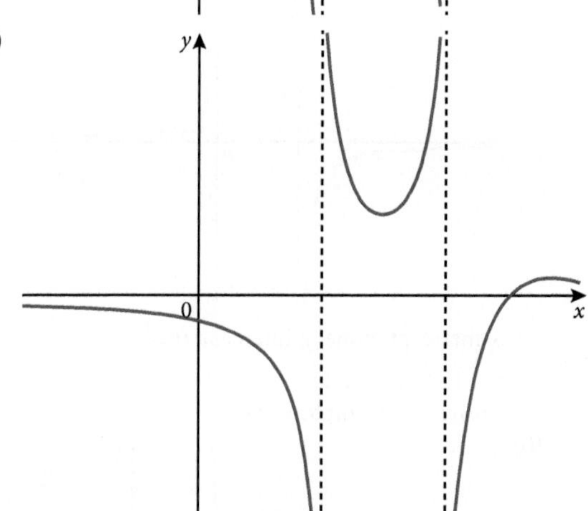

5. **(a)**

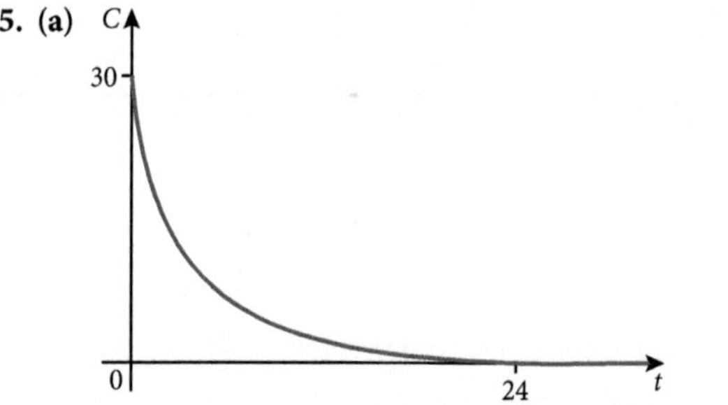

(b) at $t = 0$ hours, concentration is 30 mg/l

(c) 14.5 hours

6. $a = 4, b = \frac{1}{3}, c = 2$

Exercise 2.3

1. **(a)** $x = 3$ **(b)** $x = 9$

(c) $x = 5$ or $x = -2$ **(d)** $x = \frac{4}{5}$

(e) $x = -5$ **(f)** $x = 1$ or $x = -2$

(g) $x = 2$ or $x = -2$ **(h)** $x = \pm\sqrt{5}$

(i) $x = \dfrac{-1 \pm \sqrt{41}}{4}$ **(j)** $x = 3$ or $x = 2$

(k) $x = \frac{4}{3}$ or $x = -4$ **(l)** $x = 2$ or $x = 8$

(m) no solution **(n)** $x = \dfrac{1 + \sqrt{41}}{2}$

2. **(a)** $-2 < x < 4$ **(b)** no solution

(c) $x < -3, x > \frac{1}{2}$ **(d)** $x \leqslant \frac{1}{3}, x \geqslant 5$

3. $k < -\frac{3}{11}, k > 3$

4. (a) $p = \frac{9}{4}$ (b) $p < \frac{9}{4}$ (c) $p > \frac{9}{4}$

5. $k < -1, k > \frac{1}{3}$

6. (a) $m + \frac{1}{n} \geqslant 2 \Rightarrow mn + 1 \geqslant 2n \Rightarrow mn - 2n + 1 \geqslant 0$;
since $m \geqslant n \Rightarrow mn \geqslant n^2$ it follows that $mn - 2n + 1 \geqslant n^2 - 2n + 1$ and since $n^2 - 2n + 1 = (n-1)^2 \geqslant 0$ then $mn - 2n + 1 \geqslant 0 \Rightarrow m + \frac{1}{n} \geqslant 2$

(b) $(m+n)\left(\frac{1}{m} + \frac{1}{n}\right) > 4 \Rightarrow (m+n)\left(\frac{1}{m} + \frac{1}{n}\right)mn \geqslant 4mn \Rightarrow (m+n)(n+m) \geqslant 4mn \Rightarrow m^2 + 2mn + n^2 \geqslant 4mn \Rightarrow m^2 - 2mn + n^2 \geqslant 0 \Rightarrow (m-n)^2 \geqslant 0$
which is true for all $m \geqslant n > 0$ and is equivalent to original inequality – thus $(m+n)\left(\frac{1}{m} + \frac{1}{n}\right) \geqslant 4$ is true for all $m \geqslant n > 0$.

7. $x = \frac{-1 \pm \sqrt{13}}{2}, x = 1$ or $x = -2$

8. $x < -2, -1 < x < 1, x > 3$

Chapter 2 practice questions

1. $x = a$ or $x = 3b$

2. $c = 5$

3. (a) $x = 1$ (b) $p(x) = 4x^2 - 8x - 15$

4. $a = -1, b = -2, c = 3$

5. (a) $m > -2$ (b) $-2 < m < 0$

6. $1 \leqslant x \leqslant 3$

7. $-1 < k < 15$

8. (a) vertical asymptote: $x = -2$;
horizontal asymptote: $y = 8$

(b) x-intercept: $\left(-\frac{1}{2}, 0\right)$ (c) y-intercept: $(0, 2)$

(d)

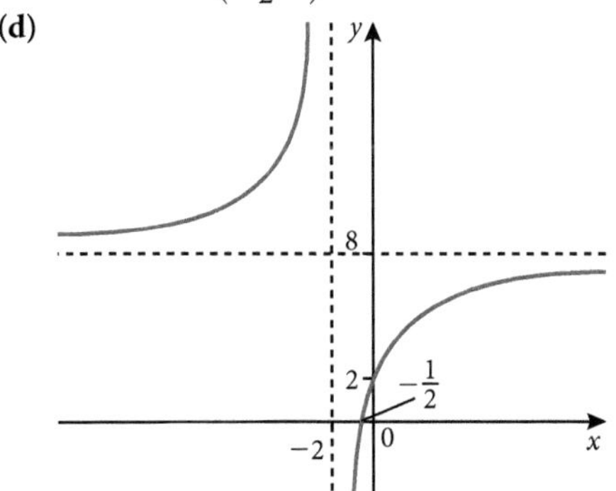

9. $k \in \mathbb{R}$

10. $\frac{-2 - \sqrt{13}}{2} < k < \frac{-2 + \sqrt{13}}{2}$

11. (a) $P\left(\frac{1}{2}, \frac{5}{2}\right)$

(b) domain: $x \in \mathbb{R}, x \neq \frac{1}{2}$; range: $y \in \mathbb{R}, y \neq \frac{5}{2}$

12. $-3 \leqslant x \leqslant \frac{1}{3}$

13. $x = \frac{19}{5}$ or $x = 7$

14. $x = \frac{-2 - \sqrt{2}}{2}$ or $x = \frac{-2 + \sqrt{2}}{2}$

15. $w = \pm\sqrt{7}$ or $w = \pm 2$

Chapter 3

Exercise 3.1

1. (a) $-1, 1, 3, 5, 7$ (b) $-1, 1, 5, 13, 29$

(c) $\frac{3}{2}, \frac{3}{4}, \frac{3}{8}, \frac{3}{16}, \frac{3}{32}$ (d) $1, 7, -5, 19, -29$

(e) $5, 8, 11, 14, 17$ (f) $3, 7, 13, 21, 31$

2. (a) $-1, 1, 3, 5, 7, 97$

(b) $2, 6, 18, 54, 162, 4.786 \times 10^{23}$

(c) $\frac{2}{3}, -\frac{2}{3}, \frac{6}{11}, -\frac{4}{9}, \frac{10}{27}, -\frac{50}{1251}$

(d) $1, 2, 9, 64, 625, 1.776 \times 10^{83}$

(e) $3, 11, 27, 59, 123, 4.50 \times 10^{15}$

(f) $0, 3, \frac{3}{7}, \frac{21}{13}, \frac{39}{55}$, approx 1

(g) $2, 6, 18, 54, 162, 4.786 \times 10^{23}$

(h) $-1, 1, 3, 5, 7, 97$

3. (a) $u_n = \frac{1}{4}u_{n-1}, u_1 = \frac{1}{3}$

(b) $u_n = \frac{4a^2}{3}u_{n-1}, u_1 = \frac{1}{2}a$

(c) $u_n = u_{n-1} + a - k, u_1 = a - 5k$

4. (a) $u_n = n^2 + 3$ (b) $u_n = 3n - 1$

(c) $u_n = \frac{2n-1}{n^2}$ (d) $u_n = \frac{2n-1}{n+3}$

5. (a) $2, \frac{3}{2}, \frac{5}{3}, \frac{8}{5}, \frac{13}{8}, \frac{21}{13}, \frac{34}{21}, \frac{55}{34}, \frac{89}{55}, \frac{144}{89}$

(b) Substitute $a_{n-1} = \frac{F_n}{F_{n-1}}$ and simplify.

Exercise 3.2

1. $3, \frac{19}{5}, \frac{23}{5}, \frac{27}{5}, \frac{31}{5}, 7$

2. (a) arithmetic, $d = 2, a_{50} = 97$

(b) arithmetic, $d = 1, a_{50} = 52$

(c) arithmetic, $d = 2, a_{50} = 97$

(d) not arithmetic, *no common difference.*

(e) not arithmetic, *no common difference.*

(f) arithmetic, $d = -7, a_{50} = -341$

3. (a) (i) 26

(ii) $a_n = -2 + 4(n-1)$

(iii) $a_1 = -2, a_n = a_{n-1} + 4$ for $n > 1$.

(b) (i) 1

(ii) $a_n = 29 - 4(n-1)$

(iii) $a_1 = 29, a_n = a_{n-1} - 4$ for $n > 1$.

(c) (i) 57

(ii) $a_n = -6 + 9(n-1)$

(iii) $a_1 = -6, a_n = a_{n-1} + 9$ for $n > 1$.

(d) (i) 9.23

(ii) $a_n = 10.07 - 0.12(n-1)$

(iii) $a_1 = 10.07, a_n = a_{n-1} - 0.12$ for $n > 1$.

(e) (i) 79

(ii) $a_n = 100 - 3(n-1)$

(iii) $a_1 = 100, a_n = a_{n-1} - 3$ for $n > 1$.

(f) (i) $-\frac{27}{4}$

(ii) $a_n = 2 - \frac{5}{4}(n-1)$

(iii) $a_1 = 2, a_n = a_{n-1} - \frac{5}{4}$ for $n > 1$.

4. $13, 7, 1, -5, -11, -17, -23$

5. 299, $299\frac{1}{4}$, $299\frac{1}{2}$, $299\frac{3}{4}$, 300

6. $a_n = \frac{40}{9} + \frac{26}{9}(n-1) = \frac{14}{9} + \frac{26}{9}n$

7. $a_n = -\frac{142}{3} + \frac{11}{3}(n-1) = -51 + \frac{11}{3}n$

8. (a) 88 (b) 36 (c) 11 (d) 16 (e) 11

9. 9, 3, −3, −9, −15

10. 99.25, 99.50, 99.75

11. $a_n = 4n - 1$

12. $a_n = \frac{19n - 277}{3}$

13. $a_n = 4n + 27$

14. Yes, 3271st term

15. Yes, 1385th term

16. No

Exercise 3.3

1. (a) Geom., $r = 3^a$, $g_{10} = 3^{9a+1}$
   (b) Arithmetic, $d = 3$, $a_{10} = 27$
   (c) Geometric, $r = 2$, $b_{10} = 4096$
   (d) Neither
   (e) Geometric, $r = 3$, $u_{10} = 78\,732$
   (f) Geometric, $r = 2.5$, $a_{10} = 7629.39453125$
   (g) Geometric, $r = -2.5$, $a_{10} = -7629.39453125$
   (h) Arithmetic, $d = 0.75$, $a_{10} = 8.75$
   (i) Geometric, $r = -\frac{2}{3}$, $a_{10} = -\frac{1024}{2187}$
   (j) Arithmetic, $d = 3$, $a_{10} = 79$
   (k) Geometric, $r = -3$, $u_{10} = 19\,683$
   (l) Geometric, $r = 2$, $u_{10} = 51.2$
   (m) Neither
   (n) Neither
   (o) Arithmetic, $d = 1.3$, $a_{10} = 14.1$

2. (a) (i) 32
   (ii) $-3 + 5(n-1)$
   (iii) $a_1 = -3$, $a_n = a_{n-1} + 5$ for $n > 1$
   (b) (i) −9
   (ii) $19 - 4(n-1)$
   (iii) $a_1 = 19$, $a_n = a_{n-1} - 4$ for $n > 1$
   (c) (i) 69
   (ii) $-8 + 11(n-1)$
   (iii) $a_1 = -8$, $a_n = a_{n-1} + 11$ for $n > 1$
   (d) (i) 9.35
   (ii) $10.05 - 0.1(n-1)$
   (iii) $a_1 = 10.05$, $a_n = a_{n-1} - 0.1$ for $n > 1$
   (e) (i) 93
   (ii) $100 - (n-1)$
   (iii) $a_1 = 100$, $a_n = a_{n-1} - 1$ for $n > 1$
   (f) (i) $-\frac{17}{2}$
   (ii) $2 - 1.5(n-1)$
   (iii) $a_1 = 2$, $a_n = a_{n-1} - 1.5$ for $n > 1$
   (g) (i) 384
   (ii) $3 \times 2^{n-1}$
   (iii) $a_1 = 3$, $a_n = 2a_{n-1}$ for $n > 1$
   (h) (i) 8748
   (ii) $4 \times 3^{n-1}$
   (iii) $a_1 = 4$, $a_n = 3a_{n-1}$ for $n > 1$
   (i) (i) −5
   (ii) $5 \times (-1)^{n-1}$
   (iii) $a_1 = 5$, $a_n = -a_{n-1}$ for $n > 1$
   (j) (i) −384
   (ii) $3 \times (-2)^{n-1}$
   (iii) $a_1 = 3$, $a_n = -2a_{n-1}$ for $n > 1$
   (k) (i) $-\frac{4}{9}$
   (ii) $972 \times \left(-\frac{1}{3}\right)^{n-1}$
   (iii) $a_1 = 972$, $a_n = \left(-\frac{1}{3}\right)a_{n-1}$ for $n > 1$
   (l) (i) $\frac{2187}{64}$
   (ii) $a_n = -2\left(-\frac{3}{2}\right)^{n-1}$
   (iii) $a_1 = -2$, $a_n = -\frac{3}{2}a_{n-1}$, $n > 1$
   (m) (i) $\frac{390\,625}{117\,649}$
   (ii) $a_n = 35\left(\frac{5}{7}\right)^{n-1}$
   (iii) $a_1 = 35$, $a_n = \frac{5}{7}a_{n-1}$, $n > 1$
   (n) (i) $-\frac{3}{64}$
   (ii) $a_n = -6\left(\frac{1}{2}\right)^{n-1}$
   (iii) $a_n = -6$, $a_n = \frac{1}{2}a_{n-1}$, $n > 1$
   (o) (i) 1216
   (ii) $9.5 \times 2^{n-1}$
   (iii) $a_1 = 9.5$, $a_n = 2a_{n-1}$, $n > 1$
   (p) (i) $69.833\,729\,609\,375 = \frac{893\,871\,739}{12\,800\,000}$
   (ii) $a_n = 100\left(\frac{19}{20}\right)^{n-1}$
   (iii) $a_1 - 100$, $a_n = \frac{19}{20}a_{n-1}$, $n > 1$
   (q) (i) $0.002\,085\,685\,73 = \frac{2187}{1\,048\,576}$
   (ii) $a_n = 2\left(\frac{3}{8}\right)^{n-1}$
   (iii) $a_1 = 2$, $a_n = \frac{3}{8}a_{n-1}$, $n > 1$

3. 6,12,24,48

4. 35,175,875

5. −36

6. 21,63,189,567

7. −24, 24

8. 1.5,$a_n = 24\left(\frac{1}{2}\right)^{n-1}$

9. $a_4 = \pm 3$, $r = \pm\frac{1}{2}$, $a_n = 24\left(\pm\frac{1}{2}\right)^{n-1}$

10. $\frac{49}{3}$

11. 10th term

12. Yes, 10th term

13. Yes, 10th term

14. €2228.92

15. £945.23

16. €2968.79

17. 7745

18. $\frac{686}{27}$

19. 10th term

20. £2921.16

Exercise 3.4

1. 11 280

2. $-\frac{105\,469}{1024}$

3. 0.7

4. $\frac{10}{7}$

5. $\frac{16 + 4\sqrt{3}}{39}$

6. (a) $\frac{52}{99}$ (b) $\frac{449}{990}$ (c) $\frac{7459}{2475}$

7. £13026.14

8. (a) 940 (b) 6578 (c) 42625

9. $\frac{n(7+3n)}{2}$ 10. 17 terms

11. 29 terms 12. $d = 4$

13. (a) 250, 125250 (b) 83501

14. $a = 1, d = 5$

15. (a) 2890 (b) 0.290 (c) −2.065

16. 11400 17. 1.191 18. 49.2

19. $\frac{6}{5}$ 20. $\frac{3+\sqrt{6}}{2}$

21. (a) $3, \frac{18}{5}, \frac{93}{25}, \frac{468}{125}; \frac{15}{4}\left(1 - \frac{1}{5^n}\right)$

(b) $\frac{1}{6}, \frac{1}{4}, \frac{3}{10}, \frac{1}{3}; \frac{n}{2n+4}$

(c) $\sqrt{2}-1, \sqrt{3}-1, 1, \sqrt{5}-1; \sqrt{n+1}-1$

22. (a) 1.945 (b) 84.2

23. (a) 127 (b) 128

24. (a) $\frac{819}{128}$ (b) $\frac{32}{5}$

25. (a) 11866 (b) 763517
(c) 14348906 (d) ~150

Exercise 3.5

1. (a) $x^5 + 10x^4y + 40x^3y^2 + 80x^2y^3 + 80xy^4 + 32y^5$
(b) $a^4 - 4a^3b + 6a^2b^2 - 4ab^3 + b^4$
(c) $x^6 - 18x^5 + 135x^4 - 540x^3 + 1215x^2 - 1458x + 729$
(d) $16 - 32x^3 + 24x^6 - 8x^9 + x^{12}$
(e) $x^7 - 21bx^6 + 189b^2x^5 - 945b^3x^4 + 2835b^4x^3 - 5103b^5x^2 + 5103b^6x - 2187b^7$
(f) $64n^6 + 192n^3 + 240 + \frac{160}{n^3} + \frac{60}{n^6} + \frac{12}{n^9} + \frac{1}{n^{12}}$
(g) $\frac{81}{x^4} - \frac{216}{x^2\sqrt{x}} + \frac{216}{x} - 96\sqrt{x} + 16x^2$

2. (a) 56 (b) 0 (c) 1225
(d) 32 (e) 0

3. (a) $x^7 - 14x^6y + 84x^5y^2 - 280x^4y^3 + 560x^3y^4 - 672x^2y^5 + 448xy^6 - 128y^7$
(b) $64a^6 - 192a^5b + 240a^4b^2 - 160a^3b^3 + 60a^2b^4 - 12ab^5 + b^6$
(c) $x^5 - 20x^4 + 160x^3 - 640x^2 + 1280x - 1024$
(d) $x^{18} + 12x^{15} + 60x^{12} + 160x^9 + 240x^6 + 192x^3 + 64$
(e) $2187x^7 - 5103bx^6 + 5103b^2x^5 - 2835b^3x^4 + 945b^4x^3 - 189b^5x^2 + 21b^6x - b^7$
(f) $64n^6 - 192n^3 + 240 - \frac{160}{n^3} + \frac{60}{n^6} - \frac{12}{n^9} + \frac{1}{n^{12}}$
(g) $\frac{16}{x^4} - \frac{96}{x^2\sqrt{x}} + \frac{216}{x} - 216\sqrt{x} + 81x^2$
(h) 112
(i) $1792\sqrt{3}$

4. (a) $x^{45} - 90x^{43} + 3960x^{41}$
(b) Does not exist as the powers of x decrease by 2 starting at 45. There is no chance for any expression to have zero exponent.
(c) $\binom{45}{43}x^2\left(\frac{-2}{x}\right)^{43} + \binom{45}{44}x\left(\frac{-2}{x}\right)^{44} + \left(\frac{-2}{x}\right)^{45}$
$= -\binom{45}{43}\frac{2^{43}}{x^{41}} + \binom{45}{44}\frac{2^{44}}{x^{43}} - \frac{2^{45}}{x^{45}}$
(d) $\binom{45}{21}x^{24}\left(\frac{-2}{x}\right)^{21} = -\binom{45}{21}\cdot 2^{21}x^3$

5. $\binom{n}{k} = \frac{n!}{k!(n-k)!} = \frac{n!}{(n-k)!k!} = \frac{n!}{(n-k)!(n-(n-k))!}$
$= \binom{n}{n-k}$

6. $(1+1)^n = \binom{n}{0} + \binom{n}{1} + \binom{n}{2} \ldots + \binom{n}{n}$
$2^n = 1 + \binom{n}{1} + \binom{n}{2} \ldots + \binom{n}{n} \Rightarrow 2^n - 1$
$= \binom{n}{1} + \binom{n}{2} \ldots + \binom{n}{n}$

7. (a) $k! = k(k-1) \cdots \times 2 \times 1 = k((k-1) \cdots \times 2 \times 1)$
(b) apply part **a**
(c) apply part **a**.

8. $\left(\frac{1}{3} + \frac{2}{3}\right)^6 = 1$ 9. $\left(\frac{2}{5} + \frac{3}{5}\right)^8 = 1$

10. $\left(\frac{1}{7} + \frac{6}{7}\right)^n = 1$ 11. 15

12. 90720 13. 16128

14. $1 + 10x + 45x^2$, 1.1045, 0.9045

15. Use definition of $\binom{n}{r}$, sum of an entry in the nth row plus twice the next entry plus the third entry is equal to the entry directly below the last entry but two rows below.

16. (a) $\frac{7}{9}$ (b) $\frac{19}{55}$ (c) $\frac{7952}{2475}$

17. −145152 18. $35a^3$ 19. 96096

20. $243n^5 - 810n^4m + 1080n^3m^2 - 720n^2m^3 + 240nm^4 - 32m^5$

21. 7838208

22. $k = 3$

Chapter 3 practice questions

1. $D = 5, n = 20$

2. $2098.63

3. (a) Nick: 20
Charlotte: 17.6
(b) Nick: 390
Charlotte: 381.3
(c) Charlotte will exceed the 40 hours during week 13.
(d) In week 11 Charlotte will catch up with Nick and exceed him

4. (a) loss for the second month = 1060 g
loss for the third month = 1123.6 g
(b) Plan A loss = 1880 g
Plan B loss = 1898.3 g
(c) (i) Loss due to plan A in all 12 months = 17280 g
(ii) Loss due to Plan B in all 12 months = 16869.9 g

5. (a) €895.42
(b) This is the future value of an annuity due = 6985.82

6. (a) $\sqrt[3]{7}, 1, \sqrt[3]{7}, 1 \ldots$ (b) 0, 2, 0, 2, …

7. (a) On the 37th day (b) 407 km

8. (a) 1.5 (b) (i) 207595 (ii) 2019
(c) 619583 (d) Market saturation

9. (a) $\sqrt{\frac{1}{4} + \frac{1}{4}} = \frac{\sqrt{2}}{2}$ (b) $\frac{1}{2}$
(c) (i) $\frac{1}{4}$ (ii) $\frac{1}{2}$ (d) (i) $\frac{1}{512}$ (ii) 2

10. **(a)** 1220 **(b)** 36920

11. **(i)** Area A = 1, Area B = $\frac{1}{9}$ **(ii)** $\frac{1}{81}$

(iii) $1 + \frac{8}{9}, 1 + \frac{8}{9} + \left(\frac{8}{9}\right)^2$ **(iv)** 0

12. **(a)** Neither, geometric converging, arithmetic, geometric diverging

(b) 6

13. **(a)** **(i)** Kell: 18 400, 18 800; YBO: 18 190, 19 463.3

(ii) Kell: 198 000; YBO: 234 879.62

(iii) Kell: 21 600; YBO: 31 253.81

(b) **(i)** After the second year

(ii) 4th year

14. **(a)** 62 **(b)** 936

15. **(a)** $7000(1 + 0.0525)^t$ **(b)** 7 years

(c) No, since $9912 < 10\,015.0$

16. **(a)** 11 **(b)** 2 **(c)** 15

17. 15, −8 **18.** $a = -2, b = -7$ **19.** 10300

20. **(a)** $a_n = 8n - 3$ **(b)** 50

21. 559 **22.** −3, 3 **23.** 9

24. **(a)** 4 **(b)** $16(4^n - 1)$

25. $|x| < \frac{5}{3}$, 10

26. **(a)** $\frac{n(3n + 1)}{2}$ **(b)** 30

27. 1275 ln2

28. **(a)** 4, 8, 16

(b) **(i)** $u_n = 2^n$ **(ii)** $2^{n+1} = 3 \times 2^n - 2 \times 2^{n-1}$

29. **(a)** $\frac{2}{3}$ **(b)** 9

30. $a = 2, b = -3$ **31.** −2, 4 **32.** $\frac{\theta}{1 - \cos\theta}$

33. **(a)** $32 + 80x + 80x^2 + 40x^3 + 10x^4 + x^5$;

(b) 32.8080401001

34. **(a)** $\$5000(1.063)^n$ **(b)** $6786.35

(c) 12

35. 7

36. $u_1 = 12, d = -1.5$

37. **(a)** $1, -nx, + \binom{n}{2}x^2, - \binom{n}{3}x^3$

(b) **(i)** $|u_3| - |u_2| = |u_4| - |u_3| \Rightarrow 3n^2 - 9n = n^3 - 6n^2 + 5n$

(ii) $n = 7$

38. **(a)** $81 + 108x + 54x^2 + 12x^3 + x^4$

(b) 92.3521

39. **(a)** 41 **(b)** $\sum_{n=1}^{41} 7 + 7n$ **(c)** 6314 **(d)** 287

40. **(a)** **(i)** $\frac{v_{n+1}}{v_n} = 2^d$ **(ii)** 2^a **(iii)** $v_n = 2^{a+(n-1)d}$

(b) **(i)** $S_n = \frac{2^a(2^{dn} - 1)}{2^d - 1}$ **(ii)** $d < 0$.

(iii) $S_\infty = \frac{2^a}{1 - 2^d}$ **(iv)** $d = -1$

Chapter 4

Exercise 4.1

1. **(a)** $y = b^x$

(b) domain $\{x : x \in \mathbb{R}\}$, range $\{y : y > 0\}$

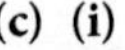

(c) **(i)**

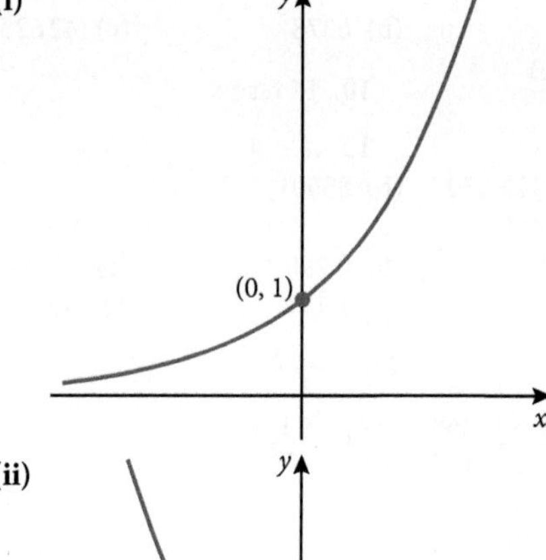

(ii)

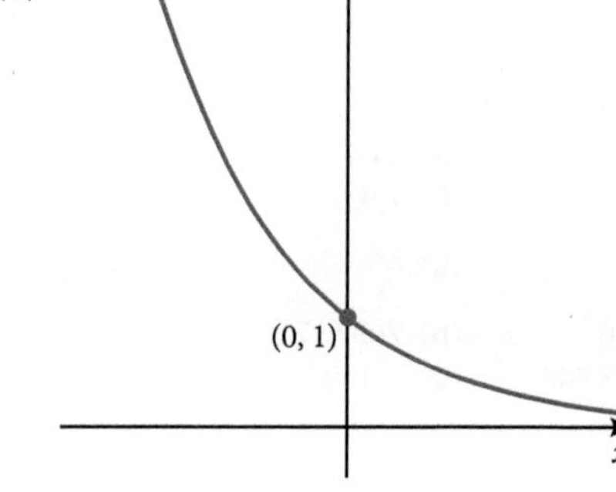

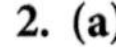

2. **(a)**

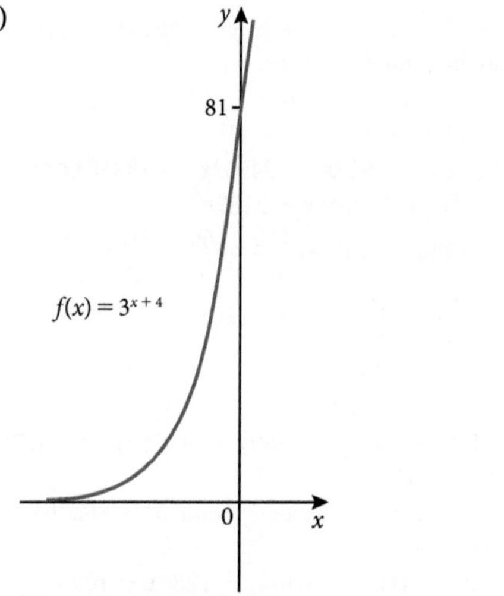

(i) y-intercept: (0, 81)

(ii) horizontal asymptote: $y = 0$ (x-axis)

(iii) domain: $x \in \mathbb{R}$; range: $y > 0$

(b)

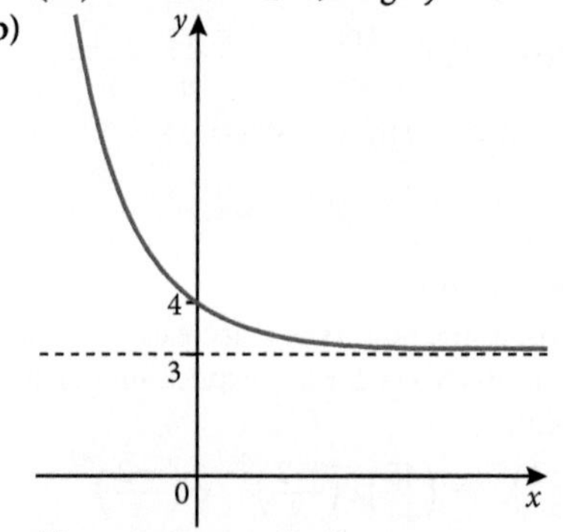

(i) y-intercept: (0, 4)

(ii) horizontal asymptote: $y = 3$

(iii) domain: $x \in \mathbb{R}$; range: $y > 3$

(c)

(i) y-intercept: $\left(0, \frac{1}{4}\right)$

(ii) horizontal asymptote: $y = 0$

(iii) domain: $x \in \mathbb{R}$; range: $y > 0$

(d)

(i) y-intercept: $(0, 1)$

(ii) horizontal asymptote: $y = 0$

(iii) domain: $x \in \mathbb{R}$; range: $y > 0$

(e)

(i) y-intercept: $(0, 1)$; x-intercept: $(-0.631, 0)$

(ii) horizontal asymptote: $y = -1$

(iii) domain: $x \in \mathbb{R}$; range: $y > -1$

3. domain: $x \in \mathbb{R}$

range: if $a > 0 \Rightarrow y > d$, if $a < 0 \Rightarrow y < d$

y-intercept: $(0, a(b)^{-c} + d)$

horizontal asymptote: $y = d$

4. $y = 2^{-x}$ $y = 4^{-x}$ $y = 8^{-x}$ $y = 8^{x}$ $y = 4^{x}$ $y = 2^{x}$

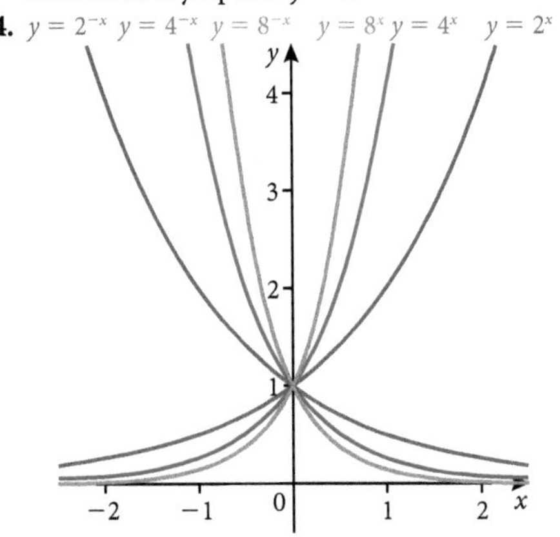

5. **(d)** $y = \left(\frac{1}{2}\right)^x$

(e) $y = \left(\frac{1}{4}\right)^x$

(f) $y = \left(\frac{1}{8}\right)^x$

6. $y = b^x$ is steeper

7. **(a)** $P(t) = 100\,000(3)^{\frac{t}{25}}$ where t is number of years

(b) **(i)** 900 000

(ii) 2 167 402

(iii) 8 100 000

8. **(a)** $N(t) = 10^4(2)^{\frac{t}{3}}$

(b) **(i)** 20 000

(ii) 80 000

(iii) 5 120 000

(iv) 10 485 760 000

9. **(a)** $A(t) = A_0(2)^{\frac{t}{10}}$

(b) 7.18%

10. **(a)** \$17 204.28

(b) \$29 598.74

(c) \$50 922.51

11. **(a)** $A(t) = 5000\left(1 + \frac{0.09}{12}\right)^{12t}$

(b)

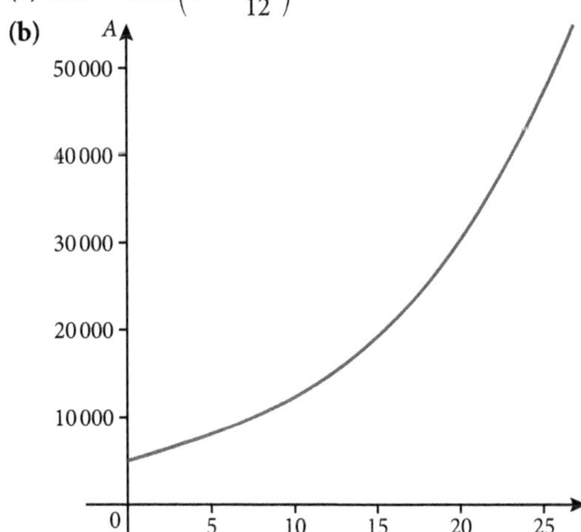

(c) minimum number of years is 16

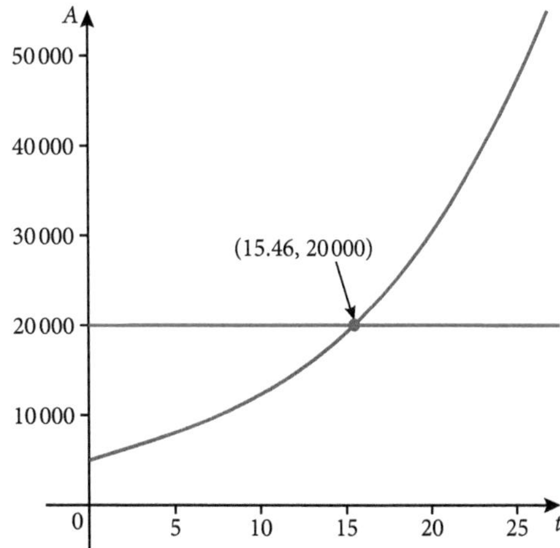

12. **(a)** \$16 850.58 **(b)** \$17 289.16

(c) \$17 331.09 **(d)** \$17 332.47

13. **(a)** 240 310 **(b)** 299 592

14. **(a)** $A(w) = 1000(0.7)^w$ **(b)** about 20 weeks

15. **(a)** Payment plan II **(b)** \$10 737 418.23

16. **(a)** $a = 2, k = 3$ **(b)** $a = \frac{1}{3}, k = 2$

(c) $a = 3, k = -4$ **(d)** $a = 10, k = \frac{3}{2}$

Exercise 4.2

1. **(a)**

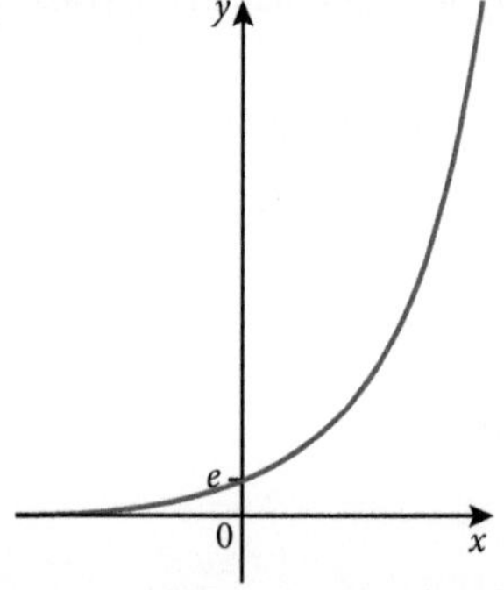

(i) x-int.: none, y-int.: $\left(0, \frac{1}{e}\right)$
(ii) horizontal asymptote: $y = 0$
(iii) domain: $x \in \mathbb{R}$; range: $y > 0$

(b)

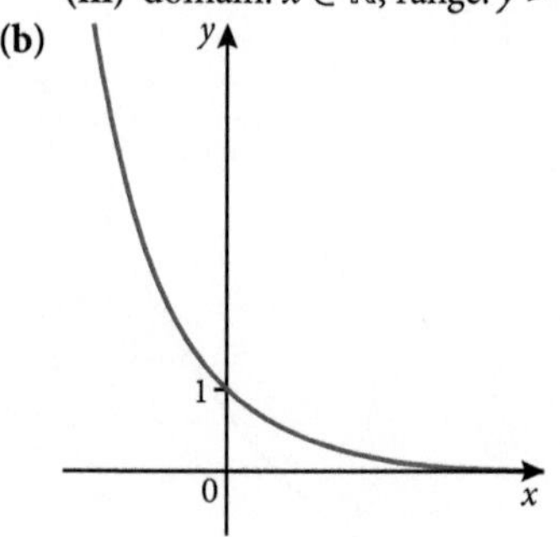

(i) x-int.: none, y-int.: (0, 1)
(ii) horizontal asymptote: $y = 0$
(iii) domain: $x \in \mathbb{R}$; range: $y > 0$

(c)

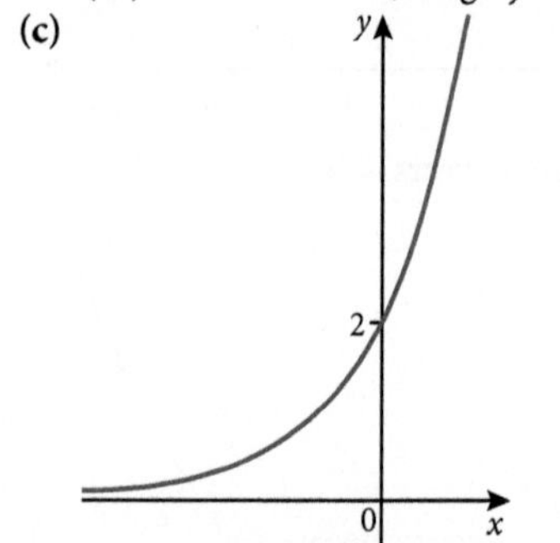

(i) x-int.: none, y-int.: (0, 2)
(ii) horizontal asymptote: $y = 0$
(iii) domain: $x \in \mathbb{R}$; range: $y > 0$

(d)

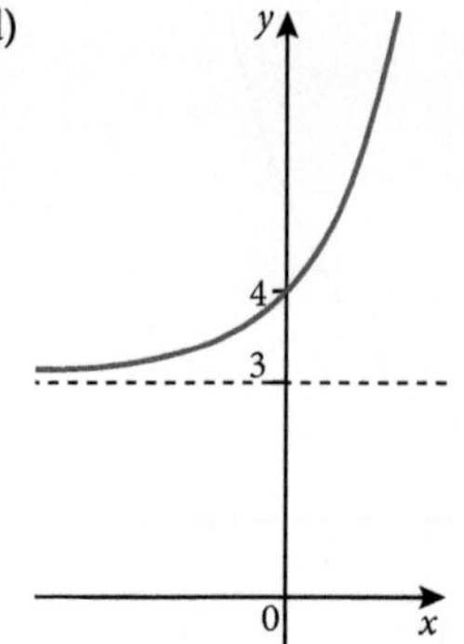

(i) x-int.: none, y-int.: (0, 4)
(ii) horizontal asymptote: $y = 3$
(iii) domain: $x \in \mathbb{R}$; range: $y > 3$

(e)

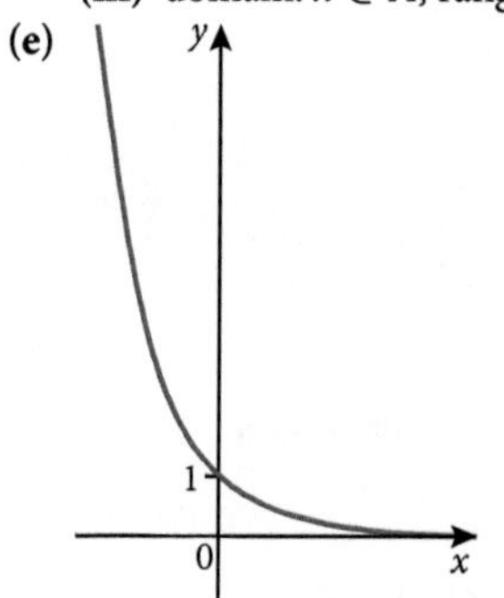

(i) x-int. : none, y-int.: (0, 1)
(ii) horizontal asymptote: $y = 0$
(iii) domain: $x \in \mathbb{R}$; range: $y > 0$

2. **(a)** Bank A: earns €113.71 in interest; Bank B: earn €113.99 in interest
(b) Bank B account earns €0.28 more in interest

3. **(a)** Blue Star
(b) \$1362.34 which is \$5.96 more than Red Star.

4. **(a)** 97.6% **(b)** 78.7%
(c) 9.16% **(d)** 0.254%

5. **(a)** 5 kg **(b)** 3.53 kg

6. **(a)** **(i)** £1568.31 **(ii)** £2459.60
(b) 15.4 years

7. **(a)** 39 **(b)** 11 minutes

Exercise 4.3

1. **(a)** $2^4 = 16$ **(b)** $e^0 = 1$ **(c)** $10^2 = 100$
(d) $10^{-2} = 0.01$ **(e)** $7^3 = 343$ **(f)** $e^{-1} = \frac{1}{e}$
(g) $10^y = 50$ **(h)** $e^{12} = x$ **(i)** $e^3 = x + 2$

2. **(a)** $\log_2 1024 = 10$ **(b)** $\log 0.0001 = -4$
(c) $\log_4\left(\frac{1}{2}\right) = -\frac{1}{2}$ **(d)** $\log_3 81 = 4$
(e) $\log 1 = 0$ **(f)** $\ln 5 = x$
(g) $\log_2 0.125 = -3$ **(h)** $\ln y = 4$
(i) $\log_{10} y = x + 1$

3. **(a)** 6 **(b)** 3 **(c)** -3 **(d)** 5 **(e)** $\frac{3}{4}$
(f) $\frac{1}{3}$ **(g)** -3 **(h)** 13 **(i)** 0 **(j)** 6
(k) -3 **(l)** $\sqrt{2}$ **(m)** 3 **(n)** $\frac{1}{2}$ **(o)** -2
(p) 88 **(q)** $\frac{1}{2}$ **(r)** 18 **(s)** $\frac{1}{3}$ **(t)** π

4. (a) 1.70 (b) 0.239 (c) 3.91
(d) 0.549 (e) 1.40 (f) 0.209
(g) 4.61 (h) 13.8

5. (a) $x > 2$ (b) $x \in \mathbb{R}\ x \neq 0$ (c) $x > 0$
(d) $x < \frac{8}{5}$ (e) $-2 \leqslant x < 3$ (f) $x \leqslant 0$

6. (a) domain $\{x{:}x > 0, x \neq 1\}$ range $\{y{:}y \in \mathbb{R}, y \neq 0\}$
(b) domain $\{x{:}x > 1\}$ range $\{y{:}y \geqslant 0\}$
(c) domain $\{x{:}x > 0, x \neq 1\}$ range $\{y{:}y < 0, y \geqslant 6.259\}$

7. (a) $f(x) = \log_4 x$ (b) $f(x) = \log_2 x$
(c) $f(x) = \log_{10} x$ (d) $f(x) = \log_3 x$

8. (a) $\log_2 2 + \log_2 m = 1 + \log_2 m$
(b) $\log 9 - \log x$
(c) $\frac{1}{5}\ln x$
(d) $\log_3 a + 3\log_3 b$
(e) $\log 10x + \log(1 + r)^t = 1 + \log x + t\log(1 + r)$
(f) $3\ln m - \ln n$

9. (a) $\log_b p + \log_b q + \log_b r$
(b) $2\log_b p + 3\log_b q - \log_b r$
(c) $\frac{\log_b p}{4} + \frac{\log_b q}{4}$
(d) $\frac{\log_b q}{2} + \frac{\log_b r}{2} - \frac{\log_b p}{2}$
(e) $\log_b p + \frac{1}{2}\log_b q - \log_b r$
(f) $3\log_b p + 3\log_b q - \frac{1}{2}\log_b r$

10. (a) $\log x$ (b) $\log_3 72$ (c) $\ln\left(\frac{y^4}{4}\right)$
(d) $\log_b 4$ (e) $\log\left(\frac{x}{yz}\right)$ (f) $\ln\left(\frac{36}{e}\right)$

11. (a) 9.97 (b) −5.32 (c) 2.06 (d) −0.179

12. (a) $\frac{\ln x}{\ln 2}$; 4.32 (b) $\frac{\ln x}{\ln 5}$; 1.86

13. $\log_b a = \frac{\log_a a}{\log_a b} = \frac{1}{\log_a b}$

14. $\log e = \frac{\ln e}{\ln 10} = \frac{1}{\ln 10}$

15. (a) $dB = 10\log\left(\frac{I}{10^{-16}}\right) = 10(\log I - \log 10^{-16})$
$= 10(\log I + 16) = 10\log I + 160$
(b) $10\log 10^{-4} + 160 = 10(-4) + 160 = 120$ decibels

Exercise 4.4

1. (a) $x = 0.699$ (b) $x = 2.50$ (c) $x = 7.97$
(d) $x = 3.64$ (e) $x = -1.92$ (f) $x = 2.71$
(g) $x = 0.434$ (h) $x = 2.12$ (i) $x = 4.42$
(j) $x = 0.225$ (k) $x = 0.642$ (l) $x = 22.0$

2. (a) $x = 3$ (b) $x = 0$ or -1

3. (a) \$6248.58 (b) $9\frac{1}{4}$ years

4. (a) 12.97 years (b) 12.92 years

5. 20 hours (≈ 19.93)

6. (a) 24 years (≈ 23.45) (b) 12 years (≈ 11.90)
(c) 9 years (≈ 8.04)

7. 6 years

8. (a) 0.997 grams (b) 127 000 years

9. (a) 37 dogs (b) 9 years

10. (a) 459 litres
(b) 8.89 minutes ≈ 8 min. 53 seconds
(c) 39 minutes

11. (a) 5 kg (b) 17.7 days

12. (a) $x = \frac{20}{3}$ (b) $x = 104$ (c) $x = \frac{1}{e^3}$ (d) $x = 4$
(e) $x = 98$ (f) $x = \pm\sqrt{e^{16}} = \pm e^8$
(g) $x = 2$ or $x = 4$ (h) $x = 9$
(i) $x = \frac{13}{5}$ (j) $x = 3$

13. (a) $x > 10^{-\frac{2}{5}}$ (b) $0 < x < 2$ (c) $0 < x < \ln 6$
(d) $0.161 < x < 1.14$ (approx. to 3 s.f.)

Chapter 4 practice questions

1. (a) $x = 2$ (b) $x = 3$ (c) $x = \frac{1}{2}$ (d) $x = 3$

2. (a) $x \approx 2.58$ (b) $x \approx 1.17$ (c) $x = 2$ (d) $x \approx 0.304$

3. (a) $\log_2(9x)$ (b) $\ln\left(\frac{3\sqrt{x-4}}{x}\right)$

4. (a) 1.89 (b) 4.85

5. (a) €2597 (b) 11 years (c) 7.18%

6. (a) \$1474.47 (b) 5.7%

7. (a) 1 (b) $\frac{3}{2}$ (c) 36

8. (a) 604 (b) 13 years

9. (a) 88% (b) \$11 610 (c) 2011

10. (a) domain: $x \in \mathbb{R}$, range: $y > 0$
(b) y-intercept: $\left(0, \frac{1}{e^2}\right)$; asymptote: $y = 0$ (x-axis)
(c) $f^{-1}(x) = 2 + \ln x$
(d) domain: $x > 0$, range: $y \in \mathbb{R}$

11. (a) 631 (b) 1270
(c) (i) $A_0 = 500$ (ii) $b = 1.06$
(d) $k = \ln 1.06 \approx$ ~0.05827

12. (a) (i) domain: $x < 0, x > 2$ (ii) domain: $x > 2$
(b) (i) $x = -\frac{2}{99}$ (ii) no solution

13. (a) $C = 5000, k \approx 0.0556$
(b) 140 753

14. (a) 8 (b) 2 (c) $\left(-\frac{2}{3}, 3\right)$

15. $x = 2$

16. $y = 16$

17. (a) $x = 3$ or $x = \frac{1}{81}$ (b) $x = 6$

18. (a) $\log\left(\frac{a^2b^3}{c}\right)$ (b) $\ln\left(\frac{ex^3}{\sqrt{y}}\right)$

19. 1900 years

20. $c = 42$

21. $x = \sqrt{e}, x = e$

22. (a) \$265.33 (b) 235 months

23. $x = 5^{\frac{5}{3}}$ or $x = 5^{\frac{-5}{3}}$

24. $k = \frac{\ln 2}{20}$

Chapter 5

Exercise 5.1

1. (a) $\frac{\pi}{3}$ (b) $\frac{5\pi}{6}$ (c) $-\frac{3\pi}{2}$
(d) $\frac{\pi}{5}$ (e) $\frac{3\pi}{4}$ (f) $\frac{5\pi}{18}$
(g) $-\frac{\pi}{4}$ (h) $\frac{20\pi}{9}$ (i) $-\frac{8\pi}{3}$

2. (a) 135° (b) −630° (c) 115°
   (d) 210° (e) −143° (f) 300°
   (g) 15° (h) 90.0° (i) 480°
3. (a) 390°, −330° (b) $\frac{7\pi}{2}, -\frac{\pi}{2}$ (c) 535°, −185°
   (d) $\frac{11\pi}{6}, -\frac{13\pi}{6}$ (e) $\frac{11\pi}{3}, -\frac{\pi}{3}$
   (f) 9.53, −3.03
4. (a) 12.6 cm (b) 14.7 cm
5. 1.5 radians, or approx. 85.9°
6. 7.16
7. (a) $13.96 \approx 14.0\text{ cm}^2$ (b) 131 cm^2
8. $\alpha = 3$ (radian measure), or $\alpha = 172°$
9. 32 cm
10. 6.77 cm
11. (a) 3π radians/second (b) 11.9 km/hr
12. 19.8 radians/second
13. $v = \omega r$
14. 28.3 cm
15. 20 944 sq metres
16. (a) $r \approx 30.6$ cm (b) difference ≈ 0.0771 cm
17. $150\sqrt{3}\text{ cm}^2$
18. area of circle $= \left(\frac{4\pi}{\pi - 2}\right)A$

Exercise 5.2

1. $t = \frac{\pi}{6}: \left(\frac{\sqrt{3}}{2}, \frac{1}{2}\right); t = \frac{\pi}{3}: \left(\frac{1}{2}, \frac{\sqrt{3}}{2}\right)$
2. (a) 0.6 (b) 1.0 (c) 0.5 (d) 0.5
   (e) 2.7 (f) 0.1 (g) 0.3 (h) 1.6
3. (a) (i) I (ii) $\left(\frac{\sqrt{3}}{2}, \frac{1}{2}\right)$
   (b) (i) IV (ii) $\left(\frac{1}{2}, -\frac{\sqrt{3}}{2}\right)$
   (c) (i) IV (ii) $\left(\frac{\sqrt{2}}{2}, -\frac{\sqrt{2}}{2}\right)$
   (d) (i) negative y-axis (ii) $(0, -1)$
   (e) (i) II (ii) $(-0.416, 0.909)$
   (f) (i) IV (ii) $\left(\frac{\sqrt{2}}{2}, -\frac{\sqrt{2}}{2}\right)$
   (g) (i) V (ii) $(0.540, -0.841)$
   (h) (i) II (ii) $\left(-\frac{\sqrt{2}}{2}, \frac{\sqrt{2}}{2}\right)$
   (i) (i) III (ii) $(-0.929, -0.369)$
4. (a) $\sin\frac{\pi}{3} = \frac{\sqrt{3}}{2}, \cos\frac{\pi}{3} = \frac{1}{2}, \tan\frac{\pi}{3} = \sqrt{3}$
   (b) $\sin\frac{5\pi}{6} = \frac{1}{2}, \cos\frac{5\pi}{6} = -\frac{\sqrt{3}}{2}, \tan\frac{5\pi}{6} = -\frac{\sqrt{3}}{3}$
   (c) $\sin\left(-\frac{3\pi}{4}\right) = -\frac{\sqrt{2}}{2}, \cos\left(-\frac{3\pi}{4}\right) = -\frac{\sqrt{2}}{2},$
   $\tan\left(-\frac{3\pi}{4}\right) = 1$
   (d) $\sin\frac{\pi}{2} = 1, \cos\frac{\pi}{2} = 0, \tan\frac{\pi}{2}$ is undefined
   (e) $\sin\left(-\frac{4\pi}{3}\right) = \frac{\sqrt{3}}{2}, \cos\left(-\frac{4\pi}{3}\right) = -\frac{1}{2},$
   $\tan\left(-\frac{4\pi}{3}\right) = -\sqrt{3}$
   (f) $\sin 3\pi = 0, \cos 3\pi = -1, \tan 3\pi = 0$
   (g) $\sin\frac{3\pi}{2} = -1, \cos\frac{3\pi}{2} = 0, \tan\frac{3\pi}{2}$ is undefined
   (h) $\sin\left(-\frac{7\pi}{6}\right) = \frac{1}{2}, \cos\left(-\frac{7\pi}{6}\right) = -\frac{\sqrt{3}}{2},$
   $\tan\left(-\frac{7\pi}{6}\right) = -\frac{\sqrt{3}}{3}$
   (i) $\sin(1.25\pi) = -\frac{\sqrt{2}}{2}, \cos(1.25\pi) = -\frac{\sqrt{2}}{2}, \tan(1.25\pi) = 1$
5. (a) $\sin\frac{13\pi}{6} = \sin\frac{\pi}{6} = \frac{1}{2}; \cos\frac{13\pi}{6} = \cos\frac{\pi}{6} = \frac{\sqrt{3}}{2}$
   (b) $\sin\frac{10\pi}{3} = \sin\frac{4\pi}{3} = -\frac{\sqrt{3}}{2}; \cos\frac{10\pi}{3} = \cos\frac{4\pi}{3} = -\frac{1}{2}$
   (c) $\sin\frac{15\pi}{4} = \sin\frac{7\pi}{4} = -\frac{\sqrt{2}}{2}; \cos\frac{15\pi}{4} = \cos\frac{7\pi}{4} = -\frac{\sqrt{2}}{2}$
   (d) $\sin\frac{17\pi}{6} = \sin\frac{5\pi}{6} = \frac{1}{2}; \cos\frac{17\pi}{6} = \cos\frac{5\pi}{6} = -\frac{\sqrt{3}}{2}$
6. (a) $-\frac{\sqrt{3}}{2}$ (b) $-\frac{\sqrt{2}}{2}$ (c) undefined
7. (a) 0.598 (b) $-\frac{\sqrt{2}}{2}$ (c) 0
8. (a) I, II (b) II (c) III
   (d) II (e) I, IV

Exercise 5.3

1. (a)

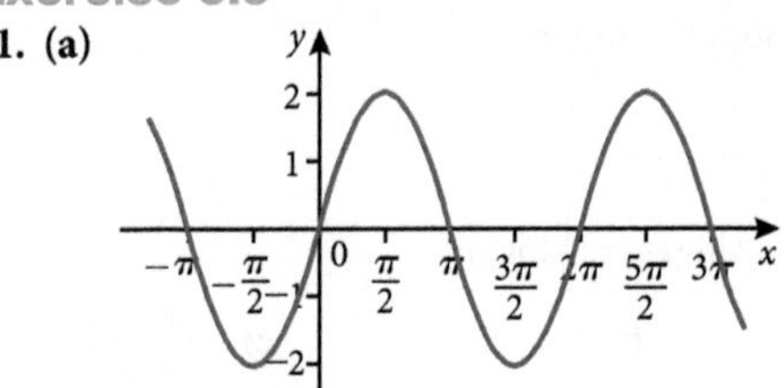

(b)

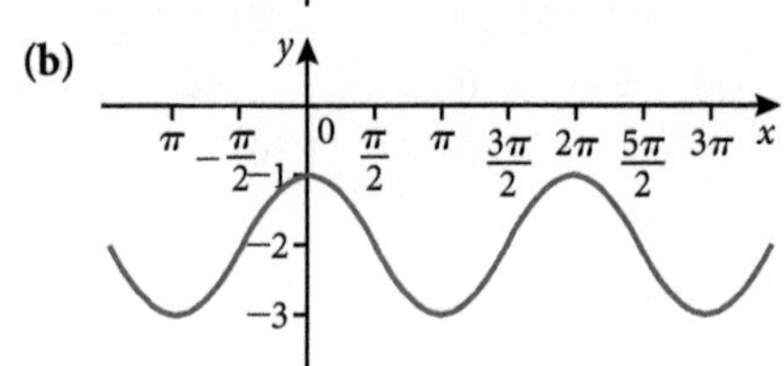

(c)

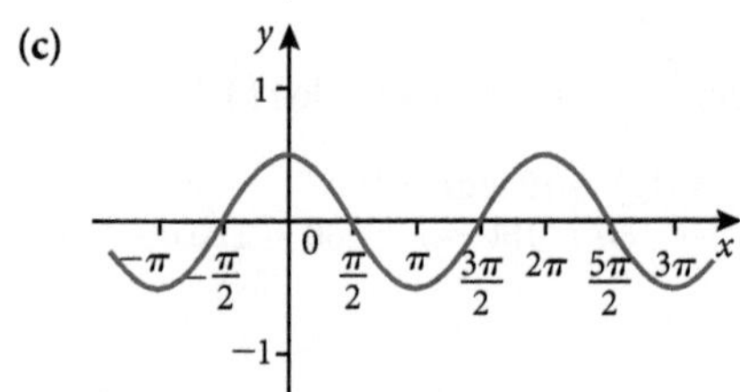

(d)

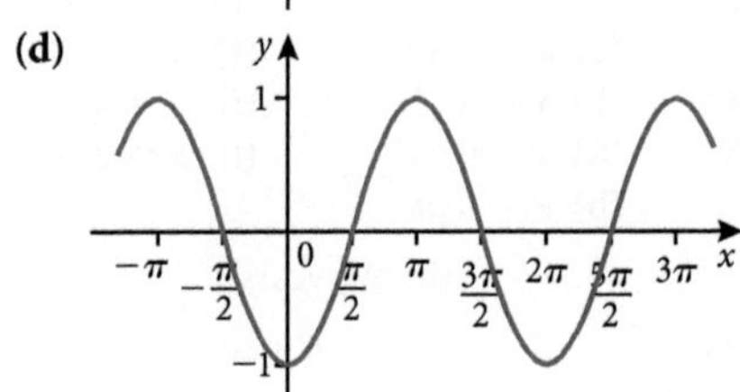

(e)

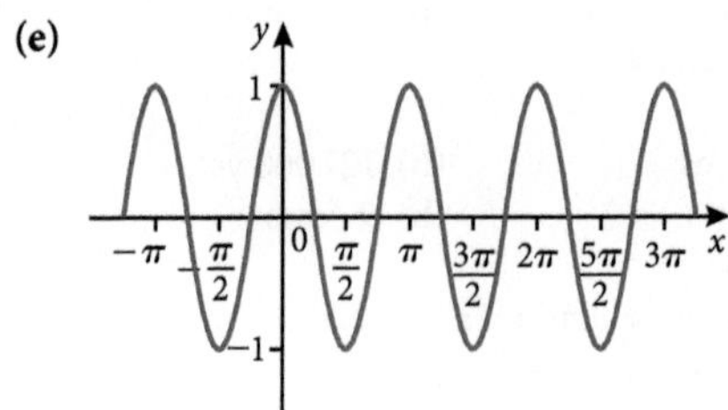

(f)

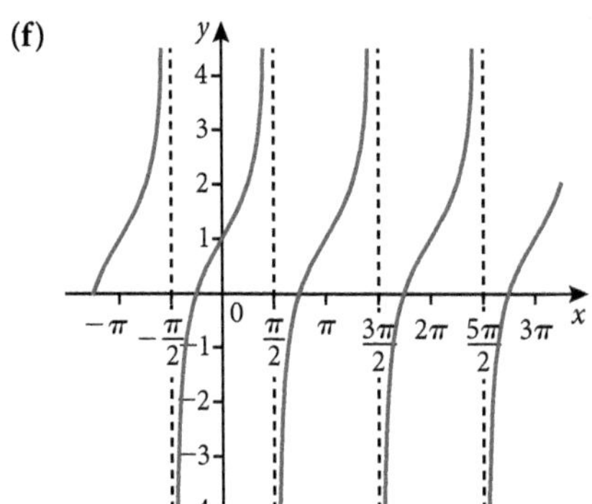

(g)

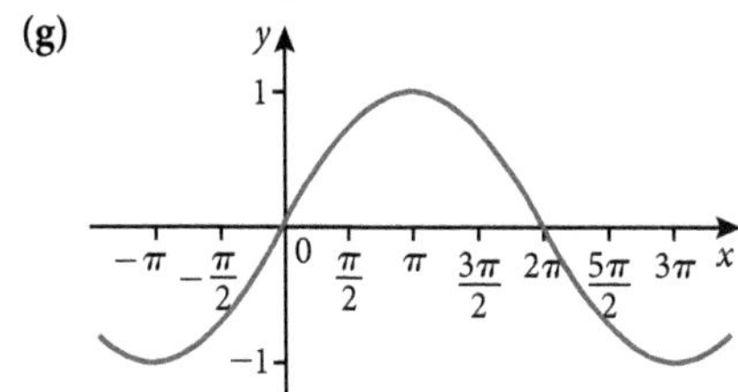

(h)

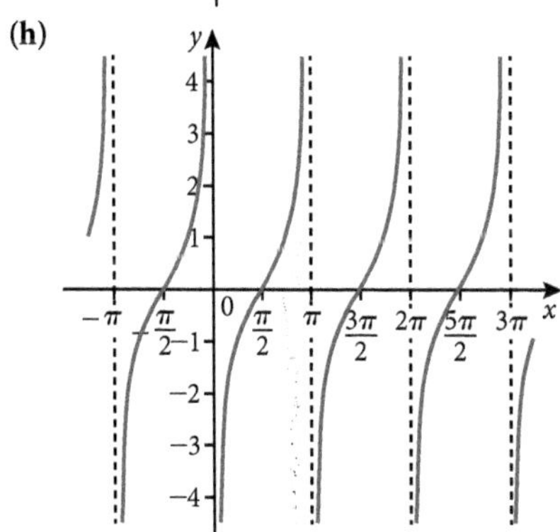

(i)

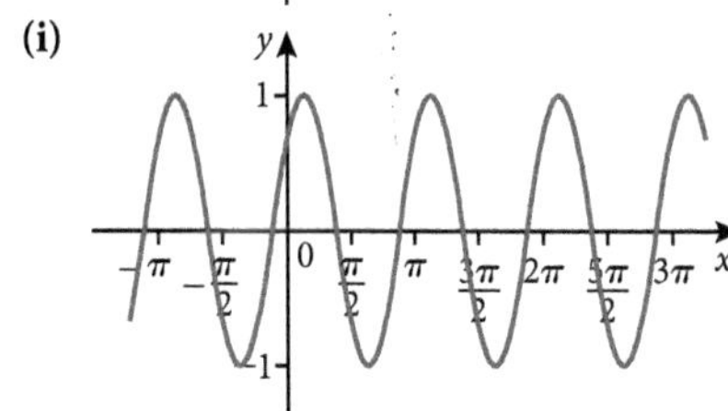

2. (a) (i)

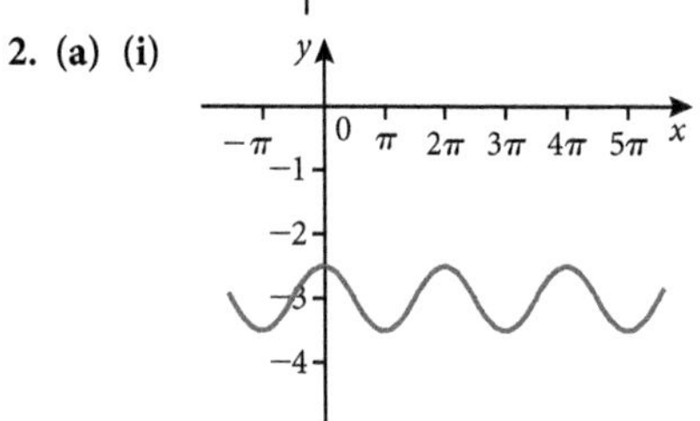

amplitude $= \frac{1}{2}$, period $= 2\pi$

(ii) domain: $x \in \mathbb{R}$, range: $-3.5 \leqslant y \leqslant -2.5$

(b) (i)

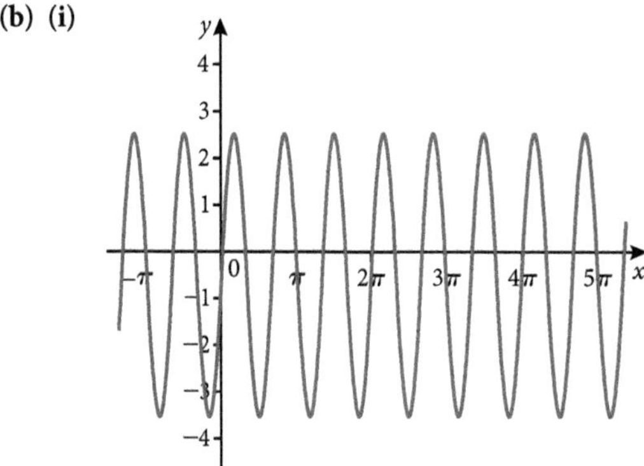

amplitude $= 3$, period $= \frac{2\pi}{3}$

(ii) domain: $x \in \mathbb{R}$, range: $-3.5 \leqslant y \leqslant 2.5$

(c) (i)

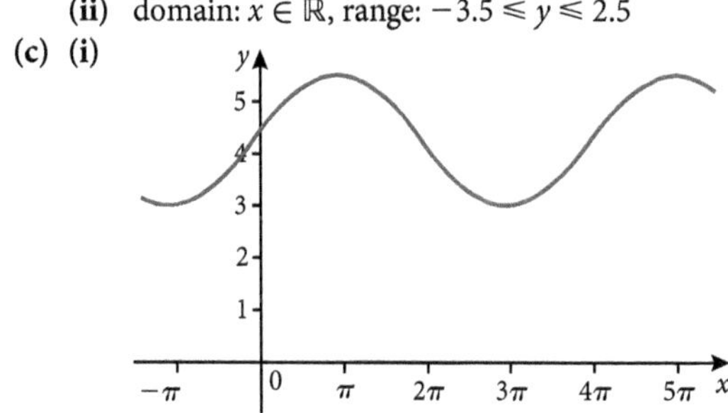

amplitude 1.2, period $= 4\pi$

(ii) domain: $x \in \mathbb{R}$, range: $3.1 \leqslant y \leqslant 5.5$

3. (a) $A = 3, B = 7$ **(b)** $A = 2.7, B = 5.9$

4. (a) $p = 8$ **(b)** $q = 6$

5. (a) $a = 2, b = 3, c = -1$

(b) $\frac{5\pi}{18}$

6. $a = 3, b = -\frac{\pi}{4}, c = -1$

Exercise 5.4

1. (a) $x = \frac{\pi}{3}, \frac{5\pi}{3}$ **(b)** $x = \frac{7\pi}{6}, \frac{11\pi}{6}$

(c) $x = \frac{\pi}{4}, \frac{5\pi}{4}$ **(d)** $x = \frac{\pi}{3}, \frac{2\pi}{3}$

(e) $x = \frac{\pi}{4}, \frac{3\pi}{4}, \frac{5\pi}{4}, \frac{7\pi}{4}$ **(f)** $x = \frac{\pi}{6}, \frac{5\pi}{6}, \frac{7\pi}{6}, \frac{11\pi}{6}$

(g) $x = \frac{\pi}{4}, \frac{3\pi}{4}, \frac{5\pi}{4}, \frac{7\pi}{4}$ **(h)** $x = \frac{\pi}{3}, \frac{2\pi}{3}, \frac{4\pi}{3}, \frac{5\pi}{3}$

(i) $x = 0, \frac{3\pi}{4}, \pi, \frac{7\pi}{4}, 2\pi$ **(j)** $x = 0, \frac{\pi}{2}, \pi, \frac{3\pi}{2}, 2\pi$

2. (a) $x \approx 0.412, 2.73$ **(b)** $x \approx 1.91, 4.37$

(c) $x \approx 1.11, 4.25$

(d) $x \approx 0.508, 1.06, 3.65, 4.20$

(e) $x \approx 2.96, 5.32$ **(f)** $x \approx 1.28, 4.42$

3. (a) $\frac{5\pi}{2}, \frac{3\pi}{2}, \frac{\pi}{2}, -\frac{\pi}{2}, -\frac{3\pi}{2}, -\frac{5\pi}{2}$

(b) $\frac{\pi}{6}, -\frac{11\pi}{6}$

(c) $\frac{7\pi}{12}, \frac{19\pi}{12}$

(d) $0, \frac{\pi}{4}, \frac{\pi}{2}, \frac{3\pi}{4}, \pi, \frac{5\pi}{4}, \frac{3\pi}{2}, \frac{7\pi}{4}, 2\pi, \frac{9\pi}{4}, \ldots, \frac{15\pi}{4}$

4. (a) $x = \frac{5\pi}{6}, \frac{3\pi}{2}$ **(b)** $x = \frac{\pi}{4}, \frac{5\pi}{4}$

(c) $x = \frac{\pi}{6}, \frac{\pi}{3}, \frac{7\pi}{6}, \frac{4\pi}{3}$ **(d)** $x = \frac{\pi}{6}, \frac{5\pi}{6}, \frac{7\pi}{6}, \frac{11\pi}{6}$

5. $t \approx 1.5$ hours
6. (a) 80th day (March 21) and approximately 263rd day (September 20)
   (b) 105th day (April 15) and approximately 238th day (August 26)
   (c) 94 days – from 125th day to 218th day
7. (a) $x = \frac{\pi}{2}, \frac{2\pi}{3}, \frac{4\pi}{3}, \frac{3\pi}{2}$ (b) $x = \frac{\pi}{2}, \frac{7\pi}{6}, \frac{11\pi}{6}$
   (c) $x = \frac{\pi}{2}, -\frac{\pi}{2}$ (d) $x \approx 0.375, 2.77$
   (e) $x \approx -0.785, 1.11$ (f) $x = \frac{\pi}{4}, \frac{3\pi}{4}$
   (g) $x = 0, \frac{\pi}{3}, \frac{5\pi}{3}, 2\pi$ (h) $x \approx 0.983, 4.12$
8. (a) $\cos x = \frac{4}{5}$ (b) $\cos 2x = \frac{7}{25}$
   (c) $\sin 2x = \frac{24}{25}$
9. (a) $\sin x = \frac{\sqrt{5}}{3}$ (b) $\sin 2x = -\frac{4\sqrt{5}}{9}$
   (c) $\cos 2x = -\frac{1}{9}$

Chapter 5 practice questions

1. (a) 135 cm (b) 85 cm
   (c) $t = 0.5$ sec. (d) 1 sec.
2. $x = \frac{\pi}{3}, \pi, \frac{5\pi}{3}$
3. $\theta \approx 2.28$ (radian measure)
4. (a) (i) -1 (ii) 4π
   (b) four
5. (a) $p = 35$ (b) $q = 29$ (c) $m = \frac{1}{2}$
6. $x = 0$ $x \approx 1.06, 2.05$
7. (a) $x = \frac{2\pi}{3}, \frac{4\pi}{3}$ (b) $x = \frac{\pi}{6}, \frac{\pi}{2}, \frac{5\pi}{6}, \frac{3\pi}{2}$
8. (a) $\sin x = \frac{1}{3}$ (b) $\cos 2x = \frac{7}{9}$
   (c) $\sin 2x = -\frac{4\sqrt{2}}{9}$
9. (a) $d = 1.6 \sin\left(\frac{2\pi}{11}\left(t - \frac{9}{4}\right)\right) + 4.2$
   (b) approximately 3.15 metres
   (c) approximately 12:27 pm to 7:33 pm
10. $x \approx 0.785, 1.89$
11. (a) 15 cm (b) area ≈ 239 cm^2
12. $k > 2.5, k < -2.5$
13. $k = 1, a = -2$
14. (a) $\frac{5}{13}$ (b) $-\frac{12}{13}$ (c) $-\frac{120}{169}$ (d) $\frac{119}{169}$

Chapter 6

Exercise 6.1

1. (a) scalene (b) equilateral
   (c) isosceles (d) equilateral
2. (0, 2, 0)
3. (a) yes (b) no
4. See Worked Solutions
5. (a) yes (b) no (c) no (d) yes
6. (a) (i) 7, (ii) $3\sqrt{5}$, (iii) $\sqrt{13}$, (iv) $2\sqrt{10}$
   (b) (i) 4, (ii) $2\sqrt{3}$, (iii) $\sqrt{13}$, (iv) $\sqrt{7}$
7. $S(2, -4, 16)$
8. (a) See Worked Solutions (b) $\frac{\sqrt{899}}{2}$
9. surface area $= 116\pi$ units^2, volume $= \frac{116\pi\sqrt{29}}{3}$ units^3
10. 78 cm
11. surface area ≈ 262cm^2, volume ≈ 330 cm^3
12. $(-1, 6, -7), (3, 4, 5), (1, 2, 3)$
13. (a) $\frac{2}{3}$ (b) $\frac{2}{3}$
14. surface area ≈ 2170 m^2, volume ≈ 6170 m^3
15. surface area $= 2052$ cm^2, volume $= 5832 + 324\sqrt{7}$ cm^3

Exercise 6.2

1. (a) (ii) $\cos\theta = \frac{4}{5}, \tan\theta = \frac{3}{4}, \cot\theta = \frac{4}{3}, \sec\theta = \frac{5}{4}$, $\csc\theta = \frac{5}{3}$
   (iii) $\theta \approx 36.9°; 53.1°$
   (b) (ii) $\sin\theta = \frac{\sqrt{39}}{8}, \tan\theta = \frac{\sqrt{39}}{5}, \cot\theta = \frac{5\sqrt{39}}{39}$, $\sec\theta = \frac{8}{5}, \csc\theta = \frac{8\sqrt{39}}{39}$
   (iii) $\theta \approx 51.3°; 38.7°$
   (c) (ii) $\sin\theta = \frac{2\sqrt{5}}{5}, \cos\theta = \frac{\sqrt{5}}{5}, \cot\theta = \frac{1}{2}, \sec\theta = \sqrt{5}$, $\csc\theta = \frac{\sqrt{5}}{2}$
   (iii) $\theta \approx 63.4°; 26.6°$
2. (a) $\theta = 60°, \frac{\pi}{3}$ (b) $\theta = 45°, \frac{\pi}{4}$ (c) $\theta = 60°, \frac{\pi}{3}$
3. (a) $x = 50\sqrt{3}, y = 100$ (b) $x \approx 8.60, y \approx 12.3$
   (c) $x \approx 20.6, y \approx 24.5$ (d) $x \approx 374, y \approx 299$
   (e) $x = 18, y = 18\sqrt{2}$ (f) $x = 200, y = 100\sqrt{3}$
4. (a) $\alpha = 60°, \beta = 30°$ (b) $\alpha \approx 67.4°, \beta \approx 22.6°$
   (c) $\alpha \approx 20.0°, \beta \approx 70.0°$ (d) $\alpha = 30°, \beta = 60°$
5. 114 metres 6. 67.4° 7. 4.05 metres
8. 4105 m 9. 44°, 68°, 68° 10. 5.76 km h^{-1}
11. 69.5 m 12. 28.7 m 13. 151 m
14. 59.2 m 15. $3\sqrt{5}$ 16. -0.6
17. $\frac{|ap + bq + c|}{\sqrt{a^2 + b^2}}$ 18. 14°

Exercise 6.3

1. (a) $\sin\theta = \frac{3}{5}, \cos\theta = \frac{4}{5}, \tan\theta = \frac{3}{4}$
   (b) $\sin\theta = \frac{12}{37}, \cos\theta = -\frac{35}{37}, \tan\theta = -\frac{12}{35}$
   (c) $\sin\theta = -\frac{\sqrt{2}}{2}, \cos\theta = \frac{\sqrt{2}}{2}, \tan\theta = -1$
   (d) $\sin\theta = -\frac{1}{2}, \cos\theta = -\frac{\sqrt{3}}{2}, \tan\theta = \frac{\sqrt{3}}{3}$
2. (a) $\sin 120° = \frac{\sqrt{3}}{2}, \cos 120° = -\frac{1}{2}, \tan 120° = -\sqrt{3}$, $\cot 120° = -\frac{\sqrt{3}}{3}, \sec 120° = -2, \csc 120° = \frac{2\sqrt{3}}{3}$
   (b) $\sin 135° = \frac{\sqrt{2}}{2}, \cos 135° = -\frac{\sqrt{2}}{2}, \tan 135° = -1$, $\cot 135° = -1, \sec 135° = -\sqrt{2}, \csc 135° = \sqrt{2}$
   (c) $\sin 330° = -\frac{1}{2}, \cos 330° = \frac{\sqrt{3}}{2}, \tan 330° = -\frac{\sqrt{3}}{3}$, $\cot 330° = -\sqrt{3}, \sec 330° = \frac{2\sqrt{3}}{3}, \csc 330° = -2$

(d) $\sin 270° = -1$, $\cos 270° = 0$, $\tan 270° =$ undef., $\cot 270° = 0$, $\sec 270° =$ undef., $\csc 270° = -1$

(e) $\sin 240° = -\frac{\sqrt{3}}{2}$, $\cos 240° = -\frac{1}{2}$, $\tan 240° = \sqrt{3}$, $\cot 240° = \frac{\sqrt{3}}{3}$, $\sec 240° = -2$, $\csc 240° = -\frac{2\sqrt{3}}{3}$

(f) $\sin\frac{5\pi}{4} = -\frac{\sqrt{2}}{2}$, $\cos\frac{5\pi}{4} = -\frac{\sqrt{2}}{2}$, $\tan\frac{5\pi}{4} = 1$, $\cot\frac{5\pi}{4} = 1$, $\sec\frac{5\pi}{4} = -\sqrt{2}$, $\csc\frac{5\pi}{4} = -\sqrt{2}$

(g) $\sin\left(-\frac{\pi}{6}\right) = -\frac{1}{2}$, $\cos\left(-\frac{\pi}{6}\right) = \frac{\sqrt{3}}{2}$, $\tan\left(-\frac{\pi}{6}\right) = -\frac{\sqrt{3}}{3}$, $\cot\left(-\frac{\pi}{6}\right) = -\sqrt{3}$, $\sec\left(-\frac{\pi}{6}\right) = \frac{2\sqrt{3}}{3}$, $\csc\left(-\frac{\pi}{6}\right) = -2$

(h) $\sin\left(\frac{7\pi}{6}\right) = -\frac{1}{2}$, $\cos\left(\frac{7\pi}{6}\right) = -\frac{\sqrt{3}}{2}$, $\tan\left(\frac{7\pi}{6}\right) = \frac{\sqrt{3}}{3}$, $\cot\left(\frac{7\pi}{6}\right) = \sqrt{3}$, $\sec\left(\frac{7\pi}{6}\right) = -\frac{2}{\sqrt{3}}$, $\csc\left(\frac{7\pi}{6}\right) = -2$

(i) $\sin(-60°) = -\frac{\sqrt{3}}{2}$, $\cos(-60°) = \frac{1}{2}$, $\tan(-60°) = -\sqrt{3}$, $\cot(-60°) = -\frac{\sqrt{3}}{3}$, $\sec(-60°) = 2$, $\csc(-60°) = -\frac{2\sqrt{3}}{3}$

(j) $\sin\left(-\frac{3\pi}{2}\right) = 1$, $\cos\left(-\frac{3\pi}{2}\right) = 0$, $\tan\left(-\frac{3\pi}{2}\right) =$ undef., $\cot\left(-\frac{3\pi}{2}\right) = 0$, $\sec\left(-\frac{3\pi}{2}\right) =$ undef., $\csc\left(-\frac{3\pi}{2}\right) = 1$

(k) $\sin\left(\frac{5\pi}{3}\right) = -\frac{\sqrt{3}}{2}$, $\cos\left(\frac{5\pi}{3}\right) = \frac{1}{2}$, $\tan\left(\frac{5\pi}{3}\right) = -\sqrt{3}$, $\cot\left(\frac{5\pi}{3}\right) = -\frac{\sqrt{3}}{3}$, $\sec\left(\frac{5\pi}{3}\right) = 2$, $\csc\left(\frac{5\pi}{3}\right) = -\frac{2\sqrt{3}}{3}$

(l) $\sin(-210°) = -\frac{1}{2}$, $\cos(-210°) = -\frac{\sqrt{3}}{2}$, $\tan(-210°) = \frac{\sqrt{3}}{3}$, $\cot(-210°) = \sqrt{3}$, $\sec(-210°) = -\frac{2\sqrt{3}}{3}$, $\csc(-210°) = -2$

(m) $\sin\left(-\frac{\pi}{4}\right) = -\frac{\sqrt{2}}{2}$, $\cos\left(-\frac{\pi}{4}\right) = \frac{\sqrt{2}}{2}$, $\tan\left(-\frac{\pi}{4}\right) = -1$, $\cot\left(-\frac{\pi}{4}\right) = -1$, $\sec\left(-\frac{\pi}{4}\right) = \sqrt{2}$, $\csc\left(-\frac{\pi}{4}\right) = -\sqrt{2}$

(n) $\sin\pi = 0$, $\cos\pi = -1$, $\tan\pi = 0$, $\cot\pi =$ undef., $\sec\pi = -1$, $\csc\pi =$ undef.

(o) $\sin 4.25\pi = \frac{\sqrt{2}}{2}$, $\cos 4.25\pi = \frac{\sqrt{2}}{2}$, $\tan 4.25\pi = 1$, $\cot 4.25\pi = 1$, $\sec 4.25\pi = \sqrt{2}$, $\csc 4.25\pi = \sqrt{2}$

3. $\sin\theta = \frac{15}{17}$, $\tan\theta = \frac{15}{8}$, $\cot\theta = \frac{8}{15}$, $\sec\theta = \frac{17}{8}$, $\csc\theta = \frac{17}{15}$

4. $\sin\theta = -\frac{6\sqrt{61}}{61}$, $\cos\theta = \frac{5\sqrt{61}}{61}$

5. $\cos\theta = -1$, $\tan\theta = 0$, $\cot\theta =$ undef., $\sec\theta = -1$, $\csc\theta =$ undef.

6. (a) (i) 30° (ii) 85°
(b) (i) 45° (ii) 7°
(c) (i) 60° (ii) 20°

7. (a) $6\sqrt{3}$ (b) 88.9 (c) $675\sqrt{2}$

8. 28.5°

9. (a) 236 cm^2 (b) 97.4 cm^2

10. (a) 9.06 cm^2 (b) 119 cm^2

11. $ab\sin\theta$

12. $x\sqrt{3}$

13. $\frac{2hf\cos\theta}{h+f}$

14. See Worked Solutions

15. (a) $A(x) = 24\sin x$
(b) $0° < x < 180°$

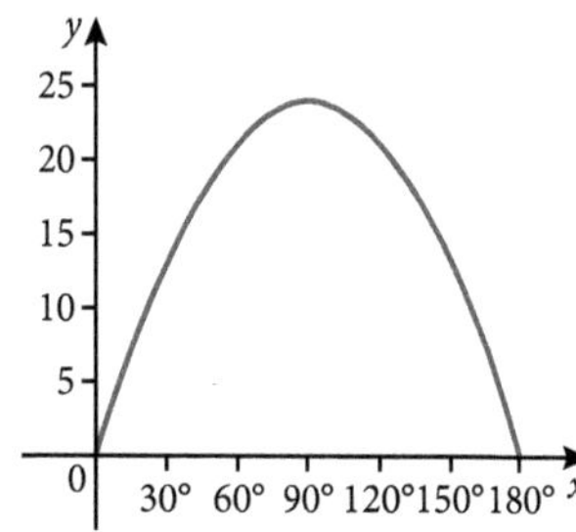

(c) (90°, 24), right-angled triangle, which will always give the maximum area because that is the maximum value of $\sin x$

16. (a) 45° (b) −33.7° (c) 60.3°

17. (a) 71.6° (b) 45°

Exercise 6.4

1. (a) infinite (b) one triangle
(c) one triangle (d) one triangle
(e) two triangles (f) one triangle

2. (a) $BC \approx 17.9$, $AB \approx 27.0$, $A\hat{C}B = 115°$
(b) $AB \approx 18.1$, $BC \approx 22.5$, $B\hat{A}C = 65°$
(c) $AB \approx 74.1$, $B\hat{A}C \approx 60.2°$, $A\hat{B}C \approx 48.8°$
(d) $B\hat{A}C \approx 81.6°$, $A\hat{B}C \approx 60.6°$, $A\hat{C}B \approx 37.8°$
(e) two possible triangles:
(1) $B\hat{A}C \approx 55.9°$, $A\hat{C}B \approx 81.1°$, $AB \approx 40.6$
(2) $B\hat{A}C \approx 124.1°$, $A\hat{C}B \approx 12.9°$, $AB \approx 9.17$
(f) two possible triangles:
(1) $A\hat{B}C \approx 72.2°$, $A\hat{C}B \approx 45.8°$, $AB \approx 0.414$
(2) $A\hat{B}C \approx 107.8°$, $A\hat{C}B \approx 10.2°$, $AB \approx 0.102$

3. 10.8 cm and 30.4 cm

4. 51.3°, 51.3°, 77.4°

5. 71.6° or 22.4°

6. 20.7°

7. area $\approx$ 151 cm^2

8. (a) (i) $BC = 5\sin 36°$, or $BC \geqslant 5$
(ii) $5\sin 36° < BC < 5$
(iii) $BC < 5\sin 36°$
(b) (i) $BC = 5\sqrt{3}$, or $BC \geqslant 10$
(ii) $5\sqrt{3} < BC < 10$
(iii) $BC < 5\sqrt{3}$

9. $x \approx 64.9$ m, $y \approx 56.9$ m

10. (a) $x = 5$
(b) See Worked Solutions
(c) $\frac{15\sqrt{3}}{14}$

11. $\frac{21\sqrt{15}}{4}$

12. (a) obtuse triangle
(b) acute triangle
(c) See Worked Solutions

13. 21.1 or 6.84
14. (a) 14 (b) $\cos\theta = \frac{3}{5}$, $WY = 2\sqrt{65}$
(c) $2\sqrt{5}$ (d) 13.9°
15. 51.3°

Chapter 6 practice questions

1. $\sin A\hat{O}B = \frac{24}{25}$
2. $\sin 2\theta = \frac{21}{29}$, $\cos 2\theta = \frac{20}{29}$
3. 101.5°
4. $\sin 2A = -\frac{120}{169}$
5. (a) 29.1 m (b) 41.9 m
6. $C\hat{A}B \approx 86.4°$
7. (a) 38.2° (b) 17.3 cm²
8. (a) $A\hat{C}B \approx 116°$ (b) 155 cm²
9. 78.5 km
10. $J\hat{K}L \approx 31°$
11. (a) 3.26 cm (b) 7.07 cm²
12. 70.5°
13. (a) 91 m (b) $1690\sqrt{3}$
(c) (i) See Worked Solutions
(ii) $A_2 = 26x$
(iii) $x = 40\sqrt{3}$
(d) (i) supplementary angles have equal sines
(ii) See Worked Solutions
14. (a) $2\sqrt{2} + 4$ (b) $2\sqrt{2} + 3\sqrt{3} + 2\sqrt{6} + 3$
15. 28.3 cm²
16. (a) $0 < \theta < 120°$
(b) See Worked Solutions
(c) 60°
17. (a) 120 cm² (b) 2.16 (c) 161 cm²
18. (a) See Worked Solutions
(b) $\sin\frac{\alpha}{2} = \pm\sqrt{\frac{1 - \cos\alpha}{2}}$
(c) See Worked Solutions
19. $\cos\theta = \frac{b}{2a}$
20. 59.5 cm
21. $\triangle ABC = 72$ cm², $\triangle ABD = 24\sqrt{3} \approx 41.6$ cm², $\triangle BCD \approx 34.6$ cm², $\triangle ACD \approx 69.3$ cm²
22. $D\hat{E}F \approx 41.9°$
23. 43.0 metres
24. 97.9°
25. (a) $y = \frac{\sqrt{3}}{3}x$ (b) 56.6°
26. length ≈ 277 km, bearing ≈ 18.5°

Chapter 7

Exercise 7.1

Note: Some answers may differ from one person to the other due to different graph accuracies.

1. (a) Student, all students in a community, random sample of few students, qualitative
(b) Exam, 10th-grade students in a country, a sample from a few schools, quantitative.
(c) Newborns, heights of newborns in a city, sample from a few hospitals, quantitative
(d) Children, eye colour of children in a city, sample of children at schools, qualitative
(e) Working persons, commuters in a city, sample of a few districts, quantitative
(f) Country leaders, all country leaders, sample of few presidents, sample of international school students, qualitative
(g) Students, all international school students, qualitative
2. Answers are not unique!
(a) Skewed to the right as few players score very high
(b) Symmetric
(c) Skewed to the right
(d) Unimodal, or bi-modal, symmetric or skewed etc.
3. (a) (b) Quantitative
(c) (d) Qualitative
4. (a) Discrete
(b) Continuous
(c) Continuous
(d) Discrete
(e) Continuous
(f) Discrete (debatable)
5.

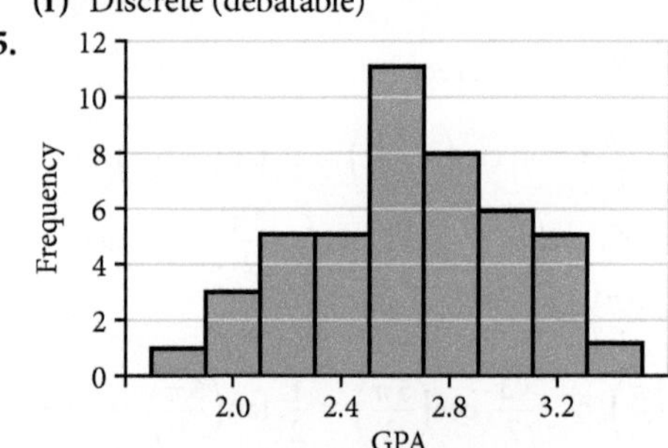

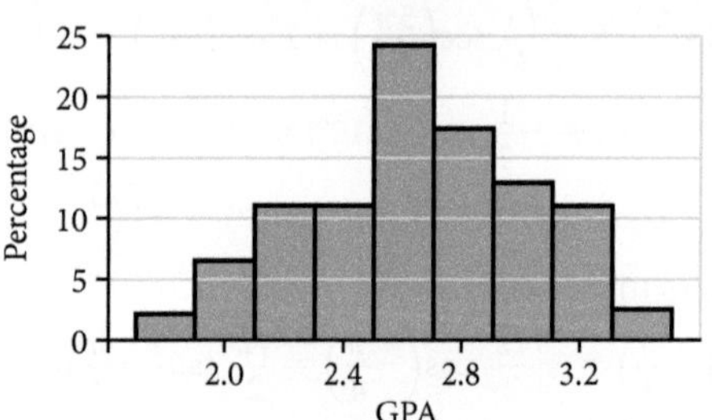

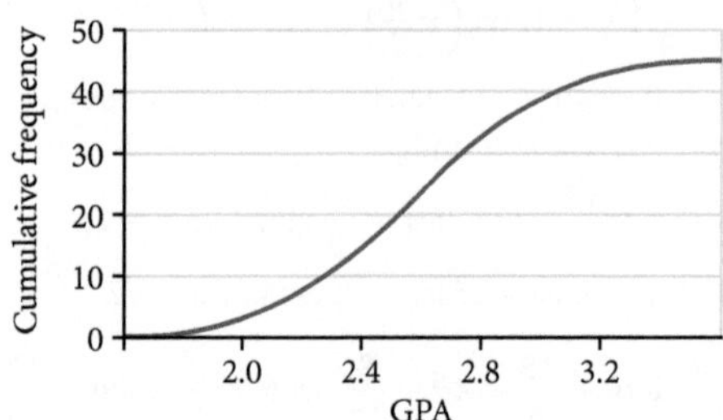

Relatively symmetric. No outliers.

6.

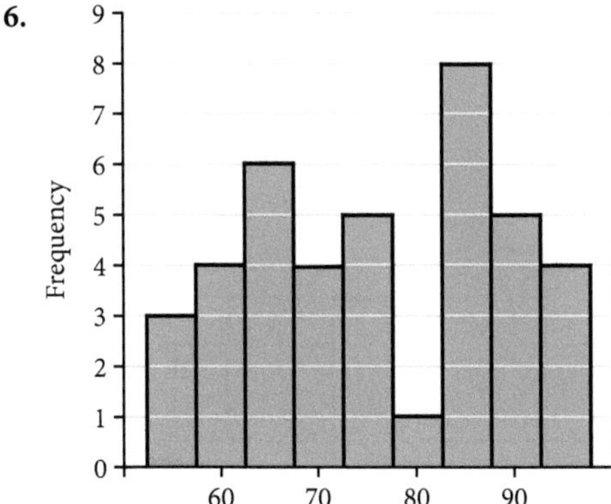

The grades appear to be divided into two groups, one with mode around 65 and the other around 85. No outliers are detected.

7. (a)

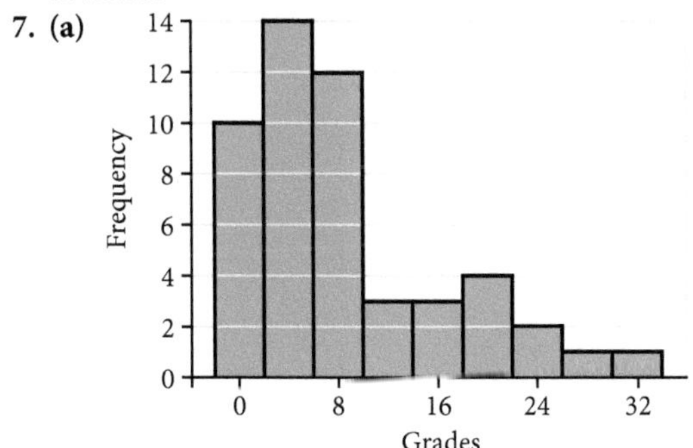

(b) The data is skewed to the right.

(c)

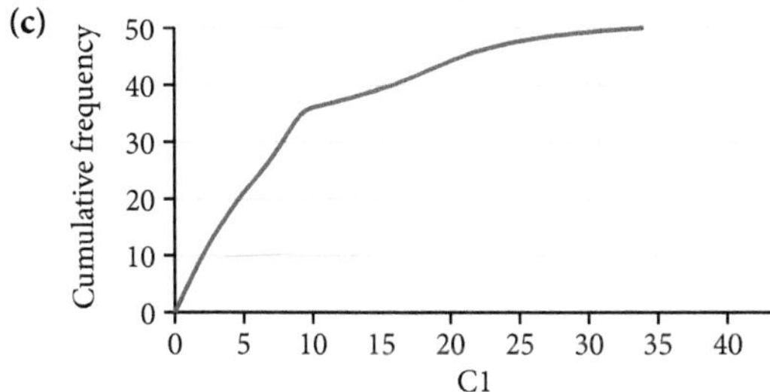

Approximately 36 out of 50 may lose their licence, about 72%.

8. (a)

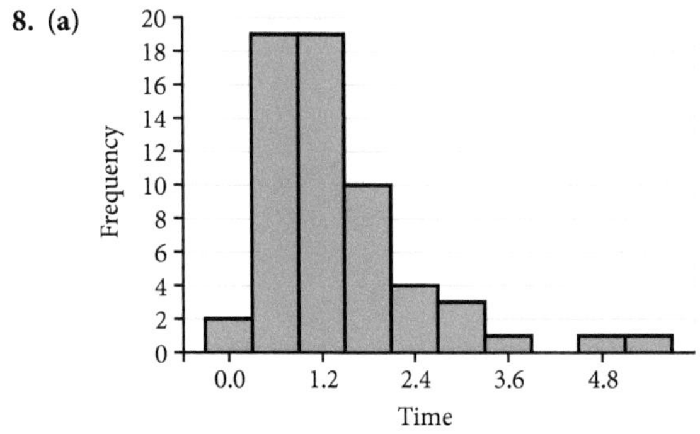

(b)

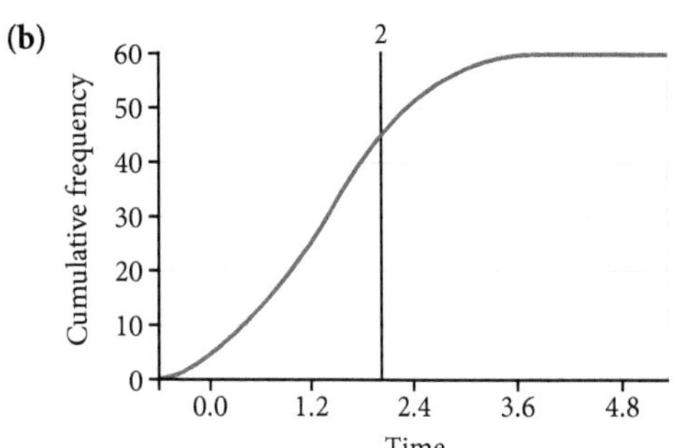

About 10 customers will have to wait more than 2 minutes.

9. (a) Skewed to the right, there is a mode at about 6 days stay, and a few extremes that stayed more than 20 days. A good proportion stayed for less than 10 days.

(b)

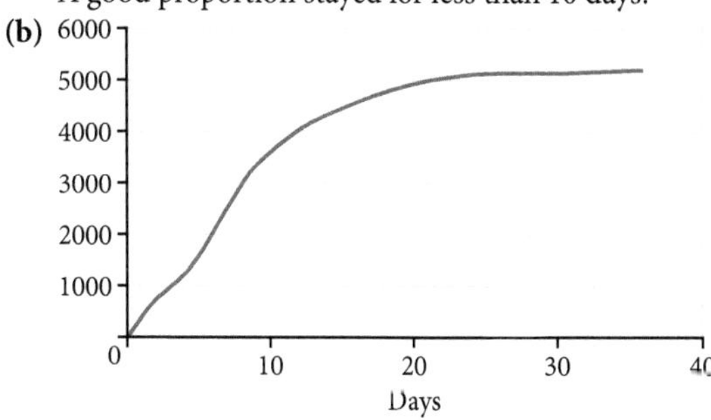

(c) Approximately % of the patients

10. (a) 40 minutes **(b)** Approximately 30%

(c)

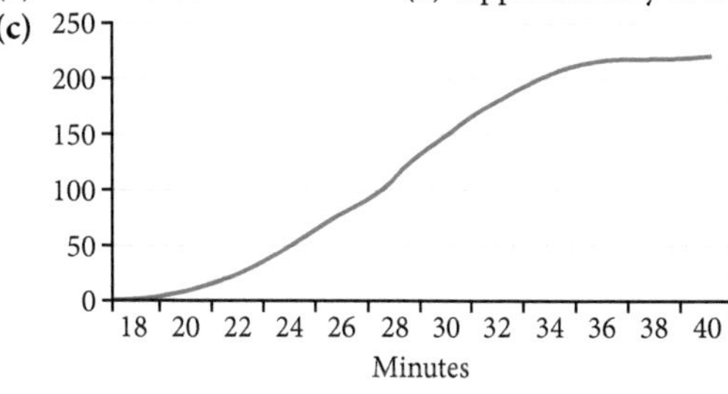

11. (a)

Speed	Frequency
$60 \leqslant$ speed < 75	20
$75 \leqslant$ speed < 90	70
$90 \leqslant$ speed < 105	110
$105 \leqslant$ speed < 120	150
$120 \leqslant$ speed < 135	40
$135 \leqslant$ speed < 150	10

(b), (c)

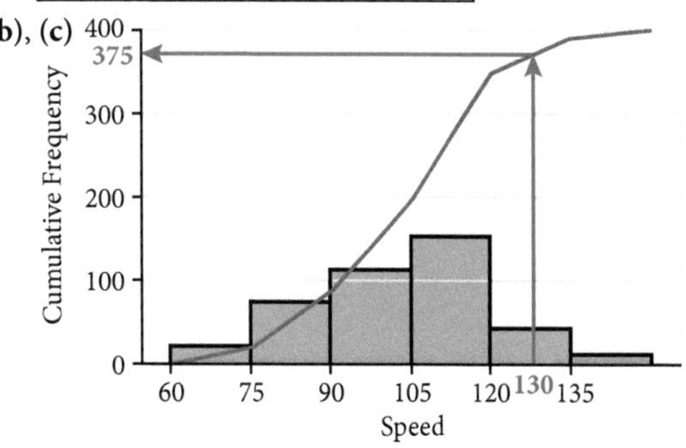

(d) $\frac{25}{400} = 6.25\%$

12. (a)

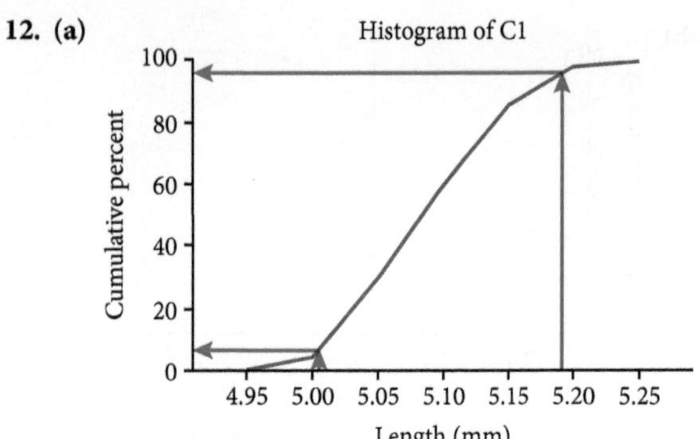

(b) about 5% at the lower end and also about 5% at the upper end.

13. (a),(b)

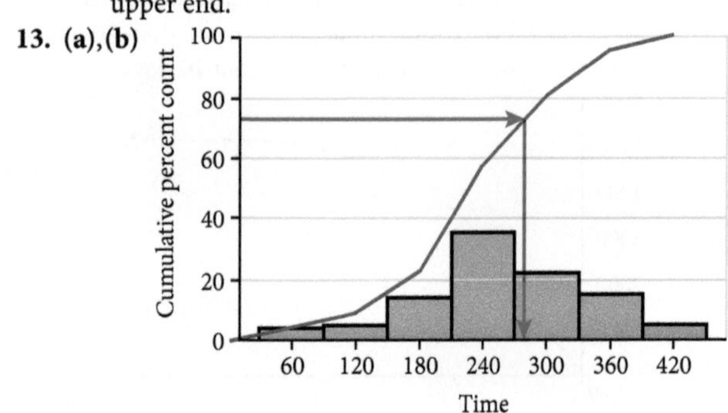

(c) As you see from diagram, about 250 seconds.

Exercise 7.2

1. **(a)** 6 **(b)** 6
   (c) It appears to be symmetric as the mean and median are the same. A histogram supports this view
2. **(a)** 7.8 **(b)** 7.5 **(c)** 7 or 8
3. Average = 1.16, median = 1. Median is more appropriate as the data is skewed to the right.
4. Mean = 307 036, median = 288 521. There are extreme values and it is skewed to the right. Median is more appropriate.
5. Mean = median = 430. It appears to be symmetric and hence either measure would be fine.
6. **(a)** €49.56
   (b) €49.93
7. 2.05 kg
8. **(a)** 29.96
   (b)

1	89
2	023344
2	5666777
3	34
3	568
4	022
4	66

Median is 27

(c)

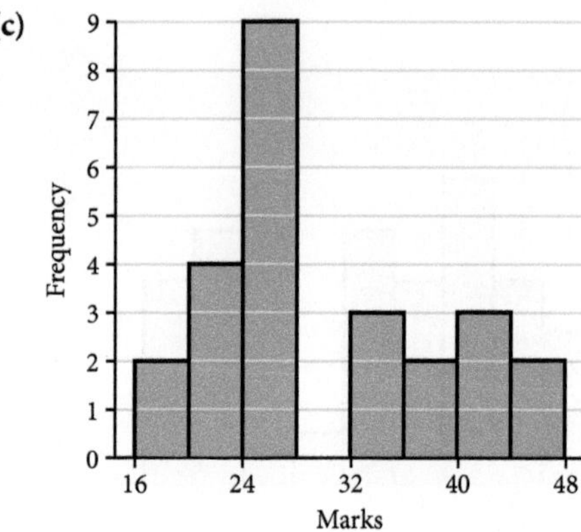

(d)

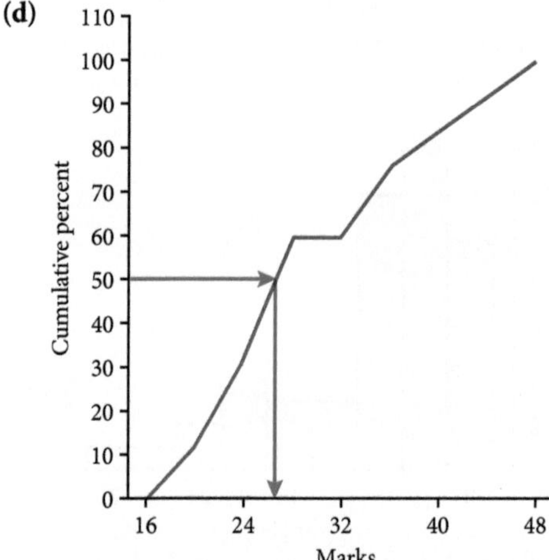

The median is approx. 27.

9. (a)

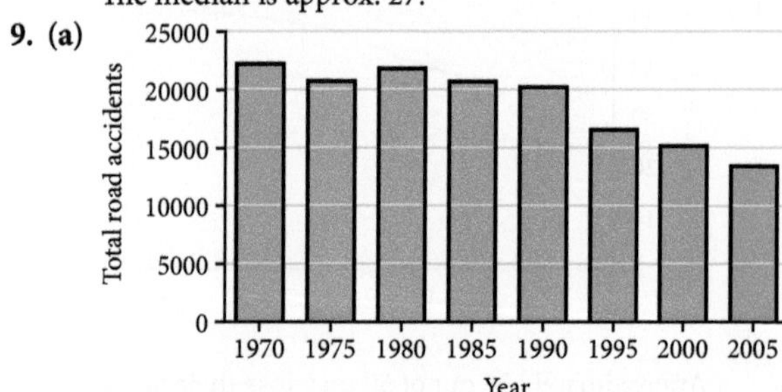

There appears to be a decline in the total number of injuries.

(b)

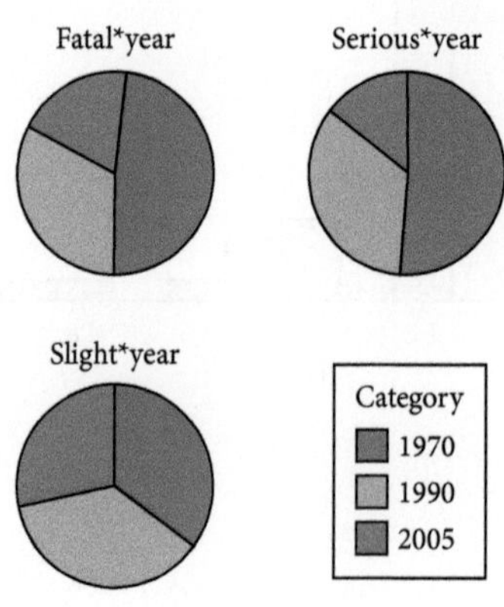

10. (a)

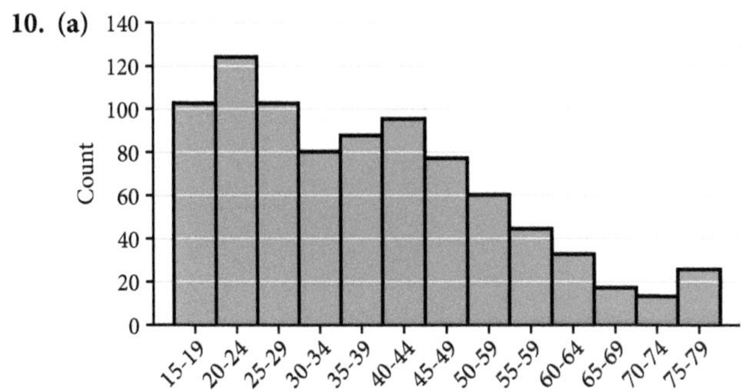

(b) 37.6

(c)

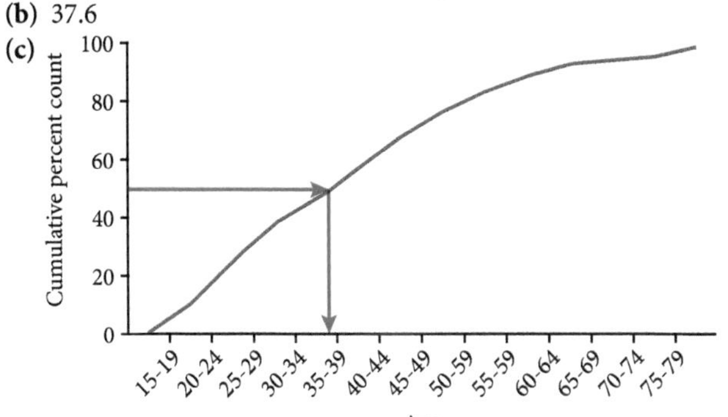

From the graph, the median is approx. at 38.

11. Median = approx. 8 days ; Mean = 9.5 days

12. Median = approx. 28 minutes; Mean = 28.7 minutes

13. Median = approx. 105 km h^{-1}; Mean = 103 km h^{-1}

14. Median = approx. 5.075 mm; Mean = 5.09 mm

15. Median = approx. 210 seconds; Mean = 228.6 seconds

16. (a) 41.6 (b) 61.6

17. (a) 61.4 (b) 63.8

Exercise 7.3

1. (a) Mean = 71.47, S_{n-1} = 7.29

(b)

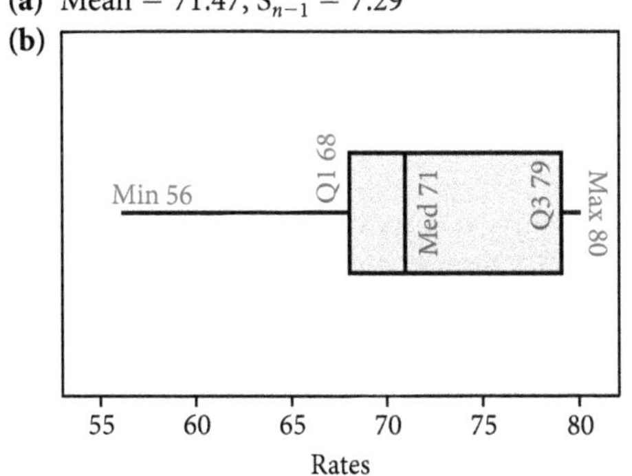

(c) No outliers

2. (a) Mean = 162.6, S_{n-1} = 23.3

(b)

11	79
12	567
13	089
14	123679
15	033445689
16	02334568
17	1344789
18	02255779

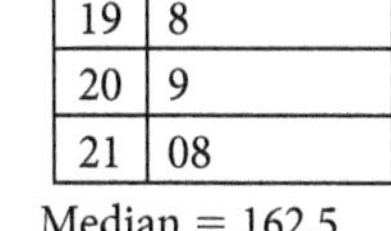

19	8
20	9
21	08

Median = 162.5

(c)

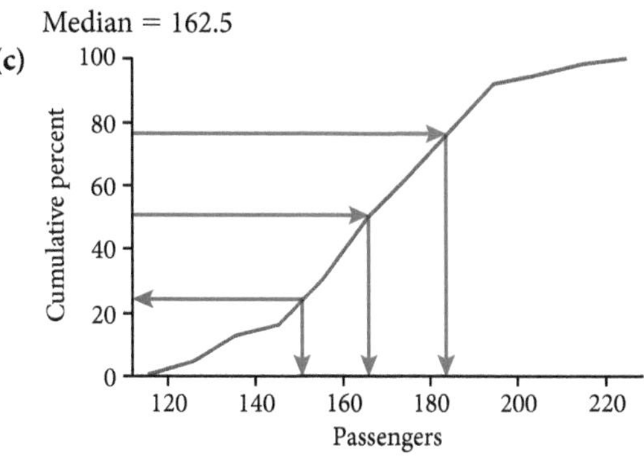

Q1 approx. 150, Median approx. 165, Q3 approx. 182

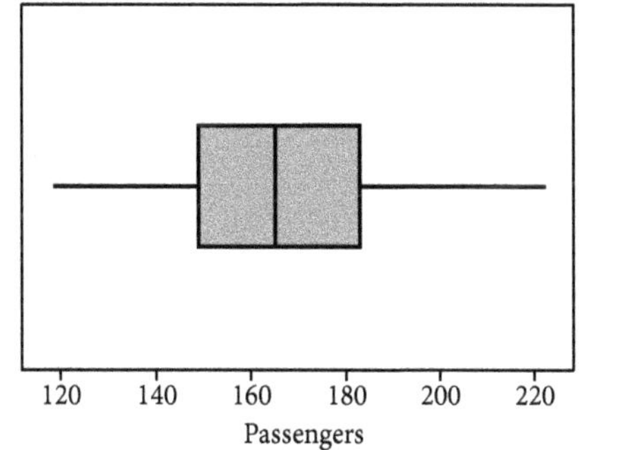

(d) Real Q1 = 146.75, Q3 = 179.25, IQR = 32.5. No outliers

(e) $\bar{x} \pm 3s_{n-1}$ = (92.55, 232.65) No outliers

3. (a) and (b)

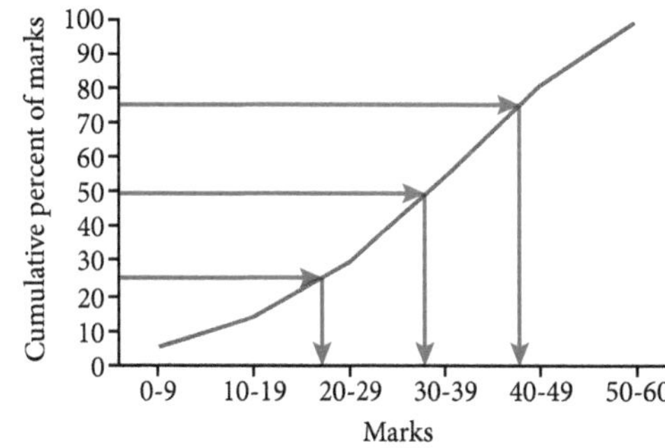

Q1 = approx. 18, Med = approx. 29, Q3 = approx. 39

4. (a)

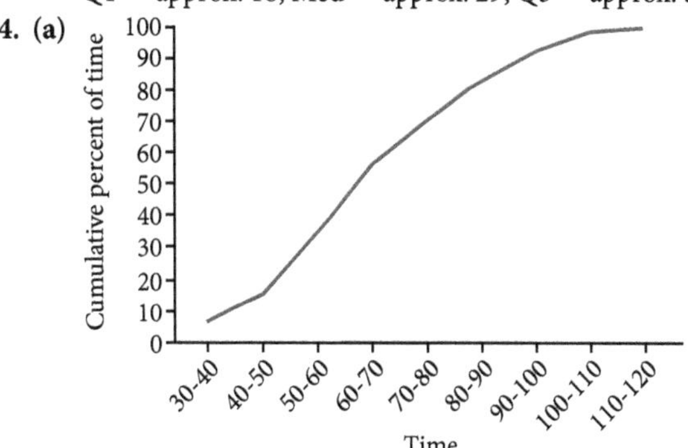

(b) approx. median = 63, IQR = 27

(c) about 68

5. 29.6
6. (a) mean = 72.1, S_{n-1} = 6.1
   (b) New mean = 85.1, S will not change.
7. (a)

$x \leqslant 10$	$x \leqslant 20$	$x \leqslant 30$	$x \leqslant 40$	$x \leqslant 50$
15	65	165	335	595

$x \leqslant 60$	$x \leqslant 70$	$x \leqslant 80$	$x \leqslant 90$	$x \leqslant 100$
815	905	950	980	1000

   (b)

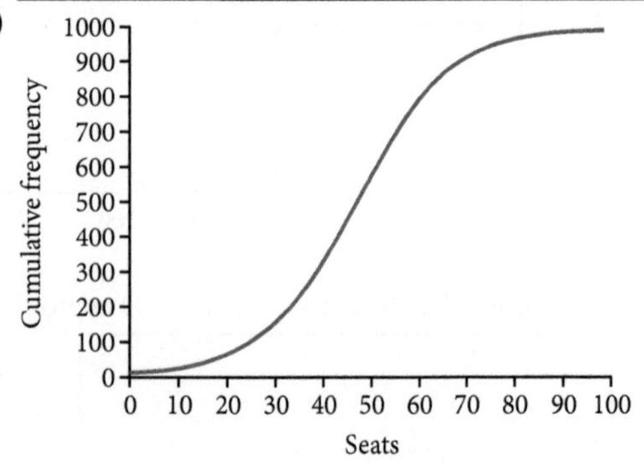

   (c) (i) Around 50
   (ii) Q1 = 40, Q3 = 60, IQR = 20
   (iii) About 170 days
   (iv) Approximately 70 seats

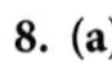

8. (a)

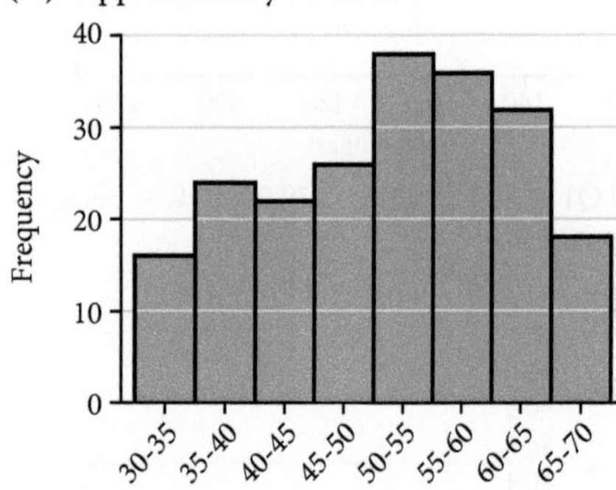

   (b)

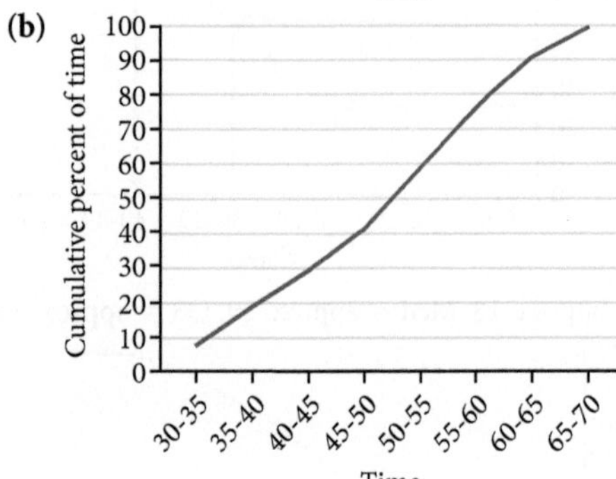

   Median = 53, IQR = 15
   (c) mean = 51.3 and S_{n-1} = 34.8
9. (a) Q1 = 165.1, median = 167.64,
   Q3 = 177.8, minimum = 152, maximum = 193
   (b)

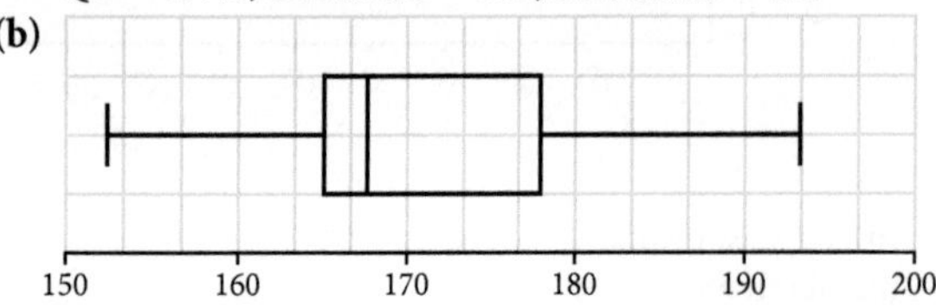

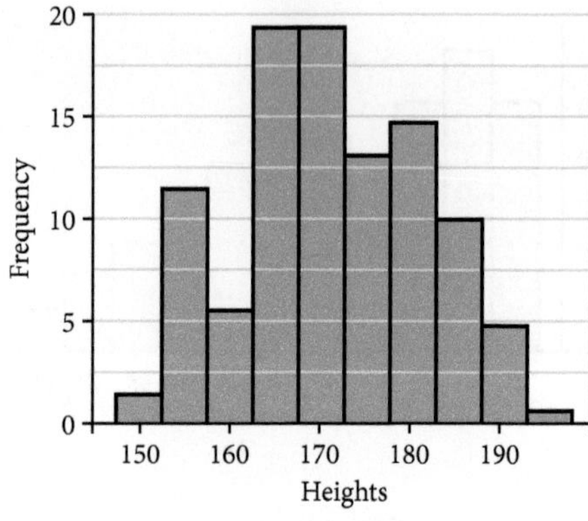

   (c) Mean = 170.5, standard deviation = 9.61
   (d) The heights are widely spread from very short to very tall players. Heights are slightly skewed to the right, bimodal at 165 and 170, no apparent outliers. The heights between the first quartile and the median are closer together than the rest of the data.
   (e)

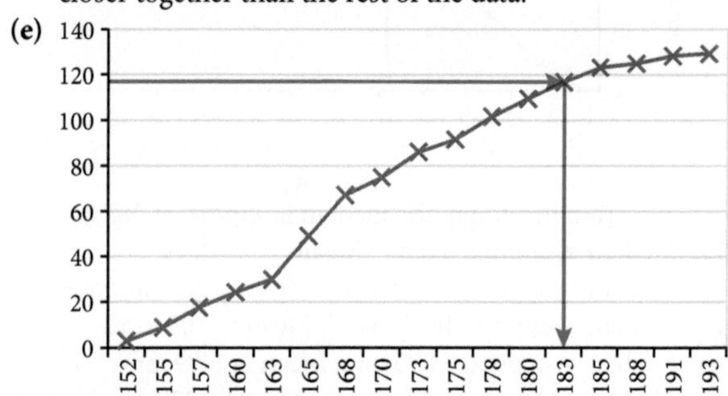

   Approx. 183 cm tall
   (f) 171.3
10. (a) 12 (b) 12 (c) 111
11. (a) 31 (b) Increase
12. 36.7
13. $x = 6, y = 11$
14. Mean = 11.12, Variance = 24.6 (calculating $\sigma^2 = 23.6$)
15. Std. dev. = 6.1, IQR ≈ 6
16. Std. dev ≈ 4.5, IQR ≈ 6
17. Std. dev ≈ 16.7, IQR ≈ 15
18. Std. dev ≈ 0.056, IQR ≈ 0.05
19. Std. dev ≈ 82.3, IQR ≈ 60

Exercise 7.4

1.

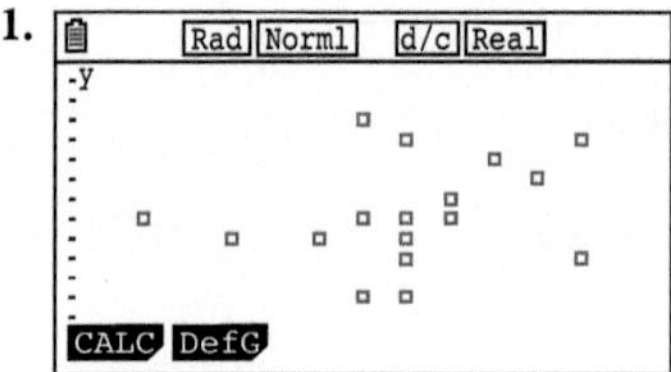

It appears that the data have a weak positive linear relationship. The correlation coefficient is 0.26 which confirms the weakness of the relationship.
The regression equation is: $y = 6.56 + 0.29x$. For every change of 1 unit in the x-values, the y-values will change, on average, by 0.29.

2. (a)

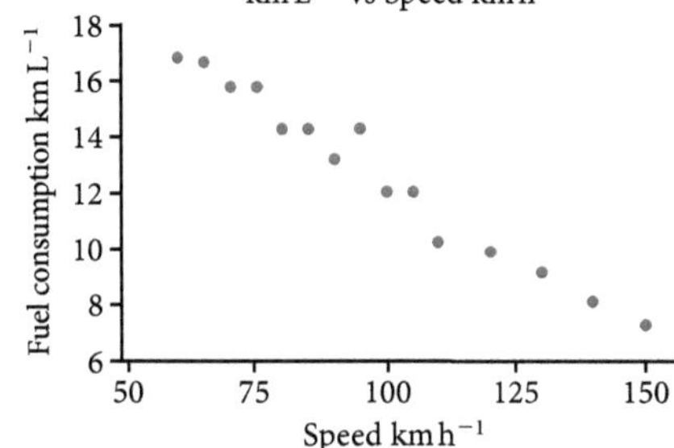

(b) We chose the speed as the explanatory variable because the car must first run to cause a fuel consumption. Hence the speed helps explain the fuel consumption. The relationship appears to be negatively sloped because the consumption is measure by the distance travelled per litre of fuel.

(c) The relationship appears to be a relatively strong negative one without any apparent outliers. The correlation coefficient is -0.986 which is very close to -1. A very strong relationship.

(d) The regression equation is:
Fuel cons.km L^{-1} = 24.1 − 0.116 Speed km h^{-1}.
For every increase of 1 km h^{-1} in speed, the average number of km per litre will decrease by 0.116 km l^{-1}. i.e. consumption will increase.

3. (a)

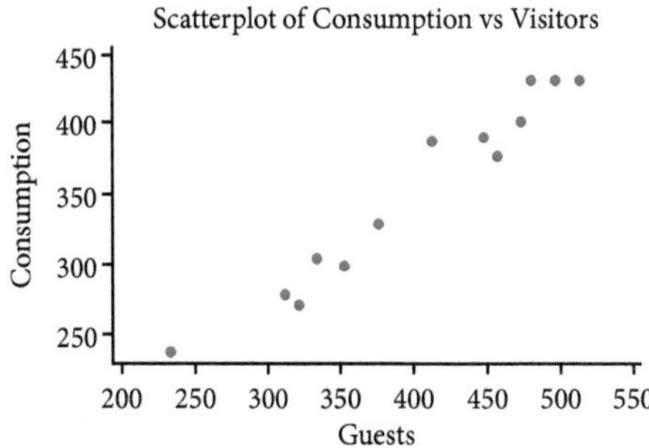

(b) The relationship appears to be a positive one except for an outlier which can be traced to be Singapore. We chose the explanatory variable to be the Income because the income level dictates how willing are people to pay for goods.

(c) The relationship is relatively strong (weakened by Singapore's numbers). The correlation coefficient is 0.621. If we remove Singapore's data, then it becomes 0.886.

(d) The regression equation is: *PPP* = 24383 + 0.351 GNI/cap. For every increase of \$1 in GNI/cap, the *PPP* will increase, on average by \$0.351.

4. (a)

(b) There is obviously a positive relationship between the number of guests and consumption. As the number of guests increases, the consumption will also increase.

(c) The relationship seems to be strong and there is an absence of outliers. The correlation coefficient is 0.978 which is very close to 1.

(d) The regression equation is:
Consumption = 40.0 + 0.777 Guests. For every increase of 1 guest, we expect, on average, that consumption will increase by 0.777.5.

5. (a)

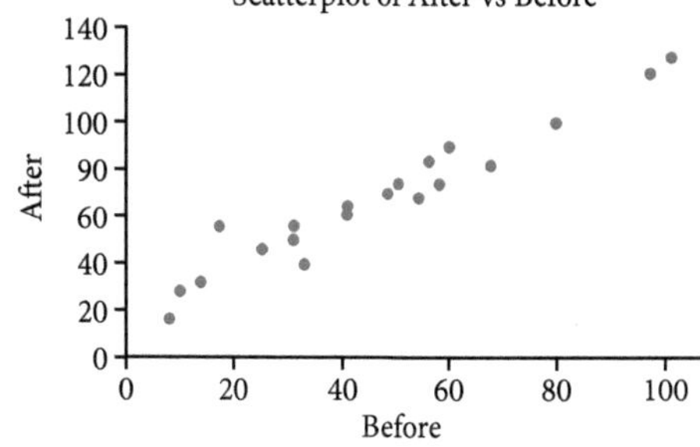

The scatter plot shows a strong positive relationship. That is, the higher the 'Before' score, the higher the 'After' score is. The regression equation is:
After = 20.2 + 1.03 Before.
This means that, on average, for every change of 1 mark on the 'Before' test, the 'After' test is expected to change by 1.03. The correlation coefficient is 0.97 indicating a very strong linear relationship. For a student with 60 score on the 'Before' test, the model predicts, on average, a score of 81.90 on the 'After' test.

6. (a)

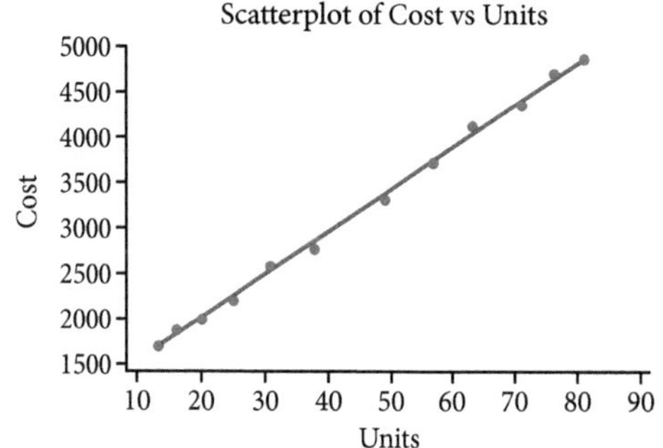

(b) The regression equation is: Cost = 1066 + 47.1 points.

(c) For every increase of 1000 units in production, the cost, on average. will increase by 47 100 Euros. The correlation coefficient is 0.999, which is almost perfect association. This is a strong linear relationship.

(d) Let number of 1000 units be x,
then: Cost = 1066 + 47.1x
$\frac{Cost}{x} = \frac{1066}{x} + 47.1$ = cost per unit. If this cost is 105,
then $105 = \frac{1066}{x} + 47.1 \Rightarrow 18.411$
Thus the number of units will be 18 40 units.

7. (a) R = 0.493. This is a relatively weak correlation between the two scores.

(b) The regression equation is: Economics = 2.07 + 0.649 Physics

(c) 4.7 (which can be rounded up to 5).

8. (a) Scatterplot of Price (€) vs Points

Appears to be a positively sloped trend.

(b) The regression equation is:
Price (€) = −2689 + 154 points.

(c) The intercept is meaningless because zero is not in the domain of the explanatory variable. On average, for every increase of 1 point, we expect the price to increase by 154 Euros.

(d) $r = 0.93$ indicating a strong association between points and price.

(e) The average price of a 63-point diamond is predicted to be 7013 Euros.

(f) Residual = 2104.

9. (a) Scatterplot of Blood Volume vs Age

(b) $r = -0.922$. There is a strong negative correlation between the stroke volume and age of patients.

(c) The regression equation is:
Blood volume = 82.5 − 0.269 Age. On average, for an increase of 1 year, we expect blood volume to be decreasing by 0.269 ml per stroke. The interpretation of the intercept of 82.5 does not make sense in this situation.

(d) On average, 45-year olds may have 70 stroke volume. Using the model to predict the 90-year old volume is not advisable as it is an extrapolation of 17 years beyond the range of collected data.

10. (a)

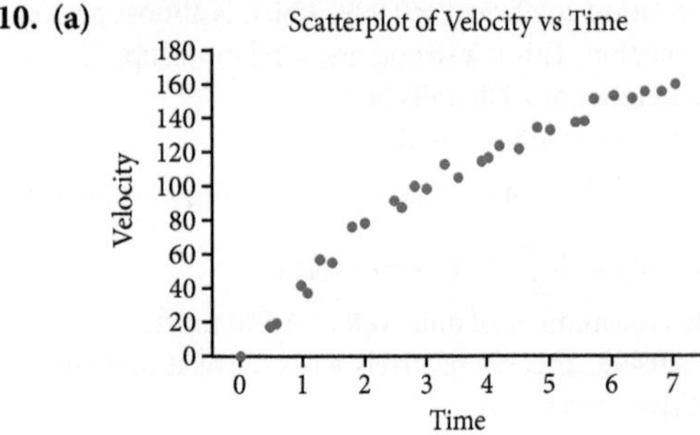

Apparently, there is a strong association between time and speed as expected. However, it appears that there is a break point around 3 seconds.

(b) The regression equation is: Velocity = 24.5 + 21.8 Time

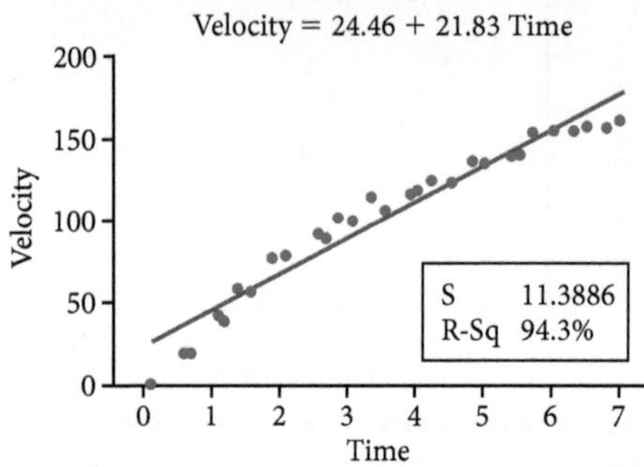

Apparently, the data do not follow a linear model through the whole range. There is a clear deviation from the line at both ends.

(c) $R = 0.97$, which is a strong association indication. However, this number may not be of great validity since the data does not appear to be linear.

(d) By splitting the data, we can clearly see that the new model fits the data better. The data clearly has two phases, one before 3 seconds and the other after.

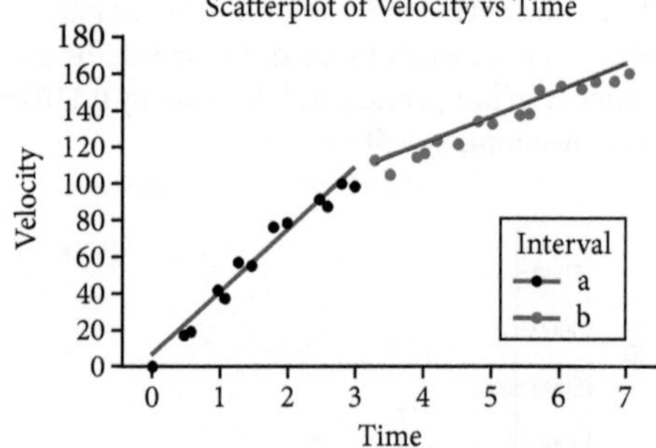

First interval: Velocity = 3.60 + 34.9 time
Second interval: Velocity = 60.12 + 14.9 time

(e) At 4 seconds, model in (b) gives a speed of 111.78 while model in (d) gives 119.78. The actual observation is 118. This shows that the error in using (b) is much larger than the error in using (d).

Chapter 7 practice questions

1. (a) 12 **(b)** $\sqrt{30.83}$

2. 4

3. (a)

Time	1.6	2.1	2.6	3.1	3.6	4.1	4.6	5.1	5.6	6.1	6.6
Frequency	2	2	6	4	11	10	5	5	3	2	0

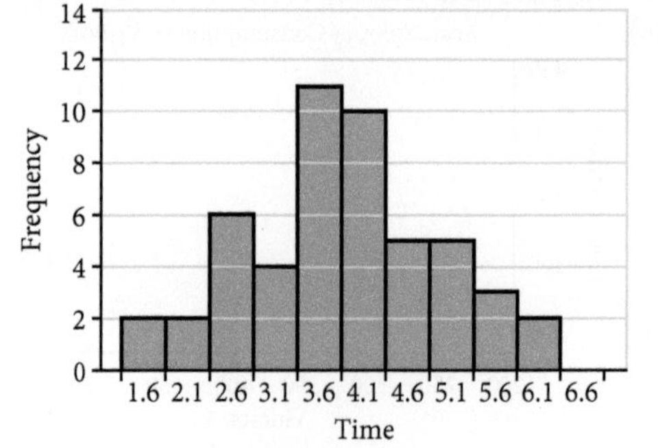

(b) 86% **(c)** approx. 4 **(d)** 3.8, 1.1

(e)

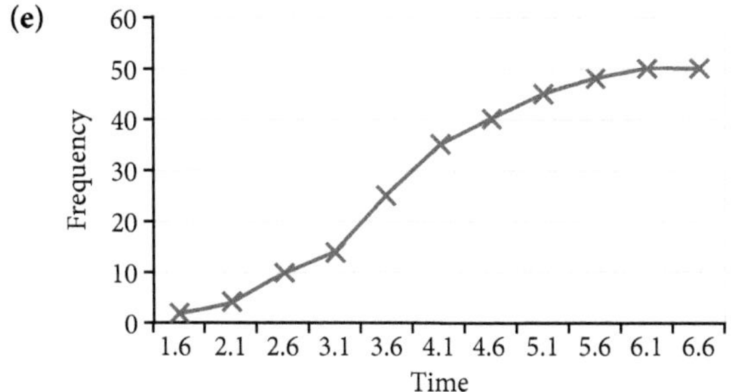

(f) Minimum = 1.6, Q1 = 3, median = 4, Q3 = 4.5, maximum = 6.2

4. (a) Median and IQR as the data is skewed with outliers.
(b) Mean = 682.6, standard deviation = 536.2
(c)

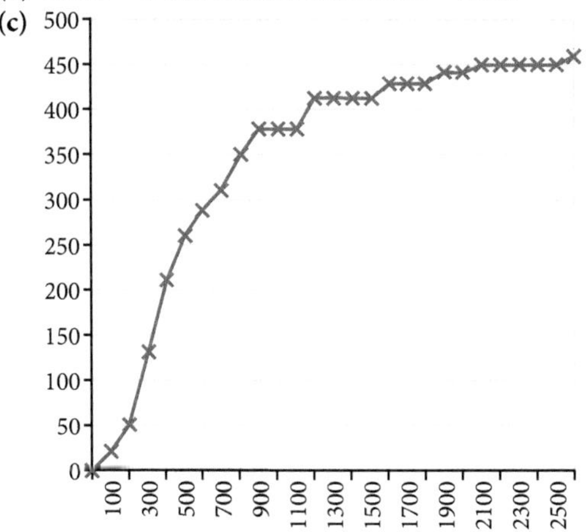

(d) Q1 = 300, median = 500, Q3 = 800, IQR = 500
(e) There are a few outliers on the right side. Outliers lie above Q3 + 1.5IQR = 1550
(f) Data is skewed to the right, with several outliers from 1600 onwards. It is bimodal at 300-400.

5. (a) Spain, Spain (b) France
(c) On average, it appears that France produces the more expensive wines as 50% of its wines are more expensive than most of the wines from the other countries. Italy's prices seem to be symmetric while Frances' prices are skewed to the left. Spain has the widest range of prices.

6. (a) Mean = 52.65, standard deviation = 7.66
(b) Median = 51.34, IQR = 2.65
(c) Apparently, the data is skewed to the right with a clear outlier of 112.72! This outlier pulled the value of the mean to the right and increased the spread of the data. The median and IQR are nor influenced by the extreme value.

7. (a) The distribution does not appear to be symmetric as the mean is less than the median, the lower whisker is longer than the upper one and the distance between Q1 and the median is larger than the distance between the median and Q3. Left skewed.
(b) There are no outliers as Q1 − 1.5IQR = 37 < 42 and Q3 + 1.5IQR = 99 > 86.
(c)

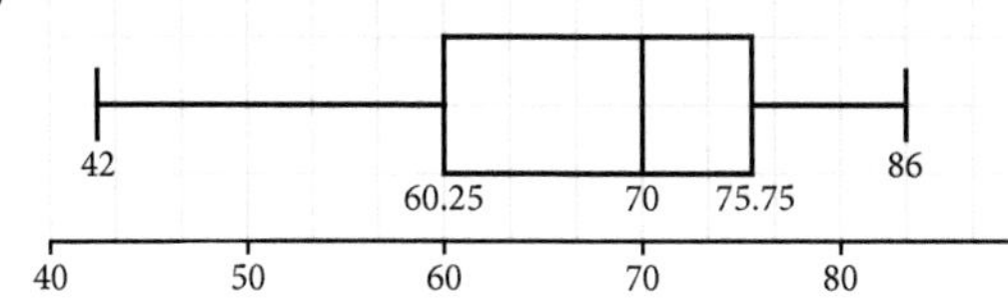

(d) See (a)

8. (a) 225
(b) Q1 = 205, Q3 = 255, 90th percentile = 300, 10th percentile = 190
(c) IQR = 50, Number in middle 50% is approx 1200
(d)

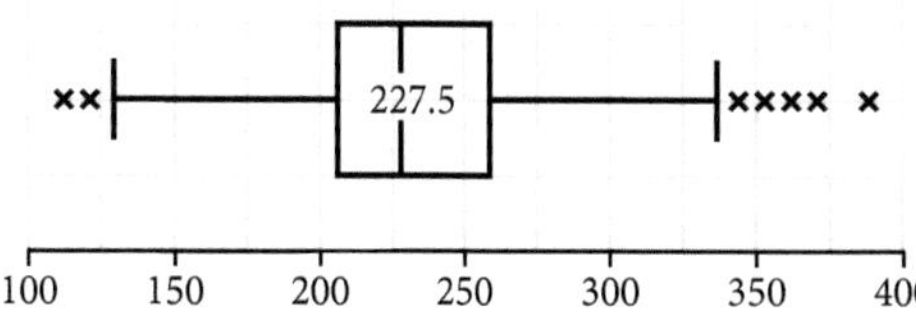

(e) The distribution has many outliers and is apparently skewed to the right with more outliers there. The middle 50% seem to be very close together, while the whiskers appear to be quite spread.

9. (a)

Speed	Frequency
26-30	9
31-34	16
35-38	31
39-42	23
43-46	12
47-50	8
51-54	1

(b)

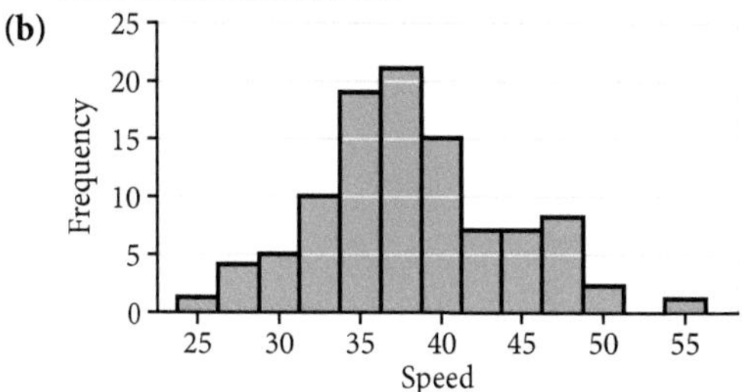

Data is relatively symmetric with possible outlier at 55.
The mode is approximately 37
Histogram created from table:

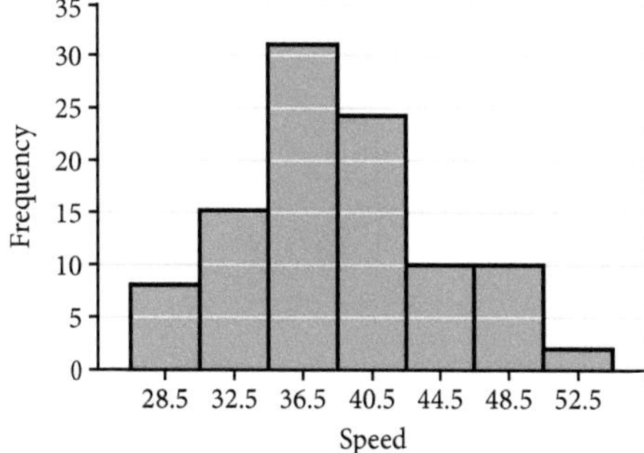

(c) Mean = 38.2, standard deviation = 5.7
(d)

Speed	Cumulative frequency
26-30	9
31-34	25
35-38	56
39-42	79
43-46	89
47-50	99
51-54	100

(e) Median = 37.6, Q1 = 34.5, Q3 = 41.3, IQR = 6.8
(f) There are outliers on the right since $Q3 + 1.5IQR = 51.5 < \text{maximum} = 54$.

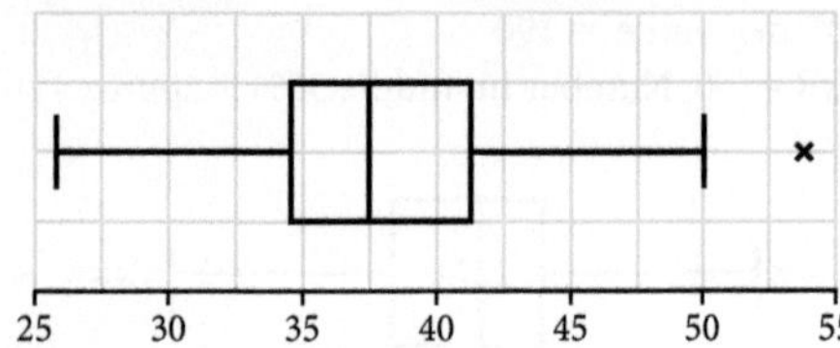

10. (a) Mean = 1846.9, media = 1898.6, standard deviation = 233.8, Q1 = 171.8, Q3 = 2031.3, IQR = 319.5
(b) $Q1 - 1.5IQR = 1232.55 >$ minimum, so there is an outlier on the left.
(c)

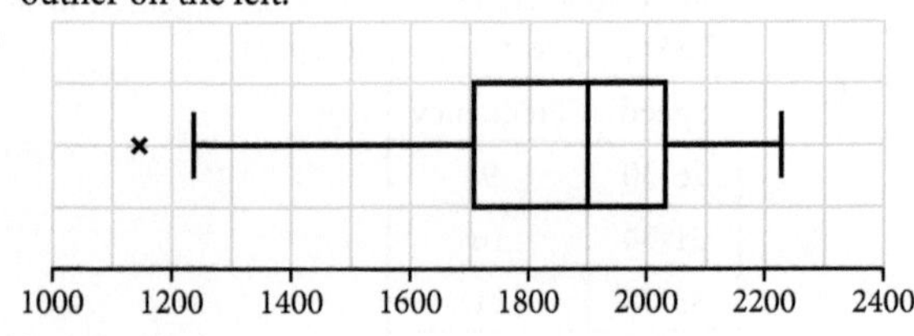

(d)]1613, 2081[
(e) The mean and standard deviation will get larger. The rest will not change much.

11. (a) 49.6 minutes (b) 48.9 minutes

12. (a)

≤10	≤20	≤30	≤40	≤50	≤60	≤70	≤80	≤90	≤100
30	130	330	670	1190	1630	1810	1900	1960	2000

(b)

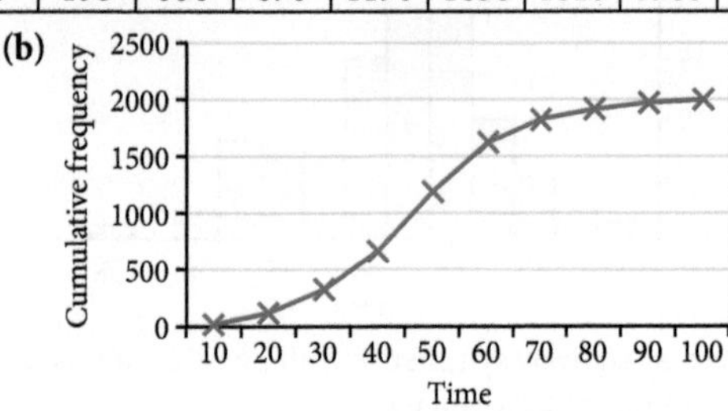

(c) (i) 47 (ii) About 500 (iii) Above 60

13. 174 cm

14. (a) $\mu = 12$ (b) Standard deviation = 5

15. $k = 4$

16. (a) 97.2 seconds
(b)

30	60	90	120	150	180	210	240
5	20	53	74	85	92	97	100

(c)

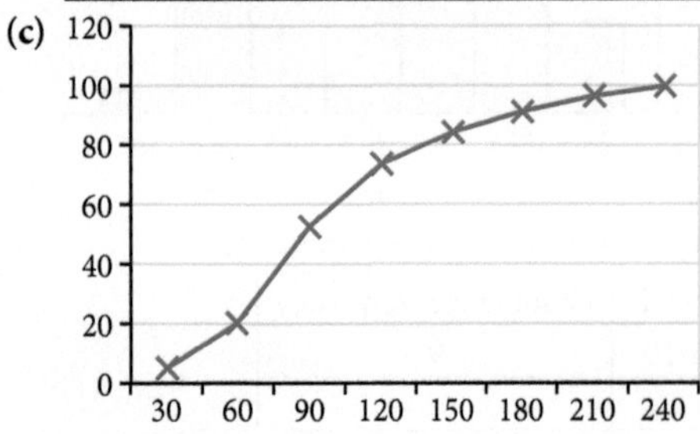

(d) All approx. median = 88
Q1 = 66
Q3 = 124

17. (a) (i) 10 (ii) 24
(b) Mean = 63, standard deviation = 20.5
(c) Skew to the left
(d) 65

18. (a) 7.41 g
(b)

Weight g	Number of packets
$w \leq 85$	5
$w \leq 90$	15
$w \leq 95$	30
$w \leq 100$	56
$w \leq 105$	69
$w \leq 110$	76
$w \leq 115$	80

(c) (i) Median = 97 g (ii) Q3 = 0 g
(d) 0
(e) 0.263

19. (a) 98.2
(b) (i) $m = 165, n = 275$
(ii)

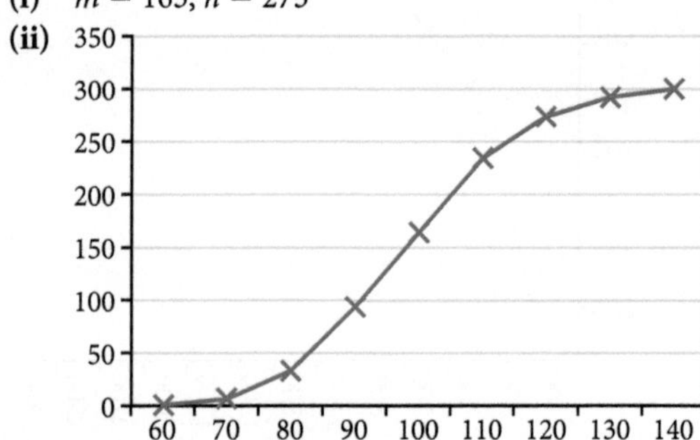

(c) (i) 20% (ii) 115 km h^{-1}

20. (a) (i) 24 (ii) 158
(b) 40
(c) 7%

21. $a = 3$

22. (a)

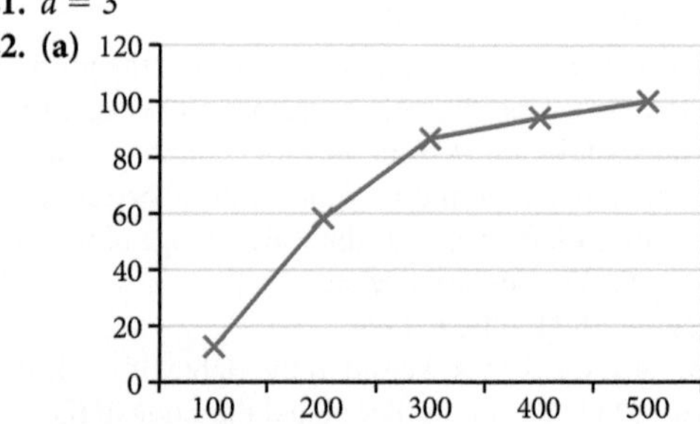

(b) IQR = approx. 120
(c) $m = 7, n = 6$
(d) $199 000
(e) (i) approx. 9 (ii) $\frac{4}{9}$

23. (a) (i) 20 mm (ii) 24 mm
(b) 10 mm

24. (a)

Mark	[0, 20[	[20, 40[	[40, 60[	[60, 80[	[80, 100[
Number of students	22	50	66	42	20

(b) Pass mark = 43%

25. (a) 183 cm (b) 14 m

26. $a = 3, b = 7, c\ 11, d = 11$

27. (a) 100 (b) $a = 55, b = 75$

28. $x = 4, y = 10$

29. **(a)** Apparently linear with two possible outliers: (7, 54) and (28, 78). It appears to be a linear relationship.

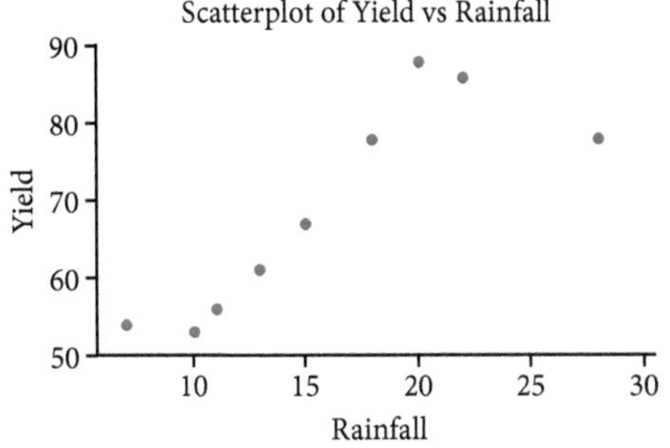

(b) 0.853. A relatively strong positive linear relationship.
(c) Yield = 40.5 + 1.78 Rainfall. On average, a change of 1 cm in rainfall corresponds to a change of 1.78 kg in crop. The intercept is not useful in this case since 0 is not in the domain of the explanatory variable.
(d) 74.3
(e) 7°.

Chapter 8

Exercise 8.1

Note: Some answers may differ from one person to the other due to different graph accuracies.

1. **(a)** {left handed, right handed}
(b) all real numbers from (say) 50 cm to 210 cm.
(c) all real numbers from 0 to 720 (say).

2. {(1,h), (2, h), …, (1, t), …, (6, t)}

3. **(a)** {(1, Hearts), …, (King, Hearts), (1, Spades), …}
(b) {[(1, hearts), (King, Diamonds)], …,[(1, Spades), (10, Diamonds)],…}
(c) a: 52, b: 1326

4. **(a)** 0.47
(b) anywhere from 0 to 20!
(c) 10 000.

5. **(a)** {(1, 1), (1,2), …, (4, 4)} **(b)** {3, 4, …, 9}

6. **(a)** {(b, b), (b, g), (b, y), (g, b), (g, g), (g, y), (y, b), (y, g), (y, y)}
(b) {(y, y), (y, b), (y, g)}
(c) {(b, b), (g, g), (y, y)}

7. **(a)** {(b, g), (b, y), (g, b), (g, y), (y, b), (y, g)}
(b) {(y, b), (y, g)}
(c) φ

8. **(a)** {(t,t,t), (t,t,h), (t,h,t), (h,t,t), (h,t,h), (h,h,t), (t,h,h), (h,h,h)}
(b) {(h,t,h), (h,h,t), (t,h,h), (h,h,h)}

9. {(I, fly), (I, dr), (I, tr), (H,dr), (H, b)}, {(I, fly)}

10. **(a)** {(1,g), (1,f), …, (0,c)}
(b) {(0,c), (0,s)}
(c) {(1,g), (1,f), (0,g), (0,f)}
(d) {(1,g), (1,f), (1,s), (1,c)}

11. **(a)** $\{(G_1, K_1, M_1), (G_1, K_2, M_1), (G_1, K_1, M_2), \ldots\}$
(b) A = all triplets containing G_2; B = all triplets not containing K_1; C = all triplets containing M_2.
(c) $A \cup B$ = All males or persons who drink; $A \cap C$ = All single males; C' = All non-single persons; $A \cap B \cap C$ = All single males who drink; $A' \cap B$ = All females who drink.

12. **(a)** {(R, L, L, S), (L, R, L, R), …}, 81
(b) {(R,R,R,R), (L,L,L,L) (S,S,S,S)}
(c) {(R,R,L,L), (R,L,R,S), …}
(d) {(R,L,R,S), (S, S, R, L), …}

13. **(a)** {(T, SY, O), (C, SN, O), …}
(b) {(T, SY, O), (T, SY, F), (B, SY, O), …}
(c) {(C, SY, O), (C, SN, O), (C, SY, F), …}
(d) $C \cap SY$ = {(C, SY, O), (C, SY, F)}
C' = {(T, …, …), (B, …, …)}
$C \cup SY$ = all triplets containing C or SY.

14. **(a)** {(1,1,1), (1,1,0), (0,1,0), …}
(b) X = {(1,1,0), (1,0,1), (0,1,1)}

Exercise 8.2

1. **(a)** $\frac{3}{10}$ **(b)** $\frac{3}{4}$

2. **(a)** 0.63 **(b)** 1

3. **(a)** $\frac{1}{52}, \frac{7}{26}, \frac{4}{13}, \frac{10}{13}$ **(b)** $\frac{1}{51}$, 13/17 **(c)** $\frac{1}{52}, \frac{10}{13}$

4. **(a)** $\frac{4}{5}$ **(b)** $\frac{11}{30}$ **(c)** 1

5. **(a)** $\frac{1}{2}$ **(b)** $\frac{1}{12}$

6. **(a)** $\frac{1}{7}$ **(b)** $\frac{4}{7}$

7. **(a)** **(i)** ((1, 1), (1, 2) , …, (6, 6))
(ii) $\frac{1}{6}$ **(iii)** $\frac{2}{9}$ **(iv)** $\frac{5}{6}$
(b) **(i)** 0 **(ii)** $\frac{1}{9}$ **(iii)** $\frac{5}{36}$ **(iv)** 0

8. **(a)** 0.04 **(b)** 0.55 **(c)** 0.1548
(d) 0.060372 **(e)** 0.104022

9. **(a)** Yes **(b)** No **(c)** No

10. **(a)** 0.06 **(b)** 0.42 **(c)** 0.3364 **(d)** 0.412

11. **(a)** 0.183 **(b)** 0.69

12. **(a)** $\frac{5}{18}$ **(b)** $\frac{1}{3}$ **(c)** $\frac{5}{18}$

13. **(a)** $\frac{3}{28} \simeq 0.107$ **(b)** $\frac{17}{190} \simeq 0.0895$ **(c)** $\frac{68}{95} = 0.716$

14. 0.75

15. $\frac{111}{400} \simeq 0.2775$

16. **(a)** 0.096 **(b)** 0.008 **(c)** 0.512

Exercise 8.3

1. $\frac{7}{20}$

2. **(a)** $\frac{1}{2}$ **(b)** $\frac{2}{5}$ **(c)** $\frac{1}{5}$
(d) $\frac{1}{10}$ **(e)** $\frac{2}{3}$

3. $P(A \cap B) = \frac{1}{9} \neq 0 \neq P(A)P(B)$

4. $\frac{29}{35}$

5. 0.9

6. **(a)** 92%
(b) **(i)** 0.64% **(ii)** 15.36% **(iii)** 14.72%
(c) 48.68%

7. **(a)** 10 000 **(b)** $\frac{9}{10}$ **(c)** 0.3439 **(d)** $\frac{1000}{3439}$

8. **(a)** $\frac{15}{16}$ **(b)** $\frac{4}{5}$ **(c)** $\frac{1}{5}$

9. **(a)** $\{(1, 1), (1, 2), \ldots, (6, 6)\}$

(b)

x	2	3	4	5	6	7	8	9	10	11	12
P(x)	$\frac{1}{36}$	$\frac{1}{18}$	$\frac{1}{12}$	$\frac{1}{9}$	$\frac{5}{36}$	$\frac{1}{6}$	$\frac{5}{36}$	$\frac{1}{9}$	$\frac{1}{12}$	$\frac{1}{18}$	$\frac{1}{36}$

(c) **(i)** $\frac{11}{36}$ **(ii)** $\frac{11}{12}$ **(iii)** $\frac{1}{3}$ **(iv)** $\frac{2}{3}$

10. **(a)** $\frac{7}{15}$ **(b)** $\frac{11}{75}$ **(c)** $\frac{9}{35}$ **(d)** $\frac{46}{75}$ **(e)** $\frac{11}{20}$

(f) No. For example, a grade 9 student is more likely to be female.

11. **(a)** **(i)** 0.56 **(ii)** 0.15

(b) $\frac{15}{56}$ **(c)** no

12.

P(A)	P(B)	Conditions for events A and B	P($A \cap B$)	P($A \cup B$)	P($A \mid B$)
0.3	0.4	Mutually exclusive	0.00	0.7	0.00
0.3	0.4	Independent	0.12	0.58	0.30
0.1	0.5	Mutually exclusive	0.00	0.60	0.00
0.2	0.5	Independent	0.10	0.60	0.20

13. **(a)** 0.30 **(b)** yes

14. **(a)** 65% **(b)** 35% **(c)** 52%

15. **(a)** 0.56; **(b)** 0.10

16. **(a)** $\frac{1}{216}$ **(b)** $\frac{91}{216}$ **(c)** $\frac{75}{216}$

17. **(a)** 0.21 **(b)** 0.441 **(c)** 0.657

18. **(a)** $\frac{23}{144}$ **(b)** $\frac{11}{144}$ **(c)** $\frac{15}{144}$ **(d)** $\frac{9}{23}$

19. **(a)** $A \cap B = \{(10, 5),(10, 4), \ldots,(10, 1),(1, 10), \ldots,(5, 10)\}$, $p = 0.069$

(b) $A \cup B = \{(1, 12), \ldots(1, 1), (2, 12), \ldots, (3, 12), \ldots, (4, 11), \ldots, (5, 10), \ldots\}p = 0.778$

(c) list, p=0.931 **(d)** list, p = 0.222

(e) same as (c) **(f)** same as (d)

(g) This is $(A \cup B) - (A \cap B)$; p = .708

20. **(a)** $\frac{1}{36}$ **(b)** $\frac{1}{28}$ **(c)** $\frac{91}{216}$ **(d)** 0.5

21. $\frac{2}{3}$

22. **(a)** 0.103 **(b)** 0.0887 **(c)** 0.537

23. **(a)** 0.10 **(b)** 0.00001

24. **(a)** 0.36 **(b)** 0.64 **(c)** 0.75

(d) 0.17 **(e)** 0.0455 **(f)** 0.682

25. **(a)** 0.8805 **(b)** 0.0471

Chapter 8 practice questions

1. **(a)** 0.30 **(b)** 0.72 **(c)** 0.7

2. **(a)** 0.0004 **(b)** 0.9996 **(c)** 0.0004

3. 0.99998

4. **(a)** **(i)** 8 **(ii)** 0.80 **(ii)** 0.15

(b) 0.083

5. **(a)** **(i)** 0.3405 **(ii)** 0.0108 **(iii)** 0.9622 **(iv)** 0.30

(b) Yes.

6. **(a)** 0.63 **(b)** 0.971

7. **(a)** 0.60

(b) Yes, P(B|A) = P(B) = 0.60

(c) 0.42

8. **(a)**

	Boys	Girls
Passed the ski test	32	16
Failed the ski test	14	12
Training, but did not take the test yet	20	16
Too young to take the test	4	6

(b) **(i)** 0.6167 **(ii)** 0.56 **(iii)** 0.1463

9. **(a)** $\frac{3}{32}$ **(b)** $\frac{3}{4}$ **(c)** $\frac{5}{32}$

10. **(a)** 0.02 **(b)** 0.64

11. **(a)** 0.4 **(b)** 0.6

12. **(a)** 0.38 **(b)** 0.283

13. **(a)**

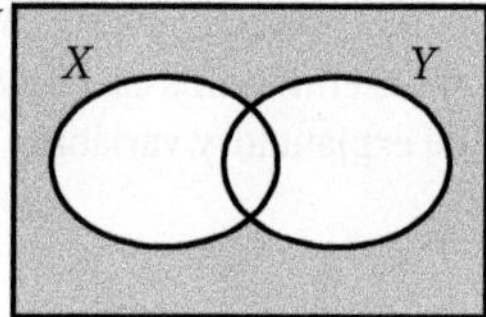

(b) **(i)** 2 **(ii)** $\frac{1}{18}$

(c) No, $n(X \cap Y) \neq 0$

14. **(a)**

	Males	Females	Totals
Unemployed	20	40	60
Employed	90	50	140
Totals	110	90	200

(b) **(i)** $\frac{1}{5}$ **(ii)** $\frac{9}{14}$

15. **(a)**

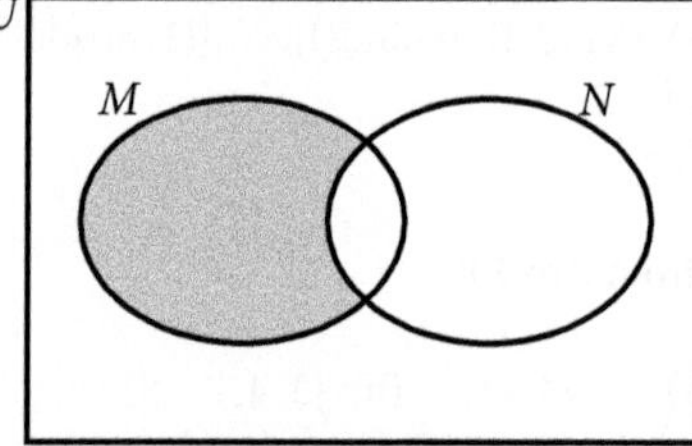

(b) 35 **(c)** 0.35

16. **(a)** $\frac{7}{12}$ **(b)** $\frac{11}{36}$ **(c)** $\frac{1}{3}$

17. **(a)** $\frac{1}{11}$ **(b)** $\frac{12}{121}$

18. **(a)** $a = 21, b = 11, c = 17$

(b) **(i)** $\frac{1}{8}$ **(ii)** $\frac{21}{32}$

(c) **(i)** 0.258 **(ii)** 0.742

19. **(a)** See Worked Solutions **(b)** $\frac{47}{160}$ **(c)** $\frac{35}{47}$

20. **(a)** $\frac{1}{6}$ **(b)** $\frac{1}{12}$ **(c)** $\frac{2}{9}$

21. **(a)** **(i)** $\frac{8}{21}$ **(ii)** $\frac{1}{6}$

(iii) No, $P(A \cap B) \neq P(A)P(B)$

(b) $\frac{10}{17}$ **(c)** $\frac{200}{399}$

22. 0.00198

23. $\frac{10}{19}$

24. $\frac{2}{5}$

25. **(a)** **(i)** $\frac{5}{36}$ **(ii)** $\frac{25}{216}$ **(iii)** $\frac{1}{6}\left(\frac{5}{6}\right)^{2n-2}$

(b) Working using **(a)(iii)** as sum to infinity of a geometric series

(c) $\frac{5}{11}$ **(d)** 0.432

26. $\frac{1}{9}$

27. **(a)** 0.80 **(b)** 0.56

28. **(a)** 0.732 **(b)** $\frac{11}{61}$

29. **(a)** $\frac{2}{3}$ **(b)** $\frac{2}{9}$ **(c)** $\frac{3}{4}$

30. **(a)** $\frac{7}{3}$

(b) **(i)** $\frac{7}{72}$ **(ii)** $\frac{2}{27}$

(c) 0.559 **(d)** $n = 8$ **(e)** 3

Chapter 9

Exercise 9.1

1. **(a)** 4 **(b)** $3x^2$ **(c)** -2 **(d)** 6

2. **(a)** 0 **(b)** $\frac{5}{2}$

(c) does not exist (increases without bound)

3. $\lim_{c \to \infty}\left(1 + \frac{1}{c}\right)^c = e$

4. $\lim_{x \to \infty} f(x) = \lim_{x \to -\infty} f(x) = 3$

5. as $x \to a$, $g(x) \to +\infty$

6. **(a)** horizontal: $y = 3$; vertical: $x = -1$

(b) horizontal: $y = 0$ (x-axis); vertical: $x = 2$

(c) horizontal: $y = b$; vertical: $x = a$

Exercise 9.2

1. **(a)** -2 **(b)** 3

(c) $\frac{1}{2}$ **(d)** -2

2. **(a)** **(i)** $y' = 6x - 4$ **(ii)** -4

(b) **(i)** $y' = -2x - 6$ **(ii)** 0

(c) **(i)** $y' = -\frac{6}{x^4}$ **(ii)** -6

(d) **(i)** $y' = 5x^4 - 3x^2 - 1$ **(ii)** 1

(e) **(i)** $y' = \cos x$ **(ii)** $\frac{1}{2}$

(f) **(i)** $y' = 2x - 4$ **(ii)** 0

(g) **(i)** $y' = 2 - \frac{1}{x^2} + \frac{9}{x^4}$ **(ii)** 10

(h) **(i)** $y' = 1 - \frac{2}{x^3}$ **(ii)** 3

(i) **(i)** $y' = -\sin x$ **(ii)** $-\frac{\sqrt{2}}{2}$

3. $a = -5, b = 2$

4. **(a)** between A and B

(b) **(i)** A, B and F **(ii)** D and E **(iii)** C

(c) pair B & D, and pair E & F

5. **(a)** $(0, 0)$ **(b)** $(2, 8)$ and $(-2, -8)$

(c) $\left(\frac{5}{2}, -\frac{21}{4}\right)$ **(d)** $(1, -2)$

6. $a = 1, b = 5$

7. $a = 1$

8. $(3, 6)$

9. **(a)** $4.\dot{6}$ degrees Celsius per hour

(b) $C'(t) = 3\sqrt{t}$

(c) $t = \frac{196}{81} \approx 2.42$ hours

Exercise 9.3

1. **(a)** $(1, -7)$ **(b)** $\left(-\frac{3}{2}, 8\right)$ **(c)** $(3, 2)$

2. **(a)** **(i)** $y' = 2x - 5$

(ii) increasing for $x > \frac{5}{2}$

(iii) decreasing for $x < \frac{5}{2}$

(b) **(i)** $y' = -6x - 4$

(ii) increasing for $x < -\frac{2}{3}$

(iii) decreasing for $x > -\frac{2}{3}$

(c) **(i)** $y' = x^2 - 1$

(ii) increasing for $x > 1, x < -1$

(iii) decreasing for $-1 < x < 1$

(d) **(i)** $y' = 4x^3 - 12x^2$

(ii) increasing for $x > 3$

(iii) decreasing for $x < 0, 0 < x < 3$

3. **(a)** **(i)** $(3, -130), (-4, 213)$

(ii) $(3, -130)$ minimum because 2nd derivative is positive at $x = 3$

$(-4, 213)$ maximum because 2nd derivative is negative at $x = -4$

(iii)

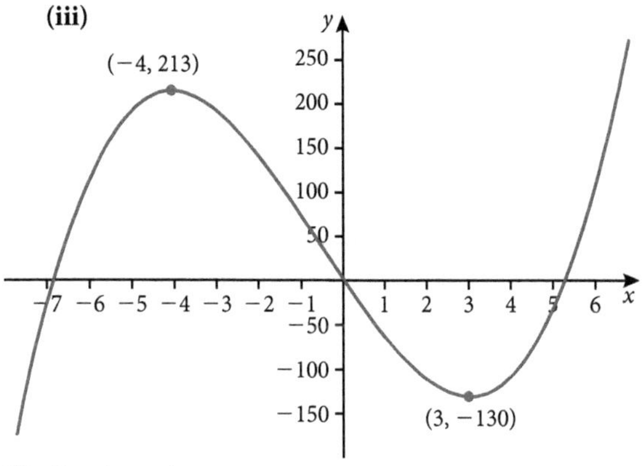

(b) **(i)** $(0, -5)$

(ii) stationary point is neither a maximum nor minimum because 1st derivative is always positive

(iii)

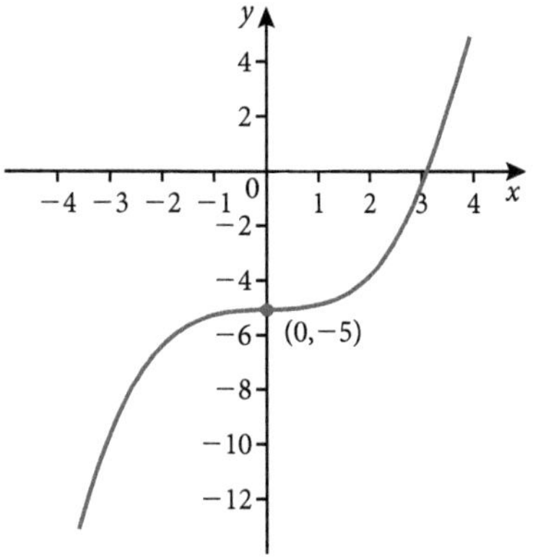

(c) (i) $(1, 4), (3, 0)$

(ii) $(1, 4)$ maximum because 2nd derivative is negative at $x = 1$

$(3, 0)$ minimum because 2nd derivative is positive at $x = 3$

(iii)

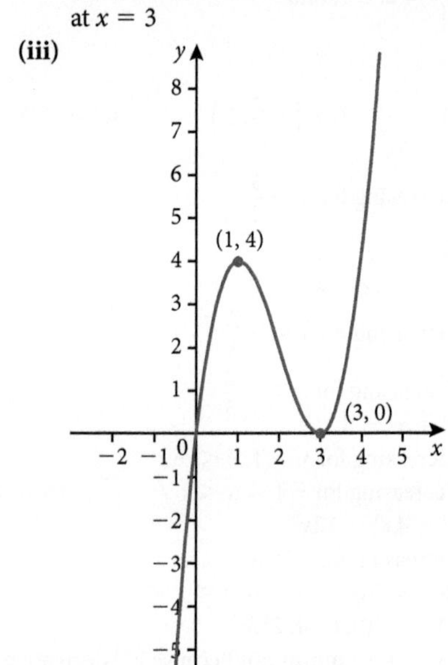

(d) (i) $(-1, 4), (0, 6), \left(\frac{5}{2}, -\frac{279}{16}\right)$

(ii) $(-1, 4)$ minimum because 2nd derivative is positive at $x = -1$

$(0, 6)$ maximum because 2nd derivative is negative at $x = 0$

$\left(\frac{5}{2}, -\frac{279}{16}\right)$ minimum because 2nd derivative is positive at $x = \frac{5}{2}$

(iii)

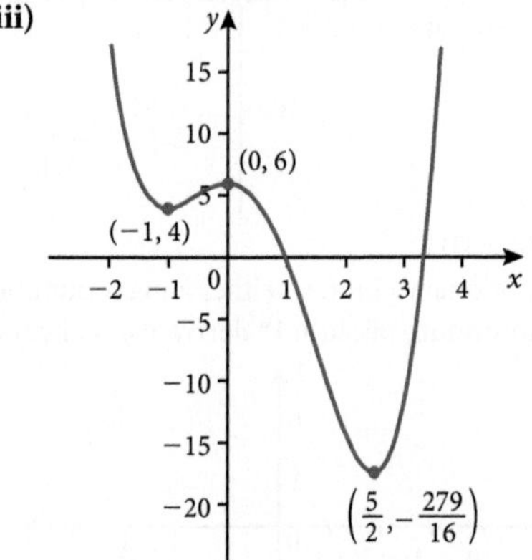

(e) (i) $(-1, 14), \left(\frac{7}{3}, -\frac{122}{27}\right)$

(ii) $(-1, 14)$ maximum because 2nd derivative is negative at $x = -1$

$\left(\frac{7}{3}, -\frac{122}{27}\right)$ minimum because 2nd derivative is positive at $x = \frac{7}{3}$

(iii)

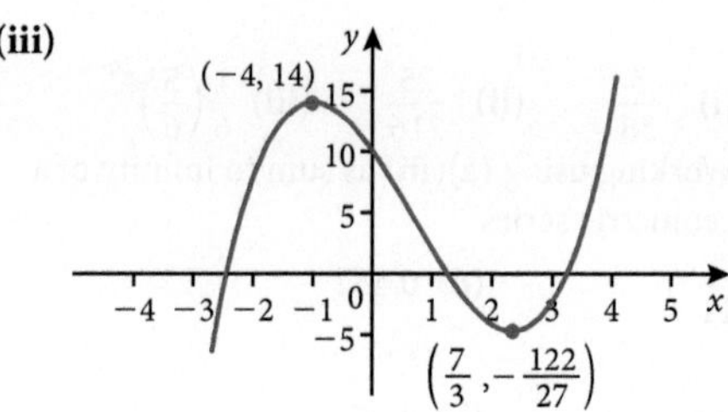

(f) (i) $\left(\frac{1}{4}, -\frac{1}{4}\right)$

(ii) $\left(\frac{1}{4}, -\frac{1}{4}\right)$ minimum because 2nd derivative is positive at $x = \frac{1}{4}$

(iii)

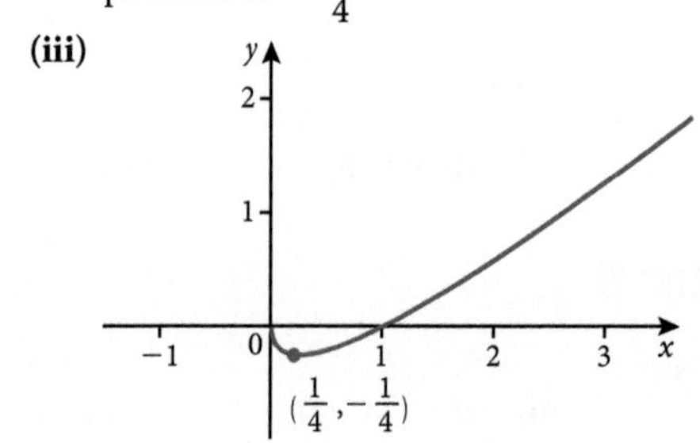

4. (a) $v(t) = 3t^2 - 8t + 1$; $a(t) = 6t - 8$

(b)

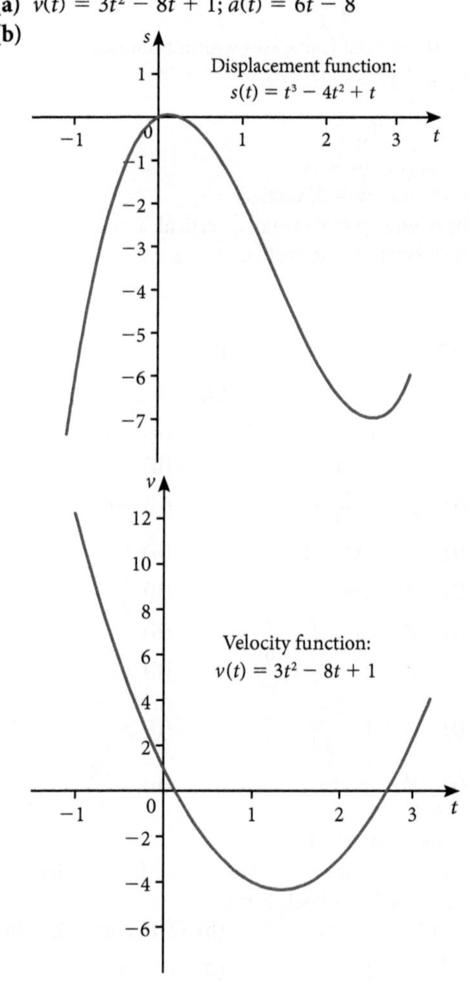

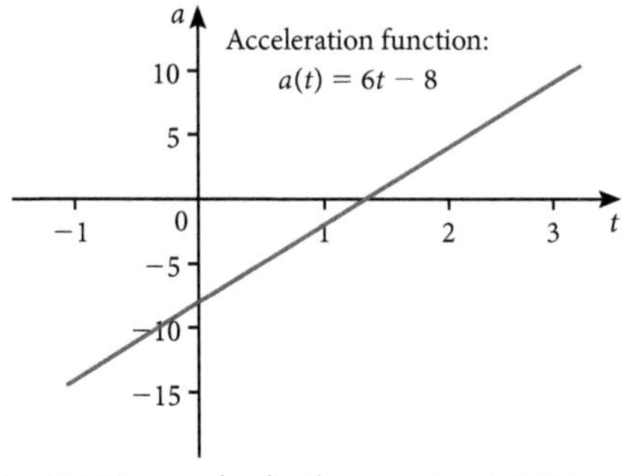

(c) $t \approx 0.131$ seconds, displacement ≈ 0.0646 m
(d) $t = 1.\dot{3}$ seconds, velocity $= -4.\dot{3}\ \text{ms}^{-1}$
(e) object moves right at a decreasing velocity then turns left with increasing velocity then slows down and turns right with increasing velocity

5. (a) relative maximum at $(-2, 16)$; relative minimum at $(2, 16)$; inflection point at $(0, 0)$
(b) absolute minima at $(-2, -4)$ and $(2, -4)$; relative maximum at $(0, 0)$; inflection points at $\left(-\frac{2\sqrt{3}}{3}, -\frac{20}{9}\right)$ and $\left(\frac{2\sqrt{3}}{3}, -\frac{20}{9}\right)$
(c) relative maximum at $(-2, -4)$; relative minimum at $(2, 4)$; no inflection points
(d) relative minimum at $(-1, -2)$; relative maximum at $(1, 2)$; inflection points at $\left(-\frac{\sqrt{2}}{2}, -\frac{7\sqrt{2}}{8}\right)$, $(0, 0)$ and $\left(\frac{\sqrt{2}}{2}, \frac{7\sqrt{2}}{8}\right)$
(e) relative minimum at $(-1, 0)$; absolute minimum at $(2, -27)$; relative maximum at $(0, 5)$; inflection points at $(1.22, -13.4)$ and $(-0.549, 2.32)$

6. (a) $x = \frac{\pi}{6}$ and $x = \frac{5\pi}{6}$
(b) maximum at $x = \frac{\pi}{6}$ because 2nd derivative is negative; minimum at $x = \frac{5\pi}{6}$ because 2nd derivative is positive

7. (a) $v(0) = 27\ \text{m s}^{-1}$, $a(0) = -66\ \text{m s}^{-2}$
(b) $v(3) = 45\ \text{m s}^{-1}$, $a(3) = 78\ \text{m s}^{-2}$
(c) $t = \frac{1}{2}$ and $t = 2\frac{1}{4}$; where displacement has a relative maximum or minimum
(d) $t = \frac{11}{8} = 1.375$; where acceleration is zero

8. $x \approx 5.77$ tonnes; $D \approx 34.6$ ($34,600); this cost is a minimum because cost decreases to this value then increases

9. $a = -3$, $b = 4$, $c = -2$

10. relative maximum at $\left(-2, -\frac{15}{4}\right)$, stationary inflection point at $(1, 3)$
$f(x) \to x$ as $x \to \pm\infty$

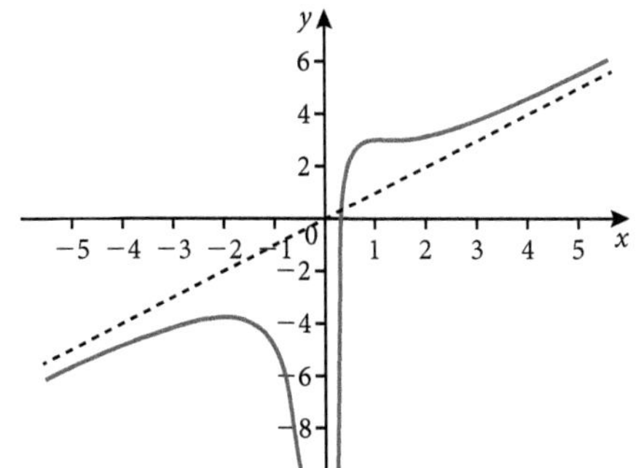

11. (a)

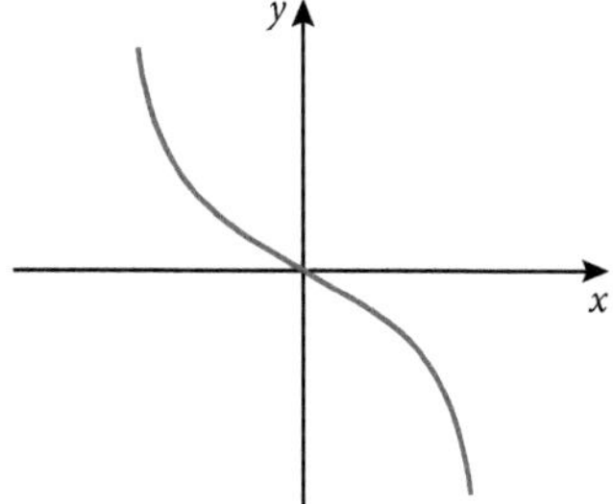

(b)

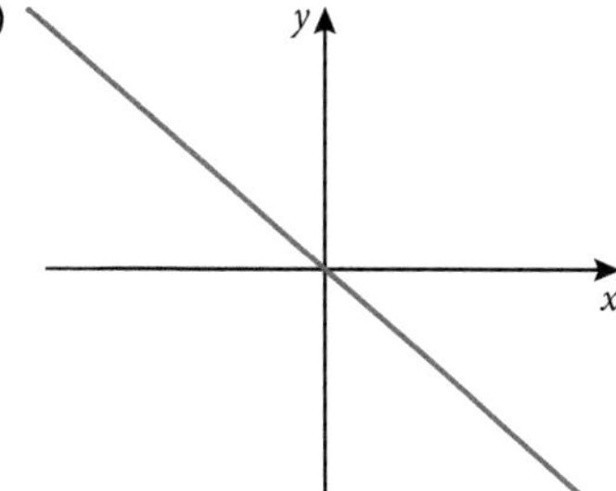

(c)

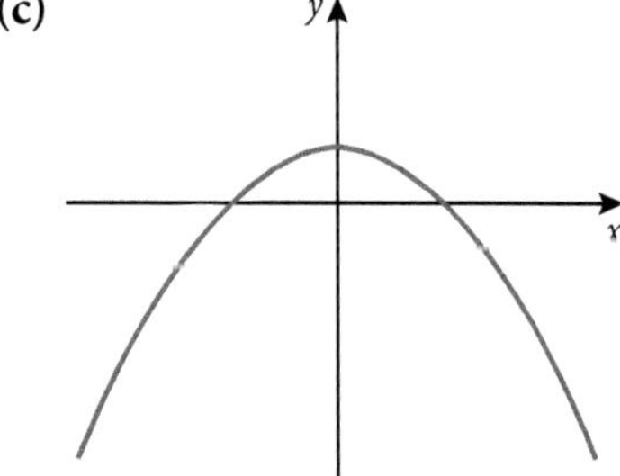

(d)

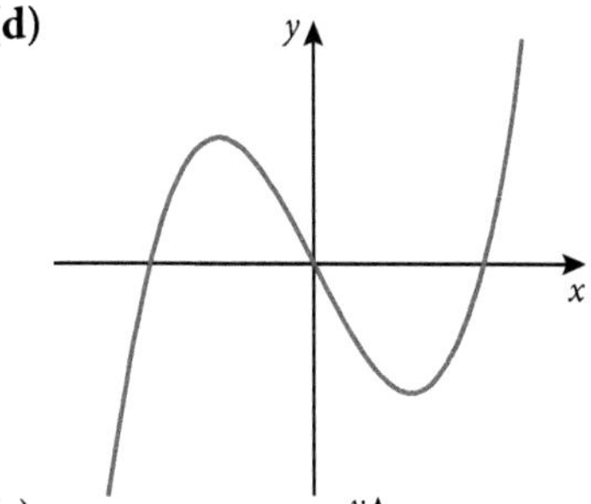

(e)

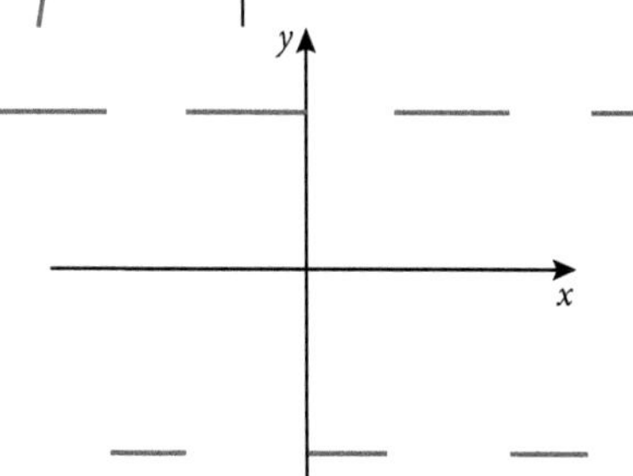

12. (a) (i) increasing on $1 < x < 5$; decreasing on $x < 1, x > 5$
(ii) min. at $x = 1$;max. at $x = 5$
(b) (i) increasing on $0 \leqslant x < 1, 3 < x < 5$; decreasing on $1 < x < 3, x > 5$
(ii) min. at $x = 3$;max. at $x = 1$ and $x = 5$

13. $x \approx 0.5$ and $x \approx 7.5$

14.

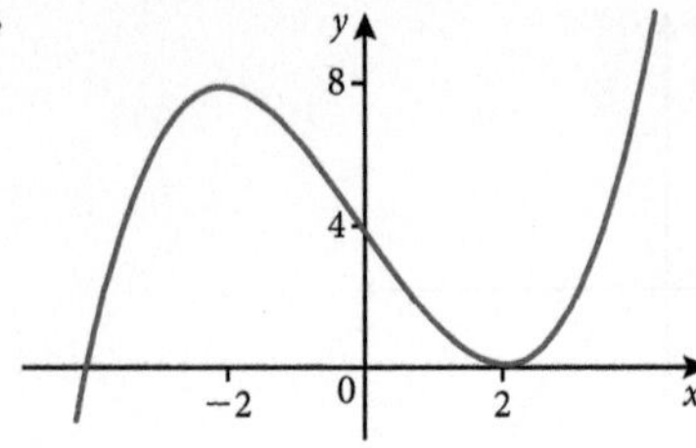

15. (a) right $1 < t < 4$; left $t < 1, t > 4$

(b) (i) $v_0 = -24$ **(ii)** $a_0 = 30$

16. (a) max. at $x \approx 6.50$, min. at $x \approx -0.215$

(b) max. is $\frac{7\pi}{4} + 1$, min. is $\frac{\pi}{4} - 1$

Exercise 9.4

1. (a) $y = -4x - 8$ **(b)** $y = \frac{4}{27}$

(c) $y = -x + 1$ **(d)** $y = -2x + 4$

2. (a) $y = \frac{1}{4}x + \frac{19}{4}$ **(b)** $x = -\frac{2}{3}$

(c) $y = x + 1$ **(d)** $y = \frac{1}{2}x + \frac{11}{4}$

3. at $(0, 0)$: $y = 2x$; at $(1, 0)$: $y = -x + 1$; at $(2, 0)$: $y = 2x - 4$

4. $y = -2x$

5. (a) $x = 1$

(b) for $y = x^2 - 6x + 20$ eq. of tangent is $y = -4x + 19$
for $y = x^3 - 3x^2 - x$ eq. of tangent is $y = -4x + 1$

6. normal: $y = \frac{1}{2}x - \frac{7}{2}$; int. pt: $\left(-\frac{1}{2}, -\frac{15}{4}\right)$

7. eq. of tangent: $y = -3x + 3$; eq. or normal: $y = \frac{1}{3}x - \frac{1}{3}$

8. $a = -4, b = 1$

9. (a) $y = 2x + \frac{5}{2}$ **(b)** $\left(\frac{2}{3}, \frac{41}{27}\right)$

10. eq. of tangent: $y = -\frac{3}{4}x + 1$; eq. or normal: $y = \frac{4}{3}x - \frac{22}{3}$

11. (a) Shows $x = 1$ is the only solution to f and $y = 12x + 4$

(b) f and $y = 9x + 5$ intersect at $(3, 32)$ which is a turning point

(c)

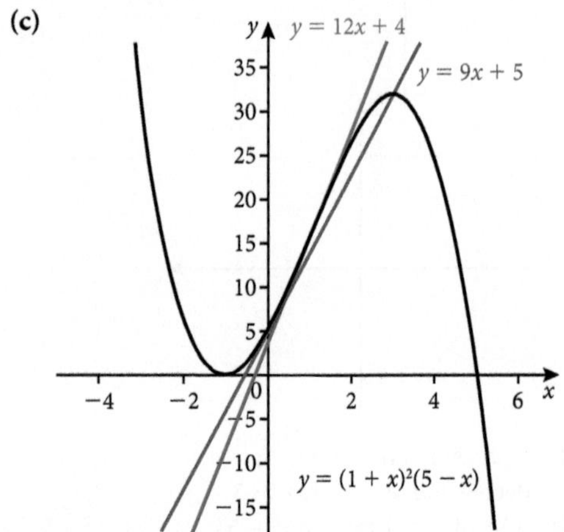

12. $y = 11x - 25$ and $y = -x - 1$

13. $y = (2\sqrt{2} - 2)x, y = -(2\sqrt{2} + 2)x$

14. (a) $y = \frac{1}{12}x + \frac{4}{3}$ **(b)** $\sqrt[3]{9} \approx 2.08$

15. $y = -\frac{1}{2\sqrt{a^3}}x + \frac{3}{2\sqrt{a}}$

16. $x_Q = -2x_P, y_Q = -8y_P$

Chapter 9 Practice questions

1. (a) gradient = 3 **(b)** $y = 3x - \frac{9}{4}$

(c)

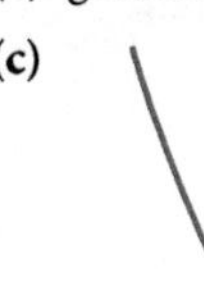

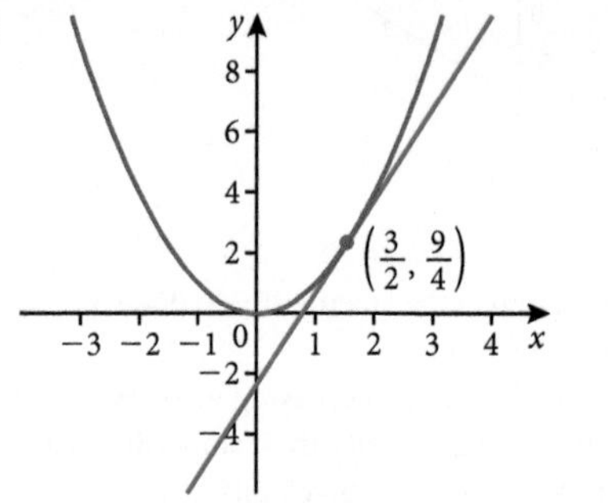

(d) $Q\left(\frac{3}{4}, 0\right), R\left(0, -\frac{9}{4}\right)$

(e) Mid-point $PR = \left(\frac{\frac{3}{2} + 0}{2}, \frac{\frac{9}{4} + \left(-\frac{9}{4}\right)}{2}\right)$
$= \left(\frac{3}{4}, 0\right) = Q$

(f) $y = 2ax - a^2$

(g) $T\left(\frac{a}{2}, 0\right), U(0, -a^2)$

(h) x-coord.: $\frac{a + 0}{2} = \frac{a}{2}$; y-coord.: $\frac{a^2 - a^2}{2} = 0$

2. $A = 1, B = 2, C = 1$

3. (a) $4x - 15x^4$ **(b)** $-\frac{1}{x^2}$

4. (a) $x = 2$ or -2; $f'(1) = -6 < 0$ (decreasing) and
$f'(3) = \frac{10}{9} > 0$ (increasing) $\therefore f(2)$ is a turning point

(b) vertical asymptote: $x = 0$ (y-axis);
oblique asymptote: $y = 2x$

5. $\left(\frac{1}{2}, 3\right)$

6. 1

7. (a) $y = 5x - 7$ **(b)** $y = -\frac{1}{5}x + \frac{17}{5}$

8. (a) $x = 1$

(b) $-3 < x < -2, 1 < x < 3$

(c) $x = -\frac{1}{2}$

(d)

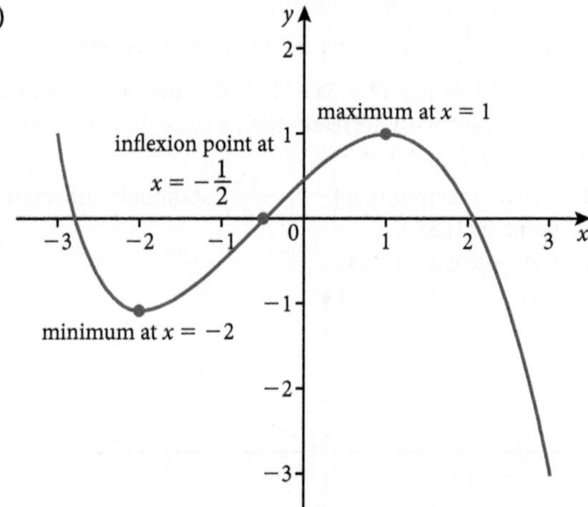

9. $b = 2, c = 3$

10.

function	diagram
f_1	d
f_2	e
f_3	b
f_4	a

11. **(a)** $\frac{2}{\pi}$ **(b)** $\frac{\sqrt{2}}{2}$ **(c)** $x \approx 0.881$

12. **(a)** **(i)** $x = 0$ **(ii)** $y = 3$
 (b) $\frac{dy}{dx} = \frac{2}{x^2}$
 (c) increasing for all x, except $x = 0$
 (d) no stationary points because $\frac{dy}{dx} = \frac{2}{x^2} \neq 0$

13. maximum at $(-1, 1)$, minimum at $(0, 0)$, maximum at $(1, 1)$

14. $a = \frac{8}{3}, b = \frac{16}{5}$

15. **(a)** $10\,\text{m s}^{-1}$ **(b)** 10 sec. **(c)** 50 metres

16. **(a)** $v = 14 - 9.8t, a = 9.8$
 (b) max height = 10 m $t \approx 1.43$ sec.
 (c) velocity = 0, acceleration $= -9.8\,\text{m s}^{-2}$

17. $(-4, 120)$

18. **(a)** $y = (-\sqrt{3})x + \frac{\pi\sqrt{3}}{3} - 2$
 (b) $y = \left(\frac{\sqrt{3}}{3}\right)x - \frac{\pi\sqrt{3}}{9} - 2$

19. $a = -2, b = 8, c = 10$

20. **(a)** $y = -7x + 1$ **(b)** $y = \frac{x}{7} + \frac{107}{7}$

21. **(a)** absolute minimum at $\left(\frac{3}{4}, -\frac{27}{256}\right)$
 (b) domain: $x \in \mathbb{R}$,range: $y \geqslant -\frac{27}{256}$
 (c) inflexion points at $(0, 0)$ and $\left(\frac{1}{2}, -\frac{1}{16}\right)$
 (d)

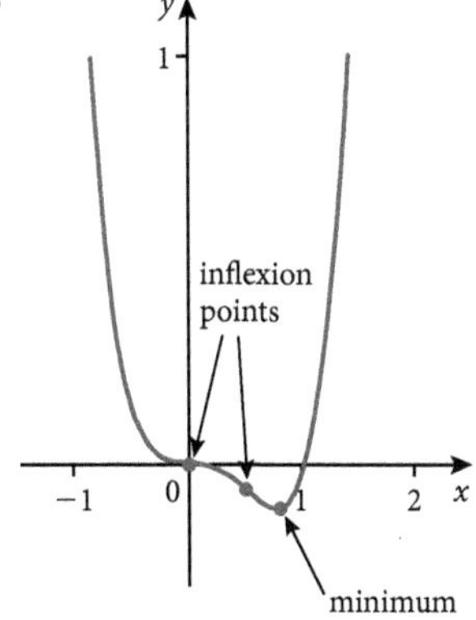

22. **(a)** $-\frac{5}{3}$ **(b)** 3

23. **(a)** $f'(x) = \frac{3x - 4}{2\sqrt{x}}$ **(b)** $f'(x) = 3x^2 - 3\cos x$
 (c) $f'(x) = -\frac{1}{x^2} + \frac{1}{2}$ **(d)** $f'(x) = -\frac{91}{3x^{14}}$

24. 3 solutions: $\left(\frac{11}{2}, \frac{1105}{8}\right)$, $(2, -15)$, and $(-2, 5)$

25. $\frac{17}{2}$ and $-\frac{47}{6}$

26. $\left(2, \frac{2}{3}\right), \left(-2, -\frac{2}{3}\right)$

27. $(-1, -2)$

28. (a) $s(4) = 16, s(2) = 20$
 (b) $s(3) = 18$

29. **(a)** stationary at $x = 0$, then $v > 0$ for $0 < x \leqslant 2\pi$ thus, there is no change in direction.
 (c) $t = 0, \pi, 2\pi$
 (d) max. value of s is 2π

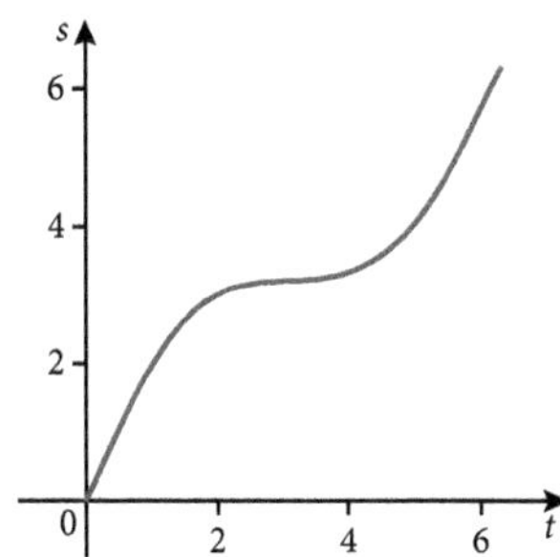

30. $a = \frac{1}{4}, b = \frac{3}{4}, c = -6, d = -\frac{5}{2}$; y-coord. is $-\frac{19}{2}$

31. absolute minimum points at $\left(-2, -\frac{1}{8}\right)$ and $\left(2, -\frac{1}{8}\right)$

32. **(a)** $y = -x + 2$ **(b)** $y = -x + \frac{\pi}{2}$
 (c) For example, deduces that $-x + \frac{\pi}{2} < -x + \pi$

33. **(a)** Show only solution is $x = 1, y = 2x$,
 (b) $(1, 2)$

34. **(a)** $v = 50 - 20t$ **(b)** $s = 1062.5$ m

35.

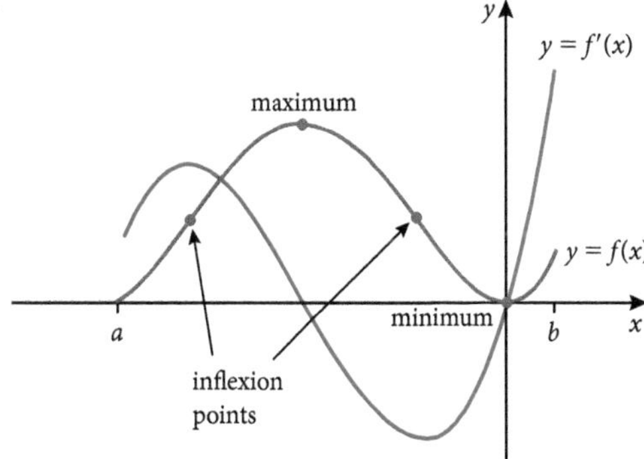

Chapter 10

Exercise 10.1

1. **(a)** $y' = -e^x$ **(b)** $y' = 1 + \frac{1}{x}$ $\left[\text{or } \frac{x+1}{x}\right]$
 (c) $y' = \frac{2e^x}{5}$ **(d)** $y' = \frac{2e}{x}$
 (e) $y' = \frac{1}{2}e^x - \sin x$ **(f)** $y' = \frac{1}{x}$

2. **(a)** $y = \frac{1}{4}x$ **(b)** $y = 2x + 1$ **(c)** $y = \frac{2}{3e}x$

3. $(0, -1)$ is an absolute maximum

4. $\frac{d^2y}{dy^2} = \frac{1}{x^2} \neq 0 \therefore$ no points of inflection

5. $x = \frac{\pi}{2}$

6. (a) $f'(x) = e^x - 3x^2; f''(x) = e^x - 6x$
   (b) $x \approx 3.73$ or $x \approx 0.910$ or $x \approx -0.459$
   (c) decreasing on $(-\infty, -0.459)$ and $(0.910, 3.73)$ increasing on $(-0.459, 0.910)$ and $(3.73, \infty)$
   (d) $x \approx -0.459$ (minimum); $x \approx 0.910$ (maximum); $x \approx 3.73$ (minimum)
   (e) $x \approx 0.204$ or $x \approx 2.83$
   (f) concave up on $(-\infty, 0.204)$ and $(2.83, \infty)$; concave down on $(0.204, 2.83)$

7. $\frac{1}{e}$

Exercise 10.2

1. (a) $y' = 12(3x - 8)^3$ (b) $y' = -\frac{1}{2\sqrt{1-x}}$
   (c) $y' = \frac{2}{x}$ (d) $y' = \cos\left(\frac{x}{2}\right)$
   (e) $y' = -\frac{4x}{(x^2+4)^3}$ (f) $y' = -3e^{-3x}$
   (g) $y' = -\frac{1}{2\sqrt{(x+2)^3}}$ $\left[\text{or} - \frac{1}{(2x+4)\sqrt{x+2}}\right]$
   (h) $y' = -2\sin x\cos x$ (i) $y' = 2xe^{x^2} - 2$
   (j) $y' = \frac{-6x+5}{(3x^2-5x+7)^2}$ (k) $y' = \frac{2}{3\sqrt[3]{(2x+5)^2}}$
   (l) $y' = \frac{2x}{x^2-9}$

2. (a) $y = -12x - 11$ (b) $y = \frac{9}{5}x - \frac{2}{5}$
   (c) $y = 2x - 2\pi$

3. (a) $v(t) = -2t\sin(t^2 - 1)$ (b) velocity $= 0$
   (c) $t = \sqrt{\pi + 1} \approx 2.04, t = 1$
   (d) Accelerates in the positive direction then slows down, turn around, accelerates in the negative direction, slows down, turn around again, then accelerates in the positive direction.

4. (a) $\frac{dy}{dx} = -1$ for $x < -1$, $\frac{dy}{dx} = 1$ for $x > -1$
   (b) $\frac{dy}{dx} = -\frac{\cos x}{\sin^2 x}$ (c) $\frac{dy}{dx} = 3(x + \sqrt{x})^2\left(1 + \frac{1}{2\sqrt{x}}\right)$
   (d) $\frac{dy}{dx} = (-\sin x)e^{\cos x}$ (e) $\frac{dy}{dx} = \frac{2\ln x}{x}$
   (f) $\frac{dy}{dx} = -\frac{3}{\sqrt{(2x+1)^3}}$ or $-\frac{3}{(2x+1)\sqrt{2x+1}}$

5. (a) (i) $y = -12x + 38$ (ii) $y = \frac{1}{12}x + \frac{7}{4}$
   (b) (i) $y = \frac{2}{3}x + \frac{5}{3}$ (ii) $y = -\frac{3}{2}x + 6$
   (c) (i) $y = 4x - 4$ (ii) $y = -\frac{1}{4}x + \frac{1}{4}$

6. (a) $\frac{dy}{dx} = 2\sin(2x); \frac{d^2y}{dx^2} = 4\cos(2x)$
   (b) $\left(\frac{\pi}{4}, 0\right)$ and $\left(\frac{3\pi}{4}, 0\right)$

Exercise 10.3

1. (a) $y' = x^2e^x + 2xe^x$
   (b) $y' = \sqrt{1-x} - \frac{x}{2\sqrt{1-x}}$ or $\frac{2-3x}{2\sqrt{1-x}}$
   (c) $y' = 1 + \ln x$
   (d) $y' = \cos^2 x - \sin^2 x$ or $2\cos^2 x - 1$
   (e) $y' = \frac{xe^x - e^x}{x^2}$
   (f) $y' = -\frac{2}{(x-1)^2}$
   (g) $y' = 2(2x-1)^2(7x^4 - 2x^3 + 3)$
   (h) $y' = \frac{x\cos x - \sin x}{x^2}$
   (i) $y' = \frac{-xe^x + e^x - 1}{(e^x - 1)^2}$
   (j) $y' = \frac{33}{(3x+2)^2}$
   (k) $y'' = 2x\ln 3x + \frac{x^2 - 1}{x}$ or $\frac{2x^2\ln 3x + x^2 - 1}{x}$
   (l) $y' = 0$

2. (a) $y = -\frac{1}{2}x + 2$
   (b) $y = \frac{1}{2}x + \frac{1}{2}$
   (c) $y = 5x - 8$

3. (a) $(-1, -2e)$ and $\left(3, \frac{6}{e^3}\right)$
   (b) $(-1, -2e)$ is a minimum, $\left(3, \frac{6}{e^3}\right)$ is a maximum
   (c) (i) $h(x) \to 0$ as $x \to \infty$
   (ii) $h(x) \to \infty$ as $x \to -\infty$
   (d) horizontal asymptote: $y = 0$ (x-axis)
   (e)

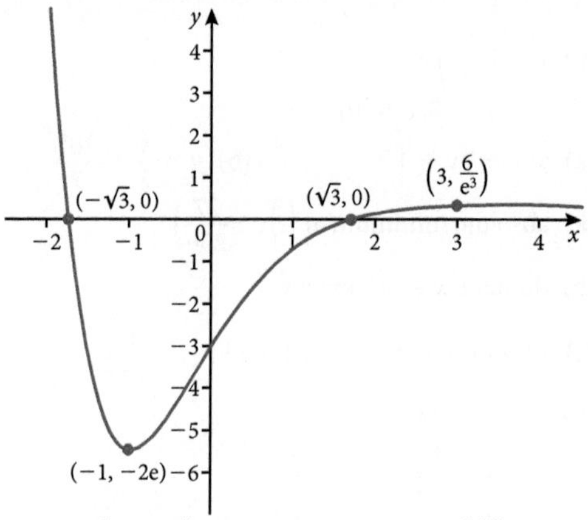

4. tangent: $y = \frac{1}{27}x + \frac{4}{9}$; normal: $y = -27x - \frac{242}{3}$

5. (a) (i) $(0, 0)$ and $(4, 0)$ (repeated)
   (ii) $\left(\frac{4}{3}, \frac{256}{27}\right)$ (iii) $\left(\frac{8}{3}, \frac{128}{27}\right)$
   (b)

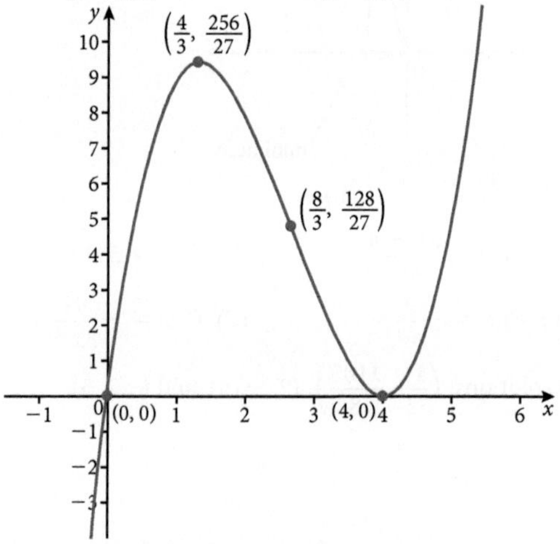

6. (a) (i) $y = -12x + 38$ (ii) $y = \frac{1}{12}x + \frac{7}{4}$
(b) (i) $y = x$ (ii) $y = -x$
(c) (i) $y = \frac{1}{4}x + \frac{1}{4}$ (ii) $y = -4x + \frac{9}{2}$

7. $A\left(\frac{16}{13}, 0\right), B(0, -16)$

8. (a) $h'(x) = \frac{-2x^2 + 8x - 6}{(x^2 - 4x + 5)^2}$
(b) (1, − 1) and (3, 1)

9. tangent: $y = \left(\frac{\pi + 2}{2}\right)x - \frac{\pi^2}{8}$;
normal: $y = \left(-\frac{2}{\pi + 2}\right)x + \frac{\pi^2 + 4\pi}{4\pi + 8}$

10. (a), (b) see worked solutions
(c) $f'(3.8) = 0$ and $f''(3) = \frac{8}{81} > 0, f''(4) = -\frac{2}{625} < 0$, therefore graph of f changes concavity from up to down at $x = 3.8$ verifying that graph of f does have an inflection point at $x = 3.8$

Exercise 10.4

1. $\left(\frac{\sqrt{6}}{2}, \frac{5}{2}\right)$ and $\left(-\frac{\sqrt{6}}{2}, \frac{5}{2}\right)$
2. $\frac{8}{\pi + 4}$ metres by $\frac{4}{\pi + 4}$ metres (or 1.12 m by 0.56 m)
3. $\sqrt{2}$ by $\frac{\sqrt{2}}{2}$
4. $13\frac{1}{3}$ cm by $6\frac{2}{3}$ cm
5. $\frac{\sqrt{5}}{2}$
6. (a) See Worked Solutions
(b) $S = 4x^2 + \frac{3000}{x}$
(c) 7.21 cm × 14.4 cm × 9.61 cm
7. $x = 5\sqrt{2\pi} \approx 12.5$ cm
8. $x \approx 3.62$ m
9. longest ladder ≈ 7.02 m
10. $d \approx 2.64$ km
11. 4 units^2
12. 6 km
13. $h = 2\frac{\sqrt{3}}{3}R$, $r = \frac{\sqrt{6}}{3}R$
14. distance of point P from point X is $\frac{ac}{\sqrt{r^2 - c^2}}$
15. $x \approx 11.5$ cm, max volume ≈ 403 cm^3
16. Students' worked solutions

Chapter 10 practice questions

1.

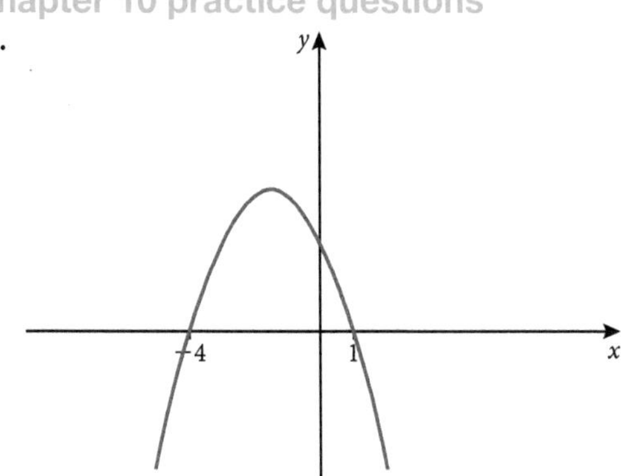

2. (a) (i) 0 and −5 (ii) $\left(-\frac{5}{3}, \frac{500}{27}\right)$
(iii) $\left(-\frac{10}{3}, \frac{250}{27}\right)$
(b)

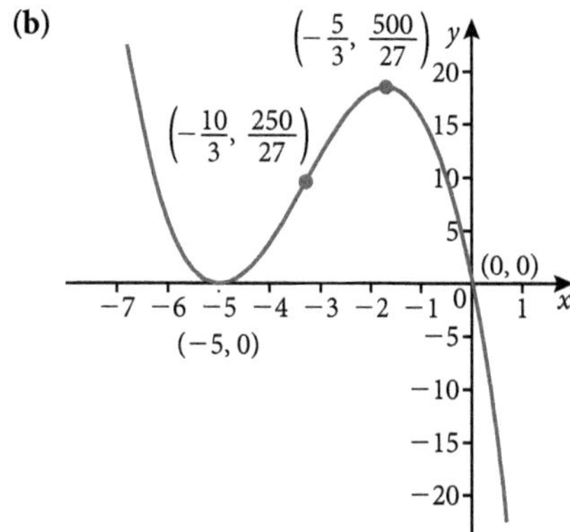

3. $\left(\frac{1}{3}, 1\right)$
4. $y = 3x + 1$
5. (a) (i) $a = -4$ (ii) $b = 2$
(b) (i) $f(x) = -3x^2 - 4x + 8$
(ii) $\frac{-2 + 2\sqrt{7}}{3}, \frac{-2 - 2\sqrt{7}}{3}$ (iii) $f(1) = 5$
(c) (i) $y = 8x$ (ii) $x = -2$
6. (a) (i) $v(0) = 0$ (ii) $v(10) \approx 51.3$
(b) (i) $a(t) = 0.99e^{-0.15t}$ (ii) $a(0) = 0.99$
(c) (i) 66 (ii) 0
(iii) as object falls it approaches terminal velocity
7. (a) $\left(-\frac{2}{3}, -\frac{149}{27}\right)$ is a minimum, $(-4, 13)$ is a maximum
(b) $\left(-\frac{7}{3}, \frac{101}{27}\right)$ is an inflection point
8. (a) (i) $g'(x) = -\frac{3}{e^{3x}}$
(ii) $e^{3x} > 0$ for all x, hence $-\frac{3}{e^{3x}} < 0$ for all x – therefore, $f(x)$ is decreasing for all x
(b) (i) $e + 2$ (ii) $g'(x) = -3e$
(c) $y = -3ex + 2$
9. (a) see worked solutions
(b) $f'(3) = 0$ and $f''(3) > 0 \Rightarrow$ stationary point at $x = 3$ and graph of f is concave up at $x = 3$, so $f(3)$ is a minimum
(c) (4, 0)
10. (a) $-\frac{4}{(2x + 3)^3}$ (b) $5\cos(5x)e^{\sin(5x)}$
11. $A = 1, B = 2, C = 1$
12. (a) $\frac{1}{x}$ (ii) $-\frac{1}{x^2}$
(b) (i) (ii) see worked solutions
(c)

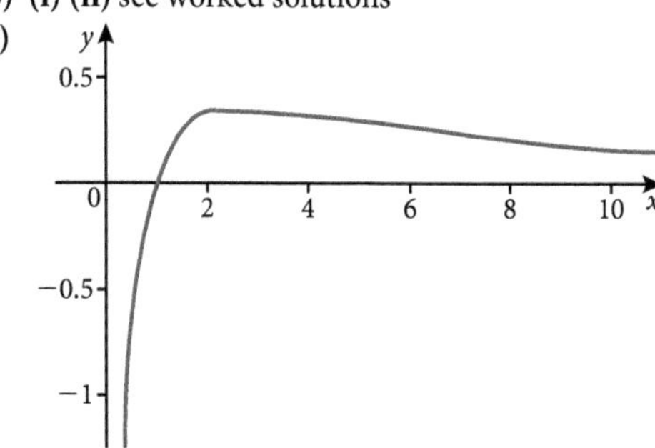

(d) $x = 4$ or $x = 2$

13. (a) $\dfrac{x^4 + 3x^2}{(x^2+1)^2}$ **(b)** $2e^x\cos(2x) + e^x\sin(2x)$

14. $a = 1$

15. (a) $x = 3$; sign of $h''(x)$ changes from negative (concave down) to positive (concave up) at $x = 3$

(b) $x = 1$; $h'(x)$ changes from positive (h increasing) to negative (h decreasing) at $x = 1$

(c)

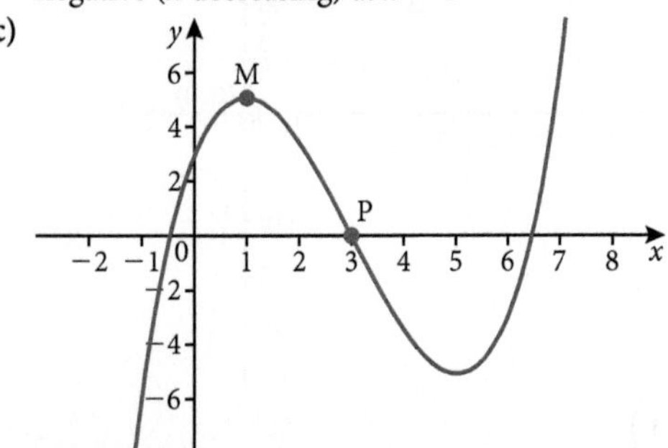

16. $y = 2ex - e$

17. $h = 8$ cm, $r = 4$ cm

18. maximum area is 32 square units; dimensions are 4 by 8

19. (a) E **(b)** A **(c)** C

20. $y = -\frac{1}{5}x + \frac{32}{5}$

Chapter 11

Exercise 11.1

1. (a) $\dfrac{x^2}{2} + 2x + c$ **(b)** $t^3 - t^2 + t + c$

(c) $\dfrac{x}{3} - \dfrac{x^4}{14} + c$ **(d)** $\dfrac{2t^3}{3} + \dfrac{t^2}{2} - 3t + c$

(e) $\dfrac{5u^{\frac{7}{5}}}{7} - u^4 + c$ **(f)** $\dfrac{4x\sqrt{x}}{3} - 3\sqrt{x} + c$

(g) $-3\cos\theta + 4\sin\theta + c$ **(h)** $t^3 + 2\cos t + c$

(i) $\dfrac{4x^2\sqrt{x}}{5} - \dfrac{10x\sqrt{x}}{3} + c$ **(j)** $3\sin\theta - 2\tan\theta + c$

(k) $\frac{1}{3}e^{3t-1} + c$ **(l)** $2\ln|t| + c$

(m) $\frac{1}{6}\ln(3t^2 + 5) + c$ **(n)** $e^{\sin\theta} + c$

(o) $\dfrac{(2x+3)^3}{6} + c$

2. (a) $-\dfrac{5x^4}{4} + \dfrac{2x^3}{3} + cx + k$

(b) $-\dfrac{x^5}{5} + \dfrac{x^4}{4} + \dfrac{x^2}{2} + 2x - \dfrac{11}{20}$

(c) $\dfrac{4t^3}{3} + \sin t + ct + k$

(d) $3x^4 - 4x^2 + 7x + 3$

(e) $2\sin\theta + \frac{1}{2}\cos 2\theta + c$

3. (a) $\dfrac{(3x^2+7)^6}{36} + c$ **(b)** $-\dfrac{1}{18(3x^2+5)^3} + c$

(c) $\dfrac{8\sqrt[4]{(5x^3+2)^5}}{75} + c$ **(d)** $\dfrac{(2\sqrt{x}+3)^6}{6} + c$

(e) $\dfrac{\sqrt{(2t^3-7)^3}}{9} + c$ **(f)** $-\dfrac{(2x+3)^6}{18x^6} + c$

(g) $-\dfrac{\cos(7x-3)}{7} + c$

(h) $-\frac{1}{2}\ln(\cos(2\theta - 1) + 3) + c$

(i) $\frac{1}{5}\tan(5\theta - 2) + c$ **(j)** $\frac{1}{\pi}\sin(\pi x + 3) + c$

(k) $\frac{1}{2}\sec 2t + c$ **(l)** $\frac{1}{2}e^{x^2+1} + c$

(m) $\frac{1}{3}e^{2t\sqrt{t}} + c$ **(n)** $\frac{2}{3}(\ln\theta)^3 + c$

4. (a) $-\frac{1}{15}\sqrt{(3-5t^2)^3} + c$ **(b)** $\frac{1}{3}\tan\theta^3 + c$

(c) $-\cos\sqrt{t} + c$ **(d)** $\frac{1}{12}\tan^6 2t + c$

(e) $2\ln(\sqrt{x} + 2) + c$ **(f)** $\frac{1}{3}e^{x^3+3x+1} + c$

(g) $\frac{1}{2}\ln(x^2 + 6x + 7) + c$

(h) $-\dfrac{k^3}{2a^4}\sqrt{a^2 - a^4x^4} + c = -\dfrac{k^3}{2|a|^3}\sqrt{1 - a^2x^4} + c$

(i) $\frac{2}{5}(3x^2 - x - 2)\sqrt{x-1} + c$

(j) $-\frac{2}{3}\sqrt{(1+\cos\theta)^3} + c$

(k) $\frac{2}{105}(15t^3 - 3t^2 - 4t - 8)\sqrt{1-t} + c$

(l) $\frac{1}{15}(3r^2 + 2r - 13)\sqrt{2r-1} + c$

(m) $\frac{1}{2}\ln(e^{x^2} + e^{-x^2}) + c$

Exercise 11.2

1. (a) 24 **(b)** 40 **(c)** $\frac{24}{25}$ **(d)** 0

(e) $\dfrac{176\sqrt{7} - 44}{5}$ **(f)** 0 **(g)** 2

(h) -268 **(i)** $\frac{64}{3}$ **(j)** 2

(k) $\ln\left(\frac{11}{3}\right)$ **(l)** $\frac{44}{3} - 8\sqrt{3}$ **(m)** $\sqrt{\pi} + 1$

(n) (i) 6 **(ii)** 6 **(iii)** 12

(o) 1 **(p)** 4 **(q)** 0

2. (a) $\dfrac{14\sqrt{17} + 2}{3}$ **(b)** $\frac{1}{\pi}$ **(c)** $\ln(2)$

(d) $16\sqrt{2} - 5\sqrt{5}$ **(e)** $\sqrt{14} - \sqrt{10}$ **(f)** $\frac{3}{2}$

(g) $-\frac{1}{2}\ln\left(\frac{37}{52}\right)$ **(h) 0** **(i)** -4

(j) $\frac{1}{6}$ **(k)** $\dfrac{e-1}{2}$ **(l)** $2\cos(1) + 2$

3. (a) $\dfrac{\sin x}{x}$ **(b)** $-\dfrac{\sin x}{x}$ **(c)** $\dfrac{\cos t}{1+t^2}$

4. (a) $\frac{1}{3}\ln\left(\frac{3k+2}{2}\right)$ **(b)** $k = \dfrac{2(e^3-1)}{3}$

5. Substitute $u = 1 - x$

6. (a) $-(1-x)^{k+1}\left(\dfrac{1}{k+1} + \dfrac{x-1}{k+2}\right)$

(b) $\dfrac{1}{(k+1)(k+2)}$

7. (a) 0; **(b)** $\sqrt{47}$; **(c)** $\dfrac{15\sqrt{47}}{47}$

Exercise 11.3

1. (a) $\frac{125}{6}$ (b) $\frac{9\pi^2}{8} + 1$ (c) $4\sqrt{3}$
   (d) $\frac{10}{3}$ (e) $\frac{8}{21}$ (f) $\frac{125}{24}$
   (g) $\frac{13}{12}$ (h) 4π (i) $\frac{59}{12}$
   (j) 362.94 (k) $3\ln 2 - \frac{63}{128}$
   (l) (between $-\frac{\pi}{2}$and $\frac{\pi}{6}$) $\sqrt{3}\ln\left(\frac{3}{4}\right) - 2\sqrt{3} + 4$
   (m) 19 (n) $\frac{37}{12}$
   (o) $\frac{1}{2}$ (p) $\frac{2\sqrt{2}}{3}$
2. $\frac{269}{54}$
3. $\frac{e}{2} - 1$
4. 25.36
5. $m = 0.973$
6. $\frac{37}{12}$

Exercise 11.4

1. (a) $\frac{70}{3}$ m, 65 m (b) 8.5 m to the left, 8.5 m.
   (c) 1 m, 1 m (d) 2 m, $2\sqrt{2}$m.
   (e) 18 m, 28.67 m (f) $\frac{4}{\pi}m, \frac{4}{\pi}m$
2. (a) $3t$, 6 m, 6 m (b) $t^2 - 4t + 3$, 0, 2.67 m
   (c) $1 - \cos t, \left(\frac{3\pi}{2} + 1\right)m, \left(\frac{3\pi}{2} + 1\right)m$
   (d) $4 - 2\sqrt{t+1}$, 2.43 m, 2.91 m
   (e) $3t^2 + \frac{1}{2(1+t)^2} + \frac{3}{2}$, 11.3 m, 11.3 m
3. (a) $4.9t^2 + 5t + 10$ (b) $16t^2 - 2t + 1$
   (c) $\frac{1}{\pi} - \frac{\cos \pi t}{\pi}$ (d) $\ln(t+2) + \frac{1}{2}$
4. (a) $e^t + 19t + 4$ (b) $4.9t^2 - 3t$
   (c) $\sin(2t) - 3$ (d) $-\cos\left(\frac{3t}{\pi}\right)$
5. (a) 12; 20 (b) $\frac{13}{2}; \frac{13}{2}$
   (c) $\frac{9}{4}; \frac{11}{4}$ (d) $2\sqrt{3} - 6; 6 - 2\sqrt{3}$
6. (a) $-\frac{10}{3}; \frac{17}{3}$ (b) $\frac{204}{25}$ (c) $-6; \frac{13}{2}$
7. (a) $\frac{166}{5}$ (b) $\frac{166}{5}$ (c) $\frac{166}{5}$
8. (a) $50 - 20t$ (b) 1187.5
9. 1.27 s
10. (a) 5 s (b) 272.5 m
    (c) 10 s (d) -49 m s^{-1}
    (e) 12.447 s (f) -73.1 m s^{-1}

Chapter 11 practice questions

1. (a) $p = 3$ (b) 3 square units
2. (a) (0, 1) (b) $2\sqrt{2} - 2$
3. $a = e^2$
4. (a) $y = \frac{x}{e}$ (b) $\ln x + 1 - 1$
   (c) See Worked solutions
5. (a) (i) 400 m
   (ii) $v = 100 - 8t$, 60 m/s
   (iii) 8 s
   (iv) 1344 m
   (b) Distance needed 625
6. (a) See Worked Solutions (b) 2.31
   (c) $-\pi\cos x - \frac{x^2}{2} + c$; 0.944
7. ln 3
8. (a) (i) See Worked Solutions
   (ii) (1.57, 0); (1.1, 0.55); (0, 0), (2, −1.66)
   (b) $x = \frac{\pi}{2}$
   (c) (i) See Worked Solutions
   (ii) $\int_0^{\frac{\pi}{2}} x^2 \cos x \,dx$
   (d) $\frac{\pi^2}{4} - 2 \approx 0.4674$
9. (a) 2π
   (b) range: $\{y \mid -0.4 < y < 0.4\}$
   (c) (i) $-3\sin^3 x + 2\sin x$ (iii) $\frac{2\sqrt{3}}{9}$
   (d) $\frac{\pi}{2}$
   (e) (i) $\frac{1}{3}\sin^3 x + c$ (ii) $\frac{1}{3}$
   (f) $\arccos\frac{\sqrt{7}}{3} \approx 0.491$
10. (a) (i) See Worked Solutions (ii) See Worked Solutions
    (b) See Worked Solutions (c) 3.69672
    (d) $\int_0^{\pi} (\pi + x\cos x)dx$ (e) $\pi^2 - 2 \approx 7.86960$
11. (a) (i) $10x + 1 - e^{2x}$ (ii) $\frac{\ln 5}{2} \approx 0.805$
    (b) (i) $f^{-1}(x) = \frac{\ln(x-1)}{2}$ (ii) See Worked Solutions
    (c) $v = \pi\int_0^{\ln 2} (1 + e^{2x})^2 dx$
12. $\pi\left(\frac{2}{15}a^5 + \frac{2}{3}a^3\right)$
13. $4\left(\frac{2}{5}\left(\frac{1}{2}x + 1\right)^{\frac{5}{2}} - \frac{2}{3}\left(\frac{1}{2}x + 1\right)^{\frac{3}{2}}\right) + c$
14. $a = -\frac{56}{27}$
15. 1
16. $k = 2$
17. 1800 m
18. $2a$ by $\frac{2}{3}a^2$
19. (a) $\ln x + 1 - k$ (b) $x > \frac{1}{e}$
    (c) (i) See Worked Solutions
    (ii) $(e^k, 0)$
    (d) $\frac{e^{2k}}{4}$ (e) $y = x - e^k$
    (f) See Worked Solutions (g) Common ratio = e

20. $a = 5$

21. $v = \sqrt{v_0^2 + \frac{4k}{m}}$

22. (a)

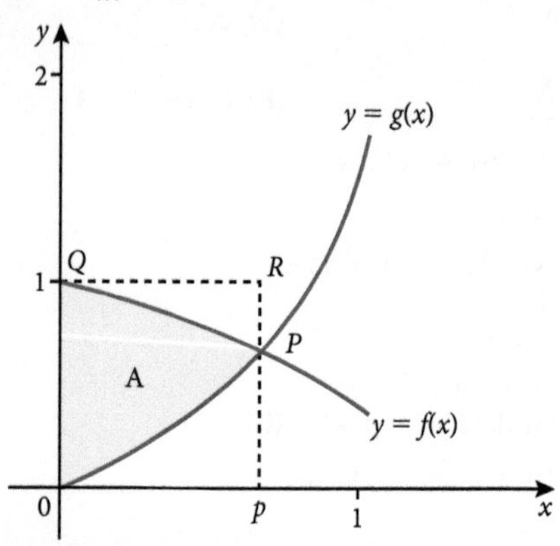

(b) See Worked Solutions

(c) 0.6937

(d) $\int_0^p (e^{-x^2} - (e^{-x^2} - 1))dx \approx \sim 0.467$

23. (a) See Worked Solutions

(b) (i) $\frac{2\pi}{9}$ **(ii)** $\frac{4\pi}{9}$ **(iii)** $\frac{6\pi}{9}$

(c) $\frac{n\pi}{9}(n + 1)$

24. (a) $t = 0, 3,$ or 6

(b) (i) $\int_0^6 \left| t\sin\left(\frac{\pi}{3}\right)t\,dt \right|$ **(ii)** 11.5 m

25. (a) 0.435

(b) $\frac{-2t}{(2 + t^2)^2}$

26. (a) $\frac{dy}{dx} = \frac{2x^2}{\sqrt{1 + x^2}} + 2\sqrt{1 + x^2}$

(b) See Worked Solutions

(c) $k = 0.918.$

27. 6 m

28. 0.852

29. (a) See Worked Solutions

(b) (i) $A = 78; k = \frac{1}{15}\ln\frac{48}{78}$ **(ii)** 45.3

30. (a)

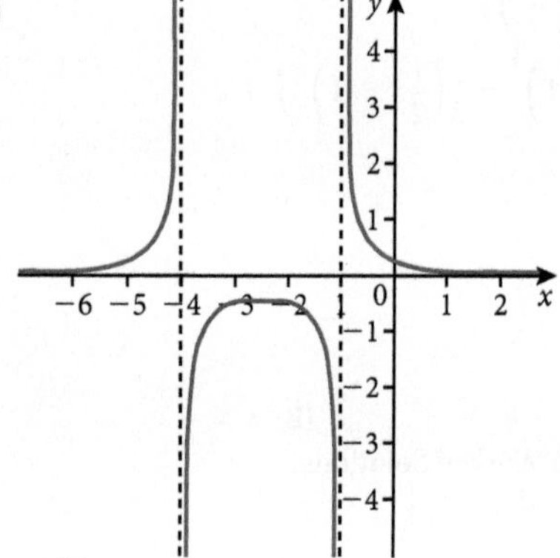

(b) $\ln\sqrt[3]{\frac{8}{5}}$

(c) area $= 2\ln\sqrt[3]{\frac{8}{5}}$

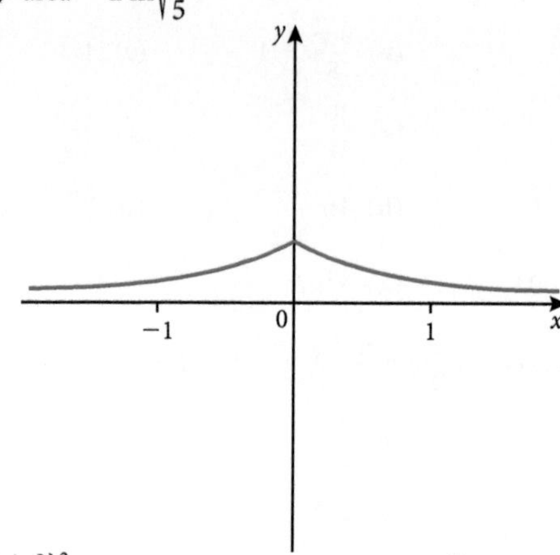

31. $\frac{(x + 2)^2}{2} - 6(x + 2) + 12\ln|x + 2| + \frac{8}{x + 2} + c$

32. (a)

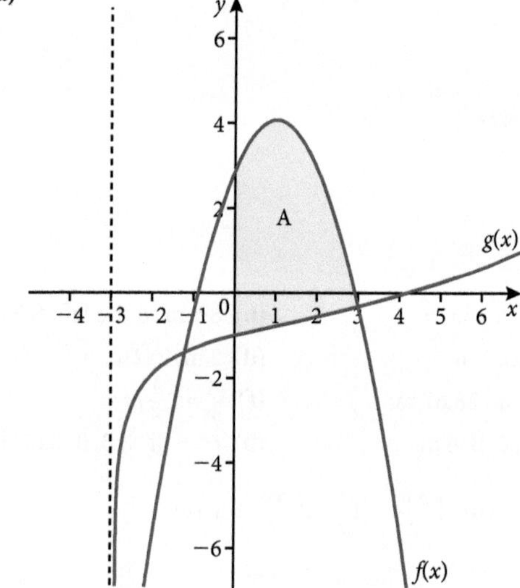

(b) (i) $x = 23$

(ii) $x - \text{int} = e^2 - 3; y - \text{int} = \ln 3 - 2$

(c) $-1.34; 3.05$

(d) (ii) $\int_0^{3.05} (4 - (1 - x)^2 - (\ln(x + 3) - 2))dx$

(iii) 10.6

(e) 4.63

33. (a)

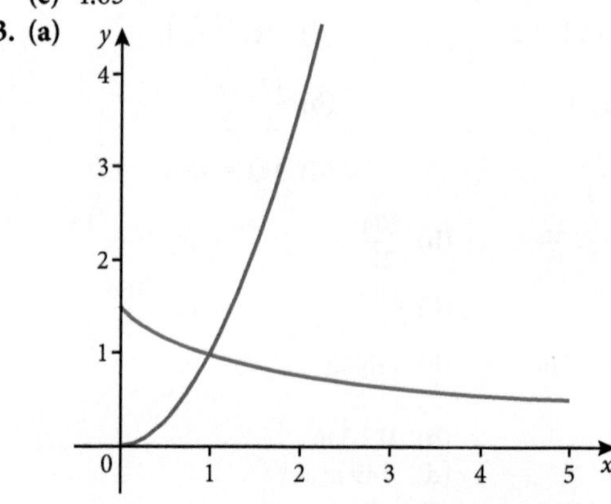

(b) (1, 1) **(c)** $y = 2x - 1; y = \left(-\frac{1}{2}\right)x + \frac{3}{2}$ **(d)** $\frac{5}{4}$

34. (a) $(1, 0), \left(e^2, \frac{4}{e^2}\right)$ **(b)** $\frac{1}{3}$

Chapter 12

Exercise 12.1

1. (a) discrete (b) continuous (c) continuous
(d) discrete (e) continuous (f) continuous
(g) discrete (h) continuous (i) continuous
(j) discrete (k) continuous (l) continuous
(m) discrete

2. (a) 0.4
(b)

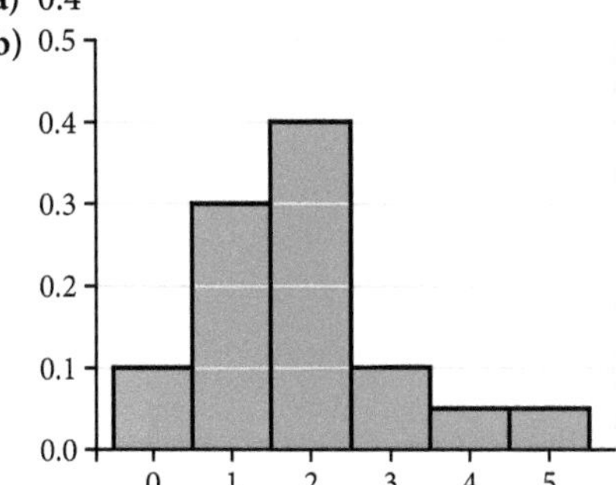

(c) 1.85, 1.19 (e) 2.85, 1.19
(f) $E(Z) = E(Y + b) = E(Y) + b$ and $V(Y) = V(Y + b) = V(Y)$

3. (a) 0.26 (b) 0.37 (c) 0.77
(d) 16.29 (e) 8.1259
(f) 4.145; 2.031475
(g) $E(aX + b) = aE(X) + b$ and $V(aX + b) = a^2V(X)$

4. (a) 0.969 (b) 0.163 (c) 3.5
(d) $\sum(x - 3.5)^2 \cdot P(x) = 1.048 \Rightarrow \sigma = \sqrt{1.048} \approx 1.02$
(e) Empirical: 0.68, 0.95; approximately 0.68, approximately 0.90

5. $k = \frac{1}{30}$

x	12	14	16	18
$P(X = x)$	$6k$	$7k$	$8k$	$9k$

6. (a) $k = \frac{1}{10}$ (b) $\frac{37}{60}$ (c) $\frac{19}{30}$
(d) $E(X) = 16$, SD= 7
(e) $E(Y) = \frac{11}{5}$; $V(Y) = \frac{49}{25}$

7. (a) $\frac{1}{50}$
(b)

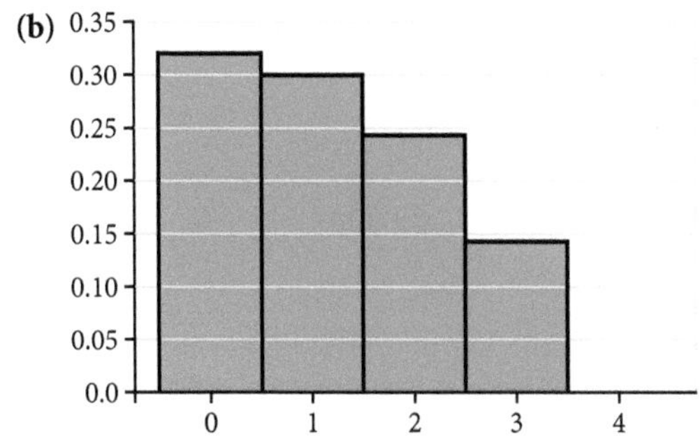

(c) $\frac{17}{25}$ (d) $\mu = 1.2$; Var = 1.08

8. (a) $P(X = 18) = 0.2$, $P(X = 19) = 0.1$, Symmetric distribution.
(b) $\mu = 17$, SD = 1.095

9. (a) $\mu = 1.9$, SD = 1.338 (b) between 0 and 5.

10. $k = 0.667$, $E(X) = 5.444$

11. (a) k =0.3 or 0.7
(b) for $k = 0.3$: $E(X) = 2.18$; for $k = 0.7$: $E(X) = 1.78$

12. (a)

x	0	1	2	3
$P(X = x)$	$\frac{1}{27}$	$\frac{2}{9}$	$\frac{4}{9}$	$\frac{8}{27}$

(b) 2

13. (a) $k = \frac{1}{10}$ (b) $\frac{1}{2}$

14. (a) See table below. (b) 0.85 (c) 0.15
(d) 48.87 (e) 2.057 (f) 0.72

x	45	46	47	48	49	50	51	52	53	54	55
CDF	0.05	0.13	0.25	0.4	0.65	0.85	0.9	0.94	0.97	0.99	1

15. (a)

x	0	1	2	3	4	5	6
CDF	0.08	0.23	0.45	0.72	0.92	0.97	1

(b) 0.72 (c) 0.97 (d) 2.63 (e) 1.440

16. (a) 0.9 (b) 0.09 (c) 0.009
(d) (i) unacceptable (ii) acceptable
(e) $P(x) = (0.1^{x-1}) \times 0.9$

17. (a) 0 (b) 0.81 (c) 0.162
(d) (i) either (ii) acceptable
(e) $(x - 1)(0.1^{x-2}) \times 0.9^2, x > 1.$

18. $n = 132$

19. (a) (i) $\frac{1}{9}$ (ii) $\frac{1}{81}$
(b) (i) $\frac{73}{648}$ (ii) $\frac{575}{1296}$
(c) (i) See Worked Solutions
(ii)

X	1	2	3	4	5	6
CDF	$\frac{1}{1296}$	$\frac{15}{1296}$	$\frac{65}{1296}$	$\frac{175}{1296}$	$\frac{369}{1296}$	$\frac{671}{1296}$

(iii) $\frac{6797}{1296}$

20. 9.3

Exercise 12.2

1. (a)

x	0	1	2	3	4	5
$P(X = x)$	0.01024	0.0768	0.2304	0.3456	0.2592	0.07776

(b)

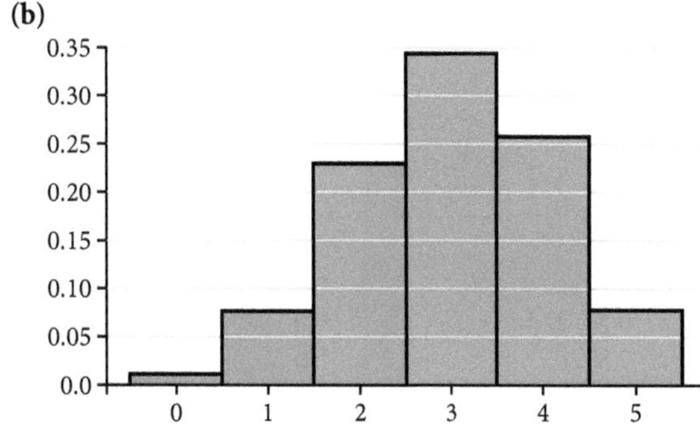

(c) (i) Mean = 3, SD= 1.095
(ii) Mean = 3, SD= 1.095
(d) between 2 and 4, and between 1 and 5.
(e) 0.8352, 0.990. Slightly more than the empirical rule.

2. (a) 0.001294494 (b) 0.000000011

(c) 0.99999999 (d) 0.99999966
(e) mean = 12, SD = 2.19

3. (a)

k	0	1	2	3	4	5
$p(x \leqslant k)$	0.11765	0.42017	0.74431	0.92953	0.98907	0.99927

(b)

Number of successes x	List the values of x	Write the probability statement	Explain it, if needed	Find the required probability
At most 3	0, 1, 2, 3	$p(x \leqslant 3)$	$p(x \leqslant 3)$	0.92953
At least 3	3, 4, 5, 6	$P(x \geqslant 3)$	$1 - p(x \leqslant 2)$	0.25569
More than 3	4, 5, 6	$p(x > 3)$	$1 - p(x \leqslant 3)$	0.07047
Fewer than 3	0, 1, 2	$p(x \leqslant 2)$	$p(x \leqslant 2)$	0.74431
Between 3 and 5 (inclusive)	3, 4, 5	$p(3 \leqslant x \leqslant 5)$	$p(x \leqslant 5) -$ $p(x \leqslant 2)$	0.25496
Exactly 3	3	$P(x = 3)$	$P(x = 3)$	0.18522

4. (a)

k	0	1	2	3
$p(x \leqslant k)$	0.02799	0.15863	0.41990	0.71021
k	4	5	6	7
$p(x \leqslant k)$	0.90374	0.98116	99836	1

(b)

Number of successes x	List the values of x	Write the probability statement	Explain it, if needed	Find the required probability
At most 3	0, 1, 2, 3	$p(x \leqslant 3)$	$p(x \leqslant 3)$	0.71021
At least 3	3, 4, 5, 6, 7	$P(x \geqslant 3)$	$1 - p(x \leqslant 2)$	0.58010
More than 3	4, 5, 6, 7	$p(x > 3)$	$1 - p(x \leqslant 3)$	0.28979
Fewer than 3	0, 1, 2	$p(x \leqslant 2)$	$p(x \leqslant 2)$	0.41990
Between 3 and 5 (inclusive)	3, 4, 5	$p(3 \leqslant x \leqslant 5)$	$p(x \leqslant 5) -$ $p(x \leqslant 2)$	0.56126
Exactly 3	3	$P(x = 3)$	$P(x = 3)$	0.290304

5. (a) p is not constant, trials are not independent.
(b) p becomes constant.
(c) $n = 3, p = \frac{5}{8}$

y	0	1	2	3
$P(Y = y)$	0.05273	0.263672	0.439453	0.244141

(d) 0.75586 (e) 1.875
(f) 0.703125 (g) 0.94727

6. (a) 0.107374 (b) 0.99363
(c) 0.89263 (d) 2

7. (a) 0.817073 (b) 1 (c) 0.0161776

8. (a) 0.033833 (b) 0.024486 (c) 0.782722

9. (a) 0.75 (b) 0.0325112 (c) 0.172678

10. (a) 0.0431745 (b) 0.997614 (c) 0.0112531
(d) 0.130567 (e) 0.956826 (f) 10
(g) 3 (h) 4, 16.

11. (a) 3 (b) 0.101308 (c) 0.000214925

12. (a)

x	0	1	2	3	4	5
$P(X)$	0.03125	0.15625	0.31250	0.31250	0.15625	0.03125

(b) 0.03125 (c) 0.03125
(d) 0.96875 (e) 0.96875
(f) (a)

x	0	1	2	3	4	5
$P(X)$	0.32768	0.40960	0.20480	0.05120	0.00640	0.00032

(b) 0.32768 (c) 0.00032
(d) 0.67232 (e) 0.99968

13. 0.91296

14. (a) 0.107 (b) 0.893 (c) $n = 14$

Exercise 12.3

(some answers are rounded)

1. (a) 0.5 (b) 0.499571 (c) 0.158655
(d) 0.682690 (e) 0.022750 (f) 0

2. (a) 0.76986 (b) 0.161514
(c) 0.656947 (d) 0.999944

3. (a) 0.008634 (b) 0.982732

4. 1.28

5. 1.96

6. (a) 0.066807 (b) 0.68269
(c) 678.16 (d) 134.90

7. (a) 1.76% (b) 509.98 (c) 5.71

8. (a) 0.9696 (b) 0.546746

9. (a) 1 day (b) 29 days (c) 112 days

10. 1.56

11. 18.9192

12. 30.81

13. 100.28

14. 29.95

15. $\mu = 21.037, \sigma = 4.252$

16. $\mu = 18.988, \sigma = 0.615$

17. $\mu = 121.935, \sigma = 34.389$

18. (a) $\mu = 6.966, \sigma = 0.324$ (b) 0.252

19. (a) 0.655422 (b) 0.008198 (c) 82 bottles

20. (a) 22.73%
(b) 0.546%
(c) 29.678
(d) 229.183

21. (a) Not likely: chance is 0.135%
(b) 15.87%
(c) 68.27% (d) 5396 km (e) 43785

22. (a) 6.817 (b) 3.4315
(c) $\mu = 64.135, \sigma = 7.545$

23. 7.3%

24. (a) 216.06
(b) 15.31

25. (a) $\mu = 111.90$, s = 17.09 (b) 0.54

26. 9.1929

27. (a) (i) $\sigma = 1.355$
(ii) $\mu = 110.37$
(b) A = 108.64; B = 112.11

Chapter 12 practice questions

1. **(a)** 34.5% **(b)** 0.416 **(c)** 3325
2. **(a)** **(i)** 0.393 **(ii)** 0.656 **(b)** 50
3. **(a)** 0.1 **(b)** 10
   (c) See Worked Solutions **(d)** 0.739
4. **(a)** $\frac{35}{128}$ **(b)** $\frac{7}{32}$
   (c) $\frac{91}{128}$
5. **(a)** **(i)** $a = -0.455, b = 0.682$
   (b) **(i)** 0.675 **(ii)** 0.428
   (c) See Worked Solutions **(d)** $t = 62.6$
6. **(a)** 69.97% **(b)** 0.00228
7. **(a)** 0.0808 **(b)** See Worked Solutions
   (c) $\mu = 25.5, s = 0.255$ **(d)** 12 500
8. **(a)** **(i)** 0.345 **(ii)** 0.115 **(iii)** 0.540
   (b) 0.119 **(c)** 737
9. **(a)** 15.9% **(b)** 227 cm
10. **(a)** 0.0912 **(b)** $a = 251, b = 369$.
11. **(a)** $a = -1, b = 0.5$
    (b) **(i)** 0.841 **(ii)** 0.533
    (c) **(i)** See Worked Solutions **(ii)** 0.647
12. **(a)** 2 **(b)** 0.182 **(c)** 0.597
13. $\mu = 66.6, \sigma = 22.6$
14. **(a)** 0.8
    (b) **(i)** See Worked Solutions
    (ii)

Y	0	1	2
$P(Y = y)$	$\frac{1}{15}$	$\frac{8}{15}$	$\frac{2}{5}$

    (c) $\frac{3}{10}$ **(d)** $\frac{1}{9}$
15. **(a)** 0.129886 **(b)** 0.676714 **(c)** 2
16. **(a)** 0.1829 **(b)** 0.3664
17. **(a)** **(i)** 0.002171 **(ii)** 0.00120
    (b) 0.8413
18. $\sigma = 0.00943\,\text{kg} \approx 9.4\,\text{g}$
19. **(a)** $x = 58.69$
    (b) $s = 3.41$
    (c) **(i)** Karl
    (ii) 0.00239
20. **(a)** $m = 1.63$
    (b) $0.44 \times 0.401 + 0.56 \times 0.242 = 0.312$
    (c) 0.434
    (d) \$6605.28

Index

Index